CHEMICAL PROCESSES IN LAKES
Werner Stumm, Editor

INTEGRATED PEST MANAGEMENT IN PINE-BARK BEETLE ECOSYSTEMS
William E. Waters, Ronald W. Stark, and David L. Wood, Editors

PALEOCLIMATE ANALYSIS AND MODELING
Alan D. Hecht, Editor

BLACK CARBON IN THE ENVIRONMENT: Properties and Distribution
E. D. Goldberg

GROUND WATER QUALITY
C. H. Ward, W. Giger, and P. L. McCarty, Editors

TOXIC SUSCEPTIBILITY: Male/Female Differences
Edward J. Calabrese

ENERGY AND RESOURCE QUALITY: The Ecology of the Economic Process
Charles A. S. Hall, Cutler J. Cleveland, and Robert Kaufmann

AGE AND SUSCEPTIBILITY TO TOXIC SUBSTANCES
Edward J. Calabrese

ECOLOGICAL THEORY AND INTEGRATED PEST MANAGEMENT PRACTICE
Marcos Kogan, Editor

AQUATIC SURFACE CHEMISTRY: Chemical Processes at the Particle Water Interface
Werner Stumm, Editor

RADON AND ITS DECAY PRODUCTS IN INDOOR AIR
William W. Nazaroff and Anthony V. Nero, Jr., Editors

PLANT STRESS–INSECT INTERACTIONS
E. A. Heinrichs, Editor

INTEGRATED PEST MANAGEMENT SYSTEMS AND COTTON PRODUCTION
Ray Frisbie, Kamal El-Zik, and L. Ted Wilson, Editors

ECOLOGICAL ENGINEERING: An Introduction to Ecotechnology
William J. Mitsch and Sven Erik Jorgensen, Editors

ARTHROPOD BIOLOGICAL CONTROL AGENTS AND PESTICIDES
Brian A. Croft

ARTHROPOD
BIOLOGICAL
CONTROL AGENTS
AND PESTICIDES

ARTHROPOD BIOLOGICAL CONTROL AGENTS AND PESTICIDES

BRIAN A. CROFT

Department of Entomology
Oregon State University
Corvallis, Oregon

A WILEY-INTERSCIENCE PUBLICATION

JOHN WILEY & SONS

New York · Chichester · Brisbane · Toronto · Singapore

Library of Congress Cataloging in Publication Data:

Croft, Brian A.

Arthropod biological control agents and pesticides/Brian A. Croft.

p. cm. — (Environmental science and technology, ISSN 0194–0287)

"A Wiley-Interscience publication."

Bibliography: p.

Includes indexes.

ISBN 0–471–81975–1

1. Entomophagous arthropods—Effect of pesticides on. 2. Insect pests—Biological control. I. Title. II. Series.

SB933.33.C76 1989

632'.7—dc 19

89–30003

CIP

Printed in the United States of America

10 9 8 7 6 5 4 3 2 1

This volume honors B. R. Bartlett of the University of California, Riverside, who pioneered the study of natural enemy–pesticide interactions and who influenced me by his ability to combine biological control research with the evaluations of "dirty pesticides."

SERIES PREFACE
Environmental Science and Technology

The Environmental Science and Technology Series of Monographs, Textbooks, and Advances is devoted to the study of the quality of the environment and to the technology of its conservation. Environmental science therefore relates to the chemical, physical, and biological changes in the environment through contamination or modification, to the physical nature and biological behavior of air, water, soil, food, and waste as they are affected by man's agricultural, industrial, and social activities, and to the application of science and technology to the control and improvement of environmental quality.

The deterioration of environmental quality, which began when all first collected into villages and utilized fire, has existed as a serious problem under the ever-increasing impacts of exponentially increasing population and of industrializing society. Environmental contamination of air, water, soil, and food has become a threat to the continued existence of many plant and animal communities of the ecosystem and may ultimately threaten the very survival of the human race.

It seems clear that if we are to preserve for future generations some semblance of the biological order of the world of the past and hope to improve on the deteriorating standards of urban public health, environmental science and technology must quickly come to play a dominant role in designing our social and industrial structure for tomorrow. Scientifically rigorous criteria of environmental quality must be developed. Based in part on these criteria, realistic standards must be established and our technological progress must be tailored to meet them. It is obvious that civilization will continue to require increasing amounts of fuel, transportation, industrial chemicals, fertilizers, pesticides, and countless other products; and that it will continue to produce waste products of all descriptions. What is urgently needed is a total systems approach to modern civilization through which the pooled talents of scientists and engineers, in cooperation with social scientists and the medical profession, can be focused on the development of order and equilibrium in the presently disparate segments of the human environment. Most of the skills and tools that are needed are already in existence. We surely have a right to hope a technology that has created such manifold environment problems is also capable of solving them. It is our hope that this Series in Environmental Sciences and Technology will not only serve to make this challenge more explicit to the established professionals, but that it also will help to stimulate the student toward the career opportunities in this vital area.

ROBERT L. METCALF
WERNER STUMM

PREFACE

The science and the practice of biological control have expanded considerably since the classic introduction of the vedalia beetle for control of the cottony cushion scale in the late 1800s to early 1900s. Recent emphasis on integrated pest management has not only taught chemical pest control specialists to appreciate and take advantage of the natural pest-regulating power of predaceous and parasitic arthropods, but it has also convinced biological control purists of the necessity of chemical pest control in many crops and managing natural enemies within the constraints of pesticide-structured agroecosystems. Major international organizations with interest in biological control are focusing attention on the efficacy of predators and parasites of pests and on the influence of pesticides on these natural control agents. Use of pesticides selectively, mostly through ecological manipulations and through physiological mechanisms, is improving. The genetic improvement, colonization, and management of pesticide-tolerant and resistant predators and parasites has come to merit a comparable footing with classical biological control and conservation/augmentation as fundamental goals of the discipline.

This book integrates research findings from numerous fields that focus on the interaction of pesticides with entomophagous arthropods. Emphasis is on those characteristics that make natural enemies unique in their responses to chemical toxins. This volume treats the history of research; susceptibility assessment; lethal, sublethal, and ecological effects of pesticides; selectivity; resistance; and resistance management. The goal is to discuss conservation of natural enemies through the use of pesticides in selective ways and the use of physiologically selective pesticides. This must be done while achieving necessary control of pests with pesticides and other pest control measures.

Appreciation is expressed to the graduate students who constructively criticized the book, including A. Knight, D. Sewell, R. Miller, K. Theiling, L. Flexner, R. Messing, S. Booth, S. Harwood, D. Carmean, and M. Arshad. I thank them for their input and blame none of them for the faults of the book. The editing of especially Karen Theiling and Russ Messing was most helpful. Kevin Currans helped me with the computer database and graphics programs. Karen Theiling developed many of the figures and Karen and my daughter Marnie worked extensively on the bibliography. Financial support came from USDA Western Regional Project W-161 on IPM, and an EPA project on Risk Assessment of Microbial Pesticides to Nontarget Organisms. Finally, I thank Candy Croft for encouragement to see this volume through to completion.

BRIAN A. CROFT

Corvallis, Oregon
July 1989

CONTENTS

PART ONE

INTRODUCTION AND SCOPE

1

NATURAL ENEMIES AND PESTICIDES: AN OVERVIEW

1.1. INTRODUCTION

Modern agriculture has come to rely extensively on synthetic chemical pesticides for pest control. Thousands of compounds from dozens of chemical classes have been developed by industry to control a wide range of pests, including insects, mites, nematodes, rodents, weeds, and bacterial, viral, and fungal pathogens. Although these toxins are targeted at plant pests, many of them are broad-spectrum biocides that have profound effects on nontarget species in agricultural ecosystems. Even the recently developed biorational pesticides, which are based on natural products and are more host- or pest-specific, can have far-reaching side effects. Biorational pesticides include agents such as microbial insecticides, insect growth regulators, chitin inhibitors, and probably many future genetically engineered pesticides.

Arthropod predators and parasitoids are the most important naturally occurring biological control agents of insect and mite pests in most crop ecosystems. Many of these entomophagous species have biologically adapted to their hosts or prey and have become efficient exploiters of herbivore populations. Even in intensively structured, highly simplified agroecosystems, they often provide complete or partial biological control of plant pests. Pesticides are often disruptive to trophic relationships involving these beneficial species. Consequently plant pest populations may increase to more damaging levels than occurred before treatment. Because of basic physiological similarities between arthropod pests and their natural enemies, pesticides often inflict severe mortality on both groups of organisms. This is especially true for those general toxins that act as nerve poisons. Nerve poisons comprise the vast majority of insecticides and acaricides in use today.

In addition to their direct impact, pesticides often disrupt trophic relationships by their toxic effects on associated species in the community, including competitors, hyperparasites, and alternate hosts or prey of natural enemies. As noted, some of the practical consequences of these disruptions are outbreaks of primary and secondary pests, increased pest control problems, and difficulties in establishing biological controls (Newsom et al. 1976, Flint and van den Bosch 1981, Velasco 1985).

Pesticides influence the biology of natural enemies in even more subtle ways. These species may experience sublethal effects on development or behavior. Fecundity, fertility, rate of development, and survivorship may be altered. Behaviors such as host or prey finding and general mobility may also change (Croft 1977; Chapter 7).

The impact of pesticides on arthropod communities may extend over long time periods and large areas. Pesticide induced perturbations in a given habitat can last for several months or years, until the delicate numerical balance needed for biological control or pest regulation is reestablished (Chapter 8). If pesticides eliminate natural enemies within a crop, pest populations may increase and subsequently emigrate to surrounding habitats. Pests may then damage crops at a considerable distance from the site where the actual chemical application took place (Vickerman and Sunderland 1977, Lopez and Morrison 1985; Chapter 8). In complex ecosystems, pesticides may cause disruptive responses in peripheral species that were not treated at all, but were sensitive to the overall ecological balance.

Because of the negative consequences associated with pesticide use, these toxins and biological control organisms have long been considered incompatible. This became particularly evident following the development of DDT and other broad-spectrum pesticides. As a result, entomologists often were split into opposing camps—those favoring chemical control methods and those favoring biological controls as the primary means to achieve pest suppression. In time it was realized that problems were associated with either of these classes of pest control methods. Often biological controls were not effective enough and yet pesticides were overly disruptive to nontarget species.

The integrated pest management (IPM) concept was developed essentially as a response to the incompatibility of pesticide and biological controls. From the beginning, its primary emphasis was to integrate these two pest control measures (Ripper et al. 1949, Ripper 1956, Bartlett 1956, 1964). Modern IPM is based on an understanding of the necessary interrelatedness of pesticides and natural enemies. Pest management seeks to exploit the more subtle interaction between biological and chemical pest control agents in the development of selective pesticides. One of the goals of this book is to highlight research on pesticide/natural enemy relationships and to show how this type of study may improve IPM.

1.2. KNOWLEDGE OF PEST/NATURAL ENEMY RESPONSES TO PESTICIDES

Much less is known about the effects of chemical pesticides on predators and parasites than on herbivorous pests (Croft and Brown 1975). This is understandable in light of the fact that pests are the primary objects of pest control activities, whereas the contributions of natural enemies to pest control and crop loss prevention are only sometimes recognized.

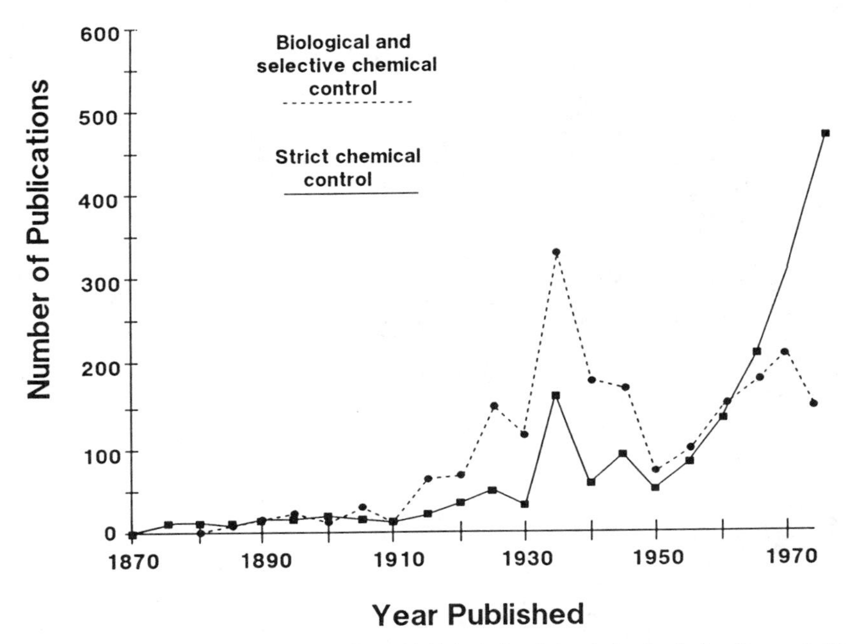

Figure 1.1. Worldwide literature citations dealing with biological and chemical control of codling moth (*Cydia pomonella*) since 1870 (taken from the bibliography of Butt 1975).

Susceptibility, resistance, sublethal effects, and other pesticide impacts on pests are commonly documented in the applied pest control literature (Brown 1978b, Metcalf 1980). In fact, during some eras this type of research almost exclusively dominated the literature, although this was not always the case. During the early years of agricultural research, a more balanced perspective between biology and chemical control of pest arthropods was maintained. However, from 1940 to 1960, pesticide impact evaluations on pests were the single most common type of literature published in applied entomology. This research was often conducted in preference to more basic study of the biology and biological control of these species.

An example of the changing emphasis in pest control is illustrated in a plot of data from the global bibliography of Butt (1975). This document contains literature on the biology (it also includes selective chemical control) and strict chemical control of the common apple pest, the codling moth *Cydia pomonella* (Fig. 1.1). From 1870 to 1910, a reasonable balance between biological and chemical control research was maintained on this species. Both areas of investigation were deemed equally necessary. Research into chemical control may have been limited by the efficacy of pesticides available during this period (e.g., heavy metal poisons, botanicals, and other inorganic products). From 1920 to 1950 almost 70% of all published research featured chemical control of the codling moth (Fig. 1.1). Since 1955, proportions have become more equal, and since the late 1960s, research with a biological or biological control emphasis has predominated.

These literature trends for *C. pomonella* are much like those cited by Metcalf (1980), who surveyed the U.S. *Journal of Economic Entomology* to obtain a similar index for a wide spectrum of economic pests. He noted that "those [publications] dealing with chemical

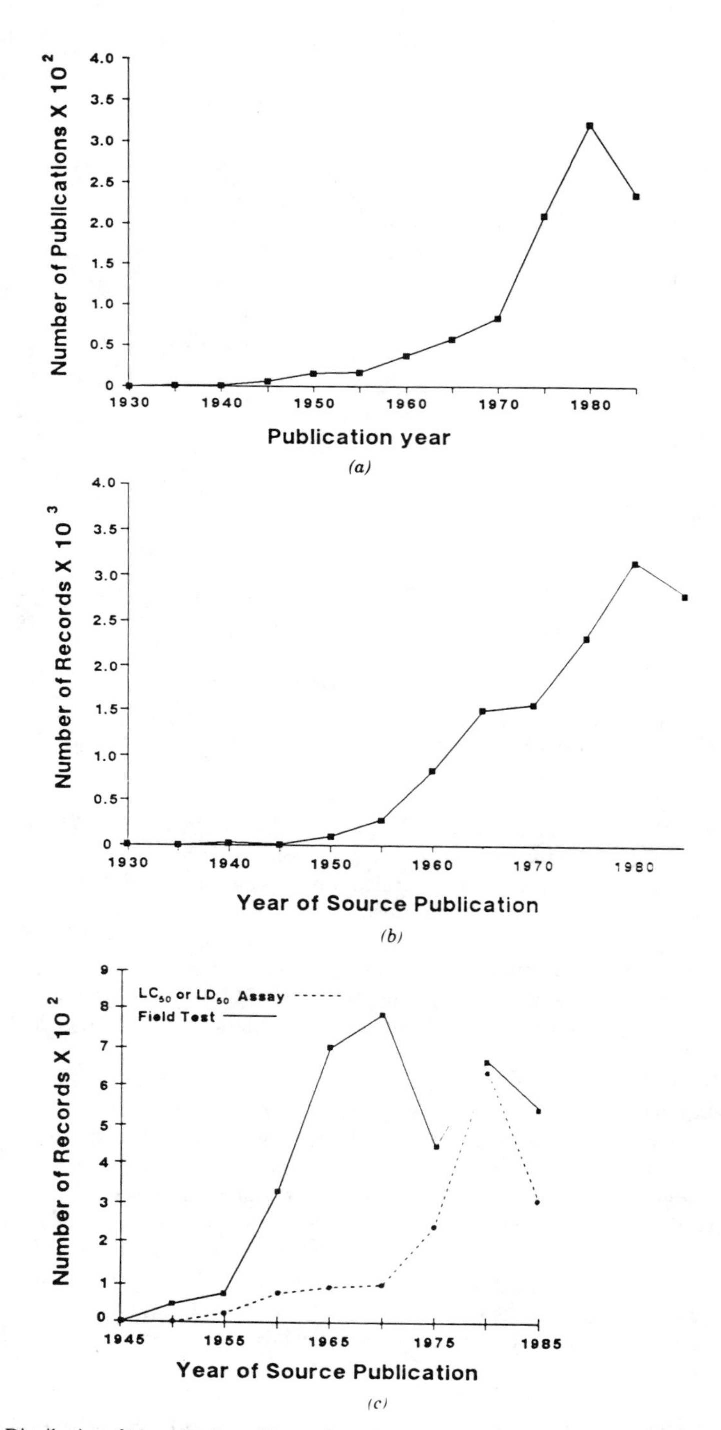

Figure 1.2. Distribution of (a) publications, (b) records, and (c) LC_{50} and LD_{50} versus field assessments of pesticide impact on arthropod natural enemies (1940–1985) (adapted from Theiling 1987).

evaluation and usage [of pesticides for pest control] comprised 76% of all the formal papers in 1950, 68% in 1960, 45% in 1970, and 43% in 1978." Presumably research in the 1980s has more heavily emphasized biological control.

In contrast to the trends for pests (Fig. 1.1), the number of publications (Fig. 1.2a) and records (Fig. 1.2b) on the impact of pesticides on natural enemies (i.e., those specific reports of pesticide side effects from a worldwide database; Chapter 2*) show an exponential increase beginning in the late 1950s. They show a particularly high rate of growth in the 1970s and 1980s (Figs. 1.2a and b). The decline in trends for the period 1980–1985 is due to incomplete sampling of the recent literature, rather than an actual decline in the numbers of records or publications (Theiling 1987; Chapter 2). The increase in pesticide impact research for natural enemies during the 1960s and 1970s occurred with a 20–30-year time lag following a similar increase for pest species (Fig. 1.1; Metcalf 1980).

Until the mid-1960s, observations of pesticide impact on natural enemies tended to be incidental. They usually were taken during pesticide evaluations on pests rather than resulting from direct studies on natural enemies (Croft and Brown 1975). However, a few in-depth field studies on the side effects of pesticides on natural enemies were conducted during this period (see reviews of Ripper 1956, Bartlett 1964). In the 1960s, a shift in perspective began that greatly expanded into the 1970s and 1980s. Studies of natural enemy responses to pesticides became more numerous and more specific. This shift is reflected in a plot of LD (lethal dosage) or LC (lethal concentration) pesticide response evaluations over time (Fig. 1.2c; Chapter 2). Median lethal tests involve more precise laboratory assessments with statistical analyses. This type of test has been increasingly employed in recent years, whereas field assessments which tend to reflect less direct and precise studies of natural enemies have leveled off (Fig. 1.2c).

Research in the 1970s and 1980s has examined in more detail the behavioral responses of natural enemies to pesticides (Croft and Brown 1975, Croft 1977, Kirknel 1974, Kurdyukov 1980). This emphasis coincides with the acceptance of IPM as the philosophy of pest control in many countries throughout the world. IPM places a high premium on the conservation of natural enemies through the development and use of selective pesticides.

The study of pesticide impact on both pests and natural enemies has been influenced by factors other than the changing emphasis in pest control. Increased understanding of basic arthropod ecology, physiology, toxicology, biochemistry, and biological control has contributed much to expand the information base available for pesticide impact research.

1.3. BOOK OBJECTIVES

This volume presents progress made in the study of the arthropod natural enemy responses to pesticides at many different levels. The primary focus is confined to predators and parasitoids of pests of agricultural crops. Discussion of pest responses to toxins is secondary. It is included to aid in understanding the complex nature of pesticide interactions between pests and natural enemies. Also, selective pesticide development, one of the primary goals of all natural enemy/pesticide research (Chapters 9 and 10), requires the pest perspective. Numerous other ecological groups of species of secondary importance may be influenced by pesticides, including hyperparasites, omnivores, and

* *Editor's note*: Chapter citations refer to chapters in our present book (Croft, 1989).

other vagrant species. The impact of pesticides on these species is discussed in Chapters 8 and 21.

Some of the more specific pesticide/natural enemy relationships that are examined are: susceptibility, sublethal and ecological effects, physiological, ecological, and integrated selectivity, and resistance development and management. Different chapters span functional levels ranging from cellular toxicology and biochemistry to regional population dynamics and community ecology of natural enemies and associated species. The goal is to integrate knowledge into a more complete understanding of natural enemy/pesticide interactions. This information may ultimately contribute to more effective IPM systems.

1.4. PERIODS OF NATURAL ENEMY/PESTICIDE RESEARCH

The development of scientific inquiry on the responses of natural enemies to pesticides can be divided into several historical periods. As discussed by Kuhn (1975), new scientific paradigms emerge from time to time due to revolutions in thinking. Certainly pest control has undergone several eras of contrasting approaches. These eras are closely associated with trends in pesticide development from the late 1800s to the present. In several cases, a new era was stimulated by an impending crisis in the current pest control system. For example, the development of resistance to pesticides often influenced the shift from one era to another (e.g., from DDT to the organophosphate and carbamate insecticides; Chapter 21). The discovery of other negative side effects such as those on birds and other wildlife sometimes precipitated movement from one chemical pest control regime to another (e.g., from lead arsenate to DDT; Chapter 21). The historical perspective of natural enemy/pesticide research presented in this chapter is undoubtedly influenced by current attitudes on pesticide use and IPM. Furthermore, developments did not proceed in as continuous a manner as the following narrative may imply.

1.4.1. The Inorganic/Natural Products Era (1870–1944)

Prior to the development of DDT in the late 1930s and early 1940s, pesticides used for agricultural pest control were derived mainly from inorganic heavy metals (e.g., lead arsenate) or they were obtained from naturally occurring organic plant toxins (e.g., nicotine, ryania, rotenone). In general, these products provided moderately effective pest control, but they did not cause such excessive disturbances in pest populations as did later synthetic organic pesticides (Chapter 2).[1] In retrospect, we have discovered that these agents were less severely toxic to natural enemies than were later pesticides (Chapters 2 and 9). Certainly, our abilities to detect their effects on natural enemies were less well developed than they are today.

The taxonomic relationship between many beneficial predators and parasitoids and pest species is fairly close—in some cases within the same family. During this era, there was little reason to suspect (nor a great deal of experimental evidence to prove) that arthropods with different feeding ecologies might respond differently, either physiologically or ecologically, to these exogenous toxins.

Some observers during this period noted that pesticide applications destroyed more natural enemies than pests. These researchers were primarily concerned about the occasional pest resurgences associated with pesticide use (Ripper 1956, Bartlett 1956). As

[1] Our knowledge of the ecology of biological control agents was not as well developed at that time, so the impact of these pesticides may have been greater than was generally indicated in the literature.

early as the late 1800s, the prominent entomologist J. H. Comstock (1880) noted that pyrethrum insecticide sprays applied for scale pest control were counterproductive due to their destructive effects on beneficial insects. He observed that the pests rather than the natural enemies returned first to treated plots, often reaching even more damaging levels than existed before treatment.

Koebele (1883), one of the early pioneers in biological control, noted that spraying California red scale [*Aonidiella aurantii* (Maskell)] also killed parasitic wasps which attacked the scale. Those wasps not killed directly left the orchard in search of food. In a few months the trees were again reinfested with red scale, natural enemies were absent, and retreatment of the orchard was required.

One early comment about insecticide selectivity was made by E. M. Green (1917). He noted:

> While wishing to kill a few individuals of species of *Lecanium* for preservation as cabinet specimens, I subjected them to the fumes of strong hydrocyanic acid gas (in an ordinary cyanide bottle) for a period of eighteen hours. In spite of this drastic treatment, large numbers of living chalcids emerged from the bodies of the coccids after their removal from the killing-bottle. This immunity of internal parasites may have a bearing upon the treatment of scale insects in the field, when they are known to be infested by useful parasites. It would now appear to be possible to fumigate scale-infested trees, with a fair probability that any chalcid parasites that may be developing within the bodies of the coccids will escape—to carry on their useful work elsewhere.

These cases typify early reports of natural enemy/pesticide interactions. They were primarily observational and usually associated with evaluations of pest control effectiveness. It is unfortunate that suggestions like Green's were not heeded to any significant degree at the time. It was not until much later that major efforts were undertaken to conserve natural enemies through the development of selective pesticides or the use of broad-spectrum pesticides in selective ways (Chapters 9–12). In the interim, the increasing incidence of pest outbreaks following pesticide treatment prompted some scientists to raise questions about the basic physiology and ecology of key natural enemies (Ripper 1956, Bartlett 1964). Were predators and parasites intrinsically more susceptible to pesticides than their hosts or prey? Could beneficial natural enemies develop resistance to pesticides as did pests? Could biological differences between pests and natural enemies be exploited to achieve selectivity?

Comparative data employing common methods of evaluation of susceptibility for both groups were limited at that time. Those few studies that were available indicated that predators and parasites were at least equally or possibly more susceptible than pests to common pesticides of that era (see summary in Ripper 1956). During this period, considerable attention was focused on the ecological effects of pesticides. Eventually, these observations would stimulate study of such factors as the modification of reproductive rates caused by pesticides (i.e., sublethal effects, hormoligosis; Chapter 7), the more rapid reinvasion of treated habitats by pests than natural enemies following pesticide application, and related disturbance in the synchrony between prey or host and natural enemy caused by pesticides (Fleschner et al. 1955, Fleschner and Scriven 1957, DeBach and Bartlett 1951, Pickett et al. 1958, Ripper 1956; see also Chapter 8).

1.4.2. The Synthetic–Organic Pesticide Era (1945–1960)

The development of DDT in 1939 and its release for civilian use in agriculture in 1945 signaled the start of a dramatic new era in insect pest control. Outstanding successes against mosquitoes, lice, and other disease vectors were followed by convincing demonstrations against the Oriental fruit moth *Grapholitha molesta*, California red scale *Aonidiella aurantii*, European corn borer *Ostrinia nubilalis*, codling moth *C. pomonella*, and other phytophagous insect pests. The success of DDT and other chlorinated hydrocarbons (e.g., BHC, aldrin, dieldrin, endrin, chlordane, etc.) was followed shortly by the commercial development of potent organophosphates, including HETP, parathion, and schradan, the first systemic insecticide. In the late 1950s, the carbamates were introduced, including the widely used compound, carbaryl. Between 1945 and 1953 about 25 new pesticides were introduced for crop protection worldwide (Perkins 1982).

This powerful arsenal of new, effective chemicals and the rapid registration of new compounds led to a shift in entomological research largely toward the evaluation of pesticides for their direct mortality against pests. The entire focus of economic entomology began to center on this approach to pest control and many other long established practices were neglected, among them biological control (see Fig. 1.1). The shift in emphasis was so dramatic that some contended (e.g., Perkins 1982) that "biological control in the U.S. nearly died as a recognizable field of research from the 1940s through the 1960s; the enormous infatuation with insecticides was clearly the major cause precipitating its near demise."

Needless to say, side-effects testing of insecticides against predators, parasitoids, and other nontarget insects was, for the most part, ignored. Despite this general trend, there were a few entomologists with foresight who continued to focus on natural enemies or on the interaction of pesticides with natural enemies. A strong contingent of biological control workers (in several laboratories) maintained an emphasis on predators and parasitoids as a primary factor in plant protection.

It is somewhat paradoxical that the importance of arthropod natural enemies in regulating or controlling some of our most troublesome pests was most conclusively proven by biological control specialists working with pesticides during this era (DeBach 1964, Huffaker 1971, Huffaker and Messenger 1976). Some of the best evidence for the role of natural enemies in curbing pest populations came from serendipitous observations of pest outbreaks resulting from the pesticide destruction of natural enemies. These observations eventually lead to the development of the insecticide check method for evaluating natural enemy effectiveness (DeBach and Bartlett 1964). This technique continues to be widely used today (e.g., Swift 1970, Edwards et al. 1979, Robertson et al. 1981, Sterling et al. 1984, Lim et al. 1986, Braun et al. 1987a, 1987b). In the insecticidal check method, paired replicated plots are used. One group is treated and one group is left untreated with a nonselective pesticide, which will exclude natural enemies while having lesser effects on phytophagous species. By comparing the treated and untreated plots, the level of biological control provided by natural enemies often can be qualitatively estimated.

The integration of biological and chemical controls was also pioneered by early IPM specialists. While pesticide use was increasing exponentially in the late 1940s and 1950s, they recorded a number of cases in which unwitting destruction of natural enemies was causing pest outbreaks. Tree fruit entomologists in Nova Scotia, Canada, were among the first to directly study conservation of beneficial predators and parasitoids by modifying the pesticide input to a crop (Pickett et al. 1958). To a large extent, their shift to an emphasis on biological controls and fewer insecticide inputs was brought on by a loss of their European fruit markets and the need to reduce production costs (Whalon and Croft 1984). After proving the efficacy of entomophagous insects, particularly against secondary pests in unsprayed apple orchards, they learned to rely on selective applications of pesticides to control codling moth, apple maggot, and other pests, thus sparing natural enemies. Another early contributor to IPM in field and orchard crops in California was A. E. Michelbacher (Michelbacher and Hitchcock 1958, Michelbacher 1962), who emphasized the maintenance of favorable natural enemy/host or prey ratios through use of nondisruptive pesticides.

Eventually, the observation that pesticides could induce pest outbreaks led to further studies of pesticide selectivity (Ripper 1956; Chapters 9–12). The need for pesticide selectivity to mammals had long been recognized because of human health side effects. However, studies of arthropod (pest) to arthropod (natural enemy) selectivity were not pursued extensively until the late 1940s and 1950s in connection with early IPM programs (e.g., Pickett et al. 1958).

In Ripper's (1956) review, he proposed the terms *ecological* and *physiological selectivity*, which are useful ways to classify pesticide selectivity. Physiological selectivity implies differences in innate activity between pests and natural enemies in response to direct contact with a toxicant. Ecological selectivity refers to differential exposure of pests and natural enemies in time and space through deliberate placement of the pesticide or exploitation of phenological or behavioral differences between these organisms, or both (Winteringham 1969). The principal type of selectivity achieved in practice during the early era of pesticide development was, and remains today, ecological rather than physiological (Newsom et al. 1976; Chapters 9 and 10).

The first documented case of pesticide resistance in arthropods was that of San Jose scale pest to lime sulfur in Washington apple orchards (Melander 1914), but it was not until the chlorinated hydrocarbons were extensively used that the number of resistant pests increased dramatically and the magnitude of the problem became widely appreciated (Brown 1971, Georghiou 1972, 1986). As with most other aspects of agricultural entomology, problems that occurred and the questions that they raised were addressed first for herbivorous pest species, and only much later for entomophagous natural enemies.

By the 1950s, some scientists questioned whether, like pests, beneficials would evolve resistant populations (Pielou and Glasser 1951, Spiller 1958, Wilkes et al. 1952). Some natural enemies were closely related to many insect and mite pests that had previously developed resistant strains (Brown 1958). Early studies to select resistant natural enemies in the laboratory produced only modest increases in resistance, which usually were unstable in both the laboratory and the field (reviewed in Croft and Brown 1975; Chapter 19). Only a few cases of naturally occurring resistance to insecticides in predators and parasites in the field were thought to have evolved during this time (1950–1970; e.g., Johansen 1957, Spiller 1958, Hoyt 1969a). None were carefully documented by using susceptible strains for comparison (Chapter 14).

Toward the end of this era, the ecologically disabling consequences of pesticides, including pest resistance, resurgence, secondary pest outbreaks, and environmental contamination, became too prevalent to ignore. These negative aspects of unilateral pesticide dependence largely contributed to and ushered in the IPM era of pest control in the late 1950s to early 1960s.

1.4.3. The IPM Era (1961–1980)

At its inception, IPM had the goal of integrating chemical and biological pest control (Bartlett 1956, Pickett et al. 1958, Perkins 1982). This concept was later expanded to include sampling, economic injury levels, economic thresholds, and various other alternative pest control measures (Stern et al. 1959). Emphasis shifted from insect eradication to ecosystem management, and a more refined ecological perspective was incorporated into plant protection. Aspects of population and community ecology became more widely appreciated in relation to such phenomena as pest and natural enemy immigration and emigration. The influence of pesticides on natural enemies was a major focus in all of these studies, as was the integrated use of these biological agents in combination with pesticides.

Early field researchers noted that systemic toxicants were often taken up from plants by phytophagous insects (Ripper 1956). In some cases, these poisons were nontoxic to entomophagous predators and parasites that fed upon pesticide-laden prey or hosts. Under these circumstances, ecological selectivity was achieved through the use of systemic pesticides. In other cases, systemic compounds were extremely toxic to beneficial species by food chain uptake (e.g., McClanahan 1967; Chapter 4), inducing pest outbreaks. Indirect effects of toxicants mediated through the food chain focused attention on the distribution and detoxification of pesticides in arthropods. Consideration was given to these modes of pesticide uptake by natural enemies (Croft 1977; Chapter 3). More specifically, this research emphasized the unique attributes of natural enemies relative to pests in terms of pesticide exposure, and led to a greater examination of the toxicology and biochemistry of pesticides in natural enemies (Chapter 6).

As noted earlier, a few physiologically selective pesticides were developed for use on major crops during the early stages of IPM development. As time went on, a few added cases were identified, but the overall rate of development of physiologically selective pesticides has been slow compared to broad-spectrum insecticides. Physiological selectivity has been most demonstrated by some of the acaricides (e.g., cyhexatin, propargite) and a few aphicides (e.g., pirimicarb, endosulfan). Few broadly selective insecticides have been identified for the management of large pest/natural enemy complexes (see Chapters 9–13). Generally, economic constraints have limited the industrial development of very specific, physiologically selective compounds for widespread commercial use (see Chapters 13 and 21).

Following the shift to a more ecological perspective in IPM, researchers began to look beyond the direct toxic effects to the more indirect and sublethal influences of pesticides on natural enemies (Fleschner and Scriven 1957, Adams 1960; Chapter 7). Whereas pest responses to pesticides had been largely measured by simple mortality studies, natural enemy responses were considered more complex, and potentially could influence populations over longer time periods. It was recognized that both direct and indirect influences of pesticides could affect the more subtle aspects of predator/prey (or parasite/host) relationships and could eliminate the impact of biological control agents in and beyond the treated environment (Croft and Brown 1975, Croft 1977). Brown (1977) proposed the phrase "harmlessness of pesticides" to describe the perspective needed in evaluating pesticide impact on natural enemies. By this he meant that one usually focused on kill potential with pests, but with natural enemies research on more subtle effects of pesticides must be considered, since the emphasis on the latter was on conservation.

Early reports of sublethal effects of pesticides on natural enemies were largely observational: for example, "the parasite was unable to locate its host in the presence of the toxicant" or "it oviposited more eggs on treated trees than on untreated trees" (e.g., Fleschner and Scriven 1957). As a result of more directed laboratory studies, effects on fecundity, fertility, development, survivorship, mobility, and predation or parasitism were measured (Croft and Brown 1975, Croft 1977; Chapter 7). These influences were difficult to relate to the population dynamics of the organism because of the complex behavioral processes involved. One could only speculate as to what the sublethal impact of pesticides in the field might be.

During the IPM era, a new emphasis in research on pesticide resistance in natural enemies began to emerge. To date less than 30 cases of resistant natural enemies are known (Croft and Strickler 1983; Chapter 14). The first well-documented cases of field-developed resistance were for predatory mites in the early 1970s (Motoyama et al. 1970, Croft and Jeppson 1970). Study of the genetics (Croft et al. 1976a) and toxicology of resistance (Motoyama et al. 1972) for these species soon followed. Within the Insecta, most resistance cases have been documented in the Coccinellidae, Chrysopidae, Cecidomyiidae, Braconidae, Aphelinidae, and Trichogrammatidae. However, genetic and toxicological study of resistance has been limited to predatory mites (Phytoseiidae) and to *Chrysoperla carnea* (Chrysopidae) (Hoy 1985; Chapters 14–16). Parasitoid species have exhibited the fewest cases of resistance and have been the least amenable to genetic improvement through hybridization and artificial selection of pesticide resistance (Chapter 20). Even relatively recently, researchers have questioned whether this group of arthropods has the potential for developing high levels of resistance to pesticides (Hoy 1987, Rosenheim and Hoy 1988; Chapter 20).

The disproportionality of resistance among pests (over 400 cases; Georghiou 1986) versus natural enemies may indicate either that there are major differences in the evolution of resistance between the two groups or that detection of resistance is far more effective for pests than for predators and parasites. The relative paucity of documented cases of resistant natural enemies is probably a result of several factors (Chapter 14). Croft (1977) has discussed the less dramatic effects of natural enemy resistance development in the field (as compared with pests), the difficulties in rearing natural enemies for resistance screening, and the absence of standard methods available for testing resistance in natural enemies as possible explanations for these differences (Chapter 14).

Historically, there have been some differences of opinion as to the number of cases of evolved resistance in arthropod predators and parasites. For example, Newsom (1967, 1974) believed that "evidence from field experience strongly suggests that resistance in natural enemies is more common [than is generally believed] and probably has evolved to the most widely used insecticides." Georghiou (1972), however, concluded that "resistance in natural enemies is severely hindered and lags far behind that of their prey or host species." Until the last decade, little definitive research has been conducted to address this question.

Recent research has conclusively proven that many beneficial predators and some parasites can develop at least low levels of resistance to most conventional pesticides. Scientific investigation has also identified some basic limitations to the development/evolution of pesticide-resistant natural enemies (Morse and Croft 1981, Tabashnik and Croft 1982, 1985). Furthermore, the biochemistry, toxicology, genetics, and ecology of resistance in predators and parasites has begun to receive increasing attention among basic and applied entomologists (Chapters 14–16).

1.4.4. Current Trends (1981–1987)

In recent years, progress has been made in a number of important areas of natural enemy/pesticide research such as susceptibility assessment, ecological impacts of pesticides, selective pesticide development, resistance evolution, and natural enemy resistance management (Chapters 8–20). These advances have aided in the continued expansion of IPM research and its implementation. Other areas of scientific progress include more complete documentation of the spatial and trophic level effects of pesticides in populations of natural enemies and pests (van den Bosch and Stern 1962, Brown 1978b, Flint and van den Bosch 1981). Basic conceptual models of these phenomena are reasonably well developed (Flint and van den Bosch 1981; Chapter 8).

Monitoring and surveillance of natural enemy susceptibility to pesticides is another area of research which received increasing attention (Figs. 1.2a and b). Beginning in the 1970s, development of standard methods for assessing side effects of pesticides on natural enemies has flourished, especially in Europe (Hassan 1985, Hassan et al. 1983, 1987; Chapter 5). Standardized procedures and pesticide screening data are now available for 30–40 major beneficial species (Chapter 5). The United States Environmental Protection Agency (USEPA) has begun to develop provisional guideline tests for assessing the impacts of microbial pesticides on several major classes of arthropod predators and parasitoids. At least two standardized resistance monitoring techniques have been proposed for natural enemies (Chapter 5). These types of monitoring tools for pests were generally established much earlier during the 1950s and 1960s.

The toxicology of selective pesticides has been stimulated by a broader interest in the ecology and effect of plant allelochemicals at different trophic levels, especially in relation to herbivory versus entomophagy (e.g., Krieger et al. 1971, Thurston and Fox 1972, Brattsten et al. 1977, Campbell and Duffey 1981, Barbosa et al. 1982, 1986). Several studies have shown that pests and natural enemies may have different types and amounts of detoxification enzymes for handling xenobiotic natural compounds and pesticides. These studies have been done with natural enemies occurring on such diverse crops as cotton (Plapp 1981), apple (Mullin et al. 1982), several deciduous fruit species (Croft and Mullin 1984), soybeans (Yu 1987), and several other crops (Mullin and Croft 1984).

Among field entomologists, attitudes toward insecticide selectivity needs in IPM programs have changed over the past decade (Metcalf 1980, 1982, Croft 1982, Hull and Beers 1985). One change has come in the degree of selectivity desired in pesticide when managing the complexes of arthropod pests and natural enemies on heavily sprayed crops such as cotton, vegetables, or apples (Chapter 9). In the early days of IPM, species-specific selectivity was often sought as an important attribute of a pesticide. Today, compounds exhibiting broader selectivity, which discriminate among major groups of species (e.g., key pest and natural enemy species), are more desirable for managing large arthropod complexes with only a few compounds (Croft 1982). The terms broad, broad-spectrum, or trophic-level selectivity convey this concept in which selectivity (i.e., in these cases both ecological and physiological) is operative at the ecological level of affected populations (Chapters 9 and 13).

Interest in more selective pesticides is demonstrated by the increased development of compounds based on models of naturally occurring toxins. Biorational pesticides are usually less broadly biocidal and more finely tuned to the physiology of the target species (Menn and Henrick 1981). These agents include natural plant metabolites, microbial pest control agents (MPCAs), insect growth regulators, and behavior-modifying chemicals (Chapters 11, 12, and 21). Increasing effort has gone into evaluating the impact of these agents on nontarget species, including predators and parasites (e.g., Beckage 1985, Flexner et al. 1986). Generally, these agents appear to be safer to arthropod natural enemies than earlier synthetic–organic products (see Chapter 2). This improved selectivity has caused the USEPA to consider modifying some types of registration requirements for the development of biorational agents as compared to conventional pesticides (Flexner et al. 1986; Chapter 11). In some cases, extreme specificity in biorational pesticides has limited their commercial development and use.

Genetic engineering technology may expand the activity spectrum of biorational pesticides for both pests and arthropod natural enemies (Kirschbaum 1985; Chapters 19, 20, and 21). The cloning and transfer of resistance genes from pests to natural enemies is another means to develop physiological selectivity in natural enemies (Hoy 1985, 1987, Berge et al. 1986). Ecological selectivity may be achieved through biotechnology by incorporating genes which produce known insect toxins into commercial crop plants (e.g., the beta endotoxin of *Bacillus thuringiensis*; Comai and Stalker 1984). Because of the potential for altering the activity of pesticides through biotechnological methods, it is believed that these tools will be the source of vast improvements in selectivity (Menn and Henrick 1981, Flexner et al. 1986).

Pesticide resistance has long been considered as a means to confer physiological selectivity to beneficials over pests. Resistant field populations of beneficials or genetically improved strains developed through hybridization, artificial selection, or biotechnological methods allow for increased biological control in treated habitats (Chapters 13–19). Management of resistance in these species is becoming an increasingly important component of IPM programs (Croft 1982, Hoy 1985). During the past two decades resistance management of natural enemy species has been widely exploited on certain tree crops such as apples and almonds and, to some extent, in greenhouses and on citrus in several countries (Chapters 17–20). Experimental and theoretical studies of factors influencing insecticide resistance (Chapter 16; Georghiou and Taylor 1976, Morse and Croft 1981, Tabashnik and Croft 1982, 1985, Tabashnik 1986a) have also contributed much to our understanding of comparative responses of pests and natural enemies to pesticides. Many physiological and ecological differences between these groups of species have been identified as being important in explaining differences in the frequency of resistance observed between them (Chapter 16).

1.5. SUMMARY

Research on natural enemy/pesticide relationships is primarily a product of the IPM era of pest control. Novel concepts and approaches to achieving selectivity in pesticide use are receiving greater attention (Hull and Beers 1985, Greenhalgh and Roberts 1987). New attitudes and policies affecting the development of selective pesticides are being developed at many public and private sector levels. Governmental regulatory agencies in several countries are giving higher priority to the inclusion of side-effect data on natural enemies as part of the requirements for new product registration (e.g., Franz 1974, USPEA provisional guidelines; see Chapter 5). Industry is providing data more frequently for pesticide impacts on selected beneficials associated with high-value crops. On a few crops, the development of new pesticides is being strongly influenced by selectivity requirements for natural enemies (e.g., spider mites and phytoseiid mites on deciduous tree fruits). Resistant beneficial species provide additional means to achieve pesticide selectivity. Much of the awareness surrounding selective pesticide development and use is being stimulated by biotechnological advances (Vaeck et al. 1987).

We are entering an exciting new era of research in which there will be a greater emphasis on selectivity to natural enemies over pests. This will include the development and greater use of physiologically selective pesticides. Improved systems of selective timing and placement with pesticides will also be involved. Future research will emphasize more basic toxicology, population genetics, and ecological responses of arthropod predators and parasitoids and their phytophagous hosts or prey.

2

PESTICIDE EFFECTS ON ARTHROPOD NATURAL ENEMIES: A DATABASE SUMMARY

2.1. INTRODUCTION

The impact of pesticides on arthropod natural enemies has been reviewed many times since IPM gained acceptance as a preferred philosophy of pest control. As might be expected, early summaries were general. They typically presented pesticide toxicity data, identified different types of side effects, and discussed ways of achieving selectivity to natural enemies (e.g., Ripper 1956, van den Bosch and Stern 1962, Bartlett 1964).

Reviews in the 1970s focused mostly on pesticide selectivity and sublethal effects (e.g.,

This chapter was coauthored by Karen M. Theiling, Department of Botany and Plant Pathology, Oregon State University, Corvallis, Oregon.

Newsom 1974, Newsom et al. 1976, Croft and Brown 1975, Croft 1977, Sukhoruchenko et al. 1977). In the 1980s, summaries were published on standardized methods of pesticide side-effect assessment (Hassan et al. 1983) and selectivity for more specific pesticide classes, including herbicides (Isaeva 1983), pyrethroid insecticides (Croft and Whalon 1982, Niemczyk et al. 1979), microbial insecticides (Flexner et al. 1986), and insect growth regulators (Beckage 1985). Other reviews focused on pesticide impact on natural enemies from the point of view of individual crops or ecological communities (e.g., apple: Croft and Whalon 1982, Niemczyk et al. 1979; soil-inhabiting predators: Basedow 1985; and regional populations of natural enemies in the USSR: Kurdyukov 1980, Novozhilov 1984).

Recent reviews of natural enemy/pesticide research have dealt with physiological and ecological selectivity (e.g., Mullin and Croft 1985, Hull and Beers 1985), standardized side-effect assessment methods and results (Hassan et al. 1987), resistance (e.g., Croft and Strickler 1983, Fournier et al. 1985a), and genetic improvement to develop resistant strains of predators and parasitoids (Hoy 1985, 1987).

Pesticide impact research on arthropod natural enemies is summarized here as a preface to the remaining chapters of this volume, which discuss susceptibility, sublethal and ecological effects, selectivity, resistance, and resistance management. This presentation documents the amount of side-effect testing that is an on-going part of IPM research. This overview was derived from analysis of a database (SELCTV), which contains information from literature published mainly from 1950 to 1986. Seldom has database technology been used to summarize pesticide side-effect data on natural enemies (e.g., see limited example in Touzeau 1986). To our knowledge, no global review of the topic has been attempted in a database format (see Theiling 1987, Theiling and Croft 1988).

2.2. THE SELCTV DATABASE STRUCTURE AND COMPOSITION

As noted in Chapter 1, literature on natural enemy/pesticide research has increased at an exponential rate since the late 1950s. It doubled from 1975 to 1984 (Fig. 1.2), and it will likely double again by 1990 (Theiling and Croft 1988). Here, and in later chapters, summaries are drawn from data for more than 600 natural enemy species and 400 agricultural chemicals, including fungicides, herbicides, insecticides, acaricides, feeding repellents, and insect and plant growth regulators. The literature used in these analyses comes from 975 investigators working in 58 countries. Many specific analyses conducted with the database cannot be cited here due to space limitations. Other comparisons and a complete listing of the literature used in the database are available in Theiling (1987).

SELCTV was developed with relational database management software. Its record structure includes 40 fields, divided into four groupings: 1) biological, 2) chemical, 3) test and summary data, and 4) reference citation information (Table 2.1). Hierarchical groupings of fields for both natural enemies and pesticides allow for different levels of searching and analysis. These fields include NE:TYPE (natural enemy type), ORDER, FAMILY, GENUS, and SPECIES and CHEM:CLASS (pesticide class), CHEM:GROUP (chemical group), and CPD: NAME (compound name).

Data were extracted from the literature, primarily from papers published in refereed journals. Minor data sources were research bulletins and dissertations. Publications were obtained from literature searches and cross-checking bibliographies from individual articles. Western countries were well represented in the collected literature, while sources from eastern Europe, the USSR, and the Far East were more limited. Information included for the latter areas was obtained largely from abstracts and translated articles.

Table 2.1. Record Structure and Field Descriptions for SELCTV Database of Documented Pesticide Side Effects on Arthropod Natural Enemies (after Theiling 1987)

SELCTV Subdivision: Biological	
GENUS:	Taxonomic genus of tested organism
SPECIES:	Taxonomic species of tested organism
FAMILY:	Taxonomic family of tested organism
NE:ATTRIBU:	Coded field of 5 natural enemy attributes: 1) ORDER 2) NE:TYPE 3) SEX 4) SOURCE 5) SUSCEPT. STATUS
STAGE:	Life stage of organism at the time of testing
PROD:UNIT:	Crop with which natural enemy is associated
LOCI:	Location where research was conducted
HOST:PREY:	Host/prey of tested natural enemy
HP:TOX:DAT:	Host/prey toxicity data in publication? (y/n)
SELCTV Subdivision: Chemical	
CPD:NAME:	Experimental or common chemical name of toxicant
FORMULATN:	Commercial formulation or trade name
CHEM:CLASS:	Type of pesticide, e.g., fungicide (f)
CHEM:GROUP:	Chemical structure classification, e.g., carbamate or pyrethroid
CHEM:RATE:	Rate or manner of application, e.g., dose, residue
CR:UNITS:	Units in which application was measured
CR:VALUE:	Numeric value associated with rate units
SC:AI:	Standard concentration in %a.i.
SELCTV Subdivision: Test and Summary	
DUR:EXP:	Duration of actual contact with toxicant
TST:METHOD:	Method used for susceptibility assessment
EVAL:TIME:	time elapsed from initial exposure to evaluation
RESP:TYPE:	Response being measured, e.g., LD_{50}, mortality
RESP:UNITS:	Response units, e.g., %+, %−, or units of assay
RESP:VALUE:	Numerical value associated with RESP:UNITS
RR:RATIO:	Resistant strain LD_{50}/susceptible strain LD_{50}
TOX:RATING:	Common response rating scale, ranging from 1 to 5
DAT:RATING:	Scale for rating precision level and statistics
SUBLETHAL:	Effects of nonlethal doses, e.g., fecundity
SLECTRATIO:	LD_{50} of host or prey/LD_{50} of natural enemy
COMMENTS:	Any other relevant features of paper
REFERENCE Subdivision: Bibliographic	
REF:NUM:	Unique identifying number assigned to each paper
AUTHOR:	Author(s) of paper from which data are extracted
PUB:DATE:	Year of publication
TITLE:	Title of publication
JOURNAL:	Scientific or technical journal of publication
VOLUME:	Volume and number of journal
PAGES:	Page numbers of publication
LANGUAGE:	Language in which paper was published
ABSTRACT:	Paper, abstract, or both?
RECORDS:	Number of records extracted from publication
KEYWORDS:	Topics of focus within publication

Each record in the database represents one screening of pesticide impact on one natural enemy species under the conditions cited. The number of records extracted per publication ranged from 1 to 600. For literature sources in which statistical analysis of data was not performed or where replication, sample size, or treatment controls were deemed inadequate, a database field was included in SELCTV for ranking the scientific rigor or precision of the information input into the database (see DAT:RATING, Table 2.1). Data of varying precision levels can be screened for in searches or analyses (however, results presented in this chapter are taken from the entire database).

Most analyses presented in this chapter involved calculations and comparisons of average toxicity ratings (TOX:RATING) and associated variance. These calculations provide a means to evaluate trends and identify possible factors affecting natural enemy responses to pesticides. Additional analyses consisted of comparing other numeric indices, such as resistance or selectivity ratios.

2.3. A SUMMARY OF SELCTV

The contents of the database are initially characterized for most of the 40 fields (see Table 2.1) across the approximately 12,600 records. Following a discussion of time trends in the literature sources (i.e., field PUB:DATE), individual fields are discussed in their approximate order of appearance in the database (Table 2.1). Time trends for other fields are characterized over 5-year increments.

2.3.1. Characterization of Database Attributes

2.3.1.1. Reference Attributes. The distribution of published research over time (PUB:DATE) as reflected in publications and records from SELCTV was referred to in Chapter 1 (see Figs. 1.2a and b). The number of records per period began to rise after 1945, with a sharp upturn after 1955. This rate of increase was maintained thereafter, with the exception of 1965–1970. The maximum records generated per year was 896 in 1983. A decrease in records per year in 1981–1985 is an artifact of incomplete literature acquisition. (This artifact applies to other time trend figures discussed later.) Trends in the publication rate over 5-year intervals were very similar to those for records (Fig. 1.2b; Theiling 1987).

2.3.1.2. Natural Enemy Attributes. Five levels of identification of natural enemies were built into the database record structure (see Table 2.1). At an ecological or functional level, natural enemies were classified as predators or parasitoids. Predators accounted for 8993 records, or just over 71% of the database. Parasitoids made up 3591 records, or approximately 29%.

Taxonomic characterization spanned 13 orders; 10 were from the Insecta and 3 from the Arachnida. The most common order represented was Hymenoptera, accounting for over 27% of records. Coleoptera made up nearly 22% of records, followed by Acari 19.2%, Hemiptera 14.1%, Neuroptera 8.7%, Diptera 4.6%, Araneae 2.5%, Thysanoptera 1.3%, and five others.

Pesticide side-effect data spanned 88 families of Arthropoda. The 20 best represented families are shown in Fig. 2.1. Three predatory families (Phytoseiidae, Coccinellidae, and Chrysopidae) and one parasitoid family (Braconidae) have been most commonly tested. Thirteen of the top 20 families represented are generalist predators or mite predators. Among those families of predators most frequently studied were many species that are well

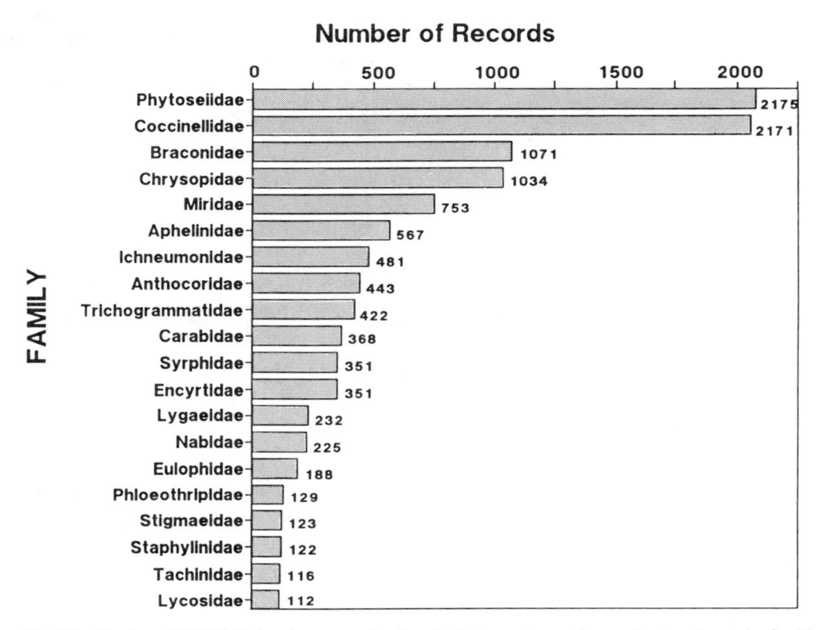

Figure 2.1. Distribution of SELCTV database records of pesticide impact on arthropod natural enemies for 20 most commonly tested families (after Theiling 1987).

documented as important biological control agents (see later identification of individual species; Huffaker and Messenger 1976, Ridgway and Vinson 1977, Hoy and Herzog 1985, Helle and Sabelis 1986). The Braconidae, Aphelinidae, Ichneumonidae, and Trichogrammatidae, in that order, were the most commonly studied parasitoid families. The value of species within these families as biological control agents of agricultural pests is also well known (e.g., *Hypera postica* on alfalfa: Dumbre and Hower 1976a, Abu and Ellis 1977, Ascerno et al. 1980; *Aonidiella aurantii* on citrus: Bartlett 1964, Bellows et al. 1985, Rosenheim and Hoy 1986; *Heliothis* spp., *Anthonomus grandis* on cotton: Plapp and Vinson 1977, Salama and Zaki 1983, O'Brien et al. 1985; several species on vegetables: Smilowitz et al. 1976, Hamilton and Attia 1977, Horn 1983).

The number of families within each order of natural enemies included in the database showed marked differences. For example, the Hymenoptera was represented by 21 families, Coleoptera by 14, and Hemiptera and Araneae by 12 each. Toxicity assessment data for eight families of Acari and seven of Diptera were also included. The remaining seven orders contained three or fewer families.

The proportion of an order accounted for by a given family has relevance when making generalizations about an order. For example, over 90% of acarine records were for the Phytoseiidae. Seven families made up the remaining 9.6%. Similarly, data for Coleoptera were 80% Coccinellidae. In contrast, none of the 21 hymenopteran families accounted for more than about 30% of records from this order.

Species ranked among the top 20 in the database and their frequencies are listed in Table 2.2. The six most commonly studied species were generalist predators or mite predators, led by *Chrysoperla carnea*. Ten of the top ranked 18 species were either phytoseiid mites or coccinellid beetles. *Encarsia formosa*, sixth ranked by species, was the

Table 2.2. Distribution of SELCTV Database Records of Pesticide Impact on Arthropod Natural Enemies for 22 Most Commonly Tested Species (after Theiling 1987)

Species	Family	No. of Records
Chrysoperla carnea	Chrysopidae	591
Amblyseius fallacis	Phytoseiidae	574
Coccinella septempunctata	Coccinellidae	440
Hippodamia convergens	Coccinellidae	375
Phytoseiulus persimilis	Phytoseiidae	323
Typhlodromus occidentalis	Phytoseiidae	268
Encarsia formosa	Aphelinidae	232
Typhlodromus pyri	Phytoseiidae	213
Trichogramma cacoeciae	Trichogrammatidae	158
Chrysopa oculata	Chrysopidae	137
Stethorus punctum	Coccinellidae	134
Metasyrphus corollae	Syrphidae	134
Phygadeuon trichops	Ichneumonidae	124
Leptomastix dactylopii	Encyrtidae	117
Cryptolaemus montrouzieri	Coccinellidae	114
Coleomegilla maculata	Coccinellidae	103
Hyaliodes harti	Miridae	102
Amblyseius hibisci	Phytoseiidae	90
Trichogramma evanescens	Trichogrammatidae	86
Opius concolor	Braconidae	85
Lycosa pseudoannulata	Lycosidae	85
Anthocoris nemorum	Anthocoridae	82

most prevalent parasitoid in the database. This aphelinid parasitoid, which attacks the greenhouse whitefly *Trialeurodes vaporariorum*, is an intensively managed species in greenhouse IPM programs. *E. formosa* has been extensively tested for susceptibility to many pesticide (e.g., Beglyarov and Maslienko 1978, Ledieu 1979, Elenkov et al. 1980, Kurdyukov 1980, Coulon and Delorme 1981, Iacob et al. 1981, Hall 1982, Hoogcarspel and Jobsen 1984). Only six of the top 22 species in the database were parasitoids, all from the order Hymenoptera. At this level of classification, only one spider (*Lycosa pseudoannulata*) appeared among the 22 species listed.

In laboratory assessments of pesticide side effects where greater precision is possible, sex of the natural enemy has occasionally been specified. In the database, this attribute was included along with four other characteristics in the NE:ATTRIBU (natural enemy attribute) field. Sex was unspecified in over 90% of database records. In 8% of records, the test organisms were female. Slightly less than 2% of records documented pesticide side effects on male natural enemies. Source of natural enemies tested was another feature recorded in the NE:ATTRIBU field. Organisms were either field collected (70%) or laboratory reared (25%). In 5% of records, the source of the natural enemies could not be ascertained.

Susceptibility status of the organism tested was designated in the NE:ATTRIBU field based on pesticide toxicity to the natural enemy. The status was assigned as *susceptible* when natural enemy mortality was greater than 10% when exposed to a pesticide at or near the recommended field rate. Approximately 71% of tests were conducted on susceptible natural enemies. *Tolerant* natural enemies were those sustaining only 10% mortality or less at these same rates. Slightly more than 26% of database records denoted the tolerant

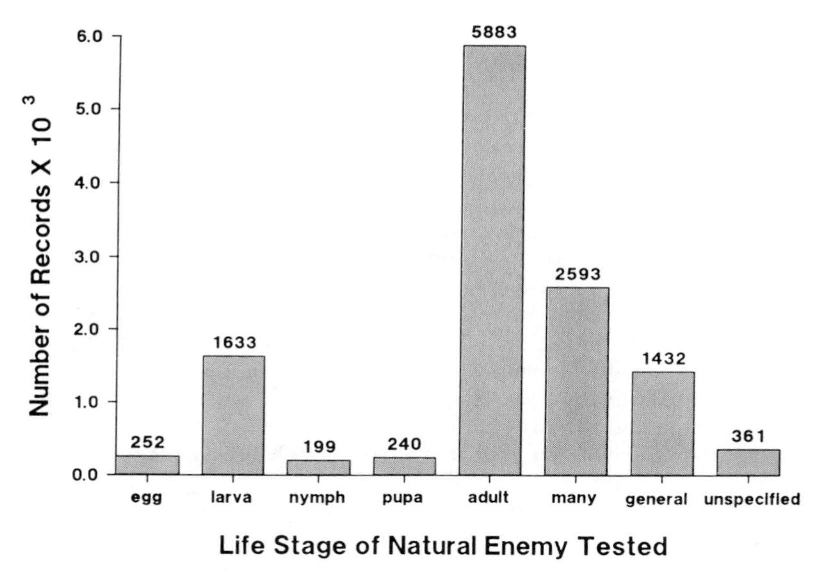

Figure 2.2. Distribution of SELCTV database records of pesticide impact on arthropod natural enemies grouped by life stage (after Theiling 1987).

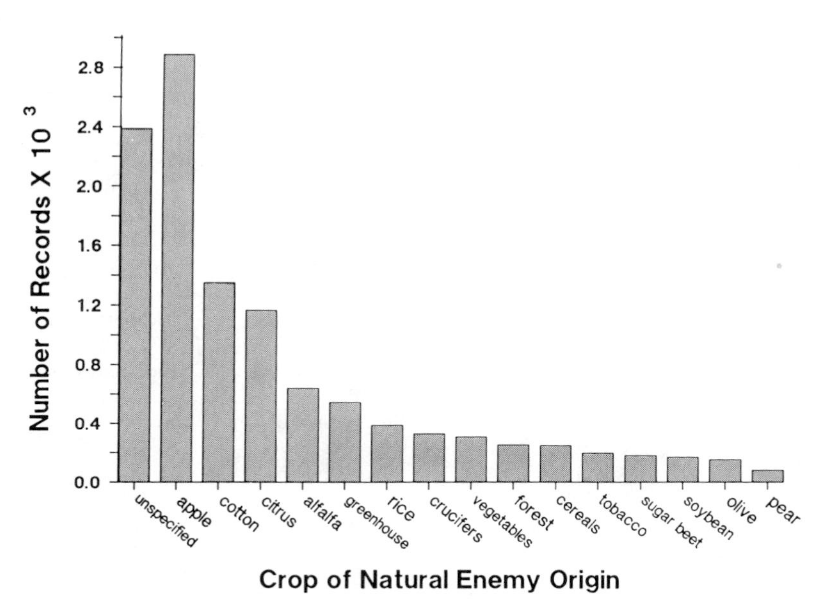

Figure 2.3. Distribution of SELCTV database records of pesticide impact on arthropod natural enemies grouped by crop of natural enemy origin (after Theiling 1987).

status. Natural enemies which exhibited a five-fold or greater survival ratio when exposed to a pesticide, as compared to a known susceptible strain, were termed *resistant*. Resistant natural enemies accounted for less than 3% of database records.

Individual life stages were tested in over 65% of recorded pesticide screenings (see Fig. 2.2). Egg and pupal stages accounted for only 2% of records each. Susceptibility of immatures (larvae and nymphs) was documented in 14.5% of records. Adults were most commonly tested at 46.7% of records. In 32% of records, the life stage was not cited specifically due to treatment of many stages. This designation most often applied to field tests of pesticide impact on natural enemies.

A two-letter code was assigned in the PROD: UNIT (production unit) field to identify the crop that the natural enemy was most commonly associated with, collected from, or on which the natural enemy was treated (Table 2.1). The database currently contains 60 crop designations. The 15 most common ones are depicted in Fig. 2.3. Natural enemies from apple, cotton, citrus, alfalfa, and greenhouse cropping systems have been the most frequent subjects of pesticide susceptibility screening. For 19% of records, largely those pertaining to laboratory testing or more basic research, the common habitat or crop of origin was not mentioned in the literature (unspecified, Fig. 2.3).

The three most commonly tested genera of natural enemies for prevalent crops are listed along with their frequencies in Table 2.3. Many are predators of indirect or induced pests in intensively sprayed cropping systems such as apple, cotton, or citrus. Pesticide side effects on phytophagous mite predators predominate in apple, citrus, and greenhouse systems. Similarly, aphid predators have been extensively studied in alfalfa, cotton, and cereal grain production units (PROD: UNIT). Spiders and a hemipteran predator predominate in rice, while in forest ecosystems, impact studies on parasitoids of

Table 2.3. Natural Enemy Genera Most Commonly Tested for Pesticide Side Effects on 10 Most Prevalent Crops from SELCTV Database (after Theiling 1987)

Crop	Predominant Genera		
Apple	Amblyseius (701)[1]	Typhlodromus (446)	Trichogramma (53)
Cotton	Chrysopa (193)	Geocoris (144)	Orius (113)
Citrus	Amblyseius (148)	Aphytis (134)	Chrysoperla (94)
Alfalfa	Chrysoperla (80)	Hippodamia (79)	Orius (71)
Greenhouse	Phytoseiulus (234)	Encarsia (202)	Trichogramma (35)
Rice	Lycosa (86)	Cyrtorhinus (51)	Oedothorax (27)
Crucifers	Apanteles (63)	Coccygomimus (39)	Brachymeria (27)
Forest	Apanteles (39)	Coccygomimus (23)	Brachymeria (17)
Cereal grains	Feronia (19)	Hippodamia (16)	Pterostichus (16)
Soybeans	Geocoris (21)	Chrysoperla (20)	Nabis (14)

[1] Number of records.

lepidopteran pests are most common. *Encarsia* and *Trichogramma* species are commonly studied in greenhouse systems, where they are released and managed for control of whiteflies and lepidopteran pests.

The geographic location where the research was conducted was specified in LOCI (location). Country was the limit of resolution for most records. Where possible, states were specified for the United States and provinces for Canada. In the database, the greatest number of studies came from the United States, namely, California, Texas, New York, and Michigan. Natural enemy testing in the United States accounted for nearly 40% of database records, followed by Canada (12%) and several western European countries: West Germany (7%), England (6%), and France (4%). Eastern European, Asian, and far eastern countries were best represented by Poland (5%), USSR (3%), and Czechoslovakia (2%).

When identified in the literature, the primary host or prey (HOST: PREY) of the tested natural enemy was recorded. The HOST: PREY field in the database presently contains 250 unique entries. The distribution of records among nine general categories of host or prey is listed in Table 2.4. Host or prey was unspecified for nearly 22% of records extracted from the literature. Another 11% listed a pest complex for a given crop. Phytophagous mites and aphids were the most comon HOST: PREY at 18.6% and 17.7% of records, respectively. The Homoptera (excluding aphids), which include scales, mealybugs, psyllids, and whiteflies, comprised another 12%. Many of the hosts or prey identified in the natural enemy/pesticide impact literature were indirect or secondary pests of crops. Generally biological control agents for these species have been more efficacious than have natural enemies for primary pests of agricultural crops (Hoyt and Simpson 1979, Hoyt and Tanigoshi 1983). Among the pest Lepidoptera, Coleoptera, and Diptera listed in the database, many are key pests whose natural enemies may provide substantial biological control (e.g., Lingren et al. 1972, Mosievskaya and Makarov 1974, Bogenschutz 1979, Brettell 1979, Garcia 1980, Somchoudhury and Dutt 1980, Ascerno et al. 1980, Andreadis 1982).

Toxicity or susceptibility data for the pesticide tested was sometimes given for the pest as well as for the natural enemy in a literature citation. The presence or absence of these

Table 2.4. Relative Distribution of SELCTV Database Records by Host or Prey Associated with Arthropod Natural Enemies Tested for Pesticide Impact (after Theiling 1987)

Host or Prey Classification		% Total Records
Phytophagous mites		18.6
Aphids		17.7
Lepidoptera—general		11.4
Homoptera—general		11.9
Mealybugs	(3.2)	
Scales	(2.7)	
Whiteflies	(2.6)	
Diptera—general		4.1
Leafminers		1.7
Coleoptera—general		1.6
Several or many		11.6
None specified		21.8

comparative data was indicated in HP:TOX:DAT (host or prey toxicity data) field. Approximately 30% of records extracted from publications provided some type of host/prey toxicity data as well as natural enemy data, while 70% did not. Comparative data were used in estimating relative toxicity relationships of natural enemies and pests. In some cases, this information was the basis for calculating selectivity ratios for predators and parasitoids (see Section 2.4.2).

2.3.1.3. Chemical Attributes. Pesticide class (CHEM:CLASS) refers to the type of pesticide or agricultural chemical tested. Some of the compounds within the eight assigned pesticide classes were not specifically pesticides, but were agricultural chemicals which can affect natural enemies. Insecticides have been most extensively studied for side effects on natural enemies (82% of records). Fungicides and acaricides were the only other classes of compounds whose side effects have been studied to any degree, at 9% and 7% of records, respectively. Herbicides accounted for only 1.4% of records. The remaining 60 records contained impact data for feeding repellents, fumigants, nutrient sprays, and plant growth regulators.

Incidence of records by pesticide class was plotted over time beginning in 1935 (Fig. 2.4). As mentioned, insecticides predominated by a large margin. Only in the past two decades have impact studies on natural enemies begun to include other pesticides and other agricultural chemicals (not included in this figure due to limited records) with any regularity. The progression of fungicide and acaricide testing over time is nearly parallel. Herbicide testing remains at a nearly negligible level. However, recent efforts made on testing these products by European researchers are not reflected in Fig. 2.4 (e.g., Boller et al. 1984, Basedow 1985, Baker et al. 1985, Hassan et al. 1987; Chapter 5).

Pesticides tested were further categorized (CHEM:GROUP) by chemical structure, origin, or taxonomy (for microbial pesticides). Twenty-eight unique pesticide groupings were represented, some of which spanned more than one CHEM:CLASS (e.g., carbamate

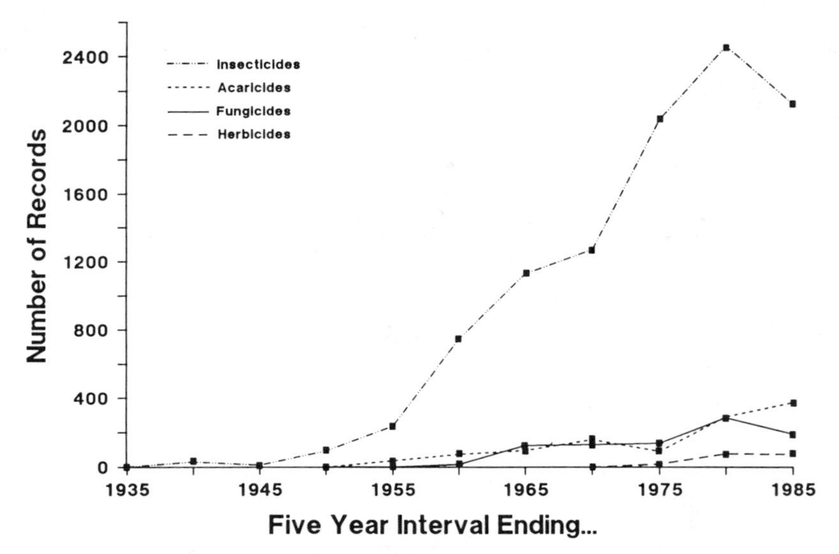

Figure 2.4. Incidence of pesticide impact research on arthropod natural enemies for each major pesticide class from SELCTV database (after Theiling 1987).

fungicides, herbicides, and insecticides). Organophosphate (40%) and carbamate (13%) pesticides comprised over half of the database. Other common chemical groupings included organochlorines and DDT derivatives (7% each), pyrethroids (6%), miscellaneous organics (5%), microbial insecticides (4%), inorganics and botanicals (3% each), and juvenile hormone-like compounds (2%). Many of the latter groupings included only insecticides.

Testing trends by CHEM:GROUP (chemical group) over time are shown in Fig. 2.5. The number of records in each chemical group was plotted over time beginning in 1935. The botanical and inorganic insecticides were relied upon more heavily before synthetic organics became available. Testing of these compounds increased, as did testing of all others, with interest in finding selective pesticides. However, after 1970 screening with botanicals and inorganics decreased to low levels. Side-effect testing with organochlorine and DDT derivatives began when nontarget side effects and control problems with these organic compounds arose during the late 1950s and early 1960s. They also leveled off and began a slow decline in the early 1970s (Fig. 2.5). As noted before, organophosphates made up the majority of insecticide impact records, representing the most widely used group of insecticides. They have continued to increase since the late 1950s to early 1960s, as have the carbamates to a lesser extent.

The curves for all of the major chemical groups depicted in Fig. 2.5 follow a similar pattern: testing on natural enemies rose rapidly with new compound development, especially after an initial delay for the older groups of pesticides; as fewer new compounds within the group were developed, testing leveled off and then decreased. The organophosphates and carbamates appear to be in the latter phase of this cycle, while the pyrethroids are still in the initial increase phase. With the newer biorational materials (e.g., the microbial and insect growth regulator pesticides), the lag between development and side-effect testing on natural enemies seems to have shortened, indicating a greater recognition of the importance of selectivity.

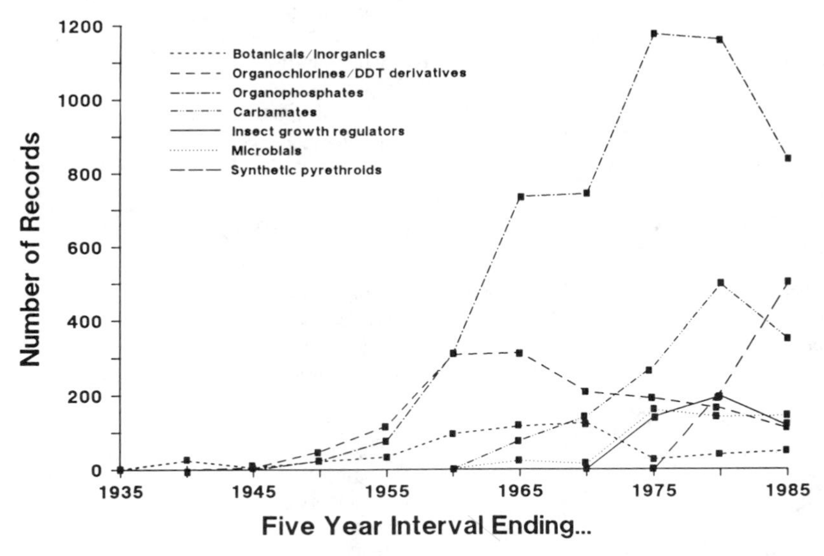

Figure 2.5. Incidence of pesticide impact research on arthropod natural enemies for each major insecticide grouping from SELCTV database (after Theiling 1987).

Common chemical names were recorded in the CPD:NAME (compound name) field for pesticides tested on natural enemies. Among 430 unique entries, most were common chemical names; however, some experimental designations and microbial species names were included. Pesticide mixtures were not entered into SELCTV. Table 2.5 contains compounds ranked within the top 30, as well as the pesticide class and group to which they belong. Interestingly, the carbamate insecticide carbaryl tops the list. This broad-spectrum insecticide has been widely used on a diversity of crops. Dicofol and cyhexatin (acaricides) and captan (a fungicide) are the only noninsecticides included in the top 30. DDT, malathion, parathion, dimethoate, and azinphosmethyl have been commonly tested for their side effects on natural enemies. Over half of the compounds in Table 2.5 are organophosphate insecticides. It is significant that 30 of the 430 entries in CPD:NAME made up over 50% of the records in the database. This indicated that while a wide diversity of pesticides were commonly tested for their effects on natural enemies, relatively few were

Table 2.5. Frequency and Classification of 30 Most Commonly Studied Pesticides in SELCTV Database of Pesticide Impact on Arthropod Natural Enemies (after Theiling 1987)

Compound Name	Pesticide Class[1]	Chemical Group	No. of Records
Carbaryl	I	Carbamate	494
DDT	I	DDT Derivative	476
Malathion	I	Organophosphate	436
Parathion	I	Organophosphate	430
Dimethoate	I	Organophosphate	390
Azinphosmethyl	I	Organophosphate	373
Diazinon	I	Organophosphate	266
Permethrin	I	Pyrethroid	239
Demeton	I	Organophosphate	237
Trichlorfon	I	Organophosphate	224
Endosulfan	I	Organochlorine	210
Oxydemeton-methyl	I	Organophosphate	204
Methyl parathion	I	Organophosphate	203
Pirimicarb	I	Carbamate	200
Fenvalerate	I	Pyrethroid	184
Methomyl	I	Carbamate	175
Toxaphene	I	Organochlorine	155
Dicofol	A	DDT Derivative	155
Fenitrothion	I	Organophosphate	154
Diflubenzuron	I	Chitin Inhibitor	143
Phosmet	I	Organophosphate	139
Carbophenothion	I	Organophosphate	129
Lindane	I	Organochlorine	129
Phosalone	I	Organophosphate	125
Phosphamidon	I	Organophosphate	125
Lead arsenate	I	Inorganic	121
Carbofuran	I	Carbamate	119
Monocrotophos	I	Organophosphate	115
Captan	F	Miscellaneous Organic	113
Cyhexatin	A	Organotin	112

[1] A = acaricide; F = fungicide; I = insecticide.

widely used exclusively—especially on the crops that were most heavily treated (e.g., cotton, apple, citrus, corn, rice).

The type of pesticide formulation and amount of active ingredient were recorded in the FORMULATN field. More than half the records in the database contained quantitative formulation data. If available, trade names were used when formulation was not specified. Among seven general categories, emulsifiable concentrates (37% of records) and wettable powders (29%) were the most widely used pesticide formulations for natural enemy impact assessments. Undoubtedly, these formulation proportions mirror their use in the field and reflect ease of use and effectiveness for pest control rather than their lack of impact on natural enemies. Minor use formulations included granulars (5%), solubles (3%), flowables (2%), and dusts (1%). Approximately 10% of records document side effects of technical grade pesticides on natural enemies. Technical grade is unformulated active ingredient, often used in more basic, laboratory research.

Chemical rate (CHEM:RATE) refers to the rate at which the pesticide was applied. Four different rate types were assigned in SELCTV. Over 50% of database records contained toxicity data for pesticides applied at a recommended field rate. This reflects that natural enemy testing was most commonly undertaken after the appropriate rate of pesticide required for pest control was established. Natural enemies were exposed to pesticides applied at a given concentration in 22% of records, to a specific dose in 2% of cases, and to a residual deposit in 6% of SELCTV. Chemical rate was not assigned to records of LC_{50} or LD_{50} bioassays.

The units in which an application was made were entered into CR:UNITS (chemical rate units). Pesticides were most often applied as percent formulation (20% of records), pounds of formulated pesticide per units of water (11%), or per cent active ingredient (a.i.) (6%). Over 150 different rate units were entered into the database. CR:UNITS apply to the numbers entered in CR:VALUE (chemical rate value). In order to make broad comparisons throughout the database based on dosage, a SC:AI (standard concentration in percent active ingredient) field was created. A limited number of conversion factors for the commonly used dosage designations have been worked out to complete this standard computation. Approximately 25% of records have values entered in this field. Further development of standardized rates will be emphasized in future updates of the database (Theiling 1987).

2.3.1.4. Testing and Summary Attributes. Ten different test methods (TST:METHD) were recognized in the database (see Fig. 2.6). Field assessments and tests in which the natural enemy was placed in contact with a fresh, dry deposit of pesticide accounted for about 30% of records each. Tests in which pesticides were sprayed. applied topically, contacted as aged residues, or used as dips for natural enemies each accounted for another 6–7% of records. Data for pesticides administered to natural enemies orally, by injection, or in tests where the mode of uptake was variable were contained in another 5% of database records. A special designation was reserved for tests in which the host of an endoparasitoid was treated with a microbial insecticide prior to or following parasitization. This designation was used for less than 1% of records.

An evaluation of the test methods reported over time indicated that laboratory contact tests have surpassed field tests as the preferred method for assessing pesticide impact on natural enemies (Fig. 1.2c; Theiling 1987). Trends for topical, residue, and dip tests showed a similar rapid increase from 1965 to 1975, but then leveled off after 1980. Reliance on spray testing has been fairly steady over the last 15 years, at a slightly lower rate than those mentioned above (Theiling 1987). Residue and contact test methods,

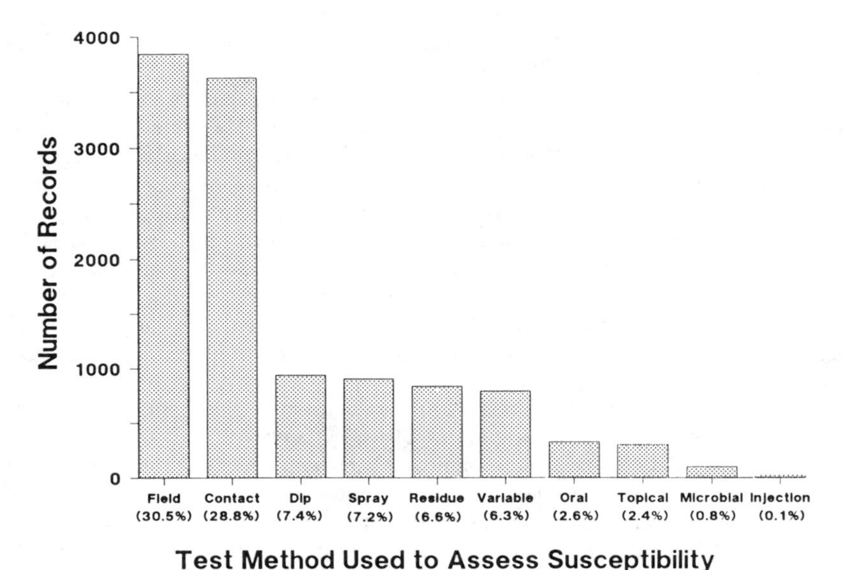

Figure 2.6. Distribution of SELCTV database records of pesticide impact on arthropod natural enemies grouped by test method (after Theiling 1987).

which accounted for approximately 60% of records, best simulate field exposure (Franz 1974, Croft 1977, Hassan et al. 1983, 1987).

The length of time that a natural enemy was in contact with the toxicant (DUR:EXP; duration of exposure) ranged from ≤ 24 hr to ≥ 1 month in database records. Nine categories were developed from 68 different exposure time entries (Theiling 1987). The most common category was ≤ 24 hr, accounting for 18% of records. Less than 10% of records were in the range of 24–48 hr, and declining percentages occurred at longer exposure intervals. Nearly 40% of database records documented susceptibility testing in which the exposure period was either variable or not specified. This condition applied, for example, to parasitoids treated at one point and evaluated at emergence.

Evaluation time (EVAL:TIME) is the elapsed time from intial pesticide exposure to evaluation of impact. There were 68 unique entries in this field, ranging from < 24 hr to > 1 month (Theiling 1987). Most natural enemies were exposed for 48 hr or less. As with duration of exposure, fewer records were found as evaluation times increased. Five percent of records documented lethal time of exposure required to cause 50% mortality (LT_{50}). Another 5% of records were assessments of parasitoids that were treated subsequent to emergence from their hosts and evaluated at emergence. Evaluation time was not specified in over 20% of records.

Response measurements encompassed three fields in the database record structure (see Table 2.1): RESP:TYPE, RESP:UNITS, and RESP:VALUE. In RESP:TYPE (response type), the measured response of a natural enemy to a pesticide was identified. Of 27 types of responses listed in SELCTV, mortality was measured in over 75% of records (Theiling 1987). Median lethal dose, concentration or time were measured in another 20% of records. The remaining responses included such sublethal responses as changes in consumption, ability to diapause, fecundity, oviposition, or developmental time

(Chapter 7). In a few cases, biochemical conversion of a pesticide to its metabolite(s) was measured as the response.

RESP:UNITS (response units) refer specifically to RESP:TYPE. Most (80%) units in the database were expressed as percentage increase or decrease of the cited response unit such as mortality or a cited sublethal response (Theiling 1987). The remaining 20% of response units applied to quantities measured in LC, LD, or LT (lethal time) assays. Examples included percent active ingredient, percent formulated pesticide, and micrograms of toxicant per insect. Most LT_{50} values were expressed in hours. RESP:VALUE (response value) is the numerical value of the measured response.

Resistance ratio (RR:RATIO) is a computation consisting of the LD_{50} or LC_{50} of a resistant natural enemy divided by that of a known susceptible strain. Resistance was much less common among natural enemies compared to pest arthropods (Chapter 14). Resistance ratios were calculated for 149 records, or just over 1% of database records. Ratios ranged from a low of 2 for a number of cases (Chapter 14) to a high of nearly 80,000 for the predaceous mite *Amblyseius fallacis* with azinphosmethyl (Motoyama et al. 1970, 1972, 1977). Most resistance ratios were calculated for phytoseiid mites (Croft and Strickler 1983, Hoy 1985). Braconid and aphelinid parasitoids together only accounted for 11% of resistance records (e.g., Pielou and Glasser 1952, Abdelrahman 1973, Schoones and Giliomee 1982, Rosenheim and Hoy 1986) and coccinellids, another 6% (Mohamad 1974, Head et al. 1977, Hull and Starner 1983). By insecticide group, resistance to organophosphates predominated among natural enemies, followed by carbamates, pyrethroids, and DDT derivatives.

A toxicity rating scale (TOX:RATING) was developed to provide a common scale for a comparison of different response types based on the percent effect. Response types include mortality, median lethal time, dose or concentration assays (which are also based on mortality), and changes in biological attributes of the natural enemy such as fertility, fecundity, developmental time, or longevity (Theiling 1987). Toxicity ratings and their frequencies in the database were 0 (no data; less than 1% of records), 1 (0% effect, 11.9% of

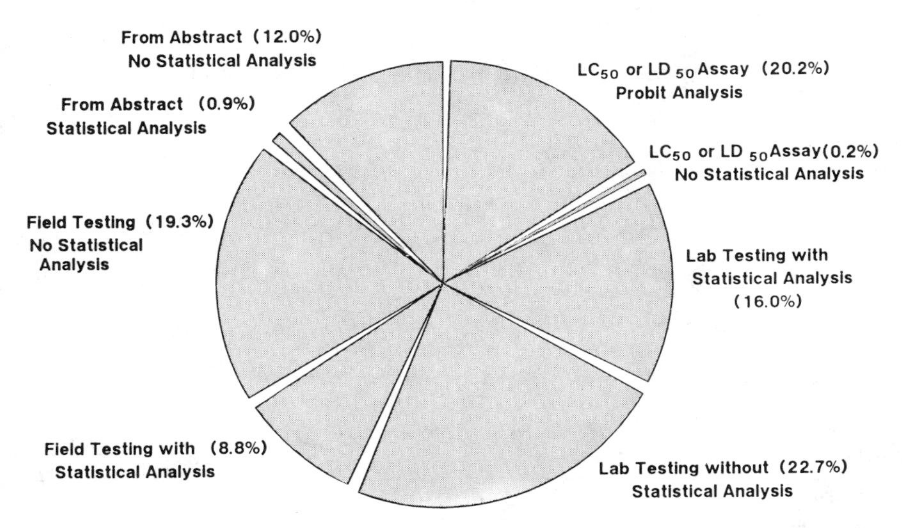

Figure 2.7. Distribution of SELCTV database records of pesticide impact on arthropod natural enemies grouped by data rating (after Theiling 1987).

records), 2 (< 10% effect, 16.2%), 3 (10–30% effect, 13.8%), 4 (31–90% effect, 28.4%), and 5 (> 90% effect, 28.7%). Thus, over half the records fell into the higher toxicity categories of 4 or 5. Zero was assigned if no response data could be extracted from an abstract or publication. Most database analyses and comparisons of toxicity were computed on the basis of toxicity ratings.

As noted earlier, ranking of pesticide impact studies on natural enemies by scientific rigor and/or level of precision was performed using a data rating scale (DAT:RATING) (see Fig. 2.7). While this procedure was somewhat qualitative, the capacity to specify a precision level for data analysis can be useful. The first digit of the two-part code indicated the relative level of precision on the response measurement, on a scale of 1–4. Presence (1) or absence (2) of summary statistics is reflected by the second digit. Figure 2.7 shows that, historically, laboratory tests without statistical analysis were the most common types of assessments reported in the literature for natural enemies on a per record basis. Next in frequency were LC or LD assays with probit analysis.

Sublethal effects of pesticides (SUBLETHAL) were recorded in 1,001 records, or 8% of the database. Some records contained measurements of more than one sublethal response, bringing the total number of reports to 1332. The sublethal effects and their frequencies in the database as well as the sublethal effects by order are listed in a two-way table

Table 2.6. Distribution of Sublethal Pesticide Side Effects among Arthropod Natural Enemy Orders from SELCTV Database (after Theiling 1987)

Sublethal Effect (No. of Records)	Order[1] HYM	ACA	COL	HEM	NEU	DIP	ARA	LEP	ORT
Fecundity (397)	133	165	28	58	11	2	0	0	0
Developmental time (150)	75	40	1	5	21	8	0	0	0
Parasitism (140)	114	0	18	0	0	8	0	0	0
Consumption (138)	0	49	39	10	18	6	14	0	2
Longevity (115)	84	5	11	9	0	2	0	4	0
Deformation (101)	30	0	22	3	29	17	0	0	0
Fertility (94)	5	10	2	47	30	0	0	0	0
Oviposition (64)	22	29	9	4	0	0	0	0	0
Repellency (63)	39	18	1	0	0	5	0	0	0
Reproduction (40)	9	30	1	0	0	0	0	0	0
Locomotion (30)	4	5	21	0	0	0	0	0	0
Order Totals	515	351	153	136	109	48	14	4	2

[1] HYM = Hymenoptera; ACA = Acari; COL = Coleoptera; HEM = Hemiptera; NEU = Neuroptera; DIP = Diptera; ARA = Araneae; LEP = Lepidoptera; ORT = Orthoptera.

(Table 2.6). Changes in fecundity were evaluated twice as often as any other sublethal effect, accounting for 30% of reports. Consumption, morphological deformation, developmental time, fertility, longevity, and parasitism were examined in about 10% of records each. Pesticide-induced sublethal effects were studied across 9 orders of natural enemies. Hymenoptera were the subject of nearly 40% of all tests, followed by Acari (26%), Coleoptera (10%), Hemiptera (10%), and Neuroptera (8%).

When records documenting sublethal effects were plotted over 5-year intervals, their numbers were seen to increase exponentially since 1960. Because the total number of records in the database is rising, sublethal effects data were also computed as percent of total records (see Fig. 2.8). The incidence of sublethal response measurements has increased slowly relative to the total records. A small peak in sublethal effects assessments can be seen for the period ending in 1945. This may reflect interest in the impact of agricultural chemicals on natural enemies, possibly related to the field failure of lead arsenate. After synthetic, organic pesticides were introduced, this type of research diminished until the 1960s. Since then, a small but steady increase in sublethal effects research has been maintained. As biorational insecticides become more widely used, measurement of sublethal effects will likely become more important in detecting the impact of these slower acting toxins on natural enemy populations (Flexner et al. 1986, Beckage 1985; Chapters 11 and 12).

Selectivity ratios (SLECTRATIO) were calculated by dividing the LC_{50} or LD_{50} of the host or prey by that of the natural enemy. Because this information was sporadically available from the literature, selectivity ratios only could be calculated for 870 records. Selectivity ratios ranged from 0.0001 to over 3000, thus spanning eight orders of magnitude. In order to compute meaningful averages on SLECTRATIO, ratios were assigned values from a logarithmic scale (the SR:INDEX) ranging from 1 to 9. A value of 5 represents equal impact on pest and natural enemy, or neutrality. Nearly 80% of selectivity ratios were computed for predators and their prey, and 20% for parasitoid host data. By

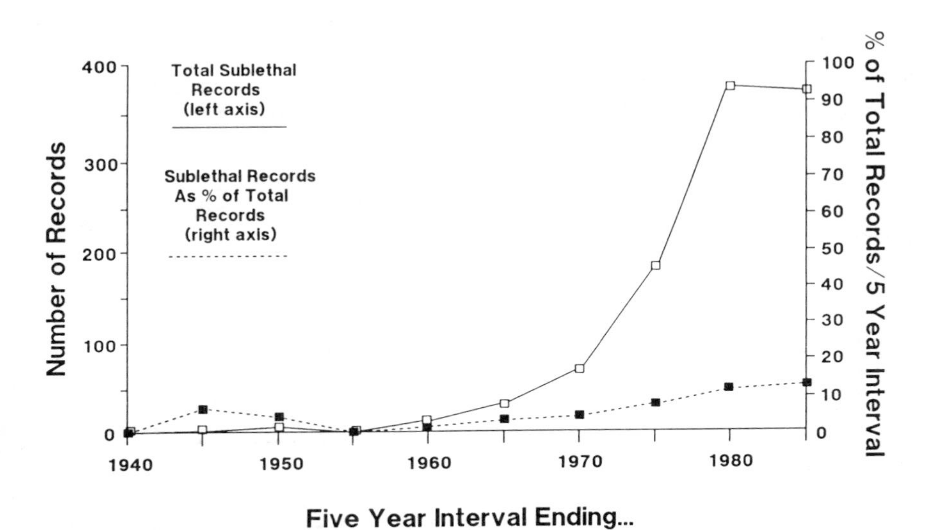

Figure 2.8. Incidence over time of documented sublethal side effects of pesticides on arthropod natural enemies from SELCTV database (after Theiling 1987).

order, selectivity ratios were calculated predominantly for Coleoptera (29%), Acari (20%), Hymenoptera (20%), and Araneae (11%). By crop, most SLECTRATIO values were from cotton, apples, rice, cereal grains, and alfalfa in descending order.

2.4. ANALYSIS OF THE DATABASE

The database analyses presented hereafter were based primarily on TOX:RATING (toxicity rating) computations. Most analysis consisted of comparing calculated average values for numeric fields based on criteria from character fields. Variances associated with toxicity rating averages were often computed.

2.4.1. Toxicity Rating Analysis

2.4.1.1. By Natural Enemy Attributes. The average toxicity of pesticide classes and of all compounds to predators, parasitoids, and all natural enemies is presented in Table 2.7. On the average, predators were less susceptible to pesticides than parasitoids. This pattern held for each pesticide group with the exception of fungicides. Susceptibility to fungicides was almost equal for both natural enemy types. In most cases, parasitoids exhibited a less variable response to pesticides than predators. Among pesticide classes, insecticides were most toxic to all natural enemies, followed by herbicides, acaricides, and fungicides (Table 2.7).

Average susceptibility values (TOX:RATING) for the 15 most common families (FAMILY) in SELCTV to all compounds were computed using the 1–5 scale. The Miridae (3.2), Ichneumonidae (3.2), and Chrysopidae (3.2) had the lowest toxicity ratings. The hemipteran families Nabidae (3.5), Lygaeidae (3.3), and Anthocoridae (3.3) were also fairly tolerant to pesticides. Most parasitoid families [e.g., Trichogrammatidae (3.7), Encyrtidae (3.7), Braconidae (3.7), and Eulophidae (3.9)] were more susceptible than families of predators. Syrphidae (3.9), a family containing important aphid predators, were highly

Table 2.7 Average Toxicity Ratings and Variances of Pesticide Classes to All Arthropod Natural Enemies, Predators, and Parasitoids from SELCTV Database (after Theiling 1987)

Pesticide Class	Predators			Parasitoids			All		
	Mean		Variance	Mean		Variance	Mean		Variance
Insecticide	3.61[1]	(7326)[2]	1.78	3.74	(2989)	1.74	3.65	(10,315)	1.77
Fungicide	2.59	(781)	1.51	2.58	(357)	1.45	2.59	(1138)	1.49
Acaricide	2.76	(747)	1.81	2.83	(144)	1.50	2.77	(891)	1.76
Herbicide	2.83	(92)	1.73	3.10	(84)	1.77	2.95	(176)	1.76
All Pesticides	3.43	(8946)	1.92	3.57	(3574)	1.86	3.47	(12,520)	****

[1] Index based on a rating scale of 1–5: 1 = 0 effect, 2 = < 10% effect, 3 = 10–30% effect, 4 = 31–90% effect, 5 = > 90% effect.

[2] Number of records.

**** = No variance calculated.

susceptible to most pesticides (e.g., Niemczyk et al. 1979, Hassan et al. 1983, Horn 1983, David and Horsburgh 1985).

The apparent tolerance of the Ichneumonidae as indicated in the above analysis is notable. It exemplifies one of the dangers of using database methods for estimating response trends of natural enemies. Close examination of individual records indicated that this value was greatly influenced by the type of tests run for this family of natural enemies. For example, many ichneumonid records in SELCTV came from tests in which pesticides were applied to parasitized hosts (e.g., Smilowitz et al. 1976, Kaya and Hotchkin 1981). These tests tended to show lower toxicity ratings than more direct exposure tests. Also, a large number of tests with ichneumonids measured side effects of insect growth regulators (e.g., Smilowitz et al. 1976, von Naton 1978, Sechser and Varty 1978, Bogenschutz 1979) and microbial pesticides (e.g., Ticehurst et al. 1982, Hotchkin and Kaya 1983a, Salama and Zaki 1983). Both of these insecticide groups were generally more selective to natural enemies than earlier, more conventional pesticides (Beckage 1985, Flexner et al. 1986). As will be discussed in more detail, there are many studies that document the high toxicity of pesticides to species from this family of natural enemies (e.g., Abu and Ellis 1977, Plapp and Vinson 1977, Otvos and Raske 1980, Rajakulendran and Plapp 1982b, Hassan et al. 1983, 1987).

Scatter plots and linear regression analysis were used to show the relationship between mean insecticide toxicity and variance to families of parasitoids (Fig. 2.9a) and predators (Fig. 2.9b) in the database. Although less evident than in other types of analysis (e.g., Table 2.7), parasitoids were more susceptible to insecticides than predators, as can be seen by examination of the mean family distribution values for both groups of natural enemies in Figs. 2.9a and b. Most obvious is that parasitoid responses to insecticides appeared to be considerably less variable than those of predators. In both figures, a linear relationship existed between susceptibility and variability in response. Negative slope values indicated decreases in variance with increases in susceptibility. The slope value for predators was lowered by the three outlying points on the left side of Fig. 2.9b. Otherwise, Figs. 2.9a and b had very similar slopes.

It appears that the variability in response of both natural enemy groups decreased similarly as the pesticide toxicity or natural enemy susceptibility increased (Figs. 2.9a and b). However, this effect alone was not of sufficient magnitude to explain the differences in variability of response between parasitoid and predator families. Other intrinsic differences in responses of natural enemies to toxins were probably involved (Chapters 6 and 16).

The 22 most common species (GENUS + SPECIES) of natural enemies included in the database were ranked in the order of their increasing susceptibility to pesticides (see Table 2.8). A spider, *Lycosa pseudoannulata*, was the most tolerant of these species. Three predator species followed, a coccinellid, a chrysopid, and a mirid. Of the phytophagous mite predators *Phytoseiulus persimilis, Typhlodromus pyri, T. occidentalis, Stethorus punctum, Amblyseius fallacis*, and *A. hibisci*, most were of intermediate susceptibility. Averages for all of these species were lower due to the inclusion of data on resistant strains (e.g., Lienk et al. 1978, Cranham and Solomon 1981, Overmeer and van Zon 1983, van de Baan et al. 1985; Chapter 14). Parasitoid species are well dispersed throughout Table 2.8, albeit mainly of moderate to high susceptibility. Again, *E. formosa* showed remarkably high tolerance for a parasitoid. This may have been due in part to testing the toxicity of diminishing residues on adults emerging from treated pupae (Hatalane and Budai 1982, Helyer 1982, Delorme and Angot 1983). These studies were commonly conducted to improve conservation of natural enemies through the timing of sprays. The high

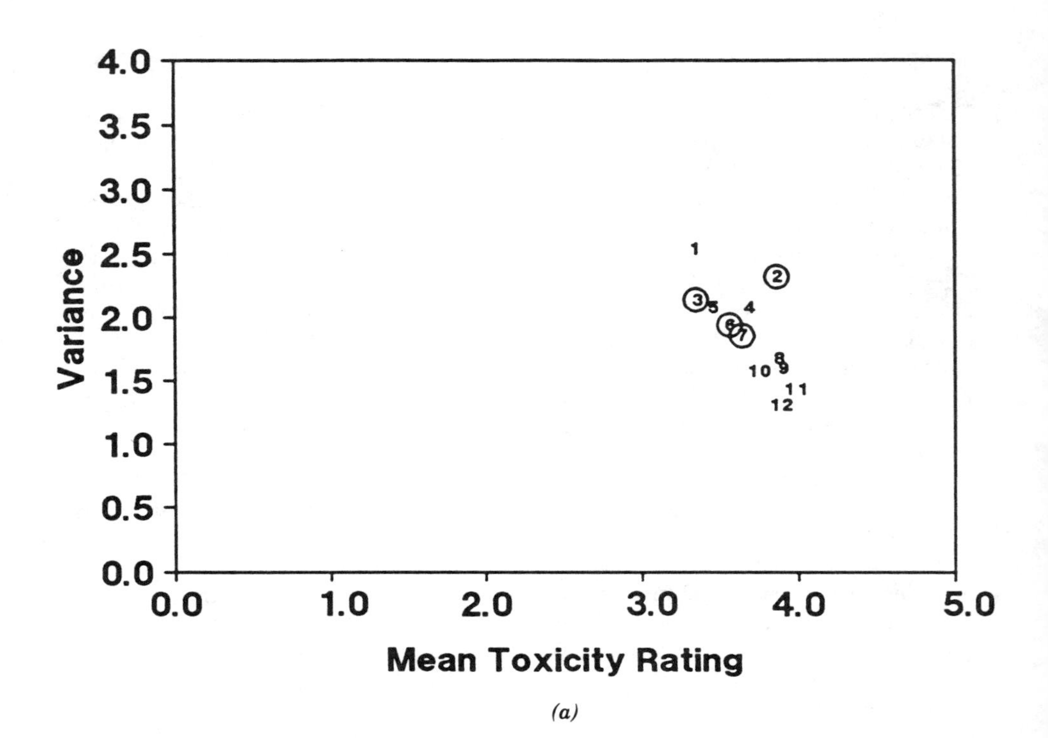
Variance
4.0
3.5
3.0
2.5
2.0
1.5
1.0
0.5
0.0
0.0
1.0
2.0
3.0
4.0
5.0
Mean Toxicity Rating
1
2
3
4
5
6
7
8
9
10
11
12

(a)

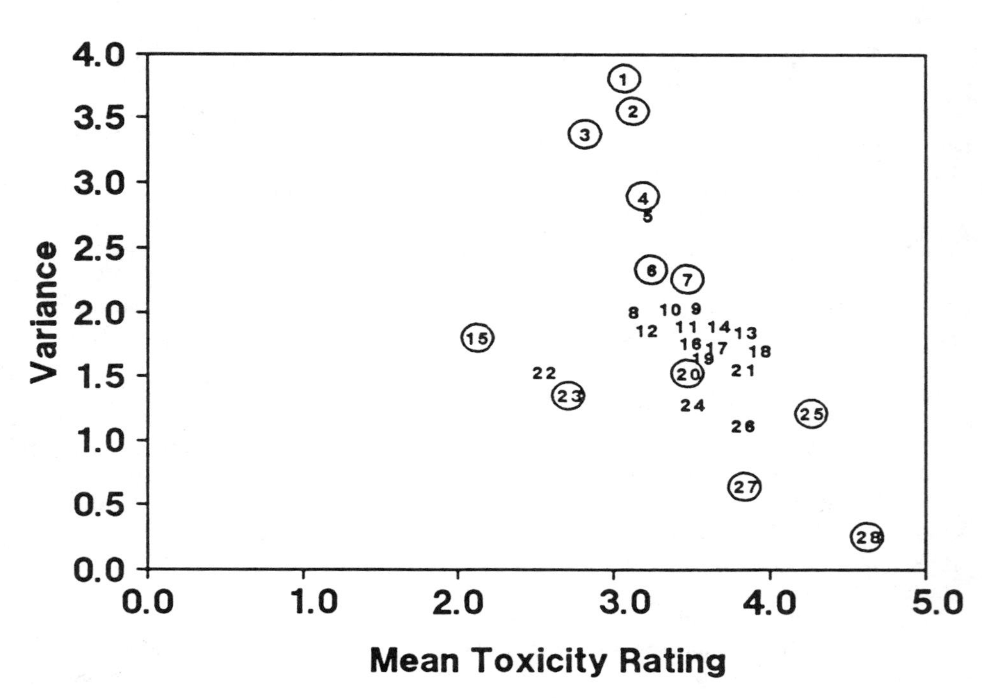
Variance
4.0
3.5
3.0
2.5
2.0
1.5
1.0
0.5
0.0
0.0
1.0
2.0
3.0
4.0
5.0
Mean Toxicity Rating
1
2
3
4
5
6
7
8
9
10
11
12
13
14
15
16
17
18
19
20
21
22
23
24
25
26
27
28

(b)

Table 2.8. Average Susceptibility of 22 Commonly Tested Natural Enemy Species to All Pesticides from SELCTV Database (after Theiling 1987)

Species	Family	No. of Records	Mean Toxicity[1]
Lycosa pseudoannulata	Lycosidae	85	3.0[2]
Cryptolaemus montrouzieri	Coccinellidae	114	3.0
Chrysopa carnea	Chrysopidae	534	3.1
Hyaliodes harti	Miridae	103	3.2
Encarsia formosa	Aphelinidae	232	3.3
Phytoseiulus persimilis	Phytoseiidae	323	3.4
Typhlodromus pyri	Phytoseiidae	268	3.4
Stethorus punctum	Coccinellidae	150	3.5
Typhlodromus occidentalis	Phytoseiidae	205	3.5
Phygadeuon trichops	Ichneumonidae	124	3.5
Chrysopa oculata	Chrysopidae	137	3.5
Amblyseius fallacis	Phytoseiidae	574	3.5
Leptomastix dactylopii	Encyrtidae	139	3.5
Hippodamia convergens	Coccinellidae	375	3.5
Anthocoris nemorum	Anthocoridae	85	3.6
Coccinella septempunctata	Coccinellidae	440	3.6
Trichogramma evanescens	Trichogrammatidae	86	3.6
Trichogramma cacoeciae	Trichogrammatidae	158	3.6
Coleomegilla maculata	Coccinellidae	102	3.7
Amblyseius hibisci	Phytoseiidae	90	3.7
Metasyrphus corollae	Syrphidae	105	4.1
Opius concolor	Braconidae	85	4.3

[1] Susceptibility is used in reference to a natural enemy, while toxicity generally refers the potency of a chemical.
[2] Index based on a rating scale of 1–5: 1 = 0 effect, 2 = < 10% effect, 3 = 10–30% effect, 4 = 31–90% effect, 5 = > 90% effect.

Figure 2.9a. Mean insecticide toxicity ratings versus variance for 12 families of parasitoids in SELCTV database (after Theiling 1987). Average toxicity ratings and variances were calculated for all insecticide records associated with each parasitoid family, using the TOX:RATING field in SELCTV database. *KEY to families*: (number in parenthesis = number of records); **1** = Tachinidae (89); **2** = Chalcididae (30); **3** = Scelionidae (42); **4** = Trichogrammatidae (326); **5** = Ichneumonidae (365); **6** = Eupelmidae (27) **7** = Platygasteridae (25); **8** = Pteromalidae (66); **9** = Aphelinidae (382); **10** = Braconidae (1016); **11** = Encyrtidae (280); **12** = Eulophidae (181). Circled numbers indicate fewer than 50 records.

Regression parameters: $y = -1.2x + 6.3$; $r = -0.72$; mean $x = 3.74$.

Figure 2.9b. Mean insecticide toxicity ratings versus variance for 28 families of predators in the SELCTV database (after Theiling 1987). Average toxicity ratings and variances were calculated for all insecticide records associated with each predator family, using the TOX:RATING field in SELCTV database. *Key to families:* **1** = Clubionidae (17); **2** = Dytiscidae (15); **3** = Hydrophilidae (11); **4** = Pentatomidae (37); **5** = Anystidae (60); **6** = Reduviidae (16); **7** = Berytidae (27); **8** = Lycosidae (110); **9** = Miridae (569); **10** = Chrysopidae (903); **11** = Carabidae (366); **12** = Malachiidae (33); **13** = Phloeothripidae (91); **14** = Hemerobiidae (64); **15** = Veliidae (33); **16** = Anthocoridae (373); **17** = Coccinellidae (1959); **18** = Syrphidae (326); **19** = Nabidae (217); **20** = Thripidae (19); **21** = Phytoseiidae (1345); **22** = Stigmaeidae (79); **23** = Micryphantidae (36); **24** = Lygaeidae (200); **25** = Macrochelidae (15); **26** = Cecidomyiidae (65); **27** = Salticidae (13); **28** = Theridiidae (16). Circled numbers indicate fewer than 50 records.

Regression parameters: $y = -0.8x + 4.6$; $r = -0.79$; mean $x = 3.61$.

susceptibility of syrphids to pesticides was exemplified again by *Metasyrphus corollae* (e.g., Grapel 1982, Hellpap 1982, Nasseh 1982).

An example of the kind of analysis that can be done with the SELCTV database, mean toxicity of pesticides to several key natural enemy species (GENUS + SPECIES) was partitioned by chemical group (CHEM:GROUP) (Table 2.9). Chemical groups can be identified which are particularly toxic or selective to a given species. For example, the spider *Lycosa pseudoannulata* had an average toxicity rating of 3.0 over all pesticides (Table 2.8). Table 2.9 shows that organophosphates are relatively harmless to this species while organochlorines are very toxic. Pyrethroids and insect growth regulators are very toxic to the coccinellid *Cryptolaemus montrouzieri*. Organophosphates and pyrethroids are highly toxic to the parasitoid *E. formosa*. While the botanicals/inorganics, organotins (acaricides), and insect growth regulators are more selective to natural enemy species, pyrethroids are particularly destructive.

The life stages (STAGE) of natural enemies differed in their susceptibility to pesticides. Table 2.10 shows the average toxicity ratings of all pesticides to natural enemies by life stage. Eggs and pupae were the most tolerant life stages to pesticides, while larvae and adults were most susceptible. When further differentiated by natural enemy type, the same trends applied to predator life stages, but eggs and adults were the more susceptible parasitoid life stages, while larvae and pupae were more tolerant. Many endoparasitoid larval stages are protected from pesticide exposure to some extent within the bodies of

Table 2.9. Average Susceptibility of 21 Commonly Tested Natural Enemy Species to Common Types of Insecticides from SELCTV Database (after Theiling 1987)

	Chemical Group[1]						
Species	BT/IO	OC/DD	OP	CA	SP	OT	IGR
Chrysopa carnea	2.6[2]	3.1	3.6	3.3	3.4	3.2	3.0
Amblyseius fallacis	3.6	3.3	3.7	4.0	4.3	2.4	2.6
Coccinella septempunctata	2.4	3.0	4.1	3.0	4.1	—	2.0
Hippodamia convergens	2.0	3.3	4.0	4.4	4.4	—	3.6
Phytoseiulus persimilis	4.0	3.2	4.2	3.3	4.4	3.6	2.8
Typhlodromus occidentalis	2.9	4.4	3.2	4.2	4.3	3.0	2.5
Encarsia formosa	3.5	3.4	4.7	3.5	4.7	2.1	2.3
Typhlodromus pyri	2.7	3.4	3.7	3.2	4.7	3.0	2.3
Trichogramma cacoeciae	3.6	4.0	4.3	4.0	4.3	—	—
Chrysopa oculata	—	3.1	3.6	3.7	5.0	4.0	1.7
Stethorus punctum	2.6	3.9	3.2	4.1	3.9	3.0	—
Metasyrphus corollae	—	4.4	4.8	4.8	—	—	1.6
Phygadeuon trichops	3.0	3.5	5.0	3.4	4.5	3.0	—
Leptomastix dactylopii	3.3	3.6	3.9	3.9	5.0	2.7	—
Cryptolaemus montrouzieri	—	3.3	4.4	4.6	4.3	—	—
Hyaliodes harti	2.0	2.9	4.1	3.7	—	—	—
Amblyseius hibisci	3.2	3.7	4.4	3.6	—	2.5	—
Trichogramma evanescens	—	2.8	3.9	4.7	—	—	—
Opius concolor	—	—	4.2	4.7	—	—	—
Lycosa pseudoannulata	—	4.4	2.3	3.7	—	—	—
Anthocoris nemorum	2.2	3.8	4.1	4.0	4.7	—	—

[1] BT/IO = botanicals and inorganics; OC/DD = organochlorines and DDT derivatives; OP = organophosphates; CA = carbamates; SP = pyrethroids; OT = organotins; IGR = insect growth regulators.

[2] All averages are based on a minimum of three data points. Index based on a rating scale of 1–5: 1 = 0 effect, 2 = < 10% effect, 3 = 10–30% effect, 4 = 31–90% effect, 5 = > 90% effect.

Table 2.10. Average Susceptibility for All Arthropod Natural Enemies, Predators, and Parasitoids Grouped by Life Stage from SELCTV Database Records (after Theiling 1987)

Life Stage	Predator		Parasitoid		All	
Egg	3.00[1]	(328)[2]	3.79	(14)	3.04	(252)
Larva/Nym	3.52	(1288)	3.22	(345)	3.46	(1623)
Pupa	3.18	(62)	3.26	(178)	3.24	(240)
Adult	3.47	(3761)	3.72	(2117)	3.56	(5884)

[1] Index based on a rating scale of 1–5: 1 = 0 effect, 2 = < 10% effect, 3 = 10–30% effect, 4 = 31–90% effect, 5 = > 90% effect.

[2] Number of records.

their hosts, as are many larval ectoparasitoids which occur with their hosts within plant tissue. These patterns of susceptibility have been noted earlier by several reviewers (e.g., Bartlett 1964, Croft and Brown 1975, Croft 1977, Hull and Beers 1985). They also have been confirmed in subsequent studies with individual species of natural enemies (Babrikova 1980, 1982, Warner and Croft 1982; Chapter 4).

2.4.1.2. By Test Attributes. Mean toxicity ratings were calculated by test method (TST:METHD) across all compounds and natural enemies (Theiling 1987). Values for contact, dip, residue, spray, and topical tests were similar, centering at about 3.50 on the 1–5 toxicity rating scale mentioned earlier. Field tests were lower at 3.39. Lower average toxicity and higher variance associated with field testing could have arisen from several sources: different modes of uptake may have occurred between test methods, refugia may have existed in field treatments and less so with other methods of assessment, and field populations of natural enemies may vary more in response to pesticides than laboratory colonies. The lowest average toxicity (2.84) by test type was for microbial insecticide tests where the active material was mediated through a host to its parasitoid.

2.4.1.3. By Pesticide Attributes. Average toxicity values across many groupings of natural enemies were calculated for each chemical grouping in the database (e.g., see summary values for insecticides, fungicides, acaricides, herbicides, and all pesticides in Table 2.7). A more detailed summary of these comparisons is given in Theiling (1987) and Theiling and Croft (1988).

One of the more meaningful comparisons of this type is for different groups of insecticides and all species of natural enemies. In Fig. 2.10, the distribution of toxicity ratings for insecticide groups is plotted chronologically in the form of a stacked bar graph. The 0% and > 10% mortality sections (i.e., those marked by solid black) for each chemical group provide a relative measure of their selectivity. Inorganic and botanical insecticides were moderately selective to natural enemies. Increasing toxicity to nontarget arthropods has prevailed since the development of DDT, and has persisted through the pyrethroids. More recent biorational insecticides appear to be increasingly selective to beneficial arthropods (Chapters 11 and 12).

A toxicity summary for major chemical groups within the insecticides, acaricides, fungicides, and herbicides is given in Table 2.11. Among insecticides, trends represented the averages of the toxicity ratings given in Fig. 2.10. Among acaricides, average trends among different chemical groups indicated that organophosphorous and carbamate types were most toxic to natural enemies. Nitrophenol derivatives, DDT-related derivatives, and organotins were intermediate, while the organochlorine and sulfur-containing

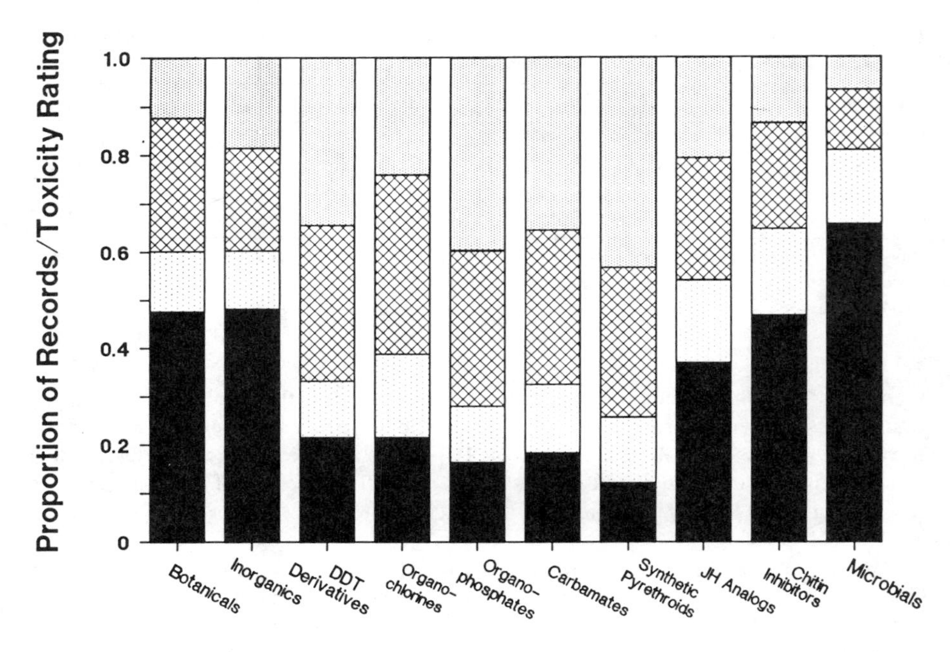

Figure 2.10. Distribution of toxicity ratings among insecticide chemical groups assessed for side effects on arthropod natural enemies from SELCTV database. Solid stacked bar: classes 1 & 2 = 0–9% effect; light stipple: class 3 = 10–30% effect; crosshatch: class 4 = 31–90% effect; heavy stipple: class 5 = > 90% effect. (After Theiling 1987.)

acaricides were fairly selective to natural enemies. Carbamate, inorganic, and miscellaneous organic fungicides were of moderate toxicity. The organotin fungicides were selective to natural enemies.

Toxicity varied greatly across the herbicide chemical groups, with no clear separation into high or low classes (Table 2.11). Values range from 3.74 for dinitrophenols (e.g., dinoseb) to 2.00 for the phenoxy herbicides (e.g., 2, 4-D). Nitrophenol derivatives, nitrogen heterocyclics, and urea derivatives were moderately toxic to natural enemies. The organometallics and phenoxy herbicides had little impact on natural enemies based on database test results.

The average toxicity values of individual compounds (CPD:NAME) to both predators and parasitoids and to all natural enemies were also of considerable interest. Results for some of the more common pesticides are shown in Table 2.12. The first three compounds listed are fungicides, of which benomyl was the most toxic. It is again evident that the fungicides were not as toxic to natural enemies as insecticides or acaricides. The center group of compounds in Table 2.12 are all insecticides. Pirimicarb, *Bacillus thuringiensis*, hydroprene, and diflubenzuron were the most selective to both predators and parasitoids, while methomyl, malathion, and permethrin were some of the most harmful compounds to these species. The last three compounds listed are acaricides. All three exhibited moderate selectivity to natural enemies.

Also tabulated in Table 2.12 is the relative susceptibility of predators versus parasitoids

Table 2.11. Average Toxicity Ratings for Insecticide, Fungicide, Acaricide, and Herbicide Chemical Groups to All Natural Enemies from SELCTV Database (after Theiling 1987)

Chemical Group	No. of Records	Average Toxicity
INSECTICIDES		
Pyrethroids	701	4.00[1]
Organophosphates	5089	3.88
Carbamates	1353	3.75
DDT derivatives	616	3.69
Organochlorines	884	3.54
JH analogues	297	3.06
Inorganics	226	2.87
Botanicals	319	2.84
Chitin inhibitors	162	2.83
Microbials	490	2.20
FUNGICIDES		
Nitrophenol derivatives	65	3.29
Carbamates	217	2.68
Inorganics	191	2.67
Miscellaneous organics	650	2.46
ACARICIDES		
Carbamates	11	4.45
Nitrophenol derivatives	262	3.18
DDT derivatives	73	2.80
Organotins	188	2.78
Organochlorines	9	2.33
Sulfur compounds	228	2.25
HERBICIDES		
Nitrophenol derivatives	15	3.74
Nitrogen heterocyclics	35	3.28
Urea derivatives	29	2.93
Carbamates	12	2.67
Miscellaneous organics	12	2.57
Organometallics	12	2.17
Phenoxy-alkyl derivatives	12	2.00

[1]Index based on a rating scale of 1–5: 1 = 0 effect, 2 = < 10% effect, 3 = 10–30% effect, 4 = 31–90% effect, 5 = > 90% effect.

to each compound. As in several types of analyses presented earlier, in most cases predators were more tolerant of pesticides than parasitoids. The same is true in this analysis. Of particular interest are compounds whose impact on predators and parasitoids differs substantially. Lead arsenate, endosulfan, carbaryl, phosalone, azinphosmethyl, chlordimeform, and permethrin were much more toxic to parasitoids than to predators. The reverse was true for benomyl and cypermethrin.

Fourteen pesticides that were highly toxic to natural enemies are listed in Table 2.13.

Table 2.12. Frequency and Average Toxicity Ratings of Common Pesticides Tested for Side Effects on Arthropod Predators and Parasitoids from the SELCTV Database (after Theiling 1987)

	Predator		Parasitoid	
Compound Name	No. of Records	Mean Toxicity	No. of Records	Mean Toxicity
Captan	91	1.98[1]	22	2.05
Bordeaux mixture*	30	2.03	10	1.90
Benomyl*	60	3.02	20	2.70
Lead arsenate	70	2.40	16	3.25
Endosulfan	143	3.37	66	3.82
DDT	342	3.66	130	3.92
Pirimicarb	166	2.98	34	3.03
Carbaryl	324	3.83	165	4.16
Methomyl	122	4.28	53	4.36
Phosalone	94	3.32	31	3.94
Trichlorfon	136	3.42	88	3.47
Azinphosmethyl*	296	3.58	69	3.96
Malathion*	311	4.23	124	4.10
Bacillus thuringiensis	100	2.06	103	2.04
Hydroprene	17	2.65	31	3.03
Diflubenzuron*	75	2.79	68	2.74
Chlordimeform	55	3.02	20	3.35
Fenvalerate*	155	3.86	29	3.79
Permethrin	178	4.03	58	4.38
Cypermethrin*	54	4.24	19	3.63
Tetradifon	69	2.13	13	2.54
Cyhexatin	96	2.79	16	3.19
Dicofol	126	2.90	29	2.97

[1] Index based on a rating scale of 1–5: 1 = 0 effect, 2 = < 10% effect, 3 = 10–30% effect, 4 = 31–90% effect, 5 = > 90% effect.
*Compounds more toxic to predators than to parasitoids on the average.

All were either carbamate, organophosphate, or pyrethroid insecticides. Compounds with the lowest average toxicity rating in the database are shown in Table 2.14. Two microbial preparations topped the list. The remaining insecticides were inorganic, naturally occurring organics, or synthetic, organic chemicals. These compounds possessed a broad range of intrinsic toxicity and span many chemical groups (CHEM:GROUP). Both ecological and physiological selectivity were represented by these insecticides. With the exception of *Beauvaria bassiana* and chlordimeform, all compounds were more toxic to parasitoids than to predators, some (e.g., schradan, ryania, nicotine sulfate, lead arsenate, tepp, and perthane) to a large extent.

2.4.2. Analysis on Other Fields

As noted earlier, averages calculated on RR:RATIO (resistance ratio) data were transformed to a log scale (RR:INDEX; log transformed resistance ratio) to allow for better comparison of a few outlying cases that were in the range of 10^5-fold. The majority of resistant natural enemies in the database were phytoseiid mites, which represent some of the highest mean RR:INDEX values that have been observed. The log

Table 2.13. Average Toxicity Ratings for Most Toxic Insecticides to Arthropod Natural Enemies from SELCTV Database (after Theiling 1987)

Name	No. of Records	Chemical Group	Mean Toxicity[1]
Formothion	45	op	4.69
Chlorthion	151	op	4.40
Parathion	418	op	4.34
Methomyl	175	ca	4.30
Mevinphos	101	op	4.28
Malathion	435	op	4.20
Methyl parathion	203	op	4.13
Permethrin	234	sp	4.12
Methidathion	89	op	4.12
Fenitrothion	154	op	4.10
Decamethrin	52	sp	4.08
Cypermethrin	73	sp	4.08
Dimethoate	389	op	4.06
Oxydemeton-methyl	204	op	4.04

[1] Index based on a rating scale of 1–5: 1 = 0 effect, 2 = < 10% effect, 3 = 10–30% effect, 4 = 31–90% effect, 5 = > 90% effect.

Table 2.14. Average Toxicity Ratings for Least Toxic Insecticides to Arthropod Natural Enemies in SELCTV Database (after Theiling 1987)

Compound Name	Chemical Group[3]	Mean Toxicity		
		All	Predators	Parasitoids
Beauvaria bassiana	MPF	1.52[1] (21)[2]	1.72 (11)	1.30 (10)
Bacillus thuringiensis	MPB	2.05 (203)	2.06 (100)	2.04 (103)
Schradan	OP	2.23 (35)	2.10 (30)	3.00 (5)
Ryania	BT	2.28 (80)	2.24 (66)	3.50 (14)
Tralomethrin	SP	2.33 (3)	— (0)	— (0)
Chlorfenvinphos	OP	2.42 (38)	2.31 (26)	2.67 (12)
Nicotine sulfate	BT	2.45 (80)	2.30 (63)	3.00 (17)
Lead arsenate	IO	2.56 (86)	2.40 (70)	3.25 (16)
Fluvalinate	SP	2.60 (5)	2.25 (4)	— (0)
Diflubenzuron	ICI	2.76 (143)	2.79 (75)	2.74 (68)
Menazon	OP	2.88 (51)	2.77 (48)	4.67 (3)
Hydroprene	JHM	2.90 (48)	2.65 (17)	3.03 (31)
Pirimicarb	CA	2.99 (200)	2.98 (166)	3.03 (34)
Vamidothion	OP	3.19 (32)	3.16 (25)	3.29 (7)
Aldicarb	CA	3.28 (65)	3.23 (59)	3.67 (6)
Tepp	OP	3.30 (33)	3.00 (21)	3.83 (12)
Disulfoton	OP	3.31 (48)	3.25 (38)	3.60 (10)
Perthane	DD	3.39 (44)	3.31 (39)	4.00 (5)
Chlordimeform	O	3.43 (108)	3.45 (88)	3.35 (20)
Trichlorfon	OP	3.44 (224)	3.42 (136)	3.47 (88)

[1] Index based on a rating scale of 1–5: 1 = 0 effect, 2 = < 10% effect, 3 = 10–30% effect, 4 = 31–90% effect, 5 = > 90% effect.

[2] Number of records.

[3] MPF = Fungus, MPB = bacteria, OP = organophosphate, BT = botanical, SP = pyrethroid, IO = inorganic, ICI = chitin inhibitor, JHM = juvenile hormone analog, CA = carbamate, DD = DDT derivative, O = other.

scale value of 1.36 for phytoseiid mites corresponds to a base-10 mean of 23-fold resistance on the average (across all compounds for all species). Resistance ratios for the remaining families of natural enemies, which included the Aphelinidae, Braconidae, Cecidomyiidae, Coccinellidae, Chrysopidae, and Lygaeidae, averaged less than a 10-fold resistance (Chapter 14).

By compound, highest average RR:INDEX values were associated with organophosphates, specifically diazinon, phosmet, azinphosmethyl, parathion, and methyl parathion. Carbaryl, permethrin, and DDT were the only non-organophosphate compounds commonly reported. RR:INDEX averages for these compounds were lower than those of the organophosphate compounds.

SLECTRATIO (selectivity ratio) analysis was performed on log-transformed SR:INDEX values as discussed previously (see Section 2.3.1.4). SR:INDEX averages were calculated for several classifications of both natural enemies and pesticide groups (Theiling 1987). Of 870 records, which included comparative pest/natural enemy toxicity data, 499 showed favorable selectivity ratios for the natural enemy. (Note: selectivity ratios were present for only about 7% of the SELCTV database records.) The mean SR:INDEX value of all of these cases was 4.61, where a value of 5.00 indicated selective survival of neither the natural enemy nor its hosts or prey.

The overall average selectivity ratio index of 4.61 could be interpreted to mean that pesticides were equally or slightly more selective to natural enemies than to pests in the literature; however, we do not believe this to be the case. Many general summaries of pesticide impact on natural enemies and pests show pests to be more tolerant to most conventional pesticides (Brattsten and Mecalf 1970, 1973, Jeppson et al. 1975, Yu 1988). Instead biases in the way selectivity data are reported are thought to be the main cause for these tendencies.

Previously cited summaries of toxicity rating distributions for all pesticides (e.g., see Section 2.3.1.4) and the data illustrated in Fig. 2.10 indicate that frequent reports of high pesticide toxicity to natural enemies are common in the literature. For example, almost 60% of all SELCTV database records had rating values of 4 or 5, which represents high mortality levels. At least half of the database records come from tests evaluating recommended fields rates of pesticides (Section 2.3.1.4). These data suggest that the

Table 2.15. Average Selectivity Ratios by Natural Enemy Family for Arthropod Pests and Natural Enemies from SELCTV Database (after Theiling 1987)

Family	No. of Records	Mean SR:Index[1]
Carabidae	10	3.70
Micryphantidae	19	4.16
Nabidae	23	4.22
Coccinellidae	239	4.27
Chrysopidae	72	4.32
Lycosidae	53	4.36
Braconidae	74	4.80
Miridae	11	4.82
Lygaeidae	12	4.83
Aphelinidae	26	4.96
Phytoseiidae	134	5.00
Ichneumonidae	33	6.21

[1] Based on log-transformed value of mean lethal mortality value for host or prey/mean value for natural enemy (see Theiling and Croft 1988).

selectivity ratios reported in the literature are biased towards compounds showing favorable selectivity to natural enemies. This is understandable since those judging publication qualifications often assign greater merit to reports of new or exceptional events rather than the norm.

By natural enemy type, average SR:INDEX values were 4.45 for predators (where pesticides are more commonly selective to these species), whereas the value for parasitoids was 5.19 (indicating the opposite trend). When these data are transformed back to actual mortality differences, they would equate to an average selectivity ratio difference of about five-fold between predators/prey and parasitoids/hosts. This ratio difference indicates that it has been easier to find selectivity pesticides that favor predators over their prey than parasitoids over their hosts. As a corollary, since parasitoids and predators often have common hosts or prey, then basic differences in susceptibility must be present between predators and parasitoids. This explanation is supported by many types of analyses reviewed in this chapter, and is favored over attributing differences in predator and parasitoid selectivity ratios to differences in the susceptibility of their prey or hosts.

Average SR:INDEX values are presented for the predominant families of natural enemies in Table 2.15. Four classes of ratio levels can be distinguished. First, carabids had the greatest advantage over their prey in a treated habitat. The next five families, including the Micryphantidae, Nabidae, Coccinellidae, Chrysopidae, and Lycosidae, were approximately 10 times more tolerant of pesticides than their prey. The Braconidae, Miridae,

Table 2.16. Average Selectivity Ratios by Compound for Arthropod Pests and Natural Enemies from SELCTV Database (after Theiling 1987)

Compound Name	No. of Records	Mean SR:Index
Pirimicarb	26	2.77
Dicofol	11	3.64
Cyhexatin	8	3.88
Thiometon	20	4.30
Demeton	27	4.33
Chlorpyrifos	13	4.38
Aldicarb	11	4.55
Dimethoate	20	4.60
Diazinon	23	4.65
Trichlorfon	9	4.67
Permethrin	44	4.75
Fenvalerate	27	4.78
Parathion	27	4.78
Azinphosmethyl	13	4.80
Carbaryl	25	4.80
Carbofuran	17	4.82
Malathion	37	4.84
Methomyl	17	4.94
Acephate	13	5.00
DDT	17	5.06
Methyl parathion	26	5.12
Lindane	14	5.14
Cypermethrin	22	5.22
Fenitrothion	17	5.47

[1] Based on log-transformed value of mean lethal mortality value for host or prey/mean value for natural enemy (see Theiling and Croft 1988).

Lygaeidae, Aphelinidae, and Phytoseiidae were of about the same susceptibility as their hosts or prey. Finally, the Ichneumonidae were most severely affected, by being nearly 100 times more susceptible to pesticides relative to their hosts. Again, these data are further evidence that the Ichneumonidae are generally quite susceptible to conventional pesticides as compared to their hosts and other natural enemies. (See earlier discussion on the overall susceptibility of this family in Section 2.4.1.1.)

By compound (CPD:NAME), pirimicarb was most selective, favoring the natural enemy by approximately 1000:1 over the host or prey (see Table 2.16). Cyhexatin and dicofol, two acaricides, followed with similar values, which were equivalent to about a 100-fold selectivity to natural enemies. Most compounds clustered about the median, neutral SR:INDEX value of 5.00. Of the different materials listed in Table 2.16, the acaricides and some aphicides probably are the compounds that show the broadest ranges of physiological selectivity to different complexes of natural enemies occurring on agricultural crops (Chapter 9).

2.5. GENERALIZATION FROM THE DATABASE ANALYSIS

At present the SELCTV database provides information on trends in the published literature *within the limits of the biases previously mentioned.* Limited data availability and the current rate of literature generation make it difficult to keep the database current and representative of the global literature. For these reasons, generalizations should be made with caution. Conclusions from the database summaries should be qualified, since the distribution of information types and sources is unknown. As evidenced by the discussion presented above for the toxicity of pesticides to ichneumonid parasitoids and the comparison of natural enemy/host or prey toxicity (selectivity ratios), literature sampling is not necessarily a random process. Aggregation of information can obscure subtle influences affecting trends. However, we believe that this method of summarization gives a more complete and quantitative overview of research trends in this field than has been possible before.

In spite of its limitations, the database approach for working with large information sets provides many other additional advantages. For example, identifying trends and characteristics of an information base is a powerful use of database technology. It can provide new perspectives and help one ask appropriate questions. This has been the case with the SELCTV database. For example, this analysis has repeatedly raised the question of whether parasitoid natural enemies are more susceptible than predators. Also, are both groups of natural enemies generally more susceptible to pesticides than pests and if so, why? The related question may be asked—are parasitoids less variable in response to toxins than predators or pests? These kinds of questions cannot be answered conclusively by interrogating the database. They need to be addressed further by experimental studies. There are many other ways that the SELCTV database can be used, many of which relate to management activities such as IPM and environmental impact assessment of pesticides on these nontarget species (Theiling 1987, Croft and Hendriks 1988).

Much of the characterization and analysis revealed by the SELCTV database summary is not discussed in greater detail in this chapter. Most, if not all, of these topic are treated later in the volume. The intention of presenting this overview is to provide an initial perspective on the types and amount of research that has been done on natural enemy/pesticide interactions and to indicate some interesting questions that can be addressed to improve possibilities for biological control and IPM in the future.

PART TWO

PESTICIDE SUSCEPTIBILITY

3

MODES OF UPTAKE

3.1. INTRODUCTION

An arthropod acquires a toxic dose of pesticide by initial contact, followed by transfer of the toxicant to the target site or sites of action within its body. Acquisition of a toxic dose involves processes that are only partially understood for any pest species (O'Brien 1967, Hartley and Grahm-Bryce 1980, Gerolt 1983). Much less is known about these events for any natural enemy.

Welling (1979), in a review of the toxico-dynamic action of pesticides, discussed intoxication in terms of uptake, translocation, activation or degradation, and interaction with the target site. Each of these steps occurs internally in the arthropod, except for some aspects of uptake. This chapter deals with the external pesticide uptake before a compound is translocated to the target site. (Pesticide transfer within a natural enemy's body is discussed in Chapter 6.) External pesticide uptake primarily involves contact at the epicuticular level, through either the exoskeleton or the gut. Types of pesticide exposure to be discussed are direct or immediate contact, residue contact, and food chain uptake. All modes of uptake may act collectively to induce mortality or sublethal responses in predators and parasites.

This chapter also includes some discussion of the behavioral responses of natural enemies to pesticides, especially those that cause them to avoid or increase pesticide uptake. Other factors affecting the susceptibility of natural enemies to pesticides are treated in Chapter 4.

Because of our limited knowledge about pesticide uptake by arthropod predators and parasites, this chapter reviews some related research on pest arthropods. In many respects, pesticide uptake by pests and natural enemies is similar. Apart from pests, the unique aspects of natural enemy biology which influence the uptake of pesticides are emphasized. Speculation about aspects of pesticide uptake by natural enemies for which relatively little is known is included to suggest needed research. Ecological selectivity is the practical application of research on pesticide uptake by natural enemies versus pests (Chapter 10).

3.2. A CONCEPTUAL MODEL OF PESTICIDE UPTAKE[1]

Figure 3.1 illustrates routes of pesticide uptake by natural enemies and associated species for 1) a predator or free-living life stage of an entomophagous parasitoid, 2) its arthropod prey or host, 3) the plant host upon which the pest arthropod feeds, and 4) an ecto- or 5) an endoparasitoid life stage associated with the plant-feeding pest arthropod. Parasitism is a specialized form of predation and is represented uniquely in Fig. 3.1 by the host-associated ecto- or endoparasite life stages.

Direct mortality or sublethal effects of pesticides (Fig. 3.1, solid lines) are caused by 1) *direct contact*, which includes immediate exposure resulting from direct interception, fallout, or vapor inhalation by the beneficial organism or associated species; 2) *residue uptake*, which includes external contact coming after the toxicant has landed on a medium, usually an inert substrate or plant surface; 3) food chain transfer, which is contact received by natural enemies via feeding on pesticides contained in or on an animal or plant host.

While the above definitions offer useful distinctions for discussion purposes, in reality they are somewhat arbitrary. There are many intermediate conditions of uptake that do not fit individual classifications. Uptake is not a series of discrete events, but is more of a continuous process including several of the subcomponents described above.

Toxicant uptake may kill a free-living form of the natural enemy (Fig. 3.1a) or any exposed ectoparasite (Fig. 3.1c). An endoparasite larva is often protected from immediate contact or residue uptake within its host. However, it may be killed (Fig. 3.1c) via food chain exposure to the toxicant received from its host (Fig. 3.1e). An endoparasite may

[1] This model attempts to represent only the gross aspects of biology of an entomophage and its host. For example, many of the unique and varied aspects of natural enemy biology such as hyperparasitism, host feeding, etc., cannot be included in this model.

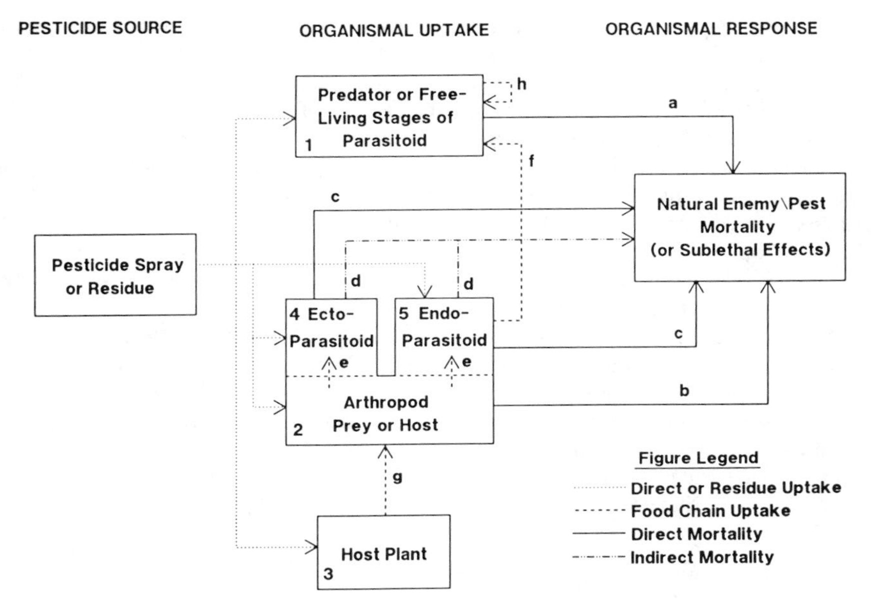

Figure 3.1. Modes of pesticide uptake by a generalized arthropod natural enemy (modified from Croft 1977).

experience a semidirect type of contact by uptake of residues through the host cuticle and trachea.[2] Pesticide uptake may cause death or sublethal effects to the host (Fig. 3.1b), which may indirectly affect associated endo- and ectoparasitoids (Fig. 3.1d). Indirect effects are not due to the pesticide itself, but to death (or some other influence) of the host or prey as it affects the entomophagous species. Indirect effects of pesticides are represented only for immature stages of parasitoids in Fig. 3.1. Immobile forms in particular and even adult predators experience indirect effects due to reductions in prey populations caused by pesticides. In fact, there is a continuum of different species forms that can be influenced by indirect effects of pesticides. They range from immobile endoparasites that are almost always killed when their hosts are killed to relatively mobile predators that are unable to find sufficient food to survive or reproduce following a pesticide treatment.

Parasitism of a host by either an ecto- or endoparasitoid can affect the susceptibility of the host to a toxicant (Fig. 3.1b). This effect is termed an *influencing factor* (see Chapter 4). Ultimately, this effect on the host can, in turn, feed back to cause indirect mortality of the natural enemy. Several other influencing factors shown in Fig. 3.1 will be discussed later in this chapter and in Chapter 4.

Toxic effects on predators or free-living parasitoid life stages (Fig. 3.1a) via pesticide transfer from other organisms can come from either external or internal sources. Pesticides can be taken up by predation and host feeding, plant and pollen feeding, cannibalism, or tropholaxis (i.e., inter- or intraspecific transfer of nourishment from one arthropod to another; Torre-Bueno 1962; Fig. 3.1h). Latent effects (Tamashiro and Sherman 1955) of

[2] Some endoparasitoids, especially certain of the Tachinidae, tap onto the tracheae of their hosts, and thus can be killed directly by volatiles. Aphid parasites, in the late larval and pupal stages, can absorb volatiles through the very thin cast skin remaining from their aphid host (Hsieh and Allen 1986). A high dose of toxicant could diffuse through host fluids and kill and endoparasitoid larva by contact rather than by a food chain pathway alone.

pesticides (Fig. 3.1f) involve transfer of pesticide from one life stage to another, resulting in delayed mortality or sublethal effects. As with immediate contact toxicity, parasitism by an ecto- or endoparasitoid of a host can affect the susceptibility of the host to food-chain mediated toxicants received from a host plant (Fig. 3.1g). This may then contribute to the mortality of the host (Fig. 3.1b) and indirect mortality of the endoparasite (Fig. 3.1d). Pesticide effects on an ecto- or endoparasitoid can be mediated by the host insect alone (Fig. 3.1e), as well as by the plant to the host and then to the natural enemy (Fig. 3.1g and e) to cause direct or indirect mortality (Fig. 3.1c or d). In some cases, food chain transfer (Fig. 3.1e) from the host to the ecto- or endoparasitoid may cause direct mortality of the natural enemy without killing the pest insect (Teague et al. 1985, Sewall and Croft 1987; see Section 3.2.3.6 on the chemotherapeutic effects of pesticides).

The multiple effects and causal factors that influence natural enemy uptake of pesticides are complex and interrelated. Few of them have been documented in the literature (Croft 1977). The toxicology, metabolism, and fate of toxins during these interactions are poorly understood for any natural enemy or chemical pesticide. Evidence for these interactions comes mostly from observations of gross responses (e.g. mortality, sublethal effects, etc.). Often, the influence of toxins can be deduced circumstantially. The relationships diagramed in Fig. 3.1 are considered below.

3.2.1. Direct Contact

Natural enemies may directly contact pesticides 1) through fallout of the toxicant while on a plant surface or other substrate, 2) by interception with aerial sprays while flying, 3) by exposure to fumigants either while flying or when on substrates, 4) from drenches or the soluble phase of soil applications of pesticides, or 5) systemically via plants and other organisms. Direct contact tends to occur shortly after application, in contrast with residue or food chain uptake that can persist for longer periods after treatment.

3.2.1.1. Direct Fallout. Basic models of factors affecting direct fallout of pesticides on pests such as houseflies, cockroaches, and mosquitoes are, in many instances, applicable to arthropod natural enemies as well (Hartley and Graham-Bryce 1980). Pests and natural enemies have many similar biological and ecological characteristics. For example, natural enemies, especially certain sedentary life stages (e.g. eggs, larvae, pupae), may be directly exposed to and temporarily immersed in a spray that is applied aerially. As with pests, the probability of a natural enemy encountering pesticide fallout is a function of the spray droplet size, its distribution, and the size of the nontarget organism. The relative proportion of a natural enemy population that is protected from a spray in a refuge may vary with species and environment (Chapter 8). For example, an organism can be located under a plant part or some other protective cover or within a host body, thereby escaping exposure to direct pesticide fallout (Sustek 1982). Hence differences in exposure to direct fallout of pesticides can occur, both among different predators or parasitoids and among different natural enemies and their hosts or prey. Differential exposure can confer ecological selectivity to a pesticide application via spatial separation (Chapter 10).

Active, adult natural enemies may be less affected by direct fallout of sprays than immatures due to their ability to escape exposure (Bartlett 1964). This may result in spatial or temporal separation of the natural enemy and pesticide (Chapter 10). Many natural enemies spend considerable time in nontreated habitats outside of the crop environment of their principal prey or hosts. They may periodically invade the treated habitat of the pest to search for food or other resources (Tabashnik and Croft 1985). Exposure to residues (as compared to direct fallout sources) of pesticides may be more common for highly mobile life stages of natural enemies than for local sedentary forms of their prey or hosts.

An example of differences in time spent in a treated versus untreated habitat among natural enemies, and between them and their prey or hosts, was reported by Talashnik and Croft (1985). In modeling resistance evolution, they estimated the time spent by 24 species of apple pests and natural enemies in a treated orchard during their life cycles. Values for these parameters were provided by scientists working pests and natural enemies of this crop. Most pests spent 85–100% of their time in the treated habitat. For alate natural enemies, values were generally 40%–98%. An extreme example was the predatory neuropteran *Chrysoperla carnea*, which spent only 49% of its life cycle in the orchard. In contrast, its prey, the European red mite *Panonychus ulmi*, the twospotted spider mite *Tetranychus urticae*, and the green apple aphid *Aphis pomi* were estimated to spend 100%, 98%, and 94% respectively, of their time in the treated habitat. Again, these differences suggest that opportunities exist to apply pesticides selectively when pests are present in the treated habitat and beneficials are outside in unexposed areas.

3.2.1.2. Aerial Exposure. Natural enemies are commonly exposed to aerial sprays of pesticides. However, this exposure is seldom if ever specifically monitored, largely because of the difficulty in studying flying arthropods (Rabb and Kennedy 1979).

Basic models of the interception of pesticide sprays by flying insects are restricted to pests. For example, the impact of droplet size and flight behavior on pesticide interception has been studied with migratory locust swarms (Hartley and Graham-Bryce 1980). These models are not very useful for natural enemies, since populations of locusts consist of fully exposed, similar individuals moving in a relatively uniform manner. Conditions would be different for a diverse population of parasitoids and predators moving within the canopy of a crop or plant ecosystem.

Achieving selectivity to natural enemies over pests would require that natural enemies be less commonly in flight during spraying or until the spray had settled or dissipated. Ecological selectivity via spatial separation would be possible if most natural enemies were sedentary and pests were actively flying; however, more often the opposite condition exists. Natural enemies are often active fliers due to their need to search for prey or hosts, especially when prey or host density is low to moderate. Under these conditions in a treated habitat, pests would be less likely to be airborne than natural enemies, and therefore aerial sprays would selectively reduce beneficial populations. Spraying may also differentially disturb pests and natural enemies, predisposing one group to exposure to aerially applied pesticides over the other (Bartlett 1966). While these scenarios of flight activity are largely conjectural, they suggest research that may help us better understand the impact of aerial sprays on pests and beneficials.

Another reason why so little research has been done on aerial uptake of pesticides by natural enemies may be because of the shortness of its duration. Normally, natural enemies and pests encounter direct fallout and aerial sprays of pesticides for only a few seconds or minutes. By comparison, residues can persist for hours or weeks in the field.

3.2.1.3. Vapor Exposure. Volatiles emanating from fumigant pesticides or from rapidly vaporizing pesticides can be extremely toxic to natural enemies (Franz 1974). This high toxicity may be because natural enemies are so much more susceptible to pesticides than their prey or hosts (Chapter 2; Theiling 1987). The size of the organism may be another important factor. Smaller organisms are generally more sensitive to pesticide vapors due to their greater surface area to volume ratio (Ripper 1956). Thus, some large natural enemies (e.g., *Chrysoperla*, spiders) may be more tolerant than their prey because of this difference in pesticide uptake. For example, coccinellid beetles may be more tolerant than aphids since vapor uptake on a surface area/weight basis is almost 10 times lower for the larger predator (Hartley and Grahm-Bryce 1980). However, many predators are of equal or smaller size than their prey. The size discrepancy between parasitoids and their hosts is even greater. These natural enemies usually are much smaller than their hosts and therefore would be discriminated against by pesticide fumes.

3.2.1.4. ***Soil Uptake.*** Soil is a complex habitat composed of aggregate particles and surrounding spaces filled with air or water. For soil-inhabiting natural enemies, this aggregate and its semiliquid environment are a principal source of pesticide uptake. As a source of pesticide exposure, soil at or near saturation is functionally similar to an aquatic habitat. When dry, it is more like a surface substrate surrounded by air (except that contact may not be confined to a single plane). Many intermediate conditions exist between these extremes. Pesticide dynamics and exposure depend on the soil type, weather conditions, and the behavior of the natural enemy (Chapter 4).

Natural enemies such as dermapterans, carabid beetles, and staphylinid beetles spend much of their time in soil habitats. Their contact with pesticides is affected by their inherent and conditioned behaviors (e.g., see discussion of sublethal effects of pesticides on carabid locomotion; Chapter 7). The factors affecting pesticide uptake and natural enemy mortality in soil habitats have been reasonably well studied (e.g., Critchley 1972a, 1972b, Kirknel 1978; Chapter 4). Those factors most commonly investigated include temperature, moisture, pH, and soil structure and type. As discussed in Chapter 4, the integration of factors influencing the susceptibility of arthropods which occur in soil habitats is probably better understood than in any other environment (including plant substrates). This is due to the relative ease in controlling exposure and measuring pesticide effects on natural enemies in this habitat (Chapter 4).

3.2.1.5. ***Systemic and Other Types of Exposure.*** Systemic pesticide transfer through a plant (or other organism) to the natural enemy is another common means of direct pesticide uptake. Systemic pesticides can be translocated to the surface of a plant or can enter a vapor phase and be toxic to a natural enemy on the plant (Cherry and Pless 1969, Cate et al. 1972). This exposure is a combination of vapor and residue uptake. Systemic transfer of pesticides to a natural enemy can also occur when the beneficial spends part of its life cycle in plant tissues. For certain hemipteran predators (e.g., mirids, anthocorids) and some hymenopteran larval ectoparasitoids and their hosts, this type of secondary association with plants is common (Clausen 1962).

Elliott and Way (1968) assessed the effects of pesticides on anthocorid species by using radioactively labeled pesticides. They evaluated the distribution of a systemic pesticide in the plant and the mortality of anthocorid eggs which were deposited in different regions of the bean plant. They found that predator nymphs suffered mortality similar to that of aphid pests. Eggs of *Anthocoris confusus* were less susceptible than eggs of *A. nemorum*, due to the lesser concentrations of the pesticide in areas of the plant that were preferred oviposition sites for *A. nemorum*.

Similarly, Aveling (1981) studied egg mortality in *A. nemorum* and *A. confusus* on hops treated with the systemic mephosfolan. Mortality of *A. confusus*, which lays its eggs in the leaves and veins, was low; however, *A. nemorum*, which lays its eggs mainly in leaf margins, had much higher mortality. Aveling suggested that mephosfolan was unevenly distributed in the leaves, and was most concentrated in the leaf margins. Behavioral avoidance of mephosfolan treated leaves was not manifested by the anthocorid adults during oviposition.

In both cases of systemic pesticide uptake by anthocorids, the effects of the toxicant on the beneficial species were not different from those observed among pests. The natural enemies exhibited behavior similar to that of the pests. The relative toxicity of the compound was likely determined by its translocation and metabolism in the plant, and the relative concentration in the plant portion where contact was made, as well as the intrinsic susceptibility of the target species.

More specialized means of direct and immediate uptake of pesticides by natural enemies probably occur in agricultural systems (e.g., with aquatic forms of natural enemies in rice ecosystems; see Dyck and Orlido 1977). Many uptake events are unique to specific natural enemy/host or prey situations. The most common types of direct uptake are those cited above.

3.2.2. Residual Contact

As alluded to previously, natural enemies most commonly take up pesticides by residual contact. While residues may be of equal or of somewhat lesser importance than direct fallout immediately following a pesticide application (See specific comparisons of contact versus residue tests of toxicity in Streibert 1981, Carruthers et al. 1985; Chapter 2), residues are more persistent phenomena. The greater susceptibility of natural enemies to low concentrations of pesticides (Chapter 2) often means that during reentry into treated habitats, predators and parasitoids may be subjected to toxic residues longer than pests. This may partially account for earlier reinvasion of treated habitats by pests than by natural enemies. Earlier pest reestablishment often leads to pest resurgence before the natural enemy can again gain regulatory control over the herbivore (Chapter 8).

The behavioral processes involved in residual uptake of pesticides by natural enemies are complex and dynamic. They include general locomotion, searching for prey or hosts, cleaning or preening, and other behavior. Pesticide deposition, redistribution, weathering, the nature of the treated substrate, and the dynamics of substrate growth or change influence residual uptake of pesticides. Several of these features are discussed hereafter, with examples of their interaction.

Understanding the dynamics of pesticide weathering in relation to natural enemy uptake has many applications to IPM and biological control. This information can help predict recolonization rates of natural enemies in treated habitats (e.g., Panis 1980; Chapter 9). The dynamics of weathering can be used to time periodic and inundative releases of natural enemies (e.g., Amaya 1982, Gupta et al. 1984, Bellows et al. 1985, Mani and Krishnamoorthy 1986). This information can help identify other means to achieve ecological selectivity with pesticides (Chapter 10).

3.2.2.1. ***Physical and Morphological Factors Affecting Residue Uptake.*** Many of the same physical processes that apply to direct contact exposure to pesticides apply to residual contact as well (e.g., distribution, particle size, nature of the treated substrate, etc.). For example, as with pests, the uptake of residual deposits during walking is affected by particle size: uptake generally increases up to a 20- to 40-μm range, but decreases at higher particle sizes (Barlow and Hadaway 1952). Particle pickup depends on the relative adhesion of the pesticide to the substrate as compared with the body part of the insect contacted. Sites of uptake of residual deposits by pests are mostly on tarsal receptors or setae (Hartley and Graham-Bryce 1980). Particles often are transferred thereafter to other parts of the body.

Gratwick (1957) measured the amounts of dye picked up by five different insect species as an index of uptake of pesticide residues. Species studied included a predatory yellowjacket wasp *Vespula vulgaris*, the carabid ground beetle *Feronia madida*, the predaceous soldier beetle *Rhagonycha fulva*, and the pests *Dysdercus fasciatus* (cotton stainer) and *Notostira erratica* (plant bug). These species were selected because of differences in size, the number of setae on their tarsi, and the number of tarsal segments, rather than for their differences in feeding habits. The actual rate of uptake *per step* was greater for the larger insects, but the smaller insects picked up more particles per unit of body weight at each step. Predaceous ground beetles, therefore, picked up more chemical, but less per unit of body weight than did pest mirids. However, as noted before, this size relationship between natural enemies and pests is not the common rule. More often, natural enemies are smaller than their prey or hosts. As with uptake of pesticide vapors (Section 3.2.1.3), this factor would lead to greater uptake of residues per unit body weight for small entomophagous forms.

Crawling (usually legless) or sluglike (e.g., vermiform) forms of natural enemies are found among many entomophagous hymenopteran and dipteran larvae of importance to biological control (Clausen 1962). These immature life stages may be exposed to pesticide residues over larger body areas and longer time periods than walking arthropods. Particles in the range of 15–20 μm are most toxic to organisms with this type of locomotion (Hartley and Graham-Bryce 1980). While these beneficials may contact more area of treated substrate than other arthropods, they also move more slowly, which could offset the potential for greater exposure. Generally, organisms with this morphology and type of locomotion are some of the most pesticide susceptible natural enemies (e.g., see extreme susceptibility of syrphid larvae; Chapter 2; Hassan et al. 1983). It is difficult to know whether their susceptibility is due to the lack of a heavily sclerotized cuticle or if crawling exposes them to higher levels of pesticide than other types of ambulatory movement (i.e., those commonly used by most other adult and immature natural enemies).

Do natural enemies take up pesticides proportionally at the same rate and through common sites of contact, the integument, tarsi, or sensillae, for example, as do pests? Intuitively, one would suspect so; however, answers to these questions are not available. Tarsal contact as a primary means of residue uptake by adults of both groups of species was mentioned earlier (Hartley and Grahm-Bryce 1980), and actual estimates of residual uptake of dyes by these anatomical parts were reported by Gratwick (1957). Detection thresholds for sugars and salts by the tarsal chemoreceptors of adult tachinid flies have been studied (e.g., Mello Filho and Batista 1983), but similar studies with insecticides have not been made. Both pests and natural enemies are highly adapted with chemoreceptors on their bodies for finding food, oviposition substrates, and mates (Askew 1971). Specialized parasitoids have highly evolved olfactory systems for host habitat and prey finding (Vinson 1975). What role these specialized organs play in pesticide detection and uptake is unknown.

3.2.2.2. ***Behavior and Residue Uptake.*** A natural enemy has little or no opportunity to escape direct fallout of pesticides, except for adults, which may take flight. Behavior of an organism plays an important role in determining the extent of residual exposure. Avoidance of residues can confer selectivity to the natural enemy (Bartlett 1966, Hislop et al. 1981, Hoy and Dahlsten 1984). Innate and modified behavioral responses of natural enemies are some of the primary determinants influencing the uptake of pesticide residues (Hartley and Graham-Bryce 1980).

Gratwick (1957) noted the importance of body-cleaning movements both in spreading particles picked up on the tarsi and in enabling pests and natural enemies to remove pesticides from their bodies. Just how these actions affect pesticide uptake and redistribution has not been studied to any great degree. Many authors have commented on the extensive preening and cleaning activities of parasitic wasps (i.e., DeBach 1964). In more specific studies, Kuhner et al. (1985) observed that by cleaning of its legs, wings, and head, adult *Diaeretiella rapae* contaminated its whole body with the moderately toxic herbicide Ramrod. Activity rates were increased and normal searching patterns greatly modified by sublethal levels of this herbicide. These studies suggest that uptake of poisons via these types of behavior may be substantially greater for parasitoid adults than for larval parasitoids or their hosts.

Since natural enemies may take up more pesticide per unit body weight per step (because of their smaller size) *and* search longer on treated substrates than pests, these factors together may multiply the overall uptake of pesticide residues. Many authors have speculated that this is a major explanation for the greater susceptibility of predators and parasitoids over their prey or hosts (e.g., Abdelrahman 1973, Franz 1974, Hoy and Dahlsten 1984, Waage et al. 1985). However, little research to prove or disprove this point has been published. Additional work is needed to identify the role that the combined influences of smaller size and greater searching activities of natural enemies have in determining their susceptibility to pesticides.

Hartley and Graham-Bryce (1980) pointed out that factors contributing to greater pesticide exposure in natural enemies must be balanced by those causing them to receive less exposure. For example, the lesser tendency for natural enemies to be present in the treated habitat than pests should be considered. Within the treated habitat, natural enemies also have a tendency to be in refugia from pesticides more often than pests. In projecting net impact, field toxicity data and other types of evaluations indicate that pesticides having persistent residues generally discriminate against natural enemies as compared with their hosts or prey (see Chapter 2). This may be a partial explanation as to why short-lived, nonpersistent compounds often show relatively selective properties to natural enemies over pests, and why more persistent compounds have such devastating effects on these smaller, more active arthropods (Chapter 10).

Another behavioral attribute of natural enemies that influences residue uptake is searching activity relative to prey density and distribution. A pest's food supply (the crop) in a treated environment is usually unlimited since it is the object of protection. However, a pesticide spray often decimates a natural enemy's food supply (the pest) and alters its distribution in space. A predator or parasite's rate of movement may increase as hunger escalates (e.g., Chiverton 1984, Hoy and Dahlsten 1984). This indirect effect associated with pesticide impact has seldom been studied directly, although several researchers have commented on its role in conferring high residual uptake by natural enemies (Abdelrahman 1973, Croft 1977, Waage et al. 1985).

Hoy and Dahlsten (1984) have demonstrated the type of research needed to evaluate the influences of pesticides on the searching behavior of natural enemies and subsequent pesticide uptake. They examined how pesticide residues influenced rates of locomotion and prey finding. Although some repellent effects were noted, sublethal doses also resulted in increased walking speeds and an overall increase in toxicant uptake (see also Chapter 7).

In summary, pesticides can directly influence the rate of locomotion of a natural enemy (Chapter 7), increasing pesticide uptake. Pesticides also may indirectly increase the beneficial's rate of locomotion through decimation of its prey or hosts, further increasing rates of pesticide uptake. More detailed study of behavioral factors and their interaction with other aspects of pesticide uptake may improve selective pesticide use in the future. It would be especially helpful to identify those compounds and application rates that do not cause both decimation of the natural enemy's food supply and increased uptake of pesticide residues by the natural enemy (see further discussion of the benefits of maintaining an adequate food supply for natural enemies which allows them to develop pesticide-resistant populations in Chapters 16 and 19).

An interesting aside to this discussion of comparative uptake of pesticides by pests and natural enemies is the differential searching behavior of primary and secondary parasitoids as impacted by pesticide residues. Horn (1983) noted that *Diaeretiella rapae*, which attacks *Myzus persicae*, is parasitized in turn by a complex of secondary parasitoids including *Aphidencyrtus aphidivorus* and *Asaphes lucens*. In field plots treated with carbaryl and malathion, he observed higher rates of primary parasitism by *D. rapae*, but lower rates of secondary parasitism by hyperparasites. Horn attributed this susceptibility differential to the searching behavior of the primary and secondary parasitoids. Secondary parasitoids tended to walk slowly over leaves searching for suitable hosts, whereas *D. rapae* spent more searching time flying among leaves. Hence, the latter species was less likely to acquire a lethal dose of insecticide. This type of study further emphasizes that differences in behavior may be a key to achieving better pesticide selectivity. More emphasis on these types of behavioral investigations is needed.

3.2.2.3. ***Pesticide Deposition.*** As implied earlier, relationships between pesticide distribution, residue levels, and selectivity in ambulatory arthropods are not well understood. Since residual deposits of pesticides are not uniform in the field, insecticide uptake should be a function of the area of insecticide contacted and the accumulated time of exposure. Generally, the more mobile an insect, the more vulnerable it should be to a highly dispersed pesticide. However, these variables do not act in solitary. For example, there may be conditions where total contact would be greater between a toxicant and a more sedentary pest. A slow moving species might remain in contact with a pesticide deposit long enough to receive greater exposure than the larger number of deposits contacted by a faster moving natural enemy. Of course, overall impact must be measured at the population level.

Unfortunately, studies of optimal residue distribution needed to achieve ecological selectivity have seldom been investigated. Identifying such distributions would require knowledge of both pest and beneficial movement. These data would have to be related to the changing residual deposit pattern of the pesticide so that some "mean free path" (Hartley and Graham-Bryce 1980) or probability of contact for each population could be predicted.

Casegrande and Haynes (1976) reported a useful study of the dynamics of pest behavior and pesticide residues. In seeking to optimize strip spraying, they evaluated alternatives for selectively controlling the cereal leaf beetle *Oulema melanopus* using as little pesticide as possible. Their simulation and experimental analysis were based on studies of beetle movement, pesticide degradation on plants, accumulation and degradation of the pesticide in the insect, and other aspects of pest behavior and mortality. Although similar comparisons were not made for the parasitoids of this pest, this kind of research could foster greater levels of selective pest control (Hall 1986; Chapter 10).

3.2.2.4. ***Weathering of Residues and Substrate Dynamics.*** As noted, the nature of the treated substrate and the weathering of residues are important variables influencing pesticide uptake by natural enemies. Reports of the toxicity of weathered pesticides to predators and parasites are common in the literature for a wide variety of crops and species (Theiling 1987; Chapter 2). Most studies, however, are only timed evaluations, with little reference to the factors influencing changes in residues. Few studies have monitored and correlated the dynamics of substrate growth, environmental factors, toxicant levels, and natural enemy behavior with the changes that occur in residue levels (e.g., Bellows et al., 1985, Morse and Bellows 1986; see below).

The most extensive studies of factors influencing the dynamics of pesticide residues on arthropod natural enemies have been done on citrus crops. Citrus grows year-round in arid regions and only sporadically experiences rainfall. The frequency and extent of rainfall has a significant influence on the need to spray for pest control.

Pioneering studies by Bartlett (1953, 1963, 1964) evaluated the effects of weathered residues on citrus on predators and parasitoids using experimental methods that were quite advanced at the time. He exposed natural enemies to residues of insecticides and acaricides at timed intervals on foliage that had been weathered under laboratory conditions. Residue losses were compared under controlled conditions, but simulation of the dynamic environmental factors that occur under field conditions was not considered. These studies provided a conservative measure of the survivorship of natural enemies in the field, since degradation of field residues would usually be more rapid.

Brandt (1982) tested the toxicity of a wide range of sprays used in commercial citrus production on the scale parasitoid *Comperiella bifasciata*. Field data were compared with laboratory data from studies where rainfall was simulated in a manner similar to the field. Results demonstrated that endosulfan residues were toxic for only two weeks during the early winter season, whereas temephos and triazophos residues were toxic for about 6 weeks.

Campbell (1975) tested the toxicity of organophosphate insecticide residues to adult *Aphytis melinus*, another scale parasite, on citrus foliage at periodic intervals after trees were sprayed in the field. Residues of 0.15% formulated malathion applied in late winter, spring, and summer were lethal for 29, 21, and 16 days, respectively. Toxicity at each seasonal interval was related to the effects of changing temperature, rainfall, and the variety of citrus leaves used as substrates.

Bellows et al. (1985) evaluated the residual activity of insecticides on the aphelinid *Aphytis melinus*, the coccinellid *Cryptolaemus montrouzieri*, and the phytoseiid mite *Euseius stipulatus*. These species attack key citrus pests. Foliage samples were collected at regular intervals from trees sprayed with acephate, dimethoate, formetanate, and sabadilla, and were analyzed for extracuticular residues. When environmental factors (e.g., rainfall) were accounted for, residue levels of pesticides coincided well with results of bioassays of acute toxicity over a 70-day interval. One exception was formetanate toxicity to *A. melinus*, which declined steadily during a period when residues remained relatively constant. They attributed this anomaly to other factors (probably formulation materials) that were conferring the changes in activity.

Morse and Bellows (1986) further reported the 48-hr residual toxicities for acephate, dimethoate, formetanate, sabadilla, chlorpyrifos, carbaryl, parathion, and methidathion to *A. melinus* and *C. montrouzieri*. By combining their results on short- and long-term residual toxicity, the impact of pesticides on these species in the field was better understood.

The increasing level of complexity and sophistication used in the above studies shows an evolution in methodology. It reflects the perceived need to more completely understand the dynamics of weathering of pesticide residues that occurs in the field. Further studies of these dynamic relationships are needed. When experiments are run in parallel with monitored real-time weather data, more timely decisions about selective chemical control of pests in the field can be made.

3.2.2.5. Exposure to Residues via Other Organisms and Sources. Contact via external residues on other organisms and on sources other than plant surfaces is a further component of pesticide uptake by natural enemies (Fig. 3.1). Residual uptake occurs through ingestion of foods, water, or other substances. Sources of external residues may be other plant parts, arthropod excrement, prey or hosts of natural enemies, members of their own species (via cannibalism), or other arthropods laden with pesticides.

With arthropod sources, there is probably little difference between uptake of external residues and internal food chain sources of pesticides (Section 3.2.3) insofar as toxicity is concerned. One difference might be in the extent of metabolism of the pesticide, when coming from an internal versus an external source. Pesticides via food chain (internal) sources would be more likely to have been metabolized than from external sources. Metabolic transformation could render a toxicant more or less toxic to a beneficial, depending on the compound and host or prey involved. Some natural enemies may primarily feed on internal fluids of their prey or hosts (e.g., piercing-sucking natural enemies), such as hemipterans, mites, and many parasitoid adults, and would therefore be exposed primarily to pesticide via the food chain. Conversely, chewing natural enemies (such as many predators) usually consume whole organisms, and would be exposed to both external and internal pesticide deposits in their hosts or prey.

There are several ways that external residues of toxicant can be transferred to a natural enemy. Adult parasitoids or predators may obtain sustenance from consumption of plant nectars, pollens, saps, free water, insect excrement (e.g., honeydew), and exposed body fluids of their hosts. Many studies have demonstrated the toxicity of pesticides administered via these sources (e.g., Brettell and Burgess 1973, Cate et al. 1972, Hegazi et al. 1982). Other studies have documented residual uptake of pesticides via predation (e.g., Ahmed 1955, Nakashima and Croft 1974, Teotia and Tiwari 1972, Kabacik-Wasylik and Jaworska 1973). Both external and internal food chain residues are usually involved. In no studies have careful distinctions been made between pesticide uptake from internal sources versus external residues. The primary goal of most research has been to determine whether or not natural enemies are conserved, rather than to identify the sources of pesticide exposure.

Cannibalism is another means of uptake of external (or internal) pesticide residues by natural enemies. Many predaceous species and immature stages of parasitoids are cannibalistic (Clausen 1962). Cannibalism among different life stages (e.g., Berendt 1973) can serve as a source of external residue uptake to a free-living natural enemy, as can feeding on fragments such as cast skins or the egg chorion (e.g., among certain coccinellids; Zeleny 1965).

To generalize, oral uptake of pesticide residues from other organisms can confer toxic doses of pesticides to natural enemies (Bartlett 1964, Hegazi et al., 1982; Chapter 2). This applies to both external residues and contamination coming from within the host's or prey's body. (See discussion of food chain sources in Section 3.2.3.) However, little study of these relationships has been made. Free-living forms of natural enemies have considerable potential for behavioral avoidance of pesticide residues via the oral route (Bartlett 1966, Jackson and Ford 1973).

3.2.2.6. Measuring Residual Effects of Pesticides. Differences among natural enemies and their hosts or prey in residual uptake versus direct fallout is an important consideration in developing methods to test side effects of pesticides (Chapter 5). Several points made above address these issues. Since natural enemies are often more susceptible to conventional pesticides than pests, most exposed forms are initially eradicated from treated habitats by sprays targeted at pests. Subsequently, they are commonly exposed to residues encountered by immigrants coming from refugia within or outside the treated area. Therefore, the dynamics of residue breakdown or redistribution are extremely important in determining the reentry period for reestablishment of biological control agents. Although direct contact tests for predators and parasites may be useful for measuring "intrinsic" differences in susceptibility, residue tests probably better estimate pesticide impact in the field (Chapter 5).

Prevailing modes of pesticide uptake also have bearing on achieving ecological selectivity with pesticides. For example, different application techniques, formulations, and adjuvants can greatly influence the dynamics of pesticide residues. These factors are discussed more in Chapters 4 and 10. In evaluating different methods of pesticide application, direct contact, residue, and food chain means of uptake are seldom quantified separately. In part, this is because the goal of most pesticide studies is net impact. Measurements are rarely taken frequently enough to separate more immediate contact from long-term residue or food chain exposure. Individual modes of uptake can best be estimated by laboratory experiments where residue levels and uptake can be carefully measured.

3.2.3. Food Chain Uptake and Transfer

Certain toxicants administered via a plant to a pest often will kill the pest, but not cause mortality to a natural enemy which subsequently feeds upon the pest. Many studies have demonstrated the selectivity of systemic pesticides (e.g., Ripper et al. 1949, Ahmed et al. 1954, Ahmed 1955, Ripper 1956; Chapters 9 and 10); however, systemic pesticides are not always selective. There have probably been more studies published which demonstrate significant mortality to natural enemies by feeding on the pesticide laden tissues of their prey or hosts. In addition, natural enemies take up pesticides from food chain sources in pest honeydews or plant nectars, pollen, and other internal sources of toxicants (Croft and Brown 1975). Hereafter, each of these sources of pesticide uptake is discussed.

3.2.3.1. Plants as Sources of Food Chain Uptake. Many natural enemies feed on plant materials at some stage in their life cycles to obtain nutrients or water (e.g., Brown and Shanks 1976). This type of uptake has been studied extensively with phytoseiid mites. Deneshvar and Rodriquez (1975), working with *Amblyseius fallacis*, observed no uptake of internal plant fluids radiolabeled with P^{32} when predators were confined to been leaves without prey. Data indicated no direct feeding on leaves by predators. When fed only on labeled prey, there were indications of transfer of systemic insecticides through the food chain. Croft and Blythe (1980), working with *A. fallacis* confined on bean leaves, observed that predators held without water took up fluids from leaves. This was confirmed through the use of translocated dyes. Porres et al. (1975), using radiolabeled phosphoric acid applied systemically, showed conclusively that *Amblyseius hibisci* accumulated tracer when held without food on avocado leaves. This predaceous mite extracted sap from the plant. Finally, Congdon and Tanigoshi (1983) observed significant mortality of *A. hibisci* when confined without prey on grapefruit leaves which were systemically treated with dimethoate. In separate experiments, they had shown that no mortality had occurred from feeding on poisoned prey (*Scirtothrips citri*).

In summary, studies with predaceous phytoseiid mites have shown that certain species directly take up systemic toxicants administered to plants. Following foliar treatment with a systemic poison which decimates their prey, predators may survive initial fallout, residue, and food chain exposures. However, they may again be exposed in a more susceptible physiological state of semistarvation by taking up water or nutrients from the treated plant. (See further discussion of the effects of starvation on susceptibility in Chapter 4, Section 4.3.2.)

Similar uptake of pesticides from facultative herbivorous/entomophagous hemipterans (Miridae, Anthocoridae, Nabidae, Pentatomidae) is probably a frequent event. This type of uptake was referred to earlier (Section 3.1.2.5) in a study by Elliott (1970). He recorded the toxicity of several systemic insecticides to anthocorid nymphs, which secondarily fed on plant fluids. In this instance, the natural enemy received a toxic dose as did the pest, except that it may have fed less extensively on the plant compared to a strictly phytophagous species.

Some interesting examples of the absence of ecological selectivity to natural enemies having both entomophagous and plant feeding habits have been documented in rice systems, where systemic poisons were administered to control planthopper and leafhopper pests (Chiu and Cheng 1976, Ku and Wang 1981). Researchers were trying to identify means to conserve parasitic wasps (*Platygaster* spp.), predaceous spiders (*Lycosa, Oedothorax*), and predaceous bugs (*Cyrtorhinus lividipennis*). Compounds were identified that did not reduce spider populations because they fed only on the hemipteran pests. However, it was much harder to find compounds that conferred a selective advantage to the mirid predator, which was both entomophagous and herbivorous (Chapter 10).

Studies demonstrating pesticide uptake from plants by parasitoids have been uncommon, probably because phytophagy is rare among them (Clausen 1962). In most cases, parasitoids take up pesticides while consuming pollen or sugars and nutrients from nectars. For example, monocrotophos and aldicarb were toxic to the ichneumonid parasitoid *Campoletes perdistinctus* when applied to cotton as a granular and translocated to the nectar of the plant (Cate et al. 1972). Residue levels in nectars containing monocrotophos were toxic for up to 6 weeks after treatment, whereas those containing disulfoton had little effect on inundative releases of adults made in the field.

3.2.3.2. Uptake via Predation. Documentation of pesticide transfer through predator–prey food chain systems is extensive. Selected studies are summarized here. Others are reviewed in Croft and Brown (1975) and Croft (1977).

McClanahan (1967) first demonstrated food chain toxicity of dimethoate, phorate, and thionazin to *Phytoseiulus persimilis* (Phytoseiidae) when fed twospotted spider mites. Binns (1971) observed both the presence and absence of food chain toxicity to *P. persimilis* when fed *T. urticae* after several systemic compounds (thionazin, methomyl, oxydemeton-methyl, aldicarb) were applied to cotton for aphid control. Nakashima and Croft (1974) observed reduced oviposition and secondary poisoning of several life stages of *A. fallacis* when adults were fed benomyl-treated *T. urticae*. Lindquist and Wolgamott (1980) showed increased mortality in *P. persimilis* when its prey, *T. urticae*, was fed lima beans drenched in phorate. Sohliesske (1979) observed high mortality when *P. persimilis* was fed on spider mites which had received aldicarb and demeton-*S*-methyl through a plant host.

Among predators of aphid pests, similar responses have been noted among the Syrphidae (Ahmed 1955), Coccinellidae (Ahmed 1955, Singh and Malhotra 1975b, Satpathy et al. 1968), and Pentatomidae (Wegorek and Pruszynski 1979). Ahmed (1955) observed that when *Coccinella undecimpunctata* were fed poisoned aphids having external residues of demeton-*S*-methyl, predator mortality was near 100%. However, when aphids were exposed to similar rates of the systemic without external contact, only 2% of the predators feeding on them were killed. Satpathy et al. (1968) observed the poisoning of coccinellid predators feeding on both live and dead aphids. Predatory carabids feeding on dead or moribund prey also sustained high levels of mortality (Kabacik-Wasylik and Jaworska 1973, Gholson et al. 1978).

A few generalizations can be made regarding the uptake of systemic poisons from prey by predators. Toxic effects apparently can be conferred from the consumption of live, moribund, or dead prey. Pesticide uptake by a predator from its prey is analogous to the uptake of a systemic insecticide from a treated plant by a pest. The extent of the effect on natural enemies likely depends on the behavior and feeding habits of the prey (amount and rate of toxicant received), the fate of the toxicant inside the prey (localization, concentration, and metabolism of the toxicant), the feeding habits of the predator (amount and rate of uptake), and the metabolic and detoxicative abilities of the predator.

Metabolism and transfer of a toxicant from plant to herbivore versus from herbivore to natural enemy may differ appreciably. The physiological systems involved in each case are quite different from each other (Chapter 6). Since a toxicant is selected for its lack of phytotoxicity, it is unlikely to be metabolized by the plant in the process of transfer to a pest; however, physiological systems of pests and natural enemies are more similar. A pest is more likely to metabolize the toxicant. Subsequent effects on the beneficial would depend on how toxic the by-products were to the natural enemy. A good example of this type of conversion among pests is that of phosphorothioates to phosphates (e.g., in the ester oxidation of parathion to paraoxon; Chapter 6). Few studies of the comparative transfer and metabolic alteration of pesticides in plant–herbivore versus herbivore–entomophage food chains have been examined. These types of studies may help explain the repercussions of systemic pesticides through the food chain, ultimately contributing to our understanding of selectivity mechanisms (Chapters 9 and 10).

3.2.3.3. Intraspecific Transfer. As with external residues (Section 3.2.2.4), food chain transfer of a toxicant can occur from one intraspecific organism to another. Cannibalism is one such source of intraspecific food chain transfer for both predators and internal parasitoids. In some species, cannibalism occurs almost independently of prey density (e.g., coccinellid and chrysopid larvae). At low prey density and high natural enemy density, many natural enemy species will engage in cannibalism and, in the process, receive a toxic dose of pesticide.

Aveling (1981) noted that young anthocorid nymphs (*Anthocoris nemorum* and *A. nemoralis*), which hatched early and were in the absence of aphid prey, often fed on unhatched eggs of their own species. Mortality among these nymphal forms was appreciable since these eggs had taken up the systemic organophosphate mephosfolan.

Another mechanism of intraspecific transfer of pesticide is tropholaxis. Pesticides in toxicant-laden food supplies may be transferred among ants and other social entomophagous insects. Dmitrienko (1979) administered chlordane and several organophosphate insecticides to adults of predaceous ant colonies. He observed high rates of mortality in immature life stages which were subsequently fed these pesticide-laden food materials. Undoubtedly, other types of intraspecific transfer occur. However, it is likely that these are only minor routes of toxicant uptake in nature.

3.2.3.4. Uptake via Honeydew and Host Feeding. It has long been known that certain pesticides, which are moved systemically through the plant–herbivore food chain via honeydew from homopteran species, can be toxic when fed upon by predators or parasitoids (Doutt and Hagan 1950, Bartlett 1964). While honeydews may be repellent to natural enemies, toxicants taken up through feeding on honeydews can be extremely toxic to natural enemies of many groups (Bartlett 1964, 1966). On the other hand, very little has been published about the systemic effects of pesticides on adult parasitoids via oviposition and host feeding or on immature endo- or ectoparasitoid stages. Major exceptions are the chemotherapeutic effects of synthetic pesticides on certain lepidopteran pests (see more detailed treatment of this phenomenon in Sections 3.2.3.5 and 3.2.3.6) and the effects of microbial pesticides on host–parasitoid systems (Chapter 11).

3.2.3.5. Food Chain Accumulation. Pesticide movement in the food chain may result in metabolic concentration of the toxicant at the predator–parasitoid level. Such pesticide biomagnification has been most commonly documented for persistent pesticides, especially among predators in aquatic and soil environments.

Wilkes and Weiss (1971) measured a 2700-fold accumulation of DDT in dragonfly nymphs (*Tetragoneuria*) due to feeding on pesticide-laden prey and direct uptake from water. Naqui and de la Cruz (1973) found higher mirex levels in predaceous dragonflies, spiders, and waterboatmen than in other aquatic herbivorous and omnivorous species. In contrast, van Halteren (1971) found no evidence of bioaccumulation in mantids fed dieldrin-treated *Drosophila.*

Gilyarov (1977) reviewed the role of small soil-inhabiting arthropods in the decomposition of DDT. He noted that certain predators, such as staphylinids (that feed on collembola), readily degraded DDT to DDE. Other predators such as spiders, myriapods, and carabids retained the DDE that entered their bodies after consuming contaminated collembola. Pesticides were not substantially concentrated by the predators.

Conversely, Korschgen (1970) observed that the carabid *Pterostichus chalcites* concentrated residues of organochlorine pesticides to high levels. Average aldrin and dieldrin levels found in these predators were 31 times those found in soils, 42 times those found in crickets, and 8.8 times the level found in an associated carabid, *Harpalus.* He attributed the abnormally high levels of pesticide to the predaceous feeding habits of the species. Humphrey and Dahm (1976) studied *P. chalcites, Harpalus pennsylvanicus,* and *Scarites substriatus* to see whether abnormally high levels of chlorinated hydrocarbon insecticides were associated with the lipids of these beetles. No consistent relationships were found.

A relatively detailed study of food chain toxicity in a plant–herbivore–natural-enemy system was conducted by Kiritani and Kawahara (1973). They examined the fate and effect of BHC passing from irrigated soil to rice, to the green rice leafhopper *Nephotettix cincticeps,* and finally to the predaceous spider *Lycosa pseudoannulata.* After feeding on rice for 2 days, the leafhoppers contained three times are much BHC as the rice plants. BHC applied to the rice plants contained 13% of the gamma isomer. The gamma isomer was found to increase and accumulate successively in plants and then in the leafhoppers. The proportion of the gamma isomer reached 35% in *N. cincticeps.* Biological concentration of gamma-BHC was particularly significant since the spider was much more susceptible to this isomer than its leafhopper prey. Similarly, in model studies of the metabolism and fate of DDT, aldrin, dieldrin, and lindane in a rice paddy ecosystem, Hsu and Hsu (1980) observed high levels of biomagnification and low biodegradation indices in the arthropod pests *Oxyahyla intricata* and *Nilaparvata lugens* and the predaceous spider *Lycosa pseudoannulata.*

Bioaccumulation of more recently developed, less persistent organophosphate, carbamate, pyrethroid, and biorational pesticides (e.g., microbials and insert growth regulators—(IGRs)) has received little study. However, there is some evidence that such bioaccumulation can occur with these pesticides. Increased concentrations of carbamate insecticide were observed in natural enemies in rice ecosystems involving plant leaf-hoppers and predaceous spiders (Choi et al. 1978). A lack of evidence for bioaccumulation was reported by Hagstrum (1970), who evaluated the concentration of carbaryl residues in the spider *Tarantula kochi* fed on topically treated houseflies.

3.2.3.6. Food Chain Uptake by Ecto- and Endoparasitoids. The interaction between a parasitoid and its host relative to toxicant transfer is complex and involves numerous feedback loops. Toxicant transfer is mediated by the physiology of the host, and is initially determined by the extent of penetration through the parasitoid cuticle. The insecticide can cause death of the host, which may or may not kill the parasitoid depending on its stage of development (Lingappa et al. 1972, Tamashiro and Sherman 1955, MacDonald and Webb 1963). If the host survives, it may detoxify the pesticide or convert it to a more toxic metabolite which can subsequently affect the internal parasitoid.

To date, most literature on host–parasite pesticide interactions has focused on the selective use of insecticides to minimize mortality of internal parasitoids. Tamashiro and Sherman (1955) evaluated the impacts of several insecticides on larval *Dacus dorsalis* parasitized by *Opius oophilus*. Reports by Lingappa et al. (1972) with the greenbug and its parasite *Lysiphlebus testaceipes*, and by Washizuka and Kuwana (1959) with rice stem borer eggs and *Trichogramma japonicum*, give insight into the dynamic interrelationships between the developmental stage of the parasitoid and the effect of the pesticide. In most cases, pesticide impact was dependent on the stage of the parasitization process when toxicant exposure occurred.

One of the most detailed studies of host–parasitoid–pesticide interactions was reported by Novozhilov et al. (1973). They conducted detailed studies with *Trissoleus grandis*, a scelionid egg parasite of the scutellerid *Eurygaster integriceps* in the USSR. After pesticide treatment, eggs were washed and pulverized to measure homogenate levels of the organophosphate trichlorfon. The chorion of the host egg absorbed from 95% to 99% of the toxicant. The amount of toxicant in the homogenate of parasitized eggs was much less (0.7–4.43%) and varied according to the developmental stage of the egg parasite at the time of treatment (Fig. 3.2). To some extent, the increasing rate of penetration was attributed to the aging of the host egg and differences in gas exchange. Differences in penetration were suggested to be due to the disruption of embryonic tissues of the host as the parasite developed. Data indicated that host embryos were to some degree involved in the active exclusion of the toxicant. Data on the mortality and emergence of parasites are shown in Fig. 3.2. Emergence of adult parasites for different treatments was highest when host eggs were treated while the parasite was in its egg stage. Emergence was lowest when host eggs were treated just prior to adult parasite emergence. Similarly, mortality of parasite adults was highest when treatments were applied to host eggs containing parasitoid pupae. Adults emerged shortly thereafter and contacted residue on host eggs.

Microbial agents as well as the newer insect growth regulators and chitin inhibitors generally show an even greater spectrum of food chain effects on endo- and ectoparasitoids and their hosts (see Chapters 11 and 12). Flexner et al. (1986) noted that for microbial insecticides, host mortality often affected the obligate endoparasitoid more than the direct toxic effect of the microbial pesticide. The complex hormonal association between a host and its parasite make application of juvenoid compounds highly destructive to developing parasites at particular stages of development (Beckage 1985; Chapter 12). At other stages

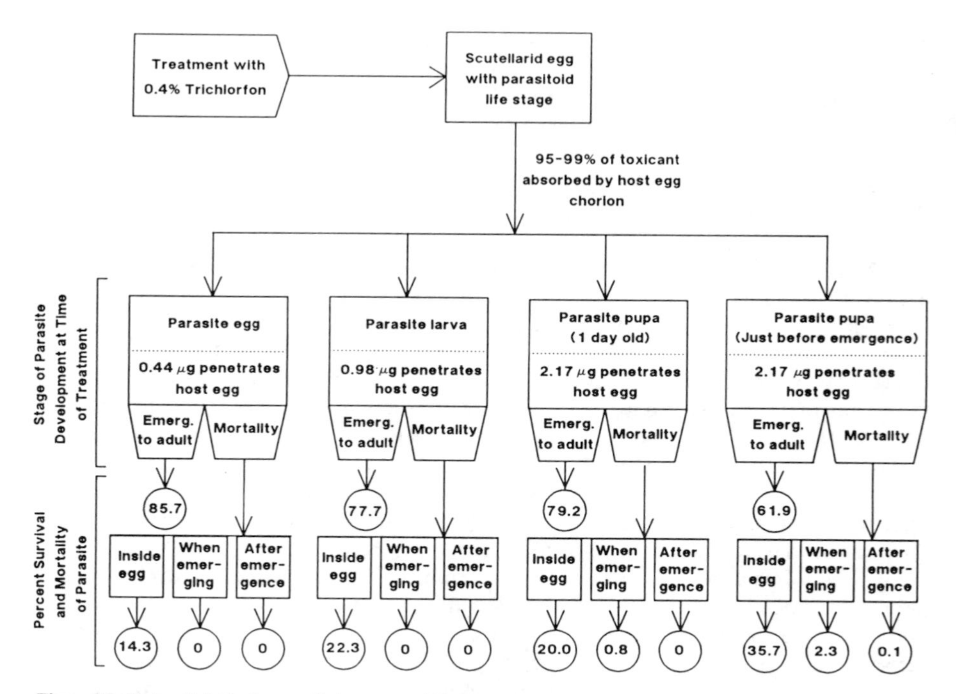

Figure 3.2. Fate of trichlorfon applied to eggs of *Eurygaster integriceps* (Scutellaridae) parasitized by *Trissoleus grandis* (Scelionidae) (after Novozhilov et al. 1973).

of development, only the host is affected. Side effects of chitin inhibitors similarly seem to be more stage-specific to natural enemies than are most conventional pesticides (Chapter 12). As with pests, late larval stages, pupae, and newly formed adults are highly sensitive to the effects of these pesticides.

Some of the most detailed physiological and toxicological studies of host–parasitoid interaction have come from evaluations of the impact of juvenile hormone-type compounds (Beckage 1985). These types of investigations are discussed further in Chapter 12.

3.2.3.7. ***Chemotherapeutic Effects on Endo- and Ectoparasitoids.*** Another very specific interaction between toxicants, parasites, and their hosts is what one might term a "chemotherapeutic" effect (Croft and Brown 1975, Sewall 1986). In these cases, a pesticide does not cause host mortality, but will kill an ecto- or endoparasitoid and may free the host from its natural enemy (Felton and Dahlman 1984, Teague et al. 1985, Horton et al. 1986). This phenomenon has been noted primarily among relatively large lepidopteran hosts and their hymenopteran parasitoids (Taylor 1954, Thurston and Fox 1972, Teague et al. 1985, Sewall and Croft 1987).

Detailed documentation of a chemotherapeutic effect was first reported by Teague and coworkers (Teague et al. 1985, D. L. Horton, unpublished studies). When incorporated into the artificial diet of *Heliothis zea* parasitized by the braconid *Microplitis croceipes*, the benzimidazol fungicide benomyl caused no effect on *H. zea*, nor did it affect parasitism of the host by adult wasps. However, subsequent emergence of parasites from the host was

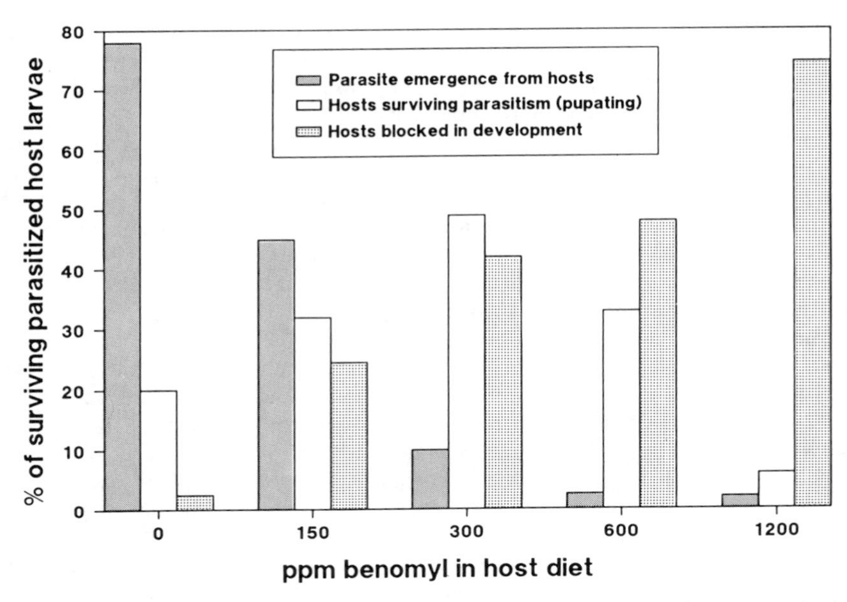

Figure 3.3. Effects of benomyl-treated artificial diet fed to third instar orange tortrix *Argyrotaenia citrana* (Tortricidae) on development of the endoparasitoid *Apanteles aristotillae* (Braconidae) (after Sewall and Croft 1987).

depressed by more than 90% at higher concentrations. Similarly, four noctuid larvae, *H. zea, Spodoptera exigua, S. ornithogalii*, and *Pseudoplusia includens*, were reared on a treated synthetic diet varying in benomyl concentrations. While hosts were not significantly affected, parasite (*Apanteles marginiventris*) emergence from the first three species was significantly depressed by increasing benomyl concentrations. Detailed dissections revealed that benomyl caused the death of parasite larvae before hosts were sufficiently damaged to cause mortality. They attributed the lack of similar parasite mortality in the host *S. ornithogalii* to unknown, unique differences in this particular host–parasite interaction. It was speculated that application of this compound in the field could subsequently affect biological control for more than one generation and contribute to increased pest survival and pest resurgence, or to secondary pest outbreaks.

In similar studies with a tortricid moth, Sewall and Croft (1987) incorporated benomyl into an artificial diet fed to third instar orange tortrix *Argyrotaenia citrana*. A concentration nontoxic to the host after parasitization was toxic via the host to the solitary endoparasitoid *Apanteles aristotillae* during development (Fig. 3.3) This same concentration, fed to hosts after parasitization, eliminated parasites from the host and increased the number of host larvae pupating as compared with controls. The effect was similar for hosts fed benomyl laced diet for only one day after parasitization versus those fed benomyl continuously. Increasing levels of benomyl caused increasing percentages of hosts blocked in larval development for up to 120 days (Fig. 3.3). Increased levels of blocked hosts were associated with parasite larvae dissected from hosts which did not pupate after one month of larval development.

3.2.3.8. Latent Transfers. Latent effects of pesticides are those expressed by a life stage of a natural enemy subsequent to the one initially exposed. As discussed in

Chapter 7, there is a fine distinction between this delayed mortality and the sublethal effect of reduced longevity. There has also been considerable debate over a proper definition of this term (Chapter 7). In this chapter, the relationship of latent effects to pesticide uptake and transfer to the natural enemy is emphasized.

Tamashiro and Sherman (1955) first described the delayed mortality from pesticides observed in adult *Opius oophilus*, a parasite of the oriental fruit fly *Dacus dorsalis*. When larval hosts containing parasitoid larvae were treated with parathion and six organochlorine compounds, emerging adults exhibited symptoms of pesticide poisoning and died shortly after eclosion. Parasites exposed to aldrin, dieldrin, endrin, and chlordane exhibited latent mortality ranging from 50% to 54% of the total emerging population at the highest dosages tested (1 to 40 μg/g body weight).

Similarly in the primary parasitoid *Diaeretiella rapae*, attacking cabbage aphids (*Brevicoryne brassicae*), Askari et al. (1984) noted that postemergent mortality of parasite adults occurred, even when camphor treatments were confined to immatures contained in aphid mummies. In the latter case, there was speculation as to whether or not the mortality observed was due to delayed exposure (as adults emerged) rather than a true latent effect.

Wiackowski (1968) investigated the effects of several organophosphate insecticides on larval *Chrysoperla carnea* and observed considerable tolerance in the stage directly treated. However, up to 80% mortality occurred in subsequently developing pupae and adults. Sell (1985) studied the effects of exposing 2- and 3-day-old larvae of *Aphidoletes aphidimyza* to several fungicides and insecticides used in greenhouse pest control. Insects surviving treatments exhibited limited effects on predation rates, duration of larval development, mature larval or adult weight, and adult egg production but adult emergence decreased with increasing concentration. These observations suggested that latent toxicity was the probably cause of the observed delay in mortality.

The expression of latent effects is probably much more common than these few cases suggest. The paucity of citations most likely reflects the tendency of scientists to concentrate on short-term effects in the laboratory evaluation of pesticides (Chapter 7). The physiological or toxicological basis for latent effects of pesticides on natural enemies is not well understood. Increased mobilization of fat body containing high pesticide residues in later life stages is a plausible explanation. Changes in detoxification enzyme levels in subsequent life stages is another possibility (Chapter 6).

3.2.3.9. Parasitism Effects on Host/Parasitoid Uptake or Transfer. Parasitism has numerous, diverse effects on both host physiology and behavior (Slansky 1986), and many of these effects can result in changes in uptake and metabolism of a pesticide by a host and parasitoid. Physiological changes leading to increased susceptibility in hosts, and indirectly in parasitoids, are discussed in Chapter 4. Behavioral changes may alter the amount of toxicant taken up by the host. This in turn feeds back to affect the physiology of the parasitoid.

Several authors have noted changes in host locomotion or movement as a result of parasitism. For example, Aphidiidae are known to cause their aphid hosts to wander away from the host-plant leaves to woody branches, stems, or lower plant surfaces (Powell 1980). The walnut aphid *Chromaphis juglandicola* seeks sheltered places such as leaf folds, angles at the base of petioles, and cracks in the bark of trees when parasitized by *Trioxys pallidus*. It is protected in these refugia from applications of insecticides (van den Bosch et al. 1962). A similar phenomenon has been noted for filbert aphid *Myzocallis coryli* when attacked by the same parasitoid (Messing 1986). Parasitoids in the Aphelinidae can also have this effect on aphids (Lykouressis and van Emden 1983). The encyrtid *Trechnites psyllae* causes parasitized pear psylla (*Psylla pyri*) nymphs to move to sheltered places on the bark of pear trees prior to mummification (Herard 1986).

While these changes in behavior among parasitized Homoptera may reduce host (and hence parasite) exposure to pesticides, parasitized Lepidoptera may behave differently. *Heliothis virescens*, when parasitized by *Microplitis croceipes*, tends to wander over plant surfaces, and thus may pick up more pesticide from previously treated surfaces (Dahlman and Vinson 1980). A parasitized nymphalid *Chlosyne harrissii* moved to the upper strata of vegetation where it might be more exposed to airborne toxicants (Shapiro 1976).

Feeding behavior is strongly unfluenced by parasitism (Slansky 1986), which may substantially alter the uptake of pesticides whose mode of entry is by ingestion. *Trichoplusia ni* larvae parasitized by *Copidosoma truncatellum* consumed 35% more than unparasitized larvae (Hunter and Stoner 1975), while *Agrotis ipsilon* parasitized by *Meteorus leviventris* showed a 24% reduction in consumption (Schoenbohm and Turpin 1977).

In coniferous forests, pesticides affect a diversity of parasitoid species and often with varying results. MacDonald (1959) first noted an increase in apparent parasitism by *Apanteles fumiferanae*. He hypothesized that parasitized spruce budworm *Choristoneura fumiferana* were less likely to be poisoned by sprays than unparasitized larvae due to a reduction in feeding on treated conifer tissues. These behavioral changes limited pesticide uptake by hosts, and as a result, overall selective conservation of parasites was achieved. Subsequent researchers have observed the same phenomenon of increasing apparent parasitism (primarily by *Apanteles fumiferanae* and *Glypta fumiferanae*) following a number of different pesticide sprays. These effects have been attributed to the differences in the dropping behavior between parasitized and unparasitized larvae (Hamel 1977, Otvos and Raske 1980). Shea et al. (1984) reasoned that if MacDonald's explanation were the sole cause of increased parasitism observed immediately after insecticide use, then it should be operative regardless of the type of insecticide used. Instead, they observed an increase in parasitism following application of only two (methomyl and permethrin) of five insecticides tested. They concluded that other factors, such as differences in host–parasitoid synchrony in different experiments and plots, may account for the differences observed in the survivorship of parasitoids.

Changes in the respiration rate of parasitized herbivorous pests may also result in changes in the amount of volatile pesticides taken up in the gaseous phase. Several researchers have demonstrated reduced respiration rates following parasitism (Jones and Lewis 1971, Sluss 1968) or injection of calyx fluid or poison gland extracts (Waller 1965).

Although these physiological and behavioral effects of parasitism on the host are well documented, their influence on pesticide uptake has not often been addressed experimentally. Further research along these lines would greatly add to our knowledge of the feedback processes which integrate host physiology, parasitism, and pesticide toxicology (see Fig. 3.1).

3.3. SUMMARY

In this chapter, a diversity of ways in which arthropod natural enemies may be exposed to pesticides are described. However, little discussion of the relative importance of each type of exposure or route of uptake can be given (see exception for the role of residue uptake). Little research quantifying these modes of uptake has been undertaken for natural enemies. It is likely that some types of uptake such as cannibalism or host feeding by parasitoids are of minor significance, whereas residue and food chain uptake are of primary importance.

Another problem arises when trying to generalize for such a diverse group of arthropods. Certainly, there are tremendous differences in pesticide exposure between individual species of natural enemies. Generalizations become difficult to make when one considers the different crops, pesticides, formulations, and application techniques that are involved.

A third factor contributing to the lack of study of the detailed dynamics of pesticide uptake by natural enemies is the methodology by which selective pesticides are evaluated. Researchers interested in this topic are primarily concerned with the overall effect of the pesticide on these beneficial species. IPM practitioners and biological control specialists are primarily interested in whether or not a natural enemy was killed. There has been very little interest in using predators or parasitoids as model systems for basic research in toxicology and the selectivity of pesticide interactions.

As more detailed research has been undertaken on ways to achieve ecological selectivity, the need to study subcomponents of pesticide exposure and uptake by natural enemies has become more apparent. It is being more widely recognized that subtle differences in the ways pests versus natural enemies are exposed to pesticides can make a large difference in the eventual impact of these toxicants in the field. Sometimes only a slight difference in susceptibility will tip the balance in favor of the entomophage, and subsequent prey or host regulation can be achieved (Chapter 10).

4

FACTORS AFFECTING SUSCEPTIBILITY

4.1. INTRODUCTION

Many factors influence the susceptibility of a natural enemy to pesticides. Some are of primary consequence, such as the nature of the organisms involved, the type of pesticide used, or the environment of exposure. Other factors are secondary and more subtle.

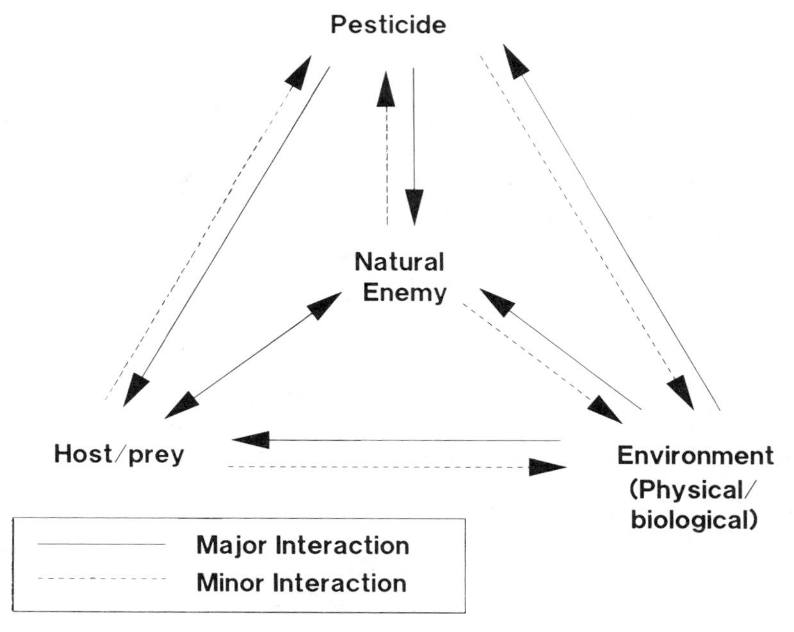

Figure 4.1. Principal classes of factors affecting natural enemy susceptibility to pesticides.

Four classes of factors affecting the susceptibility of arthropod predators and parasites to pesticides are represented in Fig. 4.1, including attributes of 1) the natural enemy, 2) its host or prey, 3) the environment of exposure, and 4) the pesticide. The scope of the environment as referred to in this chapter includes meteorological and other abiotic conditions and the substrate, medium, or habitat of exposure. Pesticide factors include type, formulation, application method, and so forth (Chapter 10). Except for pesticide-related factors (which can be controlled by the operator), each of these classes of factors may vary in nature.

Major (⟶) and minor (- - →) interactions between factors affecting natural enemy susceptibility to pesticides are indicated in Fig. 4.1. Pesticides, for example, may be directly toxic to both herbivores and natural enemies (⟶); however, natural enemies seldom affect pesticide dynamics (- - →). Minor exceptions may be when natural enemies activate pesticides for uptake by other predators—in the case of proinsecticides, i.e., those compounds which must be converted to an intermediate product to be toxic (Fukuto 1984; Chapters 3 and 5). Natural enemies may also affect pesticide dynamics during transfer from one natural enemy to another—such as by cannibalism or tropholaxis (Chapter 3). Prey or hosts may affect the susceptibility of natural enemies to pesticides by their own nutritional status and by the habitat they occur in or create. The environment (biological and physical) is seldom directly affected (⟶) by any other component of the system (although destruction of crops by pests may modify the microhabitat). Biological and physical attributes of the environment can greatly affect deposition, redistribution, and persistence of a pesticide and, therefore, uptake and toxicity to both pests and natural enemies (Chapter 3).

In chapter 3, the discussion of natural enemy/pest interactions focused mostly on *pesticide uptake*. Here, the emphasis shifts to factors influencing the *susceptibility* of predators and parasitoids to pesticides. Some factors (e.g., certain environmental and pesticide-related ones) influence both uptake and susceptibility such that discussions of these influences are difficult to separate. For this reason some overlap occurs between this chapter and Chapter 3. As with uptake, many factors affecting susceptibility to pesticides are similar for natural enemies and pests. However, many unique differences between these two groups of species may be exploited to achieve pesticide selectivity.

4.2. INTRINSIC ATTRIBUTES OF NATURAL ENEMIES

Many biological attributes of a natural enemy may influence its susceptibility to pesticides. Croft (1977) cited size, weight, sex, diapause, nutrition, and stage of development as influencing factors. Most apply to arthropods in general and are addressed in greater detail below.

4.2.1. Group, Family, Species, and Biotype

4.2.1.1. Innate Tolerance. Basic attributes which define a natural enemy's susceptibility to pesticides may be termed intrinsic characteristics. These attributes may be a direct result of the organism's short- and long-term exposure to toxins. Adaptations to previous exposure may be expressed within a single population of a species, within all populations of a species, within a taxonomic family of natural enemies, or within broader groupings, such as predators and parasitoids or entomophages and herbivores. Short-term factors influencing susceptibility tend to be expressed in individual populations, whereas long-term evolutionary adaptations may be evident across higher taxa.

Short-term factors influencing the response of an organism or a population to pesticides include natality, diet, stress, and other factors which contribute to general vigor. Under most conditions their influence on susceptibility is modest, although under extreme conditions such as starvation, their impact can be significant. Differences in vigor are most evident among resistant strains, and can be as high as 10- to 100-fold. The most extensive studies examining the intrinsic variability of natural enemy responses to pesticides have been surveys of pesticide-resistant strains (see Section 4.2.1.2., also Chapter 14). Variable responses between susceptible strains of a natural enemy species may also occur, but they are usually of a lesser magnitude.

Krukierek et al. (1975) evaluated the toxicity of oxydemeton-methyl (metasystox) to biotypes of *Trichogramma* spp. by two different techniques. The susceptibility of biotypes differed by more than three fold and was partially attributed to the test methods used. However, some variability beyond that due to methodology was present. (All strains tested were reared under nearly identical laboratory conditions and pesticide resistance was not involved.)

Evolutionary adaptation to natural toxins (Chapter 6)[1] may influence the innate tolerance of a species to pesticides. Adaptation to natural toxins tends to be interspecific, rather than intraspecific. Plant secondary compounds received through their prey or hosts or through direct feeding on plant materials may be of prime importance in predisposing natural enemy species to detoxify pesticides (Mullin and Croft 1985, Yu 1983, 1987). As

[1] Sometimes a resistance factor may operate across a long time scale (e.g., DDT and synthetic pyrethroid cross-resistance factors; see Croft et al. 1982); however, most resistance factors are of a shorter duration.

discussed more extensively in Chapters 6 and 16, some natural enemies are more capable of detoxifying pesticides than others. Natural enemies may also possess a lower detoxification potential than their hosts or prey. Detoxification capabilities of an organism often correlate with susceptibility or intrinsic tolerance to pesticides (Brattsten and Metcalf 1970, Croft and Morse 1979, Croft and Mullin 1984, Berenbaum and Neal 1987).

4.2.1.2. Pesticide Resistance. The development of resistant strains is a short-term phenomenon which influences the susceptibility of natural enemies to pesticides (Tabashnik and Croft 1982, 1985, Rosenheim and Hoy 1986). At least 30 natural enemy species have evolved resistant strains (Chapter 14).

In studies of early resistance detection, Grafton-Cardwell and Hoy (1986b) surveyed four strains of *Chrysoperla carnea* from different areas of California for their responses to six pesticides (permethrin, fenvalerate, diazinon, phosmet, methomyl, and carbaryl). Preadapted resistant strains were being sought for genetic improvement experiments (Chapters 19 and 20). The largest difference in LC_{50} values between strains was with carbaryl (2.6-fold). Measurable levels of tolerance or resistance to all six insecticides were present in populations from heavily sprayed cotton. Populations from less frequently treated sites were also variable in responses to all compounds, indicating that intrinsic tolerance as well as resistance was involved.

In a similar study with the parasitoid *Aphytis melinus*, Rosenheim and Hoy (1986) evaluated response variability among 13 field-collected strains from California citrus groves to 5 insecticides (carbaryl, chlorpyrifos, dimethoate, malathion, and methidathion). Differences in LC_{50} values ranged from 1.8- to 7.8-fold. Low LC_{50} values to carbaryl, chlorpyrifos, and dimethoate, ranging from 1.8- to 2.9-fold, indicated that innate tolerance was primarily responsible for differences among strains. Differences in LC_{50} values for malathion and methidathion were 7.8- and 7.6-fold, respectively. The magnitude of these differences indicated that resistance was present for these compounds. Patterns in LC_{50} values were correlated with the history of insecticide exposure at collection sites, both within groves and regionally. Individual population responses were best explained by an analysis that combined the influences of both grove and regional pesticide histories. Data indicated that short-term exposure to pesticides had preadapted some strains to be less susceptible to pesticides than others.

4.2.1.3. Combined Tolerance and Resistance. In the studies mentioned above, innate tolerance and evolved resistance contributed to the pesticide susceptibility of natural enemy populations. The influence of tolerance can only be separated from that of resistance by testing many individuals from different populations representing both untreated and treated habitats (Roush and McKenzie 1987).[2] Seldom, if ever, have such large-scale screening studies been attempted for natural enemies. Such studies would be helpful in addressing the intrinsic variability in the response of natural enemies to pesticides and the distribution and magnitude of resistant strains (Chapters 2, 14–16).

The predatory mite *Amblyseius fallacis* is a species in which tolerance and resistance factors have been extensively studied in the laboratory (Croft et al. 1976a, 1982). In Fig. 4.2, the responses of a susceptible and a resistant strain are plotted as LC_{50} values for DDT and

[2]Considering the widespread distribution of pesticides in presumably untreated areas, one can never be sure that previous selection of natural enemies with pesticides has not occurred, even when they are collected from relatively pristine environments.

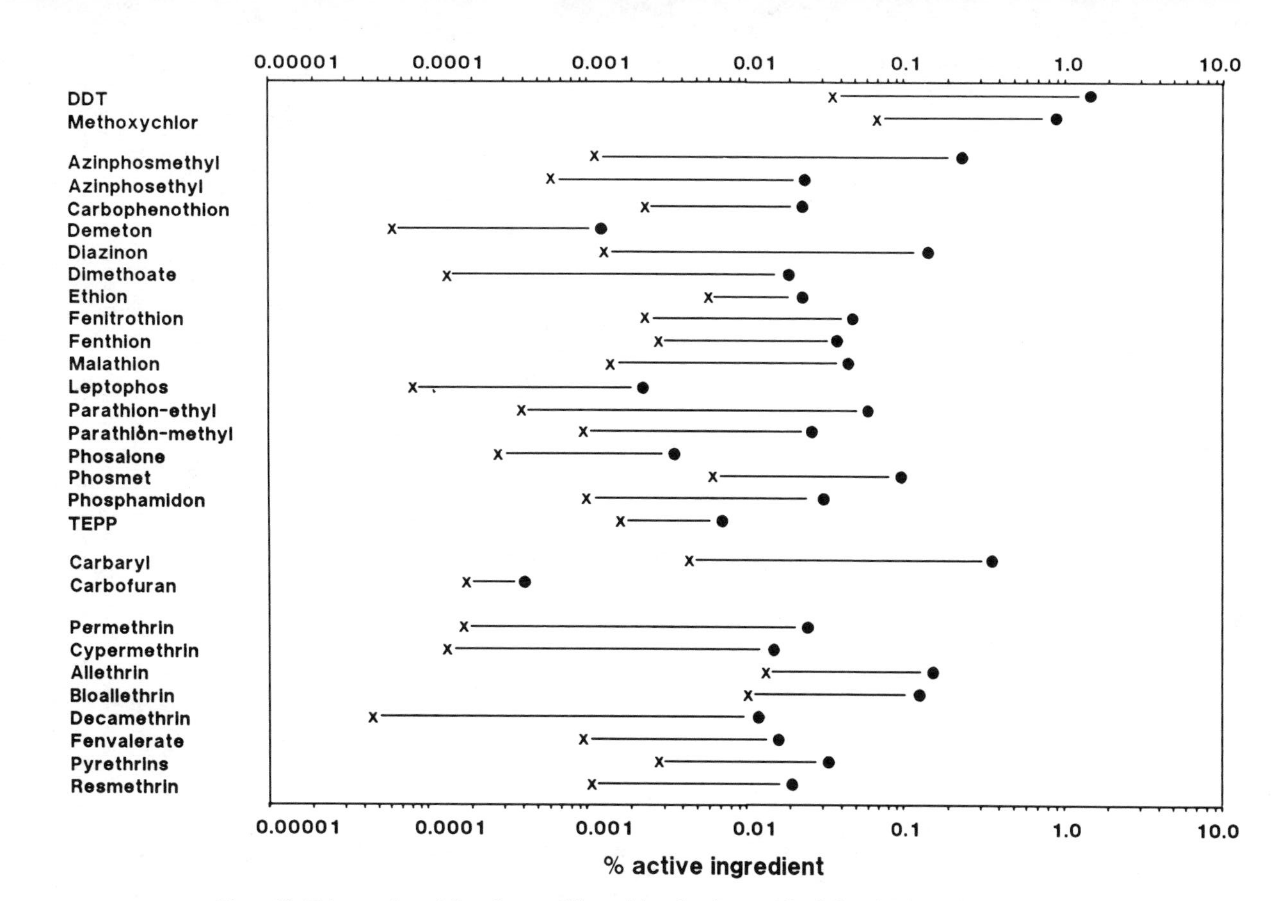

Figure 4.2. Tolerance ($\times$ = LC_{50} of susceptible strain) and resistance (● = LC_{50} of resistant strain) levels of insecticides in strains of the predaceous mite *Amblyseius fallacis* (after Croft et al. 1976a, 1982).

derivative compounds, organophosphate, carbamate, and pyrethroid insecticides. As can be seen, the greatest tolerance levels are for DDT and the related compound methoxychlor, the synthetic pyrethroid allethrin, the organophosphates ethion and phosmet, and the carbamate carbaryl. When a resistance factor was added, the highest LC_{50} values were to DDT and methoxychlor, carbaryl, the organophosphates azinphosmethyl, diazinon, parathion, and phosmet, and the pyrethroids allethrin and bioallethrin.

In Chapter 2, susceptibility ratings from the SELCTV database were presented for a wide range of beneficial species and pesticides. Summaries indicated significant differences in susceptibility among different taxa, but the degree of variability due to intrinsic tolerance versus resistance could not be determined. Few cases of resistance have been documented for natural enemies (Chapter 14); however, resistance could be more prevalent than is commonly realized (Chapters 14 and 16). If so, then resistance may be influencing the variability in the responses of natural enemies to pesticides to a greater extent than is generally recognized.

Besides aggregate data, many individual studies have reported remarkable pesticide tolerance among certain natural enemy species. *Chrysoperla carnea* (Wiackowski 1968, Miszczak 1975, Wilkinson et al. 1975, Khalil et al. 1976, Grapel 1982, Hellpap 1982, Grafton-Cardwell and Hoy 1985), the spider *Lycosa pseudoannulata* (Kiritani and Kawahara 1973, Chu et al, 1976a, 1976c, 1977, Dyck and Orlido 1977), and the stigmaeid mite *Agistemus fleschneri* (Abdel-Salam 1967, Nelson et al. 1973) are among the more notable examples. The consistent, high susceptibility of *Syrphus* spp. (Stern et al. 1959, Bonnemaison 1962, Mardzhanyan and Ust'yan 1966, Natskova 1974, Grapel 1982, Nasseh 1982, Hellpap 1982, Hassan et al. 1983), the mirid *Cyrtorhinus lividipennis* (Dyck and Orlido 1977, Ooi et al. 1979, Reissig et al. 1982), the aphidiid *Diaeretiella rapae* (Shorey 1963, Wiackowski and Dronka 1968, Wiackowski and Herman 1968, Delorme 1975, 1976), and the braconid *Opius concolor* (Monaco 1969, Liotta 1974, Maniglia 1978) is equally notable.

In the above citations, relative susceptibility was often assessed for a single species. In addition, a number of comparative tests have been published which demonstrate large differences in susceptibility between closely related species, even within the same genus. For example, among the predaceous Chrysopidae in Zimbabwe, *Chrysoperla boninensis, C. congrua*, and *C. pudica* are the predominant species which feed on important cotton pests of the region (Brettell 1979, 1982, 1984). Studies have shown that, while these species are of comparable susceptibility to many compounds, they differ appreciably in response to carbaryl and DDT (Brettell 1982). The most common species, *C. boninensis*, is more tolerant to carbaryl than *C. congrua* or *C. pudica* by a factor of several hundred fold. This high level of tolerance in *C. boninensis* is similar to that observed with *C. carnea* in other areas of the world (Wilkinson et al. 1975, Rajakulendran and Plapp 1982a, Brettell 1984).

An example of variable susceptibility among coccinellid predators associated with California citrus was presented by Bartlett (1963) (Table 4.1). With the exception of the highly tolerant *Cryptolaemus montrouzieri* (see SELCTV toxicity ratings for this species in Tables 2.8 and 2.9), generalist feeders such as *Hippodamia convergens* and *H. quinquesignata* were much less susceptible to pesticides than the specialists *Lindorus lithophane, Rodolia cardinalis*, and *Stethorus picipes*. In this case, long-term feeding habits may have preadapted the generalist species to better survive pesticide exposure (see Chapter 6).

Some families of natural enemies generally show greater tolerance to pesticides than others (Table 2.15). The Chrysopidae (Putnam 1956, Kharizanov and Babrikova 1978, Franz et al. 1980, Pape and Crowder 1981), Coccinellidae (Ahmed 1955, Bartlett 1964, Kehat and Swirski 1964, Asquith and Colburn 1971, Adbel-Aal et al. 1979), and Carabidae

Table 4.1. Average Residual Toxicity[1] of 61 Insecticides to Six Coccinellid Species Associated with California Citrus (Taken from Bartlett 1963)

		Chemical Class[3]				
Species	Host Range	DD (6)	CH (8)	OP (14)	CA (2)	Combined (includes fungicides and other pesticides)
Cryptolaemus montrouzieri	Several mealybugs	2.25	2.38	2.79	4.00	2.13
Hippodamia convergens[2]	Many aphides	1.75	1.94	3.42	4.00	2.15
Lindorus lithophane	Diaspine scale	2.33	2.94	3.42	4.00	2.44
Rodolia cardinalis	Cottony cushion scale	2.50	2.63	3.71	4.00	2.59
Stethorus picipes[2]	Spider mites	2.42	3.00	3.50	4.00	2.49
Hippodamia quinquesignata	Many aphids	1.42	1.63	3.28	4.00	2.00

[1] Values based on scale of 1 = no effect; 4 = highly toxic.
[2] Collected from overwintering sites; all other strains came from actively reproducing populations. (Bartlett 1963).
[3] DD = DDT derivatives; CH = chlorinated hydrocarbons; OP = organophosphates; CA = carbamates.

(Herne 1963, Mardzhanyan and Ust'yan 1966, Hassan 1969, Gregoire-Wibo 1980, Hagley et al. 1980) are good examples. In contrast, the parasitic Ichneumonidae (Abu and Ellis 1977, Plapp and Vinson 1977, Otvos and Raske 1980, Hassan et al. 1983, 1987; Chapter 2) and the predaceous Syrphidae (Niemczyk et al. 1979, Hassan et al. 1983, Horn 1983, David and Horsburgh 1985; Chapter 2) are very susceptible.

Natural enemies differ in susceptibility and in the variability of their responses to pesticides at even broader levels of generalization, for example, between predators and parasitoids. This point is supported by analyses presented in Chapter 2 (e.g., Tables 2.7, 2.10 and 2.12) and by data from Bartlett (1963), who tested several coccinellid predators and hymenopteran parasitoids. Bartlett's (1963) data are summarized in Table 4.2. Generally, parasitoids were more susceptible and less variable in their responses than were predatory beetles to all compounds tested except for diazinon, dieldrin, and methoxychlor. While data have been published to the contrary (e.g., Hassan et al. 1983, 1987), this author feels that the majority of research does, in fact, support the greater tolerance and variability in the response of predators to pesticides. As noted in Chapter 2, differences in the susceptibility of predators and parasitoids and their prey or hosts need to be examined more thoroughly by experimental studies.

A natural enemy's ecological characteristics may also be indicative of its relative susceptibility to pesticides. These include degree of specialization, feeding habits and alternate food sources, and the nutritional status of their prey or hosts. Among phytophagous pests having different host ranges and feeding habits, detoxification potentials and overall susceptibility to plant secondary compounds and pesticides are modified both by the degree of feeding specialization and by the types and amounts of secondary compound contained in host plants (Biernbaum 1987). Therefore, the susceptibility differential between predators and parasitoids may exemplify the influence of feeding strategy on susceptibility, since predators are considered to be more general feeders than parasitoids (DeBach 1964).

Table 4.2. Mean Toxicity of Toxicants and Variability in Response of Adults of 5 Parasitic Hymenopterans and 6 Species of Coccinellid Predators Feeding on Citrus Pests (Taken from Bartlett 1963)

Compound	Pesticide Group	Parasitoids[1]		Predators[2]	
		Mean Toxicity	±SD	Mean Toxicity	±SD
DDT	DD	4.00	0.00	2.83	1.03
Methoxychlor	DD	3.70	0.27	3.83	0.41
Dieldrin	OC	3.60	0.55	3.50	0.63
Aldrin	OC	3.20	0.45	2.00	0.63
Toxaphene	OC	4.00	0.00	2.75	0.42
Diazinon	OP	3.80	0.45	3.50	0.63
Demeton	OP	3.60	0.42	2.25	0.76
Dioxathion	OP	4.00	0.00	3.00	0.89
Chlorothion	OP	4.00	0.00	3.36	0.94

[1] Species included: *Aphytis lignanensis, Metaphycus luteolus, Spalangia drosophilae, Metaphycus helvolus, Leptomastix dactylopii.*

[2] Species included: *Cryptolaemus montrouzieri, Hippodamia convergens, Lindorus lophanthae, Rodolia cardinalis, Stethorus picipes, Hippodamia quinquesignata.*

4.2.2. Life Stage

The exposed life stages of a natural enemy may also influence their response to pesticides. The susceptibility of a particular life stage is typically studied to assess the potential for ecological selectivity (Chapter 10). A window for timing pesticide application is sought during which beneficials are most tolerant and pests are susceptible.

In Table 4.3 (top part) susceptibility data are summarized for individual life stages of natural enemies within families or orders (after Croft 1977). These comparative assessments are based on direct applications of pesticides using similar methods of evaluation. Toxicity ratings range from 1 to 3–5, depending on the number of life stages tested for each group. In the lower part of Table 4.3 toxicity rating summaries from SELCTV are given, based primarily on single life stage tests (see Table 2.10; Theiling 1987). Data from the two independent analyses presented in Table 4.3 are similar, except that values from SELCTV are less extreme. This might be expected because observations for all compounds and methods of evaluation were pooled in the database analysis.

Conclusions from the data in Table 4.3 are similar to those made by earlier researchers. Bartlett (1964) observed that adult parasitoids and predators were most severely affected by pesticides. Larval neuropterans, dipterous predators, and many predaceous coccinellids exhibited greater adult susceptibility (for an exception see Chang and Huang 1963). Holometabolous pupae and prepupae and the eggs of predators were relatively immune to pesticides even when exposed directly. There are some notable exceptions to these generalizations for the effects of microbials and other biorational pesticides on natural enemies (Chapters 11 and 12).

Greater tolerance exhibited by a particular life stage of a natural enemy may be due to physiological, behavioral, and chemical factors (Croft 1977). 1) An endoparasitoid may be associated with a particular host stage which is physiologically tolerant to pesticides (e.g., egg, pupae; Dumbre and Hower 1976a, Novozhilov et al. 1973). 2) Large glycogen or fat body storage reserves associated with certain life stages may enable a natural enemy to sequester a pesticide, thereby avoiding intoxication (larval coccinellids and chrysopids;

Table 4.3. Relative Toxicity of Pesticides to Life Stages of Arthropod Natural Enemy Groups (Adapted from Croft 1977 and Theiling 1987)

	Natural Enemy Group				
Stage	Chrysopidae (Neuroptera)	Coccinellidae (Coleoptera)	Syrphidae (Diptera)	Hymenoptera	Phytoseiidae (Acari)
SOURCE: SELECTED PAPERS[1]					
Adult	3.0[3,4]	3.3	3.5	3.9	3.2
Pupae	—	1.5	2.0	2.1	—
Late/nymph or larvae	2.0	2.3	2.7	3.1	3.3
Early/nymph or larvae	2.3	2.3	—	—	3.7
Egg	1.5	1.3	—	—	2.0
SOURCE: SELCTV[2]					
Adult	3.3[3]	3.4	3.7	3.7	3.5
Pupae	—	2.0	2.4	3.2	—
Larvae or nymph	3.2	3.7	3.9	3.3	3.1
Egg	2.4	2.8	3.4	3.7	2.9

[1]Specific studies of comparative life stage toxicity. Scales used in rating individual tests ranged from 1–3 to 1–5 (Croft 1977); number of observations per group ranged from 6 to 18.

[2]Pooled data from independent observations of life stage susceptibility. Scale used in all tests was 1–5 (Theiling 1987); number of observations per life stage ranged from 49 to 1264.

[3]1 = most tolerant to a pesticide; 4 or 5 = most susceptible to a pesticide.

[4]Comparisons between groups should be made with caution, since different scales were employed by researchers in different groups.

Takeda et al. 1965). 3) Certain parasitoid life stages may be impervious to cuticular penetration by a toxicant (e.g., eggs, pupae; Bartell et al. 1976). Croft (1977) also discussed instances of pesticide vulnerability which can be life stage-associated. 1) A natural enemy may be particularly sensitive to poisoning when progressing from one stage to the next (e.g., pupation). 2) High mortality may be experienced by an adult parasitoid emerging from its host (Dumbre and Hower 1977, Plewka et al. 1975). 3) A natural enemy may acquire a toxic dose of insecticide through unique feeding behavior (e.g., certain adult forms feed on pollens, nectars, and honeydews; Bartlett 1956).

Detoxification capabilities undoubtedly vary by life stage, influencing a natural enemy's susceptibility to pesticides. As with pests, detoxification systems may vary qualitatively and quantitatively as well (Chapter 5), involving such diverse systems as mixed function oxidases (or cytochrome p-450 monooxygenases), esterases (Ishaaya and Casida 1981), glutathione-*S*-transferases, or epoxide hydrolases (Mullin and Croft 1985, Croft and Mullin 1984, Yu 1987).

4.2.3. Age

The age of an arthropod within a life stage may affect its susceptibility to pesticides. This effect has been most commonly studied in adult and egg stages of natural enemies. Critchley (1972b) studied the toxicity of thionazin to carabid adults (*Harpalus aeneus*) by comparing newly emerged beetles and mature beetles (up to one year of age). Both sexes of newly emerged individuals were more susceptible than mature ones; however, younger, more active individuals may have taken up more pesticide than older beetles, rather than having been more inherently susceptible.

Aguayo and Villaneueva (1985) noted that young adult female *Meteorus hyphantriae*, attacking the arctiid pecan pest *Hyphantria cunea*, were more susceptible than older females to topical applications of methyl parathion. A similar tendency among the hymenopterous parasites of *Spodoptera littoralis* (*Mictopletis rufiventris*, *Chelonus inanitus*) was noted by Hegazi et al. (1982) in oral and contact tests with several organophosphates.

Moosbeckhofer (1983) evaluated the effects of diazinon, chlorpyrifos, turbufos, chlormephos, lindane, aldicarb, and carbofuran for ovicidal action on the carabids *Pterostichus cupreus* and *P. sericeus*. The toxicity of many of the compounds declined as increasingly older eggs were tested. El-Banhawy and Abou-Awad (1985) observed that older eggs of the predatory mite *Amblyseius gossypii* were more sensitive than younger eggs to flucythrinate, cyfluthrin, pyridaphenthion, and fenvalerate. Warner and Croft (1982) found that mortality to azinphosmethyl varied from 35% to 80% during the egg stage of the cecidomyiid *Aphidoletes aphidimyza*. Highest mortality occurred near eclosion, 30–50 hr after treatment. They suggested that the heightened susceptibility was due to delayed cholinesterase inhibition. During normal development, acetylcholine and cholinesterase levels increase as the embryo matures; the presence of the organophosphate inhibits cholinesterase, but acetylcholine levels do not reach lethal levels until maturation when neuromuscular activity increases.

4.2.4. Size, Weight, and Sex

Size, weight, and sex are closely related in arthropods. Relationships between these variables and the susceptibility of natural enemies to pesticides have not been extensively studied, but some generalizations can be made. In most natural enemies, the female is larger than the male (Clausen 1962). Among predatory phytoseiid mites and many parasitic hymenopterans, this size differential is associated with diploidy and different forms of arrhenotoky. Females are diploid, whereas males are often haploid.

There are many reports of male natural enemies being more susceptible to pesticides than females when no correction is made for size, surface area, or weight (Abdelrahman 1973, Adams and Cross 1967, Pielou and Glasser 1951, Lingren et al. 1972, Hukusima and Kondo 1962, Elsey and Cheatham 1976, Respicio and Forgash 1984, Aguayo and Villaneueva 1985, Lasota and Kok 1986, Scott and Rutz 1988). However, similar levels of susceptibility between sexes have also have reported (e.g., Respicio and Forgash 1984), especially when size differences are minimal. Rarely are male natural enemies less susceptible than females (e.g., Critchley 1972b).

As noted in Chapter 3, the relationship between surface area and weight has been documented for many arthropods. Generally, smaller insects have a larger ratio of surface area to body weight. Since pesticide toxicity depends on the dose per unit body weight, the organism with the greater surface area per unit body weight will be more severely affected if uptake and surface area are closely correlated. Several authors have attributed the greater susceptibility of smaller natural enemies to this factor (e.g., Krukierek et al. 1975; Chapter 3). While important, size is only a partial explanation for the greater susceptibility of natural enemies compared to their hosts or prey (Chapter 2).

Mowat and Coaker (1967) observed a negative correlation between the residual toxicity of dieldrin and the weight of six carabid beetles. A similar relationship was demonstrated for five carabids (Critchley 1972b), although thionazin toxicity was better correlated with surface area than with weight (Fig. 4.3). Critchley (1972b) related size or surface area to the sex of the beetles, males being smaller than females and also more susceptible (Fig. 4.3). Conversely, Novozhilov et al. (1973) observed that trichlorfon was more toxic to females than males of the scelionid parasitoid *Trissolcus grandis* when adult emergence was monitored from eggs of a hemipteran host; no explanation was given.

Although the association between size or weight and susceptibility among males and females of most natural enemies has been clearly documented, other factors may contribute to this relationship. Takeda et al. (1965) observed that malathion toxicity was

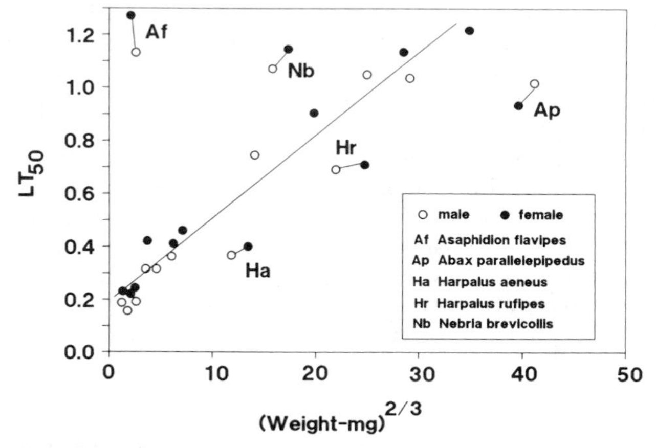

Figure 4.3. Relationship of toxicity (LT_{50}) of thionazin-treated soil to surface area (weight$^{2/3}$) of adult carabid beetles (after Critchley 1972b).

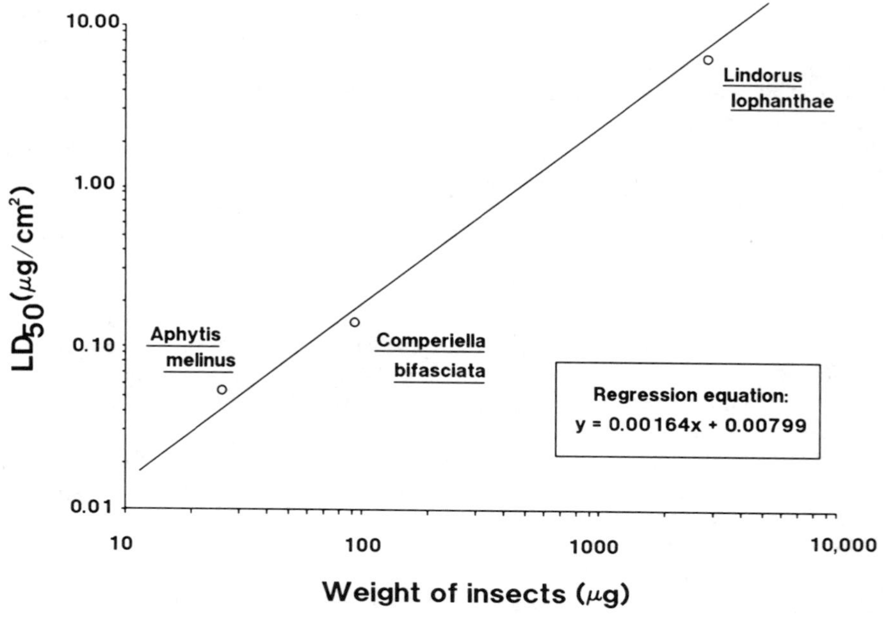

Figure 4.4. Relationship between weight of three species of insects and their LD_{50}. Data are presented on log scale. (after Abdelrahman 1973).

greater to males than to females of *Coccinella septempunctata*, but this was attributed to the greater fat body content of females.

Abdelrahman (1973) conducted an extensive study of weight and toxicity with natural enemies of the California red scale *Aonidiella aurantii*. Weight and insecticide toxicity were negatively correlated for two parasites and a coccinellid predator (Fig. 4.4). The weight of *Aphytis melinus* ranged from 8 to 47 g; and without correction, this factor accounted for 30% of the variation in malathion LD_{50} values in residue tests. To correct for this variance, the weight of a parasite was estimated based on the size of the scale host and number of parasites per scale. A pretreatment weight estimate was also made by correlating wing length with weight. A corrected weight function was thus obtained, greatly reducing variance about the regression line.

The relationship between weight or size and susceptibility is probably similar for most pests and natural enemies. Intraspecific size variation among endoparasitoids may be greater than among pests (Dahlman and Vinson 1980, Beckage and Riddiford 1982). Size variation may be multiplied when variably sized parasitoids develop in variably sized hosts. Competition for resources, resulting in greater size variability, occurs during multiple or super parasitization of a host. Resource-dependent size variation may be less acute for mobile pests and natural enemies that can more readily locate new food sources.

4.2.5. Diapause

Diapausing arthropods tend to be more pesticide tolerant than nondiapausing forms. This has been demonstrated for the aphelinid parasite *Aphelinus mali* (Schneider 1958) and the predaceous phytoseiid mite *Typhlodromus pyri* (Watve and Lienk 1976). Atallah and Newsom (1966) found both significant (toxaphene) and nonsignificant (endrin) differences in the susceptibility of diapausing versus nondiapausing forms of the coccinellid *Coleomegilla maculata*.

More detailed studies of pesticide susceptibility in diapausing natural enemies have appeared in recent years. Bartell et al. (1976) studied penetration of carbofuran, methoxychlor, methyl parathion, and phosmet into cocoons of nondiapausing and diapausing *Bathyplectes curculionis*, an endoparasitoid of the alfalfa weevil *Hypera postica*. Cocoons with diapausing parasitoids were relatively impervious to radiolabeled insecticide. Nondiapausing forms were less protected, and pupae suffered 100% mortality at comparable dosages. The authors concluded that differences in penetration and susceptibility to insecticides in *B. curculionis* were related to cocoon structure. They speculated that parasitoid physiology and time elapsed after cocoon formation affected penetration and toxicity.

4.2.6. Circadian Rhythm

Some pests exhibit a circadian rhythm in their susceptibility to insecticides (e.g., houseflies and cockroaches. Sullivan et al. 1970; pink bollworm: Ware and McComb 1970). To examine this phenomenon, Abdelrahman (1973) exposed equal-aged replicates of the parasitoid *Aphytis melinus* to 0.06 g/cm^2 of malathion residues on an hourly basis over a 3-day period. A daily rhythm of susceptibility was demonstrated, with peaks at 9 and 15 hr (Fig. 4.5). Many species have daily rhythms that affect their responses to stimuli. Factors such as feeding or enzyme production could be implicated. Abdelrahman (1973) concluded that bioassays should be conducted at a specific time of day to minimize experimental variability (Chapter 5).

4.3. ATTRIBUTES OF HOST OR PREY

As indicated in Chapter 3, the host or prey of a natural enemy can influence pesticide uptake by the beneficial, especially for closely associated organisms such as larval endo-

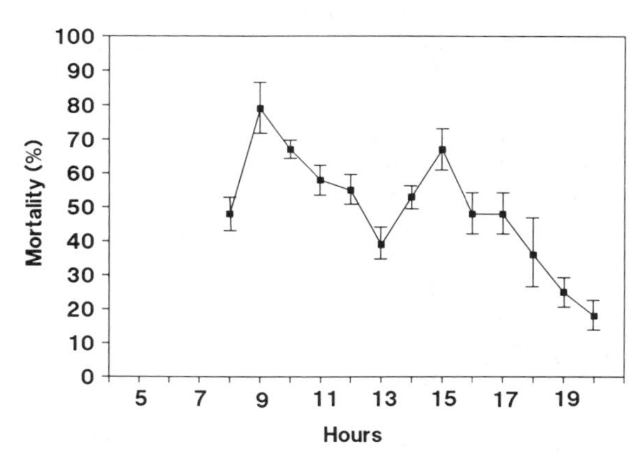

Figure 4.5. Circadian rhythm of susceptibility of *Aphytis melinus* to malathion. Each point represents average mortality plus standard deviation based on six replicate determinations (after Abdelrahman 1973).

and ectoparasitoids and their hosts. In this section, host or prey feeding strategies, density, nutrition, stress, and the feedback effects of parasitism on host/natural enemy susceptibility are discussed.

4.3.1. Host or Prey Feeding Strategies

Since herbivorous pests are the primary target of insecticides, they determine the environment of exposure for entomophagous natural enemies. The relationship between a monophagous natural enemy and its host or prey can be very specific. It may determine the natural enemy's feeding behavior and the timing of its development. These traits are strongly fixed in populations of highly specialized natural enemies after long evolutionary association with their prey or hosts.

Diet composition, especially just prior to pesticide exposure, can condition responses of polyphagous natural enemies in several ways. For example, diet can affect the susceptibility of a polyphagous parasite by altering certain of its body constituents or by causing different types of stresses (Section 4.3.3). Predaceous arthropods must be similarly affected. The influence of different hosts on an endoparasitoid exposed to benomyl through its food chain was discussed previously (Teague et al. 1985; Chapter 3). *Apanteles marginiventris* developing in *Spodoptera ornithogalii* was not affected by the fungicide, but was highly susceptible when developing in three other hosts. The observed differences in the susceptibility of the natural enemy were attributed to unique host/parasitoid interactions that varied with the different host species; however, specific factors were not identified. Perhaps *S. ornithogalii* was able to detoxify or sequester benomyl in such a way that the parasitoid was not subsequently affected.

4.3.2. Density and Starvation

A delicate balance is maintained between pest and natural enemy densities. As a result, beneficials may experience periods of cyclic abundance related to the density-dependent regulation of their hosts or prey (Chapter 8). Under these conditions, starvation stress may affect the susceptibility of natural enemies to pesticides. The influence of feeding history on the susceptibility of field-collected specimens is manifested by high variability in the results. Hence, testing of these specimens is avoided when possible (Chapter 5). The feeding history of a natural enemy should be standardized when laboratory side-effect tests are conducted (Franz 1976, Croft 1977).

The importance of food stress in conditioning a natural enemy's response to a pesticide is a function of its feeding strategy. Many natural enemies can go for long periods without feeding. In some cases, adult feeding is of minor consequence to successful oviposition and population increase (see proovigenic versus synovigenic parasites, DeBach 1964). In other cases, adult feeding is critical for survival and reproductive maturation. For the latter species, prey or host density and nutritional status can dramatically influence their susceptibility to pesticides.

Critchley (1972b) observed that *Bembidion lampros* adults, when deprived of food for 4 days and exposed to organophosphate residues, died more rapidly than well-fed individuals. Abdelrahman (1973) observed a linear relationship between starvation and the toxicity of malathion residues to *A. melinus*. Mortality differences were two fold between 0- and 6-h pre-treatment starvation periods. Kot and Plewka (1970) found more than a 20-fold difference in LC_{50} values for demeton-methyl between fed and unfed *Trichogramma evanescens* (Fig. 4.6).

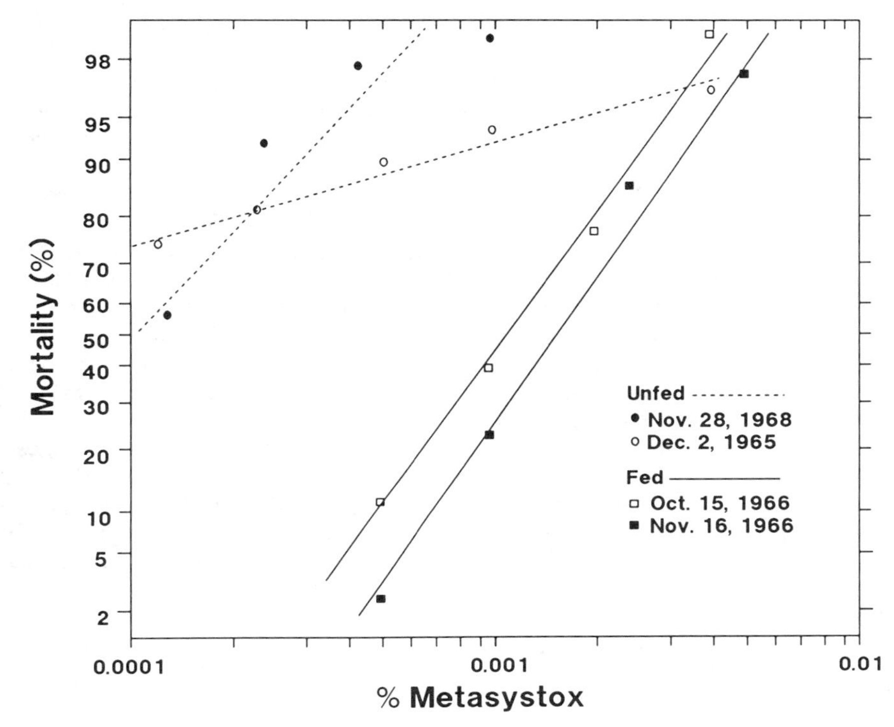

Figure 4.6. Susceptibility of unfed and fed females of *Trichogramma evanescens* to metasystox (after Kot and Plewka 1970).

4.3.3. Nutrition

A natural enemy's susceptibility to pesticides is a function of both the quantity and the quality of food ingested. Nutrition has implications for both field and laboratory evaluations of pesticide side effects (Chapter 5).

Krukierek et al. (1975) observed different mortality rates in *Trichogramma evanescens* exposed to metasystox when reared on eggs of four different lepidopteran hosts (Fig. 4.7a). Parasites reared from eggs of *Arctia caja* were least susceptible, generally larger, and exhibited greater longevity than parasitoids that developed from eggs of *Eilema* spp., *Sitotroga cerealella*, or *Pieris brassicae*. Survivorship of parasitoids also varied when developing from different generations of laboratory-reared, demeton-*S*-methyl tested *Arctia caja* (Fig. 4.7b). Over time, host quality had decreased due to laboratory rearing. Differences in parasite mortality were attributed to their vigor, which in turn was influenced by the size and nutritional value of their host.

Differences in susceptibility were observed for the braconid *Mictopletis rufiventris* when its host *Spodoptera littoralis* was fed insecticide in artificial diet versus on castor bean leaves (Hegazi et al. 1982). Although the amount of toxicant taken up by the host and parasite could not be standardized, parasites were generally less susceptible when reared on bean-plant-fed hosts. These differences in response could have been due to the induction of detoxification enzymes by compounds in the castor bean plant (in either the host or the natural enemy). Induction is common among phytophagous pests feeding on different plants (Yu et al. 1979, Berry et al. 1980).

The influence of hosts or artificial diets on natural enemy susceptibility may relate to fat body composition. The fat body is a known metabolic sink for sequestration of pesticides (Chapter 6). Polyphagous parasitoids may have different fat body constituents depending on their feeding histories (Thompson and Barlow 1983, Thompson 1986). One would also expect to find different fat body constituents among parasitoids developing in a common host which had fed on different host plants (or different artificial diets).

Several authors have discussed the relationship between natural enemy susceptibility, fat body composition, and pesticide toxicology (e.g., Hamilton and Kieckhefer 1969, Hoffman and Grosch 1971, Dumbre and Hower 1976a). Takeda et al. (1965) studied this relationship experimentally using *Coccinella septempunctata*. Although direct causation was not proven, seasonal tolerance to malathion was directly correlated with the fat content of the beetle. Early in the season, the fat body of the beetles was depleted, presumably due to a lack of feeding, and susceptibility to malathion was high. Later in summer and early autumn, presumably after extensive feeding, the fat content was higher and a correspondingly greater tolerance was observed.

Further research elucidating the effects of host or prey nutrition on the responses of natural enemies to pesticides is needed. This information will contribute to improved pesticide surveillance tests (Chapter 5). Nutritional standardization of both the pest and

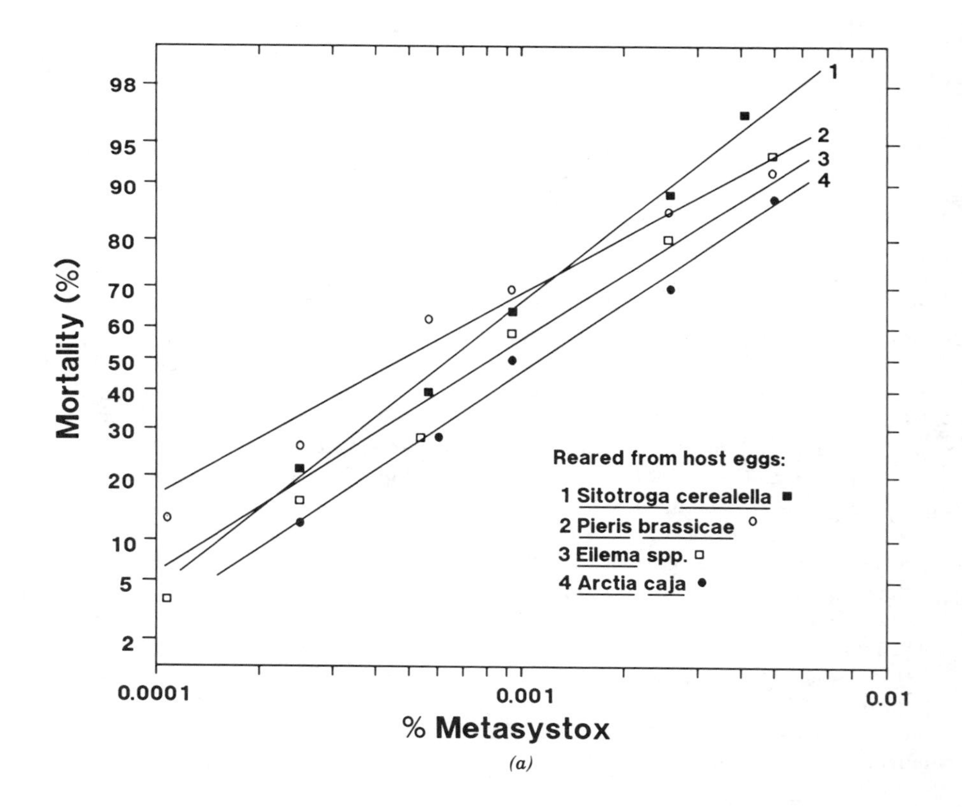

(a)

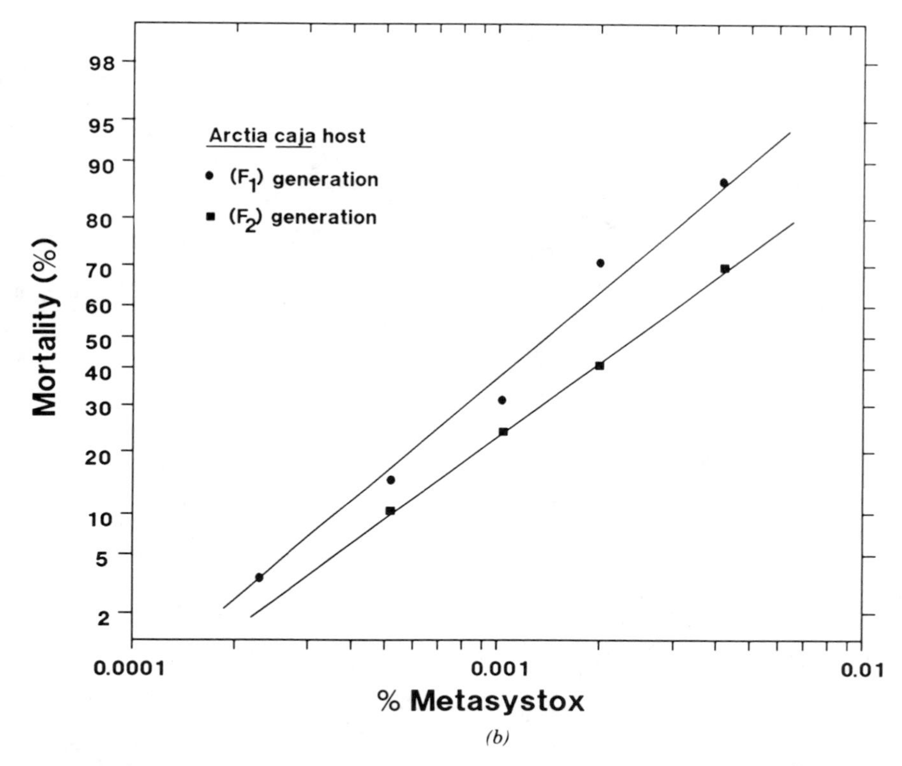

(b)

Figure 4.7. Dosage mortality for *Trichogramma evanescens* treated with metasystox. (a) Reared from different hosts (generation 1). (b) Reared from different host generations (after Kot and Plewka 1970).

the natural enemy is needed prior to toxicity testing. Pretreatment standardization should include the amount as well as the quality of food natural enemies are given (Chapter 5).

4.3.4. Feedback of Parasitism on Host/Parasitoid Susceptibility

Parasitism can have a marked effect on host susceptibility, which then influences the susceptibility of the immature parasitoids associated with the host (see also Section 3. 2, 3.6, Chapter 3). Of the host or prey attributes discussed in this section, feedback of parasitism on host susceptibility has been most thoroughly studied, usually with an emphasis on host mortality.

Abu and Ellis (1977) noted that third instar *Hypera postica*, parasitized by either *Bathyplectes curculionis* or *Microctonus aethiopoides*, were slightly more susceptible to topical applications of carbofuran, carbaryl, malathion, phosmet, and methoxychlor than were similarly treated, unparasitized host larvae. Most parasite larvae died within their susceptible hosts (Table 4.4). Dumbre and Hower (1976b) observed almost no difference in susceptibility between parasitized and unparasitized *H. postica* when the parasite *M. aethiopoides* was in the egg stage. However, an increasing effect on host susceptibility was

Table 4.4. Effects of Parasitism on Toxicity of Pesticides to Host Insect Larvae Associated with Endo- or Ectoparasitoids

Host Species (Stage)	Parasitoid Species (Stage)	Compound	Fold Difference in Host Susceptibility[1]	Reference
Hypera postica (larvae)	*Microctonus aethiopoides* (larvae)	Carbofuran	1.0	Abu & Ellis 1977
		Malathion	1.0	Abu & Ellis 1977
		Phosmet	1.2	Abu & Ellis 1977
		Carbaryl	1.0	Abu & Ellis 1977
		Methoxychlor	1.1	Abu & Ellis 1977
Hypera postica (larvae)	*Bathyplectes curculionis* (larvae)	Carbofuran	1.0	Abu & Ellis 1977
		Malathion	1.2	Abu & Ellis 1977
		Phosmet	1.2	Abu & Ellis 1977
		Carbaryl	1.1	Abu & Ellis 1977
		Methoxychlor	1.0	Abu & Ellis 1977
Hypera postica (larvae)	*Microctonus aethiopoides* (eggs)	Carbofuran	1.1	Dumbre & Hower 1976a
		Parathion-methyl	0.9	Dumbre & Hower 1976a
		Methidathion	0.9	Dumbre & Hower 1976a
		Methoxychlor	1.0	Dumbre & Hower 1976a
Hypera postica (larvae)	*Microctonus aethiopoides* (early larvae)	Carbofuran	2.5	Dumbre & Hower 1976a
		Parathion-methyl	2.0	Dumbre & Hower 1976a
		Methidathion	1.9	Dumbre & Hower 1976a
		Methoxychlor	2.0	Dumbre & Hower 1976a
Hypera postica (larvae)	*Microctonus aethiopoides* (late larvae)	Carbofuran	2.4	Dumbre & Hower 1976a
		Parathion-methyl	2.1	Dumbre & Hower 1976a
		Methidathion	2.0	Dumbre & Hower 1976a
		Methoxychlor	2.3	Dumbre & Hower 1976a
Heliothis virescens (larvae)	*Cardiochiles nigriceps* (larvae)	Parathion-methyl	14.2	Fix & Plapp 1983
		Permethrin	2.5	Fix & Plapp 1983
Spodoptera littoralis	*Microgaster rufiventris*	Methamidophos	0.8	Hegazi et al. 1982
		Phosfolan	0.8	Hegazi et al. 1982
		Prothiofos	0.8	Hegazi et al. 1982
Dacus dorsalis (larvae)	*Opius oophilus* (larvae)	Chlordane	16.6	Tamashiro & Sherman 1955
		Aldrin	14.2	Tamashiro & Sherman 1955
		Endrin	8.3	Tamashiro & Sherman 1955
		Lindane	1.6	Tamashiro & Sherman 1955
		Parathion-methyl	1.1	Tamashiro & Sherman 1955

[1] Ratio of host mortality: nonparasitized/parasitized

noted as the parasite larva developed (see Table 4.4). Parasitized hosts were about twice as susceptible as unparasitized ones. Probable influences of the parasitoid larvae were 1) release of teratocytes into the host haemocoel, 2) breakdown of host tissues by mandibular feeding, and 3) destruction of essential fluids and the fat body (Dumbre and Hower 1976b).

More pronounced effects of parasitism on host and, therefore, natural enemy susceptibility to pesticides have been widely reported for braconid endoparasitoids. Fix and Plapp (1983), working with *Heliothis virescens* and *Cardiochiles nigriceps*, observed a 14.2-fold difference in LD_{50} values to methyl parathion between parasitized and unparasitized hosts, but only a 2.3-fold difference to permethrin (Table 4.4). They suggested that permethrin had less toxic effects on parasites within their hosts, and therefore was better for field use than methyl parathion. Tamashiro and Sherman (1955) tested several organochlorines and methyl parathion on *Dacus dorsalis* parasitized by *Opius oophilus*. Parasitized hosts were six to eight times more susceptible to chlordane, aldrin, and dieldrin than unparasitized *D. dorsalis* (Table 4.4). Delayed toxicity was also evident in emerging adult parasites (Chapter 3, Section 3.2.3.7).

In contrast, Hegazi et al. (1982) found larvae of *Spodoptera littoralis* parasitized by *Microgaster rufiventris* to be less susceptible to oral or topical applications of organophosphates than unparasitized larvae. While fully developed larvae inside the host did not die, adult parasitoids which emerged later were killed. Hegazi et al. (1982) did not speculate as to the mechanism that conferred lower susceptibility to parasitized hosts.

Several hypotheses have been advanced to explain the differences in pesticide susceptibility between parasitized and unparasitized hosts. Fix and Plapp (1983) proposed that parasitism may interfere with detoxification enzyme production by the host. Obviously, in the latter stages of host consumption, most physiological functions of the host are impaired. Abu and Ellis (1977) attributed the greater susceptibility of *Hypera postica*—parasitized by *Microctonus aethiopoides*—to the host's weakened condition due to reduced feeding and to stress from parasite larval development. Studies by El-Sufty and Fuhrer (1981) showed that parasitism of *Cydia pomonella* and *Pieris brassicae* by *Apanteles glomeratus* reduced the cuticle thickness of the host by limiting incorporation of nitrogenous compounds. This influence rendered parasitized hosts and parasites more susceptible by affecting penetration of toxicants through the host integument.

Novozhilov et al. (1973) measured penetration of trichlorfon through the egg chorion of *Eurygaster integriceps* and the resulting toxicity to different life stages of the scelionid parasite *Trissolcus grandis*. Penetration of pesticide through the host egg at 24 or 72 hr was dependent on the developmental stage of the parasite (egg < larvae < pupae). Trichlorfon penetration was apparently facilitated by changes which took place in the host egg as embryonic tissues were disrupted by the developing parasite.

Insecticide treatment (demeton-*S*-methyl) of eggs of the Indian meal moth *Sitotroga cerealella* which contained developing *Trichogramma evanescens* caused mortality in adult parasites that was correlated with the stage of parasite development (Kot and Plewka 1970).

These studies point out possible mechanisms relating parasitism to host susceptibility and feedback effects to the parasitoid. Advances in understanding endoparasitoid consumption of hosts and how it affects pesticide transfer and metabolism in the natural enemy are needed. Hawlitzky and Boulag (1974) detailed the larval consumption habits of the endoparasitoid *Phaneroptoma flavitestacea* on *Ephestia kuehniella* using histological studies. Although they did not evaluate pesticide effects, a toxicological examination of pesticide transfer during this type of study would be useful in elucidating the complexities of host–parasitoid–pesticide interactions.

4.4. ENVIRONMENTAL ATTRIBUTES

Attributes of a natural enemy's environment which can influence its susceptibility to pesticide include both abiotic and biotic components. Meteorological or physical variables, for example, affect intoxication via temperature, relative humidity, moisture, and light. These abiotic factors can directly affect volatility, degradation, and penetration of the pesticide, as well as the response of the natural enemy. Biotic attributes of a natural enemy's environment include associated plants, soil, and hosts or prey. Other species in ecological association with natural enemies which influence their susceptibility to pesticides are discussed more fully in Chapter 8.

4.4.1. Temperature and Relative Humidity

The effects of temperature and relative humidity on the pesticide susceptibility of many biological organisms are well documented. However, studies of specific temperature influences on natural enemy susceptibility are limited, and studies featuring the influence

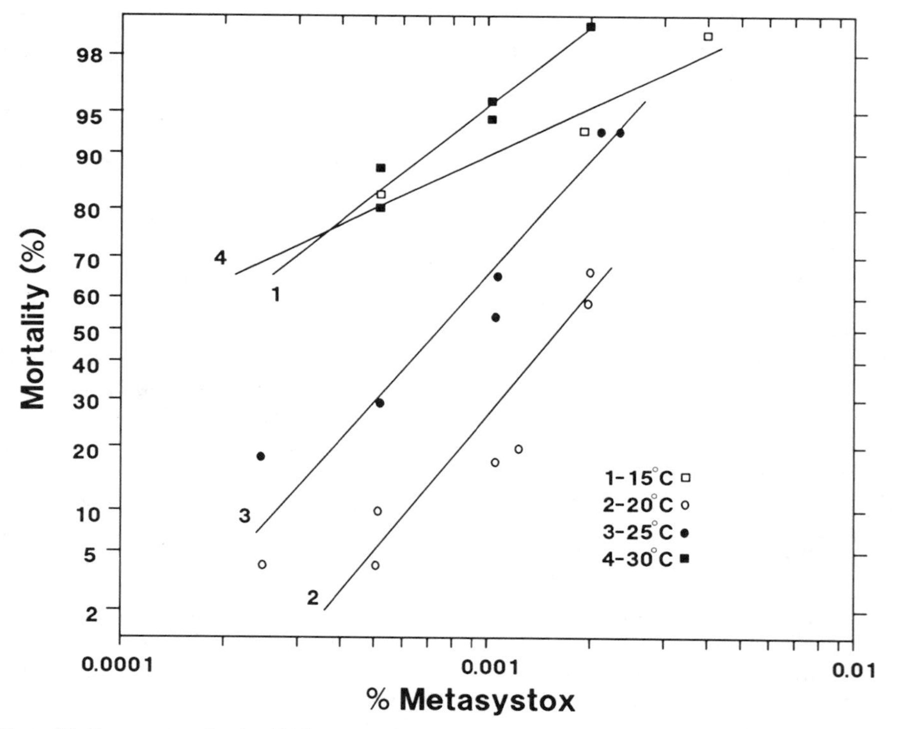

Figure 4.8. Dosage mortality for *Trichogramma evanescens* reared at different temperatures and treated with metasystox (after Kot and Plewka 1970).

of relative humidity are entirely lacking. Influences of temperature can be examined in terms of pretreatment and post-treatment regimes.

Krukierek et al. (1975) observed pronounced differences in the susceptibility of *Trichogramma evanescens* to demeton-*S*-methyl for populations reared at constant pretreatment temperatures of 15, 20, 25, and 30° C (Fig. 4.8). Insects reared at 15 or 30° C were the most susceptible (as measured by LC_{50} values), whereas those reared at 20° C were the most tolerant to the compound. Differences in LC_{50} values were near 10-fold and were attributed to the vigor condition of the reared parasites. In particular, the debilitating physiological effects of rearing parasites at extreme temperatures were cited. When rearing conditions were changed from extreme temperatures to those in the range of 20–25° C for two to three generations, mortality returned to normal in populations exposed to pesticides.

Everson and Tonks (1981) investigated post-treatment temperatures of 15, 20, 25, and 30° C on the toxicity of several acaricides to adult female *Phytoseiulus persimilis* and *Tetranychus urticae* using a slide-dip technique. Toxicities of cyhexatin, dicofol, propargite, citrazon, chlorobenzilate, and permethrin were positively correlated with temperature (Fig. 4.9). Under certain temperature regimes, pesticides were more toxic to the spider mites than to the predators, whereas others favored survivorship of the prey. Generally, higher temperatures increased susceptibility in natural enemies to a greater extent than in pests. Cooler temperatures, were more conducive to greenhouse IPM using *P. persimilis* and selective pesticides. The positive correlation of toxicity with temperature for

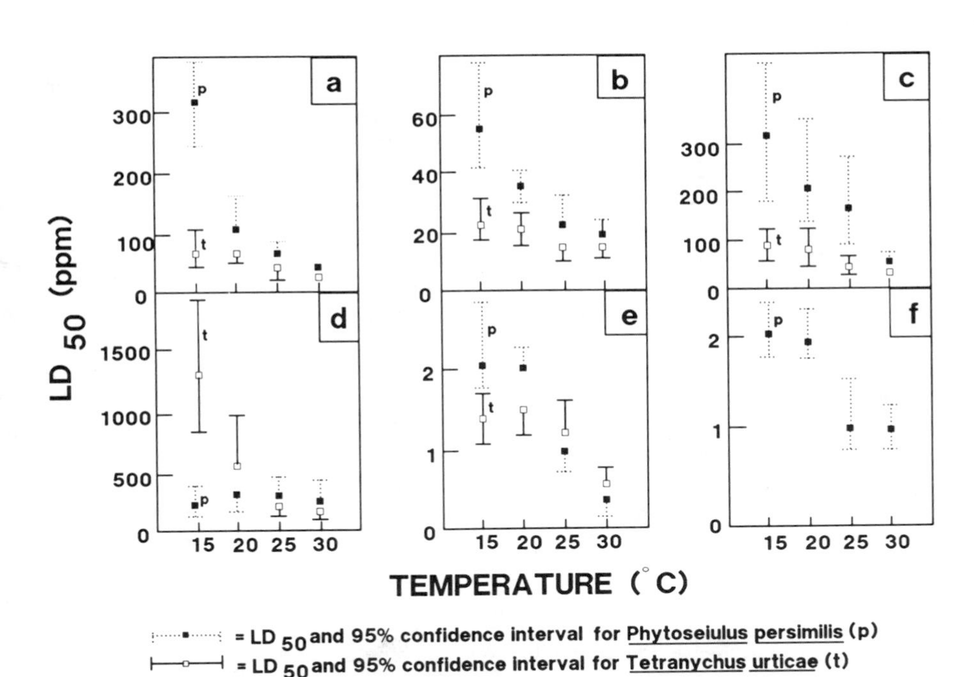

Figure 4.9. Effect of temperature on toxicity (LC_{50}) of six pesticides to *Tetranychus urticae* (t) and *Phytoseiulus persimilis* (p). The 95% fiducial limits are indicated by heavy and light horizontal bars for *T. urticae* and *P. persimilis*, respectively. Chemical pesticide designations: a = cyhexatin, b = chlorobenzilate, c = citrazon, d = propargite, e = dicofol, f = permethrin (after Everson and Tonks 1981).

permethrin was unusual. More often, synthetic pyrethroids show a negative correlation between temperature and activity as measured by mortality (e.g., Harris and Kinoshita 1977).

In the above studies, it is difficult to separate the influence of temperature on basic physiological processes influencing susceptibility from those affecting arthropod activity and, therefore, pesticide uptake. While these distinctions are unimportant in terms of net effect, they are useful to a more basic understanding of susceptibility relationships.

No data on the pretreatment effects of relative humidity on the susceptibility of natural enemies to pesticides could be found in the literature. Relative humidity probably affects intoxication via rates of uptake or adhesion to the insect (Chapter 3) and by the stress of dehydration. Many of the same points relating to temperature probably apply to relative humidity. At the extremes of relative humidity or when an organism experiences water stress, an additive or synergistic effect with temperature could also be involved.

Low humidity would probably have a greater effect on the pesticide susceptibility than would high humidity for most insects. (An exception would be for microbial pesticides, which are favored by high humidity; Chapter 11.) Many natural enemies normally occur on leaves where humidity is near saturation. Thus they are probably well adapted to these conditions. In laboratory studies, high post-treatment relative humidity should be maintained for the holding period before measuring the side effects of pesticides (Chapter 5). Otherwise control mortality is excessive and non-pesticide-related mortality confounds experimental results.

4.4.2. Moisture

The effects of moisture on natural enemy susceptibility have often been studied in soil ecosystems (Edwards 1966). The aqueous solubility of a pesticide influences its movement through the soil column, as well as its adsorption and desorption to soil particles. Pesticide residues on the soil surface degrade approximately 10 times faster (due to light-induced decomposition and a lack of moisture) than when incorporated into the soil (Edwards 1966). Soil moisture also affects soil compaction and soil texture, which in turn can affect natural enemy movement and pesticide uptake (Chapter 3).

Critchley (1972b) measured the relationship between soil moisture and the toxicity of thionazin on the predaceous carabid *Bembidion tetracolum* (Fig. 4.10a). Soil moisture was negatively correlated with carabid mortality when exposed to thionazin. Critchley attributed the differential toxicity of the compound to desorption and adsorption, which influence pesticide distribution in soil. With compaction due to the saturation of soils, beetle burrowing was greatly reduced (Fig. 4.10b), which subsequently reduced pesticide uptake. However, compaction also reduced vaporization of thionazin, slowing the depletion of residues, which then increased predator uptake of the poison. As with other physical factors, the extremes in soil moisture may cause water stress, which directly influences pesticide *susceptibility* rather than uptake.

4.4.3. Light

Photoperiod can influence diapause induction, which in turn affects the susceptibility of natural enemies to pesticides (Section 4.2.5). Light can affect the responses of natural enemies to pesticides by influencing movement and searching for prey or hosts (Chapter 3). Most studies of the effects of light on the susceptibility of natural enemies have focused on behavioral effects and on rates of pesticide uptake.

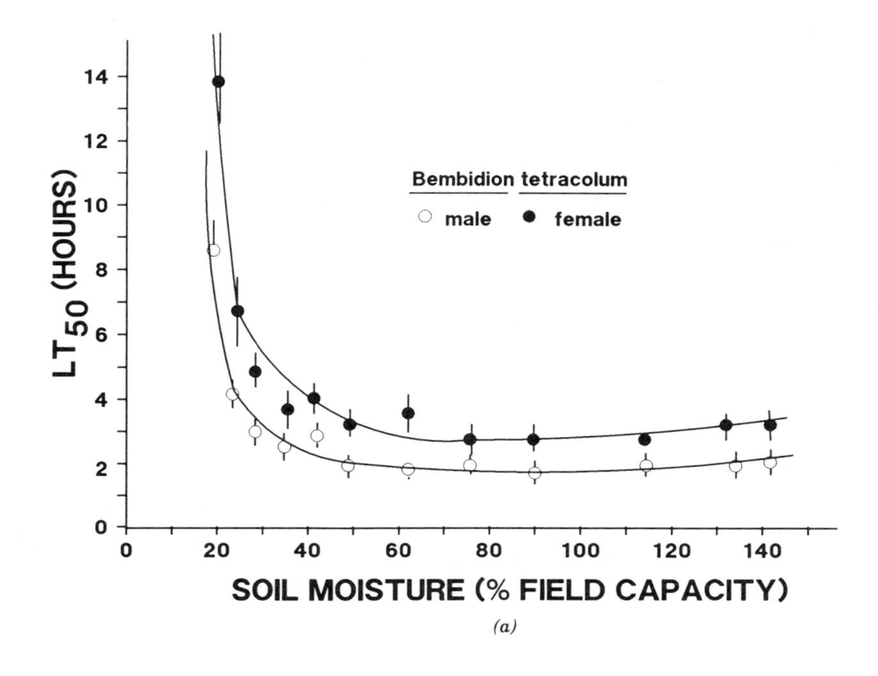

(a)

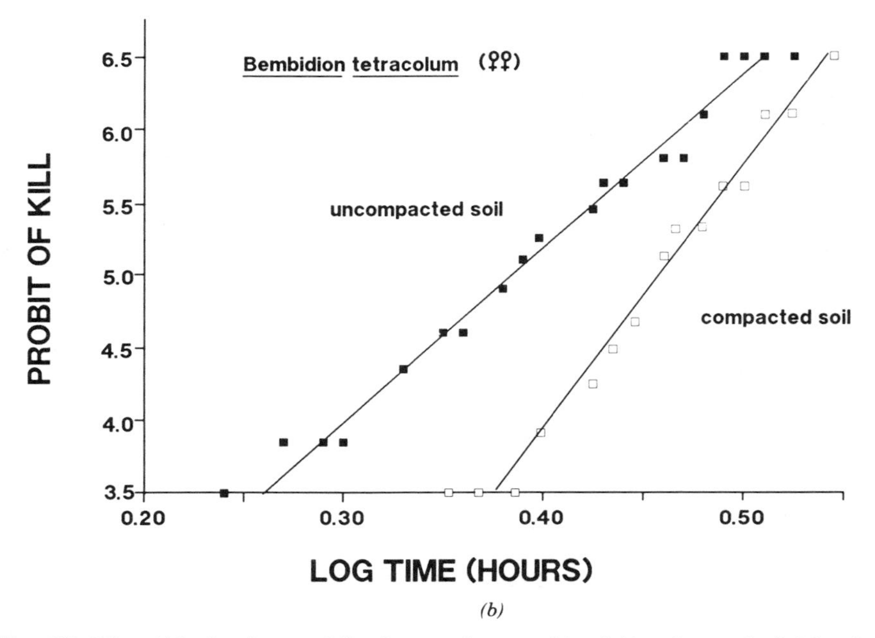

(b)

Figure 4.10. Effect of (a) soil moisture and (b) soil compaction on toxicity of thionazin-treated soil (16 ppm) to *Bembidion tetracolum* (after Critchley 1972b).

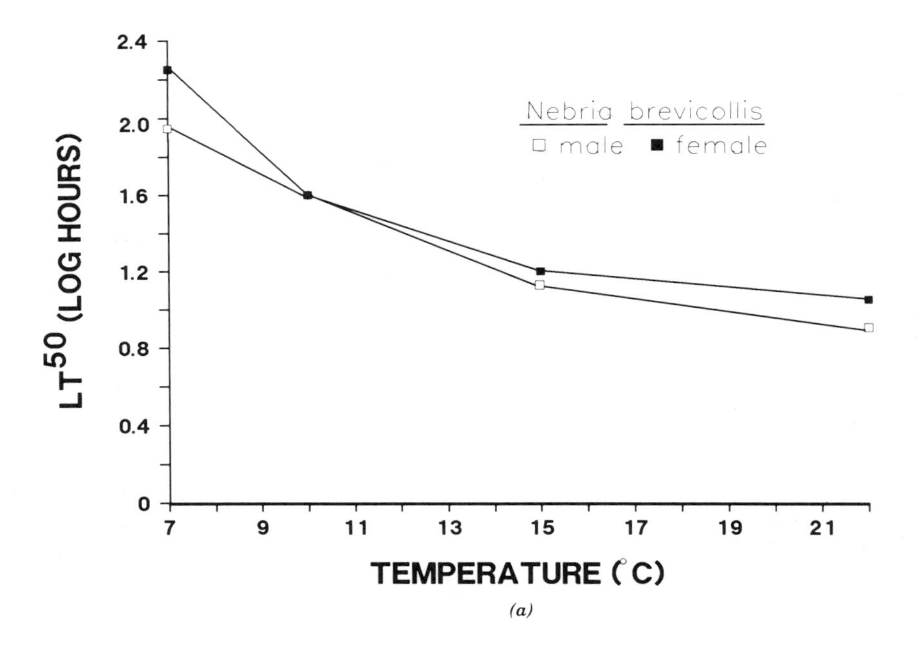

(a)

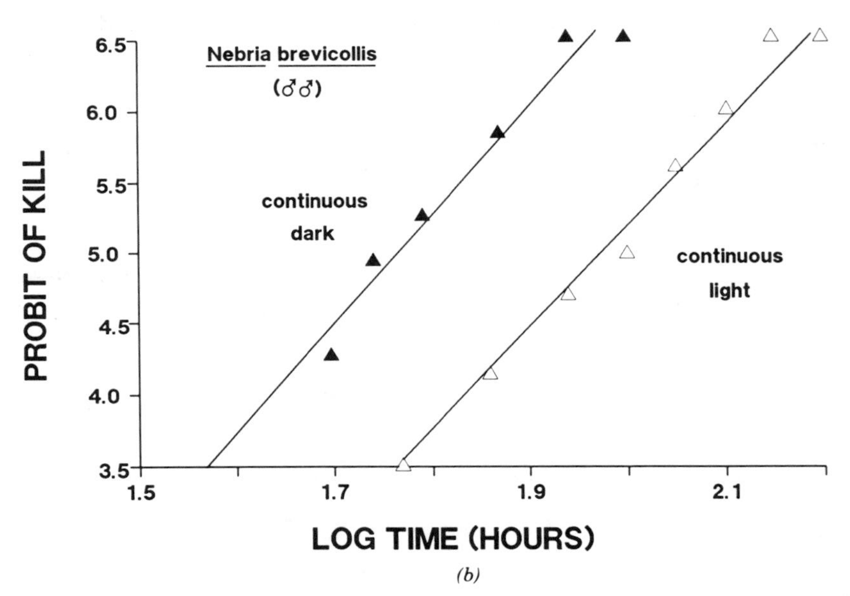

(b)

Figure 4.11. Effect of (a) temperature and (b) light on toxicity of thionazin-treated soil (16 ppm) to *Nebria brevicollis* (after Critchley 1972b).

Critchley (1972b) studied the effect of light on the toxicity of thionazin to carabid beetles in treated soil (Fig. 4.11b). He noted that because of its effect on the activity of these nocturnal predators, darkness increased the rate of pesticide uptake and subsequent mortality by 1.6-fold as compared to activity in continuous light. In relation to side-effect testing, Franze et al. (1976) discussed how light influenced natural enemy movement and therefore exposure to pesticide residues. Photoperiod and light intensity should be standardized in laboratory studies of pesticide toxicity to natural enemies in order to control protocol-related variation (Chapter 5).

4.4.4. pH

The acidity or alkalinity of the natural enemy's environment can affect pesticide toxicity and persistence, especially in aquatic or soil environments. The effectiveness of pesticides in formulations is extremely pH dependent, and considerable effort has gone into producing pesticides with optimal effectiveness in relation to pH requirements (Hartley and Graham-Bryce 1980).

In sandy loam soil where pH was altered by the addition of calcium carbonate, the onset of thionazin toxicity to carabid beetles was more rapid in highly acid soils (e.g., pH $\leqslant 5.0$), and was significantly more toxic than in slightly acid soils (pH 6.0) or slightly alkaline soils (pH 7.6; Critchley 1972b). Toxicity was directly correlated with residue levels of pesticide in the soil as determined by gas–liquid chromatography.

4.4.5. Habitat Form and Structure

Little is known about habitat form and structure relative to exposure and susceptibility of the natural enemy to pesticides. Nevertheless, it is clear that these factors often determine the extent and nature of exposure, especially in relation to pesticide distribution and natural enemy behavior (see Section 4.3.2). Microhabitat undoubtedly defines many of the physical relationships between temperature, relative humidity, and light that influence natural enemy susceptibility. In the literature, most references to habitat-related effects on natural enemy susceptibility are peripheral.

The most detailed, specific studies of habitat effects on natural enemy susceptibility address plant and soil characteristics. Plant form and structure may influence pesticide toxicity to natural enemies through characteristics of leaves, stems, bark, foliage density, and so forth. For example, natural enemy behavior can be influenced by leaf pubescence. Veinal patterns may influence natural enemy movement and the movement of systemic pesticides within the plant, variably affecting organisms within the plant habitat (Campbell 1975; Chapter 3). Studies of phytoseiid mites on apple have attributed pesticide toxicity to different searching patterns of the predatory mite and its rate of pesticide uptake (Blommers and Helsen 1986). Both leaf pubescence and veinal patterns appeared to be involved.

On a broader scale, the ways in which populations or communities of plants vary in their form and structure may also affect natural enemy exposure and susceptibility to pesticides. Crops are not of uniform size and shape, and numerous species occur in polycultures or in mixed wild and cultivated communities. Again, these features of crop architecture create the natural enemy's microhabitat environment. The influence of plant communities on natural enemy movement or host utilization could result in very different pesticide exposure episodes depending on the proportion of the habitat treated, the searching behavior of the beneficial, and other factors.

The presence of refugia or untreated reservoirs can appreciably limit the exposure of natural enemies to pesticides. A natural enemy's host or prey can modify this habitat to actually create new areas of refuge. For example, leafroller pests (Tortricidae) often curl leaves into relatively tight rolls that are not easily penetrated by pesticides. Natural enemies within such rolls may escape exposure. In this case, the host habitat serves as a refugium for the natural enemy. Aphid and eriophyiid mite galls also serve as refugia for natural enemies.

Refugia may involve plants, soil, or other structures (see further discussion in Chapters 8 and 10). Few studies have specifically attempted to identify the variety and relative importance of refugia in which natural enemies escape pesticides. Various pesticide application methods have been evaluated as a means to provide refugia for key natural enemy species, thereby achieving ecological selectivity (Newsom et al. 1976, Hull and Beers 1985). For example, lower-central foliage in fruit orchard trees can be left unsprayed to conserve predatory mites early in the season when their spider mite prey are more widely distributed (Madsen 1968).

In soil habitats, two studies have been published in which extensive efforts were made to evaluate the influences of soil habitat structure on natural enemy susceptibility to pesticides (Critchley 1972b, Kirknel 1978). These studies are discussed more extensively in Section 4.5, where the interaction of factors influencing pesticide susceptibility of natural enemies is addressed.

4.5. PESTICIDE-RELATED FACTORS

It is well known that different pesticide types, formulations, application methods, and so forth influence the susceptibility of natural enemies. They are the primary basis of pesticide impact on natural enemies, more so than other influencing factors. Pesticide-related factors are largely operational—under the control of the pest manager. In contrast, most natural enemy, host–prey, or environmental factors are uncontrollable in nature. The attributes of pesticides that influence natural enemy susceptibility are briefly discussed here. (See also Chapter 10 for ecological selectivity of pesticides.)

4.5.1. Type, Dosage, and Formulation

In the SELCTV database summary presented in Chapter 2, side effects of several types of pesticides on natural enemies were summarized. Toxicity ratings showed differences among application methods, for example, lower toxicity for field applications as compared to various laboratory methods of exposing natural enemies. These values represent aggregations of independent studies which were conducted using varying methods of exposure and evaluation.

More specific comparisons of different pesticide types, formulations, and application methods have been reported by reviewers of earlier literature. Bartlett (1964) concluded that natural enemies were usually less susceptible to stomach poisons (e.g., lead arsenate) than contact poisons. He also concluded that natural enemies were generally less affected by systemic pesticides than were plant sucking pests. He found no consistent relationship between the susceptibility of natural enemies to emulsifiable concentrates, wettable powders, or other widely used formulations.

More recently, individual studies have shown that microencapsulated, granular, and pelleted formulations are less detrimental to highly mobile natural enemies due to their more specific placement in the environment (Newsom et al. 1976; Chapter 10). Dusts are often selectively toxic to natural enemies compared to other formulations, presumably due

Table 4.5. Percent Mortality of Predaceous Spiders and Bugs Caged on Rice Plants Treated with Various Methods and Formulations of Carbofuran Insecticide (Adapted from Dyck and Orlido 1977)

	Weeks after Transplanting of Rice										
Method of Application	1	2	3	4	5	6	7	8	9	10	11
Species: Lycosa pseudoannulata											
Root coated at transplant		19			12						
Granules broadcasted at transplant		12			19			0			
Soil incorporation before transplant		38			25		8				
Hand placement of capsules[1]			66		50		0				
Granule applicator[1]			25		12			8			
Liquid applicator[1]			16		12			8			
Granules broadcasted every 2 weeks				33[2]		38[2]			12[2]		
Foliar sprays every 2 weeks[3]				33[2]		44[2]			18[2]		
Control (no spray)			6	16	25	12	0	0	0	0	0
Species: Crytorhinus lividipenris											
Root coated at transplant			88		85		50			25	
Granules broadcasted at transplant			59		40			35			7
Soil incorporation before transplant			92		50		35				16
Hand placement of capsules[1]				95	88		85			20	
Granular applicator[1]				62	75			25			14
Liquid applicator[1]				95	90			35			18
Granules broadcasted every 2 weeks				53[2]		90[2]	70		75[2]	65	32
Foliar spray				86[2]	100[2]	20			80[2]	40	28
Control (no spray)		38	20	26	30	27	20	20	30	20	12

[1] Mortality measured 3–5 days after caging. No food was provided to predators.
[2] Insecticide applied within the week indicated.
[3] Foliar spray applied at a rate of 0.5 kg a.i./ha, but all other insecticide treatments applied at 1 kg a.i./ha.

to the entomophage's tendency to search more extensively for food than does its prey or hosts (Bartlett 1964; Chapter 3). Baits are widely recognized as the safest formulations for killing pests and conserving natural enemies (Bartlett 1964; see discussion of formulations as means to achieve ecological selectivity in Chapter 10).

In specific studies, Hamlen (1975) evaluated the effect of comparable carbofuran and oxamyl rates in flowable versus granular formulations on the survival of *Encyrtus inflexus* parasitizing the hemispheric scale *Saissetia coffeae*. A consistently higher survivorship was observed when granular formulations were used as compared with flowable formulations.

Franz and Fabrietius (1971) evaluated seven commercial products or formulations of the fungicide zineb on uniparental strains of the parasitoid *Trichogramma cacoeciae* using standardized test procedures. Mortality ranging from 99.80% to 95.45% was observed at a given concentration when corrections for control mortality were made. Although their results showed no formulation-dependent patterns in zineb toxicity, their studies exemplified the type of evaluation that is needed for comparative analysis.

In a few cases, the influence of synergists as pesticide additives has been tested for their effects on natural enemies as compared to nonsynergized treatments with the same compound. Feng and Wang (1984) observed that the addition of either piperonyl butoxide or EPN (O-ethyl O-(4-nitrophenyl) phenylphosphonothioate) to the pesticides acephate, dichlorvos, profenofos, or fenvalerate increased their toxicity to the braconid *Apanteles plutellae*. Croft and Mullin (1984) reported an increase in carbaryl toxicity to the braconid wasp *Oncophanes americanus*. Synergist ratios of 14.0 and 5.5 were observed for the adult and late larval stages, respectively.

4.5.2. Application Methods

Methods of pesticide application have often been the focus of ecological selectivity studies (Chapter 10). Placement of the compound to maximally affect a pest while minimizing impact on the natural enemy is the primary objective. In studies designed to compare the impact of different application methods on pests and natural enemies, confounding factors often obscure test results. For example, in field tests it is difficult to separate density-related influences on natural enemies and their host or prey from other modifying factors (e.g., weather, alternate hosts or prey). These interactions make direct comparisons of application methods extremely difficult.

Dyck and Orlido (1977) compared application methods and formulation types for their influence on carbofuran toxicity. Eight carbofuran treatments were evaluated on the predatory spider *Lycosa pseudoannulata* and the predatory bug *Cyrtorhinus lividipennis* caged with planthopper prey or rice plants (Table 4.5). Granules broadcasted over water in rice fields or granular root zone applications were the least toxic to the spider, although mortality of the predatory bug was high.

In similar tests in South Korea, Choi et al. (1978) sampled populations of micryphantid and lycosid spiders in rice following an application of carbofuran by broadcasted granules, encapsulation, and liquid soil injection against leafhoppers and planthoppers. Spider populations were reduced by all treatments, but higher mortality was observed with encapsulation than either liquid or granular applications.

4.6. INTERACTION OF FACTORS

Although progress has been made in research on individual factors affecting natural enemy susceptibility to pesticides, complications inherent in the investigation of their interactions have generally precluded extensive study.

Critchley's (1972b) study of individual factors influencing carabid beetle susceptibility is exemplary in this regard. The influences of beetle sex and size, starvation, age, temperature, moisture, pH, light, soil compaction, and several pesticide attributes on natural enemy susceptibility were evaluated. These laboratory studies were conducted in a sandy loam soil similar to field crop soils (Critchley 1972b). This research represents the most extensive investigation to date that identifies the individual variables affecting the susceptibility of any natural enemy group.

In a related study, Critchely (1972a) studied the effect of thionazin empirically in the field. Long-term effects on carabids were investigated, particularly in relation to residual pesticide persistence at different soil depths. Thionazin granules broadcasted at 11.2 or 44.8 kg active ingredient (a.i.) per hectare (ha) considerably reduced carabids for 8 weeks following application. Lower concentrations of 2.24 and 8.96 kg/ha increased carabid activity due to sublethal effects on locomotion. Smaller, more active carabid species were most affected, whereas larger species which appeared later in the growing season were less influenced. Differences in susceptibility were also related to the burrowing habits of each carabid species. Burrowing species often were more susceptible than nonburrowing forms, presumably due to greater contact with pesticide residues.

Results of both field (Critchley 1972a) and laboratory (Critchley 1972b) studies examined together provide one of the most detailed and comprehensive analyses of factors influencing natural enemy susceptibility. While Critchley (1972a, 1972b) did not experimentally test the interactive effects of the individual variables, their net impact in the field was observed.

Kirknel (1978), however, developed an experimental design to evaluate the individual and interactive effects of several factors on the toxicity of organophosphate soil

Table 4.6. Analysis of Variance for Effects of Pesticide Compound, Dosage, Depth of Application, and Time of Application on Staphylinid Beetle *Aleochara bilineata* in Sandy Soil (after Kirknel 1978)

Source of Variation	Degrees of Freedom	Mean Square (S^2)	F-Statistic (F)	Significance (P Level)
Total	127	2742.1		
1. Dosages	3	2854.5	6.25	0.01
2. Depths	1	4947.8	10.84	0.01
3. Times	3	17839.6	39.08	0.001
4. Compounds	3	38735.8	84.86	0.001
Interactions				
1 × 2	3	657.6		
1 × 3	9	970.0		
1 × 4	9	1064.4	2.33	0.05
2 × 3	3	1163.2		
2 × 4	3	700.7		
3 × 4	9	8841.0	19.37	0.001
1 × 2 × 3	9	596.7		
1 × 2 × 4	9	488.1		
1 × 3 × 4	27	959.5		
2 × 3 × 4	9	1286.4		
1 × 2 × 3 4	27	456.5		

Standard deviation = 19.26; standard deviation in percent of mean = 71.1.

applications to the staphylinid *Aleochara bilineata*. The basic components of these studies, which included two soil types, four compounds, four concentrations of each compound, two application depths, and four time periods, are presented in Table 4.6. Variables were ranked in the order of increasing importance as they affected natural enemy susceptibility over a 305-day period. The most significant differences were observed among different compounds; however, significant differences among concentrations, depths, and exposure times were also observed. Occasionally, significant interaction effects were observed between time of application and different compounds tested, or dosages and compound used.

There are obvious limitations in the experimental studies cited above, such as lack of replication and control of all experimental variables. However, these studies emphasize the complexity and dynamic nature of variables which influence natural enemy responses to pesticides in the field. These systems require more study under experimentally controlled conditions, as well as in the field where less controlled, more realistic appraisals can be made.

It is notable that some of the best studies of factors influencing natural enemy susceptibility have been conducted in soil environments rather than aerial plant habitats. Aerial habitats are undoubtedly more variable in temperature, relative humidity, and light (see Section 4.4.5). Understanding the dynamic properties of a changing environment (Fig. 4.1) in any of these systems is indeed a challenging task. These phenomena must be better understood in order to make substantial progress in the conservation of beneficials in agricultural ecosystems.

5

STANDARDIZED ASSESSMENT METHODS

5.1. INTRODUCTION

When evaluating the impact of pesticides on natural enemies, it is extremely difficult to compare data from one test type or experimenter to another. Even when the same method and natural enemy population are employed, anomalies arise in results because of subtle differences in the practices used.

Standardization is a continual problem in testing the susceptibility of predators and parasitoids to pesticides. Quality control requires standard protocols, educational materials to train workers, and follow-up to ensure high performance. As more organisms and compounds are tested, quality control becomes increasingly critical. Standardization becomes more complicated when the susceptibilities of natural enemies and their prey or hosts are measured simultaneously (i.e., to obtain selectivity indices; Chapter 9). In spite of using standardized tests in selectivity studies, identical methods are seldom possible for both pests and natural enemies because of their morphological and biological differences.

Efforts to standardize susceptibility tests for natural enemies have been limited. However, their necessity has become increasingly clear as we have come to understand how small alterations in natural enemy behavior caused by pesticides can interfere with biological control (Chapter 8). Mortality is the extreme of pesticide impact. Methods used to detect such sublethal effects of pesticides as influences on longevity, fecundity, sex ratio, and behavior must be reliable and sensitive (Chapter 7).

While efforts to develop standardized tests of pesticide side effects are desirable for key species, costs of research must be balanced against the benefits gained from such information. Side effects testing is not feasible for all natural enemies, but it is cost effective to test some representative species. Ideally, standardized tests should maximize simplicity and replicability, while minimizing costs.

Some federal and state pesticide agencies are beginning to require that pesticide side effects on natural enemies be assessed as part of pesticide registration, especially on crops where IPM is well developed (Franz 1974, Hassan et al. 1983, EPA 1982). As these requirements become more common, the need for standardized testing will intensify.

The initial focus of this chapter is on objectives of pesticide side-effect testing. Protocols used for different natural enemies and pesticides are reviewed. In Europe, standardized tests have been developed for 22 predators and parasitoids (Hassan 1985). The United States Environmental Protection Agency (USEPA) is developing similar tests for the effects of microbial pest control agents (MPCAs) on five natural enemy groups (Lighthart 1986). These and other side-effect test methods are reviewed hereafter. Factors to consider in the design of standardized side-effect tests are also discussed.

5.2. DEFINITIONS AND OBJECTIVES OF STANDARDIZED TESTING

Test methods which have been used in the past to evaluate the impact of pesticides on arthropod natural enemies were presented in Chapter 2 (Fig. 2.6). These methods, classified by how the natural enemey was exposed to pesticides, included field, contact, dip, spray, residue, oral, topical, and injection tests. Field and contact tests have been most commonly reported in the literature (Fig. 2.6).

The test methods listed in Fig. 2.6 are the prototypes from which *standardized* susceptibility tests evolve. In reviewing these methods, one is struck by the number of

Table 5.1. Test Methods Used to Assess Side Effects of Pesticides on Arthropod Natural Enemies

Family/Group	Species	Crop	Test Level[1]	Test Type[2]	Stage Tested	Reference
Anthocoridae	*Anthocoris nemoralis*	Several	L*	—	—	Staubli unpubl., Hassan et al. 1983
Anthocoridae	*Anthocoris nemorum*	Several	L*	R	L	Firth unpubl., Hassan et al. 1983, Hassan 1985
Aphelinidae	*Encarsia formosa*	Greenhouse	L*	C	A	Ledieu 1979
			L*	C	A	Hoogcarspel & Jobsen 1984
			L*	D	P/A	Helyer 1982
			L*	R	A	Oomen unpubl., Hassan et al. 1983, Hassan 1985
			L*	C	P	Hassan 1985
			SF*	R	All	Oomen unpubl., Hassan 1985
Aphelinidae	*Cales noacki*	?	L*	?	?	Vivas unpubl., Hassan et al. 1983
Aphelinidae	*Aphytis holoxanthus*	Citrus	L	C/R	All	Havron et al. 1987a[3]
Aphidiidae	*Diaeretiella rapae*	Cabbage	L, SF*	C/R	All	Kuhner et al. 1985; Hassan 1985
Araneae	*Lynyphilid spiders*		L*	?	N/A	Shires unpubl., Hassan et al. 1983
Araneae	*Lepthyphates tenuis*	Several	L*	R	A	Inglesfied unpubl., Hassan 1985
Araneae	*Coleotes terrestris*	Forest	L*	R	A	Albert & Bogenschutz 1984
Braconidae	*Opius* spp.	Greenhouse	L*	R	A	Ledieu unpubl., Hassan et al. 1983, Hassan 1985
Carabidae	*Agonum dorsalis, Pterostichus cupreus, P. melanarius, Nebria brevicollis*	Several	L*	C/R	A	Edwards et al. 1984b, Hassan 1985
Carabidae	*Pterostichus cupreus*	Several	SF*	R	A	Edwards unpubl., Hassan 1985
Carabidae	*Agonum dorsalis*	Several	L*	C/R	A	Edwards unpubl., Hassan et al. 1983
Carabidae	*Bembidion lampros*	Several	L*	C/R	A	Chiverton unpubl., Hassan et al. 1983, Hassan 1985

Table 5.1 (*Contd.*)

Family/Group	Species	Crop	Test Level	Test Type	Stage Tested	Reference
Cecidomyiidae	*Aphidoletes aphidimyza*	Several	L	D	E/L	Warner & Croft 1982
Chrysopidae	*Chrysoperla carnea*	Several	L*	C	All	Suter 1978
		Several	L*	C/R	L	Bigler et al. 1984, Hassan 1985
Coccinellidae	*Cryptolaemus montrouzieri*	Citrus	?*	?	?	Brown unpubl., Hassan et al. 1983
Coccinellidae	*Coccinella septempunctata*	Several	?*	?	?	Pinsdorf 1977
			L*	C/R	L	Hassan 1985
Encyrtidae	*Leptomastix dactylopii*	Citrus	L*	R	All	Viggiani & Tranfaglia 1978
Ichneumonidae	*Coccygomimus turionellae*	Forest	L*	C/R	A	Bogenschutz 1975, 1979, Hassan 1985
		Forest	SF*	R	All	Bogenschutz unpubl., Hassan 1985
		Forest	F*	C/R	All	Bogenschutz 1979
Ichneumonidae	*Phygadeuon trichops*	Forest	L* SF*	R	A	Plattner & Naton 1975, Plattner 1979, Naton 1978, 1983, Hassan 1985
Lygaeidae	*Geocoris pallens*	Cotton	L	R	A	Yokoyama et al. 1984
Phytoseiidae	*Amblyseius bibens*	Several	L*	C/R	All	Overmeer & van Zon 1981, 1982
Phytoseiidae	*Amblyseius fallacis*	Apple	L	D	A/E	Croft et al. 1976a, FAO 1984[2]
		Apple	L	C	E	Nakashima & Croft 1974
		Apple	L*	?	?	Streibert 1981
		Apple	F*	C/R	All	Vanwetswinkel & Plevoets unpubl., Hassan 1985
Phytoseiidae	*Amblyseius potentillae*	Apple	L*	R	I	Overmeer & van Zon 1981, 1982, Hassan 1985

Phytoseiidae	*Phytoseiulus persimilis*	Greenhouse	L*	C	I	van Zon and Van der Geest 1980
			L	D	All	Petrushov & Zelenkova 1976
			L*	R	A/I	Samsoe-Petersen 1983, 1985, Hassan 1985
			L*	S	E/A	Helyer 1982
			L	C/R	A/I	Goodwin & Bowden 1984
			SF*	R	A/I	Vanwetswinkel unpubl., Hassan 1985
Phytoseiidae	*Amblyseius longispinosus, Phytoseiulus persimilus*	?	L	C/R	A/I	Goodwin & Bowden 1984
Phytoseiidae	*Typhlodromus occidentalis*	Apple	L	C	A	Readshaw 1977
		Apple	L	R	A	Readshaw 1977
Phytoseiidae	*Typhlodromus pyri*	Apple	L*	R	A	Overmeer & van Zon unpubl., Hassan et al. 1983, Hassan 1985
		Apple	F*	C/R	All	Englert unpubl., Hassan 1985
		Apple	F*	C/R	All	Easterbrook unpubl., Hassan 1985
Phytoseiidae	Several spp.	Grape	F*	R	All	Boness 1983
Staphylinidae	*Aleochara bilineata*	Several	?*	?	?	Naton et al. unpubl., Hassan et al. 1983
Staphylinidae	*Aleochara bilineata*	Several	L*	R	A/E/L	Samsoe-Petersen 1985, Hassan 1985
		Several	SF*	R	All	Naton unpubl., Hassan 1985
Syrphidae	*Syrphus vitripennis*	Several	L*	R	L	Riekman unpubl., Hassan et al. 1983, Hassan 1985
Trichogrammatidae	*Trichogramma cacoeciae*	Several	L*	C	A	Franz & Fabrietius 1971, Hassan 1974, 1977, 1985, Hassan et al. 1987
			L*	R	A	Hassan 1980, 1985
			SF*	R	All	Hassan 1985
Tachinidae	*Pales parva*	?	L*	C	A	Huang 1981

Table 5.1 (*Contd.*)

Family/Group	Species	Crop	Test Level	Test Type	Stage Tested	Reference
Tachinidae	*Drino inconspicua*	Forests	L*	C	A/L	Huang unpubl., Hassan et al. 1983, Hassan 1985
		Forests	SF*	R	A	Huang unpubl., Hassan 1985
General	Several spp.	Several	L*	C/R	Several	Blaisinger 1979
General	Several spp.	Orchards	F*	C/R	All	Blaisinger 1986
General	Several spp.	Orchards	F*	C/R	All	Sechser & Bathe 1978, Sechser 1981
General	Several spp.	Orchards	F*	C/R	All	Nikusch & Gernot 1986a
General	Several spp.	Several	SF*	C/R	All	Boness 1983
General	Several spp.	Pear	F*	C/R	All	Reboulet et al. 1984
General	Several spp.	Orchards	F*	C/R	All	Blanc 1986
General	Several spp.	Orchards	F*	C/R	All	Staubli 1986
General	Several spp.	Soil systems	F*	C/R	All	Koenig 1983, 1985
General	Several spp.	Arid crops	F	C/R	All	Stevenson et al. unpubl., Hassan 1985
General	Several spp.	Apple	F	C/R	All	Staubli et al. unpubl., Hassan 1985

[1] *These tests were part of the IOBC/WPRS activity as reported by Hassan 1985, Hassan et al. 1983, 1987. L = laboratory; F = field; SF = semifield; ? = unknown.
[2] C = contact; R = residue; D = dip; S = spray; C/R = contact/residue; ? = unknown.
[3] These tests have been proposed as standardized tests for resistance to pesticides for natural enemy species as well as for side effects.

factors that can influence results (Chapter 4). To some extent, the number and variety of currently available tests reflect the difficulties encountered in standardizing tests for a group of organisms as diverse as arthropod natural enemies.

Standardized methods are protocols that are developed and accepted by a peer group. A peer group usually includes personnel from public institutions, industry, governmental units, or combinations thereof, who are interested in test results. (See Sechser 1981 for an example of a standardized test developed by industry). Sometimes the peer group voluntarily follows standards for side-effect testing. In other cases, a regulatory or industry group with legal or economic power may require conformity to standards. Standardized methods are usually validated by the peer group. Validation means that results from a standard test have been compared to results of field applications of the pesticide and that the correspondence is good.

The distinction between susceptibility assessment and *standardized* susceptibility assessment is somewhat arbitrary. In some cases, a formal designation has not been made simply because the need to standardize has not been recognized. However, there may be a consensus among researchers as to what constitutes an acceptable test for a natural enemy group or species. For example, slide-dip and leaf disk spray or dip techniques for phytoseiid mites have been widely accepted (Table 5.1). Widely adopted test methods have existed for some time for *Chrysoperla carnea, Trichogramma* spp., several coccinellids, and a few other common natural enemies.

The objectives of standardized tests are to devise measures to thoroughly document the impact of pesticides on natural enemies under field conditions and to estimate the relative susceptibilities of natural enemies and their hosts or prey. The latter is applicable to basic research objectives as well. As noted, standardized tests for natural enemies should assess mortality, *as well as* other, more subtle adverse effects on predation or parasitization. Measuring the absence of pesticide effects on natural enemies is often more difficult than assessing direct mortality. The problem is analogous to measuring chronic effects of pesticides on other nontarget organisms such as wildlife or humans. Even though no obvious effect may be noted, it is difficult to prove that all possible influences have been accounted for.

5.3. GENERALIZATIONS ABOUT SUSCEPTIBILITY ASSESSMENTS

While most side-effects tests are designed to simulate field impact of pesticides, they are only models of this event. It is often necessary to employ several test types, each targeted at different components of exposure (i.e., direct contact, residue, and food chain uptake; Chapter 3). Tests should be conducted so that different factors influencing susceptibility can be controlled or manipulated experimentally (Chapter 4).

Side effect tests can be placed along a continuum ranging from laboratory experiments to field evaluations (Fig. 5.1). The relative merits of these tests have been discussed by Franz (1974), Boness (1983), Hassan (1985), and Hassan et al. (1983, 1987). Side effects tests tend to be somewhat conservative in estimating harmlessness to natural enemies so as not to underestimate pesticide impact (Naton 1983). For example, in laboratory contact tests, the test organism is completely exposed by topical or dip application. In the field, however, pesticide uptake may be lower due to irregularities in spray deposits, behavioral avoidance of exposure, or protection in refugia (Chapter 3). It has been proposed that the impact of a compound in the laboratory will seldom be exceeded in the field (Hassan et al. 1983, 1987). As implied, there are exceptions to this rule, (e.g., Nakashima and Croft 1974, Hassan et al.

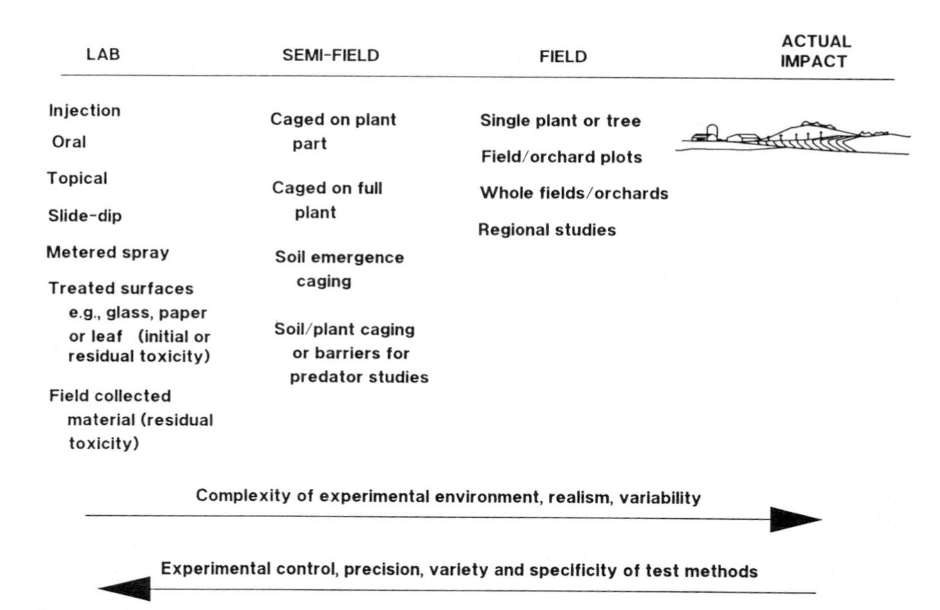

Figure 5.1. Generalized representation of the range of susceptibility assessment methods used to evaluate side effects of pesticides on arthropod natural enemies.

1983), especially with laboratory tests which may underestimate chronic, long-term effects (Chapter 8).

A review of the conditions under which side effects testing of natural enemy species occurs provides insight into some of the associated constraints. Side-effects tests are usually conducted either late in the pesticide development process or after a compound has been approved for field use. Pesticide impact on natural enemies is assessed long after the compound has been screened for pest efficacy, mammalian toxicity, and phytotoxicity. Target pests and certain other nontarget species are of greater concern than natural enemies. Therefore, the former species predetermine the pesticide formulation, dosage, and application that will be imposed upon predators and parasitoids. Usually, by the time side-effects tests on natural enemies are considered, uses of the new compound for pest control have already been established. The question then becomes, how will the new compound affect biological control by predators and parasitoids. The selectivity of new pesticides to natural enemies should be given a higher priority and should receive earlier consideration in the pesticide development process. As noted earlier, there is presently some movement in this direction.

Personnel interested in pesticide side-effects assessment often ask whether or not species-specific tests are needed to assess the effects of pesticides on any particular natural enemy. As is done with pests, can a few model species be screened for pesticide impact? To what extent can results be extrapolated to other species? Natural enemies possess unique biological characteristics and ecological associations with their hosts or prey which would seem to preclude much generalization. Those who have studied the question concur (e.g., Hassan et al. 1983, 1987). On the other hand, we may be too limited in our knowledge to know how to generalize about pesticide side effects on natural enemies.

In light of these constraints and complexities, it is realistic to question whether or not standardization of side-effect assessments for predators and parasitoids is possible. If so, what variables should be standardized? What individual species are the most appropriate models for testing and how much generalization from species to species is possible? These questions are addressed following a review of some of the available methods for standardized susceptibility assessment.

5.4. PROTOCOLS FOR STANDARDIZED TESTING

Despite the fact that the need for standardized tests to measure pesticide side effects on natural enemies has long been recognized (e.g., Ripper 1956, Bartlett 1964, Croft and Brown 1975), formal efforts to develop these tools have been limited. In the meantime, a massive amount of side-effects testing has gone on throughout the world (Chapter 2). Protocols proposed for several classes of pesticides and natural enemies are summarized hereafter. Reference to these efforts is not an endorsement, but is made simply to illustrate the type of research that has been conducted towards standardization.

Those standardized tests of pesticide side effects which have been developed for natural enemies are listed in Table 5.1. Most tests have come from a large-scale European research project (see those marked *). The preponderance of tests are laboratory contact and/or residue evaluations made on individual life stages of natural enemies. Semifield tests have been developed for several species (e.g., Ledieu 1979, Bogenschutz 1979, Naton 1978, 1983). Standardized field tests have been proposed, primarily for natural enemies attacking pests in orchart and vineyard habitats (Steiner 1977, Sechser and Bathe 1978, Blaisinger 1979; see other cases and review of these types of tests in Hassan 1985).

5.4.1. A European Program

A working party under the auspices of the Western Palearctic Regional Section of the International Organization for Biological Control (IOBC/WPRS) first organized a program to develop standardized tests in 1974. This group functions principally in western Europe (see project summaries in Franz 1974, Franz et al. 1980, Hassan et al. 1983, 1987). Overall, the effort has included 50–100 scientists working in 9 countries with at least 22 natural enemy species. (See tests marked * in Table 5.1.) As part of the program, scientists regularly exchange data and publish results summaries (e.g., Hassan et al. 1983, 1987). Initially, only laboratory tests were considered in the joint testing effort. More recently, semifield and field procedures have been developed (Hassan 1985, Hassan et al. 1987).

The IOBC/WPRS effort has been extensive in the number of species and compounds tested. Later, selected results from the IOBC/WPRS program are summarized and compared with data from the SELCTV database analysis (Section 5.6; Chapter 2). Hassan et al. (1987) recently reported pesticide side effects for 22 new pesticides and 19 natural enemies. This brings the total to 62 pesticides evaluated on seven classes of crops. More specific summaries of IOBC/WPRS research can be found in the literature cited in Table 5.1. Common principles of standardization agreed upon by the IOBC/WPRS group are discussed below.

5.4.1.1. Test Types and Combinations. The IOBC/WPRS group contends that no single test method provides sufficient information to reflect the side effects that a pesticide would have on a beneficial organism in the field. Therefore, a sequence of tests are recommended which include laboratory, semifield, and field methods. Generally, pesticides should first be examined in the laboratory on 4–6 beneficial species for a given pesticide and crop of interest (Section 5.4.1.3).

The IOBC/WPRS group contends that pesticides found to be harmless to a natural enemy in the laboratory are most likely to be harmless in the field and no further testing is required. If a pesticide is harmful in laboratory tests, semifield tests are recommended to evaluate the duration of harmful activity and the impact of residues using plant or soil substrates. Finally, field tests are recommended to assess the impact of both direct sprays and residues of pesticides.

5.4.1.2. Standard Test Types. The following test types are recommended by the IOBC/WPRS group: a) laboratory, initial toxicity, b) semifield, initial toxicity, c) semifield, persistence, and d) field tests. Standardized methods for field tests are still under development.

5.4.1.2.1. Laboratory, Initial Toxicity Test. Laboratory tests for initial toxicity involve a) exposure to a fresh, dry pesticide film, b) recommended field concentration of pesticide, c) applying compound to glass plates, plants, or soil (sand), d) an even film of pesticide, 1–2 mg fluid/cm^2, e) laboratory-reared arthropods of uniform age, f) adequate ventilation, g) water-treated controls, h) four evaluation categories (see below), and i) measurement of the reduction in biological control capacity (e.g., in percent parasitization, consumption, or mortality).

Beneficials are exposed to a fresh dry pesticide film applied at recommended field concentrations on glass plates, leaves, or soil. The choice of substrate depends on the behavior of the natural enemy. Tests are carried out under controlled temperature and humidity conditions favorable to the beneficial. Mortality in water-treated control units should not exceed 10–20%. When closed cells are used, forced ventilation is recommended to avoid the accumulation of pesticide fumes.

The pesticide solution is sprayed as an even film on the experimental surface. The amount of wet solution deposited on the experimental surface is measured by weighing the target before and immediately after spraying. At the end of the exposure period, the reduction in beneficial capacity is measured. This may include changes in parasitism, fecundity, or fertility as compared with controls. Pesticides are classified into four categories, depending on the degree of damage they cause to the beneficial: 1 = harmless (<30% of the negative effect measured), 2 = slightly harmful (30–79%), 3 = moderately harmful (80–99%), and 4 = harmful (>99%).

5.4.1.2.2. Semifield, Initial Toxicity Test. Semifield assessments of initial toxicity include: a) exposure to a fresh, dry pesticide film, b) recommended field concentration of pesticide, c) wet spraying of plants, d) field cages under field or field-simulated conditions, e) water-treated controls, f) laboratory-reared arthropods of uniform age, g) adequate contact of pesticide through dense foliage, h) food and/or host or prey near the center of treated foliage, i) adequate exposure period before evaluation, and j) four evaluation categories (see Section 5.4.1.2.1).

Plant or soil units are placed in the field under cover from rain and in partial shade or under field-simulated environmental conditions. Plants should have dense foliage to provide sufficient leaf surface for the pesticide treatment. Cages should be designed to suit the size and shape of the test plants. Plants—not the experimental cages—are sprayed with pesticide to the point of run-off. After the pesticide has dried, the plants are caged, the beneficials are released, and, if necessary, supplemental food such as sugar solution, honey, or pollen is provided. After the pesticide is applied, the hosts or prey are placed among the treated plants. Performance of the beneficials is then compared with water-treated controls using the same classification system cited for laboratory tests.

5.4.1.2.3. Semifield, Persistence Test. Semifield tests of persistence are composed of a) exposure to pesticide residues, b) recommended field concentration of pesticide, c) wet spraying of plants, d) weathering under field or field-simulated environmental conditions, e) laboratory-reared arthropods of uniform age, f) water-treated control g) monitoring for up to 1 month after treatment, and h) four evaluation categories (see Section 5.4.1.2.1).

To test the persistence (duration of harmful activity) of pesticide residues, plants or soil are treated and maintained under field or field-simulated environmental conditions. Beneficials are exposed to the treated substrate at different time intervals after application. The duration of harmful activity is the time required for the pesticide residue to diminish in effectiveness until a reduction in biological control potential of less than 30%, compared with the control, is reached. The pesticides are classified as follows: <5 days duration of harmful activity = short lived, 5–14 days = slightly persistent, 15–30 days = moderately persistent, and >30 days = persistent.

5.4.1.2.4. Field Test. Field tests are composed of a) crops and beneficials directly treated with pesticide, b) laboratory-reared or resident field populations of arthropods, c) sampling beneficials before and after treatment, d) recommended field rates and application methods reflecting agricultural practices, e) water-treated controls, f) estimation of both dead and living populations, g) collection of sufficient data for statistical analysis, h) four evaluation categories: 1 = harmless (<25% mortality), 2 = slightly harmful (25–50%), 3 = moderately harmful (51–75%), 4 = harmful (>75%).

5.4.1.3. Choice of Organism. Beneficials chosen for testing should be relevant to the crops on which the pesticide is to be used. The researcher is usually interested in certain crops, pests, and pesticides and then decides which beneficials should be tested. The alternative would be to test pesticides on as many beneficials as is practical. The IOBC/WPRS group recommends the following guidelines:

Field Crops, e.g., Wheat: One general predator (e.g., Chrysopidae, Coccinellidae), one aphid parasitoid (e.g., Aphidiinae), and one soil-inhabiting predator (e.g., Carabidae, Staphylinidae).

Vegetables, e.g., Brassicae *Crops*: One aphid predator, one aphid parasitoid, one egg parasitoid of Lepidoptera (e.g., *Trichogramma*), one larval (or pupal) parasitoid of Lepidoptera (e.g., Tachinidae, Braconidae), one soil-inhabiting predator, and one parasitoid (e.g., Cynipidae, Ichneumonidae) of soil-inhabiting pests.

Vegetables in Glasshouses, e.g., Cucumber: The predatory mite *Phytoseiulus persimilis* and the whitefly parasitoid *Encarsia formosa*, one aphid predator, and one aphid parasitoid.

Fruit Orchards, e.g., Apple: One general predator, one aphid parasitoid, one predatory mite *Typhlodromus*, one egg parasitoid of Lepidoptera, and one larval (or pupal) parasitoid of Lepidoptera.

Vineyards (Grape): The predatory mite, *Typhlodromus pyri* and one egg parasitoid of Lepidoptera.

Forests: One larval (or pupal) parasitoid of Lepidoptera, one egg parasitoid of Lepidoptera, one general predator, and one predatory mite.

5.4.1.4. Other Considerations. The IOBC/WPRS group recommends using laboratory-reared arthropods of uniform age for testing. Although changes in populations can occur during laboratory colonization and rearing (e.g., genetic bottlenecks and drift), such limitations are deemed preferable to the unknown conditions of field-collected specimens. They recommend that laboratory colonies only be supplemented with field-collected material when reduced performance has been demonstrated. Initially, a susceptible strain of a beneficial should be chosen for testing the side effects of pesticides. If a resistant strain is available, it also should be tested.

5.4.2. Interim USEPA Protocols for Microbial Pesticides

In the United States, standardized tests for assessing the effects of pesticides on arthropod natural enemies have not been established for conventional pesticides (as they have in Europe). While synthetic organic pesticides have been extensively tested in this country (see Tables 2.2 and 2.12), the importance of developing standardized tests has not been widely recognized. Neither has there been pressure by regulatory agencies or other public or private interest groups for standardized tests.

The need to develop standardized tests for microbial pest control agents (MPCAs) in the United States has arisen more recently from the impending registration of genetically engineered pesticides. Modified *Bacillus thuringiensis* and some insect viruses are the most promising early candidates (Kirschbaum 1985). In 1983, the USEPA issued a request for interim protocols for MPCAs (EPA 1982).

Side-effects tests of naturally occurring microbial pesticides have been conducted similarly to those for conventional pesticides (Chapter 11). In many cases, conventional methods are inappropriate for this class of pesticides (Flexner et al. 1986). Problems associated with conventional methods are 1) a lack of sensitivity to sublethal or chronic effects over a longer period than that normally allotted for side-effect tests, 2) inadequate quantification of the dosage administered to a natural enemy in a test, 3) inappropriate exposure of the natural enemy to the microbial (e.g., topical or residue application for an organism that must be ingested to be toxic), and 4) a lack of attention to the indirect effects of microbials on the natural enemy via its host or prey.

In 1982, the USEPA began to develop interim registration protocols for microbial pesticides, of which microbial insecticides are a subset. Many microbials being considered for registration are species-specific and are taxonomically similar to organisms that have already been registered (Tanada 1984). Should these organisms have less stringent requirements for registration than organisms from an unregistered class or family of microbials, or than those known to cause disease in nontarget beneficials? Research is needed to evaluate this question more specifically.

The USEPA has suggested that microbial pest control agents be tested on at least one natural enemy species from three of the following groups: predaceous Neuroptera, predaceous Hemiptera, predaceous Coleoptera, predaceous Acarina, and parasitic Hymenoptera. A summary of the proposed tests for each of these five groups is given in Table 5.2. More detailed descriptions of side-effect tests for a representative predator and parasitoid are presented in Section 5.4.2.2. Entomophages were selected for USEPA protocols on the basis of their importance as biological control agents on major crops. Other considerations included the effort already being made to rear and test these organisms as well as their availability from commercial sources.

Table 5.2. A Summary of Attributes of USEPA Interim Protocol Tests to Measure Impact of Microbial Pesticides on Arthropod Natural Enemies[1]

Group Test Type	Recommended Species	Microbial Pesticide Life Stage Tested				Results Monitored
		Virus	Bacterium	Fungus	Protozoan	
PHYTOSEIIDAE						
1. Contact	Af, Mo, Pp	AF, I	AF, I	AF, I	AF, I	Mortality
2. Indirect	Af, Mo, Pp	AF	AF	AF	AF	Mortality
HEMIPTERA						
1. Contact	*Geocoris*	A, N	A, N	A, N	A, N	Mort., Ovip.
2. Feeding	*punctipes*	A, N	A, N	A, N	A, N	Mort., Ovip.
CHRYSOPIDAE						
1. Contact	*Chrysoperla*	L	L	L	L	Mortality
2. Feeding	*carnea*	A, L	A, L	A, L	A, L	Mortality
COCCINELLIDAE						
1. Contact	Cm, Hc[3]	A, L	A, L	A, L	A, L	Mortality
2. Feeding	Cm, Hc	A, L	A, L	A, L	A, L	Mortality
PARASITOIDS (hymenopteran)						
1. Contact	*Trichogramma*	A	A	A	A	Parasitism
2. Indirect	*pretiosum*	E	E	E	E	Emergence

[1] Summarized from methods reported by Lighthart (1986). All tests run at dosages including 0.1, 1.0, and 10.0 times normal field rates; dose mortality data subject to Abbott's formula and probit analysis.

Key:

Af = *Amblyseius fallacis*; Mo = *Metaseiulus occidentalis*; Pp = *Phytoseiulus persimilis*; Cm = *Coleomegilla maculata*; Hc = *Hippodamia convergens*; AF = adult female; A = adult (either sex); I = immature; L = larvae; E = egg; N = nymph.

5.4.2.1. Interim Protocol Tests. Microbial pesticides have both direct and indirect effects on predatory and parasitic arthropods, as do conventional synthetic organic pesticides. They may be directly toxic to the natural enemy and they may indirectly affect them by reducing the quantity or quality of the natural enemy's food. Flexner et al. (1986) concluded that the indirect effects of MPCA's were generally more common and of greater impact than direct effects (see Chapter 11). Sometimes, microbial pesticides may have a beneficial effect by weakening the prey or host, making it more susceptible to attack by the actively pursuing predator or parasitoid.

Because microbial pesticides are often taken up orally, they are usually administered to natural enemies through contaminated food in standardized tests (Table 5.2). Fungi are exceptional because their common mode of entry is through the cuticle. Topical, contact, or residue exposure is recommended for evaluation of fungal agents. However, because there is some possibility of direct contact with all MPCAs, topical, contact, or residue tests are occasionally used for viruses, bacteria, and protozoa as well (Table 5.2). Nonoral exposure tests are conducted as a precautionary measure and should not be relied on exclusively for nonfungal microbial agents (Chapter 11).

The USEPA standardized susceptibility tests developed to date for MPCAs (Table 5.2) differ from other types of tests in several ways. Evaluation times are usually longer because microbials have a slower mode of action and may require reproduction of the microbe in the infected host or natural enemy. Control of temperature and relative humidity are critically important in these tests, since MPCAs have very different activities within the normal range of environmental conditions. The status of the host or prey of the natural enemy is extremely important in microbial tests. These organisms are often the source of microbial inoculum. Infected hosts or prey can indirectly affect the natural enemy by their status as a less than adequate food source (Chapter 3). The density of both hosts and natural enemies can also affect transfer of a microbial agent between and among hosts or prey and natural enemies (Chapter 11).

5.4.2.2. Tests for Representative Natural Enemies. Two interim standardized susceptibility tests to assess pesticide side effects on predaceous and parasitic natural enemies are considered here in greater detail. *Chrysoperla carnea* and *Trichogramma* spp. have been chosen as representative species. Both are key natural enemies of plant feeding arthropods in many crop production systems.

5.4.2.2.1. A Protocol for Chrysoperla carnea. Adult *C. carnea* are nonpredaceous nectar feeders that are active and quite mobile. The incidence of adult contact with microbial pesticides is probably minimal. Adults most likely contact MPCAs during nectar feeding or oviposition. Lacewing eggs and pupae are not easily penetrated and are less susceptible to pesticides than larvae (Chapter 4). Therefore, they are also less likely to be affected by microbial pesticides. The primary life stage affected by MPCAs through either contact or oral routes would be larval. *C. carnea* larvae rest and search on treated surfaces, and feed on infected prey (e.g., spider mites, aphids, lepidopteran eggs and larvae). Ingestion of prey involves piercing their integument and sucking the predigested liquid content. Larvae may also occasionally feed on nectar and water. Besides direct mortality, pathogenic effects on the larvae are sublethal and chronic. These side effects may include decreased predation, slower development, incomplete development, and reduced fecundity and fertility in subsequently developing adults. For *C. carnea*, five standardized tests have been proposed, two for adults and three for larvae.

TEST 1—ADULTS: Because *C. carnea* adults actively feed on plant nectar and preen extensively, oral ingestion is a likely route of uptake. The recommended test method for adults is similar to that described by Wilton and Klowden (1985). Newly emerged adults are held by the wings with forceps and offered 1 μl of solubilized or heat-inactivated (for controls) microbial pesticide. Adults drink readily upon emergence. After the drop is ingested, the test insects are placed either individually or as a treatment group in cages with food and water. Mortality is assessed after 48 and 72 hr. If no mortality occurs after 72 hr, observations should be made daily for 7 days. Equal numbers of males and females should be tested. Initial doses tested should be those recommended in the field. Serial dilutions of the initial dose may be desirable if side effects are noted in preliminary tests.

TEST 2—ADULTS: Because fungal pathogens may penetrate the adult lacewing integument, contact toxicity tests should be conducted for this MPCA. A 1-oz container (or other suitable disposable container) is sprayed or dusted with the fungal pesticide. Adults are provided food and water. Mortality is assessed after 48 and 72 hr. Other test conditions are identical to those in test 1 (above).

TEST 1—LARVAE: This test consists of dipping or spraying a test cage with MPCA and placing first instar *C. carnea* larvae on the dried residue with untreated prey. It is important to treat every surface that the larvae will contact, since larvae may be repelled by treated surfaces. After 7 days, larvae are moved to untreated petri dishes and fed untreated prey until pupation. Percent larval mortality, mean larval duration, percent successful emergence of adults, and the number and percent egg hatch from 20 mated females should be measured for a 2-week period. As noted, initial concentrations should be discriminatory, followed by serial dilutions if necessary.

TEST 2—LARVAE: This method is similar to that described by Salama et al. (1982). Two-day-old *C. carnea* larvae are fed for 7 days on prey that have been sprayed with MPCA. If a target prey is not available, another prey which is susceptible to the microbial pesticide should be used (e.g., aphids or larvae and eggs of lepidoptera). After 7 days, the neuropteran larvae are fed untreated prey until pupation. Other treatment conditions are the same as those in test 1.

TEST 3—LARVAE: The most rigorous assay for microbial side effects is to feed *C. carnea* larvae target hosts which have *ingested* MPCA (Salama et al. 1982). However, target hosts may be difficult to obtain, inoculate, rear, and feed to the lacewing larvae. Caution is recommended in using artificial diets for rearing prey because they sometimes alter the nutritional quality of the prey. This may reduce the vitality and survival of lacewing larvae and may confound experiments due to inadequate larval nutrition (Tulisalo 1984). Larvae are fed for 7 days on the target prey, which has been fed for 24 hr on an artificial diet containing the MPCA. The control *C. carnea* receive prey fed on artificial diet containing the heat-inactivated MPCA. After 7 days, lacewing larvae are fed until pupation on untreated diet. Artificial diets are not available for some insect groups; in these cases, the lacewing larvae should be fed the microbial directly in artificial diet without involvement of the target host. Test conditions and evaluation procedures similar to those reported above should be used. (For further details on test conditions, see Table 5.2; also Grafton-Cardwell and Lighthart 1986).

5.4.2.2.2. A Protocol for Trichogramma spp. The epizootiology of microbial pathogens in parasitic insects has received limited attention (Tanada 1964; Chapter 11). In general, transmission is through the spiracles, anus, mouth, and integument. Transmission may also be transovarial. Stinging the host or biting during host feeding can also result in transmission. Adult parasitoids or host-associated larval stages are most likely to be exposed to MPCAs in nature. Therefore of the 3 standardized tests proposed for the egg parasitoid *Trichogramma*, two are for adults and one is for immature parasitoids developing within infected host eggs.

TEST 1—ADULTS: Young adult males and females of known age are aspirated into (1-oz) treated plastic cups with lids. Adult *Trichogramma* are active feeders on plant nectars, honeydew, and other sugar based substances. Honey provided to adult parasitoids should be laced with the field rate of microbial pesticides added to water. Serial dilutions may be similarly provided, thereafter. Water (only) in honey should be included as a control. Parasitoids in cups should be stored at 28° C under a long-day photoperiod until the assay is completed. Mortality should be evaluated a 24-hr intervals up to 196 hr.

TEST 2—ADULTS: Microbial effects on fecundity are assessed by holding adult females treated in test 1 for 24 hr, then providing host eggs for 6 hr on days 2, 4, and 6. Host eggs are then removed and held at 28° C under long days, 65% relative humidity, for 8 days to assess percent hatch and survival to adulthood. Parasitism rates and survival of treated, developing parasites are compared to controls treated only with water.

TEST 3—LARVAE: Fresh lepidopteran eggs (e.g., *Sitotroga* sp.) are presented to adult female *Trichogramma* for 6 hr at 28° C. Equal numbers of adult parasites should be used in each replicate and treatment to obtain nearly equal rates of parasitism. After the adult parasitoids are removed, the parasitized eggs are sprayed with the recommended field rate and serial dilutions of the MPCA. Parasitized eggs are held at 28° C, 65% relative humidity, until adult emergence is complete in the water-treated controls. Normally, the adults emerging from parasitized eggs should be monitored at 8, 10, and 12 days after treatment. (For further details on this test, see Table 5.2 and Hoy and Lighthart 1986.)

5.4.3. Protocols for Natural Product Pesticides

Natural products pesticides, including the insect growth regulators (IGRs) and chitin synthesis inhibitors, are newer insecticides being developed and used for controlling insect pests (Chapter 12). To date, side-effect tests for these agents resemble those employed for conventional pesticides. Natural products pesticides, however, have modes of action that make them difficult to evaluate using conventional methods (Staubli 1986). IGRs and chitin inhibitors act slowly and can influence natural enemies over several developmental stages. They not only cause mortality, but commonly inflict sublethal effects on both the target species and natural enemies such as arrested development and morphological abnormalities (Chapters 7 and 11).

Test methods could be improved to better evaluate the side effects of IGRs. Because most IGRs act on developing immature stages, timing of application is critical and treatment of adults that have undergone melanization is inappropriate. As with microbial pesticides, the more chronic, indirect effects of IGRs on beneficials may be more important than direct effects. Therefore, evaluation periods should be lengthened to allow for the detection of both direct and indirect effects of IGRs on natural enemy populations (Beckage 1985; Chapter 12).

5.5. FACTORS TO CONSIDER IN THE DEVELOPMENT OF STANDARDIZED TESTS

There are many factors to consider in the development of standardized side-effect tests for natural enemies. Some of these factors were referred to in reviewing the IOBC/WPRS and USEPA testing protocols. Environmental factors such as temperature, relative humidity, light, and moisture were discussed in Chapter 4. Chemical, operational, and biotic factors also influence standardized susceptibility tests. The following discussion parallels that of Franz (1974), who pioneered research to develop standardized procedures.

5.5.1. Types of Tests

Field assessments often give results most similar to actual field use of a pesticide. Direct, residue, and food chain routes of exposure are combined in field tests; however, experimental control is usually limited and reproducibility may be low. Field tests are influenced by uncontrolled variables such as wind, rainfall, relative humidity, and temperature. These variables should be monitored. Field tests are quite costly to conduct and therefore are not extensively replicated. Examples of these types of tests have been reported by Blaisinger (1979), Sechser (1981), Sechser and Bathe (1978), Steiner (1977), Boness (1983), Naton (1978), and Englert (1984) for deciduous tree fruits and other crops (see Table 5.1).

At the other extreme in evaluating the side effects of pesticides on natural enemies is the laboratory test (Fig. 5.1). Laboratory tests are conducted under controlled conditions and are designed to measure more specific aspects of pesticide exposure (Chapter 3). Most standardized tests developed to date are laboratory tests (Table 5.1). Most laboratory assessments are based on direct or contact exposure of one or more life stages of the entomophage. These types of tests are usually limited to one or two specific modes of pesticide uptake, while actual field exposure may encompass a wider range of uptake routes (Chapter 3). Laboratory tests are usually conducted under specified environmental conditions, which may not represent the range of conditions encountered in the field. Lastly, although individual laboratory tests may attempt to integrate various subcomponents of actual field impact, they may fail to detect sublethal or chronic side effects (Chapters 7 and 8).

Between the extremes of field and laboratory assessments are a wide range of tests which have been termed *semifield* by Hassan et al. (1983). Examples of semifield tests have been reported by Bogenschutz (1975, 1979) and Naton (1983) for the parasitic ichneumonids *Coccygomimus turionellae* and *Phygadeuon trichops*, respectively (see also Table 5.1). In these studies, researchers focused on measuring the impact of pesticide residues on adult parasitoids. Ledieu (1979) reported a residual contact, semifield procedure for the aphelinid *Encarsia formosa*. In this case, specific methods were used to overcome some of the problems associated with both laboratory and field tests.

5.5.2. Experimental or Technical Considerations

5.5.2.1. Compound, Formulation, and Application. The pest and crop with which a particular natural enemy or natural enemy complex is associated dictates the compound or compounds that should be screened for side effects. In addition to insecticides, many fungicides, herbicides, nematocides, rodenticides, and other chemicals applied to agricultural systems are potential candidates for screening on natural enemies.

There may be considerable overlap in side-effects testing using the same pesticides and natural enemies, but from different crops. Many natural enemies are effective biological control agents in many different cropping systems. Most pesticides are initially developed on a relatively small subset of major crops (e.g., cotton, rice, soybeans, corn), but are then tested and registered on many minor crops (e.g., small fruits, most vegetables, and orchard crops). For this reason, it may be useful to generalize the effects of pesticides on natural enemies measured on one crop to a different crop from which populations have not been specifically tested. (Note: These extrapolations should be done with caution since resistance may develop among crop-specific natural enemy populations; see Chapter 4, Section 4.2.1.2, and Chapter 14.)

As with the choice of compounds for testing, the type of formulation and method of application are also usually dictated by the crop/pest system with which one is working. Occasionally, alternate formulations may be evaluated as a means to improve ecological selectivity to natural enemies (Chapter 10). Often, application methods are not identical to those used in the field, but various techniques can be used to mimic field application methods in the laboratory.

One may deliberately deviate from formulations or application techniques used in the field in order to ensure complete exposure to a compound or to make relative toxicity comparisons between species. For example, topical application or dipping the organism into a pesticide solution may give a more standardized comparison of innate susceptibility than a simulated field exposure method. Topical and dip applications provide a more uniform dose to the test organism. Technical grade pesticide is commonly used to achieve greater precision in LC_{50} or LD_{50} tests because formulated compounds may contain variable amounts of inert ingredients (Hartley and Graham-Bryce 1980).

5.5.2.2. Dosage or Concentration. The recommended field rate of a toxicant is most commonly used in standardized tests to predict probable impact in the field, but lower rates may also be used to determine selective dosages for natural enemies. Selective rates reduce, but do not eliminate pests, thus leaving a food supply for surviving beneficials. Reduced rates are also evaluated to estimate the impact of diminishing residues on immigrating beneficial species. Often comparative LD, LC, or LT values are obtained using serial dilutions of toxicant to estimate selectivity ratios between the two species groups (Chapter 2). Rather than comparing common LC_{50} or LC_{95} values, some have advocated comparing the LC_{20} of the natural enemy to the LC_{95} of the host, thus identifying a rate of pesticide that is effective against the pest while maximally conserving the beneficial natural enemy (e.g., Bartlett 1958, Hower and Davis 1984; Chapter 9).

5.5.2.3. Duration of Exposure. It is difficult to arrive at a suitable period of exposure for an experimental animal in the laboratory which simulates field conditions. Researchers usually choose a specific period of time and then use it as a standard in subsequent tests (Chapter 2). Since direct contact is an instantaneous event, dipping tests are made by immersing the insect in the pesticide solution for 1–10 seconds. Exposure to residues may last for several hours or days. By using a standard evaluation time for a given test, a good relative comparison among compounds can be obtained.

For topical applications, the time of exposure is not defined by the experimenter, since the entire amount of toxicant remains on the organism until it is absorbed or degraded. For dip tests, usually only a few seconds of exposure are sufficient to cover the entire organism. Adhering residues then remain on the organism. For spray tests, experimental animals may or may not be covered with the toxicant.

Treated animals are usually transferred to an untreated observation chamber after spraying to evaluate subsequent mortality. The same may be true for dipped animals. In spray tests, exposed animals may be treated initially and then left in contact with residues on the treated substrate. This type of treatment may best reflect field exposure, since the organism is exposed over the entire period of the chemical's persistence. This test may span an entire life cycle or multiple generations of the natural enemy. Such a long-term exposure would still fall short of estimating field impact, since pesticide applications are often repeated with some degree of overlap in remaining residue.

5.5.2.4. Test Evaluation Period. As discussed in Chapter 2 (Section 2.3.1.4), the preponderance of tests ($> 76\%$) conducted to date have been evaluated within 96 hr after treatment. This reflects the large proportion of tests that have been conducted in the laboratory. It probably also indicates the tendency of experimenters to look more at short-term effects of pesticides on natural enemies. As evaluation periods lengthen, maintaining untreated controls without mortality becomes a problem. Beyond 96 hr, it is difficult to maintain the food supply, the treated substrate, and other conditions favorable for survival of the natural enemy and its host or prey. Also, a natural enemy may transit a given life stage in a few days and in some cases a few hours. Unless evaluations are made across generations, tests are limited by the developmental time of the test organism. For evaluations made over several generations, the probability that other variables are influencing test results increases and makes pesticide impact assessment more difficult.

Tests of longer duration, lasting from 96 hr to 1 month or longer comprise less than 25% of all timed side-effect tests (Chapter 2). These are primarily field and semifield tests in which more gradual effects of environmental factors on the residual toxicity of pesticides can be measured. For some of the newer chemicals such as microbials, IGRs, and chitin inhibitors, toxic effects are not manifested until several days or weeks have elapsed (Chapters 11 and 12). The indirect effects of these agents on the host or prey of natural enemies can be more important than direct toxic effects on natural enemies. For these agents, more long-term evaluations are almost essential.

5.5.2.5. ***Substrates.*** The ideal substrate for side-effects testing is the surface upon which the predator or parasite is most often found in the treated environment. Plant substrates vary in such characteristics as pubescence and cuticular topography. Aside from variable characteristics, leaves deteriorate over time once excised from the plant. Both *Trichogramma* adults and predatory mites show variable mortality depending on the deterioration of leaf substrates in residue toxicity tests (Franz 1974, Croft unpublished). Because of variation in pubescence and other surface features, residues may not be uniformly distributed (Blommers and Overmeer 1986; Chapter 3). Different types of soil as test substrates can affect the toxicity of pesticide residues (e.g. Critchley 1972a, 1972b; Chapter 4).

In spite of these difficulties, plant or soil substrates may provide more realistic results than artificial substrates. The best artificial substrate is glass, which provides a very uniform deposition of pesticide for almost all compounds (Hartley and Graham-Bryce 1980). Other artificial substrates employed include wax and filter paper, and occasionally plastic. However, chemical interactions may occur between organic pesticides and noninert substrates such as plastic.

5.5.2.6. ***Degree of Coverage.*** In residue tests, it is difficult to achieve uniform coverage over an entire test chamber. Animals can avoid differential deposits on cage parts, plant tissues, introduced foods, and water sources. Occasionally, whether due to inherent differences in natural enemies themselves or to pesticide influences, the locomotion of natural enemies will vary on treated surfaces; greater pesticide uptake may be experienced by one species over another (Chapter 3). This factor can also influence results between tests when different procedures are used.[1] Although this behavior may occur in treated field situations, one should seek to reduce test variability and be aware of the effects that behavioral avoidance of pesticides can have on standardized test results.

5.5.2.7. ***Experimental Animals.*** Experimental animals should be as similar as possible in terms of age, feeding history, and environmental exposure to minimize

[1] Abdelrahman (1973) showed that, while earlier studies demonstrated considerable pesticide tolerance among aphelinid parasites of citrus scales, the toxicity of many compounds may have been underestimated because of methodology. Earlier authors confined parasitoids to cells with treated floors. Abdelrahman confined his insects in cages which had been entirely covered with an even deposit of pesticide. Since it is known that many parasitoids are repelled by pesticides, this explanation may account for the high LD_{50} values obtained by earlier authors as compared with those reported by Abdelrahman (1973).

variability in results not due to pesticide treatment. When working with laboratory colonies, care must be taken to avoid genetic bottlenecks during colonization or laboratory rearing. Laboratory strains will sometimes differ in their responses to pesticides due to the limited gene pool of founding individuals (Mackauer 1976). Other problems related to changing genotypes may occur when tests are replicated over appreciable time periods. Over time, laboratory population responses may vary due to random genetic drift alone.

Depending on the duration of the test, food or water provisions may be necessary to minimize control mortality and the synergistic effects of starvation or dehydration on pesticide-treated natural enemies. Numerous studies have shown that starved animals are more susceptible to pesticides than well fed ones (e.g., Kot and Plewka 1970, Abdelrahman 1973; Chapter 4). Even before testing, the previous feeding history of a test organism can influence its survival of pesticide exposure. In all tests with laboratory-reared specimens, one should standardize feeding as much as possible. During the test, care should be taken not to contaminate the natural enemy's food with pesticide, unless one is measuring food chain uptake. Pesticide uptake is difficult to standardize in tests where toxicants are administered orally. In tests where food is not treated, the food itself may represent a small refugium where the natural enemy may escape exposure, depending on the size of the natural enemy.

5.5.2.8. ***Evaluation Procedures.*** Measurement of mortality or sublethal effects of pesticides on individual life stages must be as unbiased as possible and must be relatively constant from one researcher to the next. Pesticide influences on developmental rates, fecundity, fertility, mobility, and behavior should be included in side-effect evaluations. Most researchers feel that evaluations of lethal and sublethal pesticide side effects throughout a complete generation of a natural enemy species are preferable to those made on only one or a few life stages. Life stage exposure should mimic that which occurs in the field, perhaps limited to one or two life stages. One problem with complete generation tests is that effects on individual life stages may be masked.

5.6. SUMMARY OF SELECTED RESULTS FROM STANDARDIZED TESTS

In Chapter 2, the results of many types of side-effect tests conducted on a diversity of natural enemy species were reviewed. Out of this background of tests, a number of standardized procedures have been proposed, tested, and validated. Of all the standardized tests proposed (Table 5.1), none compare in scope and integration to those efforts made by the IOBC/WPRS working group. In this section, a tabular summary is presented and selected examples from their studies are reviewed. IOBC/WPRS data are compared with output from the SELCTV database (Theiling 1987).

In Table 5.3, IOBC/WPRS data on the contact and/or residual toxicity of 20 insecticides, 12 fungicides, and 8 herbicides to 14 natural enemy species are listed. Data are presented as toxicity ratings on a 1–4 scale with means computed by compound and species. It should be noted that field rates of insecticides and fungicides are about one-tenth those of herbicides, and therefore comparisons across these major classes of pesticides are not necessarily made at the same concentration ranges. As might be expected, insecticides and acaricides are most toxic to natural enemies. Herbicides were the second most toxic group, while fungicides, in general, are the least toxic.

The 10 most toxic compounds to the entire group include four organophosphate

Table 5.3. Toxicity of 40 Pesticides to 14 Natural Enemy Species Evaluated by Standardized Test Methods (Adapted from Franz et al. 1980, Hassan et al. 1983)

Pesticide			Natural Enemy Species[1]															
Name	Class/ Type	Rate a.i.	Tc	Pp	Di	Ef	Ld	Os	Pt	Ct	Ap	Pr	Cc	Sv	An	Cm	Tc[3]	Mean
Dicofol	A/DDT	.15	3[2]	3	—	3	4	—	2	1	4	—	1	4	—	—	2	2.70
Lindane	I/OC	.10	4	—	4	4	3	3	4	1	2	1	1	4	—	—	2	2.75
Endosulfan	I/OC	.10	4	4	—	4	2	—	4	4	2	—	1	4	—	—	3	3.20
Demeton-S-methyl	I/OP	.10	4	3	—	4	4	—	4	2	4	—	4	4	—	—	3	3.60
Trichlorfon	I/OP	.10	4	—	4	4	3	3	4	4	4	4	1	4	—	1	3	2.77
Pirimiphos-methyl	I/OP	.20	4	—	3	4	4	4	4	4	4	4	4	4	—	1	4	3.69
Phosalone	I/OP	.20	4	4	—	4	4	—	4	4	4	—	1	3	—	—	4	3.60
Methidathion	I/OP	.075	4	—	4	4	4	4	4	4	4	4	4	4	4	—	4	4.00
Pirimicarb	I/CA	.10	4	4	—	4	2	—	4	1	3	—	1	3	—	—	1	2.70
Methomyl	I/CA	.10	4	—	4	4	4	4	4	4	4	3	3	4	4	—	4	3.85
Propoxur	I/CA	.15	4	—	3	4	4	3	4	4	4	3	4	4	4	—	2	3.62
Fenbutatin oxide	A/OT	.05	1	1	—	1	1	—	1	1	1	—	1	2	—	—	—	1.11
Cyhexatin	A/OT	.10	4	2	—	4	1	—	1	3	4	—	4	3	—	—	4	2.90
Azocicdotin	A/OT	.10	4	—	3	3	4	3	4	3	4	1	3	3	—	—	3	3.17
Benzoximate	A/	.15	2	—	2	1	1	2	1	1	1	1	1	4	—	—	—	1.55
Fenvalerate	I/SP	.075	4	—	4	4	4	4	4	1	4	4	3	4	4	—	4	3.69
Permethrin	I/SP	.02	4	—	4	4	4	3	4	4	4	4	4	4	4	4	4	3.93
Pyrethrum + PBO	I/SP	.10	4	3	—	4	3	—	1	1	—	—	1	4	—	—	2	2.56
Bacillus thuringiensis	I/MPB	.10	1	1	—	1	1	—	1	1	1	—	1	2	—	—	—	1.11
Diflubenzuron	I/IGR	.05	1	1	—	1	1	—	1	1	1	—	4	1	—	—	—	1.33
Summary: Insecticide/Acaricides (Average Rate of Application = .106% a.i.)																		2.89
Bupirimate	F	.04	1	1	—	1	1	—	1	1	1	—	1	1	—	—	—	1.00
Captan	F	.15	1	—	1	1	1	1	1	1	1	1	1	3	1	1	—	1.15
Captafol	F	.20	1	1	—	1	1	—	2	1	1	—	1	3	—	—	—	1.33
Triadimefon	F	.10	1	—	1	1	1	1	1	1	1	1	1	2	1	—	—	1.08

Table 5.3 (*Contd.*)

Pesticide			Natural Enemy Species[1]															
Name	Class/ Type	Rate a.i.	Tc	Pp	Oi	Ef	Ld	Os	Pt	Ct	Ap	Pr	Cc	Sv	An	Cm	Tc[3]	Mean
Dichlofluanid	F	.20	4	2	—	3	2	—	2	1	2	—	2	2	—	—	3	2.30
Vinclozolin	F	.05	1	—	1	1	1	1	1	1	1	1	1	1	—	—	—	1.00
Mancozeb	F	.20	3	1	—	3	1	—	1	1	3	—	3	3	—	—	—	2.11
Carbendazim	F	.05	1	—	1	1	1	1	1	1	4	3	1	1	—	1	—	1.42
Pyrazophos	F	.05	4	4	—	4	1	—	4	4	4	—	3	4	—	—	4	3.60
Ditalimfos	F	.075	3	—	3	3	—	1	2	1	2	1	1	1	—	1	3	1.83
Thiophanat-methyl	F	.10	1	1	—	1	1	—	1	1	4	—	1	2	—	—	—	1.44
Sulfur	F	.40	4	—	2	4	1	1	1	2	3	1	1	2	—	—	4	2.17
Summary: Fungicides (Average Rate of Application = .135% a.i.)																		1.70
Diclofop-methyl	H	.75	2	1	—	1	2	—	1	1	—	—	1	2	—	—	—	1.38
Desmetryne	H	.25	4	—	1	1	1	1	1	1	1	2	1	1	—	—	2	1.42
Dinoseb	H	1.25	4	4	—	4	4	—	4	4	4	—	4	4	—	—	—	4.00
Difenzoquat	H	1.00	4	—	3	4	4	3	1	1	4	4	1	3	—	—	2	2.83
Phenmedipham	H	2.25	1	1	—	2	1	—	1	1	4	1	2	4	—	—	—	1.89
Propachlor	H	1.00	4	—	3	4	4	3	4	1	3	3	1	3	—	—	2	3.18
Propyzamid	H	.75	3	1	—	1	2	—	1	1	—	—	1	3	—	—	—	1.63
Monolinuron	H	.75	4	—	4	4	4	3	4	1	4	3	2	3	—	—	—	3.27
Summary: Herbicides (Average Rate of Application = 1.00% a.i.)																		2.45

[1] Species and their mean toxicity ratings to all compounds: Tc = *Trichogramma cacoeciae* (2.925); Pp = *Pales pavida* (2.150); Di = *Drino inconspicua* (2.750); Ef = *Encarsia formosa* (2.775); Ld = *Leptomastix dactylopii* (2.359); Os = *Opius* spp. (2.450); Pt = *Phygadeuon trichops* (2.370); Ct = *Coccygomimus turionellae* (1.875); Ap = *Amblyseius potentillae* (2.789); Pr = *Phytoseiulus persimilis* (2.450); Cc = *Chrysoperla carnea* (1.925); Sv = *Syrphus vitripennis* (2.925); An = *Anthocoris nemorum* (3.143); Cm = *Cryptolaemus montrouzieri* (1.500); Tc = *Trichogramma cacoeciae* (persistence test) (3.000).

[2] Classes of evaluation: 1 = harmless (< 50% mortality), 2 = slightly harmful (50–79%), 3 = moderately harmful (80–90%), 4 = harmful (> 99%).

[3] Residue persistence test, classes for evaluation: 1 = short lived (< 5 days toxicity), 2 = slightly persistent (5–15 days), 3 = moderately persistent (16–30 days), 4 = persistent (> 30 days).

insecticides (methidathion, pirimiphos-methyl, phosalone, and demeton-*S*-methyl), two pyrethroid insecticides (permethrin and fenvalerate), and one carbamate insecticide (propoxur). One herbicide (dinoseb) and one fungicide (pyrazophos) are among the 10 most toxic compounds.

Among the 10 least toxic compounds are six organic fungicides (vinclozolin, bupirimate, triadimefon, captan, captafol, carbendazim), three insecticides/acaricides (one organotin, fenbutatin oxide; one microbial insecticide, *Bacillus thuringiensis*; and one IGR, diflubenzuron), and one organic herbicide (diclofop-methyl).

Of the five most tolerant natural enemies to all compounds, one is a predator and four are parasitoids: *Coccygomimus turionellae* (Ichneumonidae), *Chrysoperla carnea* (Chrysopidae), *Pales pavida* (Tachinidae), *Leptomastix dactylopii* (Encyrtidae), and *Phygadeuon trichops* (Ichneumonidae). (*Anthocoris nemorum* and *Cryptolaemus montrouzieri* were eliminated from the comparison because of too few observations.) The five most susceptible species were *Syrphus vitripennis* (Syrphidae), *Trichogramma cacoeciae* (Trichogrammatidae), *Amblyseius potentillae* (Phytoseiidae), *Encarsia formosa* (Aphelinidae), and *Drino inconspicua* (Tachinidae). Interestingly, in comparing the mean toxicity of pesticides against eight parasitoids and six predators, there were no significant differences between the two groups (parasitoids = 2.457 versus predators = 2.455, on a relative toxicity scale of 0–4 as discussed in Section 5.4.1.2; see discussion of relative susceptibility of parasitoids and predators in Chapter 2). It is interesting to note the broad range of selectivities found among some compounds. For example, diflubenzuron, cyhexatin, and mancozeb were selective to a few individual species, but were relatively toxic to others (Table 5.3). Also, the highly variable toxicity of phosalone, pyrazophos, pirimicarb, and endosulfan among natural enemy species is similarly noteworthy.

Table 5.4. Effects of Pesticides on *Pterostichus cupreus* Evaluated by 4 Different Means of Exposure (Adapted from Edwards et al. 1984b)

		% Mortality			
Compound	Equivalent Rate, g a.i./ha	Soil	Beetle	Soil, Beetle, & Barley	Glass
Gamma-HCH	400	7	7	0	100
	2000	100	7	0	—
Dimethoate	70	0	7	0	100
	350	47	100	100	100
Cypermethrin	25	20	7	0	100
	125	20	20	60	100
Pirimicarb	140	0	0	0	0
	700	0	0	0	0
Chlorfenvinphos	2000	0	40	20	100
	10000	20	100	100	—
Carbophenothion	400	7	0	40	100
	2000	20	34	56	100
	10000	20	100	—	—
Trichlorfon	1500	0	0	40	100
	7500	100	80	80	100
Carbendazim	250	20	0	0	0
	1250	0	0	0	20

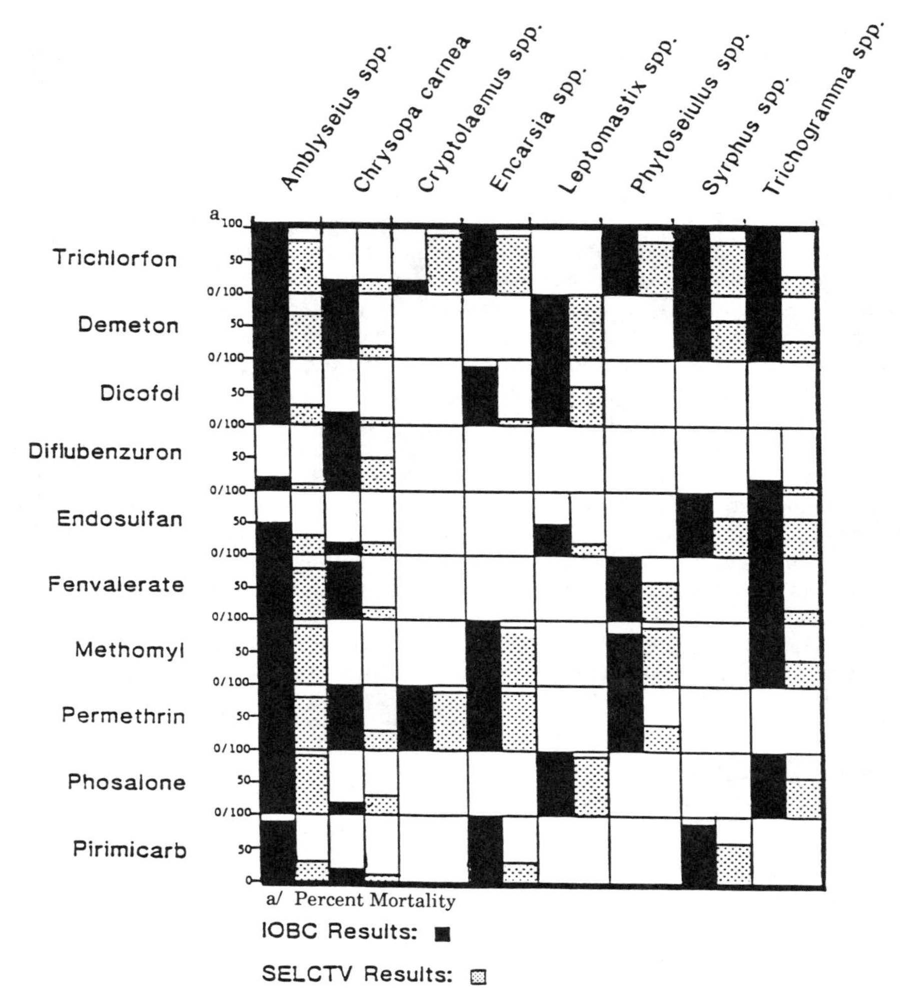

Figure 5.2. Comparison of susceptibility assessments from IOBC/WPRS standardized tests and SELCTV database records for common but independent natural enemy-insecticide pairings.

In Fig. 5.2, test results of the IOBC/WPRS (Hassan et al. 1983) and average toxicity values from the SELCTV database (Theiling 1987) are compared in histogram form. SELCTV mean values were computed from three or more data points for individual compounds and natural enemy species. Different rating scales were used in the two studies, making translation to a common mortality scale necessary. Some common differences in mortality estimates are notable between the IOBC and SELCTV data (Fig. 5.2). To some degree these may be artifacts of the translation process; more specifically they are due to using the midpoints of mortality ranges for each graphic representation. Blank boxes in Fig. 5.2 indicate insufficient data for comparison.

Overall, the IOBC data tend to show somewhat higher mortality values across all species and compounds than SELCTV estimates (Fig. 5.2). Undoubtedly, this is because the former tests are direct laboratory contact/residue bioassays run under standard conditions, while the latter include data from such a diversity of tests including field assessments. As discussed in Chapter 2, the side effects of pesticides on natural enemy species may be less severe when field tests are employed rather than direct application methods in the laboratory.

In terms of natural enemy susceptibility, IOBC and SELCTV data differ most in relation to *Trichogramma* spp. The two independent analyses show good congruence in data ratings for *Amblyseius* and *Syrphus* spp. By compound, phosalone and methomyl are similarly rated by both sources for toxicity to all natural enemies tested. The results for trichlorfon are conspicuously inconsistent for *Cryptolaemus* and *Trichogramma*. Other striking dissimilarities between IOBC and SELCTV toxicity values include pirimicarb on *Amblyseius*, demeton and fenvalerate on *Chrysoperla carnea*, dicofol and pirimicarb on *Encarsia*, and demeton and fenvalerate on *Trichogramma*. These differences are not easily explained. Most other comparisons were in fair to good agreement.

It would appear that averaging overall data for individual species (i.e., as done in SELCTV) gives useful information on toxicity trends among species and compounds that is comparable to that obtained under standardized conditions of experimental testing. This agreement lends credibility to the use of data from SELCTV in constructing susceptibility tables for individual species that could be useful to pest managers when information from specific tests is not available.

While many individual tests—and in some cases, comparisons of two or more test methods—have been run using standardized test methods (e.g., see Streibert 1981 for a comparison of a laboratory contact test versus a semifield residue test), very few comprehensive evaluations of standardized laboratory, semifield, and field techniques have been made. One of the more extensive studies is that reported by Edwards et al. (1984b; Table 5.4). They evaluated pesticide toxicity to carabids using a variety of methods to expose adult beetles: 1) spray deposits on soil, 2) direct sprays on the beetle, such as might be received by a diurnal species, 3) sprays on a vessel of soil and growing barley in which the beetle had been established for 24 hr (a simulated field exposure), 4) treated food, and 5) dry spray deposits on glass. Results were similar for eight species of carabids and eight compounds, except that beetles exposed to dry spray deposits on glass slides experienced greater mortality. Representative results for one species and four exposure methods are given in Table 5.4.

5.7. SUMMARY

The data presented in Table 5.4 and Fig. 5.2 present useful examples of the efforts needed to establish comprehensive standardized tests to measure pesticide side effects on complexes of natural enemy species. While these data give the impression that an index for pesticide impact is at hand, it must be remembered that these tests represent either responses of a single strain or mean effects on all strains tested to date. Individual investigators may be working with unique natural enemy strains which may require additional verification.

It is also appropriate to question how standardized these test results are. For example, if one looks more closely at the individual reports by scientists working on side-effect testing, one becomes cognizant of subtle or even overt differences between tests run using the same method and population of natural enemies. What *can* be said is that through present research, a greater effort is being made than ever before towards standardization of test procedures. However, there is still variation in the results of these tests. We must conclude, therefore, that while attempts at standardization can improve our ability to compare results among species (between laboratory experiments and field tests, and between one researcher and another), these tests are not definitive. To some extent, this will always be true because of the uniqueness of each organism and each experimenter involved in testing. In predicting pesticide side effects, there will always be some uncertainty and risk in extrapolating to the field. This should not deter our efforts because uncertainty is present in other aspects of the pesticide impact assessment as well (e.g., pest control efficacy, side effects on other nontargets).

Before substantial progress can be made in developing improved methods of side-effects assessment, the knowledge base of pesticide–natural enemy interactions must be expanded. A greater understanding of basic natural enemy biology is needed, as is a better grasp of natural enemy/pesticide interactions. This information may allow us to enter a more predictive mode from which we can generalize from standard tests using representative species to field effects of pesticides on other natural enemy species.

Finally, who will ultimately establish and maintain the standardized protocols for side-effect testing? Will it be a consensus developed by a scientific group such as has occurred in western Europe? Perhaps it will come from a state or federal regulatory agency as a part of refining registration protocols. Such considerations are under discussion by federal agencies in several countries, and hopefully a consensus will be forthcoming. Such guidelines should have input from industry (e.g., see Sechser 1981), who see that it is in their best interest to develop and use these procedures before they are mandated to do so. It is clear that these protocols will eventually evolve. The question is, under what institutional setting the use of these guidelines and associated policies will be implemented.

6

PHYSIOLOGY AND TOXICOLOGY

6.1. INTRODUCTION

Earlier chapters have dealt with pesticide/natural enemy relationships primarily in the external environment of the organism. However, suborganismal relationships, including physiological, biochemical, and toxicological factors, are equally important. Suborganismal factors operate near or at the target site of action. They affect the susceptibility of natural enemies and the selectivity of pesticides. Understanding these factors also provides insight into the responses of natural enemies to pesticides at the population or community levels of biological organization (Chapter 8).

In this chapter, a comparative discussion of the toxicology of pesticides to arthropod natural enemies and herbivores is presented. Review of similar information for pests adds perspective to the limited data available for natural enemies. The discussion rests on a limited number of studies conducted on predators and parasites in relation to the larger

This chapter was coauthored by C. A. Mullin, Department of Entomology, Pennsylvania State University, University Park, Pennsylvania.

database for herbivorous species and, in particular, major pests. Because mammalian selectivity is essential, much more research has been conducted on differential toxicology between pest arthropods and mammals. Mammalian toxicology is discussed to a limited extent to illustrate aspects of susceptibility and selectivity for natural enemies and pests which should be emphasized in future pesticide research.

In comparing the physiology, biochemistry, and toxicology of mammals and entomophagous and phytophagous arthropod species, it should be noted that susceptibility, resistance, and selectivity are closely related phenomena. The initial discussion focuses on basic physiology, including metabolism, nutrition, endocrinology, and excretion, with an emphasis on opportunity factors which may be exploited to achieve selectivity (Chapters 9–13).

6.2. ENTOMOPHAGE VERSUS HERBIVORE PHYSIOLOGY

6.2.1. Nutrition

Herbivorous and carnivorous arthropods consume foods that are fundamentally different in terms of nutrition; foods of arthropod origin are higher in protein and lower in carbohydrate than are plant-based diets. Animal-derived diets usually provide a balanced complement of amino acids, fatty acids (FA), and vitamins for entomophagous species. Plant-derived diets lack some of these elements. Because of these nutritional differences, entomophagous arthropods have about twice the assimilation efficiency or digestibility of herbivorous arthropods (House 1977, Slansky and Scriber 1982). Nutrition determines basic enzyme kinetics and the ultimate biochemical composition of body tissues. Therefore, the contrasting dietary intake of herbivores and their natural enemies may contribute to differences in pesticide susceptibility and resistance potential between these groups. However, these aspects of comparative natural enemy/pest physiology and toxicology have seldom been examined.

Studying diet as a clue to toxicology may offer limited insight because food quality depends on the nutrients available after digestion. The nutrients obtained must meet the consumer's physiological needs. Although dietary components are different in plant- and insect-based foods, entomophagous and herbivorous arthropods appear to have similar nutrient requirements; these include certain amino acids, vitamins, sterols, fats, and carbohydrates (Thompson 1981, 1986).

Our understanding of the dietary requirements of natural enemies has been advanced with the development of artificial diets for many parasitoids and predators (Singh and Moore 1985). For example, comprehensive work with the tachinid parasitoid *Agria housei*, the hymenopteran parasitoids *Exeristes roborator* and *Itoplectes conquisitor*, and the neuropteran predator *Chrysoperla carnea* (House 1977, Thompson 1981) has shown that larvae of these entomophages require the same amino acids, water soluble vitamins, and a sterol (e.g., cholesterol) that immature herbivores require. However, entomophages do not require the polyunsaturated fatty acids needed by plant pests (Thompson 1981, 1986). A discussion of the lipid constituents of entomophagous arthropods may help explain the apparent difference in fatty acid requirements between these two groups.

In mammals, long-chain fatty acids ($>C_{18}$ in length) with higher unsaturation are typical of the structural fats of carnivores as compared to herbivores (Crawford 1970). In insects, however, higher polyunsaturation is not correlated with increasing entomophagy (Fast 1964, 1970, Thompson 1973). Polyunsaturated fatty acids greater than C_{18} have only been studied to a limited extent in insects since they are relatively rare. In one study, Stanley-Samuelson and Dadd (1983) found $C_{20:4}$ and greater polyunsaturated fatty acids in several herbivores and a predaceous neuropteran, the owlfly. Their data did not indicate any dramatic difference between these two ecological groups.

A more general observation relating to lipid constituents is that herbivores tend to have higher fat body contents than entomophagous insects (Fast 1964). Higher lipid content can be important in the sequestration of toxicants and can confer tolerance to lipophilic compounds such as DDT and other synthetic pesticides (Section 6.3.2; Atallah and Newsom 1966).

Fatty acid composition of many dipteran and hymenopteran parasitoids is a close mimic of that of their hosts (Bracken and Barlow 1967, Barlow 1972, Thompson and Barlow 1974). Braconids, ichneumonids, pteromalids, and eulophids exhibit fatty acids almost identical, both qualitatively and quantitatively, to their hosts'. *Exeristes comstockii*, a generalist ichneumonid, duplicates the fatty acid profiles of its hosts, regardless of whether it develops on a dipteran, lepidopteran, or hymenopteran larva. Initially it was thought that ichneumonids like *E. comstockii* might lack the capability for *de novo* fatty acid synthesis. However, later studies (Thompson and Barlow 1972, Thompson and Johnson 1978, Jones et al. 1982) found enzyme pathways for both fatty acid synthesis and turnover in these species. As with host insects, parasitoids retain only monodesaturases and cannot synthesize polyunsaturated fatty acids. By comparing the lipid metabolism within parasitic Hymenoptera (and, more specifically, species not greatly influenced by the host's fatty acid composition), it appears that regulation of fatty acid content lies with triglycerides and phospholipids on enzymes that join the fatty acid to the appropriate glycerol intermediate. It seems that many parasitoids are deficient in fatty acid transferases, or that the specificity of these enzymes has been altered relative to their hosts (Jones et al. 1982). These enzymes could secondarily confer differential susceptibility to pesticides in pests versus beneficial species by the interactions between lipophilic toxicants and lipid sinks used for sequestration.

All arthropods require external sources of sterols for growth and development because of their inability to biosynthesize these compounds. Sterols function in insect membrances and as precursors to moulting hormones (ecdysteroids) and defensive secretions (Kircher 1982). Cholesterol meets the sterol needs of most insects, but the predominant sterols in plants (e.g., sitosterol, stigmasterol, and campesterol) must be converted to cholesterol to satisfy this need. Most phytophagous insects remove the methyl and ethyl groups at C_{24} in plant sterols and convert them to cholesterol. However, carnivores seem to lack this capability and require cholesterol or a similar analog in their diet (Svoboda et al. 1978).

Several entomophagous hemipterans and hymenopterans that are unable to dealkylate 24-alkylsterols and have insufficient cholesterol in their diets will utilize 24-alkyl ecdysteroids as moulting hormones (Svoboda and Lusby 1986). This differential capability to dealkylate sterols occurs within the Coccinellidae, which includes both phytophagous and carnivorous members. Hence, the plant-feeding Mexican bean beetle *Epilachna varivestris* can dealkylate phytosterols while the entomophage *Coccinella septempunctata* cannot (Svoboda et al. 1978, Svoboda and Robbins 1979). In phytophagous species, 29-fluorophytosterols undergo lethal dealkylation to the highly poisonous fluoroacetate, whereas in entomophages the lethal dealkylation does not occur. Lethal dealkylation has recently been used to manipulate insecticide activity among herbivorous pests (Prestwich et al. 1983), but it has not been exploited to achieve selectivity to entomophages.

Other nutritional and physiological differences between natural enemies and their prey or hosts may prove useful in the design of selective insecticides. For example, dietary triglycerides are of limited value as energy sources for some hymenopteran parasitoid larvae (Thompson 1977, 1981). Interestingly, Diptera and Hymenoptera, which comprise over 90% of the entomophagous parasitoids (Price 1981), rely on carbohydrates such as trehalose as energy sources for flight. Many phytophagous pests, however, use lipids for flight energy (Fast 1970). Therefore, entomophagous insects retain three- to four-fold higher levels than herbivores of key regulatory enzymes for carbohydrate catabolism, including hexokinase, glycogen phosphorylase, and phosphofructokinase (Crabtree and Newsholme 1975). Conceivably, chemicals that interfere with carbohydrate mobilization could have greater effect on entomophages than herbivores. Conversely, compounds selectively inhibiting lipid metabolism could favor beneficial species over their host or prey due to disruption of nutrient utilization. A critical question is whether or not herbivores have higher enzyme levels for lipid metabolism. Also in question is whether or not selectivity factors based on these basic systems would also confer a safety factor to mammals. Finally, the successful use of all of the above strategies would depend on the ability of either arthropod group to metabolically compensate for (or develop resistance to) these critically blocked enzyme pathways.

6.2.2. Respiration

Predaceous insects, especially adults and larvae, have higher respiratory rates than their prey, presumably due to the increased metabolic demands associated with searching behavior. Oxygen consumption in a carnivorous mite (Thurling 1980) and a coccinellid predator (Tanaka and Ito 1982) were about twice that of herbivores from the same taxonomic groups, yet carnivores survived starvation better by markedly reducing respiration. Higher rates of locomotion and other respiratory-associated activities such as energy metabolism may confer faster pesticide uptake (Chapter 3), activation, and metabolism by natural enemies, which could render them more susceptible than their prey or hosts. In contrast, entomophages could have distinct advantages over phytophages by having greater respiratory reserves to sustain them during food stress.

Interestingly, the acaricide cyhexatin is highly toxic to all phytophagous mites, but has little effect on most predaceous mites and insects (Croft 1981a). This selectivity may be explained by its mode of action. Cyhexatin inhibits mitochondrial Mg^{2+}-dependent ATPase, and at higher concentrtions, Na^{+}–K^{+} ATPase (Ahmad and Knowles 1972, Desaiah et al. 1973), and thus functions as a respiratory inhibitor. Since predaceous mites generally have greater respiratory reserves and more resiliency during metabolic stress, they would be less affected by respiratory inhibitors than herbivorous mites. In this case, physiological differences between natural enemies and pests have been successfully exploited to achieve physiological selectivity in the field as well as having commercial value for pesticide development.

6.2.3. Endocrinology

Endocrinological factors may suggest other means for achieving selectivity between pests and beneficials. For example, the juvenile hormones (JHs) and the steroidal moulting hormones (ecdysteroids) which regulate metamorphosis, moulting, reproduction, and diapause in insects have no comparable structures in vertebrates (Gilbert et al. 1980, Coudron et al. 1981, Kircher 1982). Therefore, much effort has focused on developing pesticides that impact on these hormone systems (Svoboda et al. 1978, Menn and Henrick 1981, Bowers 1982; Chapter 12). However, much less is known about differences in endocrine physiology and metabolism between herbivores and their natural enemies.

Parasitoids attacking lepidopteran hosts induce marked alterations in the electrophoretic profiles of host hemolymph proteins (Vinson and Barras 1970, Smilowitz 1973, Barras et al. 1982). Certain protein bands which are absent in parasitized larvae correlate with arrested larval to pupal ecdysis of the host (Smilowitz 1973, Smilowitz and Smith 1977). Data suggest a biochemical basis for endocrine control of the host by the parasitoid.

Parasitoid involvement with the endocrine system of its host has long been implied in the literature (Riddiford 1975, Vinson and Iwantsch 1980, Thompson 1983, Beckage 1985). Juvenile hormone esterase, a key degradative enzyme regulating the larval–pupal molt in lepidoptera, exhibits an age–activity profile very similar to ecdysis "proteins" noted above (Hammock and Quistad 1981). The possible coincidence of these biochemical events is further indicated by the abnormally depressed JH esterase at molt in parasitized insects, thereby prolonging larval stadia of the host (Beckage and Riddiford 1982, Thompson 1983). Alternately, some wasps induce precocious pupation of parasitized hosts by initiating a premature decline in JH titer. In addition, parasitoids depend on ecdysteroids of their hosts as initiation cues for development (Baronio and Sehnal 1980). One would expect that, just as herbivores have developed adaptive mechanisms to deal with toxicants encountered in their plant foods (e.g., allelochemicals), so too would natural enemies be similarly adapted to respond to the effects of hormones coming from their host or prey.

Insect bioregulators having JH action have received extensive study recently because of their safety to nontargets other than arthropods. Juvenile hormone is present in insects and probably arachnids and acari; however, it does not occur in some arthropods and higher animals. Hence a large selectivity is possible, as exemplified by the 1,700,000-fold higher toxicity of methoprene to cockroaches than to the rat (see Hollingworth 1976). Screening of over 5000 JH analogs for insect morphogenicity or toxicity has led to the commercial development of two JH mimics, methoprene and kinoprene (Menn and Henrick 1981). Juvenoids of distinct embryogenetic and metamorphogenetic activities for Lepidoptera, Diptera, and also aphids have been noted (Henrick 1982).

The selectivity of juvenoids to beneficial and injurious arthropods has been mixed. Certain diaryl juvenoids are inactive against hymenopteran parasitoids (Henrick 1982). While hydroprene has little activity on some dipteran and hymenopteran parasites (Smilowitz et al. 1976, Vinson and Iwantsch 1980), it is deleterious to others (McNeil 1975, Vinson and Iwantsch 1980, Ascerno et al. 1983) at dosages used for pest control. While endocrinological impediment of parasitoids by the indirect action of JH mimics in the hosts is expected, it is gratifying that some JH mimics have favorable selectivity to some life stage of insect (Bull et al. 1973) and acarine predators (El-Banhawy 1980). (see Chapter 12 for further discussion of the JH selectivity.)

One major difference in JH biochemistry that may be exploited to achieve selectivity between lepidopterans and their parasitoids has been discussed by Feyereisen (1987). He contends that some major insect parasitoid orders (i.e., Hymenoptera and Diptera) appear to biosynthesize only JH III, whereas major lepidopteran pests synthesize the higher homologs, JH II, JH I, JH 0, and 4-methyl-JH I instead of or in addition to JH III. Thus inhibitors of JH synthesis for the higher homologs of JH III might be more selective to the beneficials. He recommended that more research be done on the origin of propionyl-CoA in the lepidopteran corpora allata. By targeting the selective disruption of enzymes responsible for propionyl-CoA synthesis (i.e., an enzyme group presumably not needed for JH synthesis outside of lepidopteran pests), a greater likelihood of achieving selectivity would be possible. This approach was favored over trying to exploit enzymes in the isoprenoid pathway of JH synthesis, which are basically of common importance to most insects. (See further discussion of these possibilities in Chapter 12.)

There are other, more ecologically oriented ways to favor parasitoid natural enemies over their hosts using JH-type compounds. Since juvenoids prolong the larval or nymphal stages of arthropods during which plant feeding is usually most injurious, their primary application has been to control adult pests. These compounds are widely used against mosquitoes and other dipteran pests of mammals. Prolonging larval development of some hosts lengthens their exposure to parasites in the field, which can contribute to greater levels of biological control (Weseloh et al. 1983).

Considerable effort has also gone into discovering compounds that impede or disrupt JH action in insects. Juvenile hormone antagonists or the so-called anti-JHs cause precocious metamorphosis, curtail larval feeding damage, and interfere with reproduction in insects that require JH for vitellogenesis. The best studied JH antagonists are the precocenes (Bowers 1982) and fluoromevalonolactone (Menn and Henrick 1981), which interfere with the biosynthesis of JH. Their narrow activity spectrum in insects, due presumably to the evolutionary response of plants to specific pest types, limits their commercial development. Little information is available to predict their selectivity for entomophages relative to herbivores (Chapter 12).

An alternate way to obtain anti-JH effects would be to stimulate JH degradation. The two major pathways for JH catabolism are ester hydrolysis and epoxide hydration. Preliminary information, however, indicates no major differences in JH degradation between herbivorous and carnivorous arthropods (Ajami and Riddiford 1973, Hammock and Quistad 1981). Also, endocrinological control of insects may arise from inhibitors of ecdysone 20-monooxygenase, a key enzyme for ecdysteroid biosynthesis, which disrupts larval moulting (Koolman 1982). However, the use of these inhibitors is not very likely to be explored until far into the future.

6.2.4. Nonnutritive Food Components

Host and prey suitability for parasitoids and predators depends on both nutritive and nonnutritive food components. These components may include feeding stimulants or feeding deterrents, the balance of which determines the final acceptability of the food (Bernays and Simpson 1982). Physical attributes of the diet are important (Vinson and Iwantsch 1982), but perception of key chemicals is often overriding. Most studies of dietary effects on different processes in entomophages are conducted by controlled manipulation of the food of their host or prey. Nutritional studies among trophic levels have revealed that deficiencies in an herbivore's diet may seriously effect entomophage development (Zhody 1976). It is also clear that plant toxins may seriously reduce entomophage fitness (Price et al. 1980, Campbell and Duffey 1981, Barbosa et al. 1982), just as they do for herbivores (Rosenthal and Jansen 1979).

Plant secondary chemicals are usually introduced to predators and parasites via consumption of accumulated chemicals in nonessential tissues of the herbivore. This sequestration of toxicants can benefit the herbivore because the allelochemicals are often more toxic to the entomophage than to the protected host (Duffey 1980). For example, the tomato alkaloid *alpha*-tomatine, acquired by the ichneumonid *Hyposoter exiguae* from its host *Heliothis zea*, seriously disrupts the endoparasitoid's development, presumably by interfering with dietary sterol utilization (Campbell and Duffey 1979, 1981). Similarly, nicotine reduces the survival of parasitoids which attack tobacco feeding Lepidoptera, including the tobacco hornworm *Manduca sexta* and the fall armyworm *Spodoptera frugiperda* (Barbosa et al. 1982, 1986). Deleterious effects of plant toxins on natural enemies may thwart IPM practices which seek to control herbivorous pests through chemical antibiosis in plants. Support of this argument comes from field studies with cotton, where a pest-resistant nectariless variety reduced natural enemies to a greater extent than it did pest species, as compared with susceptible controls (Adjei-Maafo and Wilson 1983); it was not established whether chemical antibiosis was responsible.

More often investigators have found plant resistance based on antibiosis to be compatible with biological control. Variable distribution of defensive chemistry throughout an individual host plant may increase searching of an herbivorous pest, and possibly increase random encounters with natural enemies (Schultz 1983). Sorghum and small grains resistant to the greenbug *Schizaphis graminum* were compatible with parasitism by the aphidiid *Lysiphlebus testaceipes* (Starks et al. 1982, Boethel and Eikenbary 1986). These resistances, in part, are due to phenolic and flavonoid factors (Juneja et al. 1975, Dreyer and Jones 1981) and suggest that phenolic defenses against phloem-sucking insects may complement biological control.

While it usually is thought that toxins are obtained from the host plant of the prey, this may not necessarily be the case. Predaceous insects (Kuo 1977, Pasteels 1978) and mites (Swirski and Dorzia 1968) are also affected by toxic components developed solely in their prey. To illustrate, out of 5 species of aphids raised on broad bean, only *Megoura viciae* was found to be toxic to the coccinellid *Adalia bipunctata* (Blackman 1967). This could have indicated that other aphids may have detoxified the toxic plant compound, but a more likely interpretation is that the toxin that killed the predator was biosynthesized in *M. viciae*.

Clearly, differential physiology, biochemistry, and presumably toxicology based on nutritional, anutritional, hormonal, and behavioral chemicals exist between arthropod herbivores and their natural enemies. It is not surprising that pesticide selectivity based on these features also exists. However, multicomponent and tri-trophic interactions may complicate the implementation of selective pesticides. For example, pesticides (in particular herbicides; Komives and Casida 1983, Williams and James 1983) can increase the levels of toxic chemicals in crops and may limit the use of natural enemies.

In the future, exploitation of the unique properties of tri-trophic biochemical interactions between plants, pests, and natural enemies may occur through the incorporation of natural toxicants into crop plant tissues by genetic engineering. Already the *Bacillus thuringiensis* endotoxin has been incorporated into commercial varieties of tobacco, maize, and soybeans (Comai and Stalker 1984). This gene will likely appear in many other crops before the end of the 1990s. Unfortunately, this endotoxin is very toxic to natural enemy species (e.g., Hoy and Ouyang 1987). Very likely, more selective toxicants, antifeedants, or suppressants can be identified and incorporated into plant tissues which effectively reduce pest population development, while not affecting their natural enemies. One of the biggest challenges facing IPM specialists when using these tactics is the heavy selection pressure exerted when these selected toxicants become ubiquitous in the crop environment. Strategies must be devised to limit the development of pest strains resistant to such genetically improved crops (Croft 1986; Chapter 21).

6.2.5. Other Physiological and Behavioral Factors

Endoparasitoids often evade host immune mechanisms by preventing encapsulation. Melanization involving phenol oxidase appears to invoke encapsulation, and it has been suggested that disruption of phenol oxidase activity in the host arthropod confers this immunity (Soderhall 1982). Conflicting data, however, indicate that phenol oxidase disruption may not be the key to suppression of encapsulation in all endoparasitoids (Sroka and Vinson 1978). Herbivores consume larger quantities of phenolics in their food than natural enemies, and these plant toxicants could kill endoparasitoids by serving as substrates for melanization reactions. No attempts to manipulate these mechanisms for selective pesticide development have been reported. How these compounds affect predaceous arthropods is unknown.

Some insect defensive chemicals are semiochemicals that communicate information between organisms. They may be produced by exocrine glands, or simply sequestered internally from food sources and stored so as not to elicit autotoxicity. Many herbivores use protective chemicals to fend off predation; many predaceous and parasitoid arthropods use similar mechanisms to escape invertebrates and higher animals (Eisner 1970, Rothschild 1972, Rodriguez and Levin 1976, Blum 1981). Major distinctions between pests and natural enemies are lacking in this area of physiology, although entomophages seem to biosynthesize a higher percentage of defensive chemicals from acetate or nitrogenous precursors. This is logical considering the high protein and lipid intake of carnivores. Examples include the norsesquiterpenoids and steroids of predaceous aquatic beetles and the alkaloidal coccinellines in Coccinellidae which are polyacetate-derived (Miller et al. 1975, Blum 1981). Alternatively, many defensive toxins are sequestered intact from the food source (see Duffey 1980), and may become autotoxic during periods of stress. For example, a lipid depot of toxin would be mobilized in a time of starvation. This may generally but not always (see below) be of greater consequence to the herbivore than to predators and parasitoids because most toxins are plant-derived.

Research has recently been devoted to the development of broad-spectrum feeding suppressants for herbivores. Most antifeedants have been obtained from natural plant sources (Kubo and Nakanishi 1979, Schoonoven 1982). Azadirachtins, the potent and widely studied antifeedant of the neem tree (*Azadiracta indica*), prevent feeding in at least 35 species of pest arthropods (Jacobson 1982, Saxena 1987). One drawback of many antifeedants is that in order to block consumption of a crop, a full coverage of the plant, including the meristem, is required. However, more potent compounds are being developed that impair the herbivores' ability to perceive and accept suitable hosts (Schoonoven 1982, Norris 1986). Many synthetic pesticides at sublethal dosages also deter arthropod feeding, including pyrethroid insecticides (Ross and Brown 1982), formamidine acaricides (Hollingworth 1975), and organotin acaricides and fungicides (Hare et al. 1983). Chlordimeform has anorectic effects at about 0.1% of a lethal dose and is an antagonist of biogenic amines (Beeman and Matsumura 1973, Matsumura and Beeman 1982). Although feeding deterrents are expected to be selectively detrimental to herbivores, this does not always mean that the toxic effects will be the same. For example, the repellent chlordimeform is much more toxic to predatory mites than to the spider mite *Tetranychus kanzawai* or the cheese mite *Tyrophagus putrescentiae* (Kuwahara 1978). However, in most cases, the relationship between deterrence and selectivity is more likely to be favorable to the natural enemy because it is not the primary agent causing selection of the host plant or other deterrent source.

6.3. ENTOMOPHAGE VERSUS HERBIVORE TOXICOLOGY

Conventional pesticides are initially screened for their toxicity to herbivores and lack of toxicity to plants and mammals (Chapter 5). For this reason, it is useful to discuss natural enemy toxicology and selectivity in light of the comparative toxicology of these associated organisms. Generalizations made for mammalian and herbivore systems will be applied to arthropod predators and parasites where information is available. Extensive reviews of pesticide selectivity to mammals are available elsewhere (O'Brien 1961, 1967, Metcalf 1964, Winteringham 1969, Hollingworth 1975, 1976, Brooks 1978). In this section, we focus on comparative mammalian–insect toxicology and the defenses animals use to escape toxicoses. We emphasize features that relate most to comparative herbivore–entomophage selectivities.

Animals avoid toxicosis primarily by: 1) avoidance, for example, by sequestering the chemical in insensitive storage tissues, such as in fat or the integument; 2) presenting a penetration barrier, such as an integument or internal lipid barriers; 3) rapid excretion of the toxicant; 4) metabolic detoxification of the toxicant; and 5) developing an insensitive target site for the toxicant. These strategies, listed here and depicted in Fig. 6.1, provide the basis of physiological selectivity, or differential pesticide effects after direct exposure (Ripper et al. 1951; Chapter 9).

Four physicochemical properties of toxicants largely dictate chemical fate in an animal (Fig. 6.1). The first, water solubility, is determined by the substituents of the chemical, such as polar or nonpolar, and ionic or nonionic groupings. Water solubility determines ease of chemical transport in or excretion from a water phase like hemolymph. Secondly, the partition coefficient, or solubility in an organic solvent relative to water, is an index of the affinity of the chemical for lipophilic sinks such as membranes and fats. Most toxicants are lipophilic, perhaps since most target sites are in or around membranes. Ease of membrane

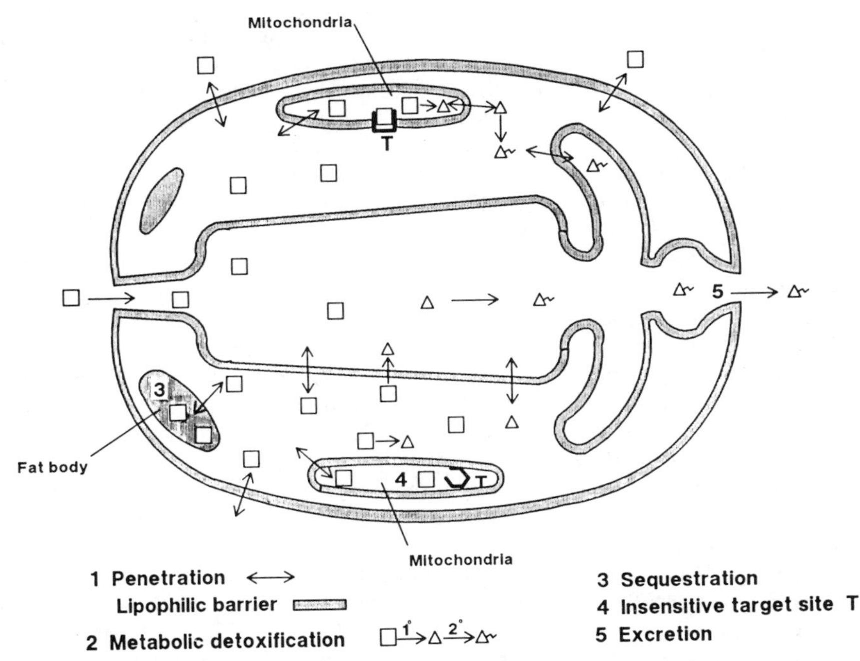

Figure 6.1. Dynamics of toxicant selectivity in an animal, including an arthropod natural enemy (after Mullin and Croft 1985).

penetration is determined mostly by the hydrophilic to lipophilic balance of the chemical. Thirdly, a steric factor dictated by its size, shape, and molecular weight influences the penetrability and fate of a chemical. Fourthly, the vapor pressure will predict if a chemical has sufficient volatility to enter an animal by an inhalation rather than by ingestion or contact.

6.3.1. Penetration

Few differences in penetration barriers have been found between arthropods and mammals. This has limited the development of highly selective pesticides based on penetration differences among these organisms. This is somewhat surprising, since the usually rigid, waxy, and chitinized cuticle of arthropods seems to contrast greatly with the often moist, flexible, keratinized skin of mammals. Nevertheless, insect and mammalian integuments contain both lipophilic and hydrophilic layers. While apolar toxicants penetrate integuments more rapidly than polar compounds, a blending of lipophilic and hydrophilic character (i.e., amphiphilicity) is required so that toxicants can penetrate into, through, and out of a multilayered barrier that is also variable in polarity.

That the integument can confer some selectivity is indicated by increases in toxicity upon direct injection of pesticide into an organism. For example, DDT is 15–30 times more acutely toxic (LD_{50}) when injected into a rat than when topically applied, whereas similar treatments of the American cockroach *Periplaneta americana* only doubles the toxicity. This suggests that DDT is selectively absorbed or passes through the insect cuticle compared with the mammalian skin, and helps account for the low hazard of DDT to humans (Metcalf 1964, O'Brien 1967, Hollingworth 1976). A parallel analysis (O'Brien 1961) with other insects demonstrates the presence of integumental barriers to DDT, such as for the milkweed bug (13-fold increase in LD_{50}, injected versus topical). Moreover, *P. americana* is not uniformly permeable to pesticidal chemicals, as exemplified by rotenone, where a 400-fold increase in acute toxicity occurs when injected beyond the cuticle (Winteringham 1969).

It is a lack of difference in penetration among herbivorous versus entomophagous arthropods that limits using integument permeability as an avenue for the design of insecticides which are safe for mammals and compatible with biological control agents, while still being effective against pests.

A limited number of studies of toxicant penetration have been made for individual natural enemy species or which compare natural enemies and pest species. Hagstrum (1970) showed that tolerance to topical applications of carbaryl in a *Tarantula* spider was due primarily to limited penetration, whereas this same compound when introduced orally via treated fruit fly prey was much more toxic. Bull and Ridgway (1969) examined comparative rates of penetration of P^{32}-labeled trichlorfon, an organophosphate insecticide, among adults of the herbivores *Heliothis virescens* and *Lygus hesperus*, and predaceous larvae of the lacewing *Chrysoperla carnea*. *C. carnea*, which is extremely tolerant to this organophosphate insecticide, showed poor penetration of trichlorfon over a 4-hr period, whereas the mirid showed a high rate of penetration and the tobacco budworm was intermediate (Fig. 6.2.a).

Comparative studies with the C^{14}-labeled organophosphate sulprofos to third instar larvae of *H. virescens*, the adult boll weevil *Anthonomus grandis*, and the lady beetle *Hippodamia convergens* showed extremely rapid penetration and loss of external dose in the highly susceptible predator. The more tolerant phytophages showed slower rates of absorption of topically applied toxicant (Fig. 6.2b).

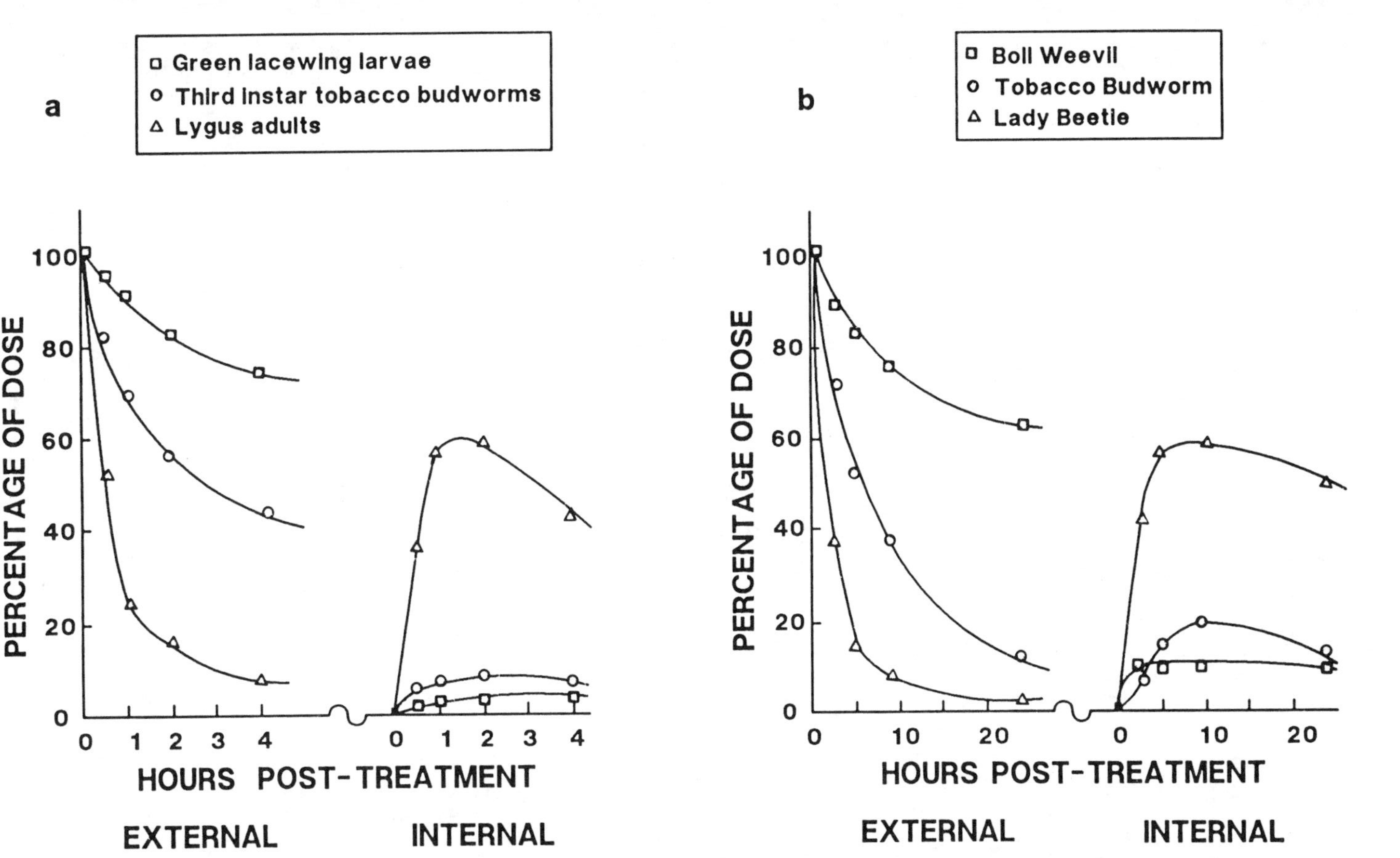

Figure 6.2. (a) Radioactivity in external and internal extracts at different times after topical treatment with P^{32}-trichlorfon (after Bull and Ridgway 1969). (b) Absorption of typically applied C^{14}-labeled sulprofos (0.1 μg/insect) by adult boll weevil *Anthonomus grandis,* adult lady beetles *Hippodamia convergens,* and third instar tobacco budworm larvae. *Heliothis virescens.* Data represent averages of three or more replicates (after Bull 1980).

Zhuravskaya et al. (1976) found that penetration of the organophosphate phosmet was faster in the cotton aphid *Aphis gossypii* than in either the adult or the third instar larvae of *Chrysoperla carnea.* These stages of the predator were 49 and 190 times more tolerant to the pesticide than the aphid. They concluded that the observed differences in rates of penetration were a significant factor in accounting for the selective action of the insecticide.

Several studies have focused on the differential penetration of pesticides through the cuticle of host insects versus the less sclerotized epicuticlar membranes of their endoparasitoids (e.g., in diapausing versus nondiapausing pupae of *Hypera postica* parasitized by *Bathyplectes curculionis*, Bartell et al. 1976; in eggs of *Eurygaster integriceps* at different developmental stages and periods of parasitization by *Trissolcus grandis*, Novozilhov et al. 1973. As noted in Chapter 3 (Section 3.2.3.6; Fig. 3.2), a range of dynamic factors, including consumption of the host, the influence of parasitism on host cuticle formation, and the stage of parasitoid development, were involved in conferring selective penetration and uptake of the pesticide by either species.

Finally, in a more integrated study of comparative toxicodynamics, Martin and Brown (1984) observed that the penetration, metabolism, and excretion of acephate in the predatory reduviid *Pristesancus papuensis* and its prey, *Pseudoplusia includens*, was much more rapid in the lepidopteran as compared to the natural enemy. Furthermore, activation to another intermediate toxicant, methamidophos, was four times greater in the predator than in the prey. They concluded that the greater activation and accumulation of toxicant in the natural enemy was primarily responsible for the ninefold greater tolerance to acephate in the pest than in the predator.

In summary, a number of cases are known where penetration barriers in natural enemies to specific insecticides have provided some basis for selectivity over their host or prey. However, no patterns are present in these cases to suggest that either group is favored by having greater penetration barriers to exogenous toxins. More research on the effects of pesticide structure on penetration and how penetration is affected by pest/natural enemy integuments is necessary before such generalizations may be possible.

6.3.2. Sequestration

Lipophilic sinks comprised of fatty tissues and intra- and extracellular lipoproteins that serve in chemical transport allow insects to sequester dietary toxicants and defend against predation (Duffey 1980, Blum 1981). In some cases, defense may be fortuitous or absent, and is not necessarily a result of sequestration. Indifferent storage of apolar and amphiphilic chemicals is much more prevalent in insects than mammals. The greater nonspecific accumulation of lipophilic toxicants in insects has been attributed to higher ingestion rates per unit of body weight, and higher ratios of lipid to water content of insects compared to mammals (Duffey 1980). It appears likely that storage of toxicants in lipid reserves can confer some pesticide tolerance and resistance in arthropods in general (O'Brien 1961, Winteringham 1969, Hollingworth 1976, Duffey 1980); however, this has not been adequately tested. Reagents that interfere with lipid deposition may selectively hinder arthropods by disrupting this mechanism of resistance.

In relation to sequestration of toxicants, the most extensive work with natural enemies has been associated with the study of fat body lipid reserves as buffers to toxicosis. Takeda et al. (1965) demonstrated in *Coccinella septempunctata* a direct correlation between low fat content in early season (presumably due to resource depletion during overwintering and limited feeding) and the severity of malathion toxicosis. In late season, summer populations which had fed extensively and had presumably obtained greater fat body

Table 6.1. Toxicity of Malathion to Adult *Coccinella septempunctata* in Relation to Seasonal Fat Body Content (Malathion = .05 a.i.) (after Takeda et al. 1965)

Date Collected	Percent Mortality	Fat per Dry Weight Basis (%)[1]
September 20–30	66	17.2
October 16	75	17.8
October 23	60	21.7
October 29	33	25.3
November 8	25	24.9

[1] Means of Males and Females

reserves, showed a lower degree of effects from the organophosphate insecticide (Table 6.1).

Although there are few other direct studies of fat body/toxicological relationships among entomophagous species, basic studies of natural enemies' fat body metabolism are beginning to receive greater attention. Olson (1980) measured the effect of age and diet on the fat body composition of adult female *Habrobracon juglandis*. Whereas the volume of fat body declined drastically within the first week after emergence and then remained relatively constant thereafter in the absence of feeding, host feeding influenced the rate of fat body loss, and sugar feeding as well as host feeding influenced the length of life.

Although few specific studies have examined fat body dynamics and toxicology, several authors have inferred causal relationships between pesticide side effects on natural enemies and the role of these organ systems in sequestration and pesticide dynamics (e.g., Hamilton and Kieckhefer 1969, Hoffman and Grosch 1971, Dumbre and Hower 1977).

6.3.3. Excretion

The excretory apparatus is primarily responsible for the transfer and elimination of harmful metabolites from internal into external fluids, where adsorption to waste solids may occur prior to defecation. Only minor differences between mammals and insects have been noted for excretion of pesticides. For example, excretion of urates or ammonia, as well as an ouabain-insensitive ion pump responsible for fluid secretion in insects (Phillips 1981), contrasts with urea excretion by an ouabain-sensitive fluid pump in vertebrates. These differences suggests that selective toxicants based on renal function may be possible. However, our understanding of the excretion of xenobiotics in insects is rudimentary compared with mammals.

Important selectivities among insects in excretory capacities are known, particularly for polar insecticides. For example, ingested nicotine is excreted rapidly in the tobacco hornworm (90% in 4 hr), whereas houseflies only excrete 10% of an applied dose in 18 hr (Self et al. 1964, O'Brien 1967). This explains, in part, the tolerance of *Manduca sexta* for its insecticidal host plant. Nicotine, a nonionic lipophilic compound like most synthetic pesticides, is rendered excretable only after degradation to polar metabolites. Tobacco feeding insects (e.g., *Manduca*) make efficient use of metabolic detoxification and rapid excretion of this toxicant.

With regard to pesticide excretion in natural enemies, the most work has been done with DDT and metabolites in entomophagous predators. Atallah and Newsom (1966) studied metabolism and excretion in the coccinellid *Coleomegilla maculata* and observed that tolerance to DDT was due, in part, to rapid metabolism of the compound to DDE and then to excretion of both the parent and the metabolized compound in feces. Dempster (1968) studied conversion of DDT to DDE in the DDT-tolerant carabid *Harpalus rufripes* and noted rapid conversion and elimination of the DDE as a nontoxic metabolite. Whereas sublethal levels of DDT markedly reduced the initial feeding rate of adult *H. rufripes* on *Pieris rapae*, conversion to DDE allowed many beetles to return to normal rates of predation. Gilyarov (1977) observed high levels of predation on collembola by staphylinids, which rapidly degraded DDT to DDE for excretion. DDE accumulation at slower rates occurred in other species such as predatory spiders, millipedes, and carabids, and mortality levels from DDT were correspondingly higher. Hagstrum (1970) studied penetration, metabolism, and excretion rates of carbaryl conjugates in the spider *Tarantula kochi* and observed total rates of excretion in the range of 1–20% of metabolized compound over a 24-hr period after treatment.

Although little is known about excretion in parasitic Hymenoptera and Diptera, endoparasitoid larvae can store nitrogenous wastes internally and then deposit them in a meconium just prior to pupation (Vinson and Iwantsch 1980). This may confer differential sensitivity to the parasite over its host to nitrogenous wastes, and perhaps nitrogenous pesticides.

In summary, excretory mechanisms have been associated with pesticide metabolism and may even account for some tolerance to pesticides in natural enemy species. However, no distinctive excretory differences that might confer differential selectivity to entomophages over their prey or hosts have been noted. The formation of meconia in parasitic wasps is the closest to being a unique excretory mechanism that could potentially confer selectivity to these species. Virtually no comparative study of these aspects of pesticide metabolism and selectivity has been investigated between pests and natural enemies.

6.3.4. Selective Metabolism

Although sequestration, penetration barriers, and excretion are important factors, metabolism and action at the target site are of greater importance in explaining species variation in selective toxicity. Generally, enzymatic detoxification is the most direct and dependable way for an animal to survive a toxicant exposure.

Metabolic transformation of lipophilic pesticides to excretable products usually proceeds by enzymatic events that can be grouped into phase I and phase II reactions. Phase I involves largely oxidative and hydrolytic reactions in which a polar group is introduced onto the toxicant. Phase II enzymes conjugate endogenous substrates such as glutathione or monosaccharides onto the polar group resulting from phase I metabolism (Kulkarni and Hodgson 1980). While these metabolic pathways function to ultimately detoxify toxicants (including pesticides and plant allelochemicals) to polar and thus excretable products, many of the initial reactions generate products that are more toxic than the parent xenobiotic. Included among these are selected oxidative reactions catalyzed by the cytochrome P-450 monooxygenases (mixed-function oxidases, polysubstrate monooxygenases) exemplified by desulfuration at the phosphorus bond (P=S to P=O), thioether oxidations, and epoxidation (Kulkarni and Hodgson 1980). A detoxification reaction working in concert can then dispose of the intermediate toxicant. An example of such a pathway is shown in Fig. 6.3. Epoxidation of olefins largely by mixed-function oxidase (MFO) can produce reactive epoxides harmful to the animal. The enzyme epoxide hydrolase (EH) catalyzes the addition of water to the epoxide, thereby detoxifying it to a more excretable 1, 2-dihydroxy metabolite. Thus the balance of the activating or

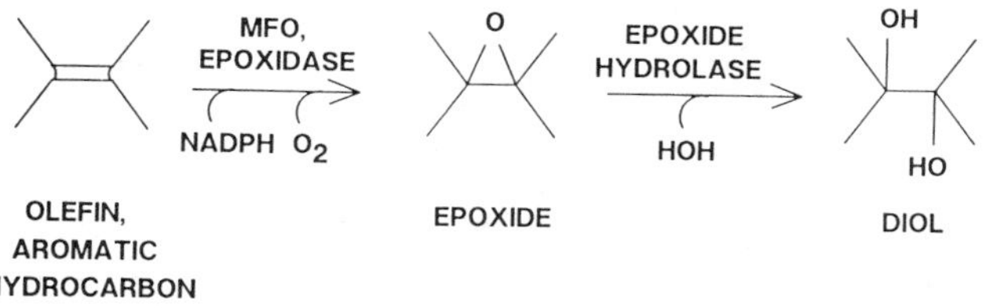

MFO, EPOXIDASE
NADPH O_2
EPOXIDE HYDROLASE
HOH
OLEFIN, AROMATIC HYDROCARBON
EPOXIDE
DIOL

Figure 6.3. Toxification–detoxification pathway for olefin and aromatic hydrocarbon metabolism (after Mullin and Croft 1985).

toxification reactions that occur with the degradatory or detoxification reactions determines the net consequence of a xenobiotic to an organism.

It is useful to consider two important concepts in pesticide selectivity, that of a selectophore and an opportunity factor. A selectophore is a chemical grouping that confers selective toxicity; it usually is associated with the target site. An opportunity factor introduces a grouping which requires activation to elicit toxicity (O'Brien 1961) and is beneficial to the organism that is better endowed to detoxify than to activate. We will call an opportunity factor a protoxiphore.

The carboxylester group is a selectophore that benefits mammals over insects since the former have greater carboxylesterase activity (RCOOR′ to RCOOH + R′OH). This is a primary reason why malathion is much more toxic to insects than mammals (O'Brien 1961, Metcalf 1964) and explains in part the relative mammalian safety of the highly insecticidal pyrethroids (Casida et al. 1983, Soderlund et al. 1983). Similarly, organophosphates are generally rendered more selective for arthropods if the ethyl esters are replaced with methyl esters. Presumably, the detoxification enzyme glutathione transferase which demethylates these organophosphates through conjugation of the methyl group to glutathione is more available to mammals than to insects, and there is a strict substrate preference for methyl over ethyl esters (Shishido 1978, Fukami 1980, Motoyama and Dauterman 1980).

Equally important for selective toxicity among the organophosphate pesticides is the application of a protoxiphore or the proinsecticide approach (Fukuto 1984). The well-known oxidative activation of the P=S to P=O in organophosphates is usually distinctly favorable to mammals. Hence the highly selective action of malathion to insects is the overall result of this rapid lethal synthesis in the insect and the more rapid detoxification in the mammal (Metcalf 1964, O'Brien 1967). While it is almost invariably true that P=S compounds are more selectively toxic to insects than mammals, exceptions (including DFP and dimefox) are known (O'Brien 1967). Another case of improved selectivity due to activative metabolism is with the amidases, which catalyze the hydrolytic opening of carboxylamides (RCONR2′ to RCOOH + HNR2′). Fluoroacetamides, such as the insecticide fluoroacetamide and the acaricide Nissol, are proinsecticides that require activation by hydrolysis to release the toxicant fluoroacetic acid, and the appropriate amidases are at higher levels in arthropods than in mammals (Matsumura and O'Brien 1963, Knowles 1975).

A third example of the use of selective activation is with *N*-derivatized analogues of *N*-methylcarbamates. Fukuto and coworkers have synthesized a large series of *N*-acetylated, phosphorylated, sulfenylated, and methoxylated derivatives of the parent *N*-methylcarbamates and found the derivatives to be almost invariably less toxic than the parent to mammals, while retaining their toxicity to insects such as mosquitoes, houseflies, and spruce budworms (Wustner et al. 1978, Fukuto and Fahmy 1981). It was later determined that insects retain higher levels of the appropriate hydrolytic enzyme to degrade the proinsecticide to the more potent *N*-methylcarbamate (see Hollingworth 1975).

From the above examples, it is clear that better selectivity is obtained if both selectophores and protoxiphores are combined in a toxicant so that both enhanced activation and decreased detoxification are exploited in the target pests. Obviously, other defensive strategies must be assessed in order to circumvent species variation that hinders the design of broadly pesticidal chemicals that are safe for nontarget species. Prominent among these are target site differences.

6.3.5. Target Site Selectivity

To gain maximal selectivity between mammals and insects requires exploiting biochemical targets that are unique to the pest. However, most conventional pesticides, partly because of their rapid action as well as our lack of knowledge concerning insect toxicology, were developed largely as nerve toxicants. Organochlorine, pyrethroid, organophosphate, and carbamate pesticides all have target sites in the nervous system. It is well known that toxicological differences between insects and mammals are more often due to morphological differences and quantitative rather than qualitative differences in physiological/neurological systems. Thus, the same neurotransmitters, neuronal and synaptic events, including transmitter release, biosynthesis, degradation, and nerve impulse propagation seem common to both insects and mammals. Only recently have sites of action unique to the insect, such as JH receptors (discussed above), been exploited for pest control. Nevertheless, there may be sufficient differences in the architecture of nerve target sites to allow for the development of significant pesticide selectivity.

Binding of a toxicant with its target site is the ultimate event prior to the harmful effect, and occurs only after penetration, alternative binding, excretory, and metabolic barriers have been circumvented. The more closely a toxicant is administered to the active site, the greater its impact. As expected, toxicity generally increases with the following routes of administration: topical, subcutaneous, oral, intraperitoneal, and intravenous. Nerve toxicants that are inhibitors of acetylcholinesterase (AChE) (including the organophosphates and carbamates) or the acetylcholine (ACh) receptor (such as nicotine) in insects must penetrate a penultimate membrane barrier prior to reaching the target site. This is necessary, since the cholinergic (ACh) junctions are limited to the central nervous system (CNS), and the CNS is ensheathed by the neural lamella and perineurium. In contrast, mammals retain peripheral cholinergic sites such as the neuromuscular junction that are unprotected by membrane sheaths.Consequently, cholinergic pesticides that are ionizable at physiological pH, and thus cannot penetrate these membranes, are selective mammalicides (O'Brien 1961, 1967). Included are cationic organophosphates, such as amiton, and nicotine alkaloids. Schradan is metabolically activated to polar metabolites that presumably fail to penetrate the nerve sheath. This organophosphate is also highly toxic to mammals, but curiously toxic to some arthropods including mites and aphids and broadly selective to many entomophagous predators and parasitoids (Chapter 9). Attempts to find ion permeability differences in the CNS of an amiton and schradan sensitive aphid and the insensitive cockroach failed. The alternative possibility that aphids have peripheral cholinergic sites has been suggested (Toppozada and O'Brien 1967, O'Brien 1967). The distinctive chemistry of aphicidal and acaricidal nerve toxicants relative to insecticides in general may be interesting to explore in this regard.

AChE hydrolytically deactivates the neurotransmitter ACh allowing for repolarization of the postsynaptic nerve membrane. Numerous cases of species differences in AChE sensitivity to organophosphates and carbamates have been noted (Metcalf 1964, O'Brien 1967, Winteringham 1969, Hollingworth 1976). The sensitivity of the AChE to these inhibitors is measured using *in vitro* kinetic methods. Satisfactory correlations of inhibitor potency (I_{50}) with *in vivo* toxicity (LD_{50}) have allowed estimation of the selective inhibition required to give a desirable toxicity ratio (mammalian LD_{50}/insect LD_{50}) of 500. Based on various studies, this would require a difference in binding between the insect and mammal AChE's of from 50 to 34,000, a level rarely observed (Hollingworth 1976). Designing selective AChE inhibitors based solely on insensitive target sites would be difficult unless combined with metabolic or alternative factors. More study on target site kinetics is necessary because of the past tendency to compare somewhat inappropriate enzyme models such as housefly head AChe and bovine erythrocyte AChE at differing states of purity.

Comparative studies of cholinergic target site differences between herbivorous pests and their entomophagous enemies have been limited. Substrate preferences for cholinesterases, the target for organophosphate and carbamate pesticides, showed no distinction in substrate–activity profiles between herbivorous, omnivorous, and carnivorous insects, including a syrphid, blowfly, and tiger beetle (Metcalf et al. 1955). Using studies of cholinesterase (AChE) activity in whole body homogenates of herbivorous and entomophagous species, Singh and Rai (1976) found much higher levels of ACh hydrolysis in the predators *Coccinella septempunctata* (252 mg ACh hydrolyzed g tissue/hr) and *Cicindella sexpunctata* (220 mg) than in the hemipteran plant bugs *Leptocosisa acuta* (30 mg) and *Dysdercus koenigii* (63 mg) and the aphid *Lipaphis erysimi* (6 mg). Similar results were obtained with AChE activity expressed on a per insect basis. The K_m (Michaelis–Menten constant) for enzyme activity indicated a general higher affinity of ACh with AChE from herbivores than entomophages. Phosphamidon, an insecticide selectivity toxic to these phytophagous pest species, showed a higher inhibition rate with AChE from *L. erysimi* and *L. acuta* than with AChE from their predators *C. septempunctata* and *C. sexpunctata*, respectively. This was due to a higher affinity between enzyme and inhibitor rather than to the rate of phosphorylation. The authors suggested that these differences in properties in AChE between the two groups might be exploited to develop selective pesticides.

More detailed comparative work on cholinesterase systems has been done with silkworm moths and their tachinid parasites than with most other host/parasitoid models (Bai Shaing et al. 1981). Enzyme activities from the head and nerve cord of the larvae of the tussah silkwork *Antheraea pernyi* and from the individual larvae and head and thorax of the adult tachinid *Blepharia tibialis* were measured in relation to the toxicity of the organophosphate insecticide dimethoate and the related compound omethoate. The bimolecular rate constants K_i for omethoate to cholinesterase of the tachinid fly were from 12 to 80 times greater than those of the silkwork. These results for *in vitro* cholinesterase inhibition were in close agreement with *in vivo* toxicity tests with dimethoate. They concluded that the mechanism of selective toxicity favoring the silkworm over its natural enemy was principally due to a difference in selective inhibition of the AChE.

Alternative nerve targets other than cholinergic systems are available in insects. Whereas cholinergic sites are centrally located in insects and thus protected by nerve sheaths, peripheral junctions are believed to have glutamate as an excitatory transmitter. In mammals, the converse is true, and glutaminergic junctions are protected behind the blood–brain barrier of the CNS. Hence glutamate antagonists should be more selective against insects. However, attempts to exploit this have failed, perhaps due to the poor understanding of glutamate deactivation (Brooks 1978, 1980).

Gamma-aminobutyric acid (GABA) is thought to be an inhibitory transmitter in the neuromuscular junctions of both insects and mammals. Action at GABA junctions is responsible in part for the effectiveness of the newer pyrethroids (Casida et al. 1983), the fermentation product avermectin, as well as the mammalicide picrotoxinin (Beeman 1982). Pyrethroids also act as axonic poisons by blocking sodium or calcium channels, presumably through inhibition of various ATPase activities (Clark and Matsumura 1982, Casida et al. 1983). More work is needed to define the target sites for these highly selective pest control agents. Promising behavior modifiers, such as the formamidines which impact on the biogenic amine systems, may also present an alternative to nonselective biocides (Beeman 1982).

The successful design of selective neurotoxicants will largely depend on a continued polyfactorial approach to the dynamics of toxicant target site interaction. The potential importance of detoxification of dieldrin (Brooks 1978) and pyrethroids (Soderlund et al. 1983) at the target site illustrates how complex the barriers to pesticide action can be, and warrants the plea by Hollingworth (1976) for future studies to be more detailed and complete so that the fundamentals of selectivity are not obscured.

6.3.6. Detoxification Enzymes

Exploration of the detoxification basis for differential pesticide selectivity between pests and natural enemies has lagged because of difficulty in rearing entomophages, as well as the insufficient biomass usually available for enzyme assay. Many natural enemies, especially parasitoids, are small, and their size precludes dissection of specific tissues where detoxification enzymes reside (i.e., the midgut, fat body, or malpighian tubules). This often necessitates the use of whole body homogenates, which may release factors that impair enzyme measurements (Wilkinson 1979). Fortunately, more sensitive enzyme assays and stabilizing additives are being developed which begin to allow for more satisfactory *in vitro* study of detoxification enzymes within whole body preparations of entomophagous arthropods. These techniques should aid in understanding biochemical events responsible for chemical selectivities.[1]

In considering the detoxification capacities of pests versus natural enemies, the discussion must begin by reviewing background about pest preadaptations to plant foods. In this regard, feeding experience as it relates to the degree of polyphagy is thought to alter pesticide susceptibility in herbivorous arthropods. It has been suggested that elevated levels of detoxification enzymes due to prolonged exposure to plant toxicants are often responsible for altered susceptibility (Gordon 1961). Early biochemical evidence for this hypothesis was provided by Krieger et al. (1971) who found increasing aldrin epoxidase (MFO) levels in lepidopteran larvae capable of consuming many plant hosts. Confirm-

[1] In discussing the comparative detoxification relationships between phytophagous and entomophagous species, it is important to point out that problems may arise in attempts to extrapolate *in vitro* detoxification capabilities to *in vivo* events. For example, the dog has enough rhodanese activity in its liver to detoxify 4 kg of cyanide in 15 min, yet cyanide is highly toxic to dogs with an oral LD_{50} of 1.6 mg/kg (Hollingworth 1976). In contrast, many insects are highly tolerant to cyanide exposure, and rhodanese activity has often been assumed to enable herbivores to readily feed on cyanogenic foliage. However, there are no apparent relationships between rhodanese activity, which is present in hymenopteran and dipteran parasitoids, as well as predaceous, herbivorous, and omnivorous insects, and the insect's feeding specialization or tolerance to cyanide (Long and Brattsten 1982, Beesley et al. 1985). Presumably, rhodanese has no prominent role in the detoxification of dietary cyanide.

ation of MFO as one explanation for tolerance of lepidopterans to pesticides comes from recent studies where up to a 45-fold enhancement of midgut MFO levels occurred in larvae which were fed various allelochemical-laden herbaceous plants relative to phytochemically benign plants (Brattsten 1979a, 1979b, Yu et al. 1979, Ahmad 1982, Yu 1982). A concomitant larval increase in the level of insecticide tolerance of up to fivefold was also noted (Yu et al. 1979). The association of enzyme induction with plant defensive chemistry was shown through the incorporation of purified allelochemicals into artificial diets of moths. These host plant influences on arthropod detoxification capabilities have been extended to hydrolytic enzymes, including esterases and epoxide hydrolases (Dowd et al. 1983, Mullin and Croft 1983), and the conjugating enzyme glutathione-*S*-transferase (Yu 1983).

Clearly, dietary plant allelochemicals may preadapt herbivores to more efficiently detoxify pesticides. However, the mitigating circumstances of the pest's preadaptation to plant defensive chemicals and related pesticides are not as simple as was originally thought (Berenbaum 1985, Berenbaum and Neal 1987). Factors such as the host specificity of plant feeding arthropods, secondary plant compound diversity among individual plant species used as food substrates, as well as unique detoxification capabilities among phytophagous insects provide many exceptions to the broad generalizations first made by Gordon (1961), Krieger et al. (1971), and others. Still the broad generalizations first developed to explain the patterns of detoxification in this group of species have proven useful and loosely applicable.

While herbivores may retain many preadaptive detoxification abilities for pesticides because they must detoxify plant allelochemicals, entomophages may be less exposed to dietary toxicants and thus have lower detoxification capabilities. This hypothesis was first elaborated to explain the generally higher susceptibility of natural enemies than herbivorous pests to pesticides (Croft and Morse 1979; Chapters 2 and 16). It was suggested that a "natural enemy may be able to express its capacity to adapt to a natural plant toxicant only after the pest had exploited its genetic plasticity in resisting the toxicant" (Croft and Brown 1975). Expression of enhanced detoxification may occur within a generation through induction to influence tolerance attributes to pesticides, or over a longer period to influence tolerance or resistance mechanisms (Chapters 5 and 16).

Evidence to evaluate detoxification differences between pests and natural enemies initially came from data reported by Brattsten and Metcalf (1970, 1973), who tested the susceptibility of a wide variety of arthropod species to carbaryl in the presence and absence of the MFO inhibitor piperonyl butoxide. Piperonyl butoxide synergizes pesticide potency by inhibiting a major detoxification system; it is well known that carbaryl is primarily detoxified by oxidative metabolism (Brattsten and Metcalf 1970). Thus a synergistic ratio (SR), which is the ratio of the LD_{50} for carbaryl in the absence versus the presence of synergism, provides an estimate of the MFO available for detoxification in the test animal.

Although Brattsten and Metcalf (1970) concluded that their survey did "not provide any obvious correlations of MFO activity to phylogenetic position, food habits, or degree of specialization," they did not specifically examine trophic groupings such as herbivores versus entomophages. Specific comparisons among these groups of species were later made by Croft and Morse (1979) and Mullin and Croft (1985). A summary of the analysis of Mullin and Croft (1985) is presented in Table 6.2. Although a high degree of variation was apparent within each trophic group, herbivores had a 10-fold higher mean tolerance to carbaryl and a two-fold higher mean synergistic ratio than carnivores. While differences in susceptibility were significant at the $P < 0.001$ level, those for synergistic ratios were not.

Table 6.2. Toxicity of Carbaryl and Synergistic Ratios to Piperonyl Butoxide for Phytophagous Arthropods Relative to Parasitic and Predaceous Natural Enemies (after Mullin and Croft 1984[1])

	Phytophages $n = 32$	Entomophages $n = 17$
LD_{50} (in μg/mg body wt.)	378 ± 198[2,3]	37 ± 18[2]
Synergistic ratio	15.2 ± 7.3	7.2 ± 1.3

[1] Based on data from Brattsten and Metcalf (1970, 1973); see later reexamination by Croft and Morse (1979) and further adaptation here.
[2] Mean $\pm$ standard error of mean.
[3] Significant difference at $P < 0.001$ between insect groups.

Croft and Morse (1979) reported earlier that mean susceptibility levels were 16.0- and 5.9-fold higher and synergistic ratios 2.1- and 1.6-fold lower, respectively, for parasitoids and predators than for phytophagous species among the data of Brattsten and Metcalf (1970, 1973). Because of the relatively indirect nature of synergist tests, they emphasized that *more direct* tests of detoxification differences between pests and natural enemies be conducted (e.g., more specific enzyme comparisons). Mullin and Croft (1985) speculated that some of the intragroup variation observed to synergized carbaryl may have been due to opposing reactions that were catalyzed by various MFOs, since both toxification and detoxification pathways are known for this compound (Oonnithan and Casida 1968). These early studies of comparative detoxification in pests and natural enemies stimulated more specific studies and a more integrated theory of comparative detoxification between entomophagous and herbivorous arthropods.

Although predators and parasitoids may lack well-developed MFOs, they may be just as adapted as herbivores in relation to nonoxidative detoxification mechanisms (Plapp 1981). Several researchers observed that oxidatively detoxified insecticides were more toxic to entomophagous arthropods than to herbivores, whereas insecticides detoxified primarily hydrolytically (e.g., by esterases) were safer for carnivores (Plapp and Vinson 1977, Plapp and Bull 1978, Ishaaya and Casida 1981, Plapp 1981, Bashir and Crowder 1983, Brown and Casida 1983).

Mullin et al. (1982) compared detoxification enzymes in the polyphagous spider mite *Tetranychus urticae* and its major acarine predator *Amblyseius fallacis*. These species were chosen as an herbivore–predator model because of their similar biologies and morphologies and because resistance was present in population cultures of both mite species (Chapter 14). Oxidative (aldrin epoxidase, MFO), hydrolytic (esterases and epoxide hydrolases), and conjugating (glutathione-*S*-transferase) enzymes that typify the metabolism of lipophilic toxicants to excretable products were measured from whole body homogenates. Major enzyme activity differences were found (Fig. 6.4) that were closely associated with patterns of susceptibility observed among strains. Comparing pesticide-susceptible strains, the herbivore had fivefold higher MFO and sixfold higher *trans*-epoxide hydrolase levels than the carnivore. Since both elevated MFO (Brattsten 1979b) and *trans*-epoxide hydrolase (Mullin and Croft 1983, 1984) have been ascribed to previous encounters with plant defensive chemicals, their results supported the allelochemical preadaptation hypothesis. Also, the herbivore and carnivore had similar esterase activities for 1-naphthyl acetate (Fig. 6.4). These data supported the hypothesis that both herbivores

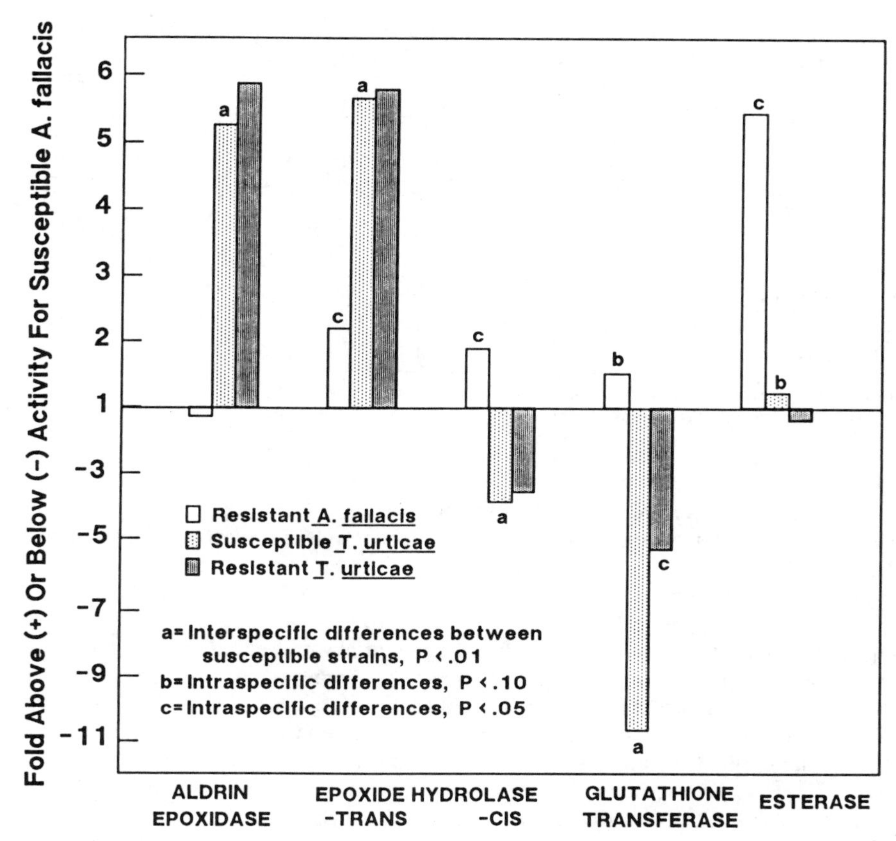

Figure 6.4. Relative fold activity of detoxification enzymes in susceptible strain of predaceous mite *Amblyseius fallacis*, compared to resistant strain and susceptible and resistant strains of its prey *Tetranychus urticae* (after Mullin et al. 1982).

and entomophages should have similarly developed hydrolytic detoxification pathways. Esteratic pathways are important in both detoxification and basal metabolism, and carboxylesterase, lipase, amidase, proteinase, cholinesterase, and thioesterase activities may all be incorporated into this general esterase measurement (Heymann 1980).

A carnivore can have an advantage over an herbivore for specific enzymes in potential detoxification capability. Thus *A. fallacis* had 11-fold higher glutathione-S-transferase and four-fold higher *cis*-epoxide hydrolase activities than the prey mite (Fig. 6.4). These enhanced levels may be due to the higher respiration rate of the predator, which can lead to autooxidative metabolites that require detoxification by these enzyme systems (Mullin et al. 1982).

These findings with a predator herbivore system led to the examination of a representative parasitoid herbivore system, that of the polyphagous tortricid *Argyrotaenia citrana* and its braconid ectoparasitoid *Oncophanes americanus* (Croft and Mullin 1984). MFO comparisons between the parasitoid versus lepidopteran larvae were hampered by the difficulty in comparing whole body homogenate activity levels for both species.

Table 6.3. Detoxification Enzyme Levels in 3 Natural Enemies and Their Respective Prey or Hosts (Adapted from Mullin et al. 1982, Croft and Mullin 1984, and Mullin 1985)

	Enzyme Activity (pmol/min-mg protein)					
Species[1]	Aldrin Epoxidase	*trans*-EH	*cis*-EH	*trans*/ *cis*-EH	Esterase ($\times 10^{-3}$)	Glutathione Transferase
Amblyseius fallacis	0.27	310	431	0.072	318	1095
Tetranychus urticae	1.44[2]	1710[2]	117[2]	14.6[2]	389	102[2]
Oncophanes americanus	0.85	407	727	0.56	307	135
Argyrotaenia citrana	26.8[2]	1340[2]	536	2.50[2]	593	288
Pediobius foveolatus	0.67	198	415	0.48	83	—
Epilachna varivestris	4.16[2]	780[2]	352	2.22[2]	94	—

[1] Composite values for preparations from both adult and immature stages.
[2] Interspecific differences between significant values for herbivore versus entomophage at $P < 0.01$ level.

However, a comparison of gut levels of the pest versus whole-body levels of the natural enemy indicated that the host had substantially higher levels of oxidative enzymes than did the natural enemy, even when calculated on a whole body weight basis. Again, *trans*-epoxide hydrolase activity was elevated in the herbivore relative to the entomophage, whereas hydrolytic esterase activitity was similar in the two species (Table 6.3), which supports the hypothesis relating detoxification potential with feeding ecology. In contrast to the predator–prey situation, the glutathione-*S*-transferase activity was similar between the parasite and its host. This may be related to the relative activity levels of the mobile mite predator *A. fallacis* compared to the more sessile ectoparasitoid *O. americanus*, hence the parasitoid might require less of a protective enzyme such as glutathione-*S*-transferase that prevents oxygen toxicity. Similar results were found for the Mexican bean beetle *Epilachna varivestris* associated with its eulophid endoparasitoid *Pediobius foveolatus* (Mullin 1985; Table 6.3).

A useful comparative index that emerged from detoxification enzyme studies with natural enemies and herbivorous pests was the ratio of *trans*- to *cis*-EH. Herbivorous prey consistently had a higher *trans*/*cis* ratio than associated entomophages. One explanation for this phenomenon may be the distribution of epoxides within plant and animal tissue. Plants biosynthesize *trans*- and higher substituted olefins, including fatty acids, phenolics, alkaloids, and terpenoids that are rare or absent in animals. Many are allelochemical defenses against herbivory. In contrast, *cis*-olefins usually have constitutive and homeostatic functions in both plants and animals (Mullin and Croft 1985, Mullin 1985). Epoxidation of olefins largely by MFO (Fig. 6.3) can produce harmful epoxides which are subsequently detoxified by the epoxide hydrolase of appropriate selectivity. Thus, herbivores in general might be expected to exhibit a high *trans*- relative to *cis*-EH level to evade plant defenses, while entomophages, due to their feeding specialization on animal tissue, would be expected to have a lower ratio.

This tendency has been confirmed for 20 species of herbivorous pests from the orders Acari, Coleoptera, Lepidoptera, and Diptera, and for seven species of natural enemies from the orders Acari, Neuroptera, Coleoptera, and Hymenoptera (Mullin and Croft 1984, 1985, Mullin 1985). Indeed, the herbivores surveyed had, on the average, 21-fold higher *trans*-EH activity and 10-fold higher total EH activity than the entomophagous arthropods considered (Table 6.4). In addition, a phytophagous coccinellid, *E. varivestris*,

Table 6.4. Epoxide Hydrolase Detoxification Enzyme Ratios in Selected Pests Relative to Their Natural Enemies (after Mullin and Croft 1984, 1985, Mullin 1985)

Species Group	*trans*-EH	*cis*-EH	*trans/cis*-EH
Herbivore pests $n = 20$	20.7[1,2]	3.0[2]	10.2[2]
Natural enemies $n = 7$	1.0	1.0	1.0

[1] Ratio of activities in nmol/min mg protein for gut preparations except for acarines and parasitoids, where whole body assays were used. Ratios calculated for mean values of pest/natural enemy.

[2] Significant difference observed in herbivore natural enemy comparison at $P < 0.05$ level.

had a much higher *trans/cis*-EH ratio (2.22 $\pm$ 0.017) than the predaceous coccinellids (0.73 $\pm$ 0.20) in the survey. These highly significant biochemical differences suggest that EH may be an appropriate biochemical target for development of a broad-spectrum insecticide against herbivorous pests that will have less impact on entomophages (i.e., a broad-spectrum, selective compound; Croft 1981a). Presumably, crop pests should be more susceptible than their natural enemies to inhibitors of *trans*-EH.

Other detoxification enzyme differences between herbivores and carnivores have been observed. Lacewing (*Chrysoperla carnea*) larvae are tolerant of the *cis*-isomers of pyrethroids, such as deltamethrin and permethrin, presumably because they retain esterases that detoxify the *cis*-isomers faster than the less insecticidal *trans*-isomers (Ishaaya and Casida 1981, Casida et al. 1983). Herbivorous insects, by contrast, only poorly hydrolyze the *cis*-isomers, but detoxify the less insecticidal *trans*-pyrethroids to a much greater extent (Soderlund et al. 1983, Jao and Casida 1984). Replacement of the ester functionality by the analogous oxime ether produces a highly toxic pyrethroid against both *C. carnea* and the coccinellid *Cryptolaemus montrouzieri* (which is also tolerant of pyrethroid esters and a number of other pesticides; Chapter 2), and again implies a key role for esterases within entomophages for pyrethroid detoxification (Brown and Casida 1983). In further investigations of permethrin toxicity to the larvae of *C. carnea*, Bashir and Crowder (1983) found that this species metabolized 70–80% of *cis*- and *trans*-isomers within 2 hr, which suggests that rapid degradation accounts for the observed permethrin tolerance in this species. Again, hydrolytic esterases were implicated as a primary mechanism.

Kharsun and Karpenko (1976) found that the organophosphates malathion and leptophos inhibited esterase activity in several predatory species, including a spider, a carabid, and a coccinellid associated with the Colorado potato beetle *Leptinotarsus decemlineata*. Gargov (1968) observed that when the coccinellid *Coccinella septempunctata* was treated with the organophosphates demeton-methyl, vamidothion, dimethoate, and trichlorfon, its cholinesterase and aliesterase activities were reduced appreciably. Feng and Wang (1984), in a comparative study of detoxification and selectivity of insecticides to the noctuid *Plutella xylostella* and its braconid parasitoid *Apanteles plutellae*, observed that synergists of oxidative (PBO) and hydrolytic (EPN) enzymes had a greater effect on the parasitoid than its host. Their data indicate that either these detoxification enzymes were more important to the metabolism of these organophosphate insecticides in the natural enemy than in the pest insect or these enzyme systems were present in greater amounts in the beneficial species.

Table 6.5. Detoxification Enzyme Activities in Spined Soldier Bug *Podisus maculiventris* and 4 Species of Its Lepidopteran Prey (Adapted from Yu 1987)

	Ratio of Relative Enzyme Activity among Species				
Detoxifying Enzymes	Spined Soldier Bug	Velvetbean Caterpillar	Tobacco Budworm	Fall Armyworm	Corn Earworm
MICROSOMAL OXIDASES					
Aldrin epoxidation	1[1]	3.8	12.0	2.0	5.7
Biphenyl hydroxylation	1	2.5	2.5	1.5	6.7
PCMA *N*-demethylation	1	3.3	1.5	1.3	1.4
PNA *O*-demethylation	1	1.9	3.0	1.6	4.1
Parathion desulfuration	1	0.2	0.02	0.2	0.1
Phorate sulfoxidation	1	29.1	4.6	44.6	79.8
GLUTATHIONE TRANSFERASES					
DCNB conjugation	1	14.0	7.3	15.0	2.7
CDNB conjugation	1	0.05	1.7	0.6	0.8
PNPA conjugation	1	5.4	7.2	5.2	3.7
HYDROLASES					
General esterase	1	2.8	4.8	4.2	4.0
Carboxylesterase	1	1.8	2.5	0.9	2.2
Epoxide hydrolase	1	1.5	2.6	1.9	0.8

[1] Ratio of pmol/min·mg protein of SSB/similar value for lepidopteran prey species.

More extensive studies of the biochemical defense capacities of an entomophagous versus several phytophagous species in relation to both secondary plant compounds and pesticides have been published by Yu (1987, 1988). He examined selectivity relationships between the pentatomid spined soldier bug *Podisus maculiventris* and its lepidopteran prey, the velvetbean caterpillar *Anticarsia gemmatalis*, the tobacco budworm *Heliothis virescens*, the fall armyworm *Spodoptera frugiperda*, and the corn earworm *Heliothis zea* using six assays for MFO, three for glutathion-*S*-transferase, and three for esterase detoxification systems (Table 6.5). The selectivity of 13 compounds from three major insecticide classes to the hemipteran predator and each prey were examined, as well the ability of the phytophages to oxidatively metabolize a number of secondary plant compounds.

The detoxification enzyme survey of Yu (1987) indicated that the predator showed more enzyme activity than its prey in only four of 24 MFO assays, three of 12 glutathione-*S*-transferase assays, and three of 12 assays for esterase (Table 6.5). Bioassay results showed that the predator was generally more susceptible to organophosphate and carbamate insecticides than the prey (Table 6.6). Yu (1988) concluded that these differences probably were due to the lesser detoxification capabilities of the beneficial species. It was not known why the predatory pentatomid was less susceptible to several pyrethroid insecticides as compared to its prey (Table 6.6). Although pyrethroids are known to be metabolized by MFO and esterases in other insects, the spined soldier bug had lower activities of these enzymes than the prey species. Yu (1988) concluded that detoxification appeared to be an unlikely factor for the observed differences in susceptibility to this group of pesticides.

Van de Baan (1988) reported another study of comparative detoxification and

Table 6.6. Susceptibility Patterns to Insecticides in Spined Soldier Bug *Podisus maculiventris* and 3 Species of Its Lepidopteran Prey (Adapted from Yu 1988)

Pesticide Class	Species LD_{50} (μg/g insect)			
	Spined Soldier Bug	Velvetbean Caterpillar	Fall Armyworm	Corn Earworm
ORGANOPHOSPHATES				
Diazinon	27.8	60.4	17.2	288.5
Methyl Parathion	0.4	1.0	7.9	138.3
Trichlorfon	16.6	31.3	87.5	120.7
Chlorpyrifos	5.0	8.5	6.4	60.5
Tetrachlorvinfos	1097.0	38.5	3.4	206.4
CARBAMATES				
Methomyl	6.4	2.3	3.1	47.3
Oxamyl	13.5	299.8	177.0	169.5
Propoxur	24.3	4.8	6.3	202.5
Carbaryl	6.0	7.9	156.8	259.7
PYRETHROIDS				
Permethrin	9.8	0.8	3.4	2.3
Cypermethrin	2.9	1.0	0.9	2.4
Fenvalerate	18.0	1.2	28.8	4.5

selectivity for the facultative predatory hemipteran *Deraeocoris brevis*, which feeds on a diversity of prey including the orchard pest pear psylla (*Psylla pyricola*). In comparing attributes of predator nymphs with susceptible and resistant adults of pear psylla (Table 6.7), near equal MFO and glutathione-*S*-transferase activity levels were observed between the predator and susceptible strains of summer form psylla; however, lower esterase levels were present in the mirid as compared to the pest psyllid. Comparisons of susceptible *D. brevis* against resistant pear psylla showed greater enzyme levels in the prey for MFOs and esterases, but not for glutathione-*S*-transferase (Table 6.7). (See further discussion of this comparison in Chapter 15, which deals with the role of detoxification enzymes in resistant natural enemies.)

In the latter three studies of comparative pest/natural enemy detoxification, the predators examined might be least likely to exhibit substantial differences in detoxification enzyme profiles as compared to their prey. More specialized entomophages such as certain parasitoids would be more likely to exhibit different detoxification enzyme profiles than their hosts. The data of Yu (1987, 1988) and van de Baan (1988) tend to support this hypothesis, although the detoxification enzyme activity in the pentatomid compared to its lepidopteran prey is perhaps lower than expected.

While studies on hemipteran predators and their prey are useful in exploring one extreme of the continuum of entomophagy versus phytophagy in detoxification potential, more data are needed from the other end of the spectrum. Comparisons between highly specialized parasitoids and their hosts are critically needed. The studies with *Oncophanes americanus* and *Pediobius foveolatus* are useful in this regard, but as noted above, they suffer from having to rely on whole body enzyme assays for the parasitoids, which obscured results to some degree. More detailed, organ-specific studies of detoxification

Table 6.7. Detoxification Enzyme Activities in Predatory Hemipteran *Deraeocoris brevis* and Resistant and Susceptible Strains of Its Prey, Pear Psylla (*Psylla pyricola*) (Adapted from van de Baan 1988)

Strain/Species Form/Date	Relative Level of Enzyme Activity[1]		
	Mixed Function Oxidase	Glutathione-S-Transferase	General Esterase
S-*Psylla*[2] Summer/June	1.00	1.00	1.00
S-*Psylla* Winter/March	—	0.38	0.14
R-*Psylla* Summer/June	1.58	1.77	3.76
R-*Psylla* Winter/October	4.08	0.79	1.67
R-*Psylla* Winter/January	2.73	0.50	2.73
R-*Psylla* Winter/March	2.18	0.21	0.27
D. brevis[3] Summer/August	1.57	1.50	0.21

[1] Based on comparisons of enzyme per unit body weight.
[2] Tests run on adult males and females.
[3] Tests run on 3rd- to 5th-instar nymphs.

enzymes between parasitoids and their hosts, with accompanying selectivity studies using secondary plant toxins and pesticides, would provide useful information in this regard. Even more desirable would be the study of long-established evolutionary relationships in a model plant/pest/natural enemy complex of species. Species representing generalist- and specialist-feeding hosts or prey as well as generalist- and specialist-feeding predators and parasitoids should be included to provide a more complete and integrated perspective on natural enemy/pest detoxification systems.

From the combined studies mentioned above, it is clear that a detailed perspective of natural enemy/pest detoxification is beginning to emerge. As with pests (Berenbaum 1985, Berenbaum and Neal 1987), natural enemies show many exceptions to the broad generalizations proposed initially to explain the patterns of susceptibility to secondary plant compounds and pesticides observed between pests and natural enemies (e.g., Croft and Morse 1979, Plapp 1981). In predicting trends in detoxification and susceptibility, one should now consider the different roles of individual detoxification enzyme systems in a carnivorous species, the diversity of secondary plant compound content of the arthropod prey or host's food, the feeding specificity of the natural enemy, and its evolutionary background and degree of specialization as an entomophage. As can be seen from this partial list, there are a variety of factors that may influence detoxification patterns in arthropod natural enemies and their hosts or prey. However, this is not to discount the possibility that there are general patterns of detoxification between these groups that can be exploited to design more selective pesticides (Chapters 9 and 21).

6.4. SUMMARY

Studies of comparative nutrition, endocrinology, excretion, sequestration, penetration, and detoxification between arthropod natural enemies and their phytophagous prey or

Table 6.8. Physiological Differences Between Herbivorous Pests and Entomophagous Natural Enemies That May Allow for Selective Pesticide Design

Feature	Herbivore	Entomophage
Lipid content to carbohydrate ratio	High	Low
Fatty acid composition	Synthesized de novo	Mostly acquired from host or prey
Sterols	Dealkylated from plant sterols	Mostly acquired from host or prey
Respiratory rate	Low	High
Sequestration potential	High	?

hosts suggest numerous possibilities for exploiting physiological, biochemical, or toxicological differences between the two groups to obtain selectivity in future pesticides. Some possibilities are shown in Table 6.8. Unfortunately, few of these factors have been studied extensively or exploited for the design of new pesticides. Based on some general detoxification differences (i.e., detoxification enzyme patterns), it may be possible to achieve broad-spectrum selectivity. Pesticides of this type would be generally toxic to a wide spectrum of herbivorous plant feeding pests, while being less toxic to natural enemies as a general class. In light of the attention being given to IPM systems on a worldwide basis, the development of such selective pesticides is long overdue. Broad-spectrum selectivity should be vigorously sought after and developed through increased study of the basic physiological, biochemical, and toxicological attributes of beneficial natural enemy species as compared to their phytophagous prey or hosts (Chapter 21).

PART THREE

PESTICIDE INFLUENCES

7

SUBLETHAL INFLUENCES

7.1. INTRODUCTION

Historically, biological evaluations of pesticide impact on arthropods have focused primarily on mortality assessments (Chapter 2). However, it is well known that sublethal doses of these compounds can affect the physiology and behavior of both target and nontarget arthropod species (Moriarty 1969, Croft 1977, Haynes 1988). In a few cases, sublethal effects have been quantified, analyzed, and their significance interpreted. However, it is apparent that only the most obvious and gross effects have been studied. Many more subtle sublethal effects have gone unnoticed. This has partly been due to the notion that local pest control (almost to the level of eradication) was a feasible and desirable objective of pesticide use, and thus only mortality was worth studying. Also,

The senior coauthor of this chapter is Russell Messing, Department of Entomology, Oregon State University, Corvallis, Oregon.

there have been conceptual, methodological, and semantic difficulties which have slowed the progress of research in this area.

This chapter examines pesticide sublethal effects on arthropod predators and parasitoids, and the relationship of these effects to other pesticide/natural enemy interactions. An understanding of sublethal pesticide side effects provides a more complete perspective of the ecological impact of pesticide usage (Chapter 8), and may contribute to improved IPM in agricultural pest control systems.

7.2. PROBLEMS IN DEFINITION AND MEASUREMENT

An example of the conceptual difficulty that arises in defining and measuring the sublethal effects of pesticides is illustrated by some of the early literature on the topic. Atallah and Newsom (1966) examined the impact of several chlorinated hydrocarbon insecticides on the reproduction and survivorship of the predaceous coccinellid *Coleomegilla maculata*, and found that higher doses of toxaphene caused higher mortality in a shorter period than lower doses, as indicated by steeper log time/probit mortality lines (Fig. 7.1). They interpreted these data as evidence of sublethal effects of pesticides on the predator's longevity. However, Moriarty (1969) contended that this was not a true sublethal effect, but merely "quicker kill" with larger doses.

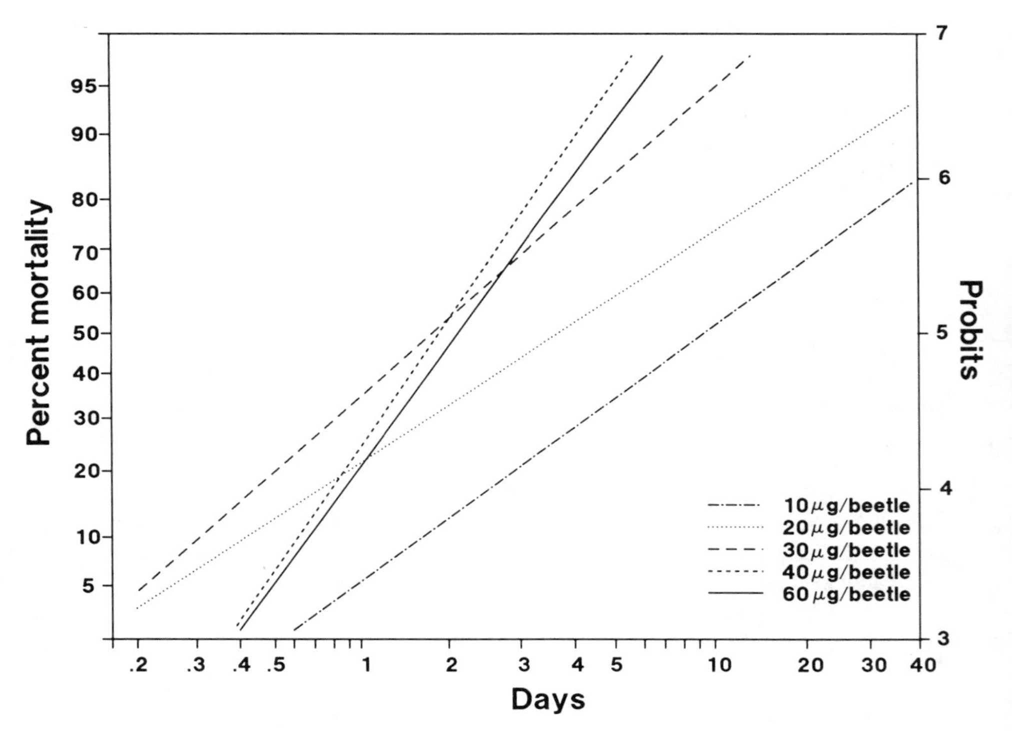

Figure 7.1. Log time-probit mortality lines for toxaphene-treated *Coleomegilla maculata* (nondiapause) in relation to longevity (after Atallah and Newsom 1966).

Tamashiro and Sherman (1955) examined the impact of several insecticides on Oriental fruit flies (*Dacus dorsalis*) and the internal braconid parasitoid *Opius oophilus*, and proposed the term "latent toxicity" to characterize this case in which larvae were exposed to a toxicant but mortality was delayed until the adult stage. However, Moriarty (1969) took issue with this terminology and defined the term more specifically. He preferred to restrict latent toxicity to situations in which mortality over time was polymodal, with the first phase defined as acute and subsequent phases as latent.

For those interested in IPM, the sublethal pesticide effects of greatest concern are those that alter the ability of entomophages to regulate the density of their hosts or prey. This ability may be affected in two ways: 1) changes in the population's intrinsic rate of increase, such as developmental rate, fecundity, longevity, and sex ratio; and/or 2) changes in the population's feeding behavior, such as general mobility, host searching, and oviposition. (Sublethal pesticide doses may also influence enzyme induction and the physiological basis for pesticide resistance; Chapters 6 and 15.)

When viewed in terms of the entomophage's capacity to regulate pest populations, the sometimes confusing semantic distinctions between sublethal, delayed lethal, and latent toxicity are less critical. It should be recognized that there is a continuum of pesticide impact ranging from immediate mortality to a slight reduction in longevity. While we may apply certain terms to intervals along this continuum, the terms are imprecise and overlap to a certain extent. Of particular interest are changes in the feeding or reproduction of natural enemy populations, and whether these are classified under one label or another becomes a moot point.

Another conceptual difficulty in the study of sublethal effects of pesticides is the distinction between actual induced changes in an individual's (or numerous individuals') characteristics, and changes in a population's characteristics due to selective mortality and genetic drift. An individual fly exposed to a sublethal dose of a toxicant may undergo physiological changes leading to increased oviposition (induction). On the other hand, a particular dose may kill 60% of a population of flies, and the survivors may be genetically predisposed for increased oviposition (selection) (Afifi and Knutson 1956). The impact on population dynamics may or may not be the same, but the distinction is important in understanding the mechanisms involved. Moriarty (1969) defined sublethal effects strictly in terms of individuals that survive toxicant contact. However, from an IPM perspective, the selection effect on populations may be just as important. For the purposes of this chapter and volume, we view the broader definition as being preferable and more useful to applied entomologists and those practicing IPM.

Research priorities for sublethal effects of pesticides have followed the typical pattern of focusing first on pests, and only much later on natural enemies and other nontarget species (Chapter 2). The time lag is even more pronounced than in other areas of pesticide/natural enemy research, because for so long the emphasis was on pest eradication, and sublethal effects even on pests were considered relatively unimportant.

There are also numerous methodological difficulties in examining sublethal effects which have slowed research progress in this area. As mentioned, the distinction between individual effects and population effects is sometimes subtle and can only be determined by careful experimentation with stringent controls. Since sublethal effects sometimes occur at pesticide doses which kill 50–90% of the target population (Dumbre and Hower 1976a), sample sizes must be increased to ensure that adequate survivors are available for research. Also, at rates below those that kill a substantial portion of the population it is possible to compare treated and untreated cohorts of individuals in order to separate induction from selection effects, but once the dose approaches the LD_{50}, this becomes impossible.

The time frame required for sublethal effect research is much more lengthy than for simple mortality studies. Synovigenic natural enemies reproduce throughout their life span, and sublethal effects on fecundity may persist for this entire period. This makes even more difficult the already challenging task of maintaining three trophic levels of organisms in a healthy and standardized condition (Chapter 5).

Dose precision requirements are usually far greater for studying sublethal effects than for mortality studies. Only 1/80 the field rate of malathion was sufficient to induce behavioral changes in *Euderophale saliens*, a eulophid parasitoid of whiteflies (Hoy and Dahlsten 1984). Only 1/1000 the LD_{100} of chlordane induced sublethal effects in house crickets (Luckey 1968).

Compared with mortality studies, where the insect is classified simply as "dead" or "alive" (or occasionally "moribund"), sublethal effects on physiology and behavior can be exceedingly subtle and multifaceted. Very few predators, parasitoids, or even pest arthropods have had their behavioral repertoires sufficiently characterized to enable researchers to analyze subtle changes in behavior. Also, pesticides can produce opposing sublethal effects, which may mask one another. For example, a toxicant may increase age-specific fecundity while decreasing longevity (Atallah and Newsom 1966), thus revealing no apparent change in overall realized fecundity. Furthermore, some sublethal effects—especially behavioral ones—are reversible, so that unless observations are made during a specific time interval, the effects may go unnoticed.

The solution to these methodological problems depends largely on the objectives of each individual research project. A population ecologist studying seasonal predator–prey population dynamics may chose to forego the distinction between induction and selection effects, and focus on changes in population density. A toxicologist, on the other hand, may be interested specifically in physiological changes in individual insects caused by sublethal doses.

Life table analysis may be one of the most effective means of teasing apart the subtle, interrelated aspects of changes in population density. Longevity, age-specific (or time-specific) fecundity, sex ratio, and generation time can be examined as they relate to the intrinsic rate of increase. Similarly, predator–prey simulation models (e.g., Gilbert et al. 1976, Waage et al. 1985) may allow researchers to focus on multiple sublethal effects and how they interact to affect population responses.

7.3. EXAMPLES OF WELL-DESIGNED TESTS OF SUBLETHAL EFFECTS

There are a few examples in the literature of well-designed tests for an analysis of sublethal pesticide effects on insects, and more specifically on arthropod natural enemies. For example, basic studies on experimental animals such as *Drosophila* (Armstrong and Bonner 1985) illustrate the type of comprehensive studies that are needed to understand the influences of sublethal effects of pesticides on predators and parasites. For natural enemy species, Jackson and Ford (1973) used a tightly defined behavioral criterion (consumption of prey eggs) to test the response of individual predaceous mites to pesticide-contaminated prey. A simultaneous presentation of uncontaminated prey served as an effective control. Bogenschutz (1979), following guidelines developed by the IOBC/WPRS working group (see Chapter 5), used "reduction in parasitization capacity" over a specified 9-day period as the criterion for testing sublethal effects of 20 pesticides on the generalist ichneumonid parasitoid *Coccygomimus turionellae*. By sidestepping detailed consideration of individual versus population effects and concentrating on a specific subsample of the total life span (in effect a form of stratified sampling), he demonstrated statistically significant sublethal side effects in rigidly controlled laboratory trials, which included relatively realistic field conditions.

A well-designed methodology for measuring the sublethal effects of pesticides on the field behavior of natural enemies was reported by Brust et al. (1986). They used tethered lepidopteran pest larvae to assess the impact of phorate on carabid predation in corn agroecosystems. In observations made over a 24-hr period, they observed a close relationship between the absolute densities of carabids in the system and the frequency of attacks on pest larvae. Although specific sublethal effects were not evaluated in their experiments, they reported on the timing of predatory activities. It appears that these methods may allow for more precise assessment of pesticide impact on carabid locomotion than has commonly been reported in the literature (e.g., Coaker 1966, Dempster 1968, Edwards et al. 1970, Chiverton 1984; see discussion in Section 7.4.2.1).

Because of the impact of pesticides on biological control organisms in cropping ecosystems, many recent workers have emphasized the need to develop standardized tests for measuring sublethal side effects (Croft 1977; Chapter 5). Franz (1974), in reviewing the development of standardized test methodology, stated that "the reduction of the performance [of natural enemies] has proven to be more sensitive and better suitable than the measuring of mortality only." This conclusion further demonstrates the importance of focusing on the total impact of pesticides on natural enemies (including sublethal effects), as opposed to more cursory evaluation of mortality

7.4. SUBLETHAL EFFECTS REPORTED IN THE LITERATURE

A variety of responses to sublethal doses of pesticides have been observed in various natural enemy species. Most have been made incidentally in connection with mortality studies, although in recent years more research concentrating specifically on sublethal effects has been conducted (e.g., Perera 1982, Hoy and Dahlsten 1984). The incidence of sublethal-effects tests on natural enemies in the literature (taken from the SELCTV database; Chapter 2) has increased dramatically since the early 1960s; however, the relative increase in these types of assessments is much less significant (see Fig. 2.8). As of the mid-1980's, 13% of pesticide side-effect tests on natural enemies included information on sublethal effects (Theiling 1987).

Specific examples of documented cases of sublethal effects on arthropod natural enemy populations are listed in Table 7.1. Cases of both induction and selection are included, as many reports are not sufficiently detailed to separate the two effects. Examples are cited as life table, behavioral, or miscellaneous aspects. It should be noted that each category is strongly associated with the others, and any given sublethal effect is likely to be mediated by elements of all three categories. A decrease in fecundity usually has a physiological or biochemical basis, and may affect feeding behavior (functional response) as well. As mentioned, the relative contribution of induction and selection to measured sublethal responses is not always clear in the published record. However, as a partial characterization, each reference is classified as a mortality study if 30% or more of treated individuals died, thus implicating selection as a possible factor in the outcome. Studies with < 30% mortality were classified as sublethal, indicating a probable induction effect.

Table 7.1. Examples of Sublethal Side Effects of Pesticides on Arthropod Natural Enemies

Class of Effect	Natural Enemy Species	Family	Pesticide	Effect	Method[1]	References
			I. Life Table Parameters			
A. FECUNDITY						
	Bracon hebetor	Braconidae	Heptachlor, carbaryl	Decreased	m	Grosch & Valcovic 1967, Grosch 1970, 1975
	Bracon hebetor	Braconidae	Permethrin	Decreased	s	Press et al. 1981
	Bracon mellitor	Braconidae	Azinphosmethyl, chlordimeform	Decreased	s	O'Brien et al. 1985
	Biosteres longicaudatus	Braconidae	Diflubenzuron	Decreased	m	Lawrence 1981
	Microctonus aethiopoides	Braconidae	Carbofuran, methidathion, parathion, methoxychlor	Decreased	m	Dumbre & Hower 1976a
	Coccinella septempunctata	Coccinellidae	Hydroprene	Increased vitellogenin	s	Gong et al. 1982
	Coccinella septempunctata	Coccinellidae	Ethiofencarb, oxydemeton-methyl, heptenophos	Decreased	m	Grapel 1982
	Coccinella septempunctata	Coccinellidae	Hydroprene	Increased	s	Chou et al. 1981
	Coleomegilla maculata	Coccinellidae	DDT	Increased	m	Attallah & Newsom 1966
	Coleomegilla maculata	Coccinellidae	Toxaphene	Decreased	m	Attallah & Newsom 1966
	Stethorus sp.	Coccinellidae	Dimethoate, demeton-methyl, phosalone	Decreased	m	Pavlova 1975

Chrysopa californica	Chrysopidae	DDT			Fleschner & Scriven 1957
Chrysopa carnea	Chrysopidae	Ethiofencarb, oxydemeton-methyl, heptenophos	Decreased	m	Grapel 1982
Chrysopa carnea	Chrysopidae	Fenvalerate, permethrin	Decreased	s	Grafton-Cardwell & Hoy 1985
Amblyseius brazilii	Phytoseiidae	Methoprene	Decreased	s	El-Banhawy 1980
Amblyseius deleoni	Phytoseiidae	Benomyl, thiophanate	Decreased	?	Kashio & Tanaka 1979
Amblyseius fallacis	Phytoseiidae	Benomyl	Decreased	s	Nakashima & Croft 1974
Amblyseius fallacis	Phytoseiidae	Azinphosmethyl, phosmet, captan, endosulfan, dikar	Decreased	s	Hislop et al. 1978
Amblyseius swirskii	Phytoseiidae	Chlorobenzilate	Decreased	s	Swirski et al. 1967
Phytoseius persimilis	Phytoseiidae	Maneb, benomyl	Decreased	s	Babikir 1978
Phytoseius persimilis	Phytoseiidae	Triforine	Increased	s	Babikir 1978
Metaseiulus occidentalis	Phytoseiidae	Avermectin	Decreased	s	Grafton-Cardwell & Hoy 1983
Telenomus sp.	Scelionidae	Trichlorfon	Decreased	m	Kamenkova 1971
Syrphus sp.	Syrphidae	DDT	Decreased	s	Way 1949
Syrphus corollae	Syrphidae	Juvenile hormones	Decreased	s	Ruzicka et al. 1978
Trichogramma evanescens	Trichogrammatidae	Metasystox, metox, DDT	Decreased	s	Plewka et al. 1975
Aphidoletes aphidimyza	Cecidomyiidae	Demeton-*S*-methyl, bromophos, triforine	Decrease in F_1	m	Sell 1984b

Table 7.1 (*Contd.*)

Class of Effect	Natural Enemy Species	Family	Pesticide	Effect	Method[1]	References
	Encarsia formosa	Aphelinidae	Sulfur, mancozeb, propineb, pyrazophos	Decreased	s	Elenkov et al. 1984
	Chrysoperla carnea	Chrysopidae	Triprene	Increased	s	Maccolini 1985
B. EGG HATCH						
	Bracon hebetor	Braconidae	Carbaryl	Decreased	s	Grosch & Hoffman 1973
	Biosteres longicaudatus	Braconidae	Hydroprene	Increased	s	Lawrence et al. 1978
	Menochilus sexmaculatus	Coccinellidae	Malathion	Decreased	s	Parker et al. 1976
	Coccinella septempunctata	Coccinellidae	Tetradifon	Decreased	m	Moghaddam 1978
	Hippodamia convergens	Coccinellidae	Diflubenzuron	Decreased	s	Ables et al. 1977
	Hippodamia convergens	Coccinellidae	Diflubenzuron	Decreased	s	Keever et al. 1977
	Geocoris pallens	Lygaeidae	Glyphosate, methomyl	Increased	s	Yokoyama & Prichard 1984
	Geocoris punctipes	Lygaeidae	Acifluorfen, benzeton	Increased	s	Farlow & Pitre 1983
	Phytoseiulus sp.	Phytoseiidae	Benomyl	Decreased	s	Bushchik & Lazurina 1982
	Syrphus corollae	Syrphidae	JH analog	Decreased	s	Ruzicka et al. 1978
	Amblyseius gossipi	Phytoseiidae	Cypermethrin, flucythrin, pyridaphenthion	Decreased	m	Abou-awad & El-Banhawy 1985

C. DEVELOPMENTAL RATE

Apanteles melanoscelus	Braconidae	JH analog	Decreased	s	Granett et al. 1975a
Apanteles mlanoscelus	Braconidae	JH analog	Decreased	s	Boone & Hammond unpubl.
Biosteres longicaudatus	Braconidae	Hydroprene	Increased	s	Lawrence et al. 1978
Cardiochiles nigriceps	Braconidae	JH analog	Decreased	s	Vinson 1974
Coccinella transversoguttata	Coccinellidae	2, 4-D	Decreased	m	Adams 1960
Coccinella septempunctata	Coccinellidae	Hydroprene	Increased	s	Chou et al. 1981
Menochilus sexmaculatus	Coccinellidae	Malathion	Increased or decreased (dose dependent)	s	Parker et al. 1976
Chrysopa rufilabris	Chrysopidae	Ethion, carbophenothion, carbaryl, azinphosmethyl	Decreased	m	Lawrence 1974
Chrysopa carnea	Chrysopidae	Parathion	Decreased	s	Bartlett 1964
Chrysopa carnea	Chrysopidae	Toxaphene	Decreased	m	Bartlett 1964
Ooencyrtus kuwanai	Encyrtidae	JH analog	Increased	m	Granett et al. 1975a
Campoletis sonorensis	Ichneumonidae	JH analog	Decreased	s	Vinson 1974
Hyposoter exiguae	Ichneumonidae	*Bacillus thuringiensis*	Increased	m	Thoms & Watson 1986
Amblyseius brazilii	Phytoseiidae	methoprene	Decreased	m	El-Banhawy 1980
Microplitis rufiventris	Braconidae	Diflubenzuron	Decreased	m	Heynen 1985
Chrysoperla carnea	Chrysopidae	Triprene	Increased	m	Maccolini 1985
Amblyseius fallacis	Phytoseiidae	Benomyl	Decreased	s	Butcher & Penman 1983

Table 7.1 (*Contd.*)

Class of Effect	Natural Enemy Species	Family	Pesticide	Effect	Method[1]	References
	Metaseiulus occidentalis	Phytoseiidae	Benomyl, thiram, demeton-*S*-methyl, azinphosmethyl	Decreased	s	Butcher & Penman 1983
D. LONGEVITY						
	Bracon hebetor	Braconidae	Carbaryl	Decreased	m	Grosch & Hoffman 1973
	Microctonus aethiopoides	Braconidae	Carbofuran, Methoxychlor	Decreased	m	Dumbre & Hower 1976a
	Opius oophilus	Braconidae	Aldrin, dieldrin, endrin, chlordane	Decreased	m	Tamashiro & Sherman 1955
	Notophilus biguttatus	Carabidae	2,4,5-T	Decreased	s	Eijsackers 1978
	Chrysopa carnea	Chrysopidae	Permethrin	Decreased	s	Shour & Crowder 1980
	Coleomegilla maculata	Coccinellidae	Endrin, toxaphene	Decreased	s	Atallah & Newsom 1966
	Menochilus sexmaculatus	Coccinellidae	Malathion	Increased or decreased (dose dependent)	s	Parker et al. 1976
	Elasmus zehntneri	Elasmidae	Phorate	Decreased	?	Singh et al. 1979
	Geocoris sp.	Lygaeidae	Methomyl	Decreased	m	Walker & Turnipseed 1976
	Amblyseius brazilii	Phytoseiidae	Methoprene	Decreased	s	El-Banhawy 1980
	Phytoseius persimilis	Phytoseiidae	Benomyl	Decreased	s	Parr & Binns 1971
	Ischiodon scutellaris	Syrphidae	Malathion	Decreased	m	Teh et al. 1983
	Trichogramma evanescens	Trichogrammatidae	Metasystox, metox, DDT	Decreased	s	Plewka et al. 1975
	Geocoris punctipes	Lygaeidae	*Bacillus thuringiensis* (beta endotoxin)	Decreased	s	Herbert & Harper 1986

E. SEX RATIO					
Cardiochiles nigriceps	Braconidae	JH analog	More females	m	Vinson 1974
Bracon mellitor	Braconidae	Azinphosmethyl, chlordimeform	More females	s	O'Brien et al. 1985
Geocoris pallens	Lygaeidae	Methomyl, methidathion	More females	m	Yokoyama & Pritchard 1984
Geocoris pallens	Lygaeidae	Propargite	More males	m	Yokoyama & Pritchard 1984
Trissoleus grandis	Scelionidae	Chlorophos	Fewer females in F_1	s	Novozhilov et al. 1973
II. Behavioral effects					
A. LOCOMOTION					
Bembidion lampros	Carabidae	Thionazin	Increased	s	Critchley 1972b
Bembidion lampros	Carabidae	DDT	Increased	s	Dempster 1968
Bembidion lampros	Carabidae	Dieldrin	Increased	s	Coaker 1966
Bembidion lampros	Carabidae	Chlorfenvinphos	Increased	s	Edwards et al. 1970
Notophilus biguttatus	Carabidae	2,4,5-T	Decreased	s	Eijsackers 1978
Poecilus cupreus	Carabidae	Diazinon, carbofuran	Increased	s	Moosbeckhofer 1983
Pterostichus melanarius, Bembidion obtusum, Trechus quadristriatus, Philonthus spp., *Tachyporus hypnorum*	Carabidae	Pirimicarb, demeton-*S*-methyl	Increased	s	Feeney 1983
Forficula auricularia	Dermaptera	Cypermethrin, deltamethrin	Decreased	s	Ffrench-Constant & Vickerman 1985
Phytoseius persimilis	Phytoseiidae	Maneb	Decreased	s	Babikir 1978

Table 7.1 (*Contd.*)

Class of Effect	Natural Enemy Species	Family	Pesticide	Effect	Method[1]	References
	Trichogramma cacoeciae	Trichogrammatidae	Benomyl	Decreased	s	Franz & Fabrietius 1971
	Trichogramma cacoeciae	Trichogrammatidae	Triazophos, endosulfan	Decreased	m	Shires et al. 1984
	Carabids	Carabidae	Fenvalerate	Increased	s	Poehling et al. 1985b
	Formicids	Formicidae	Bendiocarb	Increased	s	Cockfield & Potter 1983
B. SEARCHING						
	Encarsia formosa	Aphelinidae	Tetradifon, dichlofluanid, benomyl	Repellent	s	Irving & Wyatt 1973
	Encarsia formosa	Aphelinidae	Pirimicarb	Increased	s	Irving & Wyatt 1973
	Phytoseius macropiles	Phytoseiidae	Carbaryl	Repellent	s	Dabrowski 1969a
	Amblyseius fallacis	Phytoseiidae	Captan, phosmet, endosulfan, azinphosmethyl	Repellent	s	Hislop et al. 1981
	Amblyseius fallacis	Phytoseiidae	Fenvalerate	Repellent	s	Penman et al. 1981
	Typhlodromus occidentalis	Phytoseiidae	Fenvalerate	Repellent	s	Penman et al. 1981
	Typhlodromus occidentalis	Phytoseiidae	Fenvalerate	Repellent	s	Riedl & Hoying 1983
	Typhlodromus pyri	Phytoseiidae	Azinphosmethyl	Repellent	s	Walker & Penman 1978
	Typhlodromus pyri	Phytoseiidae	Azinphosmethyl	Repellent	s	Collyer 1976
	Episyrphus balteatus	Syrphidae	Phenmedipham	Repellent	s	Tanke & Franz 1977
	Trichogramma cacoeciae	Trichogrammatidae	Phenmedipham	Repellent	s	Tanke & Franz 1977
	Trichogramma pretiosum	Trichogrammatidae	Endosulfan, permethrin	Repellent	m	Jacobs et al. 1984
	Typhlodromus Occidentalis	Phytoseiidae	Diquat, paraquat	Repellent	m	Boller et al. 1984

Bracon mellitor	Braconidae	Chlordimeform	Increased (honey water)	s	O'Brien et al. 1985
Harpalus rufipes	Carabidae	DDT	Decreased	s	Dempster 1968
Chrysopa carnea	Chrysopidae	Pirimicarb	Decreased	s	Grapel 1982
Coccinella septempunctata	Coccinellidae	Pirimicarb	Decreased	s	Grapel 1982
Menochilus sexmaculatus	Coccinellidae	Triphenyltin acetate	Decreased	s	Abdul Kareem et al. 1977
Lycosa pseudoannulata	Lycosidae	BPMC, propoxur	Decreased	s	Chu et al. 1976c
Lycosa pseudoannulata	Lycosidae	*Gamma*-BHC	Repellent	s	Kiritani & Kawahara 1973
Oedothorax insecticeps	Lycosidae	Dicrotophos, ofunack, fenitrothion	Increased	s	Chu et al. 1977
Amblyseius fallacis	Phytoseiidae	Fenvalerate	Repellent	s	Penman et al. 1981
Phytoseius persimilis	Phytoseiidae	Captan	Repellent	s	Jackson & Ford 1973
Agistemus exsertus	Stigmaeidae	Dicofol, tetradifon	Decreased	s	Hassen et al. 1970
Xanthogramma aegyptium	Syrphidae	Phosphamidon	Repellent	m	Azab et al. 1971
Several carabids	Carabidae	Phorate	Decreased	m	Brust et al. 1986
Aphidoletes aphidimyza	Cecidomyiidae	Demeton-*S*-methyl	Decreased	m	Sell 1984a

Table 7.1 (*Contd.*)

Class of Effect	Natural Enemy Species	Family	Pesticide	Effect	Method[1]	References
D. OVIPOSITION						
	Encarsia formosa	Aphelinidae	Pynosect 30	Reduced	s	Perera 1982
	Encarsia formosa	Aphelinidae	Pirimicarb, *gamma*-BHC, tetradifon, dichlofluanid	Reduced	s	Irving & Wyatt 1973
	Aphytis holoxanthus	Aphelinidae	Oil	Reduced	s	Rosen 1967
	Coccygomimus turionellae	Ichneumonidae	20 compounds	Reduced	s	Bogenschutz 1979
	Trichogramma pretiosum	Trichogrammatidae	Savol (oil)	Reduced	s	Ables et al. 1980
	Trichogramma pretiosum	Trichogrammatidae	Methomyl,	Reduced	s	Bull & House 1983
	Trichogramma pretiosum	Trichogrammatidae	permethrin, methyl parathion			
	Trichogramma brasiliensis	Trichogrammatidae	Cypermethrin, fenvalerate	Repellent	s	Singh & Varma 1986
	Amblyseius fallacis, Metaseiulus occidentalis	Phytoseiidae	Benomyl, dichlofluanid, thiram, azinphosmethyl	Reduced	s	Butcher & Penman 1983
	Typhlodromus sp.	Phytoseiidae	Pyrethroids	Reduced	s	Sekita 1986
			III. Miscellaneous			
	Biosteres longicaudatus	Braconidae	Diflubenzuron	Deformed ovipositor	m	Lawrence 1981

Microctonus aethiopoides	Braconidae	Hydroprene	Deformed abdomens	s	Ascerno et al. 1983
Bracon hebetor	Braconidae	JH analogs	Potentiation of radiation-induced sterility	s	Grosch 1975
Bracon hebetor	Braconidae	Carbamates	Protection from radiation sterility	s	Grosch 1975
Harpalus aeneus	Carabidae	Thionazin	Ejection of formic acid	s	Critchley 1972b
Chrysopa carnea	Chrysopidae	JH analog	Diapause termination	s	Ruzicka et al. 1978
Semiadalia undecimnotata	Coccinellidae	JH analog	Diapause termination	s	Ruzicka et al. 1978
Nasonia vitripennis	Pteromalidae	Ecdysterone	Diapause termination	s	de Loof et al. 1979a
Ischiodon scutellaris	Syrphidae	Malathion	Weight loss	s	Teh et al. 1983
Gonia cinerascens	Tachinidae	Methoprene	Weight increase	s	Verenini 1984
Chrysoperla carnea	Chrysopidae	Triprene	Weight increase	s	Maccolini 1985

[1] Studies are classified as mortality studies (m) if $\geq 30\%$ mortality occurred when natural enemies were exposed to the pesticides and as sublethal-effect studies (s) if $< 30\%$ mortality occurred after pesticide exposure.

7.4.1. Life Table Parameters

As mentioned above, life table analysis may be the best way to examine pesticide sublethal effects on the population responses of natural enemies. Pesticides can alter every parameter of population growth. While these parameters have been measured individually, seldom have integrated life tables, which include all of the component elements, been constructed. Part of the reason life tables have been so rarely used is because results from these techniques have not been deemed sufficiently useful for the effort needed to obtain them. Simple mortality analyses are usually sufficient to measure the grosser aspects of pesticide impact on natural enemies. However, we contend that the more subtle aspects of sublethal effects, in some cases, can *only* be detected by such integrated, aggregate estimation techniques. Below are some examples of sublethal effects of pesticides as reflected by the subcomponents of life table analysis.

7.4.1.1. ***Fecundity.*** Sublethal doses of pesticides usually affect the fecundity or fertility of adult female natural enemies either by directly reducing oviposition or by decreasing percent egg hatch. Similar influences on sperm production probably occur, but have seldom been studied. Depending on the methodology used, it is often difficult to distinguish between sublethal effects on egg hatch via affected adults and direct ovicidal activity on eggs. Both effects are probably common in the field.

Herbicides, fungicides, and all major classes of insecticides have sublethal effects on the fecundity of natural enemies. In fact, fecundity seems to be one of the most sensitive biological characteristics to sublethal effects, and is the most important in terms of population dynamics. Although in most cases fecundity is decreased, several compounds have caused a distinct increase in both the number of eggs laid and the subsequent percent hatch (Table 7.1).

Atallah and Newsom (1966) observed that DDT increased oviposition rates of *Colemegilla maculata* by 67%; however, survival to adulthood of the larvae hatching from these eggs was reduced by 25%. Similarly, Plewka et al. (1975) found that *Trichogramma evanescens* treated with demeton-*S*-methyl showed a temporary increase in oviposition, but longevity was subsequently reduced, leading to an overall decrease in fecundity. The predaceous mite *Amblyseius swirskii*, when exposed to residues of 0.12% phenkapton on lemon leaves for 24 hr, showed an initial increase in oviposition for 3 days followed by a sharp decrease for the remainder of the trial (Swirski et al. 1967). Average fecundity for the entire trial was lower for the treated mites than for the controls, but the authors did not measure the effects of the change in age-specific fecundity on the intrinsic rate of increase. An interesting report of a sublethal effect was that of reduced fecundity in F_1 female *Aphidoletes aphidimyza* when exposed to sublethal concentrations of several pesticides (Sell 1984b). This report raises questions regarding the persistence of such effects in populations and the necessary duration of the tests run to evaluate these effects (Chapter 5).

These cases illustrate the multifaceted (and frequently opposing) sublethal effects that can be manifested in a single population. They also point out the importance of evaluation over a suitable time period, and the value of more integrated life table analysis to measure overall changes in reproductive performance.

Grosch (1975) showed that adult female *Bracon hebetor* which were topically treated with carbaryl showed a significant decrease in subsequent egg hatch only for some of the higher sublethal doses; however, the number of eggs laid was significantly lower for all sublethal doses. Parker et al. (1976) showed a similar decrease in the number of eggs laid over the life span of the coccinellid *Menochilus sexmaculatus* when topically treated with sublethal doses of malathion. In this case, the reduction in fecundity was correlated with a reduction in longevity and a reduction in the proportion of the life span during which eggs were laid.

Both Farlow and Pitre (1983) and Yokoyama and Pritchard (1984) showed significant *increases* in the percent egg hatch of predaceous big-eyed bugs (*Geocoris* sp.) after exposure to the herbicides glyphosate, acifluorfen, and benzeton. These results are interesting both in their implications for the manipulation of natural enemy populations, and also in pointing out the need to examine the effects of *all* chemicals applied to a crop, not only those with obvious insecticidal activity. Recently, more attention has been paid to both lethal and sublethal effects of herbicides, fungicides, defoliants, plant growth regulators, and other chemicals used in crop management (see Chapter 2).

7.4.1.2. ***Longevity.*** As mentioned previously, reductions in longevity may be conceived of as sublethal effects, but no clear lines distinguish direct mortality from latent mortality from slightly reduced longevity. From a more practical perspective, it is the resulting amount of feeding or reproduction that occurs between treatment and death that is important. In the study by Parker et al. (1976), the impact of reduced longevity on population increase was clearly shown in the quantitative evaluation of decreased oviposition. Other reports, however, have indicated reduced longevity as a sublethal effect without giving information on the impact on fecundity (Kot and Plewka 1970, Singh et al. 1979).

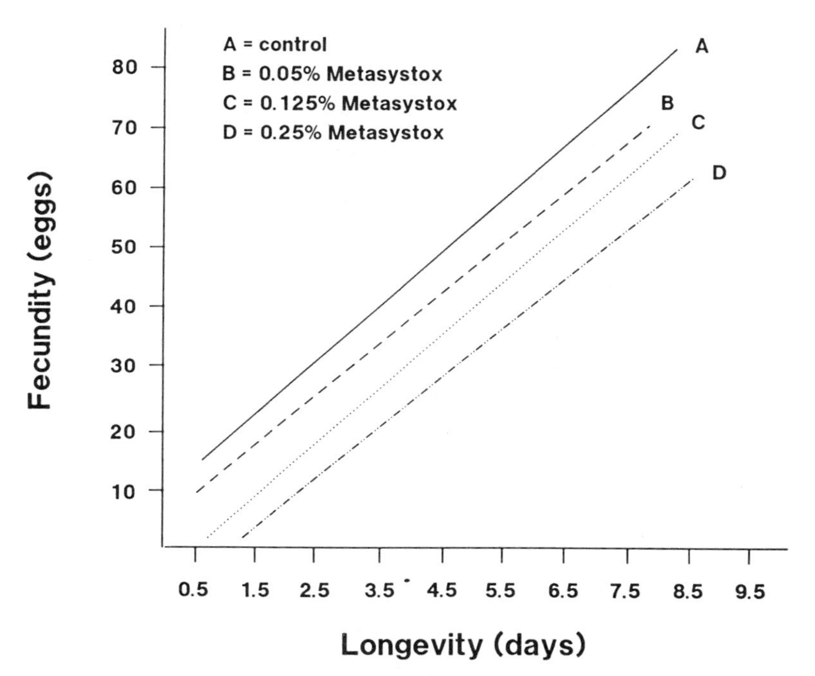

Figure 7.2. Relationship between fecundity and longevity of *Trichogramma evanescens* treated with demeton-*S*-methyl (after Plewka et al. 1975).

Plewka et al. (1975) examined the effects of sublethal doses of demeton-*S*-methyl, DDT, and metox on the parasitoid *Trichogramma evanescens*, developing within eggs of *Sitotroga cerealella*. They observed that all three toxicants reduced both the average longevity and the average fecundity of treated populations. (Data for three concentrations of demeton-*S*-methyl for all age classes of females are summarized in Fig. 7.2.) By using the regression of longevity on fecundity to estimate the decrease in fecundity compared with controls for various age classes of adults, the authors were able to separate the direct decrease in fecundity (induction) from the overall decrease due to reduced longevity. For example, at 0.25% metasystox they found a total decrease in fecundity for all age classes of 72.7%, while the average decrease for particular age groups was only 36% (i.e., there were fewer individuals in older age classes).

In cases of reduced longevity in which death occurs before the adult stage is reached, the effect on fecundity is obviously zero (except in rare cases of paedogenesis). In terms of life table parameters this effect is equivalent to instantaneous death. However, predators and parasitoids may still feed on their hosts or prey during the larval stage, and thus contribute to a functional response, if not a numerical response. A similar situation may occur when pesticides induce sterility in an arthropod. A sterilized insect may be functionally equivalent to a dead one in terms of intrinsic rate of increase, but it may continue to feed on its hosts. Sterilized insects may also influence population dynamics by mating with, and using up, the reproductive capacity of normal individuals, as in the sterile male technique used for pest control. Grosch and Hoffman (1973) showed that carbamate pesticides interfere with the mitotic apparatus of the braconid *Bracon hebetor*, resulting in poor egg hatch accompanied by only a slight decrease in egg production. Grosch (1972) suggested that such sterilized females might trap the sperm from males, which would have even greater effects on the population's reproductive performance.

7.4.1.3. ***Developmental Rate.*** The developmental rate (or generation time, time to first offspring) is a critical life table parameter which may have a greater impact on the intrinsic rate of increase than total fecundity (Mackauer 1986). Adams (1960) was one of the first to note developmental rate differences in a natural enemy species exposed to sublethal doses of a pesticide. He observed that coccinellids (*Coccinella transversoguttata*) treated with the herbicide 2,4-D showed an increase in the mean time to pupation for all age groups treated, except for 1-day-old larvae. Lawrence et al. (1973) also noted a slower developmental rate for third instar larvae and pupae of *Chrysoperla carnea* when exposed to topical treatments of azinphosmethyl, ethion, carbophenothion, and carbaryl, while chlorobenzilate had no effect on development.

Parker et al. (1976) found that sublethal doses of topically applied malathion either increased or decreased preoviposition periods of *M. sexmaculatus*, depending upon the precise dose. One-tenth of the LD_{50} resulted in a 14% decrease in the duration of the preoviposition period, while eight-tenths of the LD_{50} caused an increase of 34%.

Although a faster developmental rate often leads to a higher intrinsic rate of increase (which is often used as an index of fitness; Lewontin 1965), it must be kept in mind that predators and especially parasitoids represent a special case in which phenological synchrony with the host or prey is often of prime importance. An increase in the developmental rate could conceivably work to a parasitoid's disadvantage if it disrupts phenological synchrony with a critical window of susceptibility in the host.

7.4.1.4. ***Sex Ratio.*** A pesticide applied to a given natural enemy species may affect males and females differentially due to sex-related differences in physiology, phenology, or behavior. Disproportionate mortality of either males or females may skew the subsequent sex ratio so that the population's intrinsic rate of increase is altered. Yokoyama and Pritchard (1984) found that laboratory populations of the predator *Geocoris pallens*, exposed to methomyl or methidathion, suffered higher male than female mortality. However, the same population exposed to propargite showed the reverse effect, with females being more susceptible. While, strictly speaking, this is a lethal rather than sublethal effect, the resultant aberration in sex ratios may affect mate finding and other behaviors of unaffected individuals.

Actual sublethal effects may also alter the sex ratio of natural enemy populations, as demonstrated by Novozhilov et al. (1973). Scelionid parasites exposed to trichlorfon while developing in the eggs of their host produced a lower proportion of females in the next generation than untreated controls. O'Brien et al. (1985) demonstrated a similar but opposite effect; offspring of adult *Bracon mellitor*, treated with the LC_5 of azinphosmethyl and chlordimeform, showed a higher female: male ratio than offspring of untreated controls. This effect was more pronounced for older treated females, suggesting that untreated adults were showing normal sperm depletion with age (thus producing more males). However, treated adults which laid fewer eggs during the beginning days of exposure had more sperm available later in the test period.

7.4.2. Behavioral Effects

Numerous authors have documented changes in natural enemy behavior resulting from exposure to sublethal doses of pesticides. Studies have ranged from measurements of general mobility (Critchley 1972b) to specific alteration of prey acceptance (Jackson and Ford 1973) to even more integrated behavioral studies (Kuhner et al. 1985).

Early reports of behavioral changes were based primarily on anecdotal observations. For example, Barker (1968) noted that the predaceous mite *Blattisocius keegani* suffered an initial knockdown from malathion sprays, but that some mites recovered and exhibited distorted searching behavior before they died, possibly from lack of food.

Vinson (1975) outlined the sequence of behavioral events necessary for a parasitoid to successfully locate, attack, and complete development in its host. With the exception of physiological host regulation, this sequence is very similar for predaceous insects. Sublethal doses of pesticides can interfere with the completion of any one of these components, and thus severely reduce or eliminate the pest-regulating ability of natural enemies.

The sublethal effects on behavior listed in Table 7.1 are grouped by effects on general mobility, searching behavior, and feeding and oviposition. In many cases the pesticide acts as a repellent; when the treated substrate is repellent, the result is classified as an interference with searching behavior—particularly host-habitat location and host location. In some cases, repellency is the result of contact with treated hosts or prey. These cases are classified as a disruption of actual oviposition or host acceptance. A natural enemy may also be repelled by contacting treated surfaces, leading to reduced consumption.

7.4.2.1. ***Mobility.*** Both increases and decreases in mobility have been documented following exposure to sublethal doses of pesticides. Increased mobility can influence the rate of pesticide uptake and therefore pesticide-induced mortality (Chapter 3); it can also be beneficial in increasing predation and parasitization rates (see below). Relatively long periods of immobility may lead to increased susceptibility of an individual to, for example, predation, dessication, and other mortality factors (Ffrench-Constant and Vickerman 1985).

Eijsackers (1978) examined the effects of the herbicide 2, 4, 5-T on the predaceous soil-inhabiting carabid beetle *Notophilus biguttatus*. He noted that the rate of movement of adults was inhibited at a much lower concentration of the herbicide than the LD_{50}. Critchley (1972b) also examined the effects of pesticides on soil-inhabiting carabids, but he observed an increase in overall activity following exposure. Other investigators have also noted an increase in carabid activity following pesticide treatments (Coaker 1966, Dempster 1968, Edwards et al. 1970). Recently, Edwards et al. (1984a) suggested that sublethal pesticide influences on the searching and predation behavior of carabids may be associated with more successful pest control programs on some cereal and vegetable crops. They noted that "it may well be that those insecticides that have been most successful in controlling pests in the field are those that not only will kill pests but also increase the activity of their predators."

Sublethal effects of pesticides on carabid activity, predation, uptake of pesticides, and so forth have come under greater scrutiny by Chiverton (1984). He noted that pitfall trap catches of *Pterostichus melanarius* increased significantly in plots treated with fenitrothion and fenvalerate for several weeks after treatment compared to control plots. Significantly more female than male carabids were caught in traps in treated plots. Corresponding decreases were observed in prey population levels following treatment, presumably due primarily to chemical effects. Significantly fewer female carabids from treated plots had guts full of solid food when compared with those captured from untreated control areas. While some research has shown direct stimulation of locomotion due to sublethal influences of pesticides in the laboratory (e.g., Critchley 1972b, Moosbeckhofer 1983), Chiverton (1984) concluded that the major reason for increased catch of ground beetles following application of pesticides in his study was simply an increased searching rate due to hunger.

Thornhill and Edwards (1985) explained the initial decreases and subsequent increases in pitfall catches of staphylinids and carabids as first due to mortality effects of *gamma*-HCH and then to sublethally induced increases in locomotion in immigrant forms as residues declined. Their data concurred with the studies of Goos et al. (1974), who noted that when an insecticide initially caused significant decreases in populations of carabid and staphylinid beetles, trap catches later approached those of the untreated plots and sometimes exceeded them. These studies demonstrate that several factors are involved in the increased activity of these epigael natural enemies following exposure to insecticide treatments. More detailed studies are needed to understand these effects and their combined influence on biological control.

Many of the reported changes in predaceous or parasitic behavior concern a general repellency to the pesticide applied (Jackson and Ford 1973, Irving and Wyatt 1973, Walker and Penman 1978). Sublethal effects on overall mobility are usually expressed in broad terms, with no analysis of the subsequent impact on the numerical or functional response of the natural enemy in relation to its host or prey. One report that does attempt to integrate sublethal behavioral effects in terms of repellency and functional response attributes is that of Perera (1982). He found that untreated controls of *E. formosa* exhibited a typical type II functional response curve (Holling 1959) when parasitizing greenhouse whiteflies on potted *Phaseolus* (Fig. 7.3). However, when synergized pyrethrin (Pynosect 30 ULV) was applied as a uniform cover spray on the bean leaves, the functional response changed to a sigmoid pattern, with host searching beginning at a relatively high 80 larvae per leaf and reaching a maximum parasitism rate of only 1.5/female/day (compared with

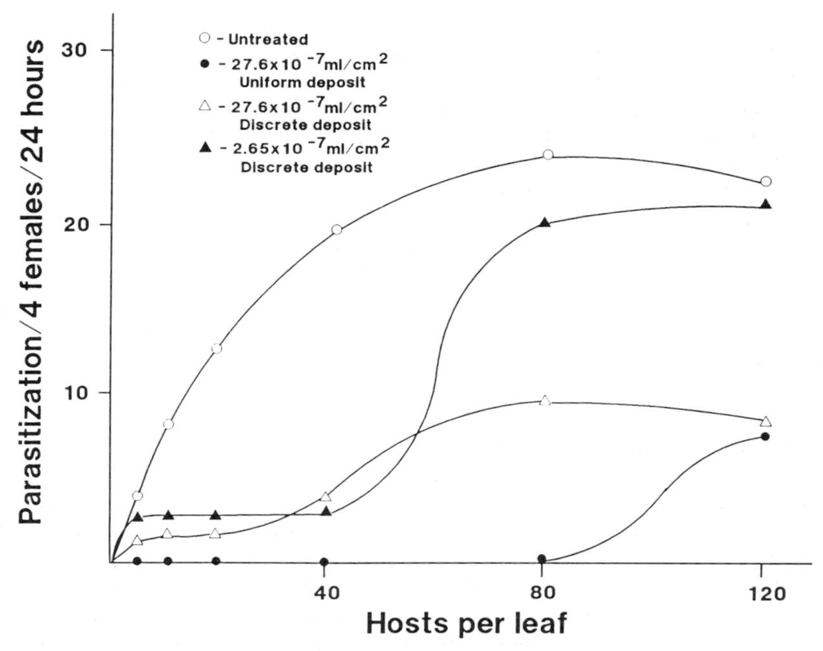

Figure 7.3. Effect of sublethal deposits of pyrethrin (Pynosect 30 ULV) on parasitism of *Trialeurodes vaporariorum* by *Encarsia formosa* (after Perera 1982).

6.6/female/day for controls). Applications of the insecticide at the same rate but in more discrete (patchy) droplets gave an intermediate curve, with host searching beginning at 40 larvae per leaf but maximum parasitism of still only 2.2/female/day. By reducing the amount of toxicant by 90% and maintaining discrete droplets, the rate of parasitism returned to near the untreated levels. Obviously in this case, significant repellent effects were modifying the functional response of the parasitoid to its host. In less detailed studies, the repellent effect of pyrethroid insecticides on a number of behavioral attributes of natural enemies has been demonstrated (e.g., Riedl and Hoying 1983, Penman et al. 1981, Jones and Parrella 1983, Jacobs et al. 1984).

Other elements of predaceous or parasitic behavior have also been shown to be affected by sublethal doses of pesticides. Irving and Wyatt (1973) examined the effects of sublethal doses of several insecticides and fungicides on the oviposition behavior of *E. formosa*. Because whitefly nymphs develop brown wound marks following oviposition by the parasitoid, stab marks were used as a measure of pesticide impact on oviposition. The authors found that the insecticides tetradifon and *gamma*-BHC, and the fungicides benomyl and dichlofluanid all decreased oviposition (due to repellency), but that pirimicarb increased the number of ovipositions by 190–226%.

Kiritani and Kawahara (1973) documented sublethal food chain toxicity to lycosid spiders feeding on contaminated planthoppers. They observed an initial decrease in prey consumption, although in some cases the spiders not only recovered but overcompensated in their feeding response, consuming more planthoppers per day than the controls.

Another detailed study of pesticide sublethal effects on parasitoid behavior was conducted by Hoy and Dahlsten (1984). By using a video camera interfaced with a microcomputer, the authors were able to record walking speed and changes in speed at intervals of 0.15 second. Three hymenopterous parasitoids (*Encyrtus saliens, Trioxys pallidus*, and *Euderauphale flavimedia*) were exposed to malathion/Staley's bait mixtures and residues. Sublethal doses of the toxicant resulted in increased walking speeds, which increased toxicant uptake sufficiently to counter the slight protection afforded by the repellency of the bait. Although the authors recommended field tests to more precisely evaluate the combined effect of these factors, this type of detailed behavioral study substantially increases our understanding of and ability to influence behavioral response.

7.4.2.2. Multiple Behavioral Components. In several of the cases listed above, more than one component of behavioral response was influenced by the sublethal effects of pesticides in the environment. These complex studies have been more frequently reported in recent years and often in association with the development of some of the more sophisticated assays developed to measure side effects of pesticides (Chapter 5).

For example, Kuhner et al. (1985) studied the influence of several herbicides on the parasitization behavior of the aphidiid *Diaeretiella rapae* attacking *Myzus persicae*. Propachlor caused 31.2% and 47.2% mortality of adults emerging from mummies and postemergent, respectively. In addition, the behavior of adult survivors was very different from controls. In Table 7.2, results of observational studies made on three adults placed in petri dishes and observed for 10-min intervals are presented. Attack related behaviors were much more common in untreated adult parasitoids, whereas certain behaviors associated with general movement were greater in pesticide contaminated adults. Also, there were

Table 7.2 Sublethal Influence of Herbicide Propachlor on Behavior of adult *Diaeretiella rapae* after Emergence from Contaminated Mummies of *Myzus persicae* (from Kuhner et al. 1985)

	Average Number of Observations/Female/ 10-Min Testing Interval	
Behavioral Response of Parasite or Attacked Aphid	Controls (No Treatment)	Herbicide-Treated
Aphid contacted without reaction	0.3	2.3
Aphid contacted with reaction	6.7	0.0
Attempted oviposition	12.3	0.0
Successful oviposition	5.3	0.0
Orientation	1.3	0.0
Defensive movement before oviposition	0.7	0.3
Defensive movement after/during oviposition	7.0	0.0
Escape after/during oviposition	1.0	0.0
Movement/searching	11.7	16.0
Resting	2.0	6.0
Cleaning of legs	9.7	14.3
Cleaning of antennae	6.7	3.0
Cleaning of abdomen	11.7	3.7
Cleaning of wings	0.3	1.7
Cleaning of head	3.3	6.0
Staggering	0.7	6.0
Falling	0.0	1.7

striking differences in the patterns of cleaning behavior in the exposed individuals compared to the controls. Finally, perhaps the most telling sign of intoxication was the virtual absence of staggering and falling in the controls and the more common observation of these activities in contaminated individuals (Table 7.2).

7.4.3. Miscellaneous Effects

Even more subtle and difficult to detect than sublethal effects on behavior are those cases in which a pesticide induces physiological or biochemical changes which have no expressed outward manifestation. For example, sublethal exposure to a toxicant might result in a decrease or overcompensatory increase in enzyme production, which would not be observable unless the affected animal were subsequently challenged with another toxicant (Ilivicky et al. 1964). Other nonvisible sublethal effects such as synergistic potentiation of natural or synthetic chemicals or of pathogens (Pristavko 1966), mortality of symbiotic microorganisms (Herschberger and Forgash 1964), and subtle changes in circulation, respiration, and excretion (reviewed in Moriarty 1969) have been shown in pest arthropods, but as yet have not been documented for any natural enemies. Grosch (1975) demonstrated sublethal effects of pesticides on radiation-induced sterility in braconid wasps (*Bracon hebetor*) by previously treating them with conventional insecticides, juvenile hormone analogs, and antibiotics. While some compounds compounded sterility, carbamates and their metabolites provided some protection from radiation damage. If selection of populations as well as individual changes in behavior are considered in a definition of sublethal effects (see Section 7.2), then the entire phenomenon of pesticide resistance may also be thought of as a sublethal effect.

A few other cases of miscellaneous effects have been noted. For example, Critchley (1972b) suggested that sublethal doses of thionazin caused adult carabids (*Harpalus aeneus*) to eject "a very potent repellent fluid" (probably formic acid). Lawrence (1981) showed that sublethal doses of diflubenzuron, applied to fruit fly larvae which contained immature stages of the parasitoid *Biosteres longicaudatus*, caused abnormally developed ovipositors in the adults. Although other studies have shown no effect on emerging parasitoids (e.g., Ables et al. 1975), the mode of action of insect growth regulating pesticides may be expected to produce similar teratogenic effects on other natural enemy species. Host regulation, which is both a physiological and a behavioral process, can be influenced in sublethal ways by the action of juvenile hormone analogs and other insect growth regulators (see Chapter 12).

7.5 MODE OF ACTION

In studying the mode of action by which sublethal doses of pesticides produce various effects, much has been inferred from the study of the lethal effects of these compounds. Also, the few direct investigations of sublethal response mechanisms that have been carried out have concentrated mainly on pest anthropods or on common laboratory model systems (e.g., *Rhodnius* or *Periplaneta*). An exception to this generalization is the series of papers by Grosch and coworkers (Grosch and Valcovic 1967, Grosch 1970, 1972, 1975, Grosch and Hoffman 1973) which utilize the braconid parasitoid *Bracon hebetor*. Sublethal doses of heptachlor altered the parasite's feeding and oviposition behavior and reduced its fecundity and development rate. It was postulated that reduced egg production was caused indirectly by a behavioral effect via nervous system toxication. In studies with sublethal doses of carbaryl, the wasps also exhibited reduced oviposition, which correlated with reabsorption of mature ova and oocytes and a more drastic decrease in fat cell size than was observed in starved individuals. In this case, reduced vitellogenesis caused by insufficient protein metabolism could not be explained by altered feeding behavior. Instead Grosch speculated that the physiology of the state of "knock-down" (i.e., increased rate of heartbeat, increased chemical activity in the ganglia, and fat body participation in carbamate metabolism) might explain the interference in egg production. In addition, carbamate breakdown products were shown to directly damage the intracellular mitotic apparatus, resulting in embryo mortality during cleavage and subsequent reductions in egg hatch.

Moriarty (1969) lists three ways in which sublethal effects can be produced: 1) direct effects of the toxicant on the nervous system, 2) indirect effects caused by chemical upset of the hormonal system and consequent changes in the titres of endocrine secretions, and 3) direct effects on nonneural target sites, such as those involved in protein synthesis. Exposure to toxicants can and probably often does result in all three types of effects simultaneously, with the outcome depending on the specific compound, dose, and physiological state of the affected organism.

Cases of latent toxicity (after Moriarty 1969) may be caused by the interaction of pesticides with normal physiological processes. *Drosophila melanogaster* sequesters sublethal doses of DDT in the fat body during the larval stage, but when histolysis and tissue reorganization occur during pupation, the toxicant is released and the pupa dies (Kalina 1950). Almost all cases of latent toxicity occur in the holometabola. Compounds which interfere with chitin formation do not show their effect, either lethal or sublethal, until the subsequent molt of the insect.

7.6. FACTORS INFLUENCING SUBLETHAL EFFECTS

The severity and duration of sublethal effects in natural enemies caused by a compound or group of compounds is a result of many chemical and ecological as well as physiological processes. For example, DDT and the organochlorine insecticides are usually much more persistent than organophosphates and carbamates, and thus may be available for a longer period of time (Moriarty 1969). Many early reports of sublethal effects targeted persistent pesticides (e.g., Way 1949, Fleschner and Scriven 1957, Atallah and Newsom 1966, Coaker 1966, Dempster 1968). Database analysis of literature reporting sublethal effects on natural enemies for each of the major classes of pesticide compounds (Theiling 1987) showed that insecticides comprised 75% of the 1001 documented cases, followed by fungicides (15%), acaricides (7%), and herbicides (3%). Among insecticides, organophosphates have been most extensively documented as causing sublethal effects, with 4% (204 out of 5093 total organophosphate records) reported cases, followed by carbamates with 8% (134 out of 1593 records). These insecticides have been most widely tested for both lethal and sublethal effects on natural enemies. Also of note is the increasing frequency of reports of repellency to natural enemies caused by pyrethroid insecticides (e.g., Penman et al. 1981, Riedl and Hoying 1983, Jones and Parrella 1983, Jacobs et al. 1984).

Modern pest management programs are placing increasing emphasis on the use of the so-called "soft" pesticides, with a view toward decreasing the high mortality usually inflicted on natural enemies by conventional toxicants. These biorational pesticides, including insect growth regulators, chitin synthesis inhibitors, anti-feedants, and microbial pest control agents, usually cause lower initial natural enemy mortality than conventional synthetic insecticides. For this very reason, they are more likely to have an effect over a longer time period and in more subtle, sublethal ways (see Chapters 11 and 12). Several field researchers have commented on the difficulty in developing standardized tests to evaluate the effects of biorational pesticides on natural enemies (e.g., Staubli 1986, Blaisinger 1986). In the SELCTV database analysis reported by Theiling (1987), juvenile-hormone-like compounds and microbial pesticides, which have been less extensively tested as compared to many synthetic organic pesticides, had high proportions of sublethal-effect reports at 50% and 26% of all records within each pesticide class, respectively (see earlier cited values of 4% and 8% for organophosphates and carbamates, respectively). These data further reinforce the need to improve methods for testing the side effects of biorational pesticides on natural enemies (see Chapters 5, 11, and 12).

Beckage (1985) reviewed the endocrine interactions of parasitoids and their hosts and discussed a number of ways in which insect growth regulators (IGRs) can influence parasites. Both by direct interaction with parasite endocrine systems and indirectly through host physiology, these compounds have been shown to cause numerous sublethal effects, including increases and decreases in fecundity, increases and decreases in developmental rate, and changes in sex ratio, diapause, and morphology (see Table 7.1 for several examples). Similarly, microbial pest control agents (MPCAs) have been implicated in causing a wide range of sublethal effects (Flexner et al. 1986). As biorational pesticides are used more extensively, increasing emphasis in side-effect testing must be placed on the more subtle aspects of natural enemy performance. Sublethal effects may actually be more important than mortality in such cases (Flexner et al. 1986).

There are basic ecological differences between arthropod pests and natural enemies which influence the degree to which each group may be exposed to sublethal doses of pesticides. For example, because natural enemies are often more susceptible to pesticides than are phytophagous pests, there may be a longer time period during the degradation of the pesticide when a dose sufficient to affect natural enemies will be present (Chapter 3). Although many natural enemies are more susceptible to pesticides than are their hosts or prey, the formulation, timing, and placement of field applications are directed as narrowly as possible at target pests. Under ideal conditions of perfect ecological selectivity, the toxicant would affect only the pests and would have no impact on beneficials. In reality, however, selectivity is imperfect and some overlap of pesticide exposure on natural enemies invariably occurs. It is in this range of overlap that sublethal effects are most likely to occur.

Predators and parasitoids are usually dependent on phytophagous pests for their food and reproduction. Thus, in most cases, the phenology of their diapause termination, developmental rates, and so forth have evolved to follow those of their host or prey with some lag time (Campbell 1975). When pesticides are applied to early generations of the pest, subsequently emerging natural enemies are likely to encounter sublethal residues. Natural enemies are usually more mobile than their hosts or prey, and appear to move more frequently into and out of treated habitats. This again makes them more likely to come into contact with sublethal residues of pesticides.

7.7. HORMOLIGOSIS

Intuitively, it seems obvious that sublethal effects of toxic substances should result in some degree of harm to the affected organism. Partial poisoning would be expected to be detrimental and result in decreased fitness. However, numerous reports in the literature have shown that sublethal doses of pesticides are capable of increasing certain aspects of the performance of natural enemies.

Luckey (1968) proposed the term *hormoligosis* (from the Greek *hormo* = excite and *oligo* = small quantities) to describe the situation in which sublethal quantities of any stress agent are "helpful" when presented to organisms in suboptimal environments. Stress agents not only include toxic chemicals, but also radiation, temperature, and minor injuries. Although "helpful" is not clearly defined, Luckey (1968) uses increase in growth rate as an easily measured factor. In his experimental work, he showed that sublethal doses of several different insecticides increased the growth rate of the cricket *Acheta domesticus*.

It is important for our purposes to define the helpful aspects of hormoligosis more clearly. We must distinguish between what is most helpful for the affected predator or parasitoid in the long run (e.g., maximum fitness) and what is helpful for the applied entomologist, who is managing natural enemy populations for pest control in the short term. These two factors are not necessarily identical.

As mentioned earlier (Section 7.4.1), a number of different life history attributes contribute to a population's intrinsic rate of increaser r_0, which is often used as an indicator of fitness. This population measurement is a result of the interaction of age-specific fecundity, developmental rate, longevity, and sex ratio, which have evolved to respond to dynamic environmental constraints. An increase in one characteristic (age-specific fecundity) may be offset by a decrease in another (longevity) to yield a net increase, decrease, or unchanged r_0.

It has been argued from the theory of natural selection that an aberration of any normal process that is induced by an insecticide is likely to be deleterious to the individual (Warner et al. 1966). However, even though a population's r_0 may decrease, a change in a specific characteristic may be taken advantage of by an applied ecologist. What is best for the natural enemy's long-term fitness is not necessarily best for the pest manager practicing IPM.

We can illustrate this point using the results of the laboratory study by Atallah and Newsom (1966). They found that sublethal doses of DDT increased the oviposition rates of the coccinellid *Coleomegilla maculata* by 67%, but that survival to the adult stage was reduced by 25%. First of all, for the reduction in survival to be meaningful it must be shown that this was indispensable mortality (i.e., not normally occurring due to cannibalism, predation, etc.). Secondly, even if the reduction in survival led to an overall decrease in r_0, the importance in the timing of predation on early season pests such as aphids may make increased oviposition in the spring more important than the subsequent decrease in predator survival. In other words, timing of attack may be more important for pest management than the predator's overall reproductive success (Mackauer and van den Bosch 1973). Sublethal pesticide stimulation of a single life table characteristic may thus be beneficial. The high mobility of coccinellids (and many other natural enemies) will ensure a continual supply of immigrants to maintain genetic diversity.

The stimulatory effect of pesticides has been noted in plants, cell cultures, fermentation microbes, and mammals as well as in insects and mites (reviewed in Luckey 1968). Factors positively affected in arthropod natural enemies have included fecundity (Fleschner and Scriven 1957, Atallah and Newsom 1966, Babikir 1978), developmental rate (Adams 1960, Lawence et al. 1978, Parker et al. 1976), mobility (Dempster 1968, Critchley 1972b), predation (Kiritani and Kawahara 1973), oviposition (Irving and Wyatt 1973), and protection against radiation-induced sterility (Grosch 1975). Induction of detoxifying enzymes and other physiological functions (Yu et al. 1979, Gong et al. 1982) may also be thought of as hormoligotic effects.

Luckey (1968) suggested that the hormoligosis principle acting on phytophagous insects may help explain many of the reported cases of arthropod pest resurgence following field applications of pesticides. In fact, this area received considerable attention by economic entomologists in the early 1970s for species such as mite and aphid pests (e.g., Huffaker et al. 1970). With modern entomologists placing increased emphasis on the relationship between pesticides and natural enemy species, it is conceivable that the principle of hormoligosis may be manipulated to the advantage of biological control agents.

7.8. CONCLUSIONS

The study of the impact of sublethal doses of pesticides on arthropod natural enemies is moving from a stage of observation and documentation to one of more rigorous scientific investigation. The objective of much of this research is to improve biological control and IPM. More basic studies are emphasizing detailed behavioral effects such as natural enemy searching capacities and biological control effectiveness. Additional basic studies on the biochemical and physiological mechanisms causing sublethal effects are needed as well. The change in emphasis in pesticide sublethal effects research comes none too soon, considering the more frequent and widespread use of biorational pesticides. These compounds seem to act more slowly and subtly on pests and beneficial species. As our understanding of sublethal pesticide influences advances, the use of sublethal doses to enhance natural enemy effectiveness might be achieved by altering rates of increase and synchrony between pests and natural enemies, and in other ways not yet envisioned.

8

ECOLOGICAL INFLUENCES

8.1. INTRODUCTION

Most chapters in this book deal with pesticide side effects on individuals or single populations of natural enemies. Here, the focus is on trophic and community levels of organization. By their direct effects on predators and parasitoids, pesticides can profoundly influence arthropod community structure. The indirect effects of pesticides acting through other species can alter the composition of the natural enemy complex responding to pests in a crop (and beyond the treated habitat). The emphasis in this

The senior coauthor of this chapter is Russell Messing, Department of Entomology, Oregon State University, Corvallis, Oregon.

chapter is on the *relationships* between populations, rather than on populations themselves.

Pimentel (1961) was one of the first to comprehensively study the effects of pesticides from a community perspective. By comparing the number of arthropod taxa and their population densities on treated and untreated crop plants, he unraveled some of the complex trophic relationships which characterize the "structure and physiology of the community." In so doing, he showed how pesticides targeted at one pest affected many species differently, and how the interdependency of these species made overall pesticide impact difficult to predict.

An example from Pimentel's study illustrates the complexity of pesticide/ecological relationships. Insecticide applications are often used to control the diamondback moth *Plutella xylostella* in crucifers. However, moth larva are heavily attacked by three hymenopterous parasitoids and several predators. The predators are highly dependent upon populations of three aphid species (*Brevicoryne brassicae, Myzus persicae*, and *Rhopalosiphum pseudobrassicae*). These aphids are themselves heavily attacked by four species of parasitoids, which is turn are attacked by several hyperparasites. Insecticide applications directly affect each of the aforementioned populaions, and each change of a link in the complex may influence diamondback moth populations.

Although the focus of pesticide studies for many years has been on the mortality of isolated species, the shift toward IPM requires that diverse and often subtle pesticide side effects be more thoroughly investigated.

8.2. THE ECOLOGY OF NATURAL ENEMY/PESTICIDE INTERACTIONS

There is a rich ecological literature and theory dealing with arthropods with which to examine specific questions of pesticide impact at the population or community levels. For example, 1) the ecology of disturbance and succession, 2) the characteristics of agroecosystems, and 3) the role of predators and parasitoids in community structure are relevant to the subject of pesticide impact. Although ecological theory in many ways lacks robust explanatory and predictive power, in the context of community dynamics, specific questions of pesticide impact may be more understandable.

The ecological effects of pesticides in crop habitats have been principally studied by two research groups. Initially, economic entomologists were interested in the practical control of economic insect and mite pests. Their studies focused on small subsets of the community, usually a single pest and its natural enemy, or at most a few associated species. More basic ecologists (and some ecological entomologists) have studied the broader ecological effects of pesticides on overall community structure (e.g., Huffaker and Rabb 1984). From the latter perspective, pesticides become models for evaluating disturbance phenomena across multiple trophic levels.

In moving from two-species interactions to multiple species, or community dynamics, detail is lost as breadth is gained. Whereas experimentation is feasible and cause and effect are more documentable in simplified predator–prey systems, generalizations become more empirical and causation less clear at the community level. As discussed later, most current research tends to focus at either end of this continuum, with relatively little attention paid to intermediate levels of ecological complexity.

Since basic ecology represents the established theory of faunal disturbance, colonization, succession, equilibrium of population regulation, and so forth, these principles can be applied to IPM systems, where pesticides are widely used. Conversely, detailed studies of pesticide impact on arthropods should provide opportunities for testing ecological theory. In this area, both ecology and IPM could benefit from an interchange of data and ideas.

8.2.1. Predators and Parasitoids in Community Structure

In nonagricultural communities, predators and parasitoids play a key role in determining the structure and organization of species populations at lower trophic levels (Paine 1966, Holt 1987). In agriculture, this role is implicit in IPM (although often on a simplified level), particularly in managing entomophagous species for purposes of biological control.

In terms of sheer numbers or biomass, predators and parasitoids comprise a large portion of an arthropod community in most agricultural systems (Table 8.1). It should be noted that beneficial natural enemies are generally only secondary consumers, while tertiary and higher level consumers such as hyperparasites are usually considered detrimental to herbivore pest control. Most surveys listed in Table 8.1 do not distinguish between these trophic levels. However, it is clear that natural enemies are major constituents of agricultural arthropod communities. Pesticide side effects, whether direct or indirect, often cause significant changes in the quantities and directions of energy and nutrient flow in the ecosystem.

In addition to biomass and species numbers of natural enemies, the diversity of relationships and trophic links in food webs are thought to contribute to the stability of

Table 8.1 Representative Examples of Arthropod Species Composition of Agricultural Cropping Systems

Crop	Total Species	% of Total Arthropods				Reference
		Herbivores	Predators	Parasites	Other	
Apple						
Apple (mites only)	29	21	63	—	5	Strickler et al. 1987
Alfalfa	591	53	37	10	—	Pimentel & Wheeler 1973
Brassicas	41	53	22	24	—	Pimentel 1961
Collards	267	33	29	38	—	Root 1973
Forest: conifers	?	22	14	13	51	Whittaker 1952
Hardwoods	?	24	18	33	24	Whittaker 1952
Grain	86	52	23	24	—	Odum 1971
Grassland	1320	41	14	38	7	Evans & Murdoch 1968
Old field	1320	68	19	9	4	Menhinick 1963

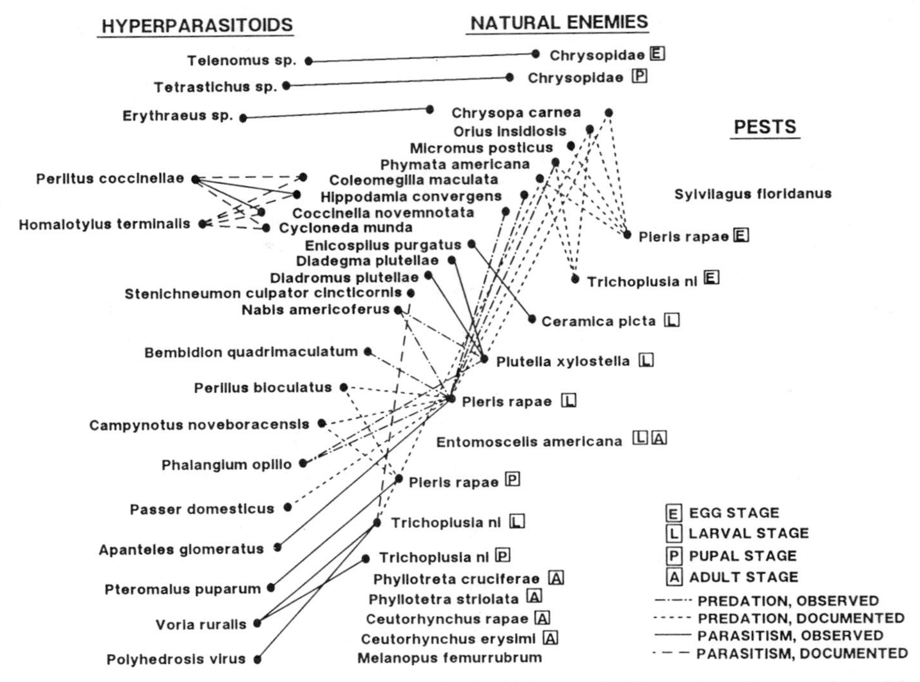

Figure 8.1. Arthropod food web relationships associated with leaves of cabbage plants (*Brassica oleracea*) in Minnesota (adapted from Weires & Chang 1973).

arthropod communities. The complexity of interactions in even small, partial food webs illustrates the importance of predators and parasitoids in community structures. Weires and Chang (1973) presented an illustration of such a food web in their description of the cabbage arthropod community. They described leaf, plant sap, root, detritus, and aphid honeydew-associated subcommunities. In Fig. 8.1 a partial food web is illustrated for pests, natural enemies, diseases, and hyperparasitoids associated with the leaves of cabbage plants in Minnesota. Although the nature of stability is difficult to characterize in a complex community representation, it is not hard to see how nutrient and energy flow in and through the secondary consumer (i.e., natural enemy) level is fundamental to the dynamics of the whole system. The corollary is that pesticide impact on these natural enemies will have multitudinous effects, both dramatic and subtle, reverberating through the food webs in many directions simultaneously.

8.2.2. The Ecology of Disturbance and Succession

Disturbance phenomena, including climatic flüctuations, are natural, recurring events in ecosystems, and arthropods have evolved ecological strategies to cope with these occurrences. Disturbance theory characterizes these events in terms of type, intensity, frequency, scale, and reliability (Schowalter 1985). Although most attention has focused on the response of herbivores to disturbance, effects on community structure, dispersal, host selection behavior, and resource quantity and quality (Schowalter et al. 1986) are equally important disturbance-related factors that influence higher trophic levels such as natural enemies.

In classifying insecticide applications as disturbance phenomena, it may be important to distinguish between pesticide classes, such as between insecticides and herbicides. Herbicides directly affect primary producers, with subsequent effects rippling through dependent trophic levels in the community (although herbicides can have significant direct effects on natural enemies; Chapter 2). Insecticides or acaricides—even broad-spectrum ones—are more selective because they have limited direct impact on primary producers. They tend to intercept the food web at the primary consumer level and above. Of course, secondary effects on vegetation via herbivores can be substantial. Also, apart from phytotoxicity, insecticides can affect plant physiology and nutrition. For example, carbaryl, a widely used carbamate insecticide, has pronounced direct effects on the movement of plant hormones in vascular tissues of apple trees (Williams and Batjer 1964), as well as having an impact on pests and natural enemies (Chapter 2).

An appreciation of the ecological factors governing the colonization and succession of disrupted habitats may allow pesticide inputs to be better managed so as to maximize biological control by natural enemies in IPM systems (see Section 8.5).

8.2.3. The Ecology of Agroecosystems

Pesticide-treated agroecosystems range from highly structured, simplified environments (such as greenhouses) to almost undisturbed and exceedingly complex biomes (e.g., coniferous forests). Most agroecosystems lie between these extremes and have attributes that make them unique in relation to other communities. Briefly, these attributes are reflected by the following generalizations.

1. They are often simplified. They objectives of agriculture are usually to maximize production of a single species (the crop) and to minimize competition or exploitation of the crop by other species. The result is a system which is greatly simplified in terms of species, architecture, microclimate, allelochemicals, and phenology. Crop breeding also leads to genetic uniformity. As a result, arthropod pest and natural enemy communities occuying these habitats tend to be more simplified than communities in more pristine environments.

2. They are regularly disturbed. Almost all crops are continually subjected to operational inputs which may severely disrupt established ecological relationships. These include cultivation, chemical treatment, fertilization, pruning, irrigation, harvest, and so forth. These inputs may reset the system to an earlier successional stage, never allowing it to reach equilibrium or what would be considered a climax community in a natural ecosystem. These factors tend to favor certain types of pests and natural enemies that are well adapted to exploit temporary ecosystem habitats.

3. They are exotic. Many of our crop plants have been introduced to this continent relatively recently from other areas of the world. Likewise, many of our pest arthropods are exotic in origin; 39% of our 600 most important pest species have been introduced accidentally in the recent past and account for 40–50% of all crop losses in the United States due to insects and mites (Sailer 1983). Many important weed species are also exotic in origin. Obviously, then, the entire floral and faunal complex associated with many of our agricultural crops represents relatively new assemblages of species which have not coevolved for long time periods, as have the species in many native habitats.

When pesticides are added to these communities which are undergoing rapid selection for adaptation to their new environments, it becomes extremely difficult to predict species shifts, competitive interactions, and resistance development. A myriad of physiological and ecological adaptations to stresses are occurring, therefore a tremendous number of factors must be taken into account in understanding the species interactions occurring in the community.

4. *They often are "islands."* Crops are most often grown in discrete land units within a heterogeneous matrix of native vegetation and altered environments. This biogeographical perspective has been used to try to understand the colonization of crop islands in relation to the theory of island biogeography developed by MacArthur and Wilson (1967). The size of crop islands and the distance from colonizing sources of arthropod species have an obvious impact on the number and diversity of arthropod species affecting the crop.

In considering these unique aspects of cropping systems, several authors have attempted to define a specific ecology of agroecosystems (see papers reviewed in Levins and Wilson 1980). Generalizations from these studies have been more relevant to IPM than those derived from the study of more pristine ecosystems. More sophisticated pest management techniques and theory must be developed concurrently with a deeper understanding of agricultural ecology.

8.3. TWO-SPECIES INTERACTIONS

In the exceedingly complex web of relationships which occur in even the simplest cropping systems, we can often do no more than empirically describe the results of a disturbance such as a broad-spectrum pesticide application. However, by sharpening the focus to the relationships between a few species (or two populations in a given habitat), we can begin to understand the mechanisms by which pesticides influence arthropods in agricultural systems, as well as the ways in which these effects reverberate through time.

8.3.1. Natural Enemy Interactions with Hosts and Prey

The dynamic interactions between predators or parasites and their hosts or prey have historically been considered important in the study of communities of agricultural arthropods. Pesticide effects on these relationships, both direct and indirect, immediate and long-term, have been at the core of IPM since its inception. However, significant developments beyond such well-established concepts as pest resurgence and outbreaks have not been pursued extensively.

8.3.1.1. Resurgence and Secondary Pest Outbreaks. Two widely accepted models depicting the effects of pesticides on pest/natural enemy relationships are shown in Fig. 8.2 (modified from Flint and van den Bosch 1981). They illustrate the phenomena of target pest resurgence and secondary pest outbreaks, both of which can be caused by pesticide applications.

For a number of reasons, pesticides are often more toxic to predators and parasitoids than to their hosts or prey (see Chapter 2). Therefore, applications of nonselective pesticides usually result in disproportionate mortality of natural enemies. By changing the numerical ratio between populations at two trophic levels, the density-related feedback mechanisms and the regulatory balance between these populations are temporarily disrupted, leading to increased survival and higher population densities of the herbivores.

Pesticide disruption of a tightly linked, two-species interaction often leads to pest

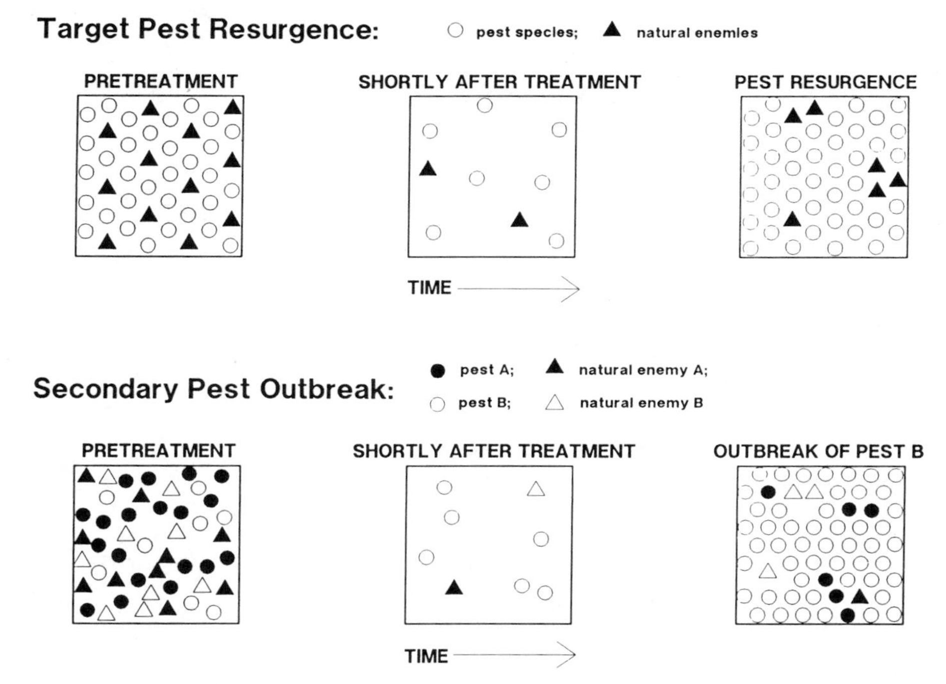

Figure 8.2. Models of target pest resurgence and secondary pest outbreak caused by pesticides (after Flint and van den Bosch 1981).

outbreaks with debilitating economic consequences, especially when coupled with pesticide resistance in the herbivore (Croft and Hoyt 1983). However, pesticide disruption also has served as a valuable research tool for demonstrating the regulating capacity of both native and imported natural enemy species (DeBach 1964). As noted in Chapter 1, the insecticide check method has been widely used to demonstrate the effectiveness of biological control in the field (e.g., Edwards et al. 1979, Sterling et al. 1984).

Even with the use of selective pesticides which are less toxic to natural enemies, the indirect effects of predator starvation or emigration can sometimes lead to subsequent pest outbreaks. Berendt (1973, 1974) showed how selective acaricides reduced prey to such low densities that predaceous mites were eliminated, even though the compound was not toxic to predators. Shires (1985) demonstrated that DDT, cypermethrin, and parathion-methyl all caused initial reductions in carabid and staphylinid populations feeding on cereal aphids in spring wheat. This initial reduction was attributed to predator mortality, and populations in all plots recovered after 4–6 weeks. However, because the cypermethrin and parathion-methyl greatly reduced aphid numbers, a secondary decrease in predaceous beetles was observed in plots several weeks later due to starvation, whereas DDT-treated plots showed little aphid mortality and no secondary decline in predators. This is a good example of the way in which pesticide effects can reverberate through a system over time, due to perturbations of trophic relationships.

A conceptual model illustrating the relationship between pesticides, natural enemies, and pests is shown in Fig. 8.3. Different degrees of pesticide toxicity to both pest and natural enemy species result in different general classes of interaction. Most compounds

Toxicity to Target Pest (columns) / **Toxicity to Natural Enemy** (rows)

	No Effect	Slightly Toxic	Moderately Toxic	Highly Toxic
No Effect	No Effect	Selective Pesticide	Selective Pesticide	Pest Resurgence
Slightly Toxic	Pest Upset	Suppression of Both Species	Selective Pesticide	Pest Resurgence
Moderately Toxic	Pest Upset	Pest Resurgence	Suppression of Both Species	Pest Resurgence
Highly Toxic	Pest Upset	Pest Resurgence	Pest Resurgence	Pest Resurgence

Figure 8.3. Generalized effects of selective and nonselective doses of pesticides on natural enemy/pest populations.

are more toxic to natural enemies than to herbivorous pests, and this often results in some degree of pest resurgence. Some physiologically selective compounds, however, show appreciably greater activity against the pest than against its predators and parasitoids. These compounds are well suited for IPM programs, provided that sufficient pest populations remain to supply a source of food for natural enemy populations. (Physiologically selective compounds include the acaricides propargite and cyhexatin, and the carbamate aphicide pirimicarb; see Chapter 10.) When a compound is highly toxic to herbivores, subsequent natural enemy starvation or emigration will often create a predator-free space in which recolonizing herbivores can expand their populations rapidly.

Besides altering absolute population densities and disrupting critical predator/prey ratios, pesticides may also affect two-species interactions by disrupting synchrony and coincidence in time and in space. Phenologically, the potential of a parasite to attack a host successfully is often a function of being present at a precise window of vulnerability, commonly a susceptible life stage of the host (e.g., Drummond et al. 1985). Actual biological control potential for multivoltine pests is often a function of predation or parasitism at the critical initiation phase of population growth (Mackauer and van den Bosch 1973). Even with comparable toxicity to natural enemy and pest populations, a pesticide may disrupt the precise phenological synchrony necessary for population regulation.

In considering colonization of an herbivorous pest and its natural enemy in a given habitat over time, generalist predators usually only colonize a crop system after prey population densities have reached a threshold level sufficient to provide adequate nutrition. Parasitoids similarly colonize (or eclose) following a lag time sufficient to ensure adequate host levels for reproduction (Campbell et al. 1974). If nonselective pesticides are applied following initial pest buildup, they will tend to kill a greater percentage of the late emerging (or colonizing) herbivores and the early emerging (or colonizing) natural enemies. Selection pressure then drives the populations apart, possibly resulting in asynchrony and reduced control.

Besides this direct effect on natural enemy/pest population synchrony, pesticides can have more subtle, indirect influences on the temporal association of predators or parasitoids and their hosts or prey. For example, it has been demonstrated in several crops that the presence of sufficient numbers of pests early in the season is necessary to enable natural enemy populations to build up to levels sufficient to exert biological control for the remainder of the season [e.g., strawberry with the cyclamen mite *Steneotarsus pallidus* and the predatory mite *Typhlodromus cucumeris* (Huffaker and Kennett 1956); apple orchards with *Tetrancyhus ulmi*, *Panonychus ulmi*, and *Aculus schlechtendali* and the phytoseiid mite *Amblyseius fallacis* (Croft and McGroarty 1977); field crops with a parasitoid of *Pieris rapae* (Parker et al. 1971)]. If early season pesticide applications reduce or eliminate the prey, predators will starve, emigrate, or fail to colonize the crop, and subsequent biological control will be poor.

Spatial coincidence of natural enemies and their hosts or prey is as important as temporal synchrony in maintaining fine-tuned feedback processes necessary for maintaining pest populations at acceptable subeconomic threshold levels. Pesticides can, at times, disrupt this spatial relationship even when they do not alter the absolute predator/prey ratios in or around the treated crop system. Based on differences in life history and the types of refugia found in the physical environment, survivors of a pesticide application may be located in disjunct micro- or macrohabitats. For example, Walker and Penman (1978) showed that azinphosmethyl applications to apple trees led to isolated, irregular bronzing due to feeding damage by *Panonychus ulmi*. The insecticide did not reduce the number of the predator *T. pyri*, but altered its spatial patterns in relation to the prey.

With many predators and almost all parasitoids, the ovipositing adult females must locate the host or prey and deposit their eggs in a suitable location. The offspring then have relatively limited searching capacity. Pesticides can disrupt spatial patterns of herbivores and require increased searching (and hence less energy available for reproduction) by natural enemy adults. (In a similar manner, disrupted distributions or reduced densities of sexually reproducing natural enemies can result in energy loss in mate finding, which contributes to reduced fitness.) Spatial disruption of immature or less mobile natural enemies and their hosts or prey can result in outright starvation or local extinction. Some pesticides have repellent effects on entomophagous species (Jackson and Ford 1973, Walker and Penman 1978; Chapter 7). When this occurs in the field, resistant or tolerant phytophagous species may have predator-free time and space in which to reproduce, until the repellent effect wears off.

In many cases, resurgence following pesticide applications is attributed to natural enemy mortality when, in fact, other effects are being expressed. Pesticides can directly alter host-plant physiology, leading to increased nutritional potential for herbivores (Chapman and Allen 1948). Sublethal doses have been shown to directly stimulate the reproductive potential of many pest species (Hodjat 1971, Stewart and Philogene 1983, Lowery and Sears 1986). Selection leading to resistant biotypes can change the genetic structure of populations, leading in some cases to increased fecundity (Eggers-Schumacher 1983).

8.3.1.2. Refugia. Even broad-spectrum, nonselective pesticides usually do not kill

Table 8.2 Examples of Known Refugia from Pesticides in Arthropod Natural Enemies

Species or Complex	Location	Crop	Chemical	Reference
Phytoseius macropilus	Under dead scale bodies, in crevices	Apple	DDT	Dabrowski 1970a
Euproctis simulis	In "winter nests" of hosts (brown-tailed moth)	Forest	DDT/ lindane	Vater 1980
Epipyrops melanoleuca	In sugar cane leaf	Sugar cane	Several organo-phosphates	Varma and Bindra 1980
Several carabids and staphylinids	Underground stones in forest litter	Forest	Pirimi-phosmethyl	Sustek 1982
Syrphid larvae	In aphid colonies on curled leaves	Beans	DDT	Way et al. 1954
Several carabids	On nearby vegetation and surrounding forest	Wheat	Several	Emel'Yanov & Yakushev 1981
Several carabids	As pupae in soil	Rape, Wheat	Deltamethrin	Basedow et al. 1985
Several parasites and coccinellids	Host and scale bodies	Ice plant	Organo-phosphates	Washburn et al. 1983
Belaustium putmani	Bark crevices	Apple	Several	Cadogan & Laing 1982

all pests and natural enemies in a treated habitat. Within most cropping systems, there are various refugia which arthropods, given their small size and mobility, can utilize to escape pesticide contact (Table 8.2). In some cases, this use of refugia may be a reaction to or evasion from toxic and repellent compounds; in many other cases the presence of at least a portion of the population in refugia at any given time is fortuitous.

Basic life history differences appear to favor different types of refuge for parasitic and predaceous species. Many endoparasitoids are protected from the effects of direct pesticide contact while within the bodies of their herbivorous hosts. This refuge may consist of living hosts (i.e., during early stage of parasitism); dead, eviscerated host bodies (i.e., many aphidiid species in aphid mummies; Hsieh and Allen 1986); or host exuviae or excretions (i.e., under scale covers; Washburn et al. 1983). (See Table 10.1 for examples of parasites escaping exposure to pesticides while in refuge within host bodies.)

In comparison with parasitoids, most true predators cannot utilize the bodies of their prey as a refuge from pesticides. In some cases, predators may be protected by prey modifications to host-plant tissues (i.e., within leaves rolled by leafrollers, or in galls or leaves curled by aphids; Way 1958). Most refugia for predators, however, occur somewhat removed from the host plant, such as under stones or leaf litter (Sustek 1982) or in surrounding vegetation (Emel'yanov and Yakushev 1981).

An interesting aspect of host regulation, which is reported among several parasitic hymenopteran families, is the ability of the developing parasitoid to alter the host's behavior in such a way as to make it leave the treated host plant and seek shelter before death occurs (Lykouressis and van Emden 1983). The selective pressure favoring such host regulation may be predation or hyperparasitization of exposed, immobile, and unprotected prey (i.e., primary parasitoids pupating within host remains). However, a by-product of this escape behavior is the refuge reached by the parasitized host which may offer protection from the pesticide to the natural enemy (van den Bosch et al. 1962).

8.3.1.3. Immigration and Recolonization. In some cases, pests and natural enemies are virtually eliminated from a treated habitat—the pests due to direct toxicity and sublethal repellency, the natural enemies from toxicity, repellency, and starvation. Although few studies have directly examined this phenomenon, there are many incidental observations indicating that there is extreme variability in the rates of recolonization of previously treated habitats by pests and natural enemies. Generally, it appears that recolonization is often much faster for plant feeding pests than for the natural enemies of these pests.

Recolonization rates of a pesticide-treated habitat are a function of many biological and physical factors regarding the species affected and the surrounding environment. Classical island biogeography theory (e.g., MacArthur and Wilson 1967) gives us some indication of what these factors might be and how they interact in general. The defaunation of oceanic islands by fumigation has often been used as a model for examining pesticide impact on crop islands and subsequent recolonization rates (Croft and Hull 1983). Size of the island (or crop patch), distance from recolonizing species pools, and vegetation diversity (or stage of succession) of the treated habitat are all important features influencing rates of recolonization and species diversity (MacArthur and Wilson 1967). These features have been examined in management contexts for a few agricultural crop islands, at least in a theoretical sense (Gut et al. 1982, Gut 1985, Liss et al. 1986). However, defaunation (Simberloff and Wilson 1969) almost completely eliminates all arthropods, whereas even with highly toxic agricultural pesticides, survivors persist in refugia within the treated crop.

The bionomic features which characterize colonization (or recolonization) rates are often discussed in terms of *r* and *K* life history strategies (Price and Waldbauer 1975). The *r* strategists, characterized by small size (easily dispersed by wind), many progeny, and rapid development, are often first to colonize a newly available, disrupted, or early successional patch. Aphids and mites are examples of early effective colonizers, and commonly present pest resurgence problems following broad-spectrum pesticide applications. In contrast, predators and parasitoids are generally slower to colonize vacant sites. There is little selective advantage for them to colonize early, as they would encounter a shortage of hosts or prey.

In Table 8.3, some observed recovery or resiliency times for a variety of natural enemies occurring in different pesticide-treated agroecosystems are listed. It is difficult to draw generalizations from this type of data, because definitions of recovery vary from first sighting of a species following a spray application to complete reestablishment of regulatory capacity. However, the potential *range* of influences that pesticides can have on natural enemies in different crops and under various environmental conditions is reflected. At one extreme, studies of aerial applications of low rates of relatively nonpersistent pesticides applied to diverse, stable ecosystems such as forests or some perennial crops show virtually no delay in natural enemy recovery (Manser and Bennett 1962). Conversely, DDT treatments of Tanzanian coffee plantations disrupted the regulatory capacity of parasitoids of coffee leafminers (*Leucoptera menricki* and *L. caffina*) for 10 years (Bigger 1973). In highly contained cereal storage bins, phosphine treatments resulted in permanent eradication of predaceous mites (Sinha et al. 1967).

Table 8.3. Comparison of Recovery Times for Natural Enemies to Recolonize Pesticide-Treated Agroecosystems

Crop	Pest	Natural Enemies	Compound	Criteria of Evaluation	Recovery Time (Pest/Natural Enemy)	Reference
Corn	*Diabrotica* sp.	Spiders, carabids	Cloethocarb	Equivalent densities	0 days	Stinner et al. 1986
Sugar cane	*Diabrotica* sp.	*Lixophaga diatraeae*	Malathion	Return to level of control plots	< 1 day	Manser and Bennett 1962
Peach	*Grapholitha molesta*	*Macrocentrus ancylivorus*	Azinphosmethyl	First survival	7 days	Pree 1979
			Phosmet		7 days	
			Permethrin		4 days	
Rape, wheat	Aphids, thrips, cecidomyiids	Carabidae	Demeton-*S*-methyl	Return to level of control plots	0 days	Inglesfield 1984
		Linyphiidae	WL 85871		8 days	
		Braconidae			15 days	
		Empidoidea			15 days	
Sugar beet	Aphids	Ichneumonids, braconids, Aphididae, mymarids, cynipids	Demeton-methyl	Return to level of controls	10–14 days	Goos & Goos 1979
Apple	*P. ulmi, E. carpinus*	*Typhlodromus arboreus*	Permetherin	Presence/ Absence	2–4 wks/1 wk?	AliNiazee 1982
Turf	*Crambus teterrellus, Pediasia trisecta*	Several ants, *Macrocheles* sp.	Chlorpyrifos	Return to level of controls	3 wks	Cockfield & Potter 1983
Cotton	H. virescens	*Diadegma* sp.	Carbaryl	Return to level of control plots	30 days	Gonzales et al. 1980
Raspberry	Two-spotted mites	*Stethorus punctum*	Parathion	First survival	1 mo	Rosenstiel 1950
Wheat	Cereal aphids	Carabids, staphylinids, linyphiid spider	Cypermethrin, dimethoate	Equivalent densities	35 days	Cole et al. 1986

Cotton	Heliothis virescens	*Trichogramma*, coccinellids, chrysopids	Diflubenzuron	Return to level of control plots	40	days	Ables et al. 1977
Soybean	*Cerotoma trifucata*, *Empoasca fabae*, *Sericothrips variabilis*	*Geocoris* spp., *Nabis* spp.	Aldicarb carbofuran	Return to level of control plots	> 44	days	Lentz et al. 1983
Grass/grains crops	Phytophagous insects	Predaceous spiders and insects	Carbaryl	Total biomass to control level	3–5	wks	Barrett 1968
Wheat, beet	Cereal pests	Linyphiid carabids	Deltamethrin	Equivalent densities	4–6	wks	Basedow et al. 1985
Apple	Phytophagous mites	*Amblyseius fallacis*	Permethrin, cypermethrin, fenvalerate	First survival	6	wks	Bostanian & Belanger 1985
Alfalfa	Not given	*Tetragnatha laboriosa*	Carbofuran	Return to level of control plots	30–90	days	Culin & Yeargan 1983
Wheat	Cereal aphids	14, carabid specise	Cypermethrin, methyl-parathion	Equivalent densities	4–6	wks	Shires 1985
Maize	*Heliothis zea*	*Harpalus pennsylvanicus*, *Pterostichus chalcites*	Turbufos	Equivalent densities	> 77	days	Lesiewicz et al. 1984
Turf	Several species	Staphylinids, eriogonids, parasitoids, eupodids	Chlorpyrifos, isofenphos	Equivalent densities	55 6	days wks	Cockfield & Potter 1983
Sugar beet	Several pests	Several carabids	Lindane	Equivalent densities	2–3	mos	Sekulic & Dedic 1983

Table 8.3. (*Contd.*)

Crop	Pest	Natural Enemies	Compound	Criteria of Evaluation	Recovery Time (Pest/Natural Enemy)	Reference
Apple	Spider mites, rust mites	*Typhlodromus pyri*	Permethrin	Equivalent densities	2–4 mos	Kapetanakis et al. 1986
Cabbage	Several pests	Several carabids	Diazinon	Return to level of control plots	11 wks	Hassan 1969
Potato	Cereal aphids	Several carabids	Thionazon	Return to level of control plots	6 mos	Critchley 1972a
Citrus	Not given	*Metaphycus luteolus*	Parathion	Biocontrol reestablished	3–7 mos	Bartlett 1951
Pear	Tetranychus urticae	*Typhlodromus occidentalis*	Fenvalerate	Presence/absence	24 wks	Riedl & Hoying 1983
Forest	Spruce budworm	Parasites	Folithion, phosphamidon		> 1 yr	Freitag & Poulter 1970
Apple	Phytophagous mites	Several predaceous mites	Parathion	Return to level of control plots	3 yrs	Dabrowski 1970a, 1970b
Coffee	Leucoptera meyricki L. caffina	Several parasitoids	DDT	Regulatory control reestablished	10 yrs	Bigger 1973
Stored products	Not given	*Cheyletus*	Phosphine	Presence/Absence (eradicated)	Indefinite	Sinha et al. 1967

[1] ng = data not given

This wide range in recovery times is not only influenced by the definition of recovery, but also reflects real biological and environmental relationships. The actual mobility, feeding biology, suitability of alternate hosts, and density dependency of different pest species and their natural enemies lead to real biological diversity in an ecological parameter as complex as recovery time. In addition, attributes of the habitats surrounding the treated crop (including vegetation diversity and species pool characteristics, topography, microclimate, prevailing winds, etc.) further contribute to this variation.

A study of the reestablishment of natural enemies of scale insects in French citrus orchards sprayed with methidathion (Panis 1980) illustrates some of the generalizations regarding recovery time that might be obtained with more careful, detailed studies addressing this question. Following spray applications, predation on scale insects by *Chrysoperla carnea* continued despite the presence of insecticide residues. Adult coccinellids (*Exochomus quadripustulatus* and *Chilocorus bipustulatus*) renewed their predation on surviving coccoid larvae about 10 days after spraying. The ectoparasitic pteromalids *Scutellista cyanea, S. nigra*, and *Moranila californica* and the endoparasitic eulophid *Tetrastichus ceroplastae* were again present in the orchard about 20 days after spraying. These parasitoids had emerged from refugia within the bodies of the host scale insects. Encyrtid parasitoids practically disappeared from the treated orchard and did not return to normal population densities for almost 3 years. The author estimated it took about 3 years for populations of most natural enemies to return to prespray levels.

These recovery times reflect toxicological as well as bionomic differences among natural enemy types that may be characteristic to some extent. For example, some chrysopid species are among the most pesticide tolerant of all natural enemies (Plapp and Bull 1978, Shour and Crowder 1980, Ishaaya and Casida 1981, Rajakulendran and Plapp 1982a, 1982b; Chapter 2) and are highly dispersive in nature (Neuenschwander et al. 1975), Croft and Whalon 1982). Therefore they are among the first natural enemies to recolonize treated habitats. Likewise coccinellids are widely occurring, highly mobile, and as generalists feed on many species of aphids, scales, and mites; they also tend to rapidly recolonize treated systems (Radwan and Lovei 1982, Gut 1985).

In a study by Shires (1985), the interaction between immigration and feedback effects of prey density on natural enemy recolonization of a soil-litter habitat was illustrated. When parathion-methyl, DDT, and cypermethrin were applied to similarly replicated, barriered plots of spring wheat, populations of carabids and three genera of staphylinids were reduced to levels 50–90% below pretreatment counts. Substantial recovery in the number of predaceous beetles occurred in all treated plots at 4–6 weeks after application, presumably due to immigration. A further fall in the number of beetles was observed 8–12 weeks after application with cypermethrin or parathion. This reduction was attributed to an indirect effect of the pesticide which eliminated the cereal aphid food supply of the predators.

In contrast to these examples with generalist predators, many parasitoids are more host-specific and limited in distribution to patches of crop habitat containing their hosts. Also, parasitoids are generally more susceptible to pesticides than predators (see Chapter 2). Therefore it is not surprising that recovery times for parasitoids tend to be longer than for predators in comparable habitats.

An exception to this generalization is the case when endoparasitic insects find refuge within the bodies of their hosts. This is an effective mechanism of pesticide tolerance for several species (see Bartlett 1958, Hsieh and Allen 1986; Chapter 10). Given some host survival and a parasitoid developmental period longer than the residual toxicity period, recolonization from within host bodies can reestablish effective biological control in the field quite rapidly following pesticide disruption (Table 8.1).

8.3.1.4. Alternate Hosts and Prey. Besides direct toxic effects of pesticides on natural enemies and indirect effects due to mortality of prey or hosts, there may also be subtle, indirect influences on alternate hosts or prey. In many agricultural systems, there is a primary predator/prey (or parasite/host) relationship critical to the biological control of a particular crop pest, and hence most of the emphasis in research and management is placed on this interaction. However, many parasitoids and almost all predators depend at least in part on alternate species of arthropods for their survival or reproduction (Clausen 1962, Dustan and Boyce 1966).

Pesticide influences on natural enemies via alternate hosts and prey are undoubtedly of widespread occurrence, although their documentation is limited. One well-researched example (largely due to management imperatives) includes predaceous mites, their spider mite prey, and their alternate prey, the rust mite. *Amblyseius fallacis* is the most effective predator of *Panonychus ulmi* in commercial apple orchards of the midwestern United States. However, effective predation requires adequate phytoseiid levels in early summer when spider mites begin to reproduce. Predator mites are dependent upon the early season apple rust mite *Aculus schlechtendali* to build up these populations. Thus, the predator/prey interaction of primary importance (*A. fallacis*/*P. ulmi*) is strongly influenced by the secondary relationship between *A. fallacis* and *A. schlechtendali*. Pesticides having little impact on spider mites or predaceous mites, but which severely impact rust mite populations (i.e., endosulfan), can profoundly influence the dynamics of the entire mite complex (Croft and McGroarty 1977).

In a similar manner, pesticides which are not significantly harmful to generalist predators such as *Chrysopa, Orius, Geocoris*, and *Nabis* may eliminate prey such as spider mites, aphids, and thrips from cotton fields treated early in the season, leading to predator emigration. Later, invading *Lygus hesperus* or *Heliothis zea* may reproduce explosively without the stabilizing influence of these generalist predators (Smith and van den Bosch 1967).

Even more subtle influences of pesticides on natural enemies via alternate prey are known to occur. Apple rust mites (*A. schlechtendali*) condition leaves and alter the acceptability of the leaf tissue for feeding by the European red mite. By decreasing rust mite density, certain pesticides can thus indirectly cause changes in species diversity and abundance of spider mites via leaf conditioning, and thus influence pest species composition and predaceous mite abundance (Croft and Hoying 1977).

Although parasitoids are generally more host-specific than predators, some rely on alternate hosts to bridge periods of primary host scarcity (Doutt and Nakata 1965). Many parasitoid (as well as predator) adults rely on honeydew to meet their nutritional requirements (van den Bosch and Telford 1964, Hagen and van den Bosch 1968). Nonselective pesticides which reduce aphid or scale populations, and hence within-crop honeydew accumulations, may force parasitoids to leave the treated habitat in search of adequate food supplies. Contamination of honeydew with pesticides may also result in direct toxic effects to those natural enemy species that rely on these exudates as a food source (Bartlett 1966).

Alternate hosts or prey may provide a nutritional refuge when pesticides drastically reduce primary host or prey population densities while having minimal impact on the natural enemies. For example, Allen (1958) discusses how DDT applied against the Oriental fruit moth *Grapholitha molesta* caused the parasitoid *Macrocentrus ancylivorus* to move to alternate hosts (i.e., the ragweed borer *Epiblema strenuana*, the strawberry leafroller *Ancylis comptana fragariae*) and subsequently to reinvade treated orchards with no apparent reduction in levels of parasitism against the target pest.

8.3.2. Hyperparasites and Generalist Predators

Just as population linkages between predators or parasites and their hosts or prey can be disrupted by pesticides, the relationships between natural enemies and *their* predators and parasites can be similarly affected. The population-regulating capacities of these higher trophic levels have not been adequately studied and remain a subject of considerable controversy (Luck et al. 1981, van den Bosch et al. 1979). However, there are cases in which the impact of secondary parasites or predators on entomophagous natural enemies has been well documented.

Semyanov (1985) reported that five species of predators and five species of parasitoids attacked the coccinellid *C. septempunctata* in the USSR, with parasitism rates reaching 80%. Frazer and van den Bosch (1973) showed that the Argentine ant *Iridomyrmex humilis* selectively preyed upon parasitized aphids, thus interfering with biological control of the walnut aphid *Chromaphis juglandicola* by the aphidiid *Trioxys pallidus*. They also reported that this parasitoid sustained hyperparasitism levels of up to 90%.

Others have noted the importance of predaceous ants in attacking natural enemies of aphid and scale pests (Way 1963, DeBach and Bartlett 1964). Inserra (1969) showed that parasitism of the red scale *Aonidiella aurantii* by *Aphytis melinus* could be increased by using lindane to control the Argentine ant, a hyperpredator of the parasitoid.

Shepard et al. (1977) suggested that insecticides applied to soybeans had enough impact on natural enemies of the predaceous earwig *Labidura riparia* to explain this species' population resurgence in a manner analogous to the resurgence of herbivorous soybean Lepidoptera. Price and Shepard (1978) offered a similar explanation for increases of the carabid *Calasoma sayi* in soybeans treated with methyl parathion, methomyl, or monocrotophos.

The predaceous mirid *Blepharidopterus angulatus* can upset the regulatory balance between the phytoseiid *Typhlodromus pyri* and the tetranychid *Panonychus ulmi* by preying on *T. pyri* eggs. Cranham (1978a) noted that diflubenzuron, which is toxic to the mirid but not to the phytoseiid, may be ideal for codling moth management in orchards where hyperpredation by mirids interfered with mite biological control.

For cases in which a parasitoid attacks a predaceous natural enemy (i.e.,*Perilitus coccinellae* on numerous coccinellids, or diplazontine ichneumonids on syrphids), pesticide applications should favor the primary biological control agent. Predators are often more tolerant of insecticides than are parasitoids which occur in the same habitat (Croft and Brown 1975; Chapter 2), and therefore selective hyperparasiticides should be easy to identify.

Only a few studies of pesticide impact on hyperparasites and primary parasitoids have been reported in the literature. Robertson (1972) reported that the carbamate mexacarbate, when applied for control of the California oakworm *Phrygaridia californica* Packard, was slightly more toxic to a hyperparasite *Dibrachys carus* (Walker) than to *Itoplectis behrumsii* (Cresson), a primary parasite of the oakworm. This may be a possible explanation for findings by Williams et al. (1969) that mexacarbate applications in Montana forests for control of the spruce budworm *Choristoneura occidentalis* resulted in higher populations of primary parasitoids and greatly increased parasitism of budworms by the braconid *Apanteles fumiferanae* Viereck and the ichneumonid *Phaeogenes hariolus* (Cresson).

Wiackowski and Herman (1968) reported that five different aphicides were more toxic to the hyperparasite *Alloxysta curvicornis* than to the primary aphid parasite *Diaeretiella rapae*. Godfrey and Root (1968) demonstrated that carbaryl applied to collards caused a decline in the ratio of hyperparasites (*Charips brassicae*) to primary parasites (*D. rapae*) of the cabbage aphid (*Brevicoryne brassicae*). Horn (1983) noted that weekly carbaryl and malathion treatments to collards resulted in lower rates of hyperparasitism than in untreated plots, although there were no differences in the rates of primary parasitism of *B. brassicae* by *D. rapae*. Horn observed that hyperparasites spent more time walking over treated leaf surfaces searching for hosts than did primary parasites, and thus may have contacted more insecticide. He raised the possibility of utilizing this behavioral difference to achieve selectivity, thereby reducing hyperparasitism and rendering biological control by primary parasites more efficient.

There are other cases in which pesticide applications have resulted in quantifiable increases in parasitism (MacDonald 1959, MacDonald and Webb 1963, Carolin and Coulter 1971), but whether or not these were due to hormoligotic effects (Chapter 7), impact on hyperparasites or other mechanisms was not determined.

8.3.3. Competition

Although it is a subject of continuing debate among animal ecologists (see references in Liss et al. 1986), competition is generally regarded as a strong element in structuring communities and relationships among populations. Pesticide impact, both direct and indirect, on the competitive balance between natural enemies (and between pests) is undoubtedly a major mechanism by which toxicants influence the ecological composition of agroecosystems.

That important competitive interactions exist among natural enemy species in agriculture has been amply demonstrated following the release of exotic biological control agents. Competitive displacement of a native or previously introduced predator or parasitoid has been documented for fruit flies in Hawaii (van den Bosch and Haramoto 1951), for citrus blackfly in Mexico (DeBach 1964), and for citrus scale insects (DeBach and Sundby 1963) and olive scale insects (Huffaker and Kennett 1966, 1969) in California. Both intrinsic and extrinsic competition among parasitoids has also received a great deal of attention in basic laboratory research (Fisher 1971).

Thus it should not be surprising that pesticides, by their differential toxicities to competing natural enemies, can alter the distribution and abundance of predator and parasitoid species, as well as species linked to natural enemies by trophic relationships. Although pesticide selectivity is usually viewed in terms of its differential effect on natural enemies and their hosts or prey, the discussion may also be relevant to competition on the same trophic level. Factors that may influence selectivity between competing natural enemies include tolerance, resistance, phenology, mobility, types of refugia utilized, and so forth (Chapter 4).

Despite the enormous literature on competitive interactions among predators and parasitoids, and the equally large database on pesticide side effects on natural enemies (Chapter 2), these two areas have not been brought together in detailed, explicit experimental or analytical research to any great extent. While there are numerous reports documenting natural enemy species shifts following pesticide applications, the exact mechanisms involved have rarely been elucidated. This may be due in part to the inherent difficulties in experimenting with basic ecological phenomena such as competition, but is probably equally attributable to the lack of collaboration between applied, pesticide-oriented economic entomologists and basic ecologists.

McMullen and Jong (1967) reported a competitive release mechanism from insecticides on the predator complex of *Psylla pyricola* in pear orchards of British Columbia. DDT applications reduced populations of the predaceous mirids *Deraeocoris brevis, D. fasciolus, Diaphnocoris prvancheri*, and *Campylomma verbasci* and moderately reduced populations of *Chrysopa carnea* and *C. oculata*. However, *Anthocoris antevolans* and *A. melanocerus*, which are tolerant to DDT, doubled in population size in the treated plots. The authors inferred that the increase in the anthocorids was due to reduced competition for food. In a similar manner, ryania-treated plots showed a drastic reduction in the populations of most predators, but large increases in the densities of the ryania-tolerant *D. brevis*.

Some insect ecologists have discussed the desirability of treating pesticide impact on arthropods as another form of "environmental resistance" (Price 1975). In examining the principles that relate disturbance phenomena to subsequent guild structure and competition among natural enemies, it becomes apparent that they clearly apply to pesticide-disrupted agricultural systems. Miller (1983) cites several host/parasitoid systems in which disturbance (due to harvesting, burning, or removal of native vegetation) led to a structured succession of recolonizing competitive parasitoids. In a similar manner, pesticide-disrupted apple orchards usually show regular, predictable patterns of recolonization by natural enemies (Croft and Hull 1983, Whalon and Croft 1986). Patterns of recolonization reflect ordering principles of niche relationships as demonstrated by the similarities observed in mite species composition in pesticide-treated and untreated apple orchards in many different geographical regions of the midwestern and eastern United States (e.g., Poe and Enns 1969, Knisley and Swift 1972, Hislop and Prokopy 1979, Farrier et al. 1980, Berkett and Forsythe 1980, Strickler et al. 1987).

Unfortunately, the actual contribution of interspecific competition to the myriad of factors which intertwine to produce species replacement patterns has rarely been documented. A few attempts to explain these relationships in terms of *r* and *K* life history strategies have been made (Miller 1980, Croft and Hull 1983), but none has provided broad explanatory or predictive power.

8.4. MULTIPLE-SPECIES INTERACTIONS

Pesticide-treated agroecosystems often demonstrate recurring, even predictable patterns of changes in the faunal structure of communities. Frequently, these patterns are noted only empirically, and the underlying mechanisms are not clearly understood. As more species and more relationships are considered, a detailed understanding of mechanisms is often lost as perception of more general patterns is gained. In some cases, this is due to our lack of knowledge about multiple-species relationships. In some well studied systems, the sheer number of interactions that may occur is constraining. At these levels, the complexity of interactions is not easily followed, even if species-to-species interactions are well understood. Several multiple-species interactions that have been documented are discussed below.

8.4.1. Species Shifts

Species shifts caused by pesticides can be short-term, caused by immediate selectivity and resistance differences (e.g., Trumble 1984, 1985), or long-term, as the interplay of chemical selection and ecological interactions is influenced through time. Many factors are involved in these complex interactions between pesticides and species populations. As mentioned in

Section 8.3.3, competition is considered one of the most important elements in determining the relative abundance of natural enemy species, although this is rarely proven. Interspecific competition may be radically altered by pesticides. Pesticide-induced changes in species composition may be fairly easy to discern, but may require much study to identify causation among factors leading to these shifts, and very few researchers have worked at this level of detail.

In a series of studies extending over a period of 10 years, Dabrowski (1968, 1969b, 1970a, 1970b) examined the mechanisms behind patterns of species shifts of tetranychid and phytoseiid mites in pesticide-treated apple orchards in Poland. Most compounds resulted in decreases in populations of *Amblyseius finlandicus* and in increases in *Typhlodromus aberrans, T. tilarium*, and *T. pyri*. Among the many factors contributing to species shifts in predaceous phytoseiids were 1) differences in intrinsic tolerance or susceptibility to the various pesticides tested, 2) differences in the degree of developed resistance, 3) changes in abundance of preferred or suitable prey, 4) changes in phytoseiid-feeding macropredators and in tetranychid-feeding competitors (i.e., *Orius, Anthocoris, Chrysopa*), 5) differences in the relative age structure of populations at the time of spraying, and 6) behavioral characteristics relative to the use of refugia.

Other studies have reported similar patterns of predictable shifts in natural enemy species following pesticide applications, although the contributing factors have not usually been determined. In some cases the overriding factor appears to be a single mechanism, such as a strongly developed pesticide resistance in one natural enemy species, while others are susceptible (e.g., among phytoseiid mites this phenomena is well documented; Croft and Hull 1983). In other cases, many factors on the order of those delineated by Dabrowski interact to cause the observed shifts.

Several researchers have documented species shifts in the natural enemy complex associated with a crop over long periods of time. These shifts are usually correlated with changing patterns of insecticide usage; however, changes in environmental conditions, cultural patterns, and host and crop genetics could be complicating factors which coincide with the pesticide/natural enemy interaction.

Chestnut and Cross (1971) analyzed results of boll weevil parasitoid surveys taken in the southern United States at periodic intervals over a period of 31 years. The six most important parasites collected in the survey were present in both 1934 and 1965 samples. *Bracon mellitor* was the most common species in both samples; in fact, parasitism by this species was higher in treated fields in 1965 than in untreated fields in 1934, lending credibility to the demonstration of pesticide resistance potential in this species by Adams and Cross (1967). The parasitoid *Aliolus curculionis*, however, went from a very minor species in 1934 to the second most abundant species in 1965. The authors concluded that developing patterns of insecticide resistance were responsible for the changes and the overall increase in parasitism over the 31-year period.

Boyce and Dustan (1955) compared parasitism rates of the Oriental fruit moth *Grapholitha molesta* for a 14-year period spanning the introduction of DDT in peach orchards of Ontario, Canada, in the mid-1940s. The use of DDT (and later parathion) appeared to significantly decrease the relative abundance of the native parasitoids *Glypta rufiscutillaris* and *Horogenes obliteratus*, while increasing parasitism by the native *Macrocentrus delicatus* and the introduced *M. ancylivorus*.

Ibrahim (1962) compared the numbers of generalist predators of cotton pests in three major growing regions of Egypt for the period before and after the widespread introduction of organic pesticides into the country. Similar sampling techniques were used in surveys in 1939–1941 and 1959–1961. In general, populations of most predators showed little change in the middle Egypt region, but showed significant decreases in the delta regions. An exception to this trend was found for certain spiders and for the anthocorid *Orius sp.*, which increased in two of the three study areas.

The value of these long-term studies lies less in pinpointing actual mechanisms of pesticide effect on natural enemy interactions than in broadly tracing regional trends which may provide valuable historical context. Long-term studies may set the stage for more detailed investigations of insecticide resistance, competition, stability, and so forth.

8.4.2. Richness, Evenness, and Diversity

Although mechanistic, cause and effect explanations of pesticide-induced species shifts have not always been forthcoming, some workers have attempted to describe these patterns in terms of ecological parameters which broadly quantify the observed changes. These quantitative studies have generally sought to conceptually characterize ecological response rather than to pinpoint details regarding particular species which may be important to the economic entomologist.

One of the earliest quantitative ecological investigations of natural enemy/pesticide interactions was by Barrett (1968). He evaluated the impact of carbaryl on a semienclosed grassland ecosystem in Georgia. Total arthropod numbers and biomass were still reduced by over 95% 5 weeks after treatment, despite the fact that insecticide residues degraded by over 90% within 14 days. An index of species diversity (Margalef 1957) for both phytophagous and predaceous species is shown as a function of time in Fig. 8.4a. Phytophagous species showed an immediate and dramatic short-term reduction in diversity following treatment, but rapidly recovered to control levels within 2 weeks. Diversity of predaceous arthropods, however, remained markedly depressed for 5 weeks following treatment. The author concluded that this prolonged effect on natural enemies was due to the reduced food supply (i.e., the phytophagous prey), resulting in a decrease in natality or reinvasion from surrounding unsprayed areas, or both. In contrast to diversity values, equitability values (Lloyd and Ghelardi 1964) for phytophagous insects "overshot" the control values and remained elevated to the end of the season (Fig. 8.4b). For predaceous insects, equitability was altered only temporarily. Thus, the species richness and evenness components of diversity reacted in opposite manners in response to pesticide disruption.

In similar experiments, Suttman and Barrett (1979) applied carbaryl to Ohio oat field and old field communities and measured changes in density, biomass, and species richness and apportionment indices for both herbivores and entomophagous arthropods. In the less complex oat field, treatments resulted in greater reductions in density and biomass of both trophic levels than in the more complex old field. However, recovery times were faster in the oat field. Within the Coleoptera, herbivore (i.e., Curculionidae, Chrysomelidae) responses were more immediate, pronounced, and short-term than were those of carnivores (i.e., Carabidae, Coccinellidae, Staphylinidae), as reflected in measures of density, diversity, and evenness (Figs. 8.5a, b, and c). The authors concluded that the major difference between the pesticide response of herbivorous and carnivorous beetles was that it took the carnivores longer to respond to a reduced food supply and they only recovered from the disruption toward the end of the growing season, while the herbivores' food supply was largely unaffected after spraying.

Culin and Yeargan (1983) used the Shannon–Weaver species diversity index to measure the effect of carbofuran, dimethoate, and azinphosmethyl on foliage-inhabiting spiders in Kentucky alfalfa fields. They observed that treated plots showed increased

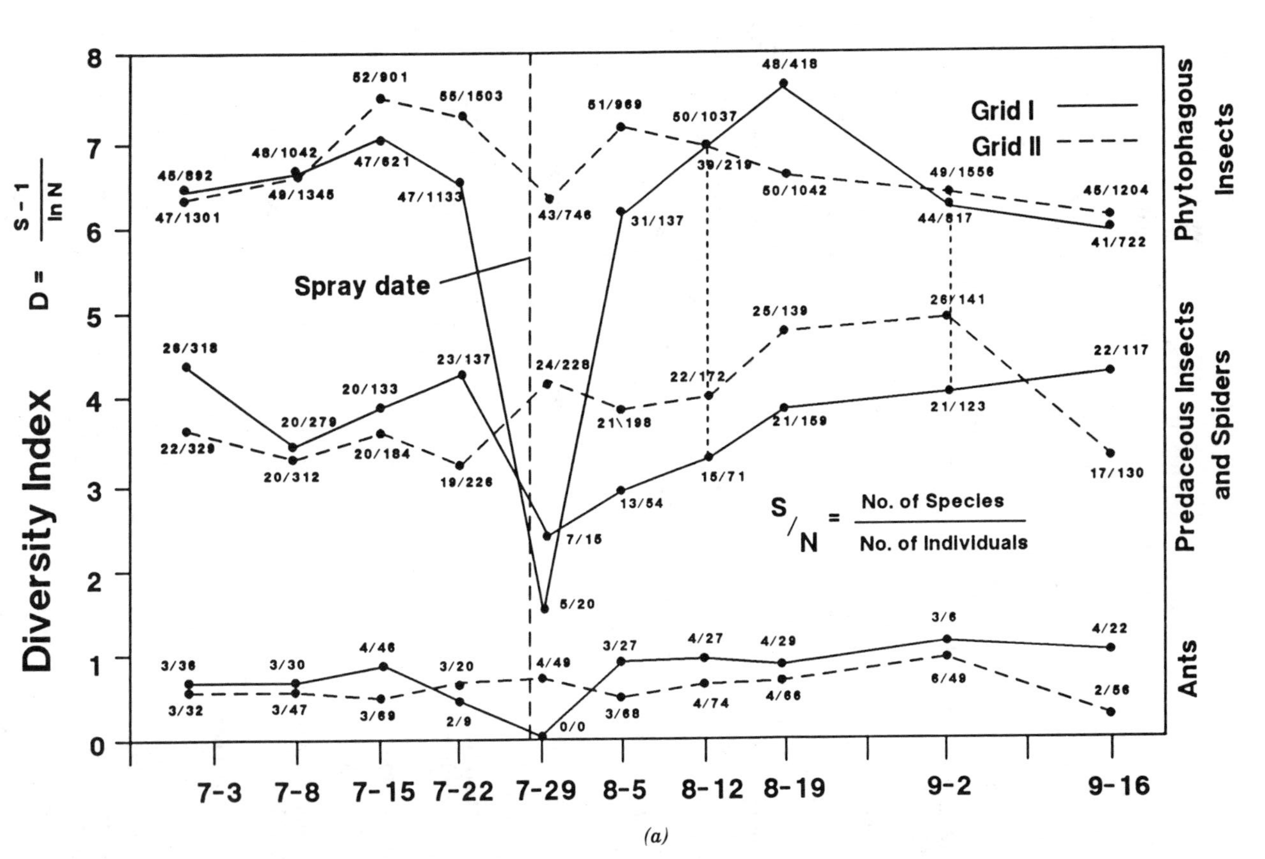

(a)

Figure 8.4a. Effects of carbaryl on the diversity of insect species for the major trophic levels in a grassland ecosystem (after Barrett 1968).

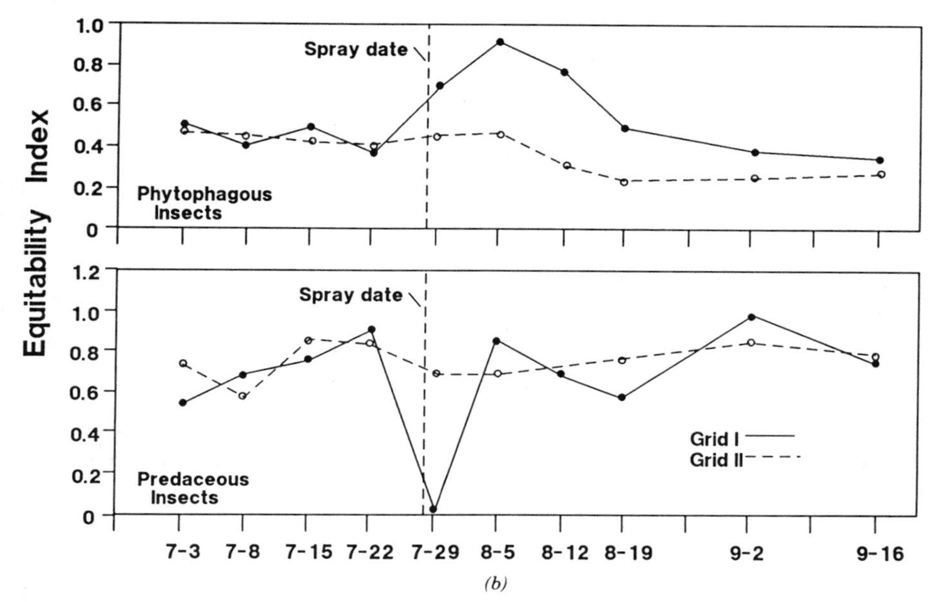

(b)

Figure 8.4b. Effects of carbaryl on the equitability index of insects from two trophic levels in a grassland ecosystem in Autumn 1968 (after Barrett 1968).

species evenness due to high mortality of the most abundant spider, *Tetragnatha laboriosa.* Gray and Coats (1983) calculated relative importance values (Pedigo 1967) for nontarget arthropods in an Iowa corn field treated with various combinations of the insecticide carbofuran and the herbicides alachlor and cyanazine. While the insecticide had little or no effect on the importance values of the nontarget species (mainly Carabidae), plots receiving herbicide applications showed lower importance values for several carabid species. Although herbicides have been shown to have direct toxic effects on some carabids (Muller 1974), the study of Gray and Coats (1983) did not distinguish between direct and indirect effects, and the herbicidal effects on weeds leading to changes in humidity, temperature, and herbivory were probably also important.

Vickerman and Sunderland (1977) compared species diversity and equitability in a winter wheat arthropod community in West Sussex, U.K., as influenced by dimethoate applications to control cereal aphids, mainly *Sitobion avenae.* Seven days after spraying, the diversity index in the treated field was 53% lower than that in untreated control plots. While both phytophagous and entomophagous species diversity was reduced by the treatment, the herbivores returned to a normal level in 6 weeks, while the carnivores required over 2 months (Fig. 8.6). The equitability of herbivorous species increased sharply after treatment and remained high for several months, while for predaceous species there was only a slight and temporary increase.

The pattern that emerges from these studies appears to indicate that pesticide applications have a more immediate, short-term effect on species diversity of phytophagous arthropods, and a more delayed, long-term influence on entomophagous species. This appears true for both commercial cropping systems (i.e., alfalfa, wheat) and more complex, mature ecosystems (i.e., old field). The explanation for this pattern lies in the food

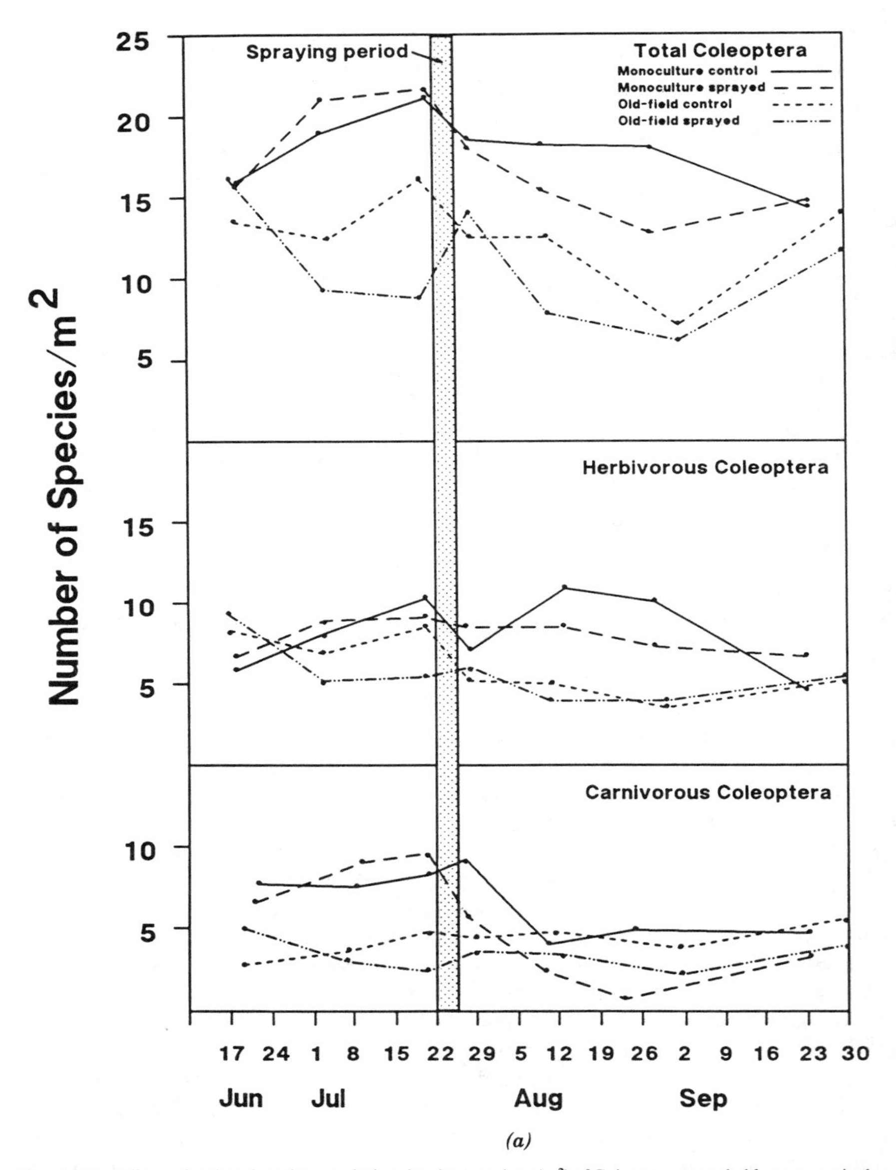

(a)

Figure 8.5a. Effects of carbaryl on the population density (numbers/m^2) of Coleoptera sampled from an agricultural monoculture and old-field plant community (after Suttman & Barrett 1979).

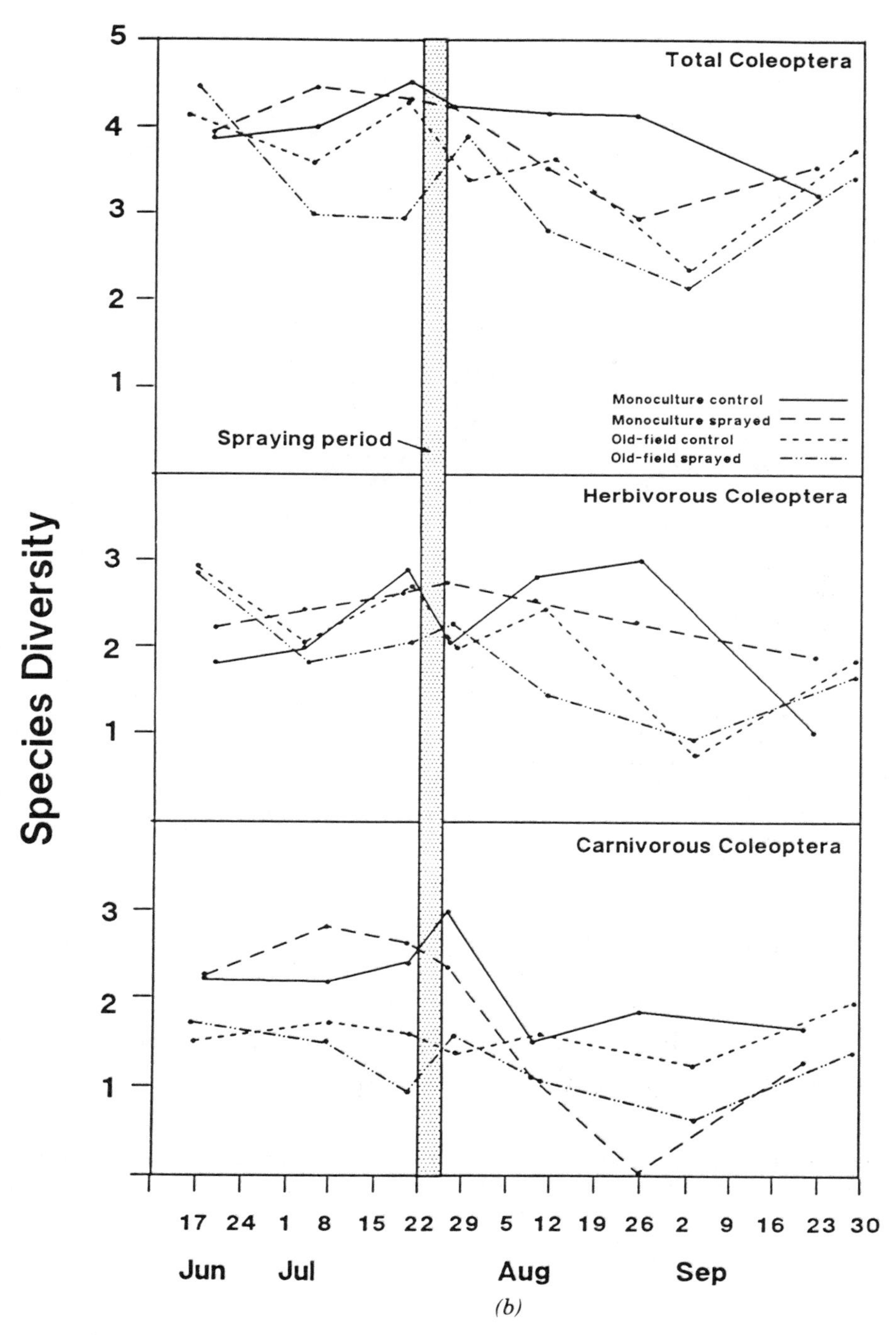

(b)

Figure 8.5b. Effects of carbaryl on the species diversity $D = S - 1/(\ln N)$ of Coleoptera sampled from an agricultural monoculture and old-field plant community (after Suttman & Barrett 1979).

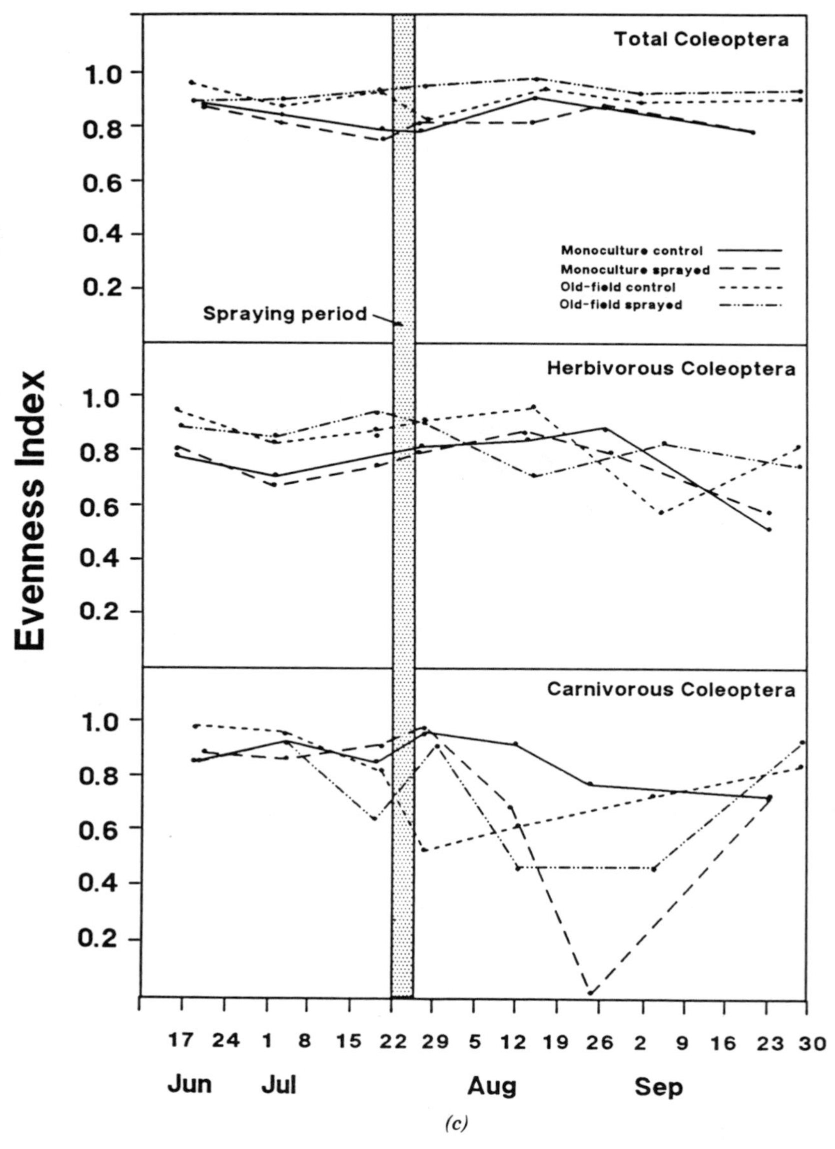

(c)

Figure 8.5c. Effects of carbaryl on evenness index values of Coleoptera sampled from an agricultural monoculture and old-field plant community (after Suttman & Barrett 1979).

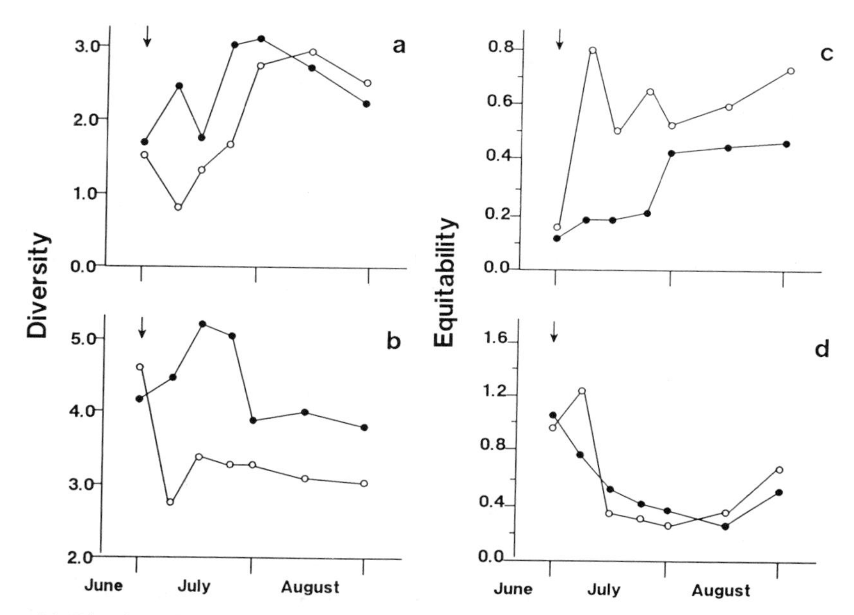

Figure 8.6. Diversity and equitability of phytophagous (*a, c*) and predaceous (*b, d*) arthropods in winter wheat before and after spraying with dimethoate (after Vickerman and Sutherland 1977).

dependency relationship between the entomophage and the phytophage. Whereas, following a major pesticide disruption, the surviving or recolonizing herbivores face an almost unlimited food supply, the carnivores are dependent upon the buildup of herbivores (and other species) for their nutritional needs.

In terms of species evenness or equitability, the indices for phytophagous arthropods tend to increase rapidly following pesticide treatment and remain elevated for a considerable time period, while values for entomophagous arthropods remain relatively unaffected. This trend may be an artifact of the entomophagous species under consideration, as most studies have focused on generalist predators rather than on the more host-specific parasitoids. Since pesticides are generally targeted against the most abundant pests that are specializing on the crop plant, these major, abundant species are drastically reduced, and herbivore evenness indices subsequently go up. A wider array of generalist (and phenologically diverse) predators is more likely to be impacted by the toxicants, however, and thus while total numbers and biomass may decrease sharply, evenness indices will be little affected. One would expect, however, that parasitoid species evenness would respond more similarly to that of herbivores, since the dominance–diversity curves would tend to parallel those of the hosts.

8.4.3. Uses of Simulation Models

To characterize natural enemy population and community response to pesticides, some researchers have turned to simulation models to mimic ecological interactions in quantitative terms. Simulation models of natural enemy/host interactions have usually incorporated the effects of insecticides as an additional mortality factor.

Sasaba et al. (1973) developed a descriptive model of pesticide effects on the green rice leafhopper *Nephotettix cincticeps* and its principal predator, the spider *Lycosa pseudoannulata*. The model predicted resurgence of leafhoppers after use of nonselective insecticides, which was corroborated by field observations. The model postulated an optimum predator/prey ratio for biological control and suggested the use of reduced applications of selective insecticides.

Kiritani (1977) and Rudd et al. (1980) used simulation models to examine pesticide effects on generalist natural enemies of rice and soybeans, respectively. In their models, natural enemy populations were considered independent of the population dynamics of the pests, and specified natural enemy equilibrium levels were depressed by insecticides with subsequent effects determined by predator and prey densities. In more theoretical studies, Barclay (1982) and Barclay and van den Driessche (1977) developed analytical models to examine pesticide impact alone and in combination with predator releases and habitat management on coupled predator and pest populations.

One of the more theoretical models of the interactions of insecticides with host parasitoid relationships was reported by Hassell (1984). He examined the equilibrium and local stability properties of these types of models in discrete time by distinguishing between different sequences of parasitism and insecticide kill in the host's life cycle. Four possibilities were considered (referred to hereafter as conditions 1–4): 1) insecticides acting before parasitism and only killing the host, 2) insecticides acting after parasitism and only killing the host, 3) insecticides acting after parasitism and also killing the parasitized host at the same rate, and 4) insecticides acting before parasitism and also killing the adult parasite. Overall, there was a distinct ranking of the different application tactics and their effects on the depression of the host equilibrium and their contribution to the stability of pest populations. Generally, condition 2 consistently led to the greatest host depression, while condition 4 usually raised the host's equilibrium. Conditions 1 and 3 were intermediate in their effects on pest depression over time. Condition 1 allowed the broadest stability conditions. In short, Hassell concluded that insecticides affecting either adult or larval parasitoids within hosts make for narrower stability conditions and greater risk of host resurgence than if only hosts were affected by pesticides.

Waage et al. (1985), in further building upon the inventory of host/parasitoid population models (see Hassell 1978 and Hassell and Waage 1984 for reviews), studied the impact of insecticides on "host depression" or "host resurgence" resulting from insecticide use. By breaking the interacting host and parasitoid life cycles into discrete components (Fig. 8.7), the timing of the insecticide application relative to parasitoid attack and the damage causing host stage were examined, while varying the parameter values simulated differences in relative toxicity. A q value was used to indicate the ratio of host population densities with and without pesticide input. (Values of $q < 1$ thus indicate host depression, $q = 1$ indicates no change, and $q > 1$ indicates an increase in host equilibrium, or pest resurgence.)

The model demonstrated, first of all, how insecticides and parasitoids complemented one another in depressing host equilibria in the case where parasitoids suffered no insecticide mortality due to a resistant life stage or refuge (i.e., pupae or within the host). It also showed that pesticide impact was greater for higher values of the pest rate of increase, and for higher values of K in the Poisson distribution (indicating greater clumping).

In the more complex interaction where pesticides kill both pest and parasitoid, the expected result of a reduction in host depression was demonstrated. It was concluded that q values can be greater than 1 (i.e., pest resurgence) when pesticides are applied at some stages in the life cycle, but less than 1 at other stages. Varying the relative susceptibilities

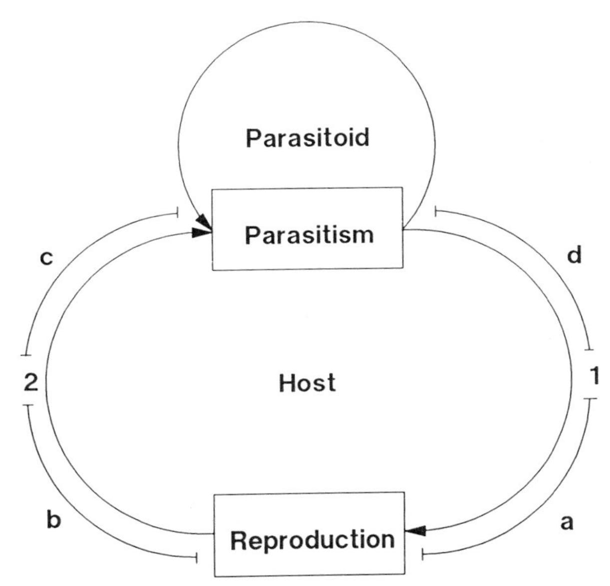

Figure 8.7. Schematic diagram of host and parasitoid life cycles. Insecticides can kill hosts either between parasitism and reproduction (1) or between reproduction and parasitism (2). Reproduction, parasitism, and the two times of insecticide action divide the host life cycles into four periods labelled *a, b, c,* and *d.* (After Waage et al. 1985.)

showed the obvious result that increased toxicity to parasitoids increased q values and led to greater probability of resurgence. The biologically realistic patterns of differential susceptibility indicated that pesticides applied before parasitism and killing parasitoid adults would be more likely to cause resurgence than applications following parasitism and killing parasitoid larvae.

The model also revealed an interesting trade-off between the relative toxicity of an insecticide to pest and parasitoid (Fig. 8.8) Any particular q value less than 1 (i.e., host depression) can be caused either by an insecticide that kills a high proportion of parasitoids and hosts, or by one that kills a low proportion of parasitoids and hosts (i.e., points A and B, respectively, on isocline for $q = 0.75$). Point A might correspond to an insecticide of high toxicity and persistence, while point B might represent one of lower toxicity, such as a microbial. Theoretically, we can accomplish the same effect in terms of host reduction with benign pesticides by tracing the isoclines out and maintaining the proper parasitoid–host ratios.

The limitations of this model, and of most simulations of this type, are that the assumptions which are made to simplify complex systems and make the mathematics tractable are often far removed from the dynamics of real populations of arthropods in field situations. For example, many host (and natural enemy) populations manifest overlapping generations, with insecticides acting on several stages at once. Also, host and parasitoid populations are probably rarely, if ever, at equilibrium, especially in highly disrupted temporary agroecosystems. Susceptibility (to both parasitism and insecticides) is not constant for different life stages of the host. And, as admitted by the authors, fine tuning of pesticide application timing may not always be compatible with other farming practices. The strength of the model, and of the modeling approach in general, is the

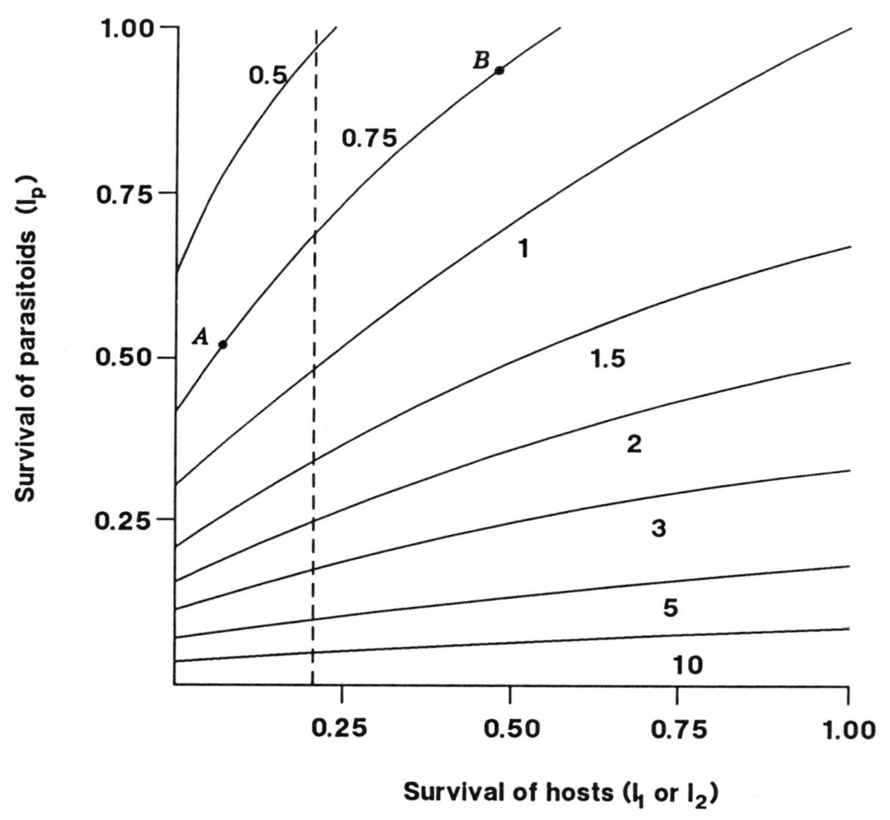

Figure 8.8. Isoclines of q values for different levels of pest I_1 or I_2 and parasitoid I_p survival following insecticide application. Values of $q > 1$ = resurgence; the host population will go extinct when host survivorship is below 0.2, or to the left of the dashed line. (After Waage et al. 1985.)

attention focused on the questions accessible to experimental study, and the conceptual organization of essential components in exceedingly complex systems.

8.5. CHARACTERISTICS OF PESTICIDE-STRUCTURED COMMUNITIES

There are a few regional cropping systems that have received such constant or regular applications of given pesticide types that they have developed what may be called *pesticide-structured arthropod communities.* In these situations, the regularity of pesticide use is such that local or even regional arthropod communities have become adapted to the toxicants as fixed components of the environment. This situation may be more common and widespread than is presently realized, but the arthropod community structures of relatively few agroecosystems have been studied in adequate detail to make these generalizations.

8.5.1. Examples from Deciduous Tree Fruits

The apple cropping system in North America is one that has received detailed and extensive study for many years in terms of the arthropod fauna and its response to certain standard pesticide regimes. In a way, the apple system is unique in that a predominant insecticide effective against the key pests in the system (azinphosmethyl) has remained widely used for over 20 years with no substantial loss of efficacy due to resistance, and hence no need to change compounds. Entomologists with lengthy experience in apple systems can predict, with reasonable accuracy, the faunal composition of orchards based on the known pesticide application history of a particular plot and region (Croft and Hull 1983).

A good example of the natural enemy component of this pesticide-structured system may be drawn from the complex of phytophagous and predaceous mites on apple foliage. In unsprayed apple orchards of eastern and midwestern North America, there is usually a diverse complex of 5–15 predatory phytoseiid species which feed on a complex of 3–5 phytophagous species (Table 8.4). Normally, population levels of the pest mite species are low and very little leaf damage is observed in unsprayed orchards. Once spraying begins, particularly susceptible species of both pests and natural enemies are displaced by moderate to highly resistant forms of mites. For example, among pests, *Bryobia* spp. are replaced by *Panonychus ulmi* and *Tetranychus ulmi*, whereas the apple rust mite *Aculus schlechtendali* seems to persist in both unsprayed and sprayed systems. Among predators, species such as *Typhlodromus pomi, T. caudiglans, Amblyseius urbraticus*, and *Phytoseius macropilus* are excluded by sprays, while resistance-prone species like *A. fallacis* soon displace them and increase to high levels in orchards sprayed with selective acaricides. Insecticide-tolerant stigmaeid mites *Zetzellia mali and Agistemus fleschneri* persist in both untreated and treated orchards. These patterns of species occurrence in relation to orchard spraying patterns are constant in several other areas of this region (Strickler et al. 1987) to the extent that they can be used as indicator species for determining the types and intensity of pesticide used and the time since trees have gone unsprayed or been abandoned (Croft and Hull 1983).

8.5.2. Examples from Soil-Inhabiting Communities

Several workers investigating the effects of regular applications of insecticides on soil-inhabiting predaceous beetles have documented similar patterns of reduced species diversity and increased abundance of what might be called indicator species. Gregoire-Wibo (1980, 1983a, 1983b) and Gregoire-Wibo and van Hoecke (1979) examined the ecological effects of aldicarb on the carabids and gasmid mites feeding on pest collembola in Belgian sugar beet fields over a 4-year period. Population densities of 21 carabid species were followed, with densities varying greatly between fields in relation to microclimate and pesticide treatments. Of the species studied, 15 reproduced in spring; these were markedly reduced by aldicarb for the entire growing season. In comparing treated and untreated fields, distinct shifts in species abundance were recognized. For example, *Metallina lampros* comprised over 50% of the specimens in untreated plots, but only 10% in treated plots, where it was replaced by the previously rare *Phylla* sp., which probably was tolerant of the carbamate pesticide. Similar trends of species displacement due to differential susceptibility were noted in carabids which reproduced mostly in summer. After several years of study, the authors were able to classify each carabid species in relation to its

Table 8.4. Principal Phytophagous and Predaceous Mites in Unsprayed versus Sprayed Apple Orchards in Midwestern and Eastern United States by Location.

Sprayed Orchards		Unsprayed Orchards		References
Pest sp.	Predatory sp.	Pest sp.	Predatory sp.	
		Michigan		
A. schlechtendali, *P. ulmi*, *T. urticae*	*A. fallacis*, *Z. mali*	*A. schlechtendali*, *Bryobia* sp, *E. uncatus*	*T. caudiglans*, *T. pomi*, *A. umbraticus*, *A. sessor*, *A. fleschneri*, *Z. mali*,	Strickler et al. 1987
		Massachusetts		
A. schlechtendali, *P. ulmi*, *T. urticae*,	*A. fallacis*, *Z. mali*, *A. fleschneri*,	*A. schlechtendali*, *Bryobia arborea*	*T. pomi*, *A. sessor*, *A. umbraticus*, *P. macropilus*, *Z. mali*	Hislop and Prokopy 1979
		Missouri		
P. ulmi, *T. urticae*	*A. fallacis*, *T. longipilus*	Unreported	*T. longipilus*, *T. pomi*, *A. sessor*, *A. ovatus*, *A. mexicanus*, *A. umbraticus*,	Poe & Enns 1969
		North Carolina		
P. ulmi, *T. urticae*, *A. schlechtendali*,	*A. fallacis*, *P. asetus*, *T. sessor*, *Z. mali*, *A. fleschneri*	*B. arboreus*, *T. urticae*, *T. schoeni*, *A. schlechtendali*	*A. sessor*, *A. nodosa*, *T. longipilus*, *A. umbraticus*, *P. macropilus*, *T. peregrinus*, *T. conspicua*, *Z. mali*, *A. fleschneri*	Farrier et al. 1980
		New Jersey		
P. ulmi, *T. urticae*, *A. schlechtendali*	*A. fallacis*, *A. fleschneri*	*T. schoeni*, *B. praetiosa*, *A. schlechtendali*	*T. pomi*, *T. longipilus*, *A. andersoni*, *A. morgani*, *A. putmani*, *P. macropilus*, *T. conspicuus*, *A. fleschneri*	Knisley & Swift 1972

pesticide sensitivity, cycle of activity, and habitat preference to the extent that the carabid complex resulting in any set of conditions was predictable.

In the studies of Dempster (1968), laboratory and field investigations were combined to examine factors governing the species abundance of carabids in crucifers in the United Kingdom following applications of DDT. Using pitfall traps, he observed a relatively diverse fauna of predaceous carabids in unsprayed plots. However, following spraying, the majority of species were greatly reduced, while two species (*Trechus quadristriatus* and *Nebria brevicollis*) became two to three times more abundant than in the unsprayed area. At least two factors appeared to be responsible for these increases. First, collembola tended to increase following DDT sprays, and both of these predators fed primarily on collembola. Secondly, these two species were more tolerant to DDT than other carabids and so were able to exploit this resource in the absence of competitors. Dempster (1968) also discussed the complex interaction of direct and indirect factors influencing community changes caused by pesticide use, including differential tolerance among species, differential exposure, recolonization potentials, and so forth. These studies, when combined with those of Critchley (1972a, 1972b, 1972c), give considerable insight into the intrinsic and extrinsic factors influencing carabid susceptibility to pesticides (Chapter 4).

8.5.3. Applications to Regional Pest Management

While the direct toxic effects of pesticides are usually limited to the field or orchard on which they are applied, several examples have been cited as to how indirect effects may extend over time and space to influence ecological systems in a much broader sphere. As IPM approaches a more ecological orientation—management rather than eradication of pests—these broader effects will receive greater emphasis and may be incorporated into comprehensive, regional IPM.

Several studies have documented the regional effects of pesticide use on natural enemies. As early as 1954, Way et al. (1954) pointed out the importance of hedgerows and other native vegetation well distributed throughout cultivated cropland as a refuge from pesticides for beneficial arthropods. Kobayashi et al. (1978) used an index of intensity of application to measure the extent of rice acreage in a region sprayed with BHC and parathion. Expressed as the percentage of the cumulative area sprayed to the total acreage of rice fields in the region, this index was well correlated with changes in both entomophages and herbivores. In general, increased regional spraying resulted in reduced levels of Hymenoptera, spiders, and other beneficial species and increased populations of planthoppers and other pests.

This regional perspective has been given due consideration in several projects examining pesticide/natural enemy interactions. Emel'yanov and Yakushev (1981) emphasized that insecticidal control of pests of winter wheat should be limited to the centers of fields, so that borders and surrounding wooded areas in the region could serve as reservoirs for predaceous carabids. Skrebtsov and Skrebtsova (1980) showed how the interspersion of flowering onions (grown for seed) and other crops attracted adult syrphids, suggesting the utility of this crop as a reservoir for protecting predators when surrounding fields were sprayed. In the USSR, microreserves have been established wherein native vegetation is preserved and beneficial arthropods are allowed to multiply. These reserves, ranging from 3 to 6 hectares each, were reported to benefit local farms (Grebennikov 1980).

From an even broader spatial perspective, Sterling et al. (1984) commented on the widespread use of insecticides to control the imported fire ant *Solenopsis invicta* in the southern United States. Although considered a pest, the fire ant is also an effective predator of the boll weevil *Anthonomus grandis* and other pests in cotton, and its widespread eradication across the southern states has severely interfered with weevil biological control. The authors recommended that in some areas throughout the region the fire ant be left unpoisoned so as to contribute to boll weevil control.

PART FOUR

PESTICIDE SELECTIVITY

9

PHYSIOLOGICAL SELECTIVITY

9.1. INTRODUCTION

Selectivity is the use of pesticides to kill plant-feeding arthropod pests while not affecting their entomophagous natural enemies (and other nontarget species). *Selectivity to natural enemies is the central theme of this volume.* More specifically, the concepts of physiological and ecological selectivity (Chapter 10) are the unifying hub around which the topics of pesticide uptake, toxicology, susceptibility, and resistance in natural enemies are interrelated (Fig. 9.1).

Pesticide selectivity may also be classified by scope or degree. Simple, species-specific selectivity refers to pesticide selectivity to one natural enemy over one prey or host species. Broad or multiple-species selectivity refers to several natural enemies being favored over one or more pests. Of course, many intermediates lie between these extremes. Broad selectivity is optimal for managing the large complexes of pests and natural enemies on agricultural crops. Generally, broad physiological selectivity has been rare in IPM.

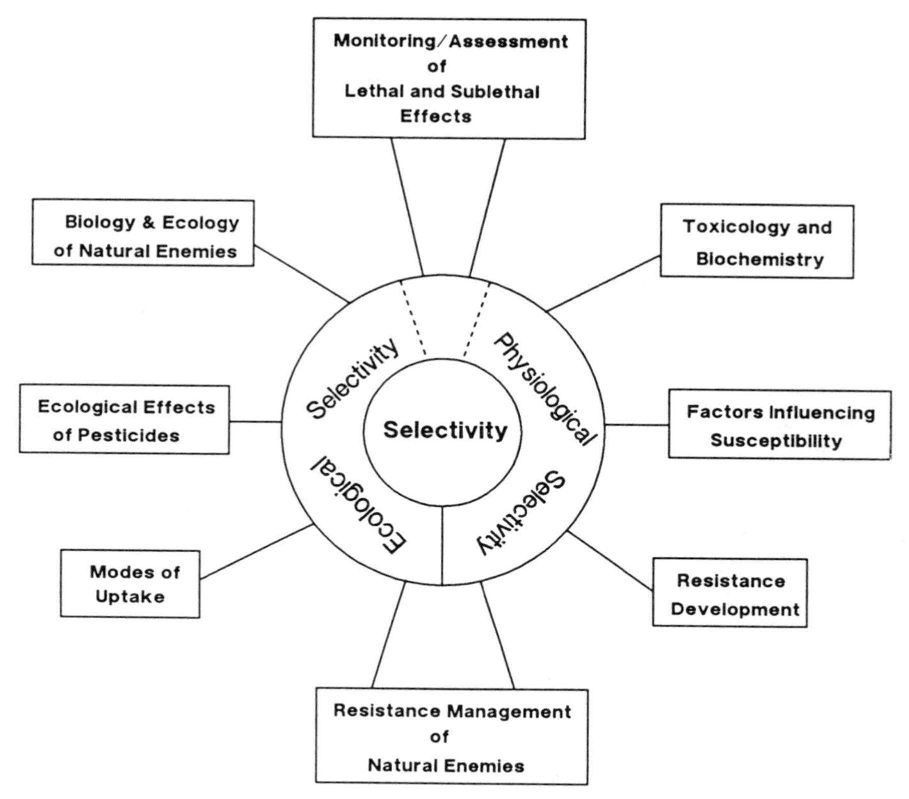

Figure 9.2. Selectivity relationship between the aphid pest *Metopolophium dirhodium* and five natural enemies to three pesticides expressed as log concentration–probit mortality lines (after Brown et al. 1983).

Physiological selectivity is addressed here before ecological selectivity (Chapter 10) for several reasons. Physiological selectivity is an inherent property of a pesticide at a particular dose, regardless of how it is used. Ecological selectivity is conditional, depending on pesticide manipulation through timing or placement. Physiological selectivity operates primarily at the physiological or organismal level, whereas ecological selectivity operates at population and community levels. In general, this volume proceeds from organismal to community levels of organization (Chapter 1).

Most pesticide selectivity to natural enemies has been ecological or a combination of physiological and ecological, rather than being strictly physiological. Even though broad physiological selectivity is rare, some degree is probably involved in many cases classified as ecological (Chapters 10 and 11). In practice, the distinction between ecological and physiological selectivity is often difficult to make. Part of the problem is methodological and is intrinsic to the measurement of the modes of uptake and action of pesticides (Chapters 3 and 6). To some degree, the problem of classifying selectivity is semantic in that we are trying to make discrete classifications when the underlying processes are continuous.

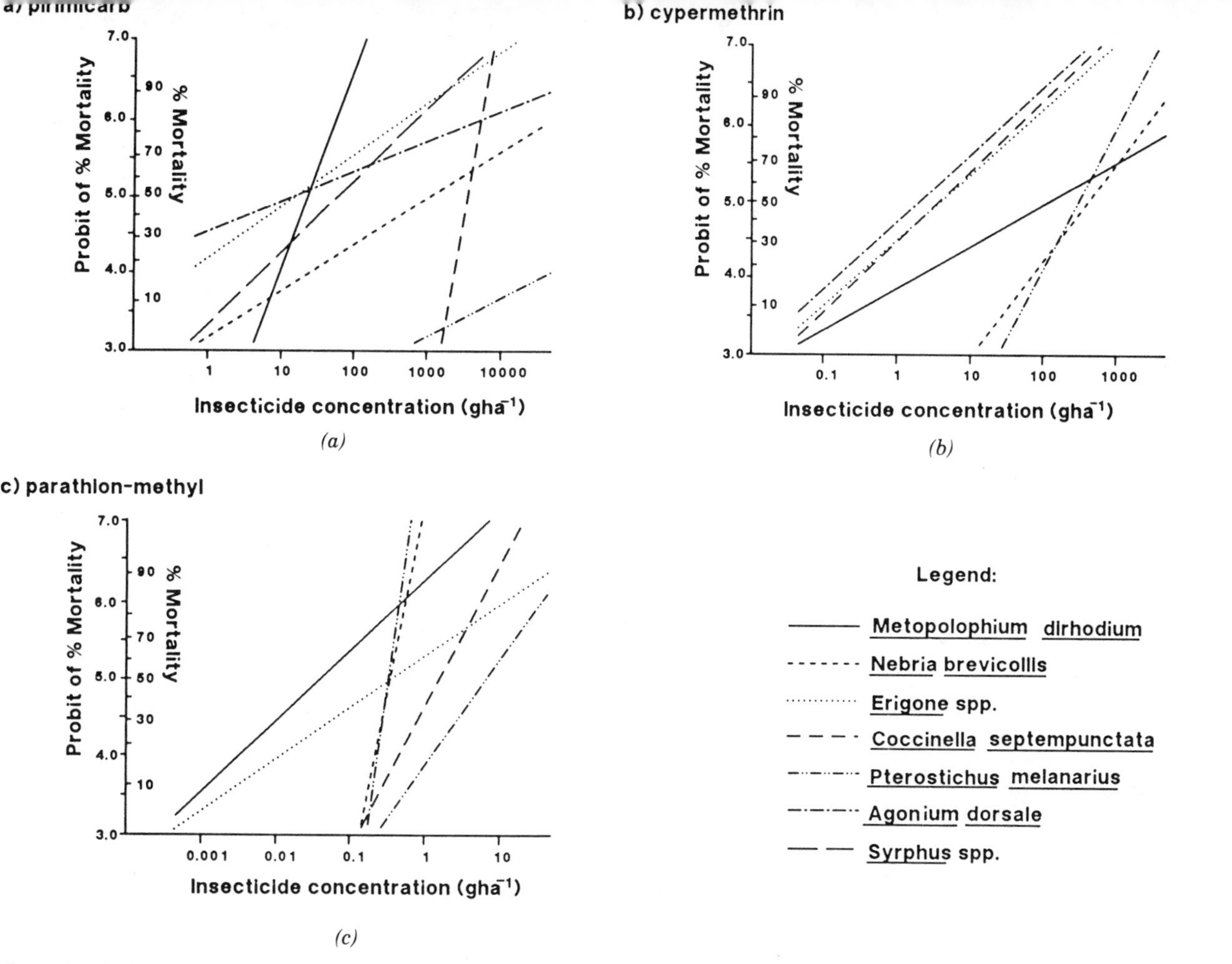

Figure 9.2. Selectivity relationships between the aphid pest *Metopolophium dirhodium* and five natural enemies to three pesticides expressed as log concentration–probit mortality lines (after Brown et al. 1983).

9.2. DEFINITION OF PHYSIOLOGICAL SELECTIVITY

Physiological selectivity refers to greater activity of a pesticide on a pest than on a natural enemy when direct contact on both species has occurred (Chapter 1). Physiological selectivity primarily involves movement of the pesticide on or within the arthropod's body and its interaction with the target site. Physiological selectivity involves such processes as penetration, translocation, activation, or degradation (Chapters 3 and 6). Other processes involved in physiological selectivity for both natural enemies and their hosts or prey are sequestration, excretion, and selective metabolism, which includes detoxification and target site insensitivity (Chapters 6 and 15).

In Fig. 9.2, different degrees of physiological selectivity (as LC_{50} curves) are illustrated for a guild of natural enemies and their prey *Metopolophium dirhodium* (Aphididae), which attacks cereal crops (Brown et al. 1983). Responses for the aphid and its predators *Coccinella septempunctata* (Coccinellidae), *Syrphus* spp. (Syrphidae), *Pterostichus melanarius* (Carabidae), *Nebria brevicollis* (Carabidae), *Agonium dorsalis* (Carabidae), and *Erigone* spp. (Araneae) to pirimicarb, cypermethrin, and parathion-methyl are illustrated. As can be seen, parathion-methyl is the most toxic compound, pirimicarb is the least, and cypermethrin is moderately toxic to the aphid and its natural enemies. In relation to selectivity at concentrations causing $>90\%$ mortality of the aphid, pirimicarb exhibits some selectivity to all six predators, parathion-methyl to three of five predators, and cypermethrin to none. The relative susceptibilities of the predators and aphid vary by insecticide, as indicated by position and slope (associated variability) of the concentration–response curves. These selectivity data are in a form commonly reported in the literature (Theiling 1987).

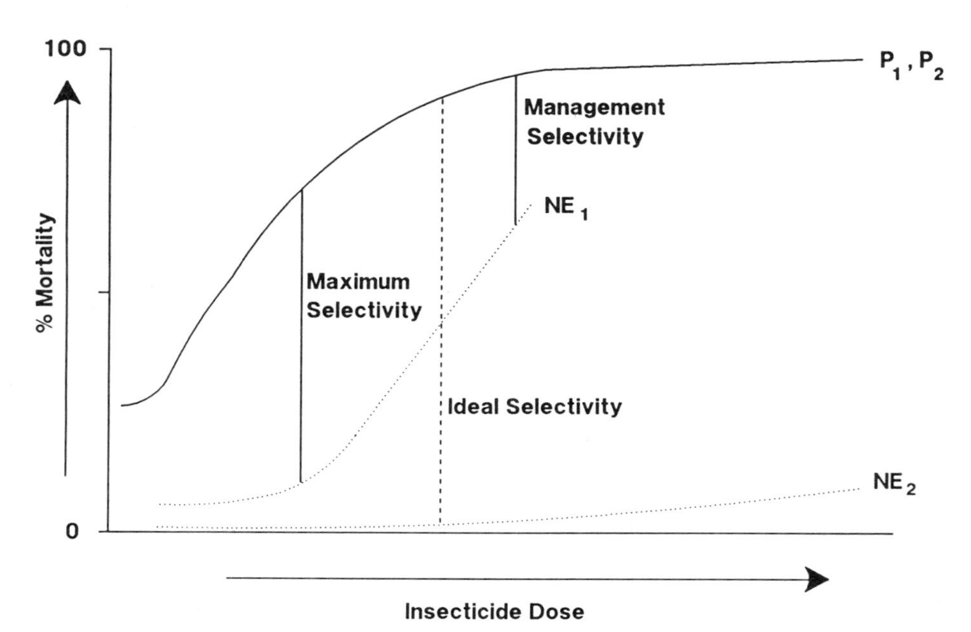

Figure 9.3. Relationship between maximum, management, and ideal physiological selectivity of an insecticide to a natural enemy and its prey or host (Modified from Hower and Davis 1984).

Rather than comparing LD, LC, or LT_{50} values or lines, it has been suggested that selectivity be defined near the point where pest mortality is high to moderate and little or no mortality occurs in the natural enemy (Bartlett 1964, Hower and Davis 1984). Several perspectives on selectivity are illustrated in Fig. 9.3 for hypothetical pests (response line P_1, P_2) and natural enemies (response lines NE_1 and NE_2). (A specific application of these concepts for the leafhopper *Empoasca fabae* and the endoparasitoid *Microctonus aethiopoides* is given in Hower and Davis 1984.) *Maximum selectivity* is the point where the difference in mortality of the pest and the natural enemy is the greatest (Fig. 9.3, see $P - NE_1$). However, *management (or realized) selectivity* may be chosen so as to achieve adequate pest control at the expense of greater natural enemy mortality. (This situation arises for many reasons, one being because few pesticides are benign to natural enemies at dosages needed for pest control; Chapter 2.) Management selectivity is determined by the economics of pest control and the benefits of control provided by the pesticide versus the biological control agent. The amount of time required for each agent to achieve pest control is also a factor. In choosing a pesticide rate for IPM, pest control is usually given precedence over selectivity to the natural enemy.

As shown in Fig. 9.3, the narrow range of difference in the dose–response curve between pest and natural enemy does not allow for adequate control of the pest while conserving the natural enemy to a very large degree. *Ideal selectivity* would be achieved if moderate to high mortality of the pest was possible, while the natural enemy was unaffected over a range of pesticide dosages (Fig. 9.3; see $P - NE_2$). With ideal selectivity, the need for chemical pest control is diminished because the surviving natural enemy contributes greater biological control compared to $P - NE_1$.

9.3. MECHANISMS OF PHYSIOLOGICAL SELECTIVITY

Several mechanisms have been identified by which a pesticide may be physiologically selective to a natural enemy over its prey or host (Chapter 6). These include penetration, sequestration, excretion, detoxification, and target site insensitivity. Only a few of these mechanisms have been examined in any detail for arthropod natural enemies and their hosts or prey.

In studies of penetration and selectivity, Bull and Ridgway (1969) observed that residues of radiolabeled trichlorfon were very slowly absorbed by tolerant *Chrysoperla carnea* larvae as compared to adult *Heliothis virescens* and *Lygus hesperus*. Highly susceptible *Hippodamia convergens*, on the other hand, showed extremely rapid penetration of radiolabeled sulprofos as compared to the more tolerant *H. virescens* and the boll weevil *Anthonomus grandis* (Bull 1980). *Tarantula kochi*, a spider, was more tolerant to carbaryl than its fruit fly prey; limited penetration and rapid excretion of the pesticide in the spider were implicated as selectivity mechanisms involved (Hagstrum 1970). Zhuravskaya et al. (1976) reported a case in which penetration and detoxification factors provide for phosmet selectivity to third instar larvae and adults of *C. carnea*. Predator larvae and adults were 49- and 190-fold more tolerant than the cotton aphid *Aphis gossypii*. Phosmet penetration was much slower for both life stages of *C. carnea* than for the prey. The more rapid entry of toxicant into the aphid was also compounded by lower enzymatic detoxification as compared to the predator.

Physiological selectivity can be influenced by factors governing pesticide penetration in natural enemies. Gordon and Cornect (1986) reported that the insect growth regulator diflubenzuron, applied against the cabbage maggot *Delia radicum*, did not affect first-instar larval *Aleochara bilineata* developing within host puparia, nor did it affect adult staphylinids when applied externally in distilled water. However, when diflubenzuron was dissolved in dimethylsulfoxide (DMSO) and applied to host puparia, emergence of staphylinid adults was suppressed by 50% compared to controls.

Sequestration or isolated storage of pesticides as a mechanism conferring physiological selectivity has never been compared between an entomophagous species and its prey or hosts. The only research on sequestration in natural enemies comes from studies of pesticide tolerance and the seasonal fat body composition of adult coccinellid beetles. Takeda et al. (1965) inferred that seasonal differences in susceptibility to malathion in beetles were due to differential sequestration in fat body tissues over time. Other studies inferring a relationship between pesticide toxicity and sequestration in natural enemies have been reported by Hamilton and Kieckhefer (1969), Hoffman and Grosch (1971), and Dumbre and Hower (1976b).

Differential excretion has been studied as a mechanism conferring tolerance to pesticides in natural enemies, but seldom for pesticide selectivity between pests and natural enemies. Tolerance studies have predominantly featured DDT and its derivatives. Atallah and Newsom (1966) reported that DDT tolerance in the coccinellid *Coleomegilla maculata* was due to its ability to rapidly convert DDT to DDE, which it excreted in its feces. In the DDT-tolerant carabid *Harpalus rufripes*, a similar rapid conversion to nontoxic metabolites was reported by Dempster (1968). Gilyarov (1977) observed rapid degradation of DDT to DDE, followed by excretion among relatively tolerant predatory staphylinids feeding on DDT-contaminated collembola. More susceptible predatory spiders, millipedes, and carabids retained the parent pesticide longer in their bodies.

By far, most research on mechanisms conferring physiological selectivity to predators and parasitoids has been conducted on selective metabolism, primarily with detoxification enzymes (Chapters 6 and 15). In contrast to the research mentioned above, detoxification studies have been reported for *both* pests and natural enemies. The roles of mixed-function oxidases, esterases, epoxide-hydrolases, and glutathione-*S*-transferases have been evaluated to some degree (Mullin and Croft 1985). However, most studies have only correlated enzyme levels and natural enemy/pest selectivity relationships.

Selectivity via differential detoxification of the pyrethroid permethrin has been compared in the highly tolerant *Chrysoperla carnea* and several of its prey. Plapp and Bull (1978) and Shour and Crowder (1980) first reported the high tolerance of the *C. carnea* larvae, which rapidly break down permethrin via high esterase activity (Ishaaya and Casida 1981, Bashir and Crowder 1983). Metabolism of permethrin via esterases was much slower in *Heliothis virescens*, which accounted for the pest's greater susceptibility (Bashir and Crowder 1983). Chang and Plapp (1983) examined the binding of DDT and *cis*-permethrin in the central nervous system of *H. virescens* and *C. carnea*. They found that both pesticides competed for the same receptors, but not at precisely the same site. Insecticide–receptor binding was stable in *H. virescens*, but was reversible in *C. carnea*. This difference in binding correlated well with the selective toxicity of permethrin. In related studies, Brown and Casida (1984) showed that a component of pyrethroid tolerance in *C. carnea* and the coccinellid *Cryptolaemus montrouzieri* was esterase detoxification. Esterase detoxification was less developed in pest larvae (*Exochomus flavipes* and *Musca domestica*), and was the primary mechanism conferring selectivity (Chapter 6).

Pesticide selectivity via detoxification enzymes can be manipulated in several ways. Feng and Wang (1984) investigated the use of synergists to alter pesticide selectivity to the diamondback moth *Plutella xylostella* and its parasitoid *Apanteles plutellae*. Results for five insecticides applied with and without piperonyl butoxide (oxidative) and EPN (o-ethyl o-(4-nitrophenyl) phenylphosphonothioate) (hydrolytic) enzyme inhibitors are listed in Table 9.1 (synergistic ratios > 1 indicate synergistic effects, and those < 1 reflect antagonistic effects). In the moth pest, antagonistic effects (i.e., lower toxicity) occurred when piperonyl butoxide was added to acephate and profenofos. EPN was antagonistic to all insecticides tested on *P. xylostella*. For the parasitoid, piperonyl butoxide was antagonistic only to EPN and highly synergistic to all other insecticides tested on *A. plutellae*. Piperonyl butoxide was especially synergistic to fenvalerate, with a synergistic ratio of over 600 (Table 9.1). In comparing the effects of the synergists for both the pest and the natural enemy, in most cases physiological selectivity to the parasitoid was decreased.

In more integrative studies of selectivity, Singh and Rai (1976) investigated acetylcholinesterase properties of the plant-sucking pests *Leptocorisa varisornis*, *Dysdercus koenigii*, and *Lipaphis erysimi* and their coleopteran predators *Coccinella septempunctata* and *Cicindella sexpunctata*. In whole-body homogenates of adults, a much higher acetylcholine hydrolysis activity per gram body tissue per hour was found in the two predators compared to the plant-feeding pests. Similar results were obtained with acetylcholinesterase (AChE) activity expressed on a per-insect basis. Michaelis–Menten constants indicated a higher average affinity of ACh with AChE from herbivores as compared to carnivores. Phosphamidon, a pesticide selective to the natural enemy species, showed a higher inhibition rate with AChE from the two aphid species as compared to their respective predators. They attributed the higher affinity to the interaction of the enzyme with the inhibitor, rather than to phosphorylation. In this case, properties of AChE were believed to confer the difference in selectivity of phosphamidon between predators and pests.

Martin and Brown (1984) investigated the lack of favorable selectivity of the systemic organophosphate acephate to a reduviid predator (*Pristhesanus papuensis*) of the soybean pest *Pseudoplusia includens*. They found that the predator had a ninefold lower LD_{50} than

Table 9.1. Synergistic Effects of Piperonyl Butoxide (PBO) and EPN (o-ethyl o-(4-nitrophenyl) phenylphosphonothioate) on the Toxicity of 5 Insecticides to *Plutella xylostella* and *Apanteles plutellae* (from Feng and Wang 1984)

	LC_{50}			Synergistic Ratio	
Species/Compound	Without Synergist	+ PBO	+ EPN	+ PBO	+ EPN
P. XYLOSTELLA					
Acephate	3.52[1]	12.1	2.38	0.29	1.48
Profenofos	1.22	1.68	1.01	0.73	1.21
Fenvalerate	2.20	1.64	2.16	1.34	1.02
Permethrin	1.00	0.87	0.79	1.15	1.27
EPN	19.3	15.6	—	1.24	—
A. PLUTELLAE					
Acephate	378.0[2]	89.8	114.0	4.21	3.32
Profenofos	0.84	0.18	0.45	4.67	1.87
Fenvalerate	0.75	0.0012	0.19	625.0	3.95
Permethrin	0.067	0.018	0.083	3.72	0.81
EPN	0.16	3.44	—	0.05	—

[1] LD_{50} expressed in mg/ml.
[2] LD_{50} expressed in μg/test tube.

the phytophagous pest. Both acephate and its activation product, methamidophos, were *in vitro* inhibitors of AChE in both species. AChE from the pest was more susceptible to inhibition by both chemicals than was AChE from the natural enemy. However, the overall rate of penetration, metabolism, and excretion of the pesticide was more rapid in the pest than in the predator; and the activation of acephate to methamidophos was four times greater in the predator than in its prey. Therefore, the authors concluded that higher phosphamidon toxicity to the natural enemy as compared to the pest was due to pesticide activation and accumulation rather than to target site sensitivity.

Miyata and Saito (1982) conducted an investigation of the mechanisms of malathion and pyridafenthion selectivity against *Nephotettix cincticeps* and *Chilo suppressalis* compared to their natural enemies, the wolf spider *Lycosa pseudoannulata* (a relatively tolerant predatory spider) and *Concephalus maculatus* (a relatively susceptible predatory spider). They observed low cuticular penetration of both compounds in the wolf spider and speculated that this accounted for its high tolerance to both organophosphate compounds. Low AChE inhibition of pyridafenthion was also involved in the tolerance exhibited by *L. pseudoannulata*. High toxicity of malathion to *C. maculatus* was thought to be due to high AChE inhibition by malaoxon, even though penetration of this pesticide was relatively slow. Miyata and Saito concluded that one mechanism of selectivity may be canceled out by another, such as was observed with *C. maculata*, where moderate penetrability was apparently overcome by high AChE inhibition.

As noted in Chapter 6 (Section 6.3.6), the most comprehensive studies comparing selectivity between a natural enemy and its hosts or prey were those of Yu (1987, 1988) for the pentatomid spined soldier bug *Podisus maculiventris* and its lepidopterous prey, the velvetbean caterpillar *Anticarsia gemmatalis*, the fall armyworm *Spodoptera fugiperda*, and the corn earworm *Heliothis zea*. Whereas the stink bug was generally more susceptible to most organophosphate and carbamate insecticides, it was more tolerant of tetrachlorvinphos, permethrin, cypermethrin, and fenvalerate compared to its prey. Yu (1988) attributed the pentatomid's high tolerance to tetrachlorvinphos to both reduced cuticular penetration and enhanced detoxication (via glutathione conjugation) of the insecticide. The predator's greater pyrethroid tolerance was less easily explained, but reduced penetration seemed to be most likely, since the natural enemy generally had lower levels of MFOs and esterase detoxification enzymes than did the prey.

9.4. PHYSIOLOGICAL SELECTIVITY VIA RESISTANCE EVOLUTION

Physiological selectivity may also be attained through development of resistant strains of natural enemies. Resistance mechanisms are of the same general types as discussed above for tolerance mechanisms (i.e., penetration, sequestration, excretion, and selective metabolism, including insensitive target sites). Target site differences among natural enemies and among pests and natural enemies have been studied more commonly in relation to resistance development and selectivity (see Chapters 14 and 15). The target site mechanisms that have been most thoroughly studied are altered acetylcholinesterase and kdr (knock down nerve insensitivity). Most cases of physiological selectivity achieved through resistance have been among predatory phytoseiid mites (Chapter 15).

9.5. CASES OF PHYSIOLOGICAL SELECTIVITY

The discussion of physiological selectivity which follows primarily concerns conventional pesticides, including the early inorganics and botanicals, DDT and organochlorines, organophosphates, carbamates, pyrethroids, acaricides, and miscellaneous insecticides. Physiological selectivity in microbial and natural products pesticides (including insect growth regulators) are treated more extensively in Chapters 11 and 12.

9.5.1. Analysis of Selectivity Literature

An extensive list of published cases of pesticide selectivity to natural enemies over their prey or hosts is presented in the Appendix. Selectivity is summarized for pesticides following the compound-structure classification system proposed by Larsen et al. (1985). Individual cases are cited where selectivity to at least one natural enemy over one host or prey species was reported. Included for each listing in the appendix are 1) chemical class (e.g., inorganics and botanicals, DDT and organochlorine derivatives, etc.), 2) crop, 3) principal host and natural enemy, 4) method of evaluation, 5) type of selectivity [i.e., physiological (P), ecological (E), both (B), or unknown (?)], 6) the ratio of selectivity between natural enemy and host or prey, and 7) the breadth of species selectivity observed.

Breadth of species selectivity refers to the number of pest and natural enemy species involved in the selectivity documentation. As noted earlier, selectivity can be simple (one species) for both natural enemy and pest (n-p), compound in the natural enemy (more than one species), but simple in the pest (nn-p), simple in the natural enemy, but compound in the pest (n-pp), or compound in both groups (nn-pp). In the Appendix, the breadth of selectivity beyond two species of natural enemy and pest was not numerically quantified as such, but these cases are apparent from the species lists associated with each case.

Simple species selectivity does not necessarily mean that other species were not affected by a particular pesticide treatment in the field. Rather this designation indicates that the documentation of selective effects only included one species of pest and natural enemy. In all cases, it should be kept in mind that the results of tests to evaluate the selectivity attributes of a particular compound are being reported. Many of these tests are based on simplified laboratory assessments, where the number and diversity of organisms and the conditions of their exposure are controlled. In the field, the actual impact of these compounds may be much greater than these studies indicate.

9.5.2. General Summary of the Appendix

The Appendix is a compilation of the published cases of selectivity *favorable* to natural enemies found in the literature described in Chapter 2 (i.e., that included in the SELCTV database). Summary analyses taken from the Appendix and SELCTV are compared hereafter (Section 9.7). While both analyses used the same literature base, the SELCTV summary (on selectivity) was based on *all* selectivity ratios (i.e., both those favoring and those not favoring natural enemies) reported as median lethal bioassays (thus including only 870, or 6.9% of the 12,600 records). Approximately 57% of these ratios (499) showed favorable pesticide selectivity to a natural enemy. The Appendix contains *only* favorable cases of selectivity (864) and was compiled from data in any form of assessment where a compound was observed to be less toxic to natural enemies than to pests.

Further explanation is needed of the criteria used to distinguish between physiological and ecological selectivity for the cases compiled in the Appendix. If a pesticide was applied directly to both the pest and the natural enemy in the laboratory, then physiological selectivity was evident. If an author stated that physiological selectivity was principally involved in a test, then this was considered evidence for physiological selectivity. If it was stated or implied that the mechanism of pesticide selectivity was through a temporal or spatial separation of the species and the pesticide, then an ecological designation was given. When both types of selectivity appeared to be involved, it was so indicated. If the type of selectivity could not be ascertained, a question mark was used. Inability to ascertain the mechanism of selectivity operating in a given test was usually due to 1) inadequate detail in reporting of test data, 2) incomplete understanding of the mode of pesticide uptake by the pest and natural enemy, or 3) a lack of a basic understanding of the compound's mode of action. The last two factors refer either to the experimenter's lack of understanding or to a general void in what was known about the compound.

The Appendix primarily consists of cases of physiological selectivity (60.1% of the total), although some instances of ecological (5.7%) and combined selectivity (6.9%) are included. In 27.4% of cases, the type of selectivity could not be distinguished. Physiological selectivity is more frequently noted in the literature because it is more novel (and therefore more publishable). However, ecological selectivity is more commonly achieved in practice as a component of most IPM programs (Hull and Beers 1985; Chapter 10).

Of the cases reported in the Appendix, 388, or almost 45% of the 864 listings, indicate simple selectivity (n-p). While useful in many pest-specific situations, such as in greenhouses where only a few pests are present, simple selectivity is often too specific to be exploited where complexes of pests and natural enemies occur. However, in many cases broader selectivity may have been found had the compound been evaluated more extensively. At least one case of simple selectivity has been reported for most widely used broad-spectrum pesticides (e.g., chlordane, methyl-parathion, methidathion, decamethrin; Appendix). However, these pesticides are generally nonselective. Over 16% of cases exhibited simple selectivity for a natural enemy and compound pest activity (n-pp). A similar 18.2% of entries exhibited compound natural enemy selectivity, but simple pest activity (nn-p). About 17% involved selectivity to multiple natural enemies and over multiple prey or hosts (nn-pp).

Selectivity from the literature listed in the Appendix is characterized by insecticide class in Table 9.2. Selectivity cases are tabulated both by the number of references (records) in the literature for a given compound and by the number of instances of selectivity to natural enemies by species. These totals may more accurately represent the amount of research conducted on different pesticide groups and their history of use rather than selectivity. For example, most cases of selectivity have been cited for organophosphates (30%), carbamates (18%), and miscellaneous insecticides and acaricides (24%) (Table 9.2). However, the

Table 9.2. A Summary of Cases of Physiological Selectivity among Insecticide Classes to Arthropod Natural Enemies (Taken from Appendix)

Insecticide Class	No. of References	No. of Natural Enemies	Species per Record Index
Botanicals/inorganics	28 (3.4%)[1]	83 (5.7%)	2.96
DDT/organochlorines	69 (8.5%)	94 (6.5%)	1.36
Organophosphates	244 (29.9%)	416 (28.7%)	1.70
Carbamates	150 (18.4%)	247 (17.1%)	1.65
Pyrethroids	49 (6.0%)	76 (5.2%)	1.55
Miscellaneous insecticides/ acaricides	196 (23.9%)	350 (24.0%)	1.79
Microbials	39 (4.8%)	87 (6.0%)	2.23
Chitin inhibitors/juvenoids	39 (5.1%)	94 (6.8%)	2.41
Total cases	814 (100%)	1447 (100%)	

[1] Percentage of total records.

number of species per reference (species per record index) may be a better index of the breadth of selectivity associated with an insecticide class. Those classes with a lower species per record index (i.e., DDT/organochlorines, organophosphates, carbamates, pyrethroids) are generally less broadly selective than those having a higher species per record index (i.e., botanicals/inorganics, miscellaneous insecticides/acaricides, microbials, chitin inhibitors/juvenoids). These conclusions will be corroborated by other analyses discussed below.

By individual compound, physiological selectivity has been most commonly reported for pirimicarb (74 cases), endosulfan (43), phosalone (35), cyhexatin (33), *Bacillus thuringiensis* (31), dicofol (30), demeton-*S*-methyl (29), diflubenzuron (28), fenbutatin oxide (25), and dimethoate (21) (Fig. 9.4). When tabulated by reports of selectivity to individual species, pirimicarb (139), *Bacillus thuringiensis* (76), phosalone (76), diflubenzuron (74), cyhexatin (66), endosulfan (59), dicofol (55), demeton-*S*-methyl (50), fenbutatin oxide (50), and trichlorfon (33) predominate. On a species per record basis, values were greatest for diflubenzuron (2.64), *Bacillus thuringiensis* (2.45), phosalone (2.17), fenbutatin oxide (2.0), cyhexatin (2.00), pirimicarb (1.88), dicofol (1.83), trichlorfon (1.83), demeton-*S*-methyl (1.72), dimethoate (1.47), and endosulfan (1.37). Any compound with ≥ 10 cases of selectivity and a species/record ratio > 2.00 can be considered broadly selective. Other compounds having a high species per record index, but having few records included are lead arsenate (6.70), fenoxycarb (4.00), ryania (3.86), phosmet (3.00), propoxur (3.00), benzomate (2.90), dinobuton (2.80), and stirophos (2.00).

From the above analyses it is clear that some of the biorational pesticides exhibit

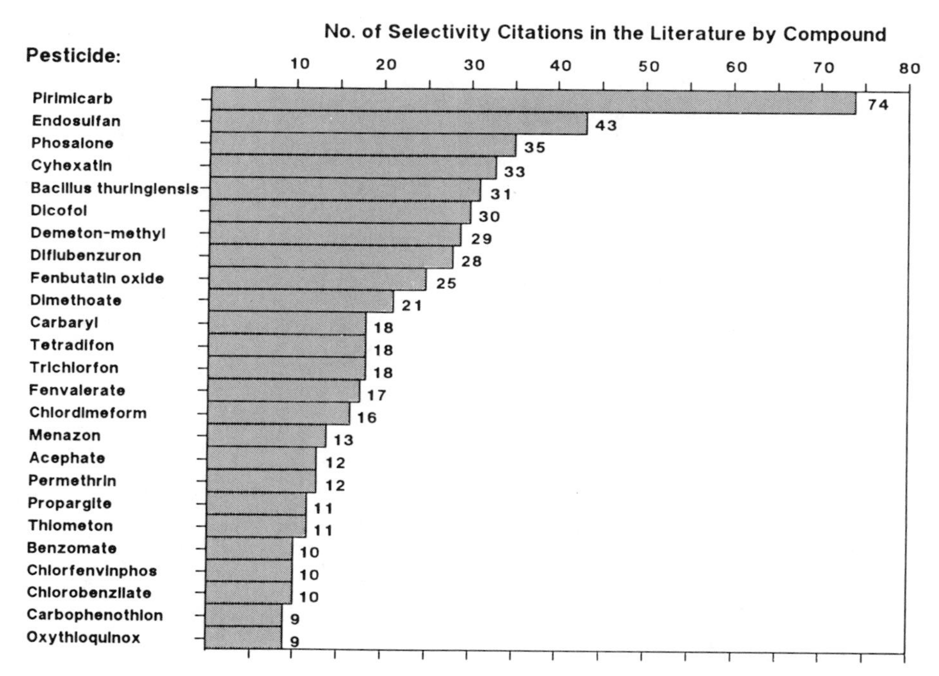

Figure 9.4. Cases of physiological selectivity among insecticides to arthropod natural enemies (Taken from Appendix).

broader selectivity than most synthetic organic pesticides, even though the former have been less thoroughly documented. Biorational pesticides are similar to the early inorganic and botanical compounds in their selectivity spectrum. However, their side effects on other nontarget species (e.g., fish, mammals, and avians) are less severe (Chapters 11 and 12).

Approximately 19% of the 113 compounds cited in the Appendix were systemic pesticides, according to Larsen et al. (1985). However, these 21 pesticides accounted for >31% of the records of physiological selectivity. Systemic pesticides are active against plant-sucking pests, such as aphids, other homopterans, hemipterans, and mites. However, the species per record index for many of the systemics is moderate to low (see data for pirimicarb, dimethoate, and demeton-*S*-methyl). These ratios may be low because selectivity studies for these compounds have often focused on relatively specific natural enemies of aphids and mites, rather than larger complexes of natural enemies.

True systemic activity *per se* is a form of ecological selectivity (Chapter 10), since spatial separation of the natural enemy and toxicant is the primary mechanism involved. The

Table 9.3. Cases of Physiological Selectivity among Insecticides from Literature Citations Grouped by Common or Taxonomic Designations of Natural Enemies (A) and of Pests (B) (Taken from Appendix)

Groups Represented	No. of Cases by Species	Percentage of Total
A. NATURAL ENEMY GROUPS		
Parasitic Hymenoptera	380	26.4
Coccinellidae	237	16.4
Phytoseiidae	231	15.9
Chrysopidae	140	9.7
Predaceous Hemiptera	103	7.1
Araneae	97	6.7
Carabidae/Staphylinidae	84	5.8
Diptera	63	4.3
Other/many/several	112	7.7
Total	1447	100.00
B. PREY OR HOST (PEST) GROUPS		
Mites	256	26.7
Aphids	252	25.4
Lepidoptera	145	14.6
Scales/mealybugs/ leafhopper/planthopper	72	7.3
Whiteflies	39	3.9
Coleoptera	23	2.3
Diptera	15	1.5
Hemiptera	3	0.3
Thripidae	2	0.2
Miscellaneous other	21	2.1
Many/several	98	9.9
?? (unknown)[1]	19	1.9
Totals	992	100.00

[1] This category applies only to pests, not natural enemies, and indicate cases where no specific hosts or prey were noted by scientific or common name.

apparent frequency of systemics showing physiological selectivity in the Appendix raises questions as to the legitimacy of these reports. The selectivity of systemics may be due to their rapid penetration into plant tissues and their rapid degradation on treated surfaces (e.g., Lelievre and Chatenet 1983). Uptake may be so rapid that it is difficult to distinguish whether selectivity is really due to a lack of toxicity to the natural enemy (physiological selectivity), or just a lack of exposure (ecological selectivity).

In many cases reported in the Appendix, direct application or other *prima facia* evidence of physiological selectivity gave conclusive evidence that many systemics such as pirimicarb and phosalone possess at least limited physiological selectivity. Fukuto (1984) pointed out that many systemics are proinsecticides. They are only active when taken up and transformed by a plant or animal. This may explain the physiological selectivity associated with these systemics.

In a related analysis of the Appendix, cases of pesticide selectivity for prominent families or orders containing natural enemies are summarized in Table 9.3. To some degree, these data reflect the values of different natural enemy groups in pesticide-treated crops or agroecosystems where IPM is practiced. Among families, the Coccinellidae were most widely favored by pesticide selectivity, followed by the Phytoseiidae and Chrysopidae. At the order level, the Hymenoptera were well represented, followed somewhat distantly by Diptera and Araneae. Beyond these major groups, percentages diminished appreciably.

In a similar analysis for pests (Table 9.3), physiological selectivity has been most commonly reported for the natural enemies of mites (26.7%), aphids (25.4%), lepidopterans (14.6%), and other homopteran pests (7.3%). These four groups together make up almost 75% of the total cases. One other category of note is that of "many/several". This listing is most often associated with field testing where specific species are less easily monitored. In some cases, it may indicate pesticides that show broad activity to many different pest groups. The latter explanation is most likely where this designation is associated with broadly selective pesticides (e.g., lead arsenate, ryania, *Bacillus thuringiensis*, diflubenzuron).

Those species of natural enemies and their prey or hosts for which cases of physiological pesticide selectivity have been most commonly reported are listed in Table 9.4. The first 28 natural enemy species make up nearly 40% of the total database. Selectivity is therefore widely dispersed among many entomophagous species. Four of the first 10 species listed are phytoseiid mites, three are coccinellid beetles; *C. carnea*, *E. formosa*, and the spider *Lycosa pseudoannulata* are other commonly listed species.

The top 25 pests cited in the Appendix make up nearly 50% of the total list, which indicates that reports of selectivity among natural enemies span a greater number of species than associated pesticide activity among their prey or hosts. Among pests, many species of mites and aphids are represented; two of the most frequently cited species are among the spider mites and six are aphid pests. Also among the top 10 pest species are the ubiquitous greenhouse pest *Trialeurodes vaporariorum* and the widely distributed and highly destructive planthopper of rice, *Nilaparvata lugens* (Table 9.4). The diamondback moth *Plutella xylostella* is notable as the only nonacarine/homopteran/hemipteran pest among the 20 most common pest species.

Those agricultural crops for which cases of selectivity to natural enemies have been most commonly reported are tree fruits (apple comprises about 90% of this crop category), cotton, greenhouse crops, crucifers, citrus, other cereals, and rice (Table 9.5). These six crops account for almost two-thirds of selectivity cases in the Appendix. Apple alone

Table 9.4. Cases of Physiological Selectivity among Insecticides from Literature Citations Grouped by Natural Enemy and Pest Species (Taken from Appendix)

Species Represented	No. of Cases by Species	Percentage of Total
A. NATURAL ENEMY SPECIES		
Chrysoperla carnea	69	4.8
Phytoseiulus persimilis	61	4.3
Coccinella septempunctata	51	3.4
Encarsia formosa	40	2.7
Typhlodromus pyri	31	2.1
Lycosa pseudoannulata	27	1.8
Amblyseius fallacis	24	1.6
Typhlodromus occidentalis	21	1.4
Hippodamia convergens	17	1.1
Menochilus sexmaculata	16	1.1
Trichogramma cacoeciae	16	1.1
Coccinella undecimpunctata	15	1.0
Scymnus includens	13	0.9
Stethorus punctillum	13	0.9
Aphelinus mali	12	0.8
Leptomastix abnormis	12	0.8
Phygadeuon trichops	12	0.8
Aphidoletes aphidimyza	11	0.8
Microplitis croceipes	11	0.8
Oedothorax insecticeps	11	0.8
Adonia variegata	10	0.7
Agistemus exsertus	10	0.7
Apanteles plutella	10	0.7
Chrysoperla formosa	10	0.7
Chrysoperla oculata	10	0.7
Cryptolaemus montrouzieri	10	0.7
Leptomastix dactylopii	10	0.7
Stethorus nigriceps	10	0.7
Total	563	38.9
B. HOST OR PREY SPECIES		
Tetranychus urticae	95	9.6
Panonychus ulmi	67	6.8
Nilaparvata lugens	35	3.5
Myzus persicae	25	2.5
Trialeurodes vaporariorum	24	2.4
Aphis gossypii	22	2.2
Brevicoryne brassicae	18	1.8
Lipaphis erysimi	17	1.7
Acyrthosiphon pisum	15	1.5
Acyrthosiphon kondoi	14	1.4
Planococcus citri	14	1.4
Aleurothrixus floccosus	13	1.3
Plutella xylostella	13	1.3

Table 9.4 (*Contd.*)

Species Represented	No. of Cases by Species	Percentage of Total
Aonidiella aurantii	11	1.1
Therioaphis maculata	11	1.1
Therioaphis trifolii	11	1.1
Microsiphum rosae	10	1.1
Aculus schlechtendali	10	1.0
Aphis craccivora	9	0.9
Nephotettix cincticeps	9	0.9
Sitobion avenae	9	0.9
Aphis fabae	9	0.9
Heliothis armigera	8	0.8
Hypera postica	8	0.8
Spodoptera littoralis	8	0.8
Total	485	48.9

contributes almost one-sixth of these, leading any other crop group by twice the number of cases. Selective acaricides and aphicides are particularly well documented on this crop (Croft 1981a, Hull and Beers 1985; Appendix). Few cases for forests, small fruits, and ornamentals may indicate lesser interest on pesticide selectivity in these crops or fewer resources available for estimating nontarget side effects.

Table 9.5. Cases of Physiological Selectivity among Insecticides from Literature Citations Grouped by Crop (Taken from Appendix)

Crop	No. of Cases	Percentage of Total
Tree fruits	153	18.8
Cotton	95	11.7
Greenhouse	84	10.3
Crucifers	52	6.4
Citrus	50	6.1
Other cereals	50	6.1
Rice	45	5.5
Other vegetables	41	5.0
Forests	19	2.3
Legumes	19	2.3
Small fruits	16	2.0
Ornamentals	13	1.6
Others	49	6.0
Several/many	48	5.0
?? (blank)	81	9.9
Total	815	100.0

9.5.3. Selectivity by Classes of Pesticides

9.5.3.1. ***Inorganics and Botanicals.*** It has long been known that the early inorganic and botanical pesticides were relatively selective to arthropod natural enemies (Ripper 1956, Bartlett 1964). In summarizing the Appendix, Table A.1, those compounds most frequently cited as being physiologically selective were lead arsenate, nicotine sulfate, ryania, and the pyrethrins. Overall, this class of compounds has shown some of the broadest selectivity to natural enemies over their prey or hosts (mean species per record index = 2.96; Table 9.6). Botanical and inorganic pesticides had the highest incidence of ecological selectivity of any insecticide group. Lead arsenate was particularly well represented in this regard (Appendix, Table A.1). The selectivity of ryania, nicotine sulfate, and pyrethrins appeared to be due more to physiological than to ecological factors.

The number of cases of selectivity for the inorganic and botanical pesticides are probably underestimated in the Appendix, Table A.1, because selectivity was less commonly reported when these compounds were widely used. Also, the literature reviewed for the Appendix emphasized the period from 1970 to the present rather than older literature.

Table 9.6. Summary of Cases of Physiological Selectivity among Botanical and Inorganic Insecticides to Arthropod Natural Enemies (Taken from Appendix, Table A.1)

Compound	No. of References	No. of Natural Enemies	Species per Record Index
Bordeaux	2	9	4.50
Lead arsenate	3	20	6.67
Nicotine	3	5	1.67
Nicotine sulfate	7	13	1.86
Pyrethrum	5	8	1.60
Ryania	7	27	3.86
Tartar emetic	1	1	1.00
Total	28	83	Mean = 2.96

Table 9.7. Summary of Cases of Physiological Selectivity among DDT Derivatives and Organochlorine Insecticides to Arthropod Natural Enemies (Taken from Appendix, Table A.2)

Compound	No. of References	No. of Natural Enemies	
Aldrin	1	1	1.00
Chlordane	1	1	1.00
Dieldrin	2	4	2.00
Endosulfan	43	59	1.37
Endrin	5	5	1.00
Lindane	7	9	1.29
Methoxychlor	4	4	1.00
Perthane	1	2	2.00
Toxaphene	5	9	1.80
Total	69	94	Mean = 1.36

9.5.3.2. DDT and Organochlorines. As a group, these chlorinated hydrocarbon pesticides were less toxic to natural enemies than the more recent organophosphates carbamates, and pyrethroids (Chapter 2). However, they were generally less toxic to herbivores as well and had a narrow range of selectivity to natural enemies (mean species per record index = 1.36; Table 9.7). Although at least one case of simple selectivity has been documented for most compounds in this class, only endosulfan has been shown to be broadly selective (Table 9.7; see recent cases for endosulfan selectivity to natural enemies in Beraldo et al. 1981, Brun et al. 1983, Bandong and Litsinger 1986). Endosulfan is particularly effective on aphids, leafhoppers, phytophagous hemipterans, and some lepidopteran pests. Endosulfan has exhibited selectivity to many natural enemy groups, including predatory mites, many aphid parasites and predators, and parasitoids associated with key pests (Appendix, Table A.2). However, the species per record index for endosulfan was surprisingly low (species per record index = 1.37, Table 9.7). At present IPM programs on several crops depend on the physiological selectivity of endosulfan while also using it in ecologically selective ways (Appendix, Table A.2).

9.5.3.3. Organophosphates. Organophosphates, more than any other pesticide class, have been widely tested for their selectivity to arthropod natural enemies (Table 9.2; Chapter 2). In the Appendix, Table A.3, the organophosphates are divided among early compounds, aliphatic derivatives, carbon cyclic derivatives, and heterocyclic derivatives. In general, organophosphates are highly toxic to most natural enemy species (Chapter 2) and are lacking in physiological selectivity. In some cases, ecological selectivity has been achieved (Chapter 10). However, one or more cases of simple physiological selectivity have been demonstrated for most individual organophosphate compounds (Appendix, Table A.3).

The organophosphates that most frequently exhibit selective properties have been systemics (Table 9.8). The first systemic organophosphate, schradan, was extensively studied for its selectivity to natural enemies and was widely heralded in this regard (Bartlett 1956, Ripper 1957). However, schradan was never widely marketed for pest control due to its high mammalian toxicity. Other selective systemic organophosphates include demeton-*S*-methyl, dimethoate, phosalone, and menazon (Appendix, Table A.3). These materials are particularly selective to the natural enemies of aphids and mites. Systemic organophosphates are commonly applied as soil drenches, side dressings, or granules. They also are applied as foliar sprays where natural enemies have been shown to survive direct exposure (Appendix, Table A.3). As indicated earlier, this suggests that these compounds possess some level of physiological selectivity.

Why physiological selectivity is common among systemic organophosphates is not clear, but it may be due to their proinsecticidal activity (Fukuto 1984). These agents may be so rapidly absorbed by plants that residues last only for a short time, but this explanation does not account for their lack of activity when directly administered to predators and parasitoids. Many systemic organophosphates can be highly toxic to natural enemies as well (see examples for phosalone in Hislop et al. 1978, Akhmedov 1981, and Hassan et al. 1983, and for demeton-*S*-methyl in Bartlett 1963, Franz et al. 1980, and Mazzone et al. 1980).

Among the organophosphates, the number of cases or levels of broad-spectrum selectivity for phosalone, phosmet, trichlorfon, demeton-methyl, phosphamidon, stirophos, dimethoate, thiometon, and acephate are noteworthy (Table 9.8). The value of these organophosphates in IPM programs on many crops is widely recognized (e.g., see summary of IPM benefits for phosalone in Lelievre and Chatenet 1983). Not only are these

Table 9.8. Summary of Cases of Physiological Selectivity among Organophosphate Insecticides to Arthropod Natural Enemies (Taken from Appendix, Table A.3)

Compound	No. of References	No. of Natural enemies	Species per Record Index
		Early OPs	
Schradan*	3	5	1.67
	Aliphatic Derivatives of Phosphorus Compounds		
Acephate*	13	24	1.85
Demeton-*S*-methyl*	29	50	1.72
Dichlorvos	4	4	1.00
Dicrotophos	2	3	1.50
Dimethoate*	21	31	1.47
Disulfoton*	4	6	1.50
Ethion	3	3	1.00
Malathion	3	4	1.25
Methamidophos	2	3	1.50
Monocrotophos*	7	10	1.43
Omethoate*	1	2	2.00
Phorate*	2	3	1.50
Phosphamidon*	5	13	2.60
Thiometon*	12	17	1.42
Trichlorfon	18	33	1.83
Vamidothion*	5	9	1.80
Subtotal	131	215	Mean = 1.64
	Carbon Cyclic (e.g., phenyl, etc.) Derivatives of Phosphorus Compounds		
Bromophos	4	4	1.00
Carbophenothion	9	15	1.67
Chlorfenvinphos	10	15	1.50
Fenitrothion	4	6	1.50
Fenthion	1	1	1.00
Heptenophos*	2	2	1.00
Parathion-methyl	1	4	4.00
Parathion-ethyl	1	2	2.00
Profenofos	1	1	1.00
Stirophos	5	10	2.00
Tetrachlorvinphos	1	1	1.00
Subtotal	39	61	Mean = 1.56
	Heterocyclic Derivatives of Phosphorus Compounds		
Chlorpyrifos	3	4	1.33
Dialifor	1	1	1.00
Diazinon	6	10	1.67
Dioxathion	2	4	2.00
Menazon*	14	18	1.29
Mephosfolan*	1	2	2.00

Table 9.8. (*Contd.*)

Compound	No. of References	No. of Natural enemies	Species per Record Index
Methidathion	1	1	1.00
Phosalone*	35	76	2.17
Phosmet	4	12	3.00
Phospholan*	1	1	1.00
Pyridafenthion	3	4	1.33
Quinalphos	1	1	1.00
Thionazin	1	1	1.00
Subtotal	73	135	Mean = 1.85
Grand total	246	416	Grand mean = 1.69

*Denotes compounds having systemic action (according to Larsen et al. 1985).

Table 9.9. Summary of Cases of Physiological Selectivity among Carbamate Insecticides to Arthropod Natural Enemies (Taken from Appendix, Table A.4)

Compound	No. of References	No. of Natural Enemies	Species per Record Index
Aldicarb*	6	10	1.67
Aminocarb	1	2	2.00
Bendiocarb	1	4	4.00
Carbaryl	18	26	1.44
Carbofuran*	8	12	1.50
Cartap	4	4	1.00
Dimetlan	1	1	1.00
Ethiofencarb	8	10	1.25
Fenobucarb	4	5	1.25
Formetanate	1	1	1.00
Isocarb	1	2	2.00
Isoprocarb	3	4	1.33
Methomyl*	7	7	1.00
MTMC	1	2	2.00
Oxamyl*	5	5	1.00
Pirimicarb*	74	139	1.88
Propoxur	3	9	3.00
Thiodicarb	4	4	1.00
Total	150	247	Mean = 1.65

*Denotes compounds having systemic action (according to Larsen et al. 1985).

organophosphates noted for their selectivity, but their rapid degradation in the environment makes them safer for use than most early persistent pesticides (Lelievre and Chatenet 1983).

9.5.3.4. ***Carbamates.*** Like the organophosphates the carbamates are generally very toxic to most natural enemies of agricultural pests (Chapter 2). Yet one of the most broadly selective products ever discovered and registered is a carbamate (pirimicarb; Table 9.9). Although highly toxic when administered directly, several systemic carbamates show moderate selectivity to natural enemies when applied as systemics (e.g., aldicarb and carbofuran; Appendix, Table A.4). Clearly, these cases represent ecological selectivity rather than physiological selectivity.

As noted, pirimicarb is one of the most selective synthetic insecticides ever registered for field use. It controls a diversity of pests, including many aphids, without causing detriment to many natural enemies (Table 9.9; Appendix, Table A.4). Pirimicarb has been most widely used for selective control of plant-sucking pests on crops such as cereal grains, greenhouse crops, and vegetables. The exact nature of its broadly selective activity is not well understood. Certainly systemic properties contribute to its high activity against pests and selectivity to natural enemies. As with some systemic organophosphates, pirimicarb is at least somewhat physiologically selective, as there are many reports of its lack of direct toxicity to natural enemies (e.g., Helgesen and Tauber 1974, Kalushkov 1982, Brown et al. 1983). This product is available for use on crops in many parts of the world, but has not been registered in the United States. (Note: This is due to concerns and costs for research on mammalian safety; Federal Register 1981.)

9.5.3.5. ***Pyrethroids.*** Of all classes of organic insecticides developed to date, the pyrethroids are probably the most toxic to natural enemies (Chapter 2). They are slightly more toxic than the organophosphates and carbamates, but there are some notable exceptions of remarkable selectivity to certain species of natural enemies (Croft and Whalon 1982; see also Chapter 13). Most of these cases, however, involve simple species selectivity or small natural enemy species complexes (species per record index = 1.55; Table 9.10 and Appendix, Table A.5).

Table 9.10. Summary of Cases of Physiological Selectivity among Pyrethroid Insecticides to Arthropod Natural Enemies (Taken from Appendix, Table A.5)

Compound	No. of References	No. of Natural Enemies	Species per Record Index
Bioresmethrin	1	1	1.00
Cypermethrin	6	12	2.00
Decamethrin	1	1	1.00
Deltamethrin	5	7	1.40
Fenvalerate	17	27	1.59
Flucythrinate	4	4	1.00
Fluvalinate	1	1	1.00
Permethrin	13	22	1.69
Tralomethrin	1	1	1.00
Total	49	76	Mean = 1.55

Pyrethroids (SPs) exhibit physiological selectivity toward a number of parasitoids and certain predators (e.g., *Chrysoperla carnea, Cryptolaemus montrouzieri, Venturia canescens*; Appendix, Table A.5; Brown and Casida 1984; Chapter 13); however, no individual compound has shown a particularly consistent pattern of physiological selectivity or broad selectivity (Table 9.10). The more numerous documentations of selectivity for permethrin, fenvalerate, and cypermethrin probably reflect their longer history of use, rather than any general selectivity properties. A few less well studied SPs such as flucythrinate, fluvalinate, and tralomethrin show greater potential for selectivity to natural enemies (e.g., Croft and Wagner 1981, Theiling 1987). Fluvalinate, developed by Zoecon Corporation, has a high margin of safety to honeybees. Apparently, the lack of side effects on natural enemies is an added bonus. This compound has not yet widely displaced the less selective SPs such as permethrin, fenvalerate, and cypermethrin in the field.

In examining the more basic properties of the pyrethroids, studies by scientists in the United Kingdom have demonstrated considerable potential for physiological selectivity with these compounds using the ichneumonid *Venturia canescens* and the Indian meal moth *Ephestia kuhniella* as a host/parasitoid model system (Elliott et al. 1983, Stevenson and Walters 1984). When adults of both the host and the parasitoid were treated topically, levels of physiological selectivity to the parasitoid varied from as high as 100-fold for bioallethrin and 69-fold for flucythrinate to near eight-fold for fenvalerate and lower for certain experimental SPs (Table 9.11). However, most SPs showed a selectivity ratio greater than eight-fold. These data demonstrate that, at least for this simple parasitoid/host system, some pyrethroids may be sufficiently selective for IPM programs.

Table 9.11. Selectivity Ratios for Indian meal moth *Ephestia kuhniella* and Its Ichneumonid Parasitoid *Venturia canescens* (Adapted from Stevenson and Walters 1984, Elliott et al. 1983)

Insecticide	*Ephestia* LD_{50} (μg/g insect)	*Venturia* LD_{50} (μg/g insect)	Ratio *Venturia*/*Ephestia*[1]
Kadelthrin	0.07	10.0	145.0
Bioallethrin	0.11	11.0	97.0
NRDC 185[2]	0.13	10.0	75.0
Flucythrinate	0.026	1.8	69.2
NRDC 108	0.16	8.0	50.0
NRDC 103	3.5	85.0	24.0
Biopermethrin	0.16	2.3	14.0
Permethrin[3]	0.13	1.8	13.8
Cismethrin	0.19	2.4	13.0
Deltamethrin	0.032	0.39	12.2
NRDC 145	0.26	3.1	12.0
NRDC 101	0.56	6.4	11.0
Bioresmethrin	0.25	2.4	9.6
Fenvalerate	0.23	1.9	8.1
NRDC 181	0.21	1.3	6.2
NRDC 100	2.4	5.5	2.3
Rotenone	160.0	> 8000.0	> 50.0
Endosulfan	2.0	47.0	24.0
Pirimicarb	10.0	77.0	7.7
Phosalone	12.0	4.7	0.4

[1] Selectivity ratio expressed as natural enemy/pest is the reverse of that used in other sections of this volume.
[2] All compounds with NRDC designation are experimental compounds.
[3] Estimated values based on results for two isomers.

Broad selectivity of SPs to the complexes of pests and natural enemies associated with apple and cotton crop production systems will be discussed later in greater detail (Chapter 13).

9.5.3.6. ***Acaricides.*** Acaricides as a group contain more physiologically selective products than any other pesticide class discussed in this chapter (Table 9.12; Appendix, Table A.6). This high margin of physiological selectivity supports the existence of major differences in the activity of acaricides on herbivorous mites and their insect and acarine natural enemies (e.g., between tetranychid and phytoseiid mites). Knowles (1975) discussed a number of biochemical and physiological differences between these two species groups which contribute to selectivity against phytophagous acarines. He concluded that differential metabolism was probably the most important.

Physiologically selective acaricides are widely employed in IPM programs and include the following compounds: dicofol, cyhexatin, fenbutatin oxide, propargite, benzomate, clofentezine, and hexythiazox (Table 9.12.; Appendix, Table A.6). Chlordimeform is a unique acaricidal compound. It is highly toxic to spider mites and to the eggs of many lepidopteran pests, yet it is not excessively toxic to many insect predators and parasites (Plapp and Vinson 1977, Plapp and Bull 1978, Rajakulendran and Plapp 1982b). These attributes account for its successful use on cotton in many areas of production. Chlordimeform was recommended for spider mite control on tree fruits for some time, but its high toxicity to many predatory mites disrupted mite management systems.

Table 9.12. Summary of Cases of Physiological Selectivity among Miscellaneous Insecticides and Acaricides to Arthropod Natural Enemies (Taken from Appendix, Table A.6)

Compound	No. of References	No. of Natural Enemies	Species per Record Index
Amitraz	1	3	3.00
Azocyclotin	2	5	2.50
Benzomate	10	29	2.90
Binapacryl	4	6	1.50
Bromopropylate	1	1	1.00
Buprofezin	3	6	2.00
Chlorobenzilate	10	13	1.30
Chlordimeform	16	30	1.88
Clofentezine	6	8	1.33
Cyhexatin	33	66	2.00
Dicofol	30	55	1.83
Dienochlor	3	3	1.00
Dinobuton	5	14	2.80
Fenbutatin oxide	25	50	2.00
Hexythiazox	3	4	1.33
Ovex	1	1	1.00
Oxythioquinox	9	10	1.11
Propargite	12	17	1.42
Soap	2	2	1.00
Tetradifon	18	25	1.39
Tetrasul	1	1	1.00
Triforine	1	1	1.00
Total	196	350	Mean = 1.79

Table 9.13. Summary of Cases of Physiological Selectivity among Microbial Insecticides and Insect Growth Regulators to Arthropod Natural Enemies (Taken from Appendix, Table A.7)

Compound	No. of References	No. of Natural Enemies	Species per Record Index
Abamectin	4	5	1.25
Bacillus thuringiensis	31	76	2.45
Beauvaria bassiana	2	3	1.50
Diflubenzuron	28	71	2.54
Epofenonane	2	5	2.50
Fenoxycarb	2	8	4.00
Hydroprene	1	1	1.00
Kinoprene	3	3	1.00
Methoprene	1	1	1.00
Nosema pyrausta	1	1	1.00
Pentafluron	1	1	1.00
Trichothecin	1	2	2.00
Triflubenzuron	1	1	2.00
Total	78	181	Mean = 2.32

9.5.3.7. ***Microbials and Insect Growth Regulators.*** The microbial pesticides and insect growth regulators (or biorational pesticides) are generally more specific in their pest activity and more selective to natural enemies than conventional pesticides (Chapters 2, 11, and 12). With the possible exception of the miscellaneous insecticides and acaricides, the biorational pesticides are more frequently physiologically selective than other insecticide classes. The basis for these unique selectivity features is discussed in greater detail in Chapters 11 and 12.

Among these pesticides, *Bacillus thuringiensis* and diflubenzuron have been widely tested for selectivity (Table 9.13; Appendix, Table A.7), extending to almost all major natural enemy groups. Diflubenzuron, in some cases, may be toxic to the immature stages of many natural enemy species (Appendix, Table A.7; Chapter 12). However, because most natural enemies are active as adults and many tend to immigrate to crop habitats rather than develop in treated systems, diflubenzuron can be used in ecologically selective ways. The principal limits to the use of both *Bacillus thuringiensis* and diflubenzuron in IPM programs are due to a lack of pest activity rather than to side effects on natural enemies (genetic engineering may expand and tailor the activity spectra of microbial pesticides; Chapter 21). Of course, it should be pointed out that most of these compounds can be highly toxic to natural enemies at particular points in their life cycles (insect viruses may be an exception; Flexner et al. 1986). However, by combining lower activity against natural enemies with precise timing and application techniques, biorational pesticides can be used with a high degree of selectivity (Chapter 12).

9.6. USING PHYSIOLOGICALLY SELECTIVE PESTICIDES

As illustrated in Fig. 9.2 and 9.3, the physiologically selective properties of a pesticide may require adjustment to match the particular response attributes of each pest/natural enemy complex of species. In some cases, physiological selectivity may be possible for all natural enemies over their prey or hosts, while in other cases, a combination of physiological and ecological selectivities must be employed for maximum natural enemy conservation.

Ecological selectivity may be achieved by separating natural enemies in time and space from the pesticide exposure of their prey or hosts (Chapter 10).

Physiological selectivity may be realized or improved through dose manipulation (Fig. 9.3). For example, many of the selective acaricides which control plant-feeding mites can be modified to conserve predatory phytoseiid mites on apple by using lower doses than normally recommended for strict pest control (Chapter 17, Section 17.2.1). Thus, a general concentration/mortality relationship can be used to adjust predator:prey ratios, thereby establishing effective biological control to augment pesticide use. Lower rates of selective acaricides can be used when a small reduction in prey will establish a predator: prey ratio sufficient to provide biological control before pests exceed economic injury levels. Higher rates of a selective acaricide may be used when predator:prey ratios are low and a greater reduction in prey is necessary (Fig. 17.1). Other specific cases of manipulating physiological selectivity in IPM and resistance management programs are discussed in Chapters 13 and 20.

9.7. SELECTIVITY IN THE SELCTV DATABASE VERSUS THE APPENDIX

As noted earlier, the selectivity data summarized from SELCTV (Chapter 2) are only a subset of that listed in the Appendix (Tables A.1–A.7). Selectivity ratios in SELCTV were based predominantly upon comparative LD_{50} or LC_{50} data. Selectivity data (in the form of prey-host/natural enemy ratios) made up only 8% of total records in the Appendix. This emphasizes the small amount of selectivity research that has been conducted using precise, median lethal assays or similar log dosage or concentration techniques to study both pests and natural enemies. Most researchers test insecticides at field rates against natural enemies (only) to evaluate selectivity because these rates or concentrations are those that natural enemies are most likely to encounter in the field. Activity on pests at the recommended field rate has usually been established earlier, during the pesticide research and development process (Chapter 1).

SELCTV data summaries by insecticide class were similar to those in the Appendix with few exceptions (Theiling 1987). Very few selectivity ratios have been published for insect growth regulators or microbial pesticides using median lethal assays; therefore selectivity information for these pesticide classes was lacking in SELCTV. However, selectivity of the biorational pesticides is evident to a limited degree in the Appendix, which draws from a broader data set. While more precise median lethal assay comparisons may be forthcoming for biorational pesticides, their side effects are less well suited for measurement using acute toxicity assays (Flexner et al. 1986; Chapters 11 and 12). Similarly, indices of selectivity for some early botanicals, inorganics, DDT derivatives, and chlorinated hydrocarbons were underrepresented in SELCTV as compared to the Appendix. Median lethal assays were seldom conducted when these chemicals were widely used. For example, while endosulfan was frequently cited as demonstrating physiological selectivity in the Appendix, it was poorly represented in the selectivity ratio analysis from SELCTV (Chapter 2). The lack of selectivity ratios in early studies may have influenced estimates for a number of other compounds in SELCTV.[1]

[1] The most numerous compounds for which selectivity ratios were reported in the SELCTV analysis were permethrin, pirimicarb, malathion, and fenvalerate. The appearance of the two pyrethroids indicated a greater reliance on high-precision testing for selectivity analysis in recent years, as was seen in general for all pesticides (Theiling and Croft 1988; Chapters 1 and 2). Again, this observation is applicable only to insecticides which cause acute toxicity, and not some of the biorational pesticides such as microbial pesticides and insect growth regulators (Chapters 11 and 12).

In comparing SELCTV and the Appendix, almost identical trends were observed for natural enemies and pests. Aphids, mites, and rice and cotton pests were predominant throughout the database summary of selectivity research, as were generalist predators and mite predators among natural enemies. The SELCTV analysis could best be characterized as summarizing the more precise selectivity information available, largely coming from laboratory studies; whereas the Appendix is drawn more broadly from all selectivity research, including a large number of field studies. However, both surveys suffer from particular biases, being nonrandom in their surveys relative to actual field use of pesticides (Theiling 1987; Chapter 2). Further research on the extent to which physiological versus ecological selectivity has been developed and employed in IPM is needed. This research should be based on more objective assessments using random samples collected from actual field implementation programs.

10

ECOLOGICAL SELECTIVITY

10.1. INTRODUCTION

The use of pesticides in ecologically selective ways, unlike physiological selectivity (Chapter 9), may involve pesticides with broad-spectrum effects on both pests and natural enemies. Selectivity is achieved by applying pesticides in ways that minimize the exposure of natural ememies while killing their hosts or prey (Untersteuholer 1970, Newsom et al. 1976, Hull and Beers 1985). These operational measures include manipulation of the pesticide formulation, timing of application, method of application, spatial distribution of treatment, as well as other means reviewed in this chapter. Pesticide uptake (Chapter 3), factors affecting susceptibility (Chapter 4), and ecological effects of pesticides (Chapter 8) are directly related to ecological selectivity (Fig. 9.1.).

Ecological selectivity as a concept was developed in association with IPM (Ripper 1959, Stern et al. 1959). Early attempts at integrating biological and chemical controls primarily involved ecological selectivity. This is probably still true today, although more physiologically selective pesticides are now available (Chapters 9 and 17). Unfortunately, the range of physiological selectivity in most compounds is limited to a few natural enemies (Chapter 9). By combining physiological and ecological selectivity, the range of selectivity can be broadened for maximum conservation of natural enemies (see integrated selectivity, Chapters 13 and 21).

Will ecological selectivity continue to be the predominant form of selectivity realized in IPM programs or will physiological selectivity become more common in the future (see Chapter 21)? Physiological selectivity has been preferred because careful monitoring and special application techniques are less critical than when implementing ecological selectivity. Conversely, because they can be applied more liberally, physiologically selective pesticides may exert greater selection pressure on pests. Thus, greater use of physiologically selective pesticides may lead to an increase in the development of pesticide resistance. Continued emphasis on *both* physiological and ecological selectivity appears to offer the most benefit and stability to IPM systems in the future.

10.2. PROBLEMS IN DEFINING ECOLOGICAL SELECTIVITY

The terms *physiological* and *ecological* selectivity are used to distinguish the means by which pesticide selectivity is achieved. However, this distinction is not easily made in some cases (see Chapter 9, Sections 9.1 and 9.5.2). When a natural enemy is exposed to less pesticide in the external environment than its prey or host, this is a clear case of ecological selecivity. When this difference in exposure occurs at the interface of the external and internal environments of the natural enemy (e.g., mouthparts), a precise definition of selectivity becomes less clear. For example, when different feeding behavior results in differential exposure to pesticides (compare piercing–sucking feeding aphids with chewing predator), this is technically ecological selectivity. Sometimes what is apparently physiological selectivity is later redefined as ecological selectivity as greater understanding of insect behavior and pesticide uptake are realized (e.g., cases involving stomach poisons and systemic insecticides; Chapters 3 and 9).

Ecological selectivity in concept and practice has changed little over the past few decades of pest control. Only modest improvements in formulation, application, and timing have been achieved. However, many researchers feel that this is a promising avenue for increasing the levels of pesticide selectivity in the field (Croft 1981a, Hull and Beers 1985, Hall 1986, 1987). Basic research on monitoring systems and computer modeling of pesticide applications (Hall 1986, 1987), pesticide residue dynamics (e.g., Reissig et al. 1985), and natural enemy and pest behavior (Hoy and Herzog 1985) may be pivotal in improving the range and degree of ecological selectivity.

10.3. A CONCEPTUAL FRAMEWORK OF ECOLOGICAL SELECTIVITY

Ecological selectivity is achieved by limiting the pesticide exposure of a natural enemy (over a pest) in its external environment. Ecological selectivity may be classified by the manner in which differential exposure of pests and natural enemies is accomplished. Cases of ecological selectivity can be classified as either separation in space or separation in time (Fig. 10.1). Spatial separation may occur within the plant unit or at the plant, field, crop, ecosystem, or even regional level. Temporal separation may occur within daily cycles of arthropod activity, single or multiple generations, or even longer time periods. Ecological selectivity can involve separation *both* in time and in space (Fig. 10.1). An alternative way to classify ecological selectivity is by methodology. This scheme includes such categories as formulation, manner of application, and so forth (Newsom et al. 1976, Hull and Beers 1985).

Ecological selectivity is founded upon exploiting differences in biology between natural enemies and their host or prey. These biological differences include aspects of life history, movement, spatial distribution, and behaviors, such as feeding, mating, and so forth. Many of the operational tactics used to achieve ecological selectivity depend on how well the biological features of pests and natural enemies are understood. The need for this type of information should reinforce the importance of fundamental research on the biology of pests and natural enemies.

10.4. SPATIAL ASYNCHRONY

Arthropod pests and natural enemies may be spatially separated at different times and life stages. Separation may range from different parts of the same plant to a scale of kilometers at the time of pesticide application. Ecological selectivity can be achieved by exploiting the

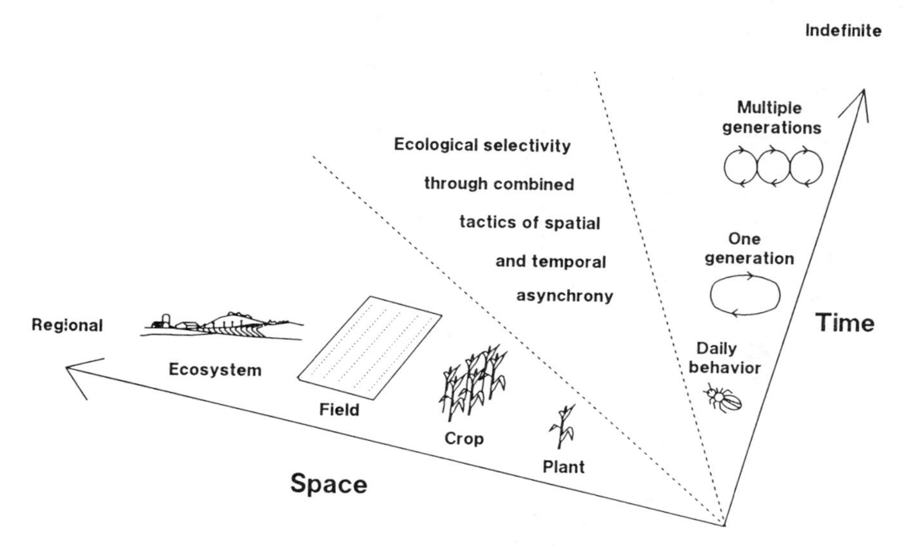

Figure 10.1. Conceptual framework of ecological selectivity in time and space.

spatial incongruence of natural enemies and pests, even with broad-spectrum pesticides (David and Somasundaram 1985).

10.4.1. Within a Plant

Spatial separation of a pest and its natural enemy may occur on a host plant when pesticide applications are made to a portion of the plant. Predatory phytoseiid mites (*Amblyseius fallacis, Typhlodromus pyri, Metaseiulus occidentalis*) often disperse to feed on pest spider mites by crawling from the groundcover or tree base where they overwinter to the lower inner central region of the apple tree. Their prey (*Panonychus ulmi, Tetranychus ulmi, T. urticae, Aculus schlechtendali*) overwinter in the tree and initially disperse more widely on the foliage than the predatory mites (Madsen 1968, Croft and McGroarty 1977). Pesticides can be applied to the outer parts of the tree by eliminating some nozzels on an airblast sprayer and by adjusting the air volume. Through selective placement of pesticides, the predators' exposure is reduced and conservation is achieved (Madsen 1968, Hoyt 1969a). A similar selective placement of white oil to outer leaves of citrus controls California red scale *Aonidiella aurantii*, yet leaves a reservoir of unaffected scales

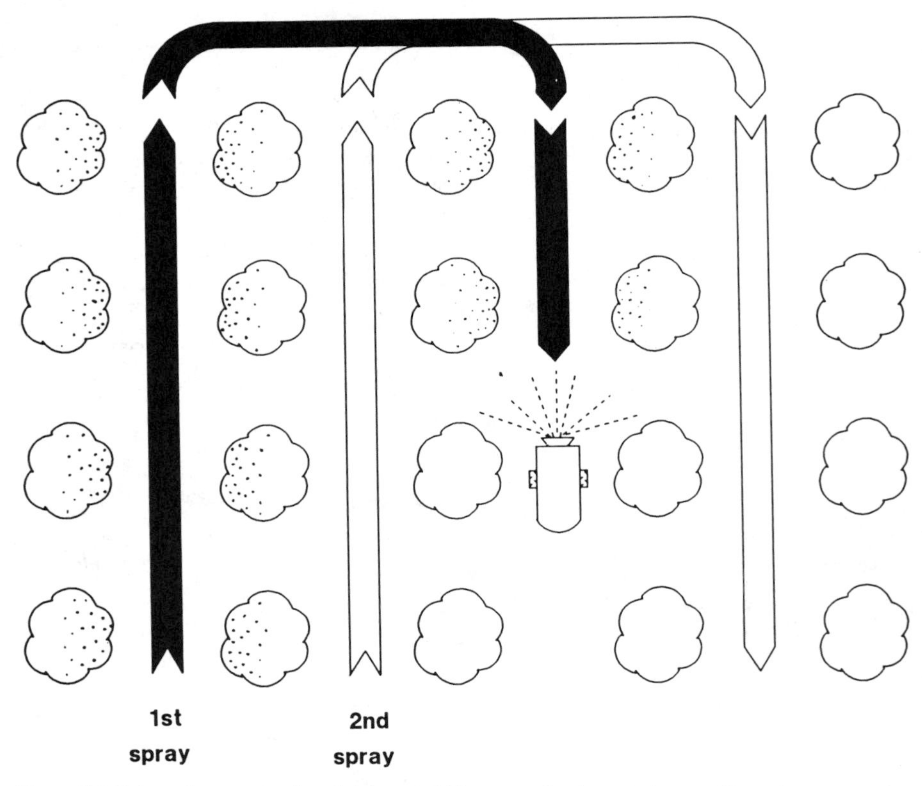

Figure 10.2. Schematic representation of alternate middle-row application technique used in apple orchards (after Hull and Beers 1985).

parasitized by *Aphytis melinus*, which survive inside the canopy of older trees (Davies 1982).

In the eastern United States, an alternate middle-row spraying technique is used to conserve predators of phytophagous mites on apple (Hull and Beers 1985; Fig. 10.2). In addition to selective placement, pesticide rates are reduced. By spraying every other row, only one half of each tree is covered and applications are made less than twice as often. Particularly mobile, alate mite predators can then develop continuously through the spraying period, whereas their less mobile spider mite prey are preferentially exposed to pesticide in the treated area. Populations of *Stethorus punctum*, a coccinellid predator of spider mites, increase faster and provide better biological control of mites on trees sprayed using the alternate middle-row technique than on conventionally treated trees. In Pennsylvania, this method is used for more than 95% of sprays applied to apples (Hull and Beers 1985).

Ecological selectivity has been achieved in Dutch apple orchards for control of *Archips rosana*. This leafroller, which overwinters under bark at the tree base, has been selectively controlled by applying methidathion, a highly toxic organophosphate to the base of the tree just before egg hatch (Gruys 1980a). Emerging larvae are killed as they move to foliage to feed and many early-season natural enemies are conserved.

Watson (1975) found that selective application of azinphosmethyl to the lower two-thirds of cotton plants adequately controlled pests, but allowed anthocorids (*Orius* sp.) to increase in the top one-third of the plant. Stam et al. (1978) reported that herbicides applied to the base of cotton conserved *Coleomegilla maculata, Scymnus louisianae, Orius insidiosus, Geocoris punctipes,* several spider species, and a parasitoid, *Eretomocerus haldemani*. However, herbicide sprays directed at both weed and crop plants greatly reduced populations of the spider *Pseudatomoscelis seriatus*.

Economic damage to tobacco by the tobacco hornworm *Manduca sexta* and the tobacco budworm *Heliothis virescens* has been prevented by treating only the upper five or six leaves of the plant (Guthrie et al. 1956, Lawson et al. 1961). Such treatment allowed for predation on the remaining larval hornworms by *Polistes*. Conservation of linyphiid spiders in Indian rice has been similarly achieved by pesticide sprays made only to aerial plant parts. Some prey and most spiders occupying the basal areas of rice hills escaped direct contact with sprays applied early in the season (Nath and Sakar 1980).

Selective pesticide placement allowed for ecological selectivity in poultry houses by treating adult flies which congregated on the ceiling at night (Axtell 1966). This contrasted with nonselective treatment of manure, which contained maggots and the acarine predator *Macrocheles muscaedomestica*. While not a case of selectivity within a plant unit, treatment of the poultry house ceiling is an analogous example of spatial separation of pesticides and natural enemies.

Partial treatment of a plant or similar production unit does not always guarantee ecological selectivity to arthropod natural enemies. The mobility and distribution of pests and natural enemies must be taken into account carefully. McClure (1977) reported that partial coverage of hemlock favored the resurgence of scale (*Fiorina externa*) rather than conservation of its predators and parasitoids. Treatment of one side of the crown virtually eliminated the parasite *Aspidiotiphagus citrinus* as well as three predaceous species, presumably due to their greater mobility over all parts of the tree. Thereafter, unchecked scale populations reached high levels in untreated portions of the tree. Preferential mortality of natural enemies due to spatially discriminating mechanisms of pesticide exposure have been widely reported in the literature (Croft and Brown 1975, Hull and Beers 1985).

10.4.2. Within a Field or Crop

On a larger scale, the distribution of pests and natural enemies in a crop or field is rarely uniform. Pests and natural enemies also differ in their patterns of movement. Spot treatments or stratified applications of pesticides across plant populations or communities may confer pesticide selectivity to predators and parasitoids.

Pruszynski et al. (1983) reported that local spot treatments of pesticides in greenhouses conserved the predator *Phytoseiulus persimilis* at the expense of its spider mite prey, *Tetranychus urticae*. When toxic pesticides had to be used to control the spider mite or *Trialeurodes vaporariorum*, applications were made only to local foci where outbreaks occurred. These foci included units as small as a single leaf or plant, but ranged up to small groups of plants. Predators survived in nearby untreated foliage from which they later emigrated after residues had dissipated.

Also in a greenhouse system, Pickford (1983) described a method for achieving ecological selectivity against *Thrips tabaci* by spraying mixtures of polybutene and deltamethrin onto plastic beneath cultures of cucumber. This treatment killed pest larvae as they descended to pupate. Natural enemies of spider mites (*Phytoseiulus persimilis*) and whiteflies (*Encarsia formosa*) on aerial plant parts were unaffected by these residues.

Using spot treatments of pesticides in the field to achieve selectivity was first demonstrated by Isley (1926) for the cotton boll weevil (*Anthonomus grandis*). He exploited the behavior of the weevil to first colonize areas near overwintering sites outside of cotton fields and in early vigorous patches of cotton within fields. By applying pesticides only to areas of aggregation, natural enemies were conserved and weevil populations lowered. Newsom et al. (1976) reported a similar tactic used against the bean leaf beetle *Ceotoma trifurcata* on soybean. In this case, control of a bean pod mottle mosaic virus vectored by the beetle and conservation of the beetle's predators and parasites were achieved.

Treatment of field perimeters has been employed to achieve ecological selectivity by spatial separation. Babenko (1980) described such a technique for the control of wheat aphids in the USSR, which initially colonized 50–100-m bands on the outside edge of fields. Natural enemies congregated in these bands and gradually became distributed throughout the field, where they provided biological control later in the growing season. In years of aphid outbreaks, field edges were sprayed early, when natural enemies were in the adult stage and highly mobile. Less mobile aphid populations were reduced by these sprays to a greater extent than more mobile natural enemies. Subsequent control by coccinellids, syrphids, and chrysopids was usually sufficient after sprays dissipated. Such treatments required only 20–25% as much insecticide as a full treatment program while achieving comparable crop yields. Basedow et al. (1976) reported similar selective control of aphids in Swedish and West German cereal crops through the treatment of field edges to conserve carabids and staphylinid predators.

Within a field, the treatment of strips or selected plots may allow for ecological selectivity to biological control agents. Mishra (1979) observed that rice sprayed in alternate strips with 0.05% parathion at *Nephotettix* and *Nilaparvata* leafhopper thresholds of 10–15 per rice hill resulted in greater survival of predatory lycosid spiders than did complete coverage. Both selective and nonselective treatments produced similar rice yields, and thus reduced control costs were realized where selective spraying was used.

In many cropping systems, trap crops are used to lure and treat pests before they are widely discovered by natural enemies (see Newsom et al. 1976). This selective placement of pesticides exploits both spatial and temporal asynchrony between pests and natural enemies, and can be used within fields as well as among fields or even across larger areas of crop and natural plant habitats.

10.4.3. Within an Agroecosystem

An agroecosystem is a mosaic of crops and surrounding environment which includes native and managed plant communities. There are several ways that ecological selectivity can be achieved at this scale of magnitude.

Deciduous tree fruit orchards include other plant communities such as groundcover and surrounding windbreak vegetation (e.g., hedgerows and nonfruit trees; Croft and Hull 1983). Pesticide applications will often conserve natural enemies if they are limited to the target crop. For example, the predatory mite *Amblyseius fallacis* occurs in the ground cover beneath trees in the early season, while pest mites are on tree foliage (Croft and McGroarty 1977). With low-volume applications of insecticide (2–10 × of dilute applications) to the apple foliage little residue will settle on the ground cover (Croft and McGroarty 1977, Pfeiffer 1985). Residues are reduced on the ground cover by modifying the spray droplet size as well as the nozzle angle. *Amblyseius fallacis* survive in the ground cover and later immigrate into orchard trees where they provide biological control of pest mite populations. In Japanese citrus orchards, the conservation of predatory spiders in the ground cover is similarly accomplished by adjusting spraying methods (Nohara and Yasumatsu 1968).

On a larger scale, Anderson (1982) obtained ecological selectivity against the turnip root fly *Hylemya floralis* in Norwegian sugar beets by using lower rates and localized treatments of isofenphos. While treatments reduced carabid beetle populations, they did not affect staphylinids. Both groups of natural enemies were highly sensitive to direct exposure. The observed selectivity was attributed to the greater ability of rove beetles to fly into treated plots from surrounding environments following treatment with the nonpersistent compound.

Similarly, Pawlizki (1984) examined the effects of different levels of fertilizer and pesticide application on the carabids *Pterostichus cupreus, P. melanarius, Carabus granulatus*, and *C. cancellatus* in German cereal and vegetable fields. He observed that extensive pesticide applications caused long-term reduction of carabids, to the extent that nearby woodlands were no longer reservoirs for these predators. This was probably due to a general suppression of species within the region and the persistence of pesticide-adapted species associated with disturbed, treated habitats (see species displacement by pesticides, Chapter 8).

Skrebtsov and Skrebtsova (1980) noted that many natural enemy species visit cultivated and nonmanaged fields adjacent to sprayed environments to obtain pollen and nectar. Specifically, they found that the syrphids *Sphaerophoria methastri, Syrphus ribesii*, and *Eristalinus aeneus* and the ant *Formica rufa* use these crops for food. Natural enemies were conserved in this refuge during spraying of nearby fields. In this instance, the untreated habitat served as a trap crop or refugium from spraying, and thus ecological selectivity was achieved. This is an interesting contrast to using a trap crop to selectively eliminate pests (Section 10.4.2.).

10.4.4. Regional Asynchrony

Cases of areawide ecological selectivity are less common than the more localized scales discussed above. Coordinating pesticide use and monitoring impact across large areas of agricultural production are difficult. However, there are probably situations where areawide ecological selectivity occurs due to neglect rather than management. Beneficials may be conserved simply because application methods do not achieve complete coverage. Many untapped opportunities probably exist for achieving ecological selectivity on a regional scale. Several of these documented cases are discussed below.

In rice-growing regions of Japan, surveys were made using sweep nets to assess the potential for biological control (Kobayashi et al. 1978). A proportionality index of the area sprayed was related to natural-enemy: pest ratios. Attempts were made to minimize the proportion of total acreage under chemical treatment and to conserve natural enemies. More detailed information on these regional pesticide management programs and how they were coordinated and implemented seasonally is unavailable.

Another large-scale application of ecological selectivity was proposed for Soviet farmers in the establishment of "microreserves" (Grebennikov 1980). These were small woodlands from 3 to 6 ha in which parasitoids and predators could multiply undisturbed by nearby agricultural operations. Conserved natural enemies then emigrated from these microreserves to crops or orchards.

Areawide ecological selectivity becomes even more difficult when the target (or non-target) organism is considered a pest as well as a natural enemy. Such is the case with the imported red fire ant *Solenopsis invicta* which attacks the cotton boll weevil *Anthonomus grandis* in the southern United States. Using a detailed analysis of mortality factors, Sterling et al. (1984) observed that the removal of ants using nonselective insecticides greatly increased the abundance of boll weevils. Biological control by the fire ant was recommended in some areas over sole reliance on areawide chemical suppression of the boll weevil. Sterling et al. (1984) recommended confining treatments to sensitive areas of human and animal activity, while leaving large areas where the natural enemy could survive and provide biological control.

10.5. TEMPORAL ASYNCHRONY

Arthropod pests and natural enemies may be asynchronous in time due to behavior, development, and population dynamics, even though they reside in the same area. Daily activities such as feeding and general movement may not be coincidental. Asynchrony within and between generations of pests and natural enemies is common. Obviously, these species must coincide at some point, or biological control would be impossible. Timing pesticide applications to take advantage of temporal asynchrony in pest–natural-enemy biology has probably been the single most common and effective means used to achieve ecological selectivity (Newsom et al. 1976).

An implementation of tactics involving temporal asynchrony requires that the phenology of both pests and natural enemies must be followed. Tools for monitoring or predicting temporal changes in populations may improve the timing of selective pesticide applications (Croft and Hoyt 1983). Considerable refinement in timing insecticide applications to maximize ecological selectivity has been one of IPMs successes on major crops during the past two decades (Newsom et al. 1976, Croft and Hoyt 1983, Hull and Beers 1985).

10.5.1. Short-Term Asynchrony

Emel'Yanov and Yakushev (1981) observed that the daily immigration of predaceous carabid predators from surrounding refugia and the occurrence of the pest *Eurygaster integriceps* in USSR wheat fields were asynchronous. They recommended that short residual compounds lasting only a few hours be used in early morning to control the pentatomid while minimizing impact on the nocturnal predator. Sustek (1982) reported selective use of pirimiphos-methyl to control forest moth pests and to conserve a complex of carabid and staphylinid predators. The short residual compound, combined with the diurnal activity of predators and their occupation of refugia during spraying were the basis of the observed selectivity.

Kalushkov (1982) studied colonization rates of pests and natural enemies on newly planted Bulgarian hops and beans. He found that methomyl could be applied for the control of *Aphis fabae* and *Phorodon humuli* before the arrival of the predators *Propylea quattuordecimpunctata, Adalia bipunctata, Coccinella septempunctata, C. quinquepunctata,* and *Cycloneda limbifer*. In this study, the selectivity mechanism involved a time scale of several days. Aphids were reduced with low rates of methomyl, but those that remained still attracted immigrating predators. When predators arrived, residues were diminished and aphids were subsequently controlled by predators.

10.5.2. Within a Generation

Temporal asynchrony of a pest and a natural enemy within one generation or reproductive cycle is commonly exploited to achieve ecological selectivity. Timing of the co-occurrence of these species is particularly critical when highly specialized natural enemies and their hosts or prey are involved. Seasonal differences in the colonization of new habitats, such as newly planted fields, also are often used to conserve highly mobile predators and parasitoids. For generalist predators or natural enemies having many overlapping generations, achieving ecological selectivity through generational asynchrony is less feasible.

Temporal differences in behavior and biology between parasitoids and their hosts have been exploited elegantly to achieve ecological selectivity. The protected status of parasitoids within hosts represents both a spatial and a temporal refuge from pesticide exposure. A great deal of research has been conducted on the exploitation of temporal asynchronies between plant-feeding pests and their parasitoids. This research is highlighted here, and many of the same principles apply to specialist predators as well.

Cases in which selectivity has been gained through the timing of sprays to kill pests while their endoparasitoids are in protected host stages are summarized in Table 10.1. The residual activity of pesticides is a key factor in achieving this type of ecological selectivity. Flanders et al. (1984) evaluated the residual toxicity of several compounds on beans for control of the Mexican bean beetle and associated pesticide infuences on various life stages of the eulophid parasitoid *Pediobius foveolatus*. Adults were primarily affected, while impact on eggs and larvae depended on the stage of the parasitized host larvae at the time of spraying. Prepupal and pupal parasitoids were unaffected within host larvae. Compounds with residues persisting less than 7–10 days (i.e., the development time of the prepupal plus pupal stages) were the most selective. These included dimethoate, acephate, and methomyl.

When several parasitoids are involved, their developmental synchrony with the host may determine the likelihood of achieving ecological selectivity through the timing of sprays. Blais (1977) reported that selectivity to the larval endoparasitoids *Apanteles fumiferanae* and *Glypta fumiferanae* of the spruce budworm in treated Canadian forests was due to the protection of pupae inside of their hosts. Apparently these two parasitoids were closely synchronized with the larval development of the host. Blais was able to distinguish ecological selectivity to these species from nonselectivity to parasitoids that were adults at the time of spraying by the high rate of mortality observed for the latter species.

Table 10.1. Cases of Ecological Selectivity Achieved through Timings of Pesticide Applications or Manipulation of Parasitoid or Host Life Cycles

Parasitoid	Host	Compound	Life Stages Conserved[a]	Mechanism of Selectivity	Reference
Trichogramma lutea	*Ephestia kuehniella*	Several	L	Refuge in host egg	Ban 1979
Trichogramma pretiosum	*Sitotroga cerealella*	Several	A	Mass releases after sprays	Amaya 1982
Trichogramma pretiosum	*Heliothis virescens*	Several	E,L,P	Refuge in host egg	Bull & House 1983
Apanteles fumiferanae	*Choristoneura fumiferana*	Several	L	Refuge in host	Blais 1977
Three Eulophid spp.	*Phytomyza atricornis*	Ethyl-parathion	L	Refuge in host	Bragg 1974
Microctonus aethiopoides	*Hypera postica*	Methyl-parathion	P	Refuge in cocoons	Dumbre & Hower 1976b
Aphelinus mali	*Eriosoma lanigerum*	DDT, BHC, diazinon, endosulfan	L,P	Refuge in host mummy	Evenhuis 1959
Aphelinus mali	*Eriosoma lanigerum*	Vamidothion pirimicarb, ethiofencarb, azinphosmethyl, dimethoate, endosulfan	L,P	Refuge in host mummy	Gloria 1978
Pediobius foveolatus	*Epilachna varivestris*	Dimethoate, acephate, methomyl	E,L,P	Refuge in host larvae	Flanders et al. 1984
Lysiphlebus testaceipes	*Schizaphis graminum*	Disulfoton, parathion	E,L,P	Refuge in host body	Lingappa et al. 1972
Trissolcus grandis, *T. simistriatus*	*Eurygaster integriceps*	Trichlorfon	E, L, P	Refuge in host egg	Malysheva & Kartavtsev 1977
Aphidius ervi	*Acyrthosiphon onobrychis*	Demeton-methyl, malathion	E, L, P	Refuge in host mummy	Orbtel 1961
Eupelmus urozonus, *Crytoptyx dacicda*, *Eurytoma martellii*, *Pnigalio mediterraneus*	*Dacus oleae*	Dimethoate, parathion	E, L, P	Refuge in host inside olive fruit	Roberti & Monaco 1967

Diglyphus intermedius, Chrysonotomyin formosa	*Liriomyza sativae*	Oxamyl, methamidophos	E, L, P	Refuge in larvae in leaf	Schuster et al. 1979
Itoplectis narangae	*Galleria mellonella*	Lindane, carbaryl, fenitrothion	E, L, P	Refuge in host pupae	Shin 1970
Trichogramma spp.	*Chilo* sp.	Several	A	Inundative releases	Sithanatham 1980b
Epipyrops melaneleuca	*Pyrilla perpusilla*	Dimethoate, vamidothion, endosulfan, BHC	E, L, P	Refuge as pupae in host plant leaves	Varma & Bindra 1980
Anabrolepis mayurai	*Melanaspis glomerata*	Malathion	A, L, P	Refuge in host	Raghunath 1983
Aphidius sp.	*Lipaphis erysimi*	Endosulfan, phosalone, demeton-*S*-methyl	A, P	Refuge in host	Singh & Rawat 1983
Telenomus chloropus, Trissolcus grandis	*Eurygaster* spp.	Several	A, L	Refuge in host egg	Rosca & Popov 1983

[a]A = adult; E = egg; L = larva; P = pupae.

A "biological window" of spray invulnerability within a generation cycle was reported by Drummond et al. (1985) for the apple blotch leafminer *Phyllonorycter crataegella* and the eulophid endoparasitoid *Sympiensis marylandensis*. Using a temperature-dependent emergence model, they showed that the difference in average emergence of the two species was only 4.5 days, with a seasonal variation of 1–8 days. Thus in some years it was possible to spray for emerging adult leafminers without eliminating the parasitoid. The mean difference in emergence was 30 days for the braconid *Photesor ornigis*, making ecological selectivity for this species even more feasible.

Pest managers often take advantage of the natural cycles of temporal asynchrony of pests and natural enemies to attain pesticide selectivity. Periodic or mass releases of natural enemies can also be manipulated to this end, often with greater flexibility (e.g., Sithanantham 1980b, Amaya 1982). These types of manipulations have been reported for several *Trichogramma* spp. (Amaya 1982), *Encarsia formosa* (Osborne 1981), and *Phytoseiulus persimilis* (Binns 1971).

10.5.3. Multigenerational or Seasonal Cycles

The timing techniques used to achieve ecological selectivity for one generation of pests and natural enemies may also be effective over multiple generations as selective chemical controls and biological controls are integrated and begin to stabilize. Natural enemy and pest population levels may then dictate the need for carefully timed applications.

Populations of the leafminer *Leucoptera meyricki* and its parasites *Zagogrammosoma variegatum*, *Achysocharella ritchiei*, and *Pharahormius leucopterae* were monitored over a 40-week period in Kenya coffee plantations to establish optimal spray timing and maximum conservation of parasitoids (Bess 1964). Sprays timed at the peak flight of adult moths did not affect parasites in leafminer mummies, thus increasing the ratio of living parasites to leafminers after spraying. Over several generations, parasitism levels in sprayed plots were higher than in unsprayed plots, primarily due to the timing of pesticide applications.

Broodryk (1980) described an ecologically selective pest control program for key pests and aphids of cotton in South Africa, which conserved the natural enemies of all pests. Pirimicarb sprays were avoided during the first eight weeks of plant growth until flowering and were applied thereafter when pests exceeded economic injury levels. Similar programs utilizing ecological selectivity have been recommended for the control of *Heliothis zea* and *H. virescens* on cotton in Texas (Croft et al. 1985).

Washburn et al. (1983) described complex timing requirements for conserving a complex of parasitoids attacking the ice plant scale *Pulvinariella mesembryanthemi* in California. They concluded that maximum parasitism was obtained when sprays were applied to mature hosts containing parasitoid immatures and that survival was minimal during periods of peak parasitoid emergence. Spraying frequency was an important factor that governed the survival and reestablishment of the parasitoids throughout the growing season.

One of the most advanced applications of ecological selectivity through timing across multiple generations was reported by Osborne (1981). He used a physiological-time model of host and parasitoid development along with a nonpersistent pyrethroid to exploit temporal asynchrony between the greenhouse whitefly *Trialeurodes vaporariorum* and *Encarsia formosa* in greenhouse studies. A 27% reduction of parasite emergence was accompanied by a 71% reduction in whitefly populations when black scales were treated with resmethrin at 0.25 lb a.i./100 gal of water after 125 degree-days (12.7°C threshold).

Under this scheme, parasitoids were able to reproduce and provide biological control.

10.6. METHODOLOGICAL CLASSIFICATION OF ECOLOGICAL SELECTIVITY

Ecological selectivity can be classified on the basis of application methodology (Newsom et al. 1976). In the ensuing discussion of ecological selectivity from a methodological standpoint, selectivity is generally achieved by spatial and temporal separation of natural enemies and pesticide exposure. The dose and the means to manipulate it were discussed at the organismal and population levels as attributes of physiological selectivity in Chapter 9. Several authors have considered dose as an ecological selectivity factor (e.g., Newsom et al. 1976, Hull and Beers 1985).

10.6.1. Formulation

Pesticide formulations can promote ecological selectivity through limiting or directing the distribution of pesticide residues. Formulation can influence pesticide application, distribution, uptake, and penetration (Chapters 4, 6, and 9). Ripper et al. (1948) were among the earliest to demonstrate that even a highly toxic pesticide could be rendered selective by formulation changes. Using DDT coated with degraded cellulose, they observed only a slight impact on the control of *Pieris brassicae* and *P. rapae*; however, mortality of the beneficials *Lucilia sericata, Syrphus* sp., *Apanteles glomeratus*, and *Mormoniella vitripennis* was significantly lessened by the new formulation as compared to a standard DDT suspension. Many subsequent studies have demonstrated ecological selectivity via formulations including baits, seed treatments, granulars, and encapsulation.

10.6.1.1. Common Formulations. Analysis from the SELCTV database indicated that most pesticides studied for side effects on natural enemies were formulated as emulsifiable concentrates (EC) (48% of all records), followed by wettable powders (WP) (37%) and granulars (G; 6%). Soluble or dispersible powders, flowables, and dusts combined comprised the remaining 9% of records. (These percentages exclude technical grade pesticides; Chapter 2.) These proportions probably reflect the relative use of different formulation types for pest control as well.

Specific products of a common formulation type may differ somewhat in certain ingredients such as the diluent, organic solvent, or other additives. Minor variations in additives can influence the toxicity of pesticides to natural enemies. To illustrate this point, Franz and Fabrietius (1971) reported the influence of the formulation of several fungicides on the parasitism response of the egg parasitoid *Trichogramma cacoeciae* attacking eggs of *Sitotroga cerealella* (Table 10.2). Different formulations of the fungicide TMTD caused only minor variations (1–13% parasitism). However, rates of parasitism in the presence of several zineb formulations having a similar range of active ingredient varied from 2–89% (Table 10.2). These data indicate that the formulations themselves may have been toxic to the natural enemies or that some type of interaction between pesticide and formulation had occurred.

In field tests that probably involved both physiological and ecological selectivity, Charlet and Oseto (1983) noted significant differences in responses of the sunflower stem weevil *Cylindrocopturus adspersus* and the braconid parasitoid *Nealiolus curculionis* to

Table 10.2 Impact of Microencapsulated Parathion on Entomophagous Insects (Modified from Dahl and Lowell 1984)

Crop	Species	Location(U.S.)	Toxicity Rating[1]	Researcher (unpubl.)
Alfalfa	*Bathyplectes curculionis*	MO, CA	*,**	Higgins, Davis
	Spiders (several species)	UT	*	Davis
	Nabids (several species)	UT, MO	**,***	Davis Higgins, Ruppel
	Coccinellids (several species)	UT, NY, MO, CA	***	Davis, Gauthier, Lamborn
Apples	*Amblyseius fallacis*	PA, NY, MI, OH	*	Hull, Lienk, Howi, Hall
		NY	***	Wieres
	Metaseiulus occidentalis	*WA, UT*	*	Hoyt, Davis
	Stigmaeid mites	MI, WA, UT	*	Howitt, Covey, Davis
	Coccinellids	WA	*	Covey
	Predaceous bugs	WA	*	Covey
	Anthocoris spp.	CA	***	Pass
	Stethorus punctum	PA	*	Hull, Asquith
	Aphelinus mali	CA	***	Pass
Corn	Chrysopids	MO	*	Keaster
	Anthocorids	MO	*	Keaster
	Coccinellids	MO	**	Keaster
	Nabids	MI	***	Ruppel
Cotton	Coccinellids	TX	*	Nemec
	Hymenoptera	TX	***	Nemec

[1]Survivorship classification: *50–100%, **30–49%, ***0–30%.

two very similar EC formulations of permethrin. For both species, Pounce 2EC was more toxic than Ambush 2EC, and the latter provided a greater selective advantage to the parasitoid over its host.

10.6.1.2. Baits and Seed Treatments. The use of baits and seed treatments can drastically decrease pesticide residues in the environment in addition to concentrating chemicals where they are effective and selective. However, in some cases natural enemies may be affected by their use. For example, researchers have found that protein hydrolysate baits attract a variety of predators. Hagen et al. (1970) found that syrphids, adult malachiid beetles, chrysopids, and coccinellids were negatively affected by bait sprays. Gholson et al. (1978) evaluated differences in the effects of spray and bait formulations of carbofuran granules and carbaryl baits on cornfield carabid predators in Iowa. They observed differences in mortality ranging from high to low among carabids exposed to direct sprays, but when exposed to granular and baits at equivalent rates, consistently low mortalities were observed.

Khalil et al. (1976) observed that seed treatments with systemic granules of aldicarb, disulfoton, and thiometon applied for control of *Heliothis* spp. in Egyptian cotton fields caused negligible reductions in populations of *Coccinella undecimpunctata, Scymnus interruptus, Paederus alfierii, Chrysoperla carnea, Orius* spp., and spiders. Foliar application of these materials at the same rates decimated natural enemies.

Rosen (1967) reported that a protein hydrolysate bait spray of malathion was highly toxic to the aphelinid parasite *Aphytis holoxanthus* under continuous exposure and confinement in the laboratory. However, in Israeli citrus groves the compound was ecologically selective to this parasitoid due to a greater uptake by its scale host.

Nasca et al. (1983), in a similar example from citrus, reported that malathion bait sprays applied to control *Ceratitis capitata* and *Anastrepha fraterculus* were selective to predaceous chrysopids and hemerobiids. Sugarcane molasses sprays with 2% malathion were applied to the entire tree crown or in confined plastic strips dipped in the bait hung in trees. Both the fruit flies and the neuropterans were attracted to the strip baits, while treatment of the entire tree crown was effective only on the pest. Surprisingly, treatment of the entire tree was more selective to the natural enemies. Apparently, predators were less attracted to the more dispersed bait on the tree than the fruit flies.

Another example of ecological selectivity achieved through the use of bait formulations was reported by Soultanopoulos and Broumas (1977) working with olive pests. Adult *Eupelmus urozonus* and its host *Dacus oleae* were confined on leaves taken from olive saplings that had been treated in the field with sprays of dimethoate, fenthion, diazinon, or bromophos or with bait sprays containing each of these compounds with protein hydrolysate. The mortality of *D. oleae* was consistently higher than that of the parasitoid in bait applications. Several of these compounds were highly toxic to the parasitoid when administered topically.

Behavioral responses of three parasitoids to a malathion bait were recorded by Hoy and Dahlsten (1984) using a video camera. The ice plant scale parasitoid *Encyrtus saliens* was extremely susceptible to the bait residues. At sublethal concentrations, avoidance responses were exhibited by *E. saliens, Trioxys pallidus* (a parasitoid of the walnut aphid), and *Euderauphale flavimedia* (a parasitoid of the iris whitefly). Pesticide avoidance by the parasitoids conferred some selectivity. They concluded that behavioral analysis could be useful for developing a bait spray program tailored to minimize pesticide impact on these species.

Bait sprays may not be selective to all natural enemies associated with a particular pest or crop. For example, ground meat baits used for fire ant control impact on nontarget ants, including predaceous forms (e.g., Apperson et al. 1984). Immature parasitoids associated with a pest may be exposed to bait spray toxicants through the food chain (Fig. 3.1). Ehler et al. (1984) followed the impact of bait sprays applied for the eradication of the Mediterranean fruit fly *Ceratitis capitata*, which also affected the nontarget gall midge *Rhopalomyia californica* by reducing its parasitoids *Torymus koebelei, Zatropos capitis, Platygaster californica*, and *Mesopolobus* sp. After bait sprays were applied, gall midge populations reached densities 90 times higher than normal. In laboratory tests, malathion bait was highly toxic to both the gall midge and its parasitoids.

10.6.1.3. Dusts. Bartlett (1951) first demonstrated that many natural and manufactured "inert" dusts affect parasitic Hymenoptera, causing up to 100% mortality when caged on dry deposits. Inert dusts were carrier components of early pesticide formulations. Subsequent studies have demonstrated that dust formulations are more toxic to natural enemies than wettable powder or emulsifiables. Yoo et al. (1984) reported that organophosphate and carbamate insecticides used for the control of several rice pests (i.e., *Nilaparvata lugens, Nephotettix cincticeps, Chilo suppressalis, Sesamia inferens*) were more toxic to the predaceous spider *Pirata subpiratus* in potted plant tests as dust than as emulsions. However, other studies have shown that diluents such as talc and diatomaceous earth can be innocuous to natural enemies such as *Itoplectis narangae*, a larval–pupal parasite of several major lepidopteran pests (Shin 1970).

While dusts may be deleterious to natural enemies, Basedow et al. (1976) developed a technique using dusts for the control of cereal pests in northern Europe that conserved predaceous carabids. Where foliar sprays of parathion reduced predators, dust applications to moist foliage conserved them. This technique was especially selective to epigeal species. Some mortality to species that climb onto plants (e.g., staphylinids and some carabids) was observed. Apparently, the selectivity mechanism involved the greater adherence of dusts to the foliage, whereas foliar sprays left more concentrated residues on the soil surface.

10.6.1.4. Granulars. Granular formulations may control pests selectively while conserving natural enemies through spatial separation of pesticide and natural enemy. Granular formulations have been most selective against soil pests using systemic pesticides (Section 10.6.2., Chapter 9). In China, Fang et al. (1984) reported control of the cotton aphid *Aphis gossypii* using either phorate or carbofuran granules applied at sowing. The coccinellids *Adonia variegata, Coccinella septempunctata,* and *Propylea japonica* and spiders that move into cotton in early summer were thus conserved. Aphids were controlled biologically in June, making subsequent chemical control unnecessary. Larval parasitism of another important pest, *Heliothis armigera*, by the ichneumonids *Campoletis chlorideae* and *Microplitis* sp. was in the range of 16–47% when systemic granule applications were used.

Saharia (1982) used granular formulations of carbofuran, disulfoton, dimethoate, and phorate against the aphid *Lisaphis erysimi* on mustard. All four insecticides were effective against the aphid but did not kill the predator *Coccinella repanda.* To further conserve the predator, it was recommended that insecticides be applied early in the season, before predators immigrated into the crop.

Newsom et al. (1976) reported ecologically selective control of the lesser cornstalk borer *Elasmopalpul lignosellus*, which feeds beneath the soil on peanuts. Growers had previously applied foliar sprays which upset secondary pests (e.g., spider mites and lepidopterans) due to natural enemy destruction. By using a granular insecticide, biological control of the borer was achieved without disrupting biological control on the foliage. As a result, the number of applications made for borer control was reduced from 8 to 10 per season to 1 or 2.

Ba-Angood and Stewart (1980) showed that granular applications of carbofuran and thiofanox for the control of cereal aphids on barley did not significantly reduce coccinellids, chrysopids, or parasitoid mummies (*Aphidius* sp.) one week following application, whereas foliar sprays of dimethoate and pirimicarb reduced coccinellids and aphid mummies appreciably (Fig. 10.3). All compounds, except pirimicarb, were known to be directly toxic to these natural enemies, but apparently not when applied as granules. Similarly, Gravena et al. (1976) compared the influence of several insecticide formulations and application methods on the predaceous coccinellids associated with cotton aphid and lepidopteran pests in Brazil. Granular aldicarb or seed dressings of carbofuran or disulfoton gave good pest control while conserving predators. Foliar applications of parathion were detrimental to these predators.

10.6.1.5. Encapsulation. Ripper et al. (1948) first demonstrated selectivity to entomophagous species by coating a broad-spectrum pesticide to confine breakdown and activation until after ingestion by pests. While encapsulation generally increases the selectivity of pesticides to natural enemies, other problems have arisen. Bees may collect and transport pesticide particles to their hives, where toxic effects on immatures have been documented (Dahl and Lowell 1984).

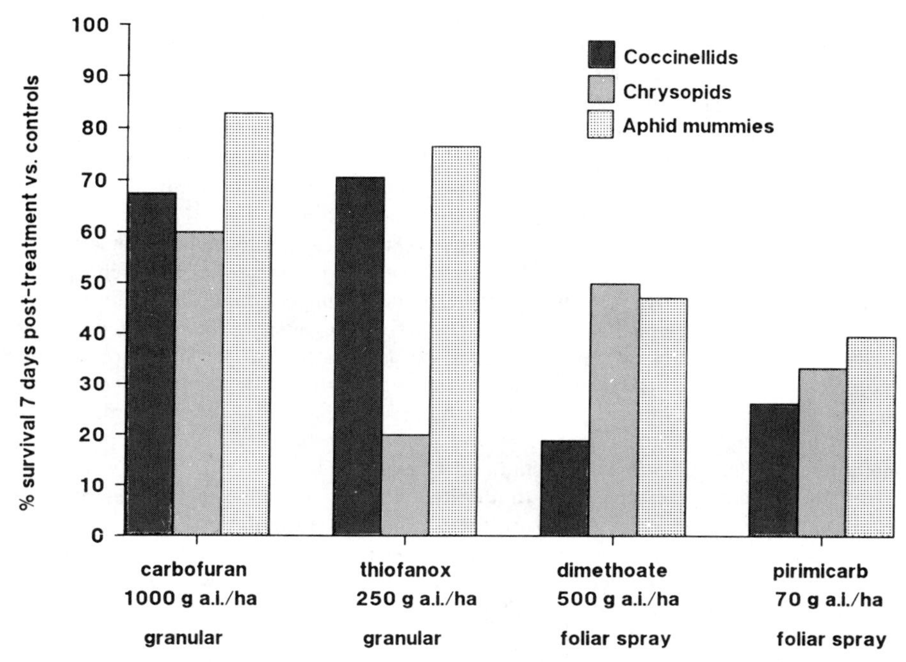

Figure 10.3. Number of predatory insects per 10 tillers of barley treated with granular (carbofuran, thiofanox) versus foliar (dimethoate, pirimicarb) insecticides (after BA-Angood and Stewart 1980).

Many field trials have been conducted to assess the impact of encapsulated pesticides on pests, honey bees, and natural enemies, particularly on alfalfa and tree fruits (Table 10.2, modified from Dahl and Lowell 1984). As can be seen, encapsulated materials show highly variable toxicity to a wide range of natural enemies. In particular, encapsulated materials were highly toxic to the predatory nabids and the few hymenopteran parasitoids that have been evaluated. Encapsulated pesticides are standard treatments on alfalfa and several tree fruit crops, and are used to a lesser degree on corn and cotton (Jackson 1984; Table 10.2).

A specific example of ecological selectivity achieved with encapsulated parathion was reported by Asquith et al. (1976) and Hull (1979a), who found this normally toxic organophosphate to be virtually nontoxic to the mite predators *Stethorus punctum* and *Amblyseius fallacis*, while showing good activity on lepidopterous pests of apple. This study typifies the literature on ecological selectivity to natural enemies with encapsulated pesticides (Dahl and Lowell 1984). Most research has involved field trials of methyl parathion. Little is known about the mechanisms by which encapsulated pesticides are selective to natural enemies.

10.6.2. Systemic Pesticides

Systemic pesticides are transported in the xylem or phloem of plant tissue to a site of uptake by plant-feeding pests. (See Appendix for a list of systemic pesticides that have selective action.) As noted in Chapter 9, many pesticides with systemic activity exhibit both physiological and ecological selectivity. Systemic insecticides are particularly effective on piercing–sucking pests (e.g., aphids and mites). Ripper (1957) first illustrated the selective use of schradan and oxydemeton-methyl, which were highly toxic to aphids, whiteflies, scales, and spider mites, but did not harm predatory coccinellids, syrphids, or the parasite *Diatreus rapae*. Many studies have since demonstrated ecological selectivity with systemics (Hull and Beers 1985, Theiling 1987; Chapters 2 and 9). Several applications of ecological selectivity in rice agroecosystems using systemic pesticides combined with other formulations and application methods are discussed below.

Pesticides with systemic properties have been extensively employed in Southeast Asian rice paddies for selective pest control. Studies have focused on systemic organophosphates and carbamates to control rice leafhoppers (especially *Nilaparvata lugens*) while conserving the spiders *Lycosa pseudoannulata* and *Oedothorox insecticeps*, the mirid *Cyrtorhinus lividipennis*, and various egg and egg-larval parasitoids (e.g., *Platygaster* spp.). While ecological selectivity to the spiders and parasitoids is commonly accomplished through the use of systemics (Chu et al. 1975, 1976a, b, c, 1977, Chang et al. 1979, Chen and Chiu 1979, Chiu et al. 1980, Chang 1981), mirid predators which feed on both plants and leafhoppers suffer high mortality (Chiu and Cheng 1976).

Further studies have been made to modify the selectivity achieved in rice systems in Southeast Asia. Choi et al. (1978) evaluated different formulations and application methods for the systemic carbamate carbofuran on spiders and leafhoppers by measuring population densities in rice fields after single root-zone applications were made (either by applying encapsulated pesticides or by injection of liquid formulations). These treatments were compared with 2–4 broadcasts of granules to irrigation water. Spiders (micryphantids and lycosids) were reduced by all treatments; however, reductions were most pronounced after placement of encapsulated pesticide. The significant reduction of spiders by root-zone applications was considered to be due to a concentration of the insecticide in the spider–leafhopper food chain. Hence, while the use of systemic pesticides may eliminate widespread residues, toxicity to natural enemies may occur through food chain uptake.

Dyck and Orlido (1977) conducted tests to examine the selective use of carbofuran in Philippine rice fields harboring several important predators, including the spider *Lycosa pseudoannulata*, the predatory bug *Cyrtorhinus lividipennis*, and the coccinellid *Micraspis crocea*. In comparing eight different application methods (Table 4.5), field trials suggested that broadcasting granules into the paddy water or root-zone applications did not kill a large percentage of the spider populations. Because the toxicant was taken up into the plant, this treatment was selective to foliage-inhabiting natural enemies. However, toxicity to *C. lividipennis* was high for any method of application of persistent pesticides.

Parallel results of selectivity with systemic insecticides have been reported on cotton. According to Ridgway et al. (1967), soil applications of aldicarb and stem applications of monocrotophos produced high mortality in adult *Nabis* spp. and *Geocoris* spp., but had very little effect on *Chrysoperla* larvae, hymenopterous parasitoids, and spiders. Again, the facultative plant-feeding species were greatly affected, compared to the more strict entomophagous feeders.

Several studies with systemics have demonstrated how pesticides can be uniquely distributed to confer ecological selectivity in the field. Sol (1961) reported that the selective application of demeton-methyl to the lower parts of sugar beets killed *Aphis fabae* while conserving syrphid, coccinellid, and chrysopid predators. Ecological selectivity was achieved on brussel sprouts in Australia with a nonsystemic (permethrin) to control the diamondback moth *Plutella xylostella* and a systemic (demeton-*S*-methyl) for the cabbage aphid *Brevicoryne brassicae* (Davies 1982). Aphids that had ceased feeding and were parasitized by *Diaeretiella rapae* moved beneath the lower leaves of the plant and escaped exposure, whereas demeton-*S*-methyl was taken up by nonparasitized aphids feeding on the plant in any location.

Newsom et al. (1976), in summarizing research with systemic insecticides, noted that the degree of selectivity to beneficials achieved by this approach had been overestimated by early researchers (e.g., Ripper 1956, Bartlett 1964, Metcalf 1964, Ridgway 1969, Cate et al. 1972). While this may have been true for early estimates of the impact of systemics, more recent results tend to confirm their selectivity to natural enemies. From the results presented in Chapter 9, it would appear that systemic pesticides have been the *most prominent group* of selective pesticides in many cropping systems. Many of these compounds are physiologically and ecologically selective.

10.6.3. Application Technology

10.6.3.1. Spray Volume and Application Techniques. Relatively little attention has been paid to the influence of application techniques on pesticide selectivity. One governing factor may be that there are so few broadly selective compounds (Chapter 9). It is generally perceived that pesticides are so much more toxic to natural enemies than to their prey or hosts that little selectivity could be achieved by manipulating application techniques.

Some comparative studies of application methods seem to support these conclusions. For example, House et al. (1983) compared the toxicity of the pyrethroid fenvalerate applied using 1) a standard hydraulic sprayer with hollow cone nozzels, 2) an electrostatic sprayer, and 3) an ultra low-volume mist blower. Predator populations (*Geocoris punctipes, G. uliginosus, Coleomegilla maculata, Hippodamia convergens, Chrysoperla* sp., *Notoxus* sp., and *Nabis* sp.) were greatly reduced by this highly toxic compound in all three treatments until plots were recolonized by immigrants seven days later.

In a study of the interaction of alfalfa plant height and insecticide application rates on beneficial arthropods, Brandenburg (1985) observed that insecticidal spray coverage, measured by water-sensitive paper at the soil surface, significantly decreased as plant height increased. However, spray volume did not have a significant effect on spray coverage. Populations of beneficial arthropods were reduced by methomyl or fenvalerate sprays applied to plants of different heights, even when the total volume of spray was reduced. Apparently these natural enemies were so mobile that they encountered a lethal dose of pesticides even though only part of the plant contained residues. He concluded that there appeared to be little potential for reducing the impact of these chemicals in alfalfa by reducing the amount of spray volume under the conditions of plant growth used in these tests.

While the impact of highly toxic insecticides has not been lessened by altering application methods (e.g., equipment, delivery systems), ecological selectivity of more selective compounds can be improved significantly through placement or timing. The use of the alternate middle-row technique in orchard ecosystems illustrated how placement alone can be a primary component of a selective pesticide program (Hull and Beers 1985). Other examples of significant benefit from application methodology can be seen in the case histories cited below for aerial, low-volume, and fumigation techniques.

10.6.3.2. Aerial versus Ground Applications. A number of studies have been conducted which compare the effects of aerial and ground application of pesticides on beneficial species. Results have been similar to pest efficacy trials with these same application methods. Generally, aerial applications have equal or less impact on natural enemies than ground applications. Often results are species-specific and depend on the distribution and mobility of the species in question (see Chapter 3).

Greene et al. (1974) evaluated the effects of several different application techniques and pesticide rates on the complex of pests and predators in Florida soybean fields. Aerial pesticide applications had little effect on *Geocoris punctipes, G. uliginosus, Nabis alternatus, N. capsiformis*, and *N. roseipennis*. Ground applications had more mixed effects on these natural enemies, with some species being almost eliminated.

10.6.3.3. Low-Volume and Electrostatic Sprays. Lingren et al. (1972) measured the responses of an ichneumonid, *Campoletis perdistinctus*, and a braconid, *Apanteles marginiventris*, to field sprays of several highly toxic pesticides (e.g., azinphosmethyl, monocrotophos, malathion, methyl parathion, toxaphene plus DDT, trichlorfon) in cotton using conventional low-volume (CLV), ultralow-volume (ULV), and low-volume (LV) applications. Residues from ULV applications were more persistent than CLV treatments. Recommendations were that the lowest rates possible be used to obtain control of cotton pests and that application methods resulting in the least persistent residues be employed to conserve parasitoids emerging from their hosts.

Research in Ontario apple orchards showed that the predaceous eurythraeid mite *Balaustium putmani* was more common in commercial orchards where low-volume sprays were used, but was usually absent in orchards where high-volume sprays were applied (Hagley and Simpson 1983). It was hypothesized that more egg masses and the sessile stage of the predator were killed by the high-volume sprays, which provided more complete coverage, inundating crevices and eliminating refugia.

In Soviet cereal grain fields, suppression of the pentatomid pest *Eurygaster integriceps* and conservation of a *Telenomus* egg parasite with the selective insecticide trichlorfon were reported by Kamenkova (1971). Mortality of the egg parasite following treatment was reduced from 35–39% when using higher volume sprays to only 4–17% when using low-volume application techniques. Timing of the spray applications to coincide with periods when about 40% of the nymphs of *Eurygaster* had reached the second instar was critical to maximizing selectivity to the natural enemy.

Electrostatically charged sprays have been developed as a means to increase the adhesion of sprays to plant foliage, thus increasing pest control effectiveness. The effect of this application technology on selectivity to natural enemies has not been widely investigated. In one recent field test in the United Kingdom, the effects of conventional and electrostatic sprays were evaluated on nontarget organisms, including carabids, staphylinids, and other pedologic species (Endacott 1983). Using pitfall traps and soil residue analysis, the study showed that lower insecticide levels were found in the soil in electrostatically sprayed plots compared to the conventionally sprayed plots. Furthermore, following initial declines in both treatments, pitfall catches of natural enemies recovered significantly faster and to higher levels after spraying with the electrostatic method than with the conventional sprayer.

10.6.3.4. Fumigation. Of all application techniques, one would not expect fumigation to favor natural enemies through lack of exposure or spatial incongruence. As discussed in Chapter 3, vapor, smoke, or aerosol exposure can be very harmful because escape from contact is rarely possible. Ordinarily, to allow survival of beneficial species, only those compounds that are physiologically selective can be applied as fumigants. Helgesen and Tauber (1974) provided a good case study with the physiologically selective carbamate pirimicarb. Equivalent quantities were applied as a spray and as a fumigant to greenhouse populations of *Myzus persicae*, the predatory species *Chrysoperla carnea* and *Phytoseiulus persimilis*, and the parasitoid *Encarsia formosa*. Natural enemies were essentially unaffected while the aphid was reduced by the combined action of the carbamate insecticide and the predatory neuropteran.

10.6.3.5. Other. One interesting selective application technique for applying permethrin, a synthetic pyrethroid insecticide, was reported by Butler and Las (1983). Gossyplure, the sex pheromone of the pink bollworm *Pectinophora gossypiella*, has been widely used to suppress populations. However, when permethrin was added to the sticker used to adhere pheromone-containing fibers to the plant, the effectiveness of the treatment was significantly increased. Selectivity to the predators *Collops* sp., several coccinellids, *Chrysoperla* sp., *Orius* sp., *Nabis* sp., *Geocoris* sp., and several mirids was observed. The pheromone-sticker plus pesticide treatment caused no measurable harm to natural enemies compared to conventional application measures, which caused from 58–100% mortality.

10.7. SUMMARY

The examples cited above illustrate how ecological selectivity can be achieved through spatial and temporal asynchrony and application methodology using broad-spectrum pesticides. Achieving ecological selectivity requires that behavioral and biological differences between arthropod pests and natural enemies be exploited. Detailed monitoring of phenology and population dynamics of pests and natural enemies is often required to meet these goals. Also, the appropriateness of techniques varies with the individual pest and the natural enemy, the crop, and the surrounding environment. Pest managers must have an appreciable understanding of the biology and behavior of arthropods in agroecosystems before these types of selectivity measures can be widely employed in the field. In many cases, too little is understood about community ecology and mobility characteristics of natural enemies and pests to implement such broadly integrated systems of pesticide selectivity (Chapters 13 and 21).

Some degree of physiological selectivity is involved in many cases of ecological selectivity. Often several operational measures must be integrated to achieve ecological or integrated selectivity in the field (Chapter 13). Furthermore, the crop, pest, natural enemy, and pesticide are being influenced by environmental factors in dynamic ways. Recent progress in modeling and monitoring research has facilitated the study of environmental effects on these components of pest management systems. Together, modeling and monitoring allow researchers to follow the dilution of pesticide residues on treated plant surfaces in more dynamic and realistic ways (e.g., Stanley 1987). Detailed studies of insect searching (e.g., Hoy and Dahlsten 1984) and oviposition behaviors provide critical input into understanding pesticide–natural-enemy interactions. Incorporating such parameters as crop growth dynamics along with improved spray delivery (Hall 1986, 1987) may hasten the development of more refined techniques for integrating pesticides and natural enemies in crop production systems.

11

MICROBIAL PESTICIDES

11.1. INTRODUCTION

The use of microbial pest control agents (MPCAs) as selective pesticides has increased in recent years because of the difficulty in achieving selectivity to arthropod predators and parasitoids with conventional pesticides (Mullin and Croft 1985; Chapters 9 and 10). Entomopathogenic viruses, bacteria, fungi, and protozoans have been widely tested for their selectivity, safety, and economic viability. Where these attributes have been found inadequate in some species, some tailoring of MPCAs may be possible with genetic

This chapter was coauthored by J. Lindsey Flexner, Oregon State University, Corvallis, Oregon.

engineering. Genetic engineering of insect pathogens is at the forefront of biotechnology. An array of engineered MPCAs with improved activity and selectivity may be registered for pest control in the near future (Kirschbaum 1985; Chapter 21).

A number of naturally occurring MPCAs have been developed commercially. Their side effects on natural enemies have largely been assessed in ways similar to conventional pesticides (Chapter 5). Many MPCAs are relatively species-specific to their hosts and occur in nature without harming natural enemies. It is a general consensus that MPCAs should be used more extensively in IPM. However, these agents can persist and replicate in nature and therefore may have substantial direct effects on arthropod predators and parasitoids. Indirect effects of MPCAs on natural enemies may be significant as well through effects on the hosts or prey of natural enemies.

The side effects of MPCAs on natural enemies are difficult to generalize. Some are probably safer and have fewer side effects on arthropod predators and parasitoids than conventional pesticides. Others may have longer term effects that cannot be assessed by the susceptibility assessment techniques used for conventional pesticides (Chapter 5). In Chapters 2 and 9, the toxicity of MPCAs compared to other pesticide groups was summarized. This chapter reviews MPCA impact on natural enemies of agricultural pests in greater detail. Methods of assessment and lethal, sublethal, and ecological side effects are considered. Current testing procedures for measuring MPCA side effects on natural enemies are critiqued. Brief comment is given to the need to assess the impact of genetically engineered MPCAs, and the role that arthropod natural enemies may play in these evaluations.

11.2. NATURAL ENEMY–MICROBIAL PEST CONTROL AGENT INTERACTIONS

MPCAs interact with arthropod natural enemies through complex biological processes. These agents are biological entities themselves. Their population parameters are dynamic and can be influenced by their host arthropod (pest or natural enemy) and other features of the biological and physical environment.

The most common MPCA–natural enemy–pest interactions can be represented in a generalized model (Fig. 11.1). Major biological components of the model are the natural enemy and its host or prey, divided into populations of susceptible and infected individuals. The other major biological component is the MPCA, which may be present as a natural population on a substrate or in an insecticide spray, or both. MPCA transmission to a natural enemy can be interspecific—from a pest to a natural enemy and vice versa (Fig. 11.1). Transmission from pest to natural enemy is most common; however, incidental transmission by natural enemies to pests via stinging or handling (and not killing) is becoming more widely recognized (Hamm et al. 1983, 1985, Levin et al. 1983). Among natural enemies and pests, intraspecific transmission of an MPCA may occur among individuals of the same generation (horizontal transmission, Fig. 11.1; Brown 1987). Infected adult females of either group can also produce infected eggs (vertical transmission, Fig. 11.1; Andreadis and Hall 1979). Other sources of infection for a natural enemy can be from sprays directed at pests or from an MPCA developing in other arthropods or natural enemy populations (Fig. 11.1).

MPCAs may indirectly impact on natural enemies by decreasing their supply of hosts or prey or by altering the quality of the arthropod natural enemy's food source (Hamm et al. 1985, Cossentine and Lewis 1986, Powell et al. 1986b). Indirect effects of MPCAs

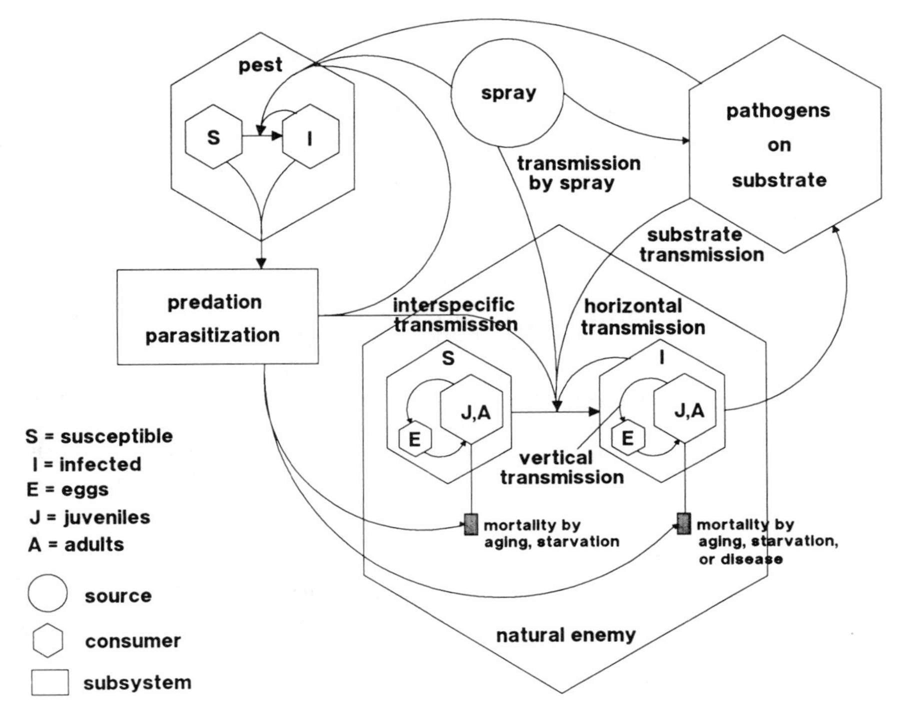

Figure 11.1. Generalized model of the interactions of microbial-pest-control agent with host–prey and natural enemey.

may be greater for parasitoids than for predators (Flexner et al. 1986), although extreme depletion of prey due to MPCA outbreaks or spray applications can drastically alter the population dynamics of predators (Chapter 8).

Generally, microbial interactions are less complex and more easily analyzed for predators than for ecto- or endoparasitoids. Transmission among predators is mostly through feeding, although for generalist pathogens like fungi, direct penetration may occur through the integument. Transmission among parasitoids can occur by contact with other parasitoids or hosts, from a spray, or from the general environment. This does not mean that microbials have less severe direct effects on predators than parasitoids, but the difference in complexity of these interactions seems to have influenced the relative amount of study on these natural enemy groups. Except for research on pathogenic fungi, which included about equal numbers of side-effects tests on predators and on parasitoids, most microbial side-effects research has been conducted on parasitoids [the percentage of cases of side-effects testing for parasitoids (as compared to predators) within each group of MPCAs is viruses (75%), bacteria (69%), and protozoans (92%); Tables 11.1–4]. In contrast, predators have more commonly been the focus of side-effects testing with conventional pesticides (Chapter 2).

Hereafter the side effects of MPCAs on natural-enemy–pest interactions are discussed in greater detail for four major classes of MPCAs. Because each of these major classes of MPCAs has a unique mode of action and side effects, they are treated individually.

Discussion includes methods of assessment, lethal effects, sublethal effects, and ecological relationships and selectivity. Throughout the discussion, occasional reference is made to MPCA side effects on the honey bee *Apis mellifera*. Many data on MPCAs are available for this species (which may be lacking for natural enemies). This information may have general application to hymenopteran parasitoids and other natural enemies.

11.3. VIRUSES

Viruses are obligate intracellular parasites. They operate only after host contact and infection. In arthropods, virus infection occurs via ingestion, integumental wounding, or by transovarial or transovum transmission. The virus then attaches to cell membrane sites in the host, where penetration of the cell membrane occurs. After penetration, the virus incorporates its genetic material with that of its host to replicate and later to release new propagules. Release may be intercellular, spreading infection within the host, or between hosts by release of free particles in fecal droppings, or by host disintegration. The time elapsed from infection to death in insects depends on the size of the inoculum and the virulence of the virus, but usually ranges from 5 to 20 days or longer (Cherwonogrodzky 1980).

The majority of economically important entomopathogenic viruses are in the genus *Baculovirus*. These include the nuclear polyhedrosis viruses (NPV), the granulosis viruses (GV), and the nonoccluded viruses. These viruses mainly infect midgut, trachea, fat, and epidermal cells and are known pathogens of a diversity of phytophagous pest arthropods.

11.3.1. Methods of Assessment

Most lethal response tests of entomopathogenic viruses have been conducted on the parasitoids and, to a lesser extent, on the predators of lepidoptera (Table 11.1). The most widely used MPCAs employed to date are relatively specific to this pest group. All data presented below or in Table 11.1 are for various strains of NPV or GV.

In judging the appropriateness of a side-effects assay for a specific natural enemy and MPCA, it is important to examine how the test organism was exposed as well as other aspects of methodology. In 35% of the tests (Table 11.1), direct toxicity of the virus was assessed. Mortality never exceeded 4% when five parasitoids and two predators of *Heliothis* were tested for contact toxicity of Viron/H (Wilkinson et al. 1975). Low contact toxicity would be expected since viruses only infect midgut epidermis or fat body tissues. Since viruses must be ingested or must penetrate the hemocoel to affect their host, contact tests for these MPCAs may be inappropriate.

The majority of the beneficial insects assayed with entomopathogenic viruses were tested for acute effects after ingestion (Table 11.1). Adults often were fed polyhedral inclusion bodies (pib) in units expressed as granules per milliliter in a contaminated food where uptake could be monitored (Hassan and Groner 1977, Cantwell and Lehnert 1979, Groner et al. 1978, Morton et al. 1975). Direct toxicity in these bioassays was usually low, indicating no acute effects on the natural enemy adults tested. Other feeding tests were conducted on larval endoparasitoids, where actual amounts of virus consumed in the host could not be quantified (Table 11.1). Tests were run either by infecting a preparasitized host larva with virus or by allowing a female parasitoid to oviposit in a preinfected host. Quantifying viral side effects on the parasitoid larvae was difficult because most effects appeared to be indirect, caused by either physiological alteration of the host or host mortality (Hotchkin and Kaya 1983a, 1985).

Table 11.1. Effects of viral[1] Pathogens on Arthropod Natural Enemies

Nontarget Insect										
Family	Species	Virus Pathogen	Virus Dosage	Method of Exposure	Treated Insect Host	Toxicity	Test Site	Reference	Sublethal Effects	Duration
Braconidae	*Chelonus blackburni*	NPV (Viron/H)	Minimum recommended field dose	Contact by adults	*Heliothis* sp.	*	Lab.	Wilkinson et al. 1975	—	—
	Chelonus insularis	HGV (Hawaiian)	Varied to cause frank infection of host before emergence	Ingested by larvae within host	*Spodoptera exigua*	*	Lab.	Hotchkin & Kaya 1983a	—	—
		HNPN (hypertrophy)			*Pseudaletia unipuncta*	****[1]			IDT	—
	Cotesia glomeratus	GV	1.0×10^5 granula/ml	Ingested by host larvae after being parasitized	*Pieris rapae*	**** (< 5 days)			—	—
				Ingested by host larvae after being parasitized		*(> 8 days)	Lab.	Levin et al. 1981		
	Cotesia marginiventris	HGV (Hawaiian)	Varied	Ingested by larvae in host	*S. exigua* *P. unipuncta*	** *	Lab.	Hotchkin & Kaya 1983a	—	—
	Cotesia marginiventris	Ascovirus (nonoccluded)	—	Ingested by noctuid host	*Spodoptera frugiperda*	**	Lab.	Hamm et al. 1985	RMW	—
	Glypta-panteles militaris	NPV (typical)	6.4×10^{10} pib/ml	Ingested by host larvae 3–8 days after being parasitized	*P. unipuncta*	*	Lab.	Hotchkin & Kaya 1983b, 1985	—	—
		HNPV (hypertrophy)	1.4×10^9 pib/ml			****			—	—
		HGV (Hawaiian)	1.8×10^{10} g/ml			****		Kaya & Tanada 1972a	—	—

Table 11.1. (*Continued*)

Nontarget Insect Family	Species	Virus Pathogen	Virus Dosage	Method of Exposure	Treated Insect Host	Toxicity	Test Site	Reference	Sublethal Effects	Duration
	Meteorus leviventris	NPV (Viron/H)	Minimum recommended field dose	Contact by adults	*Heliothis* sp.	*	Lab.	Wilkinson et al. 1975	—	1–120 hr
	Phanerotoma flavitestacea	HGV	1 μl of infected plasma	Injected into parasitised host larvae	*Anagasta kuehniella*	***	Lab.	Kaya & Tanada 1972a	—	129–216 hr
Chalcidae	*Brachymeria intermedia*	NPV (Viron/H)	Minimum recommended field dose	Contact by adults	*Heliothis* sp.	*	Lab.	Wilkinson et al. 1975	—	1–120 hr
Ichneumonidae	*Campoletis sonorensis*	HGV (Hawaiian)	Varied for frank injection	Ingested by larvae in host	*S. exigua* *P. unipuncta*	* *			IDT	—
		HNPV (hypertrophy) NPV	5×10^6 pib/ml	Ingested by larvae in host *Heliothis* exposed to virus 24, 36, 48, 60, and 72 hr after parasitization	*H. virescens*	**** (< 36 hr); *** (48 hr); **(60 hr); *(72 hr)	Lab. Lab.	Hotchkin & Kaya 1983a Irabagon & Brooks 1974	—	—
		NPV (Viron/H)	Minimum recommended field rate	Contact by adults	*Heliothis* sp.	*	Lab.	Wilkinson et al. 1975	—	1–120 hr
	Devorgilla canescens	HGV, GV	1 μl of plasma	Ingested by larvae in host	*A. kuehniella*	*	Lab.	Kaya & Tanada 1972a		

Ichneumonidae	*Hyposoter exiguae*	NPV	1×10^5 pib/ml	Ingested by larvae in host *T. ni* exposed ≤ 3 days before and 6 days after parasitization	*Trichoplusia ni*	**** (< 48 hr); *** (72–96 hr); ** (120–148 hr)	Lab.	Beegle & Oatman 1975	—	192–216 hr
		HNPV, HGV (Hawaiian)	Varied	Ingested by larvae within host	*S. exigua* *P. unipuncta*	*	Lab.	Hotchkin & Kaya 1983a	—	—
Trichogrammatidae	*Trichogramma cacoeciae*	NPV	1, 5, 10 times the LC_{100} for *M. brassicae*	Contact and ingested by adults and sprayed on eggs	*Mamestra brassicae*	*	Lab.	Hassan & Groner 1977	IDT	—
	Trichogramma euproctidis	GV	—	Sprayed on *Sitotroga cerealella*	*Xestia c-nigrum*	*	Lab.	Vybornov et al. 1982	—	—
	Trichogramma evanescens	NPV		eggs before parasitization		*			—	—
Carabidae	*Calosoma sayi*	NPV	3.3×10^8 pib	Fed to host larvae	*Spodoptera frugiperda*	*	Lab.	Young & Hamm 1985	—	24–360 hr
Coccinellidae	*Hippodamia convergens*	NPV (Viron/H)	Minimum field rate	Contact by adults	*Heliothis* sp.	*	Lab.	Wilkinson et al. 1975	—	1–120 hr
Tachinidae	*Compsilura concinnata*	HGV (Hawaiian), HPNV (hypertrophy)	Varied	Ingested by larvae in host	*P. unipuncta*	*	Lab.	Hotchkin & Kaya 1983a	—	—

Table 11.1. (*Continued*)

Nontarget Insect										
Family	Species	Virus Pathogen	Virus Dosage	Method of Exposure	Insect Host	Toxicity	Test Site	Reference	Effects	Duration
	Parasarchophaga misera	NPV	—	Ingested by larvae in host	*Spodoptera litura*	*	Lab.	Battu 1977 Battu & Dilwari 1978	—	—
	Voria ruralis	NPV (Viron/H)	Field rate	Contact by adults	*Heliothis* sp.	*	Lab.	Wilkinson et al. 1975	—	1–120 hr
		NPV	—	Ingested by larvae in host	*T. ni*	*	Lab. Fld.	Vail 1981		
Chrysopidae	*Chrysoperla carnea*	NPV	1, 5, 10 × LC_{100} for *M. brassicae*	Ingested and contacted by larvae	*M. brassicae*	*	Lab.	Hassan & Groner 1977	—	—
		NPV (Viron/H)	Field rate	Contact by adults	*Heliothis* sp.	*	Lab.	Wilkinson et al. 1975	—	1–120 hr

*(0–10% mortality) no or slight toxicity.
**(10–40% mortality) low toxicity.
***(40–80% mortality) moderate toxicity.
****(80–100% mortality) high toxicity.

[1] *Baculovirus* subgroup A—NPV; *Baculovirus* subgroup B_1—GV.
[2] Host died before larvae completed development.
‡IDT—increased developmental time; RMW—reduced mean weight.
Key: NPV—nuclear polyhedrosis virus; GV—granulosis virus.

The duration of viral tests varied. Many tests monitored the infected host until it died or the parasitoid emerged (Table 11.1). Others were run with end points ranging from 3 to 60 days.

11.3.2. Lethal Effects

Few cases of direct parasitoid or predator mortality induced by a baculovirus infecting a host have been documented (Hotchkin and Kaya 1983a, 1985). Indirect mortality of parasitoid larvae more commonly occurs (Table 11.1). Immature parasitoids may die if a virus renders their host unsuitable before the larvae mature. This usually occurs if the host dies before the parasitoid completes development (Irabagon and Brooks 1974, Beegle and Oatman 1975, Levin et al. 1981, Hamm et al. 1985). Indirect mortality may also be due to physiological alteration of the host, which can produce toxic by-products (Kaya 1970, Kaya and Tanada 1972a, 1973, Hotchkin and Kaya 1983a, 1983b).

Mortality of the ichneumonid *Campoletis sonorensis* was 96% when its host, *Heliothis virescens*, was infected with 4×10^6 polyhedral inclusion bodies (pib)/ml NPV, 24 hr after being parasitized. Mortality decreased to 49% when *H. virescens* was infected 48 hr after being parasitized and to 6% at 72 hr after oviposition (Irabagon and Brooks 1974). Similarly, when *Trichoplusia ni*, which had been infected 1–3 days earlier with 1×10^5 pib/ml NPV, were parasitized by *Hyposoter exiguae*, all parasitoids died when *T. ni* died (Beegle and Oatman 1975). Of the ichneumonids, 92% died where *T. ni* larvae were infected with the virus 48 hr after parasitoids oviposited, 56% died when hosts were infected at 96 hr postoviposition, and 20% when infected at 144 hr postoviposition. Similar results have been reported for the braconid *Cotesia glomeratus*, which experienced 100% mortality when *Pieris rapae* was infected with 10^5 GV granules/ml 96 hr after parasitization. Mortality decreased linearly and was approximately 10% when the host was infected 216 hr after oviposition (Levin et al. 1981).

To summarize, parasitoid survival and emergence in a virus-infected host can be greatly affected by the timing of host exposure to the virus. Parasitoid survival increased as the time between oviposition and virus infection increased (Levin et al. 1981, Hamm et al. 1985). Thus, the survival of parasitoids emerging from virus-infected hosts depends on the developmental rate of both the virus and the host larvae.

Another cause of indirect mortality of parasitoid larvae is host alteration. A Hawaiian strain of granulosis virus (HGV) and the hypertrophy strain of nuclear polyhedrosis virus (HNPV) both produced toxic factors that affected the development of the braconid *Glyptapanteles militaris* (Kaya 1970, Kaya and Tanada 1972a, 1972b, Hotchkin and Kaya 1983a, 1983b, 1985). When *Pseudaletia unipuncta* were fed the HGV within 24 hr of parasitization, 91% of *G. militaris* died in the first instar. Conversely, only 6% of the parasitoids died when *P. unipuncta* was fed a different GV strain (Kaya and Tanada 1972b). Kaya and Tanada (1972a) reported a similar effect on *Phanerotoma flavitestacea* when its host, *Anagasta kuehniella*, was infected with the HGV. A hypertrophy strain of NPV (HNPV) also killed 100% of the *G. militaris* in the first or second instar when fed to its host, *P. unipuncta*, yet parasitoids within *P. unipuncta* infected with a typical strain of NPV appeared normal as second instars (Kaya and Tanada 1973). When viron-free plasma from armyworms infected with the hypertrophy strain was injected into uninfected, parasitized armyworms, parasitoid development was adversely affected, suggesting that the viron of the hypertrophy strain was not the toxic factor (Kaya and Tanada 1973). Hotchkin and Kaya (1983b) tested the toxic effect of the HGV and HNPV on one tachinid, two ichneumonids, and three braconids and found that only *G. militaris* died in HNPV-infected *P. unipuncta* before the host died. The braconid *Chelonus insularis* died after its HGV- or HNPV-infected host died, probably because *P. unipuncta* died before *Chelonus* completed development within its factitious host (Hotchkin and Kaya 1983b).

The paucity of information on viral effects on predatory arthropods may reflect the less severe indirect effects of MPCAs on these species. Parasitoid survival after host exposure to an MPCA is dependent on the survival of its specific host. Predators can continue to feed on unexposed or uninfected prey following microbial treatments, barring a substantial reduction in their food supply. Because of the specificity and selectivity of MPCAs compared to conventional pesticides, predators may be able to feed on a number of diseased prey with limited effects on their survivorship.

11.3.3. Sublethal Effects

Few tests measuring side effects of viruses (Table 11.1) on adult and larval natural enemies have reported sublethal effects (Irabagon and Brooks 1974, Hassan and Groner 1977, Levin et al. 1981, Hotchkin and Kaya 1983a, Hamm et al. 1985). However, reduced developmental rates of parasitoid larvae within virus-infected hosts have been noted (Beegle and Oatman 1975, Hotchkin and Kaya 1983a). There were no adverse effects on larval feeding, adult fecundity, or egg viability of *Chrysoperla carnea* when adults were sprayed directly or when larvae were fed *Mamestra brassicae* infected with NPV (Hassan and Groner 1977). Hassan and Groner (1977) found no reduction in the parasitism rate or the development of *Trichogramma cacoeciae* exposed to NPV directly or by ingestion of infected *M. brassicae*. Beegle and Oatman (1975) found no significant difference in pupal developmental time, adult emergence, adult longevity, or percent parasitism between *H. exiguae* that emerged from infected versus uninfected *T. ni*. Adult body size and the subsequent fecundity of parasitoids were not measured. However, the developmental time of *H. exiguae* that successfully emerged from parasitized *T. ni* was significantly shorter than for those emerging from uninfected hosts. This may have been due to an actual decrease in the developmental time of the parasitoids within infected hosts, or the mean developmental time may have been reduced because parasitoid larvae that developed more slowly died when their host died from viral infection.

Conversely, Hotchkin and Kaya (1983a) found that *H. exiguae* developing within HNPV-infected *P. unipuncta* took longer to emerge from their host than parasitoids developing in uninfected hosts. They suggested that longer development was related to the nutritional quality of the host. Because viral infection reduced host size, nutrition may have been insufficient for the developing parasitoid (Hotchkin and Kaya 1983a).

11.3.4. Ecological Relationships and Selectivity

In general, baculoviruses have a narrower range of activity compared to conventional pesticides, and little evidence of direct toxicity to natural enemies has been documented (Table 11.1). Most lethal and sublethal effects were apparently indirect, and were associated with host unsuitability. All lethal effects reported were confined to immature parasitoids developing in virus-infected hosts. Mortality was greatest when parasitoids oviposited in virus-infected hosts and decreased as the period between oviposition and host infection increased (Levin et al. 1981). Thus, the ability to discriminate virus-infected hosts should provide a strong selective advantage for parasitoids of species that exhibit periodic viral epizootics.

Discrimination of virus-infected hosts has been documented for several parasitoid species. Versoi and Yendol (1982) observed that *Cotesia melanoscelus* females could discriminate between uninfected and NPV-infected gypsy moth larvae *Lymantria dispar*. The percentage of oviposition attempts in the noninfected larvae was 68.7%, significantly greater than the 32% oviposition attempts in virus-infected larvae. Levin et al. (1983) reported that 84% of *Apanteles glomeratus* that developed to maturity in GV-infected *Pieris rapae* transmitted the virus to the larvae of hosts which they subsequently attacked. Adult parasitoids did not discriminate between treated and untreated hosts, although healthy larvae were initially attacked with a greater frequency than diseased hosts. Also, adult parasitoids discriminated between parasitized and unparasitized healthy hosts, but not between parasitized and unparasitized GV-infected hosts. The two studies cited above imply that parasitoid species vary in their abilities to discriminate healthy and diseased hosts, and some species are able to avoid wasting eggs on hosts which soon will die (Hotchkin and Kaya 1983a). Thus, parasitoids that can discriminate against infected hosts in the field may experience lower impact than predicted from laboratory tests, where parasitoids have no option but to lay their eggs on infected host material. In addition, studies such as that reported by Levin et al. (1983) indicate the significant role that parasitoids can play in the dissemination of GV in the field.

11.4. BACTERIA

The most widely used commercially available MPCAs are bacteria. Commercial production, however, has been restricted primarily to varieties of only two species, *Bacillus thuringiensis* (*B.t.*) and *B. popilliae* (*B.p.*). Production of *B.t.* began in the United States after World War II, and by the mid 1960s *B.t.* was being used to protect 50% of the cole crops in southern California from *Trichoplusia ni* (Burges and Hussey 1971). *B.t.* has subsequently been registered on over 40 species of lepidopterous pests, mosquitoes, and black files, with hundreds of tons of spore products being manufctured each year in at least five different countries (Cherwonogrodzky 1980).

B.t. is a gram-positive, aerobic, spore-forming bacterium whose vegetative cell becomes a sporangium, producing a spore and a parasporal protein crystal (Burges and Hussey 1971, Cherwonogrodzky 1980). The death of insects can occur from the ingestion of the spores, crystals, or exotoxins. An insect must have a strongly alkaline midgut (pH 9.0–10.5), specific proteolytic enzymes, and suitable receptor sites to be susceptible (Cherwonogrodzky 1980). Many natural enemy species lack one or the other of these factors, and therefore are not affected by *B.t.* preparations. However, *B.t.* may induce mortality in some predators and parasitoids (Hamed 1979, Muck et al. 1981, Tanada 1984).

11.4.1. Methods of Assessment

Characterization of commercial formulations and strains of *Bacillus* spp. is important for several reasons: 1) different strains have different selectivity features; 2) different formulations have different carriers, ultraviolet screens, and inert ingredients, which may be toxic to natural enemies; 3) different varieties of *B.t.* are formulated with different amounts of active ingredients; 4) these varieties also have different production rates of exotoxins.

Few studies have analyzed the effects of carriers, ultraviolet screening agents, and inert ingredients used in commercial preparations of *B.t.* on natural enemies (Table 11.2). Haverty (1982) tested the effect of the nonaqueous oil carrier of Dipel 4L on all life stage of

Table 11.2. Effects of Bacterial Pathogens on Arthropod Natural Enemies

Nontarget insect										
Family	Species	Bacteria Pathogen	Bacteria Dosage	Method of Exposure	Treated Insect Host	Toxicity	Test Site	Reference	Sublethal Effects‡	Duration
Aphelinidae	*Encarsia formosa*	Entobakterin (*B.t.* var. *galleriae*)	—	—	*Trialeurodes vaporariorum*	*	Lab.	Beglyarov & Maslienko 1978	—	—
	Aphytis melinus	*B.t.* + surfactant	3760 ppm	Residue (on scale cover)	*Aonidiella aurantii*	*	Lab.	Davies & McLaren 1977	—	96–960 hr
Braconidae	*Agathis rufipes*	Entobakterin	0.5% a.i.	Ingested by adults	*Cydia pomonella*	*	Lab.	Mosievskaya & Makarov 1974	—	—
	Apanteles fumiferanae	Dipel WP	453 g/7.6 liter/0.4 ha	Contact and ingestion	*Choristoneura occidentalis*	*	Fld.	Hamel 1977	—	—
		Thuricide	7.6 BIU/acre; 8 BIU/acre	Ingested by larvae in host	*C. occidentalis*	*	Fld.	Thompson et al. 1977	—	—
	Ascogaster quadridentata	Entobakterin	0.5% a.i.	Ingested by adults	*C. pomonella*	*	Lab.	Mosievskaya & Makarov 1974	—	—
	Bracon brevicornis	Bakthane (*B.t.* var. *thuringiensis*)	3×10^{10} spores/ml	Ingested by larvae in host	*Sesamia cretica*	**	Lab.	Temerak 1980 1982a, b	REP, RAL, IDT	—
	Cardiochiles nigriceps	Dipel WP	16,000 IU/mg	Ingested by adults	*Heliothis virescens*	***	Lab.	Dunbar & Johnson 1975	RAL	48–192 hr
		Biotrol XK	2 mg/ml	Contact by adults		*	Lab.	Wilkinson et al. 1975	—	1–120 hr
	Cotesia melanoscelus	Dipel HG (*B.t.* var. *kurstaki*)	9.3–75.0 IU/ml	Ingested by larvae in host	*Lymantria dispar*	*	Lab.	Ahmad et al. 1978	—	186–336 hr
										—
		Dipel 4H	19.8 BIU/ha			*	Fld.	Ticehurst et al. 1982	IPP	—

		Dipel 4H	2.0×10^{10} IU/ha			*	Fld.	Weseloh et al. 1983	IPP	—
	Cotesia glomeratus	Bactospeine (*B.t.* var?)	0.05% a.i.	Ingested by adult	*Pieris brassicae*	**	Lab.	Marchal-Segault 1975a	RHC, IDT, RAL	—
		Dipel	1×10^8 spores/ml 1×10^9 spores/ml	Ingested by adult		** ****	Lab.	Muck et al. 1981	—	336 hr
	Meteorus levirentris	Thuricide HPC	Field rate	Contact by adults	—	*	Lab.	Wilkinson et al. 1975	—	1–120 hr
	Microdus rufipes	Bitoxibacillin 202	70 millard/g	—	*C. pomonella*	*	Fld.	Kazakova & Dzhunusov 1977	—	—
	Microplitis demolitor	*B.t.* var. *entomocidus*	500 μg/ml	Ingested by larvae in host	*Spodoptera littoralis*	****	Lab.	Salama et al. 1982	REP	—
	Phanerotoma flavitestacea	Bactospeine	700 IU/mg	Contact of adult and ingested by larvae in host	*Ephestia kuchniella*	**	Lab.	Marchal-Segault 1975a,b,c	RHC, IDT, RAL	—
	Rogas lymantriae	*B.t.* strain HD-1S-1980	24 IU/ml	Ingested by larvae in host	*L. dispar*	*	Fld.	Wallner et al. 1983	IPP, SSR	—
	Trioxys pallidus	*B.t.* var. *thuringiensis*	22 ppm	Topical adult, feeding larvae	*Chromaphis juglandicola*	*	Lab.	Purcell & Granett 1985	—	24–48 hr
Chalcididae	*Brachymeria intermedia*	Thuricide HPC	Recommended field rate	Contact of adult	—	*	Lab.	Wilkinson et al. 1975	—	—
Encyrtidae	*Ageniaspis fuscicollis*	β-exotoxin of *B.t.* var. *thuringiensis*	—	Contact by adult; ingested by adult	—	*	Lab.	Korostel & Kapustina 1975	—	—
		Dipel	1×10^8 spores/ml	Ingested by adults	*Yponomeuta evonymellus*	**	Lab.	Hamed 1979	—	—
	Leptomastix dactylopii	Dipel	0.10% a.i.	Contact of adult	—	*	Lab.	Franz et al. 1980	—	—
Eulophidae	*Tetrastichus evonymellae*	*B.t.* var. *kurstaki*	1×10^8 spores/ml 5×10^8 spores/ml	Ingested by adult	*Y. evonymellus*	**	Lab.	Hamed 1979	—	—

Table 11.2. (*Continued*)

Nontarget insect										
Family	Species	Bacteria Pathogen	Bacteria Dosage	Method of Exposure	Treated Insect Host	Toxicity	Test Site	Reference	Sublethal Effects‡	Duration
Formicidae	*Formica pratensis*	Bitoxibacillin 202	0.5% a.i., 1.0% a.i. 1.5% a.i.	Ingested by adults	*P. brassicae*	*	Lab.	Legotai 1980	—	—
	Pheidole megacephata	Dipel Bitoxibacillin	Recommended field rate	Ingested by larvae and adult	*Cylas formicarius*	*	Lab.	Castineiras & Calderon 1982	—	72 hr
	Solenopsis sp.	*B.t.*	0.032 kg/ha	Contact adult	*Perileucoptera coffeda*	*	Fld.	Gravena 1984	—	24–168 hr
Ichneumonidae	*Agrypon* sp.	*B.t.* var. *kurstaki*	1×10^8 spores/ml 5×10^8 spores/ml	Ingested by adults	*Y. evonymellus*	*	Lab.	Hamed 1979	—	1–120 hr
	Campoletis sonorensis	Thuricide HPC	Recommended field rates	Contact by adults	—	*	Lab.	Wilkinson et al. 1975	—	—
	Diadegma armillata	*B.t.* var. *kurstaki*	1×10^8 spores/ml	Ingested by adults	*Y. evonymellus*	**	Lab.	Hamed 1979	—	—
	Exeristes comstockii	*Serratia marcescens*	2–20% infection of host	Ingested by larvae in host	*Galleria mellonella*	****	Lab.	Bracken & Bucher 1967	—	—
	Glypta fumiferanae	Dipel	453 g/7.6 liter/ 0.4 ha	Contact by adults	*C. occidentalis*	*	Fld.	Hamel 1977	—	—
		Dipel Thuricide	7.6 BIU/acre 8 BIU/acre	Ingested by larvae in host	*C. occidentalis*	*	Fld.	Thompson et al. 1977	—	—
	Hyposoter exiguae	Dipel	1130 μg/g diet	Ingested by noctuid host	*Heliothis virescens*	****	Lab.	Thoms & Watson 1986	—	1–312 hr
	Phobocampe unicincta	Dipel 4H	19.8 BIU/ha	Ingested by larvae?	*L. dispar*	*	Fld.	Ticehurst et al. 1982	—	—
	Phygadeuon trichops	Dipel	0.10% a.i.	Contact by adults	—	*	Lab.	Franz et al. 1980	—	—
		Dipel	—	Contact by adults	—	*	Lab./ Fld.	Naton 1978	—	—

	Pimpla instigator	*B.t.* var. *thuringiensis*	—	Ingested by adults and larvae in host	*P. brassicae*	*	Lab.	Biache 1975	—	432 hr
	Pimpla turionellae	Dipel	1×10^7, 1×10^9 spores/g	Ingested by adults	—	* **	Lab.	Muck et al. 1981, Hamed 1979	—	—
	Zele chlorophthalma	*B.t.* var. *entomocidus*	500 μg/ml	Ingested by larvae in host	*S. littoralis*	***	Lab.	Salama & Zaki 1983	RPE, RPP, IDT	—
Scelionidae	*Telenomus alsophilae*	Biotrol XKW® Dipel WP Thuricide HPC	0.32, 0.64 lb/ga 0.10, 0.20 lb/ga 0.40, 0.80 lb/gc	Contact from exposed egg masses	*Ennomos subsignarius*	*	Lab / Fld.	Kaya & Dunbar 1972	—	240 hr
Trichogrammatidae	*Trichogramma cacoeciae*	Dipel	0.10% a.i.	Contact by adult	—	*	Lab.	Franz et al. 1980	—	—
		Dipel Thuricide Bactospeine	54.7×10^9 spores/g 50.7×10^9 spores/g 1.0×10^9 spores/g	Contact and ingested by adults and larvae	*Sitotroga cerealella*	*	Lab.	Hassan & Krieg 1975 —	REP	—
		Entobakterin	1.0% a.i.	Contact by adults	—	*	Fld.	Kapustina 1975	—	—
	T. embryophagum	Entobakterin	1.0% a.i.	Contact by adults	—	*	Fld.	Kapustina 1975	—	—
		Entobakterin	0.6% a.i.	Contact by adults	*C. pomonella*	*	Lab.	Kostadinov 1979	—	24 hr
	T. evanescens	Entobakterin	1.0% a.i.	Contact by adults	—	*	Fld.	Kapustina 1975	—	—
		Bitoxibacillin 202	0.5, 1.0, 1.5% a.i.	Ingested by adults	*P. brassicae*	*	Lab.	Legotai 1980	—	—
	T. pallidum	Dipel Entobakterin Toxobakterin® Exobakterin®	1.0% a.i. 1.0% a.i. 1.0% a.i. 1.0% a.i.	Contact of larvae and ingested by larvae	*S. cerealella*	*	Lab.	Kiselek 1975	—	—

Table 11.2. *(Continued)*

Nontarget insect										
Family	Species	Bacteria Pathogen	Bacteria Dosage	Method of Exposure	Treated Insect Host	Toxicity	Test Site	Reference	Sublethal Effects‡	Duration
	Trichogramma sp.	Bitoxibacillin 202	70 milliard/g	—	*C. pomonella*	****	Lab.	Kazakova & Dzhunusov 1977	—	—
	Trichogramma sp.	*B.t.* var. *thuringiensis*	—	Contact by adults	—	*	Lab.	Korostel & Kapustina 1975	REP	24 hr
	Trichogramma sp.	*B.t.* var. *thuringiensis*	—	Ingested by adults		****	Lab.	Korostel & Kapustina 1975	—	—
	Trichogramma sp.	Dendrobacillin®	—	Adults exposed to residue	—	****	Lab.	Tolstova & Ionova 1976	—	24 hr
	T. nubilale	Thuricide	9.8 BIU	Field test on egg masses	*Ostrinia nubilalis*	*	Fld.	Tipping & Burbutis 1983	—	24–408 hr
Carabidae	*Bembidion lampros*	Dipel	1.12 kg/ha five times during season	Contact by adults	*Hylemya brassicae*	*	Fld.	Obadofin & Finlayson 1977	—	24–168 hr
			0.1% a.i.	Contact by adults		*	Lab.			
			0.5% a.i.	Contact by adults		**	Lab.			
Coccinellidae	*Adonia variegata*	Dendrobacillin	2.0% a.i.	Contact by adults	*Heliothis armigera*	*	Fld.	Umarov et al. 1975	—	1–360 hr
		Entobakterin	2.0% a.i.	Ingested by adults		*				
	Coccinella septempunctata	Bitoxibacillin 202	70 millard/g	—	*C. pomonella*	****	Lab.	Kazakova & Dzhunusov 1977	—	—

		Dipel	1.0% a.i.	Ingested by larvae	*Aphis* sp.	*	Lab.	Kiselek 1975	—	—
		Entobakterin	1.0% a.i.			**				
		Ecotoxin®	1.0% a.i.			****				
		Toxobakterin	1.0% a.i.			*				
		Bitoxibacillin 202	0.5, 1.0 1.5% a.i.	Ingested by adults	*P. brassicae*	*	Lab.	Legotai 1980	—	—
		Dendrobacillin	2.0% a.i.	Contact by adults	*H. armigera*	*	Fld.	Umarov et al. 1975	—	1–360 hr
		Entobakterin	2.0% a.i.	Ingested by adults						—
	Coccinella undecim-punctata	*B.t.* var. *entomocidus*	500 μg/ml	Ingested by adults	*Aphis punicae*	*	Lab.	Salama et al. 1982	REP, RHC	—
		Dendrobacillin	2.0% a.i.	Contact by adults	*H. armigera*	*	Fld.	Umarov et al. 1975	—	1–360 hr
		Entobakterin	2.0% a.i.	Ingested by adults					—	1–360 hr
	Cryptolaemus montrouzieri	Entobakterin	1.0% a.i.	Ingested by adults	—	*	Lab.	Kiselek 1975		
	Hippodamia convergens	Thuricide HPC	Recommended field rate	Contact by adults	—	*	Lab.	Wilkinson et al. 1975	—	1–120 hr
	Stethorus loxoni	*B.t.*	Recommended field rate	Contact by adults	—	*	Lab.	Walters 1976a	—	—
	S. nigripes	*B.t.*	Recommended field rate	Contact by adults	—	*	Lab.	Walters 1976a	—	—
		B.t.	0.375 g/1 of 85% WP	Contact by adults	—	*	Lab.	Edwards & Hodgson 1973	—	—
		B.t.	Recommended field rate	Contact by adults	—	*	Lab.	Walters 1976a	—	1 and 18 hr
Staphylinidae	*Paederus alfierii*	*B.t.* var. *entomocidus*	500 μg/ml	Ingested by adults	*S. littoralis*	*	Lab.	Salama & Zaki 1983	—	

Table 11.2. (*Continued*)

Nontarget insect										
Family	Species	Bacteria Pathogen	Bacteria Dosage	Method of Exposure	Treated Insect Host	Toxicity	Test Site	Reference	Sublethal Effects‡	Duration
Chrysopidae	*Chrysoperla carnea*	*B.t.* var. *galleriae*	—	Ingested by larvae	—	**	Lab.	Babrikova et al. 1982	—	—
		B.t. var. *kurstaki*	—	Ingested by adults	—	***	Lab.		—	—
		B.t. var. *thuringiensis*	—						—	—
		Dipel	0.10% a.i.	Contact by adults	—	*	Lab.	Franz et al. 1980	—	—
		Dipel	1.0% a.i.	Ingested by larvae	*S. cerealella* eggs	**	Lab.	Kiselek 1975	—	—
		Entobakterin	1.0% a.i.			*				
		Exotoxin	1.0% a.i.			**				
		Toxobakterin	1.0% a.i.			*				
		Bitoxibacillin 202	0.5, 1.0, 1.5% a.i.	Ingested by larvae and adults	—	*	Lab.	Legotai 1980	—	—
		B.t. var. *entomocidus*	500 μg/ml	Ingested by larvae	*S. littoralis*	*	Lab.	Salama et al. 1982	RHC	—
		Dendro-bacillin	2.0% a.i.	Contact and ingested	*H. armigera*	*	Fld.	Umarov et al. 1975	—	1–360 hr
		Thuricide HPC	Recommended field rates	Contact by adults and larvae	—	*	Lab.	Wilkinson et al. 1975	—	1–360 hr
	Chrysoperla formosa	Entobakterin, Bactospeine	—	Contact by adults	—	**	Lab.	Babrikova & Kuzmanova 1984	—	—
	Chrysoperla perla	Dipel	—	Ingested by larvae		***				

Anthrocoridae	*Orius niger*	Dendro-bacillin	2.0% a.i.	Contact and ingested	*H. armigera*	*	Fld.	Umarov et al. 1975	—	1–360 hr
		Entobakterin	2.0% a.i.						—	1–360 hr
Berytidae	*Jalysus spinosus*	Dipel	16,000 IU/mg 2 mg/ml	Contact by adults	*H. virescens*	*	Lab.	Dunbar & Johnson 1975	—	48–480 hr
	Picromerus bidens	*B.t.* var. *kurstki*	1×10^8 spores/ml 5×10^8 spores/ml	Ingested by adults	*Y. evonymellus*	*	Lab.	Hamed 1979	—	—
Lygaeidae	*Geocoris punctipes*	*B.t.* var. *thuringiensis* (β-exotoxin)	0.08–9.5 μg/insect	Topical, feeding residue	*Heliothis zea*	**	Lab.	Herbert & Harper 1986	—	1–624 hr
Miridae	*Campylomma diversicornis*	Dendro-bacillin	2.0% a.i.	Contact and ingested by adults	*H. armigera*	*	Fld.	Umarov et al. 1975	—	1–360 hr
		Entobacillin	2.0% a.i.							
	Campylomma verbasci	Dendro-bacillin	2.0% a.i.	Contact and ingested by adults	*H. armigera*	*	Fld.	Umarov et al. 1975	—	1–360 hr
		Entobakterin	2.0% a.i.							
Syrphidae	*Sphaerophoria scripta*	Bitoxibacillin 202	70 milliard/g	—	*C. pomonella*	****	Lab.	Kazakova Dzhunusov 1977	—	—
Tachinidae	*Bessa fugax*	*B.t.* var. *kurstaki*	1×10^8 spores/ml	Ingested by adults	*Y. evonymellus*	*	Lab.	Hamed 1979	—	—
	Blepharipa pratensis	Dipel 4L	19.8 BIU/ha		*L. dispar*	**	Fld.	Ticehurst et al. 1982	RPP	—
	Compsilura concinnata	Dipel 4L	19.8 BIU/ha		*L. dispar*	**	Fld.	Ticehurst et al. 1982	RPP	—
	Pales pavida	Dipel	0.10% a.i.	Contact by adults	—	*	Lab.	Franz et al. 1980	—	—
	Parasetigena silvestris	Dipel 4L	19.8 BIU/ha		*L. dispar*	*	Fld.	Ticehurst et al. 1982	—	—

Table 11.2. (*Continued*)

Nontarget insect										
Family	Species	Bacteria Pathogen	Bacteria Dosage	Method of Exposure	Treated Insect Host	Toxicity	Test Site	Reference	Sublethal Effects‡	Duration
	Voria ruralis	Thuricide HPC	Recommended field rates	Contact by adults	—	*	Lab.	Wilkinson et al. 1975	—	1–120 h
	Zenillia dolosa	*B.t.* var. *kurstaki*	1×10^8 spores/ml 5×10^8 spores/ml	Ingested by adults	*Y. evonymellus*	*	Lab.	Hamed 1979	—	—
Aeolothripidae	*Aeolothrips intermedius*	Dendro-bacillin Entobakterin	2.0% a.i. 2.0% a.i.	Contact and ingested by adults	*H. armigera*	*	Lab.	Umarov et al. 1975	—	1–360 hr
Thripidae	*Scolothrips acariphagus*	Dendro-bacillin Entobakterin	2.0% a.i. 2.0% a.i.	Contact and ingested by adults	*H. armigera*	*	Lab.	Umarov et al. 1975	—	1–360 hr
Mantidae	*Tenodera aridifolia sinensis*	*B.t.* var. *kurstaki*	18,000 IU/mg; 150 μg/ml	Ingested by adults	*Trichoplusia ni*	*	Lab.	Yousten 1973	—	168 hr
Lycosidae	*Pardosa agrestis*	Bitoxibacillin 202	0.5, 1.0; 1.5% a.i.	Ingested by adult	*P. brassicae*	*	Lab.	Legotai 1980	—	—
Phytoseiidae	*Metaseiulus occidentalis*	*B.t.* var. *thuringiensis* (β exotoxin)	Proposed field rate	Treated leaves, eggs, and larvae	*Tetranychus pacificus*	****	Lab.	Hoy & Ouyang 1987	—	48 hr

*(0–10% mortality) no or little toxicity.
**(10–40% mortality) low toxicity.
***(40–80% mortality) moderate toxicity.
****(80–100% mortality) high toxicity.

‡REP—reduced egg production; RHC—reduced host consumption; RPP—reduced percent parasitism; RAL—reduced adult longevity; IPP—increased percent parasitism; IDT—increased developmental time; SSR—skewed sex ratios.

Aphytis melinus, Hippodamia convergens, and *Chrysoperla carnea* and found that mortality never exceeded 13.4%, even at a high application rate. Marchal-Segault (1975b) tested the spores, crystals, and bentonite carrier of Bactospeine separately after noting that adult life spans of *Apanteles glomeratus* and *Phanerotoma flavitestacea* were reduced by contact or consumption of the commercial formulation. The abrasive bentonite had an adverse effect on the adult life span. Different commercial formulations of *B.t.* have different numbers of viable spores per gram (Forsberg et al. 1976). Concentrations range from 70 billion spores per gram for Agritol to 2.5 billion spores per gram for Biotrol BTB 2.5D (Forsberg et al. 1976). Formulations of Thuricide range from 3 billion spores per gram in Thuricide to 65 billion spores per gram in Thuricide 65B (Forsberg et al. 1976). Dulmadge et al. (1976) detailed a standardized procedure for expressing rates of active spores of *B.t.* that quantified the problem of formulation variation. Often in the literature, information on the commercial preparation and formulation is incomplete or lacking altogether.

The variety of *B.t.* used is very important because serotypes and entomopathogenic components can be quite different. Varieties containing the heat-stable beta exotoxin may be the most directly detrimental to many beneficial insects (Tanada 1984; Section 11.4.2). Varieties of *B.t.* known to produce the beta exotoxin are *thuringiensis, dendrolimus, morisoni, tolworthi*, and *darmstadiensis*. Cantwell et al. (1966) reported that Kreig and Herfs (1963, 1964) found a significant difference between five subspecies of *B.t.* on honey bees when tested at a dose of 1×10^8 spores/ml. Three subspecies had little effect, but *thuringiensis* and *dendrolimus* killed 100% and 60% of the bees, respectively, in 9 days. The differences in mortality may have reflected the differences in the amount of beta exotoxin produced by each strain (Forsberg et al. 1976). *B.t.* formulations containing the beta exotoxin caused greater mortality of *Coccinella septempunctata* larvae than those not containing the beta exotoxin (Kiselek 1975).

The actual dosage of pathogen applied to or ingested by a natural enemy is often not clearly stated. With the exception of Dulmadge et al. (1976), no attempt has been made in the literature to standardize applications into international units as recommended some 20 years ago (Forsberg et al. 1976). Thus, it is difficult to compare or contrast different tests using different formulations and different dosage measurements.

Most lethal response tests of bacterial pathogens have been conducted on predators and parasitoids of lepidopteran pests (Table 11.2). Of the natural enemies tested with bacterial pathogens in laboratory experiments (see Table 11.2), 36% were exposed by direct contact (e.g., spray, dip, or residue). Of the 25 species tested by direct or residue exposure, only 3 species showed greater than 10% mortality. Intuitively, these results seem logical because *B.t.* usually must be ingested by an insect to initiate toxicosis. However, Tolstova and Ionova (1976) reported high mortality (100%) of adult *Trichogramma* sp. within 24 hr when exposed to dried deposits of Dendrobacillin. Details of their tests were not provided. The carabid *Bembidion lampros* experienced moderate mortality (15%) after 72 hr of residual exposure to Dipel (Obadofin and Finlayson 1977).

In the cases cited, natural enemies may have been adversely affected by an inert ingredient within the formulation or internally contaminated during direct exposure testing. Internal contamination can occur if spores 1) are ingested while the insect is grooming, 2) are consumed in a contaminated food source, or 3) directly penetrate the body cavity through an integumental wound. It is highly unlikely that the spores, crystals, or exotoxins of *B.t.* would cause significant mortality of natural enemies when applied directly to their intact integuments.

The majority of the bioassays conducted on arthropod natural enemies with *B.t.* tested for acute pathological effects after spores had been ingested by the insects (Table 11.2).

Most predators and adult parasitoids were tested by contaminating the prey or host with a known amount of the pathogen and monitoring uptake. Effects on parasitoid larvae were measured by allowing female parasitoids to oviposit in contaminated hosts or by feeding parasitized hosts a contaminated food source. Quantifying the effect on parasitoid larvae presents a special problem. As with viruses, lethal or sublethal effects may not be directly induced by the bacteria, but may be indirectly caused by altering the available food source. Marchal-Segault (1975b, 1975c) observed significant indirect mortality to the larvae of two braconid parasitoids when their hosts died before the parasitoids completed their development. Parasitoid mortality was believed to be due to inferior food quality rather than to the direct poisoning of the host-feeding larvae (Marchal-Segault 1975b). A high percentage of indirect mortality was also reported for *Microplitis demolitor* when *Spodoptera littoralis* died before the parasitoid completed its development (Salama et al. 1982). Much of the literature fails to specify whether reduction in parasitoid emergence from *B.t.*-infected hosts was due to direct or indirect mortality (Table 11.2).

Quantifying the amount of bacteria ingested by parasitoid larvae is also difficult. Some parasitoid larvae do not consume their entire host, feeding only on a specific organ or group of organs (e.g., the fat body of lepidopterous larvae). Thus, a parasitoid might be able to complete development within an infected host without ingesting toxins within the host tissue of hemolymph.

The particular life stage of the organism being tested may also affect bioassay results (Table 11.2). Considerable literature has been published in which life stage is identified as a key factor in determining a natural enemy's susceptibility to a chemical pesticide. The adult stage is generally considered to be most susceptible to chemical insecticides, although a few species of larval coccinellids are an exception to this generalization (Croft 1977).

The opposite may be true when parasitoids and predators are exposed to high dosages of *B.t.* In general, larvae of natural enemies are more susceptible to *B.t.* than adults (Tanada 1984). Kiselek (1975) found that Dipel and Exotoxin caused 20% and 18% mortality of *Chrysoperla carnea* larvae, respectively, but both compounds were harmless to adults. Kiselek (1975) also reported that 28% and 90% of *Coccinella septempunctata* larvae were killed by Entobakterin and Exotoxin, respectively, yet neither formulation had an effect on adults. Conversely, Babrikova et al. (1982) found that three varieties of *B.t.* caused 75% mortality in adult *C. carnea*, but only 30% mortality in the larvae. Babrikova and Kuzmanova (1984) similarly reported that these same preparations caused only 20–40% mortality to larvae, but 60–90% mortality in adult *Coccinella septempunctata, Chrysoperla formosa*, and *Chrysoperla perla*. Kazakova and Dzhunusov (1977) reported 100% mortality in adult *Trichogramma* sp. but no mortality if larvae were exposed to Bitoxibacillin 202.

Thus, the effects of *B.t.* on different life stages of natural enemies are highly variable. This may be related to the different ways tests were administered or to the limited data that are available. It would appear that more extensive studies are needed before patterns in susceptibility among individual life stages of natural enemies can be determined.

The effects of other types of bacteria on different life stages have been even less thoroughly investigated. Bracken and Bucher (1967) found that the bacterial pathogen *Serratia marcescens* was highly toxic to the adult ichneumonid *Exeristes comstockii*, yet had no noticeable effect on the ichneumonid larva. In this situation, parasitoid larvae were contaminated while feeding upon their infected host *Galleria mellonella*, but mortality did not occur until the adult stage. Survival of parasitoid larvae was normal (over 90%), suggesting that considerable ingestion of *S. marcescens* was tolerable, but the bacteria remained within the parasitoids causing 100% mortality of adults due to septicemia (Bracken and Bucher 1967).

An additional explanation for the variability of results between different bioassays when testing similar varieties of *B.t.* on the same organisms (Table 11.2) is the amount of time allotted for the evaluation of a lethal response. Evaluation times for *B.t.* assays are very similar to those used for conventional pesticides (Chapter 2). However, many tests have shown that direct lethal effects of *B.t.* on arthropods increase when assayed over a longer period of time. For example, the LD_{50} of Thuricide HP for honey bee larvae was 9×10^8 spores per gram if measured after 48 hr and only 0.2×10^8 spores per gram if measured at 96 hr (Ali et al. 1973). Thus, a reduction in dosage of over 95% gave the same lethal response when measured after a longer incubation period. Therefore, it would be advisable to lengthen lethal time response measurements in bioassays for these agents to allow for pathogen reproduction and infection. Obviously, the duration of assays is limited by the length of time that an insect can persist under the particular test conditions.

11.4.2. Lethal Effects

Both direct and indirect lethal effects of bacterial pesticides on nontarget beneficial insects have been observed with various formulations of *B.t.* (Table 11.2). Direct mortality of parasitoids from infection and septicemia has been documented (Hamed 1979, Muck et al. 1981, Thompson et al. 1977, Niwa 1985). Hamed tested seven species of adult parasitoids and one predator of *Yponomeuta evonymellus.* He found that two species of tachinids, *Bessa fugax and Zenillia dolosa,* as well as one ichneumonid *Agrypon* sp. and the hemipteran *Picromerus bidens* were not susceptible to *B.t.* var. *kurstaki* when evaluated by exposure to contaminated food sources. However, two ichneumonids, *Diadegma armillata* and *Pimpla turionellae,* one encyrtid *Ageniaspis fusicollis,* and the eulophid *Tetrastichus evonymellae* were found to be susceptible after ingesting spores of *B.t.* with food sap. Large numbers of vegetative cells of *B.t.* were detected in the hemolymph and gut of the dead parasitoids.

Muck et al. (1981) also tested the ichneumonid parasitoid *Pimpla turionellae* and the braconid parasitoid *Cotesia glomeratus* with *B.t.* var. *kurstaki* (Dipel). They found that high concentrations applied orally (e.g., 10^8 and 10^9 spores/ml) significantly increased the mortality of *C. glomeratus* after two weeks (39% and 100% mortality, respectively, compared to 9% in the control). *P. turionellae* was less affected than *C. glomeratus.* Eighteen days after the continuous application of 10^7, 10^8, 10^9 spores/ml, *P. turionellae* mortality reached 13%, 18%, and 26%, respectively, compared to 2% in the control. Histopathological examination of *P. turionellae* after ingestion of Dipel at the rate of 10^9 spores/ml indicated that damage to the midgut was caused by the *B.t.* endotoxin.

Direct mortality of parasitoids has been observed by Thompson et al. (1977) in field tests using *B.t.* against the western spruce budworm *Choristoneura occidentalis* and the douglas fir tussock moth *Orgyia pseudotsugata.* Two important budworm parasitoids, *Glypta fumiferanae* and *Apanteles fumiferanae,* acquired lethal *B.t.* infections from their budworm host. Fully developed parasitoid larvae would emerge from their host only to die before they could pupate. Upon inspection, dead parasitoids were found to contain all stages of the bacillus and were presumed to have died from infection. In almost all cases the host budworm remained alive until after the parasitoid emerged.

Predators may also be diectly affected by *B.t.* Direct mortality of *Coccinella septempunctata* has been reported by Kazakova and Dzhunusov (1977) and Kiselek (1975), using Bitoxibacillin, Entobakterin, and Exotoxin. However, other *B.t.* formulations tested on *C. septempunctata* and other coccinellids have had little or no effect (see Table 11.2). Similar results have been obtained for the predaceous lacewing *Chrysoperla carnea.* Babrikova (1982) and Kiselek (1975) measured significant mortality in lacewings when the larvae and adults were fed several varieties of *B.t.* (see Table 11.2). Several others have reported that feeding *C. carnea* other formulations and varieties of *B.t.* had little or no measurable effect (Umarov et al. 1975, Legotai 1980, Salama and Zaki 1983).

Recently direct tests of the toxicity of the beta exotoxin of *B.t.* var. thuringiensis have been evaluated on several natural enemies. Herbert and Harper (1986) reported no toxicity to adult *Geocoris punctipes*, but nymphs fed treated *Heliothis zea* experienced mortality from 0 to 50%. A topical LD_{50} for fourth instar nymphs was 0.25 μg per insect. The mean adult life span was also reduced from 26 to as low as 4 days by higher rates of treated prey. Hoy and Ouyang (1987) reported the beta exotoxin to be toxic to adult *Metaseiulus occidentalis* when applied to plant leaves. Egg production by treated females declined within 48 hr after treatment. Treated eggs hatched, but larvae did not develop further, and treated larvae were unable to develop to adults. Purcell and Granett (1985) evaluated the toxicity of the beta exotoxin against adult and late third instar larvae of *Trioxys pallidus*, a parasitoid of the walnut aphid *Chromaphis juglandicola.* When applied topically or by ingestion, this toxin had no adverse effects on either stage of the parasitoid.

11.4.3. Sublethal Effects

Effects from the consumption of sublethal dosages of *B.t.* have been reported for several predators and parasitoids (Table 11.2). Documented sublethal effects include reduced egg production, reduced host consumption, reduced or increased percent parasitism, reduced adult longevity, increased parasitoid development rate, and skewed sex ratios. Often it is not apparent whether these effects are direct (e.g., sublethal response of the natural enemy) or indirect (e.g., sublethal response of the host, rendering it more or less suitable as a food source).

Total egg production was reduced by 27% when the braconid parasitoid *Microplitis demolitor* and ichneumonid parasitoid *Zele chlorophthalma* were reared on *Spodoptera littoralis* infected with *B.t.* var. *entomocidus* (Salama et al. 1982, Salama and Zaki 1983). Egg production by the braconid *Bracon brevicornis* was reduced by 36% and 69% when it was reared on *Sesamia cretica* injected with 3×10^{10} spores/ml and 3×10^{16} spores/ml *B.t.*, respectively (Temerak 1980). The predator *Coccinella undecimpunctata* produced 30% fewer eggs after being fed aphids sprayed with a 500 μg/ml suspension of *B.t.* (Salama et al. 1982). Korostel and Kapustina (1975) reported that adult *Trichogramma* sp. that survived consumption of *B.t.* had reduced fecundity.

The host consumption rate for the predator *Chrysoperla carnea* was reduced by 34% and 26% when fed infected *S. littoralis* and contaminated *Aphis durantae*, respectively (Salama et al. 1982). Consumption of aphids by *C. undecimpunctata* was reduced 30% when *B.t.*-contaminated *A. durantae* were fed to the coccinellid larvae (Salama et al. 1982).

Parasitism of *S. littoralis* by the ichneumonid parasitoid *Zele chlorophthalma* was reduced by 75% compared to the control when the host larvae were fed on a diet containing *B.t.* A decrease in percent parasitism has also been reported for two tachinid parasitoids of the gypsy moth, *Compsilura concinnata* and *Blepharipa pratensis*, in field plots sprayed with *B.t.* (Ticehurst et al. 1982).

Surprisingly, field studies also show that parasitism of host species is often greater in plots treated with *B.t.* than in untreated plots. Parasitism of the gypsy moth (*Lymantria dispar*) larvae by *Cotesia melanoscelus* and the density of the parasitoid's cocoons were greater in hardwood plots treated with *B.t.* than in untreated plots (Ticehurst et al. 1982, Weseloh et al. 1983). These field plots showed a strong correlation between percent parasitism and gypsy moth larval size. *C. melanoscelus* parasitized the small early instars of *L. dispar*. Developmental rates were retarded in gypsy moth larvae that survived *B.t.* doses (Weseloh and Andreadis 1982, Weseloh 1984). Thus, the retarding effect of *B.t.* infection kept *L. dispar* larvae smaller for a longer period compared to untreated larvae, and this permitted females of *C. melanoscelus* to parasitize larger numbers of hosts (Weseloh et al. 1983).

This same phenomenon has been documented for the introduced braconid parasitoid of the gypsy moth *Rogas lymantriae* (Wallner et al. 1983). In addition, the sex ratio of *R. lymantriae* was significantly skewed in *B.t.*-treated plots because the female parasitoid oviposited a higher percentage of male progeny in the smaller hosts (Wallner et al. 1983).

The rate of parasitoid development was shown to be affected when parasitoids were reared on infected hosts. When *Ephestia kuehniella* and *Pieris brassicae*, the respective hosts of the parasitoids *Phanerotoma flavitestacea* and *Cotesia glomeratus*, were fed sublethal doses of Bactospeine, normal food consumption was reduced for both host species, and the larval period was lengthened for both the hosts and their parasitoids (Marchal-Segault 1975b). Ahmad et al. (1978) found that treating gypsy moth larvae with Dipel HG caused an increase of approximately 3 days in the total development time of its braconid parasitoid *C. melanoscelus*. Similar results were obtained when *Zele chlorophthdlma* parasitized *B.t.*-infected *Spodoptera littoralis* (Salama and Zaki 1983).

Adult longevity has also been shown to be adversely affected when adult parasitoids ingested or were exposed to *B.t.* (Table 11.2). The longevity of adult *P. flavitestacea* was reduced from 70 days to 34 and 20 days, respectively, when adult parasitoids were infected by contact with *B.t.* residues of 1% and 5% a.i. (Marchal-Segault 1975a). Adult longevity of *C. glomeratus* was reduced from 24 days to 13 and 10 days, respectively, when they were fed a sugar solution containing 1% and 5% a.i. of *B.t.* (Marchal-Segault 1975a). Adult *Bracon brevicornis* also had significantly shorter life spans when they were kept in vials with host larvae that had been infected with *B.t.* (Temerak 1980).

11.4.4. Ecological Relationships and Selectivity

Often the physiological or ecological characteristics of an organism allow it to escape a potentially toxic pesticide. If an organism is in a less susceptible or protected life stage (e.g., diapause, quiescence, or refuge) or is not residing in the habitat during application of the toxin, then the impact on that organism in the field may be less than predicted from laboratory experiments (ecological selectivity). *B.t.* is highly selective to many natural enemies due to the specific physiological mechanisms required to initiate infection. Ingestion of spores by adult predators and parasitoids occurs only if the insect 1) consumes the bacillus by host feeding or drinking water, 2) ingests spores while grooming or cleaning sensory organs that were contaminated with the bacteria, or 3) ingests spores while nectar, pollen, or leaf feeding.

Most predaceous Hemiptera and Coleoptera are entomophagous both as larvae and adults. However, predatory syrphids, chrysopids, the majority of the parasitic diptera, and many parasitic Hymenoptera do not host feed as adults. The question then becomes how many viable *B.t.* spores do adult insects contact during their daily activity within a treated habitat. At the recommended field rate, the average number of bacillus spores is 11×10^9 spores/m^2 (1.1×10^4 spores/mm^2) (Forsberg et al. 1976). This field rate is 1000-fold lower than most of those used to initiate high mortality of adult parasitoids in the laboratory by contamination of food sap (i.e., 10^9 spores/ml to 10^7 spores/ml, Muck et al. 1981). Thus, while scientific investigation is lacking, the probability that an adult parasitoid would consume lethal doses of *B.t.* while searching for hosts and nectar feeding within a *B.t.*-sprayed field might be very low. The amount of *B.t.* ingested by host-feeding parasitoids, likewise, has not been investigated. The likelihood of lethal *B.t.* consumption becomes even more remote if the parasitoids' nectar feeding is restricted to an area outside the treated habitat.

Parasitoid larvae may also escape exposure to *B.t.* by altering the behavior of their host, making it less likely to consume the bacillus. For example, Hamel (1977) found that populations of two parasitoids of the western spruce budworm *Choristoneura occidentalis* increased in Dipel-treated field plots over a 21-day period compared to untreated plots. Three other parasitoid species decreased over the same time period. *Apanteles fumiferanae* and *Glypta fumiferanae* both attack prediapausing, first instar budworm larvae. The parasitized budworm larvae were negatively phototropic, which reduced their feeding stimulus in the spring when *B.t.* was applied. Most other parasitoids of the budworm attack later instars, which contact and ingest the bacillus (Hamel 1977). Thus, even if *A. fumiferanae* and *G. fumiferanae* were shown to be susceptible to *B.t.* in the laboratory, the probability of field infection may be low. *Phanerotoma flavitestacea* and *Cotesia glomeratus* within their respective lepidopterous hosts, *Pieris brassica* and *Ephestia kuehniella*, actually delayed or hindered the rate of the lepidopterans' bacterial infection because their feeding rates on MPCA-treated food was reduced (Marchal-Segault 1975c). Thus, parasitized larvae may have a higher probability of avoiding or delaying a lethal *B.t.* dosage than unparasitized larvae.

Adult parasitoids can prevent direct or indirect mortality from a pathogen to their progeny by discriminating between infected and uninfected hosts. *C. melanoscelus* females discriminated between uninfected and NPV-infected gypsy moth larvae (Versoi and Yendol 1982). This may also apply to *B.t.*-infected larvae. The ichneumonid *Zele chlorophthalma* parasitized 79% of *Spodoptera littoralis* fed on an untreated artificial diet, but only 20% of the *S. littoralis* fed on a diet treated with *B.t.* (Salama and Zaki 1983). These results suggested that *Zele chlorophthalma* found infected *S. littoralis* less suitable than uninfected hosts. The extent to which hymenopterous parasitoids can discriminate between *B.t.*-infected and uninfected hosts has not been investigated directly.

11.5. FUNGI

Entomopathogenic fungi are an important factor in natural insect control worldwide (Cherwonogrodzky 1980). All major taxonomic groups of fungi contain arthropod attacking species, but the insect mycoses discussed in the literature are primarily from the genera *Beauveria, Metarrhizium*, and *Aspergillus* (Burges and Hussey 1971). Species most commonly used in biological control programs are *Fungi Imperfecti*, including *Beauveria bassiana, Metarrhizium anisopliae, Nomuraea rileyi*, and *Paecilomyces farinosus.*

Fungal infection generally involves cuticular penetration by the conidial germ tube, although it may sometimes occur through the respiratory or alimentary tract (Burges and Hussey 1971). After germination, mycelia penetrate tissue by the release of enzymes and by mechanical pressure (Cherwonogrodzky 1980). Free-floating "yeastlike" mycelia multiply in the hemocoel. The death of hosts occurs from the release of fungal toxins or from tissue destruction. Hyphal growth continues until the insect is filled with mycelia. Fruiting bodies erupt through the cuticle, releasing spores to infect other susceptible insects (Cherwonogrodzky 1980).

Fungi are highly dependent on the microclimate surrounding their host. High humidity is critical for spore germination and to increase spore production on fruiting bodies. Light affects sporulation on the host after it is dead as well as longevity of spore viability (Burges and Hussey 1971).

Currently, fungal pathogens are not being extensively pursued as commercial products in the United States (Cherwonogrodzky 1980). *Beauveria bassiana* is mass-produced in the USSR under the registered name of Boverin. In China, *B. bassiana* is also produced and widely used for insect control. Although some of the entomopathogenic fungi are host-specific, others such as *B. bassiana* are broad-spectrum.

11.5.1. Methods of Assessment

The effects of assays conducted with entomopathogenic fungi on nontarget beneficial insects are listed in Table 11.3. The majority of this research has been done in the USSR, where most arthropod natural enemies were inoculated by allowing the organism to contact the fungal spores directly (Table 11.3). Toxicity was low and mortality never exceeded 10% in most tests. Information on the microenvironment surrounding the host was usually lacking, so it is not known whether these fungi were tested under optimal or suboptimal conditions for their growth and population increase. Surprisingly, the only tests showing greater than 10% mortality were those in which the natural enemies consumed the fungal spores. Ingestion of fungal spores by natural enemies could be important if large areas were to be sprayed with entomopathogenic fungi.

11.5.2. Lethal Effects

Kiselek (1975) reported that 50% of *Cryptolaemus montrouzieri* larvae died when they consumed mealybugs sprayed with a 1% a.i. solution of Boverin. Boverin was harmless to the adults of this predaceous coccinellid. Most remaining tests reported little or no mortality of natural enemies exposed to entomopathogenic fungal spores (Table 11.3). Indirect mortality of the braconid *Microplitis croceipes* was reported to occur when its larval host *Heliothis zea* was infected with *Nomuraea rileyi* before oviposition, or when fungal infection occurred within 24 hr of parasitization (King and Bell 1978). When *H. zea* was infected with *Nomuraea rileyi* 24 hr after parasitization, the parasitoid completed its development and emerged before the *H. zea* died.

Overwintering mortality could increase substantially for many natural enemies if large quantities of entomopathogenic fungi were used commercially. In England, overwintering mortality of the coccinellid *Adalia bipunctata* was 36%; the primary cause of death was the fungal pathogen *Beauveria bassiana* (Mills 1981). Hibernating adult *Coccinella septempunctata* were also attacked by *B. bassiana* and infection rates of 50% were reported in overwintering populations in southeastern France (Hodek 1973). Thus, by increasing the amount of inoculum residue in the environment as a result of commercial entomopathogenic fungal application, the natural overwintering mycoses of many inactive nontarget insects might be increased, particularly if the fungal pathogen had a cosmopolitan host range.

Mortality of the carabid beetle *Harpalus rufipes* was slightly increased when *Paecilomyces farinosus* and *Paecilomyces fumosoroseus* cultures were incorporated into soils in which the adult beetles overwintered. In the same tests, *B. bassiana* did not increase overwintering mortality (Kabacik-Wasylik and Kmitowa 1973).

Table 11.3. Effects of Fungal Pathogens on Arthropod Natural Enemies

Nontarget Insect										
Family	Species	Fungus Pathogen	Fungus Dosage	Method of Exposure	Treated Insect Host	Toxicity	Test Site	Reference	Sublethal Effects[‡]	Duration
Aphelinidae	*Encarsia formosa*	Boverin® (*Beauveria bassiana*)	Recommended rate	Contact by adults	*Trialeurodes vaporariorum*	*	Lab.	Beglyarov & Maslienko 1978	—	240 hr
		Verticillium lecanii	3.6×10^7 blastospores/ml	Contact by adults	*T. vaporariorum*	*	Lab.	Hall 1982	—	—
Braconidae	*Cotesia glomeratus*	*Beauveria bassiana*	—	Infects host after female oviposits	*Pieris brassicae*	*	Lab.	Fuhrer & El-Sufty 1979	—	—
	Microplitis croceipes	*Nomuraea rileyi*	1×10^9 spores/larva	Brushed on bollworm	*Heliothis zea*	*	Lab.	King & Bell 1978	—	—
Tricho-grammatidae	*Trichogramma cacoeciae; T. embryophagum; T. evanescens*	Boverin	0.5% a.i.	Contact by adults	—	*	Fld.	Kapustina 1975	—	—
	T. pallidum	Boverin	1% a.i.	Sprayed on host eggs	—	*	Lab.	Kiselek 1975	—	—
Carabidae	*Broscus cephalotes*	*Beauveria bassiana*	—	Contact by adults	—	*	Lab.	Kabacik-Wasylik & Kmitowa 1973	—	—
	Harpalus rufipes	*Paecilomyces farinosus*	—		—	*	Fld.	Kmitowa & Kabacik-Wasylik 1971	IOM	— —
	Pterostichus vulgaris	*Paecilomyces fumosoroeus*	—		—	*				

	Calosoma sayi	*Variamorpha* sp.	2.5×10^6 spores/ml diet	Adults fed diet	*Spodoptera frugiperda*	*	Lab.	Young & Hamm 1985	—	24–360 hr
Coccinellidae	*Coccinella septempunctata*	*Beauveria bassiana*	1%	Ingested by adults and larvae	—	*	Lab.	Kiselek 1975	—	—
	Cryptolaemus montrouzieri	Boverin	1%	Ingested by adults and larvae	*Sitotroga cerealella*	*** (larvae)	Lab.	Kiselek 1975	—	—
Chrysopidae	*Chrysoperla carnea*	*Beauveria bassiana*	—	Sprayed on fields	*Leptinotarsa decemlineata*	*	Fld.	Gusev et al. 1975	—	—
		Boverin	1%	Ingested by adults and larvae	*S. cerealella*	*	Lab.	Kiselek 1975	—	—
	Pheidole megacephala	*Beauveria bassiana*	Recommended field rate	Ingested by adults and larvae	*Cylas formicarius*	*	Lab.	Castineiras & Calderon 1982	—	72 hr
Pentatomidae	*Perilloides bioculatus*	*Beauveria bassiana*	—	Sprayed on field	*L. decemlineata*	*	Fld.	Gusev et al. 1975	—	—

*(0–10% mortality) no or slight toxicity.
**(10–40% mortality) low toxicity.
***(40–80% mortality) low toxicity.
****(80–100% mortality) high toxicity.
‡IOM—increased overwintering mortality.

Few studies have investigated the sublethal effects of entomopathogenic fungi on natural enemies and beneficial insects.

11.5.3. Ecological Relationships and Selectivity

Despite the broad host range of many entomopathogenic fungi, assays conducted on arthropod natural enemies using fungal pathogens have shown this group to be relatively selective (Table 11.3). However, this statement is based on a small number of tests. One interesting ecological relationship that has been documented is that parasitism may predispose some hosts to fungal infection. King and Bell (1978) found that the rate of *N. rileyi* infection of *Heliothis zea* was extremely low if applied prior to parasitization by *M. croceipes*. However, when the fungus was applied either at the same time or after oviposition, the infection rate was significantly higher, indicating that parasitism predisposed the host to infection by the fungus. They also found that if the parasitoid and fungus were introduced into the host at the same time, then *Microplitis croceipes* required a significantly longer period of time (1.1 days) to complete development than was required for *N. rileyi* to kill *H. zea*. Indirect parasitoid mortality from the pathogen would then occur (King and Bell 1978).

Parasitoid-induced host susceptibility to the fungus *Beauveria bassiana* has been compensated for by the braconid *Cotesia glomeratus* (Fuhrer and El-Sufty 1979). This parasitoid reduced fungal competition by the production of a fungistatic substance that has been shown to inhibit the development of *B. bassiana* in the hemolymph of *Pieris brassicae*. This substance was produced in the teratocytes of immature *C. glomeratus*. The derivatives of the parasitoid serosa were suspended in the host hemolymph, producing a fungistatic material that compensated for the enhanced host susceptibility (Fuhrer and El-Sufty 1979). This may be one mechanism evolved by braconid parasitoids to reduce fungal mycosis that is triggered by oviposition.

In a similar study with the braconid *Ascogaster quadridentatus* parasitizing the codling moth *Cydia pomonella*, El-Sufty and Fuhrer (1985) observed that stinging of the host promoted cuticular penetration of *Beauveria bassiana*. While initial penetration was promoted, further growth of the fungus in the endocuticle was checked to considerable extent, as were dissemination of the disease in the hemocoel, destruction of tissues, and development of fungal spores. These modified effects on the development of the fungus were thought to be induced by the parasitization process of *A. quadridentata*.

Another extensive study of interference with fungal pathogens was reported by Powell et al. (1986b), involving the aphidiid parasitoids of cereal aphids. In field populations of aphids, parasitism levels declined through the season as fungal infections increased. In laboratory tests, the fungus *Erynia neoaphidis* took 3–4 days to kill the rose-grain aphid *Metopolophium dirhodum*, whereas the parasitoid *Aphidius rhopalosiphi* took 8–9 days at 20° C. When aphids were infected by the fungus less than 4 days after being parasitized, parasitoids were prevented from completing their development. Conversely, when infection occurred more than 4 days after parasitism, development of the fungus was significantly impaired. There was no histological evidence that the fungus invaded the tissues of the parasitoid when both attacked the same aphid.

In summary the complex interrelationships involved in the interaction of fungi and arthropod parasitoids within their hosts is reasonably well documented for this group of MPCAs. Similar studies of these associations in other groups of MPCAs are needed to more fully understand the wide range of adaptations that both of these groups of biological control agents have evolved to exclude or coexist with each other.

11.6. PROTOZOANS

Entomopathogenic protozoans include flagellates, ciliates, amoebae, coccidia, haplospo-rida, neogregarines, and microsporida (Burges and Hussey 1971). Protozoans with the most commercial potential are the microsporidia, particularly those from the genus *Nosema*. The primary mode of infection of protozoans is by ingestion, although transovum and transovarial transmission have also been documented. Protozoans seldom cause mortality, but reduce fitness by lowering vitality, progeny number, life span, and stimuli responses. Once thought to be highly host-specific, most protozoans are now known to be cosmopolitan, with the ability to cross-infect several insect orders.

Currently, most protozoans are not being pursued as commercial insecticides. However, *Nosema locustae*, produced under the trade name Nolo, is used to reduce grasshopper infestations. Most protozoans are not commercially attractive because they do not act fast enough to prevent crop damage, unlike some viruses or *B.t.* (Burges and Hussey 1971). If damage is to be circumvented using protozoans, the pathogens must be introduced seasonally prior to an expected pest outbreak (Burges and Hussey 1971). Commercial manufacture is labor-intensive and costly because protozoans must be produced *in vivo*. However, as advances are made in pest outbreak prediction and mass production is improved, the commercial potential of protozoans may increase.

11.6.1. Methods of Assessment

Results of assays conducted with protozoans on arthropod predators and parasitoids are listed in Table 11.4. The majority of tests have been conducted on hymenopterous parasitoids. Beneficial insects ingested the protozoans in over 80% of assays (Table 11.4). The toxicity to parasitoids ranged from high (Huger and Neuffer 1978, Bell and McGovern 1975) to low (Kaya 1979), depending on the protozoan and the beneficial tested. Tests for contact toxicity were very low, as expected, because the protozoan must enter the alimentary canal to have a toxic effect. Duration of tests was often longer than for most of the other MPCA assays, with one test terminated after 100 days (McNeil and Brooks 1974). Protozoan tests should be longer than other assays because seldom are vital functions of the host disrupted quickly; rather a slow debilitation generally occurs (Burges and Hussey 1971).

11.6.2. Lethal Effects

Mass rearing of the codling moth parasitoid *Ascogaster quadridentata* failed when the parasitoid became infected with *Nosema carpocapsae*, a microsporidian parasite of the codling moth (Huger and Neuffer 1978). Thus, the parasitoid colony crashed due to this infection. A similar phenomenon occurred when Bell and McGovern (1975) reared *Bracon mellitor* on boll weevil, which they had purposefully infected with the microsporidian *Glugea gasti*. No parasitoids emerged from microsporidia-infected weevil larvae. Imma-ture *B. mellitor* dissected from infected hosts contained *G. gasti* spores, suggesting that mortality was directly induced by the protozoan (Bell and McGovern 1975). *B. mellitor* was also found to be susceptible to a neogregarine, *Mattesia grandis*, after ingesting the protozoan while feeding on infected boll weevils (McLaughlin and Adams 1966). Andreadis (1982) found an inverse correlation between the infection of the European corn borer *Ostrinia nubilalis* with the microsporidian *Nosema pyraustae* and the incidence of parasitism by the braconid *Macrocentrus grandii* when *N. pyraustae* infections exceeded

Table 11.4. Effects of Protozoans on Arthropod Natural Enemies

Nontarget Insect										
Family	Species	Protozoan Pathogen	Protozoan Dosage	Method of Exposure	Treated Insect host	Toxicity	Test Site	Reference	Sublethal Effects[‡]	Duration
Braconidae	*Ascogaster quadridentata*	*Nosema carpocapsae*	—	Ingested by larvae	*Cydia pomonella*	****	Lab.	Huger & Neuffer 1978	—	—
	Bracon mellitor	*Glugea gasti*	100% infected hosts	Ingested by larvae	*Anthonomus grandis*	****	Lab.	Bell & McGovern 1975	—	1–360 hr
		1069 (unidentified)				*			—	1–360 hr
		Mattesia grandis	1×10^7 spores/mm^{-2}	Ingested by larvae and adults	*A. grandis*	**	Lab.	McLaughlin & Adams 1966	RPP	1–168 hr
	Cardiochiles nigriceps	*Nosema campoletidis*	—	Contact and ingested by parasitoid larvae	*Heliothis zea*	**	Fld.	Brooks & Cranford 1972	—	1–336 hr
		Nosema heliothidis	—		*H. virescens*	*			—	1–336 hr
	Macrocentrus grandii	*Nosema pyrausta*	—	Ingested by larvae	*Ostrinia nubilalis*	**	Fld.	Andreadis 1980, 1982	RAL	—
	Cotesia glomeratus	*Nosema polyvora*	—	Ingested by larvae	*P. brassicae*	***	Lab.	Issi & Maslennikova 1964	RNW, RF	—
		Nosema mesnili	—	Ingested by larvae or direct penetration of larvae	*P. brassicae*	***	Lab.	Larsson 1979	—	1–2400 hr

Chalcidae	*Spilochalcis side*	*Nosema campoletidis*	—	Contact	*Campoletis sonorensis*	*	Lab.	McNeil & Brooks 1974	—	1–2400 hr
		Nosema heliothidis	—	Contact	*Campoletis sonorensis*	*		McNeil & Brooks 1974	—	1–2400 hr
Ichneumonidae	*Campoletis sonorensis*	*Nosema campoletidis*	—	Exposed to host larvae	*H. zea*	*			—	1–240 hr
		Nosema heliothidis	—	Exposed to host larvae	*H. virescens*	*			REP, RAE	—
Tricho-grammatidae	*Trichogramma evanescens*	*Nosema pyrausta*	—	Ingested by larvae	*O. nubilalis*	***	Lab.	Huger 1984	—	—
Braconidae	*Microgaster croceipes; Apanteles marginiventris*	*Vairimorpha* sp.	1.6 × 10^5 spores/ml in diet	Ingested by host	*Helothis zea*	**	Lab.	Hamm et al. 1983	IDT	—
Pteromalidae	*Catolaccus aeneoviridis*	*N. campoletidis*	—	Contact	*C. sonorensis*	*	Lab.	McNeil & Brooks 1974	—	—
		N. heliothidis	—							
Tachinidae	*Bonnetia comta*	*Vairimorpha necatrix; Vairimorpha* sp.		Ingested by by noctuid host	*Agrotis ipsilon*	***	Lab.	Cossentine & Lewis 1986	IDT, RMW	1–720 hr
Reduviidae	*Zelus exsanguis*	*Vairimorpha necatrix*	300 spores/mm^2	Ingested by saltmarsh caterpillar	*Estigneme acrea*	*	Lab.	Kaya 1979	—	—
		Pleistrophora sp.	500 spores/mm^2			*				

*(0–10% mortality) no or slight toxicity.
**(10–40% mortality) low toxicity.
***(40–80% mortality) moderate toxicity.
****(80–100% mortality) high toxicity.
‡REP—reduced egg production; RPP—reduced percent parasitism; RAL—reduced adult longevity; RMW—reduced mean weight; RF—reduced fecundity; IDT—Increased developmental time.

45% of corn borer population. Although parasitoid discrimination of infected hosts may have been involved, these findings suggest that the presence of *N. pyraustae* had a significant adverse effect on field populations of *M. grandii* and may help explain the diminishing role of this and other introduced parasitoids as natural control agents of the European corn borer in the United States (Andreadis 1982).

11.6.3. Sublethal Effects

Many sublethal effects have been documented for parasitoids infected by a protozoan associated with the host of the parasitoid. Reduced adult longevity, reduced time to adult emergence, reduced percent parasitism, reduced fecundity, and reduced mean adult weight have been cited (Table 11.4). Andreadis (1980) found that when *Macrocentrus grandii* developed within larvae of *Ostrinia nubilalis* which were infected with *Nosema pyraustae*, the parasitoids had significantly shorter adult life spans. The life span of adult males was reduced from 14.3 days for uninfected individuals to 2.3 days for infected individuals. The life span of adult females was reduced from 16.6 days for uninfected individuals to 1.9 days for infected individuals. *Nosema pyraustae* also adversely affected the pupal development of *Macrocentrus grandii*. Cossentine and Lewis (1986) observed that microsporidia from the genus *Variamorpha* caused an increase in time to adult emergence and reduced pupal weight in the tachinid parasitoid *Bonnetia comta* when developing in the noctuid host *Agrotis ipsilon*. In both of the above cases, these effects may have been indirect—due to a reduced food supply available to the parasitoid resulting from pathogenesis.

Huger (1984) observed that parasitism rates were normal when the egg parasitoid *T. evanescens* was initially released into the field. However, an incidence of *N. pyraustae* in *O. nubilalis* eggs reduced parasitism levels by 50% in the infected F_1 generation of the parasitoids. Reduced fecundity was also observed in *Cotesia glomeratus* that parasitized *Pieris brassicae* infected with *Nosema polyvora*. The mean adult weight of the parasitoid was reduced, and the smaller specimens also suffered a greater proportion of overwintering mortality (Issi and Maslennikova 1964). Percent parasitism was reduced by 40% when *Bracon mellitor* was infected with the protozoan *Mattesia grandis* from its host, the boll weevil (McLaughlin and Adams 1966). Whether all these effects were a direct result of the protozoan or indirectly caused by reduced food for the parasitoids was not always clear.

11.6.4. Ecological Relationships and Selectivity

Unlike the other MPCAs, most of the protozoans tested in these assays were capable of infecting the parasitoids of their hosts. Subsequently, protozoans had some effect on parasitoid fecundity or caused a slow reduction in fitness (Table 11.4). This type of effect on natural enemies is not easily quantifiable, especially in assays of short duration, because indirect impacts on population dynamics are often difficult to detect.

Cossentine and Lewis (1986) found that the microsporidia *Variamorpha necatrix* and *Variamorpha* sp. decreased the number of tachinids (*Bonnetia comta*) which pupated from noctuid hosts. *V. necatrix* had a more significant impact. Close inspection revealed that detriment to the parasitoid was related to nutritional deficiencies caused by the accumulation of undigestible spores in the parasitoid's gut lumen rather than a toxic or disease reaction. Issi and Maslennikova (1964) found that when *Cotesia glomeratus* was infected by *Nosema polyvora* transmitted from its host *Pieris brassicae*, the percent of individuals that went into diapause was greatly reduced. Such infections could significantly increase overwintering mortality of *C. glomeratus*. Subtle effects of this type can only be discovered by carefully scrutinizing host–pathogen–parasitoid interactions.

With regard to the transmission of protozoan pathogens by parasitoids attacking their hosts, Hamm et al. (1983) observed that *Microgaster croceipes*, produced in larvae of *Heliothis zea* that were infected with the microsporidian *Variamorpha* sp., transmitted the pathogen to about 10% of the larvae that it attacked subsequently. Similarly exposed *Apanteles marginiventris* did not transmit the microsporidian from host to host. These studies emphasize the importance of interspecific transmission of MPCAs from natural enemies to their hosts (Fig. 11.1) and the species specificity of these relationships.

11.7. SUMMARY

Studies of MPCA impact on arthropod natural enemies have focused more on ecto- and endoparasitoids than on arthropod predators. In general, parasitoids have more complex interactions with MPCAs than predators. However, this does not imply that an MPCA does not directly or indirectly affect both groups of natural enemies significantly.

Direct mortality to arthropod predators and parasitoids can be caused by bacterial and protozoan MPCAs. Few cases of direct mortality from viruses of pests have been noted for arthropod natural enemies in the literature. The direct effect of fungi on arthropod natural enemies has not been well studied and has probably been underestimated. Future studies with fungi should focus on direct side effects that occur when the insect is in a favorable environment for fungal development (e.g., in overwintering hibernacula).

Sublethal side effects of MPCAs or latent mortality may be more important with MPCAs than more direct, acute mortality for most natural enemies tested (particularly with protozoans). Mortality from indirect effects of natural enemies consuming their MPCA-infected hosts is probably more significant than direct effects for any MPCA. In pest management, indirect effects from competition between MPCAs and parasitoids may be significantly reduced if MPCAs are applied after parasitoids infect their hosts.

There are probably many more subtle interactions between natural enemies and entomopathogens than have been documented to date, especially compared with known interactions between beneficials and chemical pesticides. Most entomopathogens are naturally occurring. Natural enemies have been evolving mechanisms such as discrimination of infected hosts and fungistatic properties in teratocytes to reduce competition with them over time. However, competition between arthropod natural enemies and microbial pathogens of insect pests can be substantial.

Standardized methods need to be developed for MPCA assays that adequately monitor the modes of action and side effects of these biological agents. Standards for expressing more precisely the dosage administered to the host or natural enemy are needed. Choosing the appropriate method of exposure is important in standardizing assays for MPCAs, and, with the exception of fungi, high mortality is generally encountered only when beneficial insects ingest MPCAs. In general, the duration of side-effects tests for MPCAs should be longer than those used in standard bioassays for conventional pesticides. Adequate assessment of MPCA side effects may require testing periods of a total life cycle or over several generations.

Extensive efforts have gone into studying basic interactions between entomophagous insects and pests of agricultural crops in IPM systems research over the past few decades. Basic understanding of species interactions at multiple trophic levels that involve MPCAs has been gained. These well-studied systems may be excellent models for evaluating the impact of genetically engineered MPCAs (Flexner et al. 1986).

12

INSECT GROWTH REGULATORS

12.1. INTRODUCTION

Insect growth regulators (IGRs) are a diverse group of insecticides that disrupt the development of insects and other arthropods. IGRs interfere with specific metabolic pathways of arthropod pests and therefore are thought to be more selective to natural enemies than conventional insecticides. While IGRs demonstrate improved selectivity, they can have harmful effects on arthropod predators and parasitoids. IGRs may cause

This chapter was coauthored by Hugo. E. van de Baan, Department of Entomology, Michigan State University, East Lansing, Michigan.

Juvenile Hormone Analogs (JHA's)

JH I

Hydroprene

Methoprene

Epofenonane

Anti-Juvenile Hormone Analogs (AJHA's)

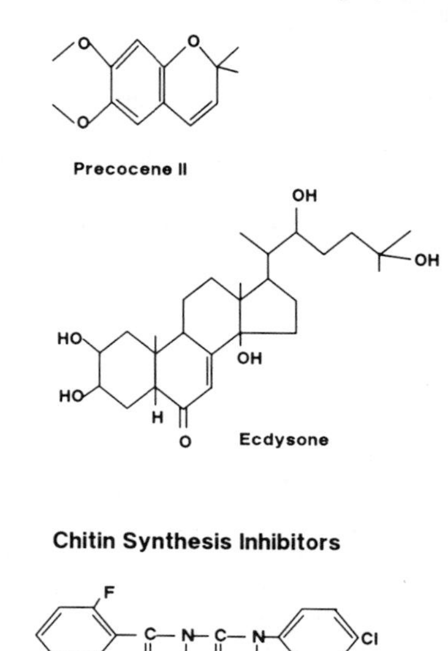

Chitin Synthesis Inhibitors

Diflubenzuron

Figure 12.1. Representative structures for major classes of insect growth regulators.

mortality, but often they interfere only with natural enemy development or their biological control capacity.

IGRs can be divided into two major groups based on their mode of action. Several types of IGRs interfere with the endocrine system, including juvenile hormone analogs (JHAs), anti-juvenile hormone analogs (AJHAs), and ecdysteroids. The benzoylphenyl ureas inhibit chitin synthesis. Representative chemical structures for these groups of compounds are illustrated in Fig. 12.1.

This chapter focuses on the effects of IGRs on beneficial arthropods and the factors affecting their toxicity and selectivity to natural enemies over their prey or hosts. For a review of the chemistry, toxicology, and pharmacodynamics of IGRs the reader is referred to Retnakaran et al. (1985). For an extensive discussion of the effects of IGRs on endocrine interactions between endoparasitic insects and their hosts, see Beckage (1985).

12.2. EFFECTS OF IGRs ON NATURAL ENEMIES

A summary of the effects of IGRs on parasitic and predaceous arthropods is presented in Table 12.1. Of the 111 cases cited, over 73% involved parasitoids, 23% involved predaceous species, and 4% involved studies of both predators and parasitoids. This is in marked contrast to the trends in research with conventional pesticides, which were elaborated using the SELCTV database. In SELCTV analysis of conventional pesticides only, predators were studied more frequently than parasitoids by almost the same ratio (73:27; Chapter 2).

The reasons why IGR side-effects data are skewed toward studies on parasitoids are probably several. IGRs are commonly directed at larval lepidopteran pests, which often have effective parasitoids. The development of parasitoids in hosts represents a unique window of vulnerability that can be highly sensitive to IGRs. There are no analogous events in predator life cycles during which physiological association with the target pest organism is as close. The influences of IGRs on parasitoids have also been studied to identify ways in which these compounds might be used to improve biological control, such as by altering rates of development or modifying diapause characteristics (e.g., Ascerno 1975, Ascerno et al. 1980).

While many compounds have insect growth regulating properties, relatively few have been substantially tested on natural enemies and their effects documented. The most common JHAs studied to date have been hydroprene (Altozar), methoprene (Altosid), kinoprene (Entsar), triprene, epofenonane, and fenoxycarb (Table 12.1). Among the exogenous juvenile hormones (JHs) tested, JH I, JH II, and JH III have been evaluated to a limited degree and almost exclusively on parasitoids. Similarly, several anti-juvenile hormone analogs including ponasterone A, several benzodioxoles, ETB, Fmev, precocenes, and ecdysterone, have been tested on the parasitoids of lepidopteran pests. By far the most widely tested IGR is diflubenzuron (Dimilin), a chitin synthesis inhibitor. Its impact has been evaluated on both predators and parasitoids. This product is used commercially on many agricultural crops. A number of experimental compounds have also been evaluated (see Table 12.1).

Examination of the published research (Table 12.1) indicates that the following factors influence the effect of IGRs on the development of parasitoids and predators:

1. *Test procedure.* Methods used to assess the side effects of IGRs include topical application to parasitized hosts, spraying or dipping of plant material infested with parasitized insects, feeding treated diet to parasitized hosts, topical application to parasitoids or predators, and field application to plants harboring pests and beneficial species. Different test methods result in different modes of pesticide uptake (Chapter 3), which may cause different effects of the IGR on a natural enemy. For example, direct treatment of a parasitoid is usually more harmful than treatment of a parasitized host, in which the parasitoid is protected to a greater extent. However, topical treatment of a parasitized host is more detrimental to an internal parasitoid than oral uptake of the IGR by the host. Passage of the IGR through the digestive tract may lessen the amount reaching the nontarget species, it may alter the compound, or in some other way it may reduce the IGR's impact on the endoparasitoid.

2. *Life stage treated.* Different life stages of predators and parasitoids may differ in their susceptibility to IGRs. Due to their unique modes of action, variable effects on different life stages are even greater than observed with conventional pesticides (Chapter 4). For example, the application of JHAs to parasitized hosts during early larval development of the parasitoid, a period when the endogenous JH titer is high, may not disrupt the parasitoid's development. The reverse is true for AJHAs and ecdysteroids. Application of AJHAs later in the process of natural enemy development may be the least disruptive. Differences in susceptibility of individual life stages of natural enemies to benzoylphenyl ureas compared to their prey or hosts may allow for reduced side effects through careful timing of applications.

3. *Dose–response relationship.* Increasing the dose generally increases effects on natural enemies, manifested by lower emergence or higher mortality (Outram 1974, Riviere 1975, Smilowitz et al. 1976, Lema and Poe 1978, El-Banhawy 1980, Mellini and

Table 12.1. Effects of Insect Growth Regulators on Nontarget Arthropod Natural Enemies

Natural Enemy/Pest	Type of Test	Effects on Natural Enemy	Reference
		Juvenile Hormone Analogues	
HYDROPRENE			
Microctonus aethiopoides/Hypera postica	Topical treatment of parasitized host	• Termination of diapause • Mortality, morphological aberrations • Interaction between parasitoid age and treatment timing	Ascerno 1975
Microctonus aethiopoides/Hypera postica	Topical treatment of parasitized host, parasitoids in diapause	• Termination of diapause • 1 μg = 8% mortality; 5 μg = 31% mortality; 10 μg = reduced eclosion, 82% adult deformation in F_1, fewer parasitized weevils; 50 μg = decrease in pupation, 100% pupal mortality	Ascerno et al. 1980
Microctonus aethiopoides/Hypera postica	Topical treatment of parasitized host, nondiapausing parasitoids	• Increased emergence, no effect on % pupation, sensitivity increased as parasitoid larvae aged	Ascerno et al. 1980
Microplitis croceipes/Heliothis virescens	Topical treatment of parasitized host	• Decrease in pupation, adult emergence • Effect dependent on timing of application	Boone, unpubl.
Nabis roseipennis, Chrysoperla carnea	Topical treatment of last instar of predator	• No preimaginal abnormalities in *C.c.*; 2.5% in *N.r.* • Adult emergence unaffected in *N.r.*; 5.9% decrease in *C.c.* • No adult abnormalities in *N.r.*	Boone, unpubl.
Aphytis melinus/Aonidiella aurantii	Parasitized scales dipped in IGR	• Mortality in pupal stage, no emergence of adults	Davies & McLaren 1977
Ooencyrtus kuwanaii/Porthetria dispar	Microapplication to parasitized egg	• 70% mortality to late larval stage at 100 ppm; 30% mortality of pupae • Decreased developmental time	Granett et al. 1975a
Apanteles melanoscelus/Porthetria dispar	Egg mass dipped in IGR	• Increased developmental time	Granett et al. 1975a

Encyrtus infelix/Saissetia coffeae	Parasitized host dipped in IGR	• Reduced emergence by 39%	Hamlen 1975
Coccinella septempunctata	Treatment of eggs	• 55% reduction of egg hatch	Kismali & Erkin 1984a
Coccinella septempunctata/ Acyrthosiphusm pisum	Topical treatment of larvae	• 3rd instar susceptible, 30% died after treatment, 10% after molting • Ingestion of treated prey had no effect	Kismali & Erkin 1984b
Gonia cinerascens/ Galleria mellonella	Topical treatment of last instar of parasitized host	• Number of puparia decreased as dose was increased	Mellini & Cesari 1982
Gonia cinerascens/ Galleria mellonella	Topical treatment of last instar of parasitized host	• Parasitoid emergence was lower in hosts treated 1–7 days before final larval instar	Mellini & Boninsegni 1983
	Topical treatment of host prepupae or pupae	• These treatments had no effect on parasitoid emergence	
Aphidius nigripes/ Macrosiphum euphorbiae	Spraying of plants infested with parasitized host	• 100% mortality of parasitoids in surviving hosts even at lowest concentration	McNeil 1975
Hyposoter exiguae/ Trichoplusia ni	Parasitized host fed treated diet	• Treated diet had no effect on development and emergence	Smilowitz et al. 1976
	Topical treatment of parasitized host	• Topical treatment caused delayed development and 30% mortality after 6.5 days at 2–20 ng/host	
Deraeocoris brevis, Chrysoperla spp./ *Psylla pyricola*	Field treatment	• No reduction in number of predators	Westigard 1979
Habrobracon juglandis	Topical treatment of parasitoid	• Decrease in egg production	Wissinger & Grosch 1975
Biosteres longicaudatus/ Anastrepha suspensa	Topical treatment of preparasitized host	• Increase in parasitoid emergence	Lawrence et al. 1978
Geocoris punctipes, Hippodamia convergens, Chrysoperla carnea	Topical treatment of immatures	• Minor deformation in *G.p.* • Arrested development in *H.c.* • 100% development in *C.c.*	Bull et al. 1973
METHOPRENE			
Nasonia vitripennis/ Sarcophaga bullata	Topical treatment of parasitoids	• Treatment of parasitoid larvae or pupae caused arrested development, incomplete coloration, and disrupted emergence	de Loof et al. 1979b
	Parasitized host puparium	• Parasitoid tolerated 80–5000× higher doses inside host puparium	

Table 12.1. *(Continued)*

Natural Enemy/Pest	Type of Test	Effects on Natural Enemy	Reference
Eucelatorias sp./ *Heliothis armigera*	Topical treatment of parasitized host	• Successful pupation, but no adult emergence	Divakar 1980
Amblyseius brazilli	Direct sprays to leaves bearing predators	• Direct spray: eggs more sensitive than adults, larvae killed after emergence	El-Banhawy 1980
	Ingestion of pollen soaked in 30 ppm IGR	• Pollen: little effect on immatures, decreased reproduction and increased mortality with increased exposure	
Nasonia vitripennis/ *Sarcophaga bullata*	Direct treatment of parasitoid prepupae	• Direct: 100% mortality in pupal stage	Fashing & Sagan 1979
	Incorporation of IGR in host diet	• Host diet: parasitoid emergence not affected, fecundity normal	
Encyrtus infelix/ *Saissetia coffeae*	Dipping of leaves with parasitized hosts	• 39% reduction in emergence	Hamlen 1975
Coccinella septempunctata	Treatment of eggs	• 36% decrease in egg hatch	Kismali & Erkin 1984a
Oenonogastra microrphopulae/ *Liriomyza trifolii*	Host placed on soil drenched with IGR	• Decrease in % parasitism	Oetting 1985
Aphytis holoxanthus/ *Chrysomphalus aonidum* *Coccophagus pulvinariae*/ *Saissetia oleae* *Tetrastichus ceroplastae*/ *Ceroplastes floridensis*	Spraying of plants	• No reduction in emergence at concentrations to 0.1% a.i.	Peleg & Gothilf 1980
Pales pavida/ *Galleria mellonella*	Topical treatment of parasitized host	• Decrease in % adult emergence with increased dose (1–10 ng/pupae emergence = 0–7%)	Riviere 1975
Deraeocoris brevis, *Chrysoperla* spp.	Field treatment	• No reduction in number	Westigard 1979
Habrobracon juglandis	Topical treatment of adult parasitoid	• Decrease in egg production	Wissinger & Grosch 1975
Nasonia vitripennis	Topical treatment of parasitoid	• No diapause induction when applied to maternal generation, eggs, or larvae	de Loof et al. 1979b

Species	Treatment	Effects	Reference
Metasyrphus corollae	Topical treatment of predator larvae	• Effect on adult male genitalia	Ruzicka et al. 1974
KINOPRENE			
Nasonia vitripennis/ *Sarcophaga bullata*	Topical treatment of parasitoid or parasitized host	• Direct treatment of parasitoid larvae or pupae caused arrested development, incomplete coloration, and disrupted eclosion	de Loof et al. 1979b
Encyrtus infelix/ *Saissetia coffeae*	Dipping leaves bearing parasitized hosts	• 57% decrease in emergence	Hamlen 1975
Coccinella septempunctata	Treated eggs	• No adverse effects	Kismali & Erkin 1984a
Opius dimidiatus/ *Liriomyza sativae*	Spraying of plants bearing parasitized hosts	• Decreased emergence of 61–76% at 1000–3000 ppm; rates needed to disrupt host were detrimental to parasitoid	Lema & Poe 1978
Chrysoperla carnea	Fed larvae artificial diet treated with IGR	• Prolonged 3rd instar, increased cocoon weight, caused mortality to all immatures • Adult survival decreased with increased rates	Maccolini 1985
Aphidius nigripes/ *Macrosiphum euphorbiae*	Spraying of plants bearing parasitized hosts	• High mortality of parasitoids in surviving hosts, 55–100% mortality 1–8 days after parasitism	McNeil 1975
Opius dimidiatus, *Opius* sp./*Liriomyza munda*	Spraying parasitized puparia Place parasitized puparia in sprayed soil	• Decreased emergence from 14% to 0% • Decreased emergence from 27% to 0%	Poe 1974a
Apanteles dignus, *A. scutellaris*/*Keiferia lycopersicella*	Spraying parasitized pupae Place parasitized puparia on sprayed sand	• No parasitoid emergence compared with 61% in check	Poe 1974b
Sympherobius barberi	Topical treatment of predator adults	• Low mortality (less than 33%)	Reeve & French 1978
TRIPRENE			
Encyrtus infelix/ *Saissetia coffeae*	Dipping leaves bearing parasitized hosts	• 32% decrease in emergence	Hamlen 1975
Chrysoperla carnea	Fed larvae artificial diet treated with IGR	• Prolonged 3rd instar, increased cocoon weight, caused mortality to all immatures	Maccolini 1985

Table 12.1. (*Continued*)

Natural Enemy/Pest	Type of Test	Effects on Natural Enemy	Reference
Gonia cinerascens/ *Galleria mellonella*	Topical treatment of host just before or after parasitism or treated host diet	• Effects depended on host development; increased concentration caused increased size, delayed development, and slight increase in mortality	Mellini & Gironi 1980
Aphidius nigripes/ *Macrosiphum euphorbiae*	Spraying of plants bearing parasitized hosts	• 100% parasitoid mortality in surviving hosts	McNeil 1975
Opius dimidiatus, *Opius* sp./ *Liriomyza munda*	Spraying parasitized puparia Place parasitized puparia in sprayed soil	• Decreased emergence from 14% to 0% • Decreased emergence from 27% to 0%	Poe 1974a
Apanteles dignus, *A. scutellaris*/ *Keiferia lycopersicella*	Spraying parasitized pupae Place parasitized pupae on sprayed sand	• No parasitoid emergence compared with 61% in check	Poe 1974b
RO = 20-3600 = ENT 70357			
Microplitis croceipes/ *Heliothis virescens*	Topical treatment of parasitoids and hosts	• 10 μg applied to host larvae caused decrease in parasitoid pupation • Effect of IGR depended on timing of application	Boone, unpubl.
Nabis roseipennis, *Chrysoperla carnea*	Topical treatment of last instar	• No preimaginal abnormalities in *C.c.*, 47.5% in *N.r.* • 10% adult emergence in *C.c.*, 40% in *N.r.* • No adult abnormalities in *N.r.*	Boone, unpubl.
Apanteles rubecula/ *Pieris rapae*	Topical treatment of parasitoid	• No effect on parasitism, emergence, or sex ratio	Wilkinson & Ignoffo 1973
Muscidifurax raptor/*Stomoxys calcitrans*	Topical treatment of parasitized host pupae	• No effect on oviposition, development, or reproduction in F_1 generation	Wright & Spates 1972
Habrobracon juglandis	Topical treatment of adult parasitoids	• Decrease in egg production	Wissinger & Grosch 1975
Geocoris punctipes, *Hippodamia convergens*, *Chrysoperla carnea*	Topical treatment of immatures	• Major deformations in *G.p.* • 0% development in *H.c.* • 90% development in *C.c.*	Bull et al. 1973

EPOFENONANE = RO–10–3108			
Chelonus sp./Trichoplusia ni	Field trials	• No effect	Buehler et al. 1985
Apanteles atar and Colpoclypeus florus/Adoxophyes orana and Pandemis heparana	Field trials	• Parasitoids less susceptible than hosts	Reede et al. 1984
Prospaltella perniciosis/ Quadraspidiotus perniciosus	Spraying of parasitized host on pumpkins	• 81% reduction in parasitism	Dorn et al. 1981
Natural enemies of Psylla pyri	Field trials	• No effect on natural enemies	Frischknecht & Jucker 1978
Sympherobius barberi	Topical treatment of adult predators	• Low mortality (less than 33%)	Reeve & French 1978
FENOXYCARB = RO–13–5223			
Apanteles atar and *Colpoclypeus florus/ Adoxophyes orana* and *Pandemis heparana*	Field trials	• Parasitoids less susceptible than hosts	Reede et al. 1984
Prospaltella perniciosis/ Quadraspidiotus perniciosus	Spraying of parasitized hosts on pumpkin	• No decrease in parasitism	Dorn et al. 1981
Metaphycus bartletti/ Saissetia oleae, Aphytis holoxanthus/Chrysomphalus aonidum	Spraying of parasitized hosts	• No effect on development of immature stages	Peleg 1983
Aphytis chrysomphali and *Comperiella bifasciata/ Aonidiella aurantii, Aphytis hispanicus, Encarsia inquirenda/Parlatoria pergandii*	Field tests	• Parasitoids less susceptible than hosts	Peleg 1983
	Other Miscellaneous Juvenile Hormone Analogues		
RO-20458			
Ooencyrtus kuwanaii/Lymantria dispar	Microapplication to parasitoid eggs Surface treatment of egg-masses	• 100 ppm, 70% mortality during late larval stage; 30% during pupal stage • Decreased developmental time	Granett et al. 1975a

Table 12.1. (*Continued*)

Natural Enemy/Pest	Type of Test	Effects on Natural Enemy	Reference
Muscidifurax raptor/ *Stomoxys calcitrans*	Topical treatment of parasitized pupae	• Oviposition, development, and reproduction unaffected	Wright & Spates 1972
Cardiochiles nigriceps, *Campoletis sonorensis*/ *Heliothis virescens*	Topical treatment of parasitized host	• Delayed parasitoid emergence • Skewed sex ratio in *Cardiochiles*, with fewer female adults produced	Vinson 1974
RO-20-3562			
Habrobracon juglandis	Topical treatment of parasitoid	• Decrease in egg production	Wissinger & Grosch 1975
ENT-70119, ENT-70357			
Nabis roseipennis, *Chrysoperla carnea*	Topical treatment of last instar	• No preimaginal abnormalities in *C.c.*; 72.5% for Ent 70119 and 47% for Ent 70357 in *N.r.* • 5–10% adult emergence in *C.c.*; 34–40% in *N.r.* • No adult abnormalities in *N.r.*	Boone unpubl.
AI 3-33972, AI 3-70119, AI-3-70348, AI 3-70350, AI 3-70356			
Trichogramma spp./ *Heliothis virescens*	Spraying cotton plants with parasitized host or during parasitization	• No effect on parasitism	Guerra et al. 1977
AY-22, 342			
Glypta fumiferanae, *Meteorus trachynotus*, *Phryxe pecosensis*/*Choristoneura fumiferana*	Topical treatment of field-collected, parasitized host larva	• Decrease in emergence from 85% in control to 57% for 0.2 μl/larva and to 31% for 0.5 μl/larva	Outram 1974
CGA 13353, CGA 34301, CGA 34302			
Apanteles glomeratus, *A. fulvipes*/*Nythobia fenestralis*, *Pteromalus puparum*, several hosts	Spraying plants with parasitized hosts	• No effect on emergence in *N.f.* or *A.g.* • 0–50% decrease in emergence in *P.p.* for 0.05% of all 3 IGRs	Scheurer et al. 1975

CGA 13353			
Apanteles sp., *Glypta fumiferanae, Actia interrupta, Phryxe pecosensis, Meteorus trachynotus, Macrocentrus iridescens/Choristoneura fumiferana*	Field trial	• No significant decrease in parasitism	Sechser & Varty 1978
JH 388 (EXPERIMENTAL OVICIDE)			
Phytoseiulus persimilis/ Tetranychus urticae	Exposure on treated leaf diseases in lab test	• Good selectivity, varying with the life stage tested	Caprioli et al. 1983
MPEP			
Aphytis holoxanthus/ Aonidiella aurantii, Ceroplastes floridensis	Sprayed plants bearing parasitized scales	• No adverse effects on larval or pupal development or on adult emergence or fecundity	Peleg 1988
		Juvenile Hormones	
JH1			
Microplitis croceipes/ Heliothis virescens	Topical treatment of parasitized host	• Decrease in pupation • Effect of IGR depended on timing of application	Boone unpubl.
Nasonia vitripennis/Sarcophaga bullata	Topical treatment of parasitoid or parasitized host puparium	• Direct: ovicidal effect; arrested development, incomplete development, incomplete coloration, and disrupted eclosion when applied to pupae • Indirect: inside puparium, parasitoid tolerated 80–5000 times higher dose	de Loof et al. 1979b
Nasonia vitripennis/Sarcophaga bullata	Topical treatment of parasitoid eggs or larvae	• No diapause induction when applied to maternal generation, eggs, or larvae	de Loof et al. 1979b
JH I, JH II, JH III			
Chelonus sp./*Trichoplusia ni*	Topical treatment of parasitized host	• Disrupted development, timing of eclosion by JH I and JH II, but not JH III	Buehler et al. 1985

Table 12.1. *(Continued)*

Natural Enemy/Pest	Type of Test	Effects on Natural Enemy	References
JH II, JH III			
Nasonia vitripennis	Topical treatment of parasitoid	• No diapause induction when applied to maternal generation, eggs, or larvae	de Loof et al. 1979a
		Anti-Juvenile Hormone Analogues	
PONASTERONE A			
Chelonus munakatae	Dipping of host	• Pupation accelerated	Sato 1968
BENZODIOXOLES (J-2710, J-3370, J-2581)			
Apanteles congregatus/ *Manduca sexta*	Topical treatment of parasitized host	• Frequent failure to emerge from host with J-2710; J-3370 and J-2581 less disruptive	Beckage et al. 1987
ETB			
Apanteles congregatus/ *Manduca sexta*	Topical treatment of parasitized host	• No effect on host or parasitoid	Beckage & Riddiford 1983
FMEV			
Apanteles congregatus/ *Manduca sexta*	Topical treatment of parasitized host	• Treatment of predetermined hosts decreased adult emergence, and effective parasitism • Treatment of terminal stage host decreased emergence by 33%	Beckage & Riddiford 1983
Microplitis croceipes/ *Heliothis virescens*	Injection of parasitized hosts	• Effects depended on host development • Early 5th instars = no effect; prewandering 5th instars = premature emergence; postwandering 5th instars = inhibited emergence	Webb & Dahlman 1986
PRECOCENE II AND ECDYSTERONE			
Nasonia vitripennis	Topical treatment of parasitoid	• Precocene II: no diapause induction if applied to maternal generation, eggs, or larvae • Ecdysterone: termination of diapause	de Loof et al. 1979a

DIFLUBENZURON			
Chrysopa oculata, Acholla multispinosa, Macrocentrus ancylivorus/Grapholitha molesta	Topical treatment of predators or contact with treated prey Treatment of parasitized hosts	• Inhibited 1st instar molt, mortality • No effect on 1st instar *A.m.* • Decreased emergence in *M.a.*	Broadbent & Pree 1984
Sturmia scutellata, Carcelia lucorum, Exorista larvarum, E. sassiata, Drino inconspicua/Lymantria dispar	Feeding parasitized hosts on treated leaves	• *E.l.* developed normally • *S.s.* and *C.l.* developed malformed pupae and no adults eclosed • *D.i.* and *E.s.* not found	Demolin 1978
Amblyseius brazilli	Sprayed leaves bearing predators Fed pollen soaked in 30 ppm IGR	• Spray: females more sensitive than eggs • Pollen: little effect on immatures, decreased reproduction, and increased mortality with increased exposure	El-Banhawy 1980
Aleochara bilineata/ Delia radicum	Residue for eggs, first instars and adults Treatment of host for parasitic phase	• No effects • Suppression when dissolved in DMSO and applied to exoskeleton of host puparium	Gordon & Cornect 1986
Apanteles melanoscelus/ Lymantria dispar	Spraying infested apple leaves in the field	• Treatment of host instars 1–2 decreased parasitoid emergence; treatment of 2–3 and 3–4 host instars did not decrease emergence	Granett et al. 1976
Apanteles melanoscelus/ Lymantria dispar	Treated diet fed to parasitized host	• Low rates: parasitoid died during larval–pupal molt; high rates: died as larvae within hosts • Parasitoids in stadia 2–3 were less affected than earlier stages	Granett & Weseloh 1975
Apanteles melanoscelus/ Lymantria dispar	Treated diet fed to parasitized hosts	• No effect on reproduction and late larval stages, 2–3 instars were affected	Granett & Weseloh 1975
Hippodamia convergens/ Trichoplusia ni	Residue exposure in petri dishes Fed treated foliage to *T. ni* and fed to *H. convergens*	• No adult mortality, 2nd instars more susceptible than *T. ni* larvae • No adult mortality, but higher mortality in 2nd instars	

Table 12.1. (*Continued*)

Natural Enemy/Pest	Type of Test	Effects on Natural Enemy	Reference
Apanteles melanoscelus, Exorista larvarum, Parasetigena silvestris/*Lymantria dispar*	Aerial spray of forest first gen. *A.m.* in host, other spp. attacking host	• 2 weeks after treatment, 80% mortality in *A. m.* and 100% in tachinids	Madrid & Stewart 1981
Cryptolaemus montrouzieri	Prey sprayed on potato leaves and fed to predator	• No adults developed from young larvae	Mazzone & Viggiani 1980
Pediobius foveolatus/*Epilachna varivestris*	Fed dipped leaves to host after parasitization	• No completion of development even at lowest dosage (3 ppm)	McWhorter & Shepard 1977
Anthocoris nemorum, Chrysoperla carnea, Episyrphus balteatus, Trichogramma cacoeciae	Residual contact	• No effect on eggs and late instars of *A.n.* and *E.b.* and 7-day-old larvae of *T.c.* inside *L. pomonella* • Detrimental to late instars of all these beneficials as well as eggs of *C.c.*	Niemczyk et al. 1985
Stethorus punctillum/*Panonychus ulmi*	Field trial	• Selective for predator	Pasqualini & Malavolta 1985
Trioxys pallidus/*Chromaphis juglandicola*	Contact, topical	• No effect on adults and larvae, no effect on reproduction	Purcell & Granett 1985
Aphelinus mali/*Eriosoma lanigerum*	Fed to parasitoids before oviposition Spraying parasitized hosts	• No reduction in number of offspring with 0.01% a.i.	Ravensberg 1981
Sympherobius barberi	Topical treatment of adults	• Low mortality (<33%)	Reeve & French 1978
Parasierola nephantidis/*Nephantis serinopa*	Parasitoid given hosts sprayed in field	• Indirect: increased dose = decreased parasitism from 100 to 0%, oviposition from 60 to 0%, number of adults produced/leaf from 8 to 0%	Sundaramurthy 1980
	Dipping parasitoids	• Direct: 40% larvae failed to spin cocoon, 25% failed to pupate, forming pupal–adult intermediates, 20% of adults survived	
Doryphorophaga doryphorae/*Leptinotarsa decemlineata*	Fed treated plants to parasitized hosts	• No effect on survival and emergence from late instar hosts	Tamaki et al. 1984

Ooencyrtus pityocampae, *Tetrastichus servadeii*, *Anastatus bifasciatus*, *Trichogramma embryophagum*/*Thaumatopoea pityocampa*	Field trials	• No effect on survival from third instar hosts at low dose, mortality at higher dose • In later instar hosts, normal fertility and ability to parasitize • No effects on hibernating hosts or parasitoids	Tsankov & Mirchev 1983
Mirids, chryspopids, coccinellids, *Trichnites insidiosus*	Field trials	• No decrease in number of predators and parasitoids at 0.14 kg a.i./ha	Westigard 1979
Pediobius foveolatus/*Epilachna varivestris*	Topical treatment of adult parasitoids or parasitized host	• No effect when adults or host treated 10 days after oviposition • Host treated just before oviposition and 4 days after oviposition had decreased parasitoid emergence from 66% to 0% and from 16% to 2%	Zungoli et al. 1983
	Field trial	• In field trial, no effect on parasitism	
CHLORFLUAZURON			
Coccinellidae, Phytoseiidae, *Orius* spp., *Nabis* spp./Lepidoptera	Field trial	• No effects	Collins et al. 1984
Trioxys pallidus/*Chromaphis juglandicola*	Contact, topical, ingestion	• No effects on adults and larvae, no effect on reproduction	Purcell & Granett 1985
BUPROFEZIN			
Phytoseiidae, *Encarsia formosa*, *Cales noacki*, *Lycosa pseudoannulata*, *Paracentrobia andoi*, *Anagrus* sp.	Field trial	• No effects	Collins et al. 1984

Table 12.1. *(Continued)*

Natural Enemy/Pest	Type of Test	Effects on Natural Enemy	Reference
Lycosa pseudoannulata, Cyrtorhinus lividipennis, Microvelia atrolineata/ Nilaparvata lugens, Nephotettix virescens, Sogatella furiafera	Direct treatment of adults or nymphs	• No effects	Heinrichs et al. 1984
PENFLURON			
Hippodamia convergens/ Trichoplusia ni	Residue in petri plate	• No adult mortality, 2nd instars more sensitive than *T. ni*	Jones et al. 1983
	Also *T. ni* fed treated leaves and given to predator	• No adult mortality, 2nd instars more sensitive than in residue test alone	
TEFLUBENZURON			
Encarsia formosa	Residue test adults	• Adults and pupae unaffected	Oomen & Wiegers 1984
	Sprayed prepupae, parasitized larvae	• Young larvae killed	
ALSYSTIN (TRIFLURON)			
Trioxys pallidus/Chromaphis juglandicola	Contact, topical ingestion	• No effects on adults or larvae, no effects on reproduction	Purcell & Granett 1985

Cesari 1982). However, natural enemies exhibit greater variability in their dose-dependent responses to IGRs than to conventional pesticides.

4. *Species and compound tested.* Different species of insects are likely to be affected differently by IGRs based on intrinsic differences in absorption, degradation, and excretion of certain xenobiotics, resulting in tolerant and sensitive species. Effects also depend on the type of IGR tested.

Each of these factors as well as the biochemistry and general mode of action will be discussed for the JHAs (including exogenous JHs), AJHAs, and benzoylphenyl ureas.

12.2.1. Juvenile Hormone Analogs

12.2.1.1. Biochemistry and Mode of Action. In holometabolous insects, juvenile hormone (JH) secreted by the corpora allata occurs at high levels in the hemolymph during the larval development. The level of JH present in the hemolymph, however, declines abruptly in the last larval instar prior to metamorphosis. Release of the molting hormone alpha-ecdysone by the prothoracic glands when the JH titer is low results in pupal transformation of the larva; in the absence of JH, transformation of pupa to adult occurs. The secretion of both hormones is controlled by neuropeptides produced in the brain. Allatotropin and allatostatin control the secretion of JH and prothoracicotropic hormone controls the secretion of alpha-ecdysone. Exposing arthropods to compounds that interfere with these complex endocrinological interactions can be highly disruptive to metamorphosis and development.

JHAs mimic the function of naturally occurring JHs in many insects (Staal 1972, 1975, Menn and Henrick 1981, Henrick 1982). Exposure to exogenous JH or JHAs during periods when endogenous JH levels are low can be highly disruptive for the larva, causing morphogenetic effects, such as the formation of larval–pupal intermediates or supernumerary larval instars. JHAs can block embryonic development in eggs and interfere with diapause and reproduction in adults.

The activity of JHs and JHAs depends on interactions with several types of proteins in the hemolymph or at the target site: 1) binding or carrier proteins, 2) degradative enzymes, and 3) receptor proteins at the target site (Fig. 12.2). These biochemical interactions will be discussed in relation to selectivity (Section 12.3). Thus, the target site activity may vary between arthropod pests and natural enemies based on intrinsic differences in their hemolymph and receptor proteins and on the structure of the JHA used.

12.2.1.2. Test Procedures. The test procedures used to administer JHAs can dramatically affect a natural enemy's response to these agents (Table 12.1), perhaps to a greater degree than with conventional pesticides (Chapter 5). The observed variation in the side effect of JHAs using different test procedures probably reflects our lack of understanding of how these pesticides influence the physiology and development of natural enemies (Beckage 1985). It may indicate the inappropriate use of methods that were originally developed to evaluate the side effects of conventional pesticides. Few methods have been developed specifically to assay the effects of IGRs on natural enemies (Chapter 5).

In generalizing about test procedures, an indirect treatment of parasitoids (via their hosts) with JHAs appears to be less detrimental than a direct treatment (Table 12.1). Among indirect treatments, host ingestion of JHAs is less harmful to developing parasitoids than when the host is exposed to topical application. For example, direct

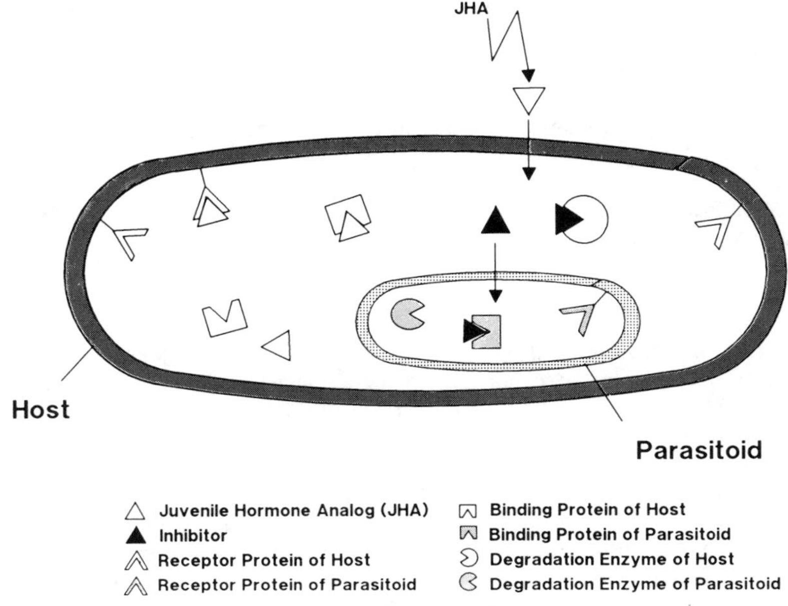

Figure 12.2. Interactions of juvenile hormone analogues and inhibitors with hemolymph and receptor proteins of arthropod hosts and endoparasitoids.

application of methoprene to prepupae and pupae of *Nasonia vitripennis* resulted in 100% mortality during the pupal stage, whereas indirect exposure via the diet of its dipteran host *Sarcophaga bullata* had no effect (Fashing and Sagan 1979). Similar results were reported by de Loof et al. (1979b) for methoprene, kinoprene, and JH I applied to this same hymenopteran parasitoid–dipteran host system. Topical applications to larvae or pupae of *N. vitripennis* caused disruption of normal eclosion and incomplete or abnormal coloration of the adult cuticle, and it arrested development in the last larval instar, pupal, or pharate stage. Parasitoids developing inside puparia of *S. bullata* tolerated doses 80- to 5000-fold higher than those tolerated from direct treatment, although the effect depended upon the compound used and the time between parasitization and treatment of the host. Artificial diet laced with hydroprene and fed to parasitized *Trichoplusia ni* did not affect the development and emergence of *Hyposoter exiguae* (Smilowitz et al. 1976). However, topical application to the parasitized host delayed development and emergence of the wasps and increased their mortality.

While laboratory bioassays with JHAs have demonstrated detrimental effects on the development of natural enemies, a less severe impact has been reported from the field. Field populations of the natural enemy may contain life stages that are insensitive to JHAs, or the immigration of adult natural enemies may confound field side-effects tests. Few direct comparisons of laboratory and field exposure to JHAs have been reported. Most evidence for the lesser impacts of JHAs on natural enemies in the field comes from assessments independent of those conducted in the laboratory.

Field application of epofenonane for the control of the pear psylla *Psylla pyri* did not affect populations of its natural enemies, which included many highly mobile adult forms

(Frischknecht and Jucker 1978; Table 12.1). Westigard (1974) observed that hydroprene and methoprene applied for the control of the pear psylla *Psylla pyricola* reduced levels of psylla nymphs without affecting adult *Deraeocoris brevis* and lacewings, *Chrysoperla* spp. The use of epofenonane and fenoxycarb in Dutch apple orchards for the control of *Adoxophyes orana* and *Pandemis heparana* showed that the parasitoids *Apanteles atar* and *Colpoclypeus florus* were less susceptible to these compounds than their hosts (Reede et al. 1984). No adverse effects on parasitoid species were observed in field trials with methoprene or fenoxycarb in citrus (Peleg and Gothilf 1980, Peleg 1983). Spraying cotton plants infested by the tobacco budworm *Heliothis virescens* with different experimental JHAs did not affect parasitization by *Trichogramma* sp. (Guerra et al. 1977). The experimental JHA, MPEP, had no adverse effects on the larval or pupal development or on the successful adult emergence and female fecundity of *Aphytis holoxanthus*, a parasitoid of the California red scale *Aonidiella aurantii* or the Florida wax scale *Ceroplastes floridensis* when plant material infested with parasitized scale was treated (Peleg 1988). No reduction in the emergence of various parasitoid species was observed when different experimental JHAs were sprayed against lepidopterous pests (Scheurer et al. 1975).

These studies indicate that JHAs often have only minor short-term effects on beneficial natural enemies. However, longer term influences on these species may be masked by factors such as the variation in life stage susceptibility and the impact of immigrating forms. Accurately assessing the effects of JHAs in the field is confounded by the difficulty of monitoring the survival of natural enemies in their most sensitive immature life stages.

Several studies using different methods of evaluation have been conducted in an effort to measure the relative importance of factors influencing the impact of JHAs, but with limited success. Staubli (1986) conducted an extensive investigation of the effect of fenoxycarb on beneficial organisms of pear pests under laboratory conditions, on individual pear trees, and under practical conditions in a pear orchard. In general, fenoxycarb was selective to predaceous anthocorids, chrysopids, hemerobiids, mites, and parasitic hymenopterans. He concluded that none of the three methods alone was sufficient to show the effects of an IGR on beneficial arthropods in the field, but in combination, a more complete understanding of IGR impact was obtained.

In a few cases, similar results from one application method of JHAs to another have been observed (Table 12.1). These examples, however, are the result of specific comparative tests of different application methods, rather than being of any sort of broad generalization on the impacts of JHAs. Definitive studies of comparative application methods have yet to be conducted on this relatively recent group of compounds.

Spraying puparia of the dipterous leafminer *Liriomyza munda* parasitized by *Opius dimidiatus* with kinoprene or triprene, or placing parasitized pupae in treated soil similarly showed the same inhibitory effect on adult parasitoid emergence from the soil (Poe 1974a). The same test protocols yielded similar results with *Cotesia dignus* and *A. scutellaris*, which parasitize *Keiferia lycopersicella* (Poe 1974b). Both topical treatment of parasitized *Galleria mellonella* and incorporation of triprene into the host diet delayed the development of *Gonia cinerascens*, increased the size of parasitoid larvae, and increased parasitoid mortality (Mellini and Cesari 1982). Such consistencies in results among test protocols are the exception rather than the rule. As pointed out by de Loof et al. (1979b), one should be very careful in extrapolating conclusions from laboratory experiments to the myriad of complex conditions that exist in the field. Furthermore, extrapolating to other host–parasitoid systems or other IGRs is inappropriate.

12.2.1.3. Life Staged Treated and Timing. The direct or, more commonly, the indirect (i.e., through its host or prey) exposure of a particular life stage of natural enemy can significantly influence the side effects of JHAs. Life-stage-associated susceptibility to JHAs has been most exclusively studied for parasitoids and their hosts, although some studies are available for predators.

Treatment of eggs of the gypsy moth *L. dispar* parasitized by *Ooencyrtus kuwanai* with hydroprene or RO-20458 during the parasitoid's late larval and meconium-cast stages caused 70% mortality of the wasps (Granett et al. 1975a). Only 30% mortality occurred when parasitoid pupae were treated and 10% when host eggs containing young larval instars or eggs were treated. Therefore, the most susceptible period for *O. kuwanai* was just prior to the larval-pupal molt.

JH I, methoprene, and kinoprene inhibited eclosion most severely when applied to hosts containing larvae and pupae of the hymenopteran parasitoid *Nasonia vitripennis*, but kinoprene was able to affect eclosion when applied as late as the black pharate adult stage (de Loof et al. 1979b). The experimental ovicide JH 388, applied to control the spider mite *Tetranychus urticae*, was selective to the phytoseiid predator *Phytoseiulus persimilis*, depending on the life stage treated (Caprioli et al. 1983).

Hydroprene substantially reduced adult emergence of *Microplitis croceipes* when the parasitized host *Heliothis virescens* was treated 7 days after parasitism, compared with treatment 3 days after parasitism. In contrast, JH I, RO-20-3600, and ethyl RO-20-3600 did not affect adult emergence, irrespective of whether the host was treated early or late after parasitism had occurred (Boone unpubl.). Emergence of *Pseudogonia rufifrons* (= *Gonia cinerascens*) was reduced when *Galleria mellonella* was treated with hydroprene during days 1–7 of the final instar (Mellini and Boninsegni 1983). No difference in parasitoid emergence was observed when pupae, prepupae, or other stages of the host were treated. When triprene was administered to *G. mellonella* parasitized by *G. cinerascens* during a period when the host was sensitive (i.e., first larval instar) and the parasitoid was insensitive (i.e., first larval instar), a slight increase in mortality of the parasitoid was observed that was proportional to that of its host. However, an increase in the weight of the parasitoid was also noted. This indicated that triprene was probably influencing the parasitoid indirectly by modifying the host (Mellini and Gironi 1980).

In summary, these life-stage sensitivity studies reflect the considerable variations in side effects that occur between parasitoid species and among compounds. Similar effects on predaceous natural enemies have also been observed, although considerably less data are available for this group (Table 12.1).

12.2.1.4. Dose—Response Relationship. Arthropod natural enemies usually respond to JHAs in a dose-dependent manner. However, some species responses may be nonlinear and fail to correlate with the levels of toxicant administered to the natural enemy or its host (Table 12.1). These responses are undoubtedly attributable to the complex interrelationships between arthropod parasitoids and their hosts.

Topical treatment of parasitized *Galleria mellonella* with methoprene caused reduced emergence of the parasitoid *Pales pavida*, which became more pronounced as the dose was increased (Riviere 1975). Similar results were obtained with hydroprene for *Gonia cinerascens* (Mellini and Cesari 1982). When hydroprene was topically applied to alfalfa weevil larvae *Hypera postica* parasitized by *Microctonus aethiopoides*, morphologically deformed adult parasitoids were obtained at rates of 1–5 μg per host (Ascerno et al. 1980). At 10 μg per host, pupation of parasitoids was inhibited and 100% pupal mortality occurred. Topical treatment of the field-collected spruce budworm with the experimental

JHA AY-22,342 resulted in reduced emergence of different parasitoid species, which declined for each as the dose was increased (Outram 1974).

Although high doses of JHAs have been observed to disrupt natural enemy development, low rates of these materials may be less deleterious. Methoprene at concentrations of up to 0.1% a.i. had no adverse effect on the various developmental stages of the hymenopterous parasitoids *Aphytis holoxanthus*, *Coccophagus pulvinariae*, and *Tetrastichus ceroplastae*. Its effectiveness in controlling soft scales made this JHA suitable for integrated control programs in citrus (Peleg and Gothilf 1980). On the other hand, some natural enemies appear to be intrinsically intolerant to various compounds regardless of dose, rendering them unsuitable as selective control agents. Mortality of *Aphidius nigriceps* in surviving hosts was high even at the lowest concentration of hydroprene, triprene, or kinoprene tested (McWhorter and Shepard 1977, McNeil 1975). Kinoprene significantly reduced adult emergence of *Opius dimidiatus*, a parasitoid of *Liriomyza sativae*, at all rates tested. Only at the highest rate (3000 ppm) did a reduction in emergence of *L. sativae* occur, indicating that dosages of kinoprene sufficient to disrupt host development were also detrimental to the natural enemy (Lema and Poe 1978).

12.2.1.5. Species and Compound Tested. Natural enemies differ in intrinsic tolerance to different JHAs, hence compounds may be selective to one species but harmful to another. One species may also respond differently to several closely related JHAs. For example, eggs of the coccinellid *Coccinella septempunctata* were sensitive to hydroprene, less sensitive to methoprene, and unaffected by kinoprene (Kismali and Erkin 1984a). The timing of eclosion was disrupted in *Chelonus* sp. when treated with epofenonane, JH I, and JH II, but not when JH III was applied (Buehler et al. 1985).

In a comparison of predator responses, a topical application of hydroprene to final instars of *Nabis roseipennis* and the green lacewing *Chrysoperla carnea* caused no preimaginal abnormalities in *C. carnea* and minimal effects on *N. roseipennis*. However, while adult emergence was unaffected in *N. roseipennis*, only 6% emergence occurred in *C. carnea* (Boone unpubl.). Treatment of the same species with RO-20-3600 caused no preimaginal abnormalities in *C. carnea*; however, 48% abnormalities occurred in *N. roseipennis*. Adult emergence in *C. carnea* and *N. roseipennis* was 10% and 40%, respectively. Topical treatment of the larvae of the beneficial predators *C. carnea, Geocoris punctipes*, and *Hippodamia convergens* with hydroprene caused minor deformation in *G. punctipes*, complete mortality in *H. convergens*, and no deleterious effects in *C. carnea* (Bull et al. 1973). RO-20-3600 had the same effect, except that adult *G. punctipes* were more severely deformed. RO-20-3600 did not affect the parasitoids *Apanteles rubecula* and *Muscidifurax raptor* when applied topically to their hosts *Pieris rapae* and *Stomoxys calcitrans* (Wilkinson and Ignoffo 1973, Wright and Spates 1972). No reduction in parasitism of *Quadraspidiotus perniciosi* by *Prospaltella perniciosus* was observed for fenoxycarb, whereas 81% reduction in parasitism was observed for epofenonane (Dorn et al. 1981).

Topical treatment of parasitized *H. virescens* with RO-20458 caused a delay in adult emergence of the braconids *Cardiochiles nigriceps* and *Campoletis sonorensis*. *C. nigriceps* adults emerged from only 33% of the hosts and the male–female ratio was reduced, whereas no negative effects were observed in *C. sonorensis* (Vinson 1974). The experimental compounds CGA 13353, 34301, and 34302 had no effect on adult emergence of the parasitoids, *Nythobia fenestralis* and *Apanteles glomeratus* (Scheurer et al. 1975), but caused a 50% reduction in the emergence of *Pteromalus puparum* after treatment of hosts with CGA 34301. Hydroprene caused high mortality in *Ooencyrtus kuwanai* when parasitized eggs of the gypsy moth *L. dispar* were treated, whereas *Apanteles melanoscelus* exhibited only increased developmental time (Granett and Weseloh 1975).

In general, however, JHAs often cause a reduction in the emergence of parasitoids when parasitized hosts are treated. Thus, topical treatments of parasitized *Hypera postica, Heliothis virescens*, or *Galleria mellonella* with hydroprene reduced the emergence of *Microctonus aethiopoides, Microplitis croceipes*, and *Gonia cinerascens*, respectively (Ascerno et al. 1980, Boone unpubl., Mellini and Cesari 1982). Similar general observations were made for other JHAs (Table 12.1).

12.2.2. Anti-Juvenile Hormone Analogs and Ecdysteroids

Several compounds with anti-juvenile-hormone activity have been reported and the most well studied are the precocenes (Retnakaran et al. 1985). Other compounds with anti-juvenile-hormone activity are fluoromevalonolactone (Fmev), benzodioxoles, and ethyl-4-[2-tert-(butylcarbonyloxy)-butoxy] benzoate (ETB). Compounds such as the precocenes are of value mainly as experimental growth regulators in the laboratory rather than as control agents of commercial value, because of their lack of activity against most Holometabola and the high dosage required to affect other insect groups.

AJHAs overcome or antagonize the action of JH, inducing precocious development of treated larvae or nymphs into miniature pupae or sterile adults. The mode of action of AJHAs such as the precocenes is based on the alkylation of proteins in the corpora allata by active epoxides of these compounds (Bowers 1981), thus resulting in cellular death and reduction of JH synthesis. The precocenes are active primarily in hemimetabolous insects, but a few holometabolous species are sensitive as well (Retnakaran et al. 1985).

A few studies have been conducted which demonstrate the detrimental or nondetrimental effects of AJHAs on parasitoids. When ETB was topically applied to third instar *Manduca sexta* parasitized by *Apanteles congregatus*, the parasitoid's development was unaffected (Beckage and Riddiford 1983). However, the host's development was also unaffected, making ETB a poor choice for pest control. Fmev decreased emergence of the parasitoid from its host. Among the benzodioxoles, J-2710 frequently suppressed emergence of *A. congregatus* from *M. sexta*, whereas J-3370 and J-2581 were less disruptive (Beckage et al. 1987).

Ecdysteroids also possess anti-juvenile-hormone activity and cause smilar side effects on natural enemies as those described for JHAs. For example, the effects of ecdysteroids injected into parasitized *Heliothis virescens* depended on the developmental phase of the host (Webb and Dahlman 1986). No effects on *Microplitis croceipes* were observed when early fifth instar *H. virescens* were treated. Premature parasitoid emergence occurred when prewandering fifth instars were treated, and emergence was inhibited when postwandering fifth instars were treated. The topical application of ecdysterone to larvae of *Nasonia vitripennis* caused diapause termination (de Loof et al. 1979a). Sato (1968) noted that pupation of the parasitoid *Chelonus munakatae* within overwintering larvae of *Chilo suppressalis* was accelerated by treatment with the phytoecdysone ponasterone A.

12.2.3. Chitin Synthesis Inhibitors

This group of biorational pesticides includes several compounds that have been widely tested and employed in the field. Their impact and selectivity have been documented on both predators and parasitoids, which are the components of large pest–natural enemy complexes on a number of commercial crops. Diflubenzuron is one of the most broadly selective insecticides discovered to date (Appendix and Section 9.5.3.6).

12.2.3.1. ***Biochemistry and Mode of Action.*** Benzoylphenyl ureas inhibit chitin synthesis by interfering with the synthesis of the enzyme chitin synthetase in the final step of chitin formation (Leighton et al. 1981). General effects of the benzoylphenyl ureas include disruption of ecdysis during molting, failure to feed due to malformed mandibles or labrum or due to blockage of the gut, delayed mortality due to a failure of adults to eclose, and ovicidal effects on eggs and after treatment of adults (Retnakaran et al. 1985). Also, changes in mating behavior have been observed, although these effects have been less precisely measured compared to those listed.

12.2.3.2. ***Test Procedures.*** As with JHAs, the side effects of chitin inhibitors have been evaluated using a diversity of test methods. However, the comparison between field and laboratory evaluations is probably more extensive for this group than for the JHAs. In most cases, little impact on natural enemies has been observed when diflubenzuron was directly or indirectly applied in laboratory tests (Table 12.1).

Ravensberg (1981) observed no reduction in the number of offspring of *Aphelinus mali*, a parasitoid of the woolly apple aphid *Eriosoma lanigerum*, when diflubenzuron was fed to female parasitoids prior to oviposition or when parasitized host nymphs were sprayed. No detriment to the parasitoid was observed following topical treatment with diflubenzuron of either adult *Pediobius foveolatus* or its host, the Mexican bean beetle *Epilachna varivestris*, 10 days after oviposition (Zungoli et al. 1983). Field trials gave comparable results and showed no difference in parasitism. No effect on adults or larvae were observed when *Trioxys pallidus*, a parasitoid of the walnut aphid *Chromaphis juglandicola*, was exposed to diflubenzuron using contact, topical, or ingestion tests (Purcell and Granett 1985). Similar results were obtained for teflubenzuron and *Encarsia formosa*, a parasitoid of the white fly *Trialeurodes vaporariorum* (Oomen and Wiegers 1984). Diflubenzuron had no detrimental effects on the staphylinid *Aleochara bilineata* when eggs, first instars, or adults were exposed to residues or when parasitized hosts were treated (Gordon and Cornect 1986).

In a few cases, both direct and indirect exposure to diflubenzuron have had a negative impact on the development of natural enemies. When larvae of *Parasierola nephantidis*, a parasitoid of the coconut blackheaded caterpillar *Nephantis serinopa*, were dipped in a solution of diflubenzuron, successful cocoon formation, pupation, and adult survival were reduced and pupal–adult intermediates were formed (Sundaramurthy 1980). Providing adult female parasitoids with hosts that were previously sprayed in the field with diflubenzuron decreased the proportion of paralyzed hosts, oviposition rates, and fecundity depending on the dose used.

The effects of chitin synthesis inhibitors appears to depend both on the test procedure and on the life stage treated (see Section 12.2.3.3). Thus, when adult *E. formosa* were exposed to residues of teflubenzuron or when pupae of this parasitoid were sprayed, no harmful effect was evident. Spraying parasitized host whiteflies, however, killed young parasitoid larvae. Feeding diflubenzuron-treated larvae of the oriental fruit moth *Grapholitha molesta* to *Chrysoperla oculata* and *Acholla multispinosa* did not affect these predators (Broadbent and Pree 1984). However, topical treatment or exposure to treated leaves increased mortality and inhibited molting of first instars of *C. oculata*, whereas *A. multispinosa* was not affected. Treatment by either feeding or topical application of oriental fruit moth larvae parasitized by *Macrocentus ancylivorus* reduced emergence of the parasitoid (Broadbent and Pree 1984). Adult females of the predatory mite *Amblyseius brazilli* were more tolerant than eggs and larvae when placed on leaf disks with residues of diflubenzuron (El-Banhawy 1980). IGR-treated pollen fed to *A. brazilli* had little effect on larvae, but adults which fed on treated pollen were less fecund. Adults of the coccinellid *Hippodamia convergens* experienced no mortality when exposed to residues of diflubenzuron or penfluron or when its prey *Trichoplusia ni* was treated with these materials (Jones et al. 1983). Second instar *H. convergens*, however, were more sensitive to treated prey than to residues of these materials.

Although laboratory assays have shown that chitin synthesis inhibitors can have negative effects on natural enemies, most field tests reported no such detrimental effects. No significant decrease was noted in predator species when pear orchards were sprayed with diflubenzuron for control of the codling moth *Laspeyresia pomonella* or the pear psylla *Psylla pyricola* (Westigard 1979). Diflubenzuron was also selective to *Stethorus punctillum*, a predator of the European red mite *Panonychus ulmi* (Pasqualini and Malavolta 1985). In rice, buprofezin did not affect beneficial arthropods such as *Lycosa pseudoannulata* (Collins et al. 1984, Heinrichs et al. 1984), and no negative effects on natural enemies were observed for this compound when used in citrus (Collins et al. 1984). The same was the case for chlorfluazuron used in cotton (Collins et al. 1984). No effects were observed on parasitoids of the pine processionary *Thaumatopoea pityocampa* following diflubenzuron sprays (Tsankov and Mirchev 1983). However, aerial sprays of diflubenzuron against *Lymantria dispar* caused high mortality of tachinid parasitoids and *Apanteles melanoscelus* (Madrid and Stewart 1981).

12.2.3.3. Life Stage Treated and Timing. Diflubenzuron was shown to be detrimental to early instars of the predators *Anthocoris nemorum*, *Chrysoperla carnea*, and *Episyrphus balteatus* and the parasitoid *Trichogramma cacoeciae* (Niemczyk et al. 1985). Late instars of these insects were not affected. Survival and emergence of Doryphorophaga doryphorae from fourth instar *Leptinotarsa decemlineata* were normal when parasitized hosts were fed on treated plants, regardless of the dose tested (Tamaki et al. 1984). However, no parasitoids emerged from third instar hosts when treated with high doses of diflubenzuron. Adults of *Hippodamia convergens* were not sensitive to diflubenzuron or penfluron, whereas the mortality of second instar larvae depended on the type of test (Jones et al. 1983). The parasitoid *Apanteles melanoscelus* was less affected by diflubenzuron when its host, the gypsy moth *Lymantria dispar*, was treated during the parasitoid's second or third larval instar than when treatment occurred during the parasitoid's egg or first larval stage (Granett et al. 1975b). Similar results were obtained in field trials when apple trees infested by gypsy moths were sprayed with diflubenzuron (Granett et al. 1976). Treatment of early larval instars of *L. dispar* reduced emergence of *A. melanoscelus*, whereas treatment of late larval instars of the host did not affect parasitoid emergence. Thus, delaying treatments until parasitoids are nearly ready to emerge from their hosts may reduce deleterious side effects.

Diflubenzuron had no significant effect on parasitoid oviposition, fecundity, emergence, or the rate of development when topically applied to either the adult parasitoid *Pediobius foveolatus* or the parasitized Mexican bean beetle larvae *Epilachna varivestris* just prior to parasitization or within 4 days of the onset of parasitism (Zungoli et al. 1983). Again, detrimental effects on the parasitoid can be minimized by timing applications to coincide with later phases of parasitoid development.

12.2.3.4. Dose–response Relationship. As noted earlier, dose–response relationships for insect growth regulating compounds can sometimes show considerable variability. A linear dose–response relationship was observed for *Parasierola nephantidis* when the parasitoid was offered *Nephantis serinopa* sprayed at different rates of diflubenzuron in the field (Sundaramurthy 1980). Increasing the dose caused reduced oviposition, host paralysis, and a decrease in the number of adults emerging from the host. When parasitized *Lymantria dispar* were fed on diet treated with low rates of diflubenzuron, the parasitoid *Apanteles melanoscelus* died during its pupal–larval molt (Granett and Weseloh 1975). Parasitoids died as larvae within the host when *L. dispar* was treated with high rates of diflubenzuron. In some cases, 100% mortality of natural enemies was observed even at the lowest dosages of the IGR, indicating low intrinsic tolerance to the compounds tested. The parasitoid *P. foveolatus* was unable to complete its development even at the lowest dose of diflubenzuron tested (3 ppm) (McWhorter and Shepard 1977).

For optimum selectivity in the field, IGRs selected for the target species should have minimal detrimental effects on beneficials. Low rates of IGRs can be effective in obtaining selectivity. Diflubenzuron used at 0.14 kg a.i./ha for control of the codling moth *Laspeyresia pomonella* caused no reduction in natural enemies of pear psylla (Westigard 1979).

12.2.3.5. Species and Compound Tested. Benzoylphenyl ureas have been shown to be detrimental to some natural enemy species and not others, using the same test procedure. Such tests indicate that natural enemies differ in intrinsic tolerance to these compounds, both as a class and among individual compounds (Table 12.1). Thus, while the tachinid endoparasitoid *Exorista larvarum* developed normally when its host *L. dispar* was fed diflubenzuron-treated leaves, *Blepharipa scutellata* and *Carcelia lucorum* formed malformed pupae and adults failed to eclose (Demolin 1978). Diflubenzuron was shown to be deterimental to eggs of *Chrysoperla carnea*, but was not toxic to eggs and late instars of *Anthocoris nemorum* and *Episyrphus balteatus* (Niemczyk et al. 1985). Mortality and inhibition of molting occurred when first instar *Chrysoperla oculata* were treated with diflubenzuron, whereas *Acholla multispinosa* was not affected (Broadbent and Pree 1984). No adult progeny were obtained from young larvae of the predator *Cryptolaemus montrouzieri* treated with diflubenzuron, whereas the impact on *Leptomastix dactylopii* was minimal (Mazzone and Viggiani 1980).

12.3. SELECTIVITY OF IGRs

IGRs are more selective to natural enemies than are most conventional insecticides (Chapter 2). Many appear to possess properties of physiological selectivity, but because IGRs act on hormone or enzyme systems common to many arthropods, physiological selectivity must be evaluated on a case by case basis. As has been discussed previously, IGRs are well suited to timing and dose manipulations, and ecological selectivity can be obtained based on those principles discussed in Chapter 10. However, complex interactions of different ecological and biological factors can interfere with achieving such selectivity. In the field, for instance, selectivity based on timing might be hindered due to overlapping life stages in the beneficial species, thus exposing susceptible life stages.

Laboratory studies have shown that *A. melanoscelus* is less affected by diflubenzuron when its host, the gypsy moth, is treated when the parasitoids are second or third instar (Granett et al. 1975b). However, when aerial spraying with diflubenzuron was conducted over an area infested with gypsy moths that were parasitized by largely third instar *A. melanoscelus*, parasitoid mortality still reached 80% (Madrid and Stewart 1981). Conversely, topical treatment of the predator *C. carnea* with 1 μg of hydroprene resulted in only 6% adult eclosion from the pupal cuticle (Boone unpubl.), but field applications at 0.4 lb a.i./100 gal in pear orchards for the control of pear psylla had minimal effect on this natural enemy (Westigard 1974). Cases like these indicate that although these compounds cause certain effects under laboratory conditions, their influence in the field can be quite different.

Physiological selectivity to natural enemies could be improved (Chapter 9) by basing the mode of action on biochemical and physiological differences between target insects and beneficials. Factors that could be considered in tailoring IGRs to increase physiological selectivity at a biochemical level include the length and shape of the molecule, the relative balance between hydrophobic and hydrophilic characteristics, and the location of double bonds and functional groups (Henrick 1982, Slama and Jarolim 1981). These factors are discussed here for the JHAs to illustrate the types of research that might lead to the development of more physiologically selective IGRs.

At the present time, few useful generalizations have been developed to explain differences in the activity of IGRs such as the JHAs in different insects (Staal 1972, 1975, Feyereisen 1987). Almost all research on the biochemical mechanisms of physiological selectivity has focused on pest arthropods rather than beneficial predators and parasitoids. Vogel et al. (1979) discussed the effects of 10 JH-active IGRs on 40 pest species placed in continuous contact with a high dose of the growth regulators. Four types of effects were recognized.

1. *Inhibition of ecdysis and metamorphosis.* JH I, epofenonane, and R-20458 caused death in the first larval ecdysis of cockroaches and scale insects. JH I caused permanent last larval stages and subsequent recovery to fertile adults in uncontaminated environments among holometabolous insects represented mainly by stored product pests. JH I and epofenonane prolonged the last larval stage followed by death as larval–pupal intermediates in uncontaminated environments in insects represented by holometabolous crop pests such as *Pieris brassicae* and *L. pomonella.*
2. *Inhibition of metamorphosis.* This effect was caused mainly by epofenonane, resulting in supernumerary larval molts. Eventual development to fertile adults in an untreated medium was observed in holometabolous stored product pests. Defective metamorphosis in the last larval stages and death as larval–adult intermediates occurred in hemimetabolous insects. Differentiation of abnormal ovaries and sterile females occasionally occurred.
3. *Inhibition of adult emergence.* JH I, RO-20458, RO-20-3600, and methoprene caused defective adult emergence and death as pharate adults in Diptera tested.
4. *Inhibition of embryogenesis.* This occurs in many insect species treated with a variety of substances.

The factors influencing selectivity among pest species may have some bearing on selectivity relationships between pests and natural enemies. For example, a biochemical mechanism enabling physiological selectivity for JHAs might be realistic for lepidopteran pests and their hymenopteran and dipteran parasitoids, because of differences in endogenous JHs in these taxonomic groups. Lepidoptera produce only JH I and JH II, whereas JH III, a more polar compound, is the principal hormone in most other species (de Kort et al. 1984, Feyereisen pers. comm.). JHAs could conceivably be developed to maximize these structure–activity differences and enhance selectivity.

As noted earlier (Fig. 12.2), the activity of JHs and JHAs depends on interactions with the following proteins in the hemolymph and target site: 1) binding or carrier proteins, 2) degradative enzymes, and 3) receptor proteins at the target site. These interactions may vary between host and parasitoid based on the structure of the JHA used and the intrinsic differences in proteins of host and parasitoid. These differences, therefore, will determine the physiological selectivity of JHAs at the biochemical level.

12.3.1. Binding Proteins

Most JH is bound to binding proteins under physiological conditions (Goodman and Gilbert 1978). Although the function of binding proteins is not well understood, possible functions include the prevention of specific uptake and absorption of JH, hormone transport, protection against enzymatic degradation, and storage of the hormone in an inert form (de Kort and Granger 1981). Two classes of JH binding proteins have been identified: low molecular weight proteins binding JH I or JH II and high molecular weight proteins binding JH III (de Kort and Granger 1981, de Kort et al. 1984).

One would expect binding proteins of a particular species to have the greatest affinity for their associated natural homologues. In *Leptinotarsa decemlineata, Locusta migratoria, Periplaneta americana*, and *Apis mellifera*, species which contain only JH III as the natural homologue, effective and high affinity binding of JH III to hemolymph proteins occurs. In *Pieris brassicae* and *Oncopeltus fasciatus*, species in which JH III is not the natural homologue, binding of JH III to hemolymph proteins is negligible (de Kort et al. 1984). These results suggest that the principal JH homologue is correlated with the characteristics of its binding protein. Other studies support this theory. In the cockroach *Leucophaea maderae*, binding proteins are found in hemolymph and ovaries with high affinity for JH III, and lesser affinity for JH I and JH 0 (Kovalick and Koeppe 1983). The JHAs methoprene and hydroprene can compete for binding sites. In *Manduca sexta*, a low molecular weight binding protein has been found with a higher affinity for JH I and JH II than for JH III (Peterson et al. 1982). JH I and JH II have been shown to be the principal homologues in this species (Peter et al. 1976, Schooley et al. 1976).

However, there appear to be large differences in affinity and specificity among JH binding proteins. JH I can effectively compete with JH III for binding sites in *L. migratoria* and *P. americana*. However, JH I is a weak competitor in hemolymph of *L. decemlineata*, indicating that the binding sites in the hemolymph of *L. decemlineata* are much more specific for JH III than those from the locust or cockroach. The cause of these differences is unclear, although large differences in hemolymph JH III esterases have been measured (de Kort et al. 1979); *L. decemlineata* contains high levels of JH III esterases compared with the locust or cockroach. This favors the protection hypothesis which states that binding proteins protect JH from degradation by JH esterases (JHE) (de Kort et al. 1983).

The different JHs of lepidopteran hosts and parasitoids and their associated binding proteins may represent potential targets for physiological selectivity. Selectivity of a JHA would be improved if the substance was bound by binding proteins in the lepidopteran host and not by binding proteins in the parasitoid. This would make a JHA more vulnerable for attack by degradation enzymes in natural enemies, whereas high levels of exogenous JH in the host would interfere with its development.

12.3.2. Degradation Enzymes

Juvenile hormones (JHs) are degraded by juvenile hormone esterases (JHEs), epoxide hydrases, and/or cytochrome P-450 monooxygenases (de Kort and Granger 1981). JHEs are mainly found in the hemolymph, although tissue-specific esterases are involved in the metabolism of JH as found in several dipteran species (Terriere and Yu 1977). Epoxide hydrases and cytochrome P-450 monooxygenases are also found in tissues (Yu and Terriere 1978). The latter type of enzymes, however, are less specific for JH degradation than are JHEs. Two classes of esterases have been distinguished in the hemolymph of insects (Sandburg et al. 1975, de Kort and Granger 1981): general carboxylesterases, which hydrolyze only free JH, and specific JHEs, which can hydrolyze both free and bound JH. There is a correlation between JHE activity and JH titer, suggesting that these enzymes play an important role in regulating the JH titer (de Kort and Granger 1981). The activity of JH-specific esterases varies among different species (de Kort et al. 1979). The honeybee, housefly, and cockroach show minimal degradation of JH in their hemolymph, compared with the Colorado potato beetle. Because rapid JH degradation does occur in the honeybee, housefly, and cockroach, it is suggested that inactivation of JH occurs in tissues.

Comparative inhibition studies of JHEs by Sparks and Hammock (1980) showed that JHEs are less important in dipteran JH metabolism and regulation than they are in Lepidoptera and Coleoptera. Degradation of JH in *T. ni* (Lepidoptera) depends on a unique carboxylic ester hydrolase. However, JHEs are sensitive to a number of phosphoramidates such as *O*-ethyl-S-phenylphosphoramidothiolate (EPPAT). Trifluoromethylketone, or TFT (1, 1, 1-trifluorotetradecan-2-one) appears to be an even more potent and selective inhibitor of JHEs in *T. ni*. This compound has been described as a transition state analog for the JHE; that is, TFT mimics the transition state of the substrate and may, therefore, be more tightly bound by JHE than the substrate, making TFT a very effective enzyme inhibitor (Hammock et al. 1982). Other compounds such as 3-octylthio-1, 1, 1-trifluoro-2-propane and 3-{(E)-4, 8-dimethyl-3, 7-nonadienylthio}-1, 1, 1-trifluoro-2-propane are also good inhibitors of JHE in *T. ni*, effectively delaying pupation when applied to prewandering larvae (Hammock et al. 1984, Prestwich et al. 1984).

These findings suggest that the JH degradation system is an excellent target for the development of new and selective insect control agents. Based on differences in JH degradation enzymes, specific compounds could be developed to selectively inhibit JHEs in Lepidoptera without inhibiting degradation enzymes in Hymenoptera. A high degree of selectivity could be incorporated into such molecules (Hammock et al. 1982). It is suggested that the potency of these inhibitors could be further increased by the synthesis of more exact substrate mimics or by modifying polarities in the molecule.

12.3.3. Receptor Proteins

Little is known about JH receptor proteins. Klages et al. (1980) found high-affinity, highly specific JH I binding sites in the cytosol of *Drosophila hydei* integument. These receptors are macromolecules with a high specificity for the natural hormone. Competition studies showed preferential binding of JH I and JH III and poor competition by the JHA methoprene (ZR 515) for the JH binding sites. Alterations in the structure of the JH III molecule markedly reduced its competitive efficiency. More basic studies are needed to understand the role of receptor proteins and their importance conferring physiological selectivity.

12.4. CONCLUSIONS

IGRs are generally less disruptive to beneficial arthropods than conventional insecticides (Chapters 2 and 9). This is due to both physiological and ecological selectivity mechanisms. Although laboratory bioassays have shown that IGRs are frequently detrimental to natural enemies, less pronounced side effects have been documented in field evaluations, probably because many natural enemies immigrate as adults and populations vary widely in the life stages that are present in field populations. Selectivity of IGRs in these instances is primarily ecological and can be greatly affected by the timing of applications.

JHAs show potential for physiological selectivity based on intrinsic differences of JHs in lepidopteran pests and their hymenopteran and dipteran parasitoids. Developing JHAs that exploit the structural differences of JHs between these taxonomic groups may provide for broader physiological selectivity in the future. However, JHAs retard or inhibit maturation, keeping pests in immature stages and thus possibly prolonging periods of crop damage. In many cases, this limits their effective use in situations where immediate control is required. While AJHAs do not have this drawback and have the potential to be selective to natural enemies, their efficacy at this time is limited.

Benzoylphenyl ureas seem to be the most promising, least disruptive IGRs under development at the present time. Limited ingestion of residues by natural enemies may be the primary mechanism of their selectivity. The ovicidal or larvicidal action of benzoylphenyl ureas against target species makes these compounds useful in IPM programs on many of the major crops. The most promising IGRs currently being used or tested in IPM programs are the chitin synthesis inhibitor diflubenzuron and the JHAs fenoxycarb, epofenonane, hydroprene, and methoprene. Their integration into integrated pest management programs should significantly improve the possibilities for achieving broad-scale selectivity in the near future (Chapters 9 and 21).

13

PESTICIDE SELECTIVITY: PYRETHROIDS

13.1. INTRODUCTION

Integrated selectivity is achieved by combining physiological and ecological selectivity to manage arthropod natural enemy–pest complexes with pesticides and to subsequently conserve biological control agents. If necessary, integrated selectivity also involves practicing pesticide resistance management. *Pesticide resistance management* means minimizing resistance development in pests while maximally exploiting resistant natural enemies (Chapter 17).

Principles and examples of ecological and physiological selectivity were presented in Chapters 9–12. Physiological selectivity is generally considered to be an inherent property of a given pesticide. However, physiological selectivity can also be achieved if natural enemies develop resistance to a previously nonselective pesticide. Given the absence of resistant strains of natural enemies, physiological selectivity can be lost if pests develop

This chapter was coauthored by Karen M. Theiling, Department of Botany & Plant Pathology, Oregon State University, Corvallis, Oregon.

resistant strains and their predators and parasitoids do not.[1] The detection, monitoring, and management of insecticide resistant natural enemies and their hosts or prey to achieve integrated selectivity and as a component of pesticide resistance management will be discussed in Chapters 14–20. This chapter develops the link between selectivity and resistance management, since selective pesticide use (ecological) and use of selective pesticides (physiological) are important tactics for limiting resistance in pests.

All available means of pesticide selectivity are usually employed in IPM programs to manage pest–natural enemy complexes in the field. However, broad selectivity is limited among current pesticides (Chapter 9). Also, the initial selectivity attributes of an insecticide may change due to resistance development. Therefore, monitoring of susceptibility in both pests and natural enemies is necessary. Achieving integrated selectivity with pesticides presents a challenge to IPM practitioners managing large pest–natural enemy complexes (Croft 1981a).

In this chapter, existing and potential means of achieving broad, integrated selectivity in the field are discussed for the pyrethroid insecticides (SPs). SPs are particularly effective on pests, but are quite toxic to natural enemies (Chapter 2). Physiological selectivity and ecological selectivity of SPs are initially considered for natural enemies associated with all crops. Thereafter, examples of integrated selectivity among the pest–natural-enemy complexes of cotton and apple in North America are discussed.

13.2. PYRETHROID SELECTIVITY TO NATURAL ENEMIES ON ALL CROPS

SPs are broad-spectrum insecticides which were initially developed in the mid to late 1970s to control a variety of arthropod pests (Elliott 1977). Their environmental stability and high insect toxicity make SPs highly attractive for use on many crops (Croft and Whalon 1982). As with beneficial predators and parasitoids, the negative impact of SPs on microorganisms, aquatic invertebrates, pollinators, fish, birds, and mammals can be significant (Smith and Stratton 1986). However, they are generally considered to be less harmful to the environment than pesticides used in the past (Elliott 1977).

The growing market for SPs on major crops around the world makes a discussion of their selectivity attributes most appropriate (Dover and Croft 1984). While the overall insecticide market has been growing at a rate of about 5% per year, the SP market has been increasing much faster (e.g., from 10 to 20% in 1984; Dover and Croft 1984). However, should widespread resistance develop among pests, the effectiveness of the SPs would be diminished (Dover and Croft 1984, NAS/NRC 1986). The potential for resistance development to the SPs has already been convincingly demonstrated: in 1976, only 6 arthropod species were recorded as resistant to these compounds; by 1980, the number had increased to 17; and by 1986, the total was more than 50 (Dover and Croft 1984, Georghiou 1986). SP cross resistance with DDT has been established in several pests, which suggests that numerous species may soon become resistant to the SPs.

As the pyrethroids become widely used, their side effects should be evaluated on a wide range of natural enemies. Selectivity research on SPs is essential to IPM. Early side-effects studies demonstrated that physiological selectivity to natural enemies is limited among the SPs (Croft and Whalon 1982). Attempts to achieve selectivity with compounds such as

[1] For example, the development of cyhexatin resistance in spider mites has recently limited IPM programs on apple in western North America. Previously these programs had relied on biological control of spider mites and use of this selective acaricide which conserves predatory mites (see Croft et al. 1987; also further discussion in Chapter 20).

fenvalerate and permethrin in the field have been mixed (e.g., Hull and Starner 1983, Croft and Whalon 1983; see also Hu 1981, Chiverton 1984, Charlet and Oseto 1983, Hsieh 1984, Nikusch and Gernot 1986a, 1986b, Poehling et al. 1985b). Mechanisms of physiological and ecological selectivity of SPs to natural enemies are not well understood. These compounds are relatively new, and little comparative research at a biochemical level has been published (Chapter 6). Biochemical mechanisms of physiological selectivity have not been high-priority research in the industrial development of the SPs (see Section 13.5. and Chapter 21).

13.2.1. Selectivity Analysis from the SELCTV Database

Attributes of SP selectivity to natural enemy complexes associated with major crops can be inferred by examining toxicity trends in the SELCTV database (Theiling and Croft 1988; Chapter 2). Pest–natural-enemy selectivity data are contained within SELCTV as selectivity ratios, although susceptibility data for natural enemies alone are much more numerous. The effectiveness of SPs on pests has long been demonstrated.

Mean susceptibility values and standard errors for major orders and families of natural enemies to the SPs are summarized in Tables 13.1 and 13.2. The greatest susceptibility and least variable responses have been recorded for predatory mites, of which the Phytoseiidae make up over 90%. Entomophagous dipterans are also highly susceptible to the SPs. Lowest mean mortalities have been computed for predatory Hemiptera and Neuroptera, the latter including the highly SP-tolerant species *Chrysoperla carnea* (Wilkinson et al. 1975, Shour and Crowder 1980). The intermediate susceptibility shown for parasitic Hymenoptera is considerably lower than expected. This group typically exhibits high susceptibility to most conventional pesticides (see values for organophosphates and carbamates in Chapter 2).

Among families (Table 13.2), similar trends in SP susceptibility were evident. Again, the high susceptibility of the Phytoseiidae was notable (mean value of 4.44). Among the Hemiptera, the Miridae, Nabidae, and Lygaeidae were somewhat tolerant, whereas the Anthocoridae were more susceptible. As expected, the Chrysopidae were more tolerant than the other neuropteran family, Hemerobiidae. The coleopteran Staphylinidae and Coccinellidae were, respectively, highly and moderately susceptible to the SPs, but the Carabidae were more tolerant. Among the Hymenoptera, the Aphelinidae, Encyrtidae,

Table 13.1. Mean Toxicity of Pyrethroid Insecticides to Major Orders or Other Taxonomic Divisions of Arthropod Natural Enemies (Taken from SELCTV database; Theiling and Croft, 1988. Values from both field and laboratory tests)

Order	Observations	Mean Toxicity (Scale 1–5)[1]	Standard Error
Acarina	110	4.30	±0.072
Diptera	24	4.29	0.153
Araneae	7	4.14	0.261
Coleoptera	150	4.06	0.087
Hymenoptera	135	4.01	0.109
Hemiptera	99	3.65	0.126
Neuroptera	130	3.66	0.121
Total	655		

[1] 1 = 0% mortality; 2 = < 10%; 3 = 10–30%; 4 = > 30–90%; 5 = > 90%.

Table 13.2. Mean Toxicity of Pyrethroid Insecticides to Major Families of Arthropod Natural Enemies (Taken from SELCTV database; Theiling and Croft 1988)

Family	Observations	Mean Toxicity (Scale 1–5)[1]	Standard Error
Anthocoridae	31	4.35	±0.210
Aphelinidae	19	4.68	0.136
Braconidae	52	3.38	0.213
Carabidae	20	3.55	0.266
Cecidomyiidae	7	4.42	0.202
Chrysopidae	118	3.53	0.122
Coccinellidae	125	4.14	0.093
Encyrtidae	11	4.91	0.091
Eulophidae	8	4.00	0.378
Hemerobiidae	7	4.00	0.309
Ichneumonidae	19	4.42	0.176
Lygaeidae	16	3.19	0.277
Miridae	21	3.86	0.242
Nabidae	16	3.88	0.272
Phytoseiidae	158	4.44	0.062
Scelionidae	6	4.17	0.401
Staphylinidae	11	4.50	0.393
Syrphidae	11	3.82	0.226
Tachinidae	6	5.00	0.000
Trichogrammatidae	15	3.87	0.291

[1] 1 = 0% mortality; 2 = < 10%; 3 = 10–30%; 4 = > 30–90%; 5 = 90%.

and Ichneumonidae showed the highest mean susceptibility responses, while both the Braconidae and the Trichogrammatidae were less susceptible.

The number of observations of SP impact on Diptera were few (Table 13.2). Predaceous Cecidomyiidae were highly susceptible, while the normally susceptible Syrphidae were less affected by the SPs. Tachinid parasitoids were highly susceptible to the SPs (Table 13.2), as has been demonstrated for most pesticide groups.

It is not clear why so much variability exists among the responses of various entomophagous groups to SPs. In general, the considerable tolerance of the Hemiptera and *C. carnea* might be expected, based on other susceptibility studies (Chapter 2). The genus *Chrysoperla*, and especially *C. carnea*, are generally tolerant to many pesticide groups. However, high susceptibility and the lack of variability in the response of the Phytoseiidae to the SPs and the apparent tolerance in certain hymenopterans are not consistent with their responses to other pesticides. Phytoseiid mites are variably tolerant to many pesticides, and resistant strains of these species are widely known; conversely, the parasitic Hymenoptera includes some of the most susceptible species to most pesticides (Chapters 2 and 14).

Plapp (1981) has speculated—and a few studies have confirmed—that parasitoids may be well adapted to detoxify pesticides hydrolytically, but not oxidatively (Croft and Mullin 1984; Chapter 6). SPs are known to be detoxified readily by esterases and other hydrolytic enzymes (Chapter 6). As with *C. carnea* (Plapp and Bull 1978, Rajakulendran and Plapp 1982a, 1982b, Chang and Plapp 1983), it is possible that highly active esterase detoxification mechanisms are the explanation for high tolerance to SPs among some parasitoid species (Feng and Wang 1984).

Table 13.3. Mean Toxicity of Pyrethroid Insecticides to Arthropod Natural Enemies Inhabiting Major Crops (Taken from SELCTV database; Theiling and Croft 1988)

Crop	Observations	% of Total	Mean Toxicity (Scale 1–5)[1]	Standard Error
Greenhouse crops	25	4	4.60	±0.15
Pear	14	2	4.36	0.17
Other crops	74	10	4.34	0.09
Apple	147	21	4.31	0.09
Cereal grains	41	6	3.88	0.15
Cotton	190	27	3.79	0.09
Vegetables	42	6	3.21	0.24
Alfalfa	18	3	3.10	0.25
Unidentified	150	21	4.09	0.09

[1] 1 = 0% mortality; 2 = < 10%; 3 = 10–30%; 4 = > 30–90%; 5 = > 90%.

Most susceptibility and selectivity research on SPs has been conducted on natural enemies associated with cotton, apple, vegetables, cereal grains, greenhouse crops, alfalfa, and pear cropping systems (Table 13.3). That nearly 50% of total SP records in the Appendix list responses for cotton and apple inhabiting species is one reason why a more detailed selectivity analysis will be presented for species associated with these crops. The average toxicity of SPs computed for crop-associated groups of natural enemies showed considerable differences in mean susceptibility and range of standard error (Table 13.3). For example, the large difference between cotton (mean = 3.79) and apple (mean = 4.31) species is meaningful from both biological and management perspectives (see Section 13.3). Mortality values for natural enemies associated with alfalfa, vegetable, and cereal grain were among the lowest for any crop, whereas species associated with greenhouses and pear were most susceptible to SPs.

Some of the trends in natural enemy responses from a particular crop to SPs can be explained. Vegetable data were heavily skewed by the many reports of high tolerance to SPs in the braconid *Apanteles plutellae*, which attacks the diamondback moth *Plutellae xylostella* (Ooi and Sudderuddin 1978, Mani and Krishnamoorthy 1984; Appendix). (This may explain, in part, the high overall tolerance of SPs observed among Hymenoptera; Table 13.1.) On cereals, several predators and parasitoids that attack aphids were quite tolerant to SPs (e.g., Coats et al. 1979, Chiverton 1984, Suss 1983, Brown et al. 1983, Ffrench-Constant and Vickerman 1985). The two most important natural enemies in greenhouses, *Phytoseiulus persimilis* and *Encarsia formosa*, were highly susceptible to most SPs (Hassan et al. 1983).

The mean toxicity of individual pyrethroids to all natural enemies was high for most compounds (Fig. 13.1). Fenvalerate and bioresmethrin were slightly less toxic than the average for all SPs, while permethrin and deltamethrin were two of the more toxic compounds in the group. The less well studied fluvalinate and tralomethrin were apparently much less toxic (Fig. 13.1). While fluvalinate has exhibited low toxicity to honeybees, its efficacy on pests may also be low (see Table 13.6). The lower toxicity of fluvalinate and tralomethrin suggests that there is potential for selectivity within this group of insecticides. However, those SPs, which are widely used, show no such levels of selectivity, partly because research on the selectivity of SPs to natural enemies has not been pursued extensively.

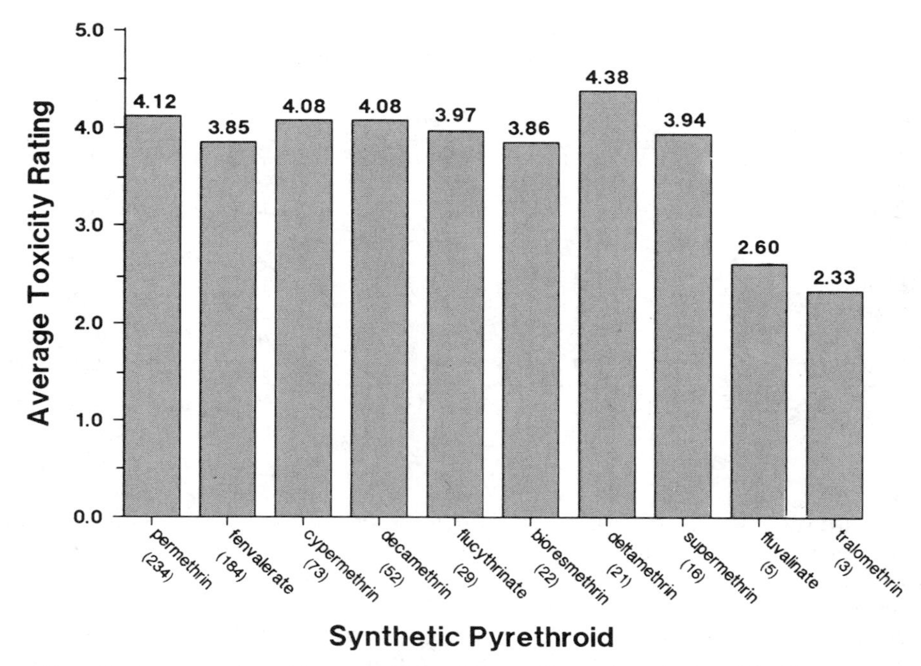

Figure 13.1. Average toxicity of 10 pyrethroid insecticides to all natural enemies from SELCTV database.

13.2.2. Physiological Selectivity

The mortality data summarized from SELCTV include many different methods of side-effects assessment, ranging from laboratory to field tests (e.g., Tables 13.1 and 13.2; Chapter 5). Generally, these data are thought to be reasonable indicators of overall trends in physiological selectivity (Theiling 1987, Theiling and Croft 1988). However, one must exercise caution in generalizing from such a broad toxicity summary. In field tests, exposure is not limited to direct contact, therefore physiological selectivity is not assured. Ecological selectivity could be involved in a number of cases included in SELCTV.

In studies where direct toxicity and physiological selectivity were specifically evaluated (i.e., only 40% of the total SELCTV database), trends similar to those reported for the entire database were found (compare Tables 13.1 and 13.2). Reports of LC, LT, LD_{50} assays, and other direct exposure tests for the major orders or related taxonomic groups of natural enemies are summarized in Fig. 13.2. The more inclusive mean toxicity values from Table 13.1 are also included for graphic comparison. On the average, direct toxicity values (Fig. 13.1) were greater than or equal to means computed for all test types combined (Table 13.1).

Apart from data aggregates from the SELCTV database, a number of individual studies of physiological selectivity for natural enemy species and their hosts or prey can be cited. Cases of both simple and broad selectivity are included (Chapter 9). Evident in these cases are the more modest levels of selectivity exhibited by the SPs compared to other pesticide classes.

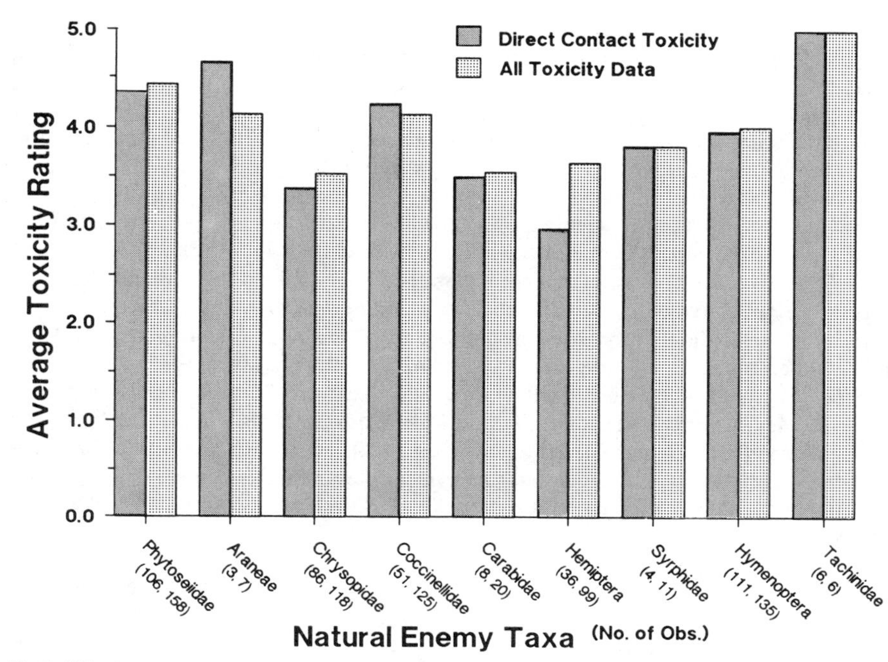

Figure 13.2. Average toxicity of pyrethroid insecticides by direct contact methods and by all exposure methods to natural enemy taxa from SELCTV database.

13.2.2.1. Simple Selectivity. Stevenson and coworkers evaluated a number of SPs, many of which had been identified and synthesized at the Rothamsted Experimental Station in the United Kingdom (Elliott et al. 1983, Stevenson and Walters 1984). They reported selectivity ratios ranging from 8- to 100-fold in favor of the ichneumonid parasite *Venturia canescens* over its host *Ephestia kuhniella* (see also Table 9.11). Their research demonstrated the potential for SP selectivity in this model parasitoid–host system. However, none of the compounds tested has effectively controlled a broad range of pests. Additional barriers to commercial development might include aspects of high mammalian toxicity, environmental persistence, or a lack of environmental stability (see also Chapter 21).

Other cases of simple physiological selectivity for the SPs are listed in the Appendix. Over 80% of the cases listed represent simple selectivity to natural enemies as compared to 20% for broad selectivity. Only in 5 of 50 cases listed did selectivity include more than three natural enemy species, and two of these were reported from field trials where physiological selectivity was not verified. The majority of SP selectivity cases in the Appendix cited fenvalerate or permethrin, although deltamethrin and cypermethrin were also well represented. The natural enemies best represented in these documentations were from the Carabidae, Chrysopidae, Coccinellidae, and Braconidae.

Abdel-Aal et al. (1979) measured the relative toxicity (LD_{50} values) of several insecticides to *Spodoptera littoralis* and the predator *Coccinella undecimpunctata*. Selectivity ratios for cypermethrin, fenvalerate, and permethrin favored the predator by 156, 37, and 2.4-fold, respectively. Almost all organophosphates, carbamates, and chlorinated hydrocarbons tested were more toxic to the predator than to the pest. In a study of other lepidopteran pests, the velvet bean caterpillar *Anticarsia gemmatalis*, the fall armyworm *Spodoptera fugiperda*, and the corn earworm *Heliothis zea* were all more susceptible to permethrin, cypermethrin, and fenvalerate than their predator, the pentatomid spined soldier bug *Podisus maculiventris* (Yu 1988). Reduced penetration in the beneficial was thought to be primarily responsible for physiological selectivity (Chapter 6).

Simple selectivity has been demonstrated for several parasitoids over their hosts using SPs (Plapp and Vinson 1977, Wilkinson et al. 1979, Suss 1983, Bogenschutz 1984, Powell et al. 1986a, Mani and Krishnamoorthy 1986, Scott and Rutz 1988). As noted earlier, reports of SP selectivity for *Apanteles plutella*, a primary parasitoid of the vegetable pest *Plutella xylostella* (diamondback moth), are numerous (Ooi and Sudderuddin 1978, Mani and Krishnamoorthy 1984, Feng and Wang 1984). Two other parasitoids of cabbage pests to which permethrin has exhibited simple selectivity are *Diaeretiella rapae*, which parasitizes *Myzus persicae* (Hsieh 1984), and *Pteromalus puparum*, which develops on the cabbage butterfly *Pieris rapae* (Lasota and Kok 1986). *Microplitis croceipes*, a braconid which attacks *Heliothis virescens*, is another hymenopteran that has shown high tolerance to SPs, which confer selectivity to it over its host (Powell and Scott 1985, Powell et al. 1986a).

13.2.2.2. Broad Selectivity. Cases of broad selectivity to several natural enemies over their prey or hosts are much more limited. The most extensive work in this regard has probably been done in cotton IPM systems, where *C. carnea* and several hemipteran, coccinellid, and parasitoid species are only moderately susceptible to the SPs, while a wide range of pests are effectively suppressed by these toxins (Plapp and Bull 1978, Wilkinson et al. 1979, Brettell 1979, 1982, 1984, Kismir and Sengonca 1980, Shour and Crowder 1980, Press et al. 1981, Ishaaya and Casida 1981, Hu 1981, Rajakulendran and Plapp 1982a, 1982b, Sukhoruchenko et al. 1982, Bashir and Crowder 1983, Powell et al. 1986a). More specific attributes of SP selectivity on cotton species are discussed in Sections 13.3 and 13.4.

Broad SP selectivity was reported for a complex of coccinellid species and two cereal pests, *Diabrotica longicornis* and *Oulema melanopus*, by Coats et al. (1979). Among eight coccinellids, a eulophid parasitoid, and the two chrysomelids, both permethrin and fenvalerate were selective to most of the coccinellids, while cypermethrin was more toxic to the predatory beetles than the chrysomelid beetles. All three SPs were considerably more toxic to the parasitoid than to the chrysomelids. This case is typical of many studies of compound selectivity, where selectivity attributes have been mixed, making simultaneous conservation of all beneficial species extremely difficult.

Among parasitoids, Waddill (1978) observed that fenvalerate was less toxic to adults of *Diglyphus intermedius* (Eulophidae), *Opius bruneipes* (Braconidae), *Apanteles* spp. (Braconidae), *Copidosoma truncatellum* (Encyrtidae), and *Telenomus remus* (Scelionidae) than to the vegetable pests *Liriomyza sativae* (vegetable leafminer) and *Keiferia lycopersicella* (tomato pinworm). Certainly the observed selectivity to a complex of parasitoids from several families must indicate that, at least for adult forms, broad selectivity is a property of this compound in certain situations. Waddill (1978) cautioned that since *L. sativae* is known to develop rapid resistance to fenvalerate, maximum biological control should be encouraged and use of this product should be parsimonious.

In alfalfa, Syrett and Penman (1980) reported LD_{50} values for the two aphid species *Acyrthosiphon kondoi* and *A. pisum* and two natural enemies, the chrysopid *Austromicromus tasmaniae* and the coccinellid *Coccinella undecimpunctata*. They found that both predators were favored over their prey by fenvalerate. Selectivity ratios were particularly high for the lacewing adults and larvae and moderate for the coccinellid adults. Activity of the pyrethroids has been shown to be inversely related to temperature (Elliott 1977, Vijverberg et al. 1983), and a marked increase in fenvalerate toxicity to the aphids and their predators was noted at lower temperatures. Therefore, it was recommended that lower concentrations of this compound should be used during cooler weather to obtain selective control of the aphids.

13.2.3. Ecological Selectivity

Several examples of ecological selectivity via temporal or spatial separation of natural enemies from pesticides were reviewed in Chapter 10. However, in the few cases cited for SPs, most were of simple ecological selectivity. In part, the paucity of cases of ecological selectivity may be due to the recent development of SPs. However, the extreme toxicity of SPs to most natural enemies may also contribute to the low number of documented cases. For example, residues of SPs are toxic to predatory mites for 30–90 days (AliNiazee and Cranham 1980, Riedl and Hoying 1983). Similarly high residual toxicities to many other natural enemies have been documented (Chapters 2 and 3). To achieve selectivity via temporal separation, a compound must have a residual time that is shorter than the life cycle of the natural enemy, otherwise there is no period for the beneficial to operate in a nontoxic environment. The persistence of most SPs leaves little room to manipulate spray applications to conserve beneficials through spatial asynchrony. Highly mobile predators or parasitoids will encounter deposits of pesticides that normally would not be contacted using less persistent compounds.

Ecological selectivity using SPs has been demonstrated in greenhouses. Osborne (1981), working with tomatoes, reported that populations of *Encarsia formosa* were conserved by carefully timing applications of the short residual compound resmethrin. He observed only a 27% reduction in parasitoid emergence when parasitized scales were treated when they were 125 degree-days old. Mean emergence of the whitefly *Trialeurodes vaporariorum* was reduced by over 70% by the same application. Parasitoids that survived treatments were able to find their hosts and reproduce, thus sustaining the parasite–host interaction through time (Chapter 10).

Ecological selectivity favoring *E. formosa* over *T. vaporariorum* has also been achieved in French greenhouses by using nonpersistent residues of bioresmethrin (Delorme and Angot 1983, Delorme et al. 1985). After evaluating this material on plants for all stages of host and parasitoid development, they found that occasional use of the compound was possible. Applications were timed to coincide with periods when few parasitoid adults were present and even then, only spot treatments were used. To further exploit the partial ecological selectivity obtained with SPs in this closed ecosystem, Delorme et al. (1985) recommended that the selection of a resistant strain of parasitoid to this compound and to deltamethrin be attempted (see Results, Chapter 20). Other cases of ecological selectivity of SP insecticides to the natural-enemy–pest complexes of cotton and apple are cited in Section 13.3.

13.3. SELECTIVITY TO SPECIES COMPLEXES OF COTTON AND APPLE

Selectivity needs for the arthropod species complexes which occur on cotton and apple are relevant to the design of SP insecticides as well as for future pesticides (Croft and Whalon

1982). Cotton represents the largest world market for new products. Requirements for pesticides on cotton dictate specifications for pesticides registered on many other crops. On a per acre basis, apple is even more intensively sprayed than cotton, although the acreage committed to apple production is much less (Croft and Hoyt 1983). In both cropping systems, the need exists for broadly selective pesticides which allow chemical and biological control of insects and mites to operate simultaneously. IPM systems on cotton and apple in North America are relatively well developed, and the conservation of natural enemies on these crops via selective pesticides is a high priority of IPM personnel (Chapter 2; Croft and Whalon 1982, Croft and Hoyt 1983, Frisbie and Adkisson 1986).

Table 13.4. Toxicity of Single Field Applications of Pyrethroids[1] to Arthropod Pest and Natural Enemy Complexes Associated with Cotton Ecosystems in North America (Updated from table in Croft and Whalon 1982[2])

	Toxicity[3]			
Species	None = 0.0	Low = 1.0	Mod. = 2.0	High = 3.0
PRIMARY PESTS				
Heliothis spp.				2.9
Boll weevil				3.0
Pink bollworm				3.0
Armyworm			1.8	
Lygus spp.			2.5	
SECONDARY PESTS				
Cotton leaf perforator			3.0	
Whitefly			2.2	
Thrips			2.3	
Aphids		0.8		
Mites	0.3			
NATURAL ENEMIES				
Chrysoperla spp.		1.5		
Coccinellids			1.8	
Orius spp.			2.4	
Nabids			2.2	
Pentatomids		1.5		
Geocoris spp.			1.7	
Spiders		1.4		
Ichneumonids				2.4
Braconids			1.8	
Trichogrammatids			2.0	
Tachinids				2.9

[1] Compounds primarily tested include permethrin and fenvalerate at field rates of 0.05–0.20 lb a.i./acre.
[2] Based in part on summaries reported in "Insecticide and Acaricide Tests 1977–1986" (ESA 1977–1986) and in part on discussions with field researchers working on insecticide application on this crop. These values are for susceptible strains.
[3] Average values based on 5–15 data sets per insect group.

13.3.1. Integrated Selectivity

In characterizing the selectivity of SPs to the pests and natural enemies of cotton and apple in North America, the species complexes associated with these crops must be described in greater detail. A partial list of pests and natural enemies associated with each crop is presented in Tables 13.4 and 13.5 along with toxicity ratings to the SPs permethrin and fenvalerate (combined values). In comparing the two crops, there are some striking similarities and several cases of what might be termed ecologically equivalent species or species complexes.

Table 13.5. Toxicity of Single Field Applications of Pyrethroids[1] to Arthropod Pest and Natural Enemy Complexes Associated with Apple Ecosystems in North America (Updated from table in Croft and Whalon 1982[2])

	Toxicity[3]			
Species	None = 0.0	Low = 1.0	Mod. = 2.0	High = 3.0
PRIMARY PESTS				
Codling moth				2.8
Apple maggot				2.7
Plum curculio				2.6
Leafrollers				3.0
Fruitworms				2.9
Lygus spp.			2.5	
SECONDARY PESTS				
Scales			2.1	
White apple leafhopper				3.0
Leafminers				3.0
Aphids		1.3		
Mites	0.2			
NATURAL ENEMIES				
Phytoseiids				3.0
Stethorus spp.				2.6
Aphidoletes spp.				2.8
Mirids			1.8	
Anthocorids				2.5
Chrysoperla spp.			2.2	
Syrphids				2.8
Spiders			1.9	
Aphelinids				2.9
Mymarids				3.0
Ichneumonids				2.8
Braconids				
Eulophids				
Tachinids				3.0

[1] Compounds primarily tested include permethrin and fenvalerate at field rates of 0.05–0.20 lb a.i./acre.

[2] Based in part on summaries reported in "Insecticide and Acaricide Tests 1977–1986" (ESA 1977–1986) and in part on discussions with field researchers working on insecticide application on this crop. These values are for susceptible strains.

[3] Average values based on 5–15 data sets per insect group.

Both crops harbor key lepidopteran pests. These include *Heliothis zea, H. virescens,* and *Pectinophora gossypiella* (pink boll worm) on cotton, while *Cydia pomonella* (codling moth) and *Grapholitha prunivora* (lesser appleworm) are found on apple. These pests directly attack the fruit (boll or pome) and are the primary targets of control programs. Fruit-feeding curculionid weevils attack both crops—boll weevil (*Anthonomus grandis*) on cotton and plum curculio (*Conotrachelus nenuphar*) on apple. Both crops harbor more sporadic lepidopteran pests. These species usually attack the foliage and occasionally the fruit, for example, armyworms and loopers on cotton versus leafrollers and budmoths on apple. Both crops harbor *Lygus* or plant bugs. Each crop has similar secondary pests such as mites, aphids, whiteflies/scales, leafhoppers, and leafminers/leaf perforators, which are usually induced to pest status by broad-spectrum insecticides. Secondary pests are often controlled by natural enemies when selective insecticides are used or when left untreated.

Apple and cotton have somewhat similar complexes of beneficials (Tables 13.4 and 13.5). However, the relative importance of each group of natural enemies differs markedly between the two crops. For example, the biological control of *Heliothis* and other lepidopterans is of primary importance on cotton. This makes the conservation of generalist predators such as *Chrysoperla*, hemipterans, coccinellids, and spiders that feed on the eggs as well as on the immature stages of these pests high priority. Conservation of the parasites of *Heliothis*, pink bollworm, and other lepidopterans is a priority for the same reasons. The most important family of parasitoids on cotton is the Braconidae (Wilkinson et al. 1979, Press et al. 1981, Powell et al. 1986a). In contrast, biological control of secondary pests such as mites, aphids, whiteflies, thrips, and cotton leaf perforators is considered less important or is often assumed.

On apple, the situation is quite different. The greatest emphasis in biological control is on spider mites. Predatory mites, particularly the Phytoseiidae, are the key regulators of spider mites on this crop. Although phytoseiids also occur on cotton, they are almost ignored as biological control agents. Conservation of other natural enemies on apple include other predators of spider mites, such as the coccinellid *Stethorus* spp., mirids, anthocorids, *Chrysoperla* spp., and spiders. Relatively little effort is made to conserve the parasitoids of key pests such as the hymenopteran braconids, ichneumonids, and dipteran tachinids. Key pests of apple are less effectively controlled by these parasitoids at the low levels required to eliminate damage to the harvested fruit (Croft and AliNiazee 1988). To some degree, there is interest in the aphelinid, mymarid, dryinid, eulophid, and braconid parasitoids of scales, leafhoppers, and leafminers, but their conservation is usually secondary to the conservation of mite and aphid predators.

The susceptibility of the pest and natural enemy complexes of cotton and apple for single applications of the SPs permethrin and fenvalerate is indicated in Tables 13.4 and 13.5. These data were obtained from seasonal field assessment tests taken in several areas of production in North America (adapted from Croft and Whalon 1982). The high toxicity of the SPs to the key pests of cotton (including *Heliothis, Plectinophora*, and *Anthonomus*) and apple (*Cydia, Rhagoletis*, and *Conotrachelus*) demonstrates why the SPs are considered so effective when viewed from a pest control perspective. Permethrin and fenvalerate were variably toxic to the secondary pests of both crops. For example, they were effective on cotton species such as leaf perforator and *Lygus*, but were considerably weaker on aphids and spider mites (Table 13.4). On secondary pests of apple, these SPs had similar mixed efficacy, being very effective on leafhoppers and leafminers and much less effective on mites and aphids.

Among natural enemies of cotton and apple pests there are major differences in the toxicity of SPs. These are due not so much to variable susceptibility among common species that occur on both crops (e.g., *Chrysoperla* spp., spiders, and hemipterans), but rather to the complement of species that make up the most important biological control agents on each crop. For example, on cotton, SPs are of low or moderate toxicity to spiders, some hemipteran species, coccinellids, *Chrysoperla* spp., and several braconid parasitoids. As mentioned, these natural enemies of key cotton pests are the primary focus of conservation efforts. However, on apple, SPs are highly toxic to phytoseiid mites and most other important predators of mites and aphids (Table 13.5), which are the primary focus of biological control programs on apple.

Since permethrin and fenvalerate were first developed and registered for use, a variety of newer SPs have been tesed for their activity on the species complexes of apple and cotton. In Table 13.6, a summary of these tests is presented for the complex of species associated with apple. Additional information on the safety to nontarget species such as honey bees and mammalian species is included. Compared to permethrin and fenvalerate, a number of the newer compounds show increased activity on apple aphids (e.g., especially the woolly apple aphid and the apple aphid). Compounds like bifenthrin and fenpropathrin have fairly good activity on spider mites. While an increase in the activity on secondary pests of apple is desirable, there is concern that too much activity may upset the predator–prey balance. Elimination of the food supply of mite predators by pyrethroid applications could increase mite problems rather than improve them due to a disruption of biological control. Resistance to these newer SPs by spider mites could develop rapidly if selection pressures were intense. (In fact, spider mite populations intensively selected either in the laboratory or in the field with some of the newer acaricidal pyrethroids have developed resistant populations rapidly, while their phytoseiid mite predators have remained relatively susceptible.)

Very little improvement in selectivity to the natural enemies of apple pests (Table 13.6) has been achieved with the newer SPs (although some improvement in honey bee conservation with flucythrinate and fluvalinate is seen in Table 13.6). All pyrethroids are still highly toxic to predatory mites and a single field application will eliminate these mites for up to 1–3 months in the field (e.g., AliNiazee and Cranham 1980, Riedl and Hoying 1983, Hull and Starner 1983, Jones and Parrella 1983, Bostanian and Belanger 1985, Bostanian et al. 1985). The only exception to this generalization involves either evolved or genetically improved SP-resistant predators (e.g., Hoy and Knop 1981, Whalon et al. 1982a; Chapters 19 and 20).

The selectivity of pyrethroids to the coccinellid predator *Stethorus punctum* is beginning to show some improvement after use in experimental plots for several years. This is a result of the development of permethrin and fenvalerate resistance in these strains (Hull and Starner 1983; see Table 13.6). However, some of the newer SPs have been shown to be less active on this predator of spider mites (e.g., bifenthrin and fenpropathrin; Table 13.6). Very little improvement in the selectivity of the newer SPs to the parasitoids of apple pests has been achieved to date (Table 13.6), although tolerance in some parasitoids associated with leafminer pests has been reported (Hagley et al. 1981).

13.3.2. Physiological Selectivity

The best defined cases of physiological selectivity among cotton natural enemies were those mentioned earlier for *C. carnea* and the hymenopteran parasitoids *Microplitis croceipes* (Powell and Scott 1985, Powell et al. 1986a), *Apanteles marginiventris* (Wilkinson et al. 1979), and *Bracon hebetor* (Press et al. 1981). Based on LD, LC, or LT_{50} comparisons, selectivity of SPs to each of these species over their prey or hosts have ranged from 10- to

Table 13.6. Toxicity and Selectivity of Pyrethroid Insecticides Registered or Proposed for Registration on Pome and Stone Fruits in the United States for 1985

	Activity Spectrum on Arthropod Species[1]																
	Pests[2]											Nontargets[3]					
Compound	Pc	Am	Cm	Lr	Stlm	Walh	Ap	Sjs	Tpb	Ps	Sm	Pm	Sp	Pr	Pa	Hb	M
Permethrin	G[4]	F	G	G	G	G	F*	P	G	G	P*	HT[5]	MT	HT	HT	HT	LT
Fenvalerate	G	F	G	F	G	G	F*	P	G	G	P*	HT	HT	HT	HT	MT	LT
Flucythrinate	F	G	G	G	G	G	F*	P	G	G	P*	HT	HT	HT	HT	LT	LT
Fluvalinate	F	F	F	F	G	G	F	P	G	P	P*	HT	HT	HT	HT	LT	LT
Fyfluthrin	F	G	G	G	G	G	F	P	G	G	P*	HT	HT	HT	HT	—	LT
Bifenthrin	F	G	G	G	G	G	F	P	G	P	F*	HT	MT	HT	HT	—	LT
Fenpropathrin	F	G	G	G	G	G	G	P	G	P	F*	HT	HT	HT	HT	—	LT
Cypermethrin	F	G	G	G	G	G	F	P	G	G	P*	HT	HT	HT	HT	—	LT

[1] Activity spectrum includes both toxic and repellent effects.
[2] Pc = plum curculio; Am = apple maggot; Cm = codling moth; Lr = leaf rollers; Stlm = spotted tentiform leafminer; Walh = white apple leafhoppers; Ap = aphids; Sjs = San Jose scale; Tpb = Tarnished plant bug; Ps = pear psylla; Sm = spider mites.
[3] pm = predaceous mites; Sp = *Stethorus punctum*; Pr = predaceous insects; Pa = parasitoid species; Hb = honey bee; M = mammalian toxicity.
[4] Activity: G = good; F = fair; P = poor.
[5] Toxicity: HT = high toxicity to nontarget; MT = moderate toxicity; LT = low toxicity.

*Compound causes upsets of this particular pest.

1000-fold. Among apple species, the most conclusive evidence for high physiological selectivity involves the mirid *Hyaluriodes vitripennis* in Pennsylvania orchards. This species readily tolerates multiple applications of SPs each season, becoming the predominant predator of secondary pests such as aphids and mites (Hull and Starner 1983), while its competitors are eliminated by these treatments. Other documented cases of physiological selectivity of SPs to orchard inhabiting natural enemies involve several carabid predators (Hagley et al. 1980), *Macrocentrus ancylivorus* parasitizing the oriental fruit moth *Grapholitha molesta* (Pree 1979), *Chrysopa oculata* (Pree and Hagley 1985), and *Apanteles ornigis*, a parasitoid of the spotted tentiform leafminer *Phyllonoryctor blancardella* (Hagley et al. 1981).

13.3.3. Ecological Selectivity

Ecological selectivity with SPs is accomplished in cotton by combining their use with the planting of short-season varieties and applications of the microbial insecticide *Bacillus thuringiensis* (Croft et al. 1985). In Texas, short-season cotton sheds over 50% of all squares and young bolls grown in the absence of insect damage. Considerable loss of fruit may cause no drop in yield if the loss occurs early in the fruiting season. During most seasons, the plant essentially runs a race to set more fruits than insects remove. The most important insect pests attacking the fruit in Texas are the boll weevil, the pink bollworm, the cotton bollworm, and the tobacco budworm. Early season sprays of *B.t.* are preferred for the control of budworms and bollworms because natural enemies are at a very sensitive stage of establishment in cotton fields at this time. If this selective microbial is not effective enough to moderately suppress these moth pests, then limited applications of SPs may be used. These are less selective to natural enemies but more effective than the microbial in suppressing the moth pests. SPs are used only when necessary, but as early as possible so as not to disrupt the biological control of *Heliothis spp.* later in the season when the plant cannot compensate for further fruit loss.

A somewhat similar approach has been experimentally evaluated on apple to attain ecological selectivity using pyrethroids (Hull and Starner 1983). The greatest damage to the crop and the most critical period of biological control of secondary pests such as mites and aphids occurs primarily later in the growing season when temperatures are hotter. Applications of SPs are restricted in early season to control such troublesome species as the *Lygus* bug, the spotted tentiform leafminer, and some tortricids. These sprays may initially decimate the small predatory insect and mite populations present. However, recovery of some species and subsequent immigration, particularly by the coccinellid *S. punctum* and certain predatory hemipteran species (e.g., *Hyaluriodes vitripennis*), is often sufficient to control these pests biologically later in the growing season in Pennsylvania orchards. While this approach to ecological selectivity with SPs to achieve temporal asynchrony between pests and natural enemies has been demonstrated experimentally, multiple applications of SPs in apple IPM systems in North America are not generally recommended (Croft and Hull 1987).

13.4. INSECTICIDE RESISTANCE AND PYRETHROID SELECTIVITY

A major factor contributing to the slow introduction and limited use of SPs for IPM on apple has been the lack of selectivity provided by these products. Another factor has been the unique patterns of resistance developed to earlier pesticides, especially the organophosphates (OPs) over the past 35 years. The scenario of resistance development (and lack thereof) to azinphosmethyl in the entire pest–natural-enemy complex on apple has facilitated selective use of this OP pesticide (Croft 1982, Whalon and Croft 1984). This resistance pattern allowed for enhanced biological control of certain groups of secondary pests, while sustaining effective chemical control of a number of key pest species.

Azinphosmethyl was first employed on apple as a nonselective broad-spectrum insecticide which had excellent activity on most key pests and relatively good activity on many secondary pests. It was somewhat weak on aphids and mites and highly toxic to most of the natural enemies of secondary apple pests (Whalon and Croft 1984). Thereafter, secondary pests, including mites, aphids, leafhoppers, leafminers, and others, developed high levels of resistance to this compound. Fortunately, no key pest of apple has developed high levels of resistance to this compound. However, a variety of natural enemies have developed azinphosmethyl resistance, including several phytoseiid mites, the coccinellid *Stethorus punctum*, and the predatory cecidomyiid *Aphidoletes aphidimyza* (Chapter 14). There is good circumstantial evidence that low-level resistance to this insecticide may have evolved in several parasitoids associated with leafhoppers, aphids, and scales on apple after more than 30 years of exposure (Croft and Strickler 1983, Whalon and Croft 1984; Chapter 20).

Most of the pest–natural-enemy complex of apple has begun to adapt to long-term chemical selection with OPs. In effect, azinphosmethyl has gone from a nonselective pesticide to a much more selective one due to resistance development in natural enemies. In reaching this state, there were very difficult periods of pest control (or lack thereof) when several secondary pests had developed resistances to the OPs, but their natural enemies had not. Later, natural enemies also developed resistant strains while key pests were still susceptible. Selective IPM programs were then implemented and biological control of certain secondary pests was attained for mites and aphids.

Because of the balance that has occurred in OP resistance in the pest–natural-enemy complex of apple in North America, researchers have been very cautious about introducing SP insecticides for widespread use. This is mostly due to their lack of selectivity to the natural enemies of mites and aphids. There is additional concern over a high propensity for resistance development to the SPs in key pests due to their known cross resistance to DDT. This compound was intensively used on apple and several pests had previously developed resistance to it (Croft 1982).

Pyrethroid insecticides have been registered for field use on cotton since the mid 1970s and on apple since the early 1980s. They have been used extensively on cotton because SPs are not as disruptive to the pest–natural-enemy complex on this crop as they are on apple. The use of SPs on apple has been very limited, except in a few local regions. As a result of these differences in the use of SPs, the patterns of SP resistance have been very different for pest species occurring on cotton and apple.

Since pyrethroids have been heavily used on cotton, concern about resistance is industrywide. For some time, resistance development in key lepidopteran pests such as *Heliothis* spp. has been of concern in certain areas of the Far East, Turkey, and Australia, where use has been most intensive (Daily and McKenzie 1987). There is some evidence that resistance to older pesticides (e.g., DDT) may have preadapted *Heliothis* populations for cross resistance to the SPs. SP-resistant strains of *Heliothis* have been reported from several areas of the southern United States (Roush and Luttrell 1987). SP resistance has also been developing in the pink bollworm populations in Texas and parts of the southwestern United States. Most critical is whether resistance will develop to SPs as a class and to the geographic extent that these products will be rendered ineffective on the vast cotton acreage on the North American continent. If this should occur, the development of more selective SPs or operational tactics for improving selective use of these compounds will become a moot point. This will affect the development of selective products not only on cotton, but on other crops as well.

In contrast to cotton, the use of SPs for apple pest control has been limited. However, SP use is increasing due to the ineffectiveness of OPs on certain leafrollers and other sporadic or secondary pests (Pree et al. 1980, Croft and Hull 1987). In spite of very conservative SP use, resistant leafminer strains have developed (Pree et al. 1987) in areas of most intensive use in North America. Interestingly, there is some evidence that predatory mites, *Stethorus punctum*, and possibly a leafminer parasitoid have also acquired some resistance to these insecticides (Strickler and Croft 1981, Hull and Starner 1983, Pree et al. 1987). Extreme caution is recommended if SPs are to be more widely employed on apple. Research on the development of new more selective compounds and ecological selectivity should be a high priority of industry scientists who work with the pest–natural-enemy complexes on this crop.

13.5. IMPROVING PHYSIOLOGICAL SELECTIVITY OF PYRETHROIDS

Analysis from the SELCTV database showed that most SPs were highly toxic to natural enemies. However, the exceptions demonstrate that the potential exists for developing more physiologically selective products (Fig. 13.1). Fluvalinate and tralomethrin were much less toxic to natural enemies than most other SPs (Fig. 13.1), but they have not been widely used for commercial pest control. A more concentrated effort by industry is needed to develop and register these types of SPs. A compound exhibiting physiological selectivity need not be the best compound for pest suppression if it allows for appreciable biological control by natural enemies. In many cases, the conserved beneficial can compensate for less effective chemical control (see examples in Fig. 9.2).

In further improving the physiological selectivity of pyrethroid compounds, more specific studies comparing chemical structure–activity relationships on pests and natural enemy species are needed. As noted earlier, the work by Stevenson and Walters (1984) comparing the relative activity of several pyrethroids in a natural enemy–pest model system (i.e., Indian meal moth *Ephestia kuhniella* and its ichneumonid parasite *Venturia canescens*) is a good example of needed research.

In similar screening studies with acarines, Croft and Wagner (1981) evaluated the relative toxicity of four experimental pyrethroids on OP-resistant strains of the spider mite *Tetranychus urticae* and permethrin- and OP-resistant strains of its predator *Amblyseius fallacis* (Fig. 13.3). Small differences in LC_{50} values were observed between resistant and susceptible strains of *T. urticae*, indicating that there was little cross resistance to the pyrethroids from previous OP selection. Spider mite LC_{50} values from LDP lines for the four compounds were also similar, each showing a flat response curve in the concentration range of 0.008–0.05% a.i. (Fig. 13.3). LDP lines for *A. fallacis* had consistently higher slope values and LC_{50} values for resistant and susceptible strains ranging from 0.0013 to 0.032% a.i. Cross resistance to each experimental SP in the permethrin resistant strains of predators ranged from 2- to 46-fold.

In the same study, fluvalinate and NCI 85913 were the least selective to *A. fallacis*. However, compound SD 57706 showed that the predator suffered slightly less mortality than its prey at lower concentrations of 0.001–0.025% a.i. Data shown in Fig. 13.3

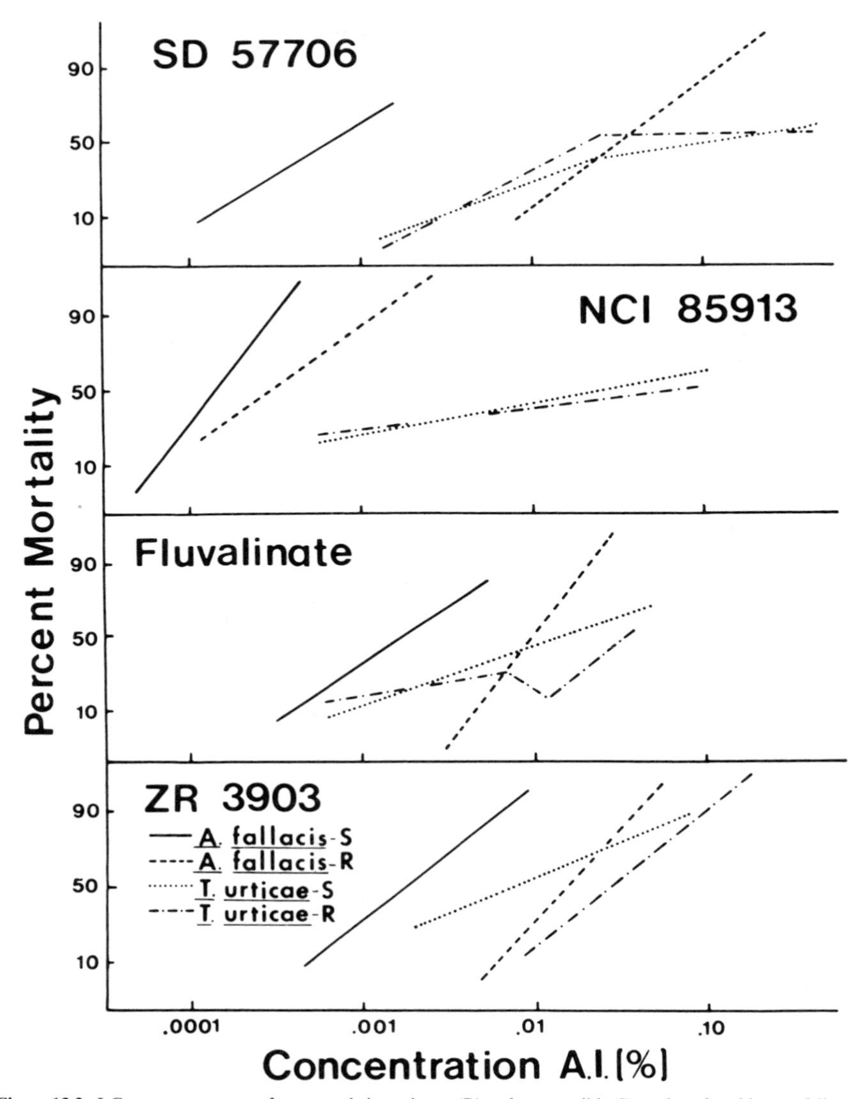

Figure 13.3. LC_{50} response curves for permethrin-resistant (R) and -susceptible (S) strains of *Amblyseius fallacis* and OP-resistant (R) and -susceptible (S) strains of *Tetranychus urticae* to four pyrethroid insecticides (from Croft and Wagner 1981).

demonstrate in principle how a more selective acaricidal pyrethroid might be integrated into this predator–prey system. Croft and Wagner (1981) noted that further screening with other compounds was necessary since it was unlikely that selective rates for these two species would be useful for suppressing other orchard pest such as codling moth or apple maggot. They recommended that an acaricidal SP more active on *T. urticae* should be found to which the predator has sufficient cross resistance to confer greater selectivity.

While the above studies illustrate the types of research necessary to develop more selective SP insecticides, they still stress simple selectivity. Once selectivity to one or more key natural enemies is established, broader selectivity should be sought for complexes of pests and natural enemies in these crop production systems. Field evaluations are often the only feasible way to carry out such diversified tests (Chapter 5). Broad selectivity will be difficult to find within the SPs based on previous experience (Appendix). However, such efforts are critical if IPM systems that have been developed over the past several decades are to be maintained.

PART FIVE

PESTICIDE RESISTANCE

14

PESTICIDE RESISTANCE: DOCUMENTATION

14.1. INTRODUCTION

Pesticide resistance has been observed in many types of arthropods, including a diversity of pest and beneficial insects, mites, and ticks (Georghiou and Saito 1983, NAS/NRC 1986). However, resistance was discovered much earlier among pests than among natural enemies of the phylum Arthropoda (Croft and Brown 1975, Croft and Strickler 1983, Hoy 1985).

The first widely recognized case of pesticide resistance among agricultural arthropod pests was reported in the early 1900s, when lime sulfur failed to control the San Jose scale on Washington State apples (Melander 1914). Since then, more than 500 species of arthropod pests are known to have evolved resistant (R) strains to one or more pesticides (Georghiou and Saito 1983, NAS/NRC 1986). This total includes about 375 cases among agricultural pests and 125 pests of medical importance to humans or animals.

Table 14.1. Documented Cases of Resistance to Pesticides in Arthropod Predators and Parasitoids

Species	Family	Crop	Pesticide	Fold-R	Conditions[1]	Locality	Reference
Aphytis africanus	Aphelinidae	Citrus	Methidathion	6	F	S. Africa	Schoones & Giliomee 1982
Aphytis coheni	Aphelinidae	Citrus	Azinphosmethyl	12	L*	Israel	Havron & Rosen 1988
			Permethrin	2	L*		
Aphytis holoxanthus	Aphelinidae	Citrus	Azinphosmethyl	4	F	Israel	Havron & Rosen 1988
				30	L*		
Aphytis melinus	Aphelinidae	Citrus	Malathion	4	L*	Australia	Abdelrahman 1973
	Aphelinidae	Citrus	Parathion	>10	F	CA, USA	Strawn 1978
	Aphelinidae	Citrus	Carbaryl	2	F	CA, USA	Rosenheim & Hoy 1986
			Chlorpyrifos	2	F		
			Dimethoate	3	F		
			Malathion	4	F		
			Methidathion	8	F		
	Aphelinidae	Citrus	Carbaryl	5–20	L*	CA, USA	Rosenheim & Hoy 1988
			Dimethoate	2	L*		
			Methidathion	2	L*		
Bracon mellitor	Braconidae	Cotton	DDT/toxaphene	9	L*	TX, USA	Adams & Cross 1967
			DDT	4			
			Carbaryl	4			
			Methyl parathion	4			
Comperiella bifasciata	Aphelinidae	Citrus	Methidathion	66	F	S. Africa	Schoones & Giliomee 1982
Encarsia formosa	Eulophidae	Greenhouse	Lindane	3	F, L*	UK	Walker & Thurling 1984
			Permethrin	15			
Macrocentrus ancylivorus	Braconidae	Peach	DDT	12	L*	Canada	Pielou & Glaser 1952, Robertson 1957
Pholetesor ornigis	Braconidae	Apple	Permethrin	3	F	Canada	Trimble & Pree 1987
			Methomyl	4			
			Fenvalerate	2			

Trichogramma evanescens	Trichogrammatidae	?	Oxydemeton-methyl	22	L*	Poland	Kot et al. 1971 Kot et al. 1975
Trioxys pallidus	Aphidiidae	Walnut	Azinphosmethyl	8	L*	CA, USA	Hoy & Cave 1988
Geocoris pallens	Anthocoridae	Alfalfa	Trichlorfon	>10	F	WA, USA	Johansen & Eves 1973
Chiracathium mildei	Clubionidae (Araneae)	Citrus, cotton	Malathion	3	F	Israel	Mansour et al. 1981
Aphidoletes aphidimyza	Cecidomyiidae	Apple	Azinphosmethyl	15	F	MI, USA	Adams & Prokopy 1977, Warner & Croft 1982
Chrysoperla carnea	Chrysopidae	Alfalfa	Carbaryl	3	L*	CA, USA	Grafton-Cardwell & Hoy 1985
			Diazinon	4	L*		
			Fenvalerate	2	L*		
			Methomyl	3	L*		
			Permethrin	11	L*		
			Phosmet	16	L*		
	Chrysopidae	Apple	Permethrin	34	F	Ont., Canada	Pree et al. 1988
			Fenvalerate	4	F		
			Cypermethrin	9	F		
			Deltamethrin	31	F		
			DDT	11	F		
			Azinphosmethyl	33	F		
			Phosmet	63	F		
			Parathion-ethyl	19	F		
			Malathion	5	F		
			Carbaryl	5	F		
			Methomyl	20	F		
		Cotton	Permethrin	46	F	CA, USA	Pree et al. 1988
			Azinphosmethyl	17	F		
			Carbaryl	6	F		
Coleomegilla maculata	Coccinellidae	Cotton	DDT	6	F	LA, USA	Atallah & Newsom 1966
	Coccinellidae	Cotton	DDT	3	F	MS, USA	Mohamad 1974
			DDT/toxaphene	4	F		
	Coccinellidae	Cotton	Methyl parathion	10–35	F	MS, USA	Chambers 1973
	Coccinellidae	Cotton	Methyl parathion	11–29	F	MS, USA	Mohamad 1974
	Coccinellidae	Cotton	Methyl parathion	11	F	MS, USA	Head et al. 1977
	Coccinellidae	Cotton	Monocrotophos	7–12	F	MS, USA	Mohamad 1974

Table 14.1. (*Continued*)

Species	Family	Crop	Pesticide	Fold-R	Conditions[1]	Locality	Reference
Stethorus punctum	Coccinellidae	Apple	Azinphosmethyl	—	F	PA, USA	Hull & Starner 1983
Stethorus punctillum	Coccinellidae	Apple	Azinphosmethyl	—	F	Italy	Pasqualini & Malavolta 1985
Labidura riparia	Labiduridae	Cotton	DDT/toxaphene	—	F	Australia	Gross & Schuster 1969, Bishop & Blood 1980
Nabis sp.	Nabidae	Cotton	Trichlorfon	>10	F	WA, USA	Johansen & Eves 1973
Amblyseius chilenensis	Phytoseiidae	Apple	Azinphosmethyl	2	F	Uruguay	Croft et al. 1976b
			Phosmet	10			
Amblyseius fallacis	Phytoseiidae	Apple	Azinphosmethyl	100–1000	F	MI, NC, USA	Motoyama et al. 1970, Croft & Nelson 1972
			Carbaryl	25–77	F, L*	IL, USA	Croft & Meyer 1973
			Diazinon	119	F	MI, USA	Croft et al. 1976a
			Parathion	103–152	F	MI, NC, USA	Motoyama et al. 1970, Croft & Nelson 1972
			Permethrin	60	L*	MI, USA	Strickler & Croft 1981
			Permethrin/fenvalerate	—	F	PA, USA	Hull & Starner 1983
Amblyseius hibisci	Phytoseiidae	Citrus	Carbaryl	—	F	CA, USA	Tanigoshi & Fagerlund 1984
			Dimethoate	22	F	CA, USA	Tanigoshi & Congdon 1983
			Parathion	—	F	CA, USA	Kennett 1970
				50–70	F	CA, USA	Tanigoshi & Fagerlund 1984
Amblyseius longispinosus	Phytoseiidae	Citrus	Carbaryl	17	L*	Taiwan	Lo et al. 1984, Lo 1986
			Carbaryl/dimethoate	15			
			Malathion	38			

Amblyseius nicholsi	Phytoseiidae	?	Fenvalerate	—	L	China	Ding et al. 1983
			Deltamethrin	—			
			Phosmet	19	L*	China	Huang et al. 1987
			Phoxim	29			
			Optunal	9			
			Dimethoate	5			
			Vapona	4			
			Trichlorfon	3			
Amblyseius potentillae (= *andersoni*)	Phytoseiidae	Apple	Azinphosmethyl	40	F	Italy	Gambaro 1975
			Phenthoate	—			Caccia et al. 1985
			Tetrachlorvinphos	33			Gambaro 1986
			Azinphosmethyl tetrachlorvinphos	—	F	Switzerland	Baillod 1986
			Parathion	100	F	Switzerland	Anber & Oppenoorth 1986
			Propoxur	2300			
			Azinphosmethyl	19			Anber & Overmeer 1988
			Paraoxon	311			
			Omethoate	21			
			Propoxur	781			
			Tetrachlorvinphos	61			
Amblyseius pseudolongispinosus	Phytoseiidae	Cotton	Dimethoate	7–13	L*	China	Ke 1987
Metaseiulus occidentalis	Phytoseiidae	Apple	Azinphosmethyl	101–104	F	WA, USA	Croft & Jeppson 1970, Ahlstrom & Rock 1973
		Grape	Dimethoate	11	F, L*	CA, USA	Hoy et al. 1979
			Methomyl	3	F		
		Almond	Methomyl	12	L*	CA, USA	Roush & Hoy 1981a
			Benomyl	—	F	CA, USA	Hoy et al. 1980
			Bayleton	—	F		
			Propoxur	>100	F	CA, USA	Roush & Plapp 1982
			Abamectin	4	L*	CA, USA	Ouyang & Hoy 1988
			Permethrin	13	L*		Ouyang & Hoy 1988
		Apple	Permethrin	10	L*	WA, USA	Hoy & Knop 1981
			Bioresmethrin	—	L*	Australia	van de Klashorst 1984
		Almond	Carbaryl/OP/permethrin	—	L*	Several	Hoy 1984a
			Carbaryl/OP/sulfur	—	L*	Several	Hoy 1984a

Table 14.1. (*Continued*)

Species	Family	Crop	Pesticide	Fold-R	Conditions[1]	Locality	Reference
Phytoseiulus persimilis	Phytoseiidae	Greenhouse	Diazinon	292	F, L*	Netherlands	Schulten et al. 1976
			Mevinphos	23			
			Malathion	6			Schulten & van de Klashorst 1974
			Oxydemeton-methyl	33			
			Parathion	>292			
			Dichlofluanid	—	F	Netherlands	Hassan 1982a
			Pirimicarb	>10	F, L*		
			Pyrazophos	>10	F		
			Fenitrothion carbaryl	—	?	Germany	Mori & Gotoh 1986
			Chinomethionate				
			Pyrazophos	—	L*	Germany	Koenig & Hassan 1986
			Pirimiphos-methyl	—	L*		
			DDVP	5	F, L*	France	Avella et al. 1985
			Deltamethrin	10	F, L*	France	Fournier et al. 1987a, 1987b
			Methidathion	200			Pralavorio et al. 1988
			Malathion	—	F	USSR	Storozhkov et al. 1977
		Apple	SP	—	L*	New Zealand	Markwick 1986
		Greenhouse	Bioresmethrin	50	L*	Finland	Manttari 1980
Typhlodromus aberrans	Phytoseiidae	Grape	Parathion	—	F	Italy/ Switzerland	Corino et al. 1986
Typhlodromus arboreus	Phytoseiidae	Apple	Azinphosmethyl	7	F	OR, USA	Croft & AliNiazee 1983
			Carbaryl	3–5	F		
Typhlodromus pyri	Phytoseiidae	Apple	Azinphosmethyl	10	F	New Zealand	Hoyt 1972
				11–42	F	New Zealand	Penman et al. 1976
				14	F	New Zealand	Wong & Chapman 1979
				12–20	F	NY, USA	Watve & Lienk 1976
				4–6	F	UK	Kapetanakis & Cranham 1983

	Carbaryl	20	F		
	Parathion	50	F		
	Bromophos	7	F	Netherlands	Overmeer & van Zon 1983
	Carbaryl	50	F		
	Parathion	100	F		
	Paraoxon	1000	F	Netherlands	van de Baan et al. 1985
	Parathion	100	F		
	Cypermethrin	3	L*	New Zealand	Markwick 1986
		14	L*	New Zealand	Markwick 1988
		35	F*		
	Azinphosmethyl	—	F	Switzerland	Baillod et al. 1985
Grape	Phosmet	—	F		
	Diazinon	—	F		
	Methidathion	—	F		
	Mevinphos	—	F		

[1] F = field; L = laboratory.

*Represents genetic selection or genetic improvement attempts; see further discussion in Chapter 20.

Pesticide resistance in an arthropod natural enemy, based on comparisons with a susceptible (S) strain, was not proven until the early 1970s (Motoyama et al. 1970, Croft and Jeppson 1970). To date, only 31 cases of resistance in individual predators or parasitoid species have been documented (Table 14.1; Croft and Brown 1975, Croft 1977, Roslavtseva 1980, Croft and Strickler 1983, Hoy 1984a). When present in natural enemy populations, pesticide resistance can be of great benefit to IPM (rather than a detriment when present in pests). Resistance can confer physiological selectivity to a predator or parasitoid over its prey or host for that compound to which resistance has developed, and sometimes for related compounds (Chapters 9 and 13).

14.2. RESISTANCE DETECTION IN NATURAL ENEMIES

Several authors have speculated on the less frequent documentation of resistance among entomophagous arthropods compared to herbivorous pests. (See reviews of this subject in Newsom 1974, Georghiou 1972, Croft and Brown 1975, Croft 1977, Croft and Strickler 1983, Tabashnik and Johnson 1988.) Historical research patterns as well as biological factors may be involved.

Research on almost all aspects of natural enemy biology, including pesticide resistance, has lagged behind that of pest species (Chapters 1 and 2). The lower number of reported cases of resistance among natural enemies may, in part, be an artifact of limited research time and resources. To illustrate the importance of this factor, Tabashnik and Johnson (1988) pointed out that an unbiased survey of resistance among both species groups could be undertaken, although no such survey has been conducted to date. Such a survey would be extremely costly. Also, it would be difficult to verify that intrinsically susceptible strains of these groups were used to document resistance.

The presence of resistance in entomophagous species may be inherently more difficult to detect than in phytophagous pests. Several factors probably contribute to this disparity. Resistant natural enemies generally are less abundant than pests. A pest developing resistance has an almost unlimited food supply (the crop plant) to exploit following pesticide treatment. When susceptible predators and parasitoids are killed by pesticides, pest population densities often increase dramatically, thus dramatizing their resistant status. In contrast, when a natural enemy first develops resistance, its food supply (the pest) is often severely reduced following a pesticide treatment. The natural enemy then either emigrates or dies, and resistance usually goes unnoticed. Resistance in entomophagous species has been most commonly documented where natural enemies and their hosts or prey have simultaneously or sequentially developed R strains, and the pesticide to which they are resistant continues to be used for the control of other pests (Croft and Morse 1979, Morse and Croft 1981; see Section 16.5.1). In these instances, the natural enemy's food supply remains abundant and resistance may be detected in both the pest and the natural enemy.

The relative paucity of documented resistance in natural enemies may reflect biological reality as well as being an artifact of research constraints. Recent comparative research on the physiological and ecological factors influencing resistance in pests and beneficials suggests several biological explanations as to why resistance may actually be less common in the latter group. Natural enemies have been shown to exhibit lower detoxification capacities and may have a lower genetic variablity in response to pesticides; natural enemies depend on the resistance status of the target pest in order to exploit their own selected resistance features; and natural enemies have a greater tendency to reside outside of treated habitats in susceptible populations as compared to pests (Croft and Morse 1979, Croft 1982, May and Dobson 1986, Tabashnik and Croft 1985, Tabashnik 1986a). These factors and others contributing to the infrequent documentation of resistance in natural enemies are discussed further in Chapter 16.

14.3. RESISTANCE DEVELOPMENT IN NATURAL ENEMIES

Pesticide resistance is an evolutionary phenomenon. It occurs when an organism has a mechanism or mechanisms to better survive and reproduce in the presence of pesticides than an S or "wild" strain (WHO 1957). After selection, a resistance trait is passed on to the next generation. That trait then occurs in greater proportions in subsequent generations until, eventually, almost all individuals in the population are resistant. Resistance mechanisms are similar to those conferring pesticide tolerance (Chapter 6) and include reduced penetration, increased detoxification, alteration of target sites, or modification of behaviors leading to reduced exposure to the toxicant (Chapter 15).

In many cases, resistance mechanisms which evolve after selection with a particular pesticide also confer resistance to related compounds with similar modes of action—a phenomenon known as cross resistance. In this way, a population can become resistant to a pesticide with which it has had no previous contact. Multiple resistance occurs when exposure to several different pesticides has led to the presence of several resistance mechanisms in the same population.

Pesticide resistance should be distinguished from pesticide tolerance. The latter is an innate lack of sensitivity to a pesticide, presumably due to previous exposure to indigenous chemicals in the natural environment. Many natural enemies are tolerant to certain pesticides (Chapters 2 and 5). Resistance to pesticides should also be distinguished from a vigor tolerance, a condition of decreased pesticide susceptibility caused by differences in nutrition, conditions of birth, and development among populations (Chapter 4). Whereas resistance is a short-term response to pesticide exposure, innate tolerance is a more long-term response to natural toxins in an organism's evolutionary past.[1] While different in their origins, innate tolerance and developed resistance can have the same net effect on a population of natural enemies: a selective advantage is gained over their prey or hosts when these organisms are exposed to pesticides.

In practice, resistance is usually measured by comparing R and S strains of a species. Thus, 12-fold resistance means that the R strain requires 12 times as much toxicant to reach an LD_{50} as the S strain. In some cases, S strains of a natural enemy are difficult to find, but researchers may be able to infer resistance based on historical data indicating susceptibility (e.g., see discussion of resistance verification in the coccinellid *Stethorus punctum*, Section 14.4.3.; Asquith and Hull 1973).

What level of resistance is significant from a genetic or biological perspective as well as from a practical management viewpoint? This has been debated from a biological perspective by scientists studying resistance in pest arthropods for years. Generally, response differences between R and S strains in the range of 3- to 7-fold indicate that a genetically determined resistance mechanism has developed, although the methods used

[1] The terms *innate tolerance* and *developed resistance* are useful distinctions in explaining why a natural enemy species may not be susceptible to a particular pesticide. Because toxicants have become so widespread and have been used for over four decades, it is becoming more difficult to be sure that what appears to be innate tolerance is not a case of cross resistance to a pesticide used in the past.

to test for resistance can influence this guideline appreciably (Rosenheim and Hoy 1988). From an IPM perspective, resistance levels in natural enemies often must be substantial—in the range of 10- to 100-fold—before significant selectivity benefits can be obtained in the field (Chapter 20). High levels of resistance in natural enemies must be obtained to be of practical value because natural enemies often are more intrinsically susceptible to pesticides than are pests (Chapter 2). Lower levels of resistance may be of some value in achieving pesticide selectivity when combined with ecological selectivity, and in allowing the reentry of natural enemies from refugia into treated habitats as pesticide residues are dissipating (Rosenheim and Hoy 1988, Hoy and Cave 1988).

This initial chapter on resistance in natural enemies focuses on the documentation of resistance and cross resistance, particularly where developed under field conditions (as opposed to genetic improvement projects; Chapter 19). Some discussion of laboratory-selected resistance is included for historical perspective. In Chapters 15 and 16, the genetics, physiology, and factors influencing resistance in natural enemies are reviewed. The second half of this volume is devoted to the management of resistance in the field and to the improvement of IPM. An approach to resistance management is presented whereby resistance in pests is minimized while maximum exploitation of resistance in beneficial predators and parasitoids is sought in order to achieve physiological selectivity with a pesticide (Croft 1981b; see also Chapter 19).

14.4. DOCUMENTED CASES OF PESTICIDE RESISTANCE

Documented cases of resistance to pesticides in 20 predaceous and 11 parasitic arthropods are listed by family, species, and pesticide in Table 14.1. Rather than treat these cases taxonomically, each case is reviewed in relation to several historical eras of research, followed by a comparison of resistance in predators versus parasitoids.

14.4.1. Early Selection Studies with Parasitoids

The first studies of pesticide resistance in natural enemies began in the late 1940s, just as resistance to DDT was being reported in a wide range of plant-feeding pests. Researchers logically questioned whether or not natural enemies of pests would respond similarly to pesticide selection pressure (e.g., Pielou 1950, Spiller 1958). Early resistance experiments with natural enemy species were specifically designed to answer this question.

Pielou and coworkers (Pielou 1950, Pielou and Glasser 1952, Robertson 1957) first demonstrated that the potential for resistance was actually present in a natural enemy. They selected *Macrocentrus ancylivorus*, a braconid parasitoid of the oriental fruit moth, in the laboratory with DDT. Selections were made by exposing newly emerged adults to crystalline residues of pesticide. Mortality levels of populations were maintained near 70% throughout the tests. A 12-fold resistance factor was initially produced following 19 generations of selection; however, no significant increase beyond this level was achieved thereafter. After 6 years of research and treatment of 3 million insects, the test was discontinued at generation 72. Afterward, reversion of resistance to near the original susceptibility level occurred after only 13 generations. This early attempt to produce a laboratory-selected, highly pesticide-resistant strain of parasitoid was typical of other early selection attempts made from 1950 through the early 1970s.

A similar attempt to select for resistance was made with *Bracon mellitor*, a braconid parasitoid of the boll weevil *Anthonomus grandis*, using DDT, methyl-parathion, toxaphene, carbaryl, or a DDT + toxaphene mixture (Adams and Cross 1967). Adult parasitoids 2–10 days old were topically treated with toxicant. Mortality in selection was maintained near 50% to ensure adequate stock in ensuing generations. Resistance levels ranging from only 4- to 9-fold were achieved after 5–10 generations of laboratory selection. Adams and Cross (1967) concluded that while the resistance levels achieved were not as spectacular as those developed by some pest insects at the time, a mechanism for resistance was clearly demonstrated in this species.

Kot et al. (1971) selected the egg parasitoid *Trichogramma evanescens*, which attacks the host *Sitotroga cerealella* by dipping parasitized host eggs into the toxicant. Mortality in selections was maintained near 50% throughout the tests to ensure sufficient stock colonies for subsequent generations. A maximum 22-fold resistance to metasystox was observed after 70 generations and 34 treatments; however, resistance was unstable. When selection was discontinued, it regressed to a level only 7.5 times greater than an S strain in just a few generations. The authors concluded that their inability to obtain a genetically stable R strain was due either to the recessive character of the resistance (which they did not determine specifically; see Chapter 15) or to an ineffective method of selecting the organism.

Studies with the California red scale parasitoid *Aphytis melinus* involved exposure of adults confined in treatment cages to pesticide sprays in laboratory tests (Abdelrahman 1973). In selection experiments, parasitoids were exposed to concentrations chosen to produce 50% mortality. A 3.4-fold increase in resistance to malathion was observed after 8 generations of selection. Abdelrahman concluded that this species possessed a mechanism for resistance because the F_8 generation had a significantly lower slope value in the log-dosage–probit mortality line than did the original colony. He felt that resistance had not peaked and that the culture had the potential for resistance to increase further. However, since its primary host *Aonidiella aurantii* was over 700 times more tolerant or resistant to malathion, he doubted that further selection of a strain would provide a sufficient increase in resistance to be of any practical significance in the field.

During this era, there were several less quantitative reports that resistance was developing in field populations of parasitoids. They included resistance to DDT in the Pacific Northwest of the United States (Johansen 1957) and to lindane in Europe (Spiller 1958) in *Aphelinus mali*, an aphelinid parasitoid which attacks the wooly apple aphid *Eriosoma lanigerum* of deciduous tree fruits. Another report stated that *Bracon mellitor*, a parasitoid of the cotton boll weevil, was resistant to insecticides applied to cotton across the southern United States (Chesnut and Cross 1971). However, no experimental verification of these field observations was made, except in the laboratory selection studies with *B. mellitor* by Adams and Cross (1967).

To summarize, early laboratory studies on the selection of insecticide resistance in parasitoids produced mixed results; however, in no case was a high, stable level of resistance produced that resulted in appreciable, practical benefits to pest control (e.g., the release or management of R strains in the field). These studies did demonstrate the potential for resistance in entomophagous species.

These early studies were disappointing to researchers at the time, who were expecting higher levels of resistance that would be useful to biological control. As a result, a general attitude about resistance in natural enemies developed. While parasitoids and predators seemed to have sufficient genetic heterogeneity to develop resistance to insecticides when selected in the laboratory, they did not develop high levels of stable resistance as had previously been shown for the house fly, spider mites, mosquitoes, *Heliothis*, and other pest species after similar selection with pesticides in the field and laboratory. (See later discussion of laboratory versus field selection of natural enemies with pesticides; Chapter 16.) Furthermore, resistance development in field populations of predators and parasitoids was uncommon. Some went so far as to speculate that natural enemies might be limited in their potential to develop highly resistant strains.

14.4.2. Resistance Documentation in Phytoseiid Mites

During the late 1960s and early 1970s, resistance in field populations of phytoseiid mites and predaceous insects associated with spider mites in apple orchards was first noted. As a result of these findings, scientists again began to consider the possibility of resistance in beneficial predators and parasitoids. Furthermore, major benefits were obtained for IPM and for pesticide selectivity in the 1970s through the manipulation and management of insecticide-R strains of natural enemies (Chapters 15–20).

A renaissance in resistance research on arthropod natural enemies began with observations of high survivorship in the predatory phytoseiid mites *Metaseiulus occidentalis* and *Amblyseius fallacis* following field application of several organophosphate (OP) insecticides in a few apple orchards in the western and eastern United States (Hoyt 1969a, Swift 1970). These resistant natural enemies appeared only a few years after their spider mite prey had developed resistance to these same compounds, which continued to be used for the control of other tree fruit pests.

Shortly thereafter, documentation studies confirming azinphosmethyl (and possibly parathion) resistance in laboratory experiments were reported almost simultaneously for *M. occidentalis* (Motoyama et al. 1970) and *A. fallacis* (Croft and Jeppson 1970). Other researchers were then stimulated to look for resistance in phytoseiid species on tree fruits around the world and for resistant natural enemies in other intensively sprayed crops.

In retrospect, the first observation of tolerance or resistance to a pesticide in a phytoseiid mite may have come long before apple researchers noticed the high survivorship of *M. occidentalis* and *A. fallacis*. Huffaker and Kennett (1953) observed that *M. occidentalis* was much more tolerant to parathion than was its relative, *T. reticulatus*. At the time they were not sure whether this lack of susceptibility was due to inherent tolerance or to resistance. Because of the lack of convincing results in earlier selection experiments with parasitoids and the lack of other supporting cases, little attention was paid to this report.

In the 1970s and 1980s, researchers at many other locations across the United States surveyed for OP resistance in local strains of *M. occidentalis* and *A. fallacis*. With few exceptions, high levels of resistance in both predators were found (see reviews in Croft and Brown 1975, Croft and Hoyt 1983; Table 14.1). Following early confirmations of geographically widespread resistance in the Phytoseiidae, the number of documented, resistant species in this group of mesostigmatid mites has grown to at least nine (Table 14.1). Among species having R strains, most research has been conducted on *Amblyseius fallacis, Phytoseiulus persimilis, Metaseiulus occidentalis*, and *Typhlodromus pyri*, with somewhat lesser amounts for *A. hibisci* and *A. andersoni* (Table 14.1). Resistance profiles for these and other related phytoseiid species have been developed.

14.4.2.1. Amblyseius hibisci. *Amblyseius hibisci* is an important predator of citrus and avocado spider mites in California. An early report by Kennett (1970) suggested that resistance to parathion had long been present in this species, although precise documentation was provided only recently (Tanigoshi and Fagerlund 1984). OP resistance (Table 14.1) or cross resistance (Table 14.2) has been documented for dimethoate,

Table 14.2. Cross Resistance to Pesticides in Predaceous Phytoseiid Mite Species

Species	Primary Selection Agent	Cross-Resistance Spectrum[1,2]																							Reference
		D	M	P	m	A	p	p	d	d	T	m	C	*M*	*P*	B	*p*	F	*D*	c	R	PP	OP	S	
Amblyseius hibisci	Parathion			X	1	4			1			2	2	0										2	Tanigoshi & Congdon 1983
																									Tanigoshi & Fagerlund 1984
Amblyseius fallacis	Azinphosmethyl	0	0	4	2	X	4	0	2	1	0		0	0	0	0	0	0	0	0	0	0	0		Croft et al. 1976a
	Permethrin	2	2														X	2	4	4	2	2	2		Croft et al. 1982
Phytoseiulus persimilis	Parathion			X	2					4					4							0			Schulten et al. 1976
Metaseiulus occidentalis	Azinphosmethyl			4		X	4	4	2	2	4		0				0	0							Hoyt, unpubl.
	Carbaryl			0		0			0				X	0	2	2	0								Roush & Plapp 1982
	Permethrin																X	2					2		Hoy et al. 1980
	Bioresmethrin																2	2		X			2		
Typhlodromus pyri	Parathion			X		0							4			4									van de Baan et al. 1985

[1] D = DDT; M = methoxychlor; P = parathion; m = malathion; A = azinphosmethyl; p = phosmet; p = phosalone; d = dimethoate; d = diazinon; T = TEPP; m = methidathion; C = carbaryl; *M* = methomyl; *P* = propoxur; B = benomyl; *p* = permethrin; F = fenvalerate; *D* = decamethrin; c = cypermethrin; R = resmethrin; PP = pyrethrins; OP = other pyrethroids; S = sulfur.

[2] X = primary selection agent; O = tested, but no cross resistance; 1 = slight cross resistance; 2 = moderate cross resistance; 3 = high cross resistance; 4 = very high cross resistance.

azinphosmethyl, malathion, and methidathion, but not for acephate or chlorpyrifos (Tanigoshi and Congdon 1983, Tanigoshi and Fagerlund 1984). A slight cross resistance to carbaryl is present in this species as well, and may be related to OP resistance due to a modified acetycholinesterase mechanism as was found in *T. pyri* (van de Baan et al. 1985; Section 14.4.2.4). Little cross resistance to other carbamates such as methomyl and no resistance to chlorinated hydrocarbons has been reported.

14.4.2.2. ***Amblyseius fallacis.*** *Amblyseius fallacis* is one of the most widely studied resistant phytoseiid mites (Tables 14.1 and 14.2). *A. fallacis* is principally a predator of tetranychid spider mites, especially *Tetranychus urticae.* It occurs on a wide variety of crops, including deciduous tree fruits, several berry species, field crops, and other vegetation in the midwestern and eastern regions of North America.

The initial inherent susceptibility of *A. fallacis* to compounds such as parathion, DDT, and diazinon was apparent in early studies by Clancy and McAlister (1958). Multiple resistance (Table 14.1) or cross resistance (Table 14.2)[2] in this species after 30 years of exposure to synthetic-organic pesticides extends to several insecticide classes including DDT and its derivatives, OPs, carbamates, and SPs. This broad resistance profile is common to many individual strains of this mite (Croft 1983; Chapter 15).

The range of resistance or cross resistance in one multiple-R strain extends to at least 29 compounds from several insecticide classes (Fig. 4.3; see Scott et al. 1983 and Chapter 15 for a more detailed history of this strain). Originally, DDT and its derivatives were the least toxic insecticides to the S strain of *A. fallacis*, yet only moderate levels of resistance to these compounds have been found in field populations. Intrinsic susceptibility and developed resistance levels to OPs, carbamates, and SPs are highly variable (Fig. 4.3). Whereas resistance to DDT, carbaryl, and permethrin is local in distribution, OP resistance occurs throughout midwestern and eastern fruit-growing areas of the United States and on many other crops (see map in Croft and Brown 1975, Brown 1977). Resistance has become so widespread that it is difficult to find susceptible populations for experimental testing (Croft et al. 1976a).

14.4.2.3. ***Metaseiulus occidentalis.*** *Metaseiulus occidentalis* has been as extensively studied for resistance as has *A. fallacis*. It is also principally a predator of spider mites, including *Tetranychus* spp. and eriophyid rust mites, which occur on several crops in the more arid regions of western North America. *M. occidentalis* is a very important biological control agent on crops such as deciduous tree fruits, berries, hops, and a number of field crops.

Organophosphate resistance is as ubiquitous for *M. occidentalis* as it is for *A. fallacis* (Croft and Brown 1975, Hoy 1984a). As noted, this cosmopolitan species was first reported to be resistant or tolerant to OP compounds as early as 1952 (Huffaker and Kennett 1953). In the 1970s, populations highly resistant or tolerant to azinphosmethyl, diazinon, phosmet, parathion, TEPP, and phosalone were observed throughout the semiarid fruit-growing regions of the United States. High levels of OP resistance have apparently developed in the USSR, where this species (*T. longipilus* of Europe = *M. occidentalis*; Hoying and Croft 1977) inhabits the forest steppe areas of the Ukraine (Matvievskii 1979).

[2] Resistance and cross resistance in this species can be determined only by examining resistance levels to the principal selection agents used in pest control and then inferring how resistance to compounds that have never or seldom been used may have arisen. The specific selection agents responsible for the observed cross resistance can only be determined by making comparative resistance studies among several strains having different resistance profiles.

The distinction between resistance and innate tolerance has not always been clear for *M. occidentalis,* because base-line susceptibility to many OP compounds has not been measured in populations collected from natural, untreated environments. Nevertheless, there is good indication that what has been termed resistance to phosalone and TEPP may be due to natural tolerance; field populations from many sources are almost uniformly immune to these compounds at all but extremely high doses.

The question of whether the lack of susceptibility to parathion in this species reflects tolerance or resistance is more puzzling. Tolerance to this compound was first indicated by Huffaker and Kennett (1953), who showed that predators were unaffected by parathion almost immediately after it was first registered for use. Further evidence of innate tolerance was provided by Croft and Jeppson (1970), who reported similar parathion LC_{50} values

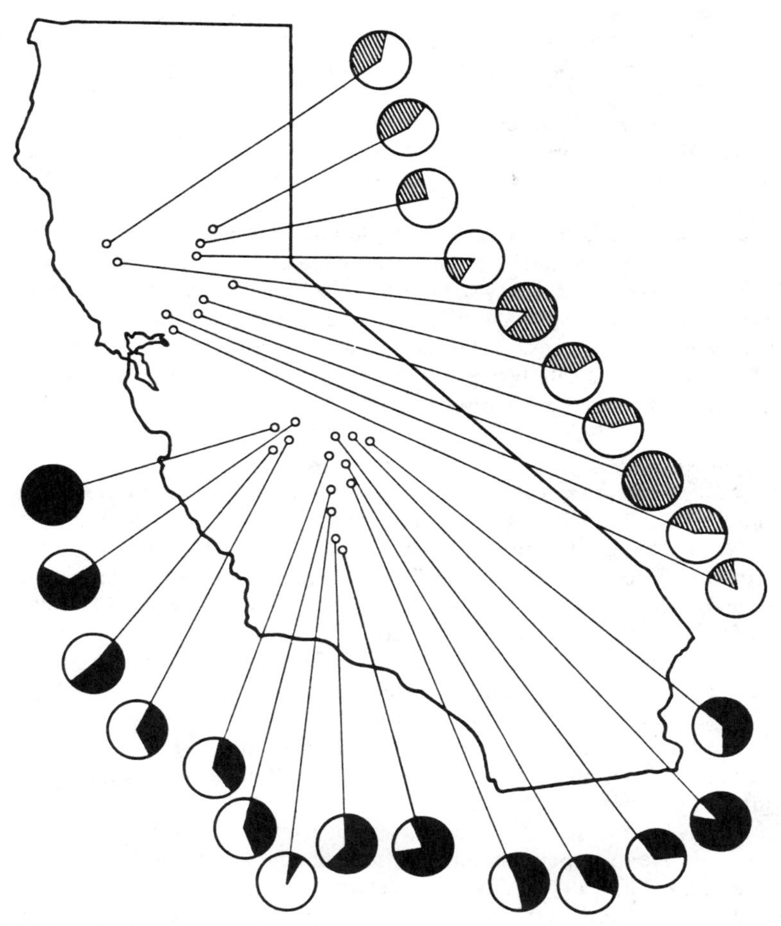

Figure 14.1. Variability in response to OP insecticides in *Metaseiulus occidentalis* in the San Joaquin valley, California. Populations sampled from pears (circles with striped shading) were tested with azinphosmethyl; predators from vineyards were tested with dimethoate. Completely shaded circles in each group represent the most resistant populations sampled as determined by LC_{50} values; partially shaded circles indicate relatively lower LC_{50} values (adapted from Hoy and Knop 1981, Hoy et al. 1980).

for two intensively selected strains from Washington and Utah commercial orchards and for two strains from untreated habitats in California. The inherent parathion tolerance of *M. occidentalis* was also indicated in an Australian study, where populations were accidentally introduced at some unknown earlier period (probably on tree stocks from the western United States). Field applications of parathion were readily tolerated by these populations, but high susceptibility to azinphosmethyl was present. (Azinphosmethyl-R strains were subsequently introduced into Australia to overcome this problem; see Chapter 18.) In a study of morphological differences in various *M. occidentalis* strains from Australia, Schicha (1978) made a brief reference to a native population that was still susceptible to parathion. Further study of this strain has not yet been reported in the literature. The question of this predator's apparently innate lack of sensitivity to parathion remains unresolved.

More recently, the range, degree, and spatial distribution of resistance and cross resistance in *M. occidentalis* populations on grape, almond, and pear crops in California have been studied by Hoy and coworkers (Table 14.2). Organophosphate resistance is widespread in *M. occidentalis* on these crops throughout the state; however, patterns of resistance are highly variable, even at adjacent orchard sites (Hoy 1985; Fig. 14.1).

Beyond OPs, resistance in native *M. occidentalis* populations to carbaryl (Hoyt unpublished) and sulfur (Hoy and Standow 1981) have been found (Table 14.1). Interestingly, sulfur resistance is a case of a natural enemy acquiring resistance to a material that is principally used as a fungicide. Also, artificial laboratory selections for resistance to carbaryl and the SP permethrin have been successful in this species (Roush and Hoy 1981a, Hoy and Knop 1981, Ouyang and Hoy 1988). Multiple resistance to OPs, carbamates, SPs, sulfur, and abamectin occur in several strains of this species (Table 14.1), which are being studied intensively for exploitation in agricultural habitats throughout the world (see Chapters 17–20). It is probably only a matter of time before most of these resistance features will be incorporated in a common superresistant strain of *M. occidentalis*.

14.4.2.4. Typhlodromus pyri. *Typhlodromus pyri* was slower to develop widespread azinphosmethyl resistance throughout its geographic range in northwestern and northeastern North America, Europe, and New Zealand (Table 14.1) than *A. fallacis* and *M. occidentalis*. In New Zealand, Hoyt (1972) first reported 10-fold resistance to azinphosmethyl, and Collyer and Geldermalsen (1975) later found populations in New Zealand with 14-fold resistance. In New York, Watve and Lienk (1976) reported nearly 20-fold resistance compared with the original data of Hoyt (1972). At the latter resistance level, predators readily tolerated field application rates, providing appreciable biological control of spider mites even where OPs were widely used. Penman et al. (1976) found resistance in New Zealand populations of *T. pyri* after 10 years of exposure to azinphosmethyl. They observed complete survival in the field at recommended field rates used for pest control.

T. pyri may be cross resistant to carbaryl as a result of OP selection (Table 14.2). This appears to be due to a common modified acetylcholinesterase mechanism (van de Vrie 1962, Watve and Lienk 1976, Collyer and Geldermalsen 1975, Kapetanakis and Cranham 1983, van de Baan et al. 1985, Hadam et al. 1986; Table 14.2). However, tolerance or resistance to carbaryl was noted much earlier than OP resistance (van de Vrie 1962); it may, in fact, be the original source of OP cross resistance rather than the reverse. Recently, higher levels of resistance to parathion ranging from 50- to 100-fold and to carbaryl and the related carbamate propoxur ranging from 50- to 2300-fold have been observed in

European populations (Kapetanakis and Cranham 1983, van de Baan et al. 1985, Hadam et al. 1986; Table 14.1). Laboratory selection of SP-resistant strains of this species are at a similar stage as compared to studies on *A. fallacis* and *M. occidentalis* (Markwick 1986; Table 14.1). Recently, an SP-selected field strain has exhibited a resistance 4 times higher than the recommended field rate of cypermethrin used in New Zealand apple orchards (Markwick 1988; Chapter 20). Effective control of phytophagous mites by *T. pyri* where multiple SP applications are used seasonally has been demonstrated (Markwick 1988).

14.4.2.5. Phytoseiulus persimilis. Resistance to OP insecticides in *Phytoseiulus persimilis* was first reported by Schulten et al. (1976) in deliberately selected Dutch glasshouse strains, and similar, independent observations have been made in several other geographic areas of Europe (Table 14.1). Parathion resistance in the Dutch strain conferred cross resistance to several OPs including diazinon, demeton-*S*-methyl, malathion, and mevinphos (Schulten et al. 1976; Table 14.2). There was even some indication of cross resistance to the carbamate pirimicarb (Table 14.2). A commercial colony used in glasshouses which was never purposefully selected for resistance also developed a high level of resistance to parathion (Schulten et al. 1976).

More recently, the spectrum of resistance in *P. persimilis* has been expanded to include moderate levels of resistance to pyrazophos, pirimicarb, DDVP, and deltamethrin (Table 14.1). Multiple resistance has been found in field-collected European populations from The Netherlands, England, France, and Germany (Hassan 1980), as well as in those selected in laboratory experiments (Table 14.1). Beyond parathion and carbaryl resistance, high levels of resistance were found to methidathion (200-fold) in a greenhouse-selected strain. This strain was further selected with several compounds, including the pyrethroid deltamethrin, with only modest success. Genetic improvement of SP resistance has been more difficult to achieve in *P. persimilis* and other predators from this family. Even after using radiation to increase genetic variation in laboratory selection trials, stable pyrethroid resistance was not obtained (Manttari 1980; see also genetic improvement of resistance in Chapter 19). Field-release studies of multiple-R strains of *P. persimilis* have undoubtedly been undertaken in greenhouses, but have not yet been widely reported in the literature. Nonetheless, significant progress has been made in the development of a multiple-R strain of this predator for use in controlling spider mites (e.g., *Tetranychus* spp.) in greenhouse systems throughout the world.

14.4.2.6. Amblyseius andersoni. Italian populations of *A. andersoni* (= *A. potentillae*; McMurtry 1977) from Po valley peach orchards were initially reported to have developed resistance to azinphosmethyl and other OPs in the early 1970s (Gambaro 1975, 1986; Table 14.1). Resistance was attributed to earlier selection with the OPs parathion and phenthoate, which had been heavily used for some 5–6 years prior to the introduction of azinphosmethyl. Gambaro (1975) reported that resistance increased very rapidly, from a few predators which survived initial azinphosmethyl applications in 1974 to survival of almost the entire population by the end of the growing season. Field data confirmed that high levels of resistance were present in predators after multiple sprays of azinphosmethyl. From 1975 to 1977, the resistant populations naturally spread to pear and apple orchards located 20–50 m away (Gambaro 1986). In 1984, these OP-resistant populations were introduced into trees inhabited by susceptible *A. andersoni*. The susceptible population was replaced by the resistant population in the course of only two seasons and kept phytophagous mites in check for the entire season despite five treatments with azinphosmethyl (Gambaro 1986). Only recently has verification of resistance to azinphosmethyl and other pesticides been published using S and R strains (Caccia et al. 1985).

A. andersoni has been reported by European researchers to be resistant to several other insecticides and fungicides (Caccia et al. 1985, Baillod 1986; Table 14.1). Currently, this natural enemy is widely relied upon for the control of plant-feeding mites (especially *P. ulmi*) throughout vineyards and orchards in southern and central Europe (Gambaro 1986, Baillod 1986; Chapters 17 and 18).

14.4.2.7. Other Phytoseiid Species. In addition to species that have developed multiple resistance over wide geographical regions, localized resistance to one or more compounds has been observed in *Amblyseius chilenensis, A. longispinosus, A. nicholsi, A. pseudolongispinosus*, and *Typhlodromus arboreus* (Table 14.1). Resistance in several of these species may be more widespread than is currently known; however, published reports of these cases are presently limited to a single or a few sites at best.

Resistance to azinphosmethyl and phosmet is present in *A. chilenensis* populations from Uruguayan apple orchards (Croft et al. 1976b). Selection of field populations continues since recommended rates for the control of other pests still cause some predator mortality, but biological control of spider mites is already partially successful in the area (Croft et al. 1976b, Carbonell and Briozzo 1981). In Chile, azinphosmethyl and diazinon R strains of *A. chilenensis* provide appreciable biological control of the spider mites *P. ulmi* and *T. urticae* on several tree fruit crops (Gonzalez 1981).

In China, resistance to both OP and SP insecticides has been either detected in the field or selected as a result of genetic improvement experiments for citrus-inhabiting populations of *Amblyseius nicholsi* (Ding et al. 1983, Huang et al. 1987). OP resistance in the range of 20- to 30-fold occurs to phosmet and phoxim, and cross resistance at lesser levels extends to several other compounds from this group of insecticides (Table 14.1).

In Taiwan, cases of laboratory-selected resistance to dimethoate, carbaryl, and dimethoate-carbaryl in *A. longispinosus* have been reported by Lo et al. (1984). Laboratory selection of *A. pseudolongispinosus*, which occurs in Chinese cotton fields, resulted in 7- to 13-fold resistance to dimethoate. This resistant strain is being tested in greenhouse studies for efficacy as a predator of spider mites (Ke 1987).

In the United States, an Oregon strain of *Typhlodromus arboreus* shows moderate levels of multiple resistance to azinphosmethyl and carbaryl (and may be cross resistant to many other compounds) in tree fruit orchards (Croft and AliNiazee 1983). This species attacks both *P. ulmi* and *T. urticae* on a diversity of crops, including apple, prune, blackberry, and filberts (Hadam et al. 1986).

14.4.2.8. Resistance Patterns in the Phytoseiidae. Despite considerable research that has been conducted on pesticide resistance in the Phytoseiidae, the prediction of resistance mechanisms (Chapter 15) or even the propensity for resistance development in specific instances of a species or pesticide is still limited. For example, whereas stable OP resistance is easily selected for in *M. occidentalis* and *A. fallacis*, it is more difficult to select for and is less stable in *T. pyri*. Conversely, carbaryl resistance in *T. pyri* is easily developed and relatively stable, but is more difficult to select for in *M. occidentalis* and *A. fallacis*. And, despite the apparent similarity of *T. occidentalis* and *A. fallacis*, these species exhibit distinctly different cross-resistance profiles to OPs (Table 14.2). For example, *M. occidentalis* is cross tolerant or develops cross resistance to phosalone, while *A. fallacis* does not.

Inconsistencies in resistance patterns at this taxonomic level have been observed among tetranychid spider mites, which have been much more extensively studied than phytoseiids (Helle and van de Vrie 1975, Herne et al. 1979). Whether these inconsistencies among both predatory and phytophagous mites represent present inadequacies in methodology and discrimination, or whether they reflect the stochastic nature of toxicological and ecological events remains unknown. In either case, patterns of resistance development to specific compounds cannot yet be predicted based on an understanding of these few model species.

In contrast, resistance prediction is possible at a more general level among families of natural enemies or among natural enemies associated with specific agroecosystems. For example, there are ecological characteristics shared by spider mites and predaceous mites that make both of these groups more likely to develop resistance than other types of pests and natural enemies (Tabashnik and Croft 1985, Croft and van de Baan 1988). Agronomic and management characteristics of different crops (e.g., the pesticide levels used) correlate to some extent with the propensity for resistance development in both pests and natural enemies (Tabashnik and Croft 1985). Detailed consideration of factors favoring resistance development in phytoseiid mites and in natural enemies associated with certain crops is further developed in Chapter 16.

14.4.3. Resistance Documentation in Predaceous Insects

Pesticide resistance among eight species of predaceous insects is listed in Table 14.1. Resistance has also been reported for other predaceous insects in the literature (e.g., Georghiou 1967, for predators of poultry flies; Redmond and Brazzel 1968, for predators attacking cotton pests; Naqui 1970, for aquatic forms of predaceous insects), but only at low incipient levels. Cases cited in Table 14.1 predominantly involve resistance levels exceeding 5- to 10-fold. In several cases, the resistance factor was high enough to allow the predator to contribute significantly to biological control in the particular agricultural crop where the resistance was found. Resistant insect predators have been found among the Anthocoridae, Cecidomyiidae, Chrysopidae, Coccinellidae, Labiduridae, and Nabidae, with 9 out of the 14 reports coming from the Coccinellidae (Table 14.1).

The earliest reference to the resistance potential of entomophagous coccinellids was made by Quayle (1943), who studied the convergent ladybird beetle *Hippodamia convergens*. He observed populations that showed higher survival rates following multiple HCN treatments than after a single treatment.

Another predaceous coccinellid showing a high propensity for resistance is the polyphagous species *Coleomegilla maculata*, which occurs on cotton and many other crops in the southern United States (Table 14.1). Atallah and Newsom (1966) first measured 6-fold resistance to DDT in populations collected in Louisiana. Subsequent samples from Louisiana cotton exhibited 10- to 29-fold resistance to methyl-parathion and 7- to 12-fold resistance to monocrotophos (Mohamad 1974, Head et al. 1977). In Mississippi, levels of methyl-parathion resistance up to 35-fold have been measured in *C. maculata* in areas where this compound has been applied extensively for three decades or more (Chambers 1973). It appears likely that regionally resistant populations of this species have developed in areas where cotton has been intensively grown and repeatedly sprayed with the same compound for several consecutive years.

In a case that closely parallels the development of resistance in the Phytoseiidae, the coccinellid *Stethorus punctum* has developed high levels of OP resistance and lower levels of SP resistance in populations from apple orchards in the eastern United States. This spider mite predator first began to survive azinphosmethyl treatments in the early 1970s, at about the same time predatory mites were beginning to develop R strains (Asquith and Hull 1973). At present *S. punctum* readily tolerates field rates of most OPs used in IPM programs in Pennsylvania and the surrounding states (Hull and Starner 1983; Table 14.1). Furthermore, after 3–5 years of intensive selection with permethrin and fenvalerate, an initial resistance response to SPs in experimental field plots was observed (Hull and Starner 1983). Specific comparisons of S and R strains have not been made with *S. punctum* due to a difficulty in finding S strains. Historical susceptibility data from field plots clearly show that resistance levels in this species are comparable to those found in the phytoseiids *A. fallacis*, *M. occidentalis*, and *T. pyri*.

In a similar though less thoroughly documented situation, *Stethorus punctillum* has developed resistance to azinphosmethyl in Italian apple orchards. Pasqualini and Malavolta (1985) reported that this predator partially survives treatments on farms where azinphosmethyl has been in use for several years.

Are the conditions fostering resistance in *S. punctum* and *S. punctillum* in some way related to the conditions present on apple that favored the widespread development of resistance among phytoseiid mites and other predators, such as *Aphidoletes aphidimyza* (Warner and Croft 1982)? A number of specific characteristics of the apple crop production system and features of the biology of *S. punctum* and *S. punctillum* that favor resistance development are discussed in Chapters 16 and 20. On apple, the intensity of pesticide selection pressure and the continuous use of azinphosmethyl and certain other OPs have contributed to resistance development in natural enemies. Also important is the unique pattern of resistance among apple pests: many secondary pests of apple have become resistant to pesticides that continued to be used for the control of other key pests, such as the codling moth, apple maggot, and plum curculio. These factors have undoubtedly contributed to the rapid and conspicuous development of resistance in *Stethorus* spp.

In both the United States and Australia, the dermapteran earwig *Labidura riparia* has developed widespread resistance to DDT–toxaphene mixtures applied to cotton (Table 14.1). To date, resistance has only been detected by efficacy field trials. This species is an important *Heliothis* predator, although it is a pest under certain conditions. In pest management field trials, Bishop and Blood (1980) observed a resurgence in earwig populations after competitors and secondary predators, including ants, carabids, and spiders, were removed from fields by DDT-toxaphene treatments. *L. riparia* was the only *Heliothis* predator that was relatively unaffected by sprays, and thereafter it was observed in large numbers feeding on prepupal and pupal stages of the pest in eastern Queensland cotton fields.

The predaceous cecidomyiid *Aphidoletes aphidimyza* has developed resistance to azinphosmethyl in apple orchards in several areas of the midwestern and eastern United States (Adams and Prokopy 1977, Warner and Croft 1982). This species is of particular interest because of its potential as a biological control agent on pests of other crops, especially greenhouse aphid pests. *A. aphidimyza* is known to feed on several hundred aphid species and is widely employed in mass release or management programs on glasshouse crops in many areas of western and eastern Europe, apparently without the benefit of resistance (Asyakin 1973, Markkula et al. 1979).

The first evidence of resistance in apple orchard strains of *A. aphidimyza* was reported by Adams and Prokopy (1977) in screening field rates of azinphosmethyl against two strains of this dipteran predator. Warner and Croft (1982) later confirmed resistance levels up to 15-fold in eggs of this species. Resistance was also appreciable in larval forms. In apple orchards, this species readily tolerates applications of many broad-spectrum insecticides, including azinphosmethyl. *A. aphidimyza* provides appreciable biological control of such troublesome aphid pests as the apple aphid *Aphis pomi* and the rosy apple aphid *Anuraphis roseus*, which previously had developed resistance to this OP (Prokopy et al. 1981, Morse and Croft 1987).

In alfalfa agroecosystems, two cases of region wide insecticide resistance in predaceous insects are of significance to IPM. In Washington State, Johansen and Eves (1973) reported trichlorfon resistance in two hemipteran predators. Resistant strains of *Nabis alternatus* and *Geocoris pallens* (Table 14.1) were documented when mortality comparisons were made against a strain without previous exposure to this OP. The resistant predators were better able to hold lygus bug (*Lygus hesperis*) populations in check, and trichlorfon was effective against the pest. Selective applications of trichlorfon were used only when the lygus bug became more abundant than the predators. This is one of the few cases in which entomophagous species became resistant without simultaneous or previous resistance in their host or prey. An important factor in this case is that these predator species are generalists and could easily turn to alternate prey following the application of sprays.

In California's San Joaquin valley, where cultivated crops are nearly contiguous and some areas are treated with pesticides almost continuously, the common lacewing *Chrysoperla carnea* has developed low-level resistance to a wide range of conventional insecticides, including several carbamates, OPs, and the more recently introduced SP permethrin (Grafton-Cardwell and Hoy 1985; Table 14.1). This is another of the relatively rare cases in which a generalist predator has developed resistance without previous or concurrent development of resistance in its principal prey. The particular significance of this case is that it represents the first documentation in which a more mobile natural enemy has acquired resistance on a regional scale. Resistance in vagrant species like *Chrysoperla* may be probable only in widely treated environments and where immigration is limited; however, a greenhouse or other confined habitat where pesticide use is intensive would also foster resistance in such vagrant species. There is a tendency for S strains to rapidly flood out resistant populations where cropland treated with pesticides represents only a part of an ecosystem (see also Chapter 16).

One indication that the immigration in *C. carnea* may not be as extensive as previously thought has recently come from the breadth of resistance found in Ontario, Canada, populations from apple (Pree et al. 1988; Table 14.1). In comparing populations from unsprayed and sprayed orchards within 0.5 km of each other, they found moderate to high levels of resistance to such diverse pesticides as DDT, parathion, azinphosmethyl, phosmet, carbaryl, methomyl, and several pyrethroids. The extent of resistance in the Ontario strain suggests that on a par with several phytoseiid mite species occurring on apple, *C. carnea* has widely adapted to the pesticide regimes used on apple. Furthermore, these resistance attributes have persisted over time, as indicated by the cross-resistance relationship between DDT and the SP insecticides. One other interesting aspect of the study of Pree et al. (1988) was the suggestion that the high SP tolerance reported for *C. carnea* may be due to cross resistances developed to other pesticides, such as DDT and the OPs.

14.4.4. Resistance Documentation in Insect Parasitoids

Parasitoids of pest arthropods have not developed resistance to insecticides in the field to the extent that predaceous forms have (Table 14.1). As noted earlier, most cases involve laboratory-selected strains, and resistance thus attained has not been high or stable in the absence of repeated selection (see Section 14.4.1).

Among field populations of parasitoids, Strawn (1978) was the first to report relatively low incipient levels of parathion resistance in the aphelinid *Aphytis melinus* from California citrus (Table 14.1). Strawn noted a similar level of resistance as that achieved in earlier artificial selection trials with malathion (Abdelrahman 1973). Havron and Rosen (1988) reported modest levels of near 4-fold resistance to azinphosmethyl in field populations of *Aphytis coheni*, which had been heavily sprayed on citrus in Israel. The best documented and most conclusive case of field-developed OP resistance comes from South African citrus orchards. After many years of continuous methidathion use, Schoones and Giliomee (1982) observed 66- and 6-fold resistance (Fig. 14.2), respectively, in two species of aphelinids, *Aphytis africanus* and *Comperiella bifasciata*. Resistance in *C. bifasciata* is a significant indication that resistance can develop to relatively high levels in field populations of parasitoids.

Another incidence of relatively high resistance in field populations of parasitoids was reported by Walker and Thurling (1984), who tested the susceptibility of three strains of *Encarsia formosa* from Great Britain, Holland, and New Zealand to lindane and permethrin. Resistance to lindane was only about 3-fold among strains, even after genetic improvement trials were conducted with the British strain for nine generations. Permethrin resistance was about 15-fold in the Dutch strain. In this case, resistance was not due to selection alone, but was influenced by cross resistance to other compounds.

One of the most extensive field surveys for resistance in parasitoids was conducted by Rosenheim and Hoy (1986), who surveyed 13 populations of *A. melinus* from throughout

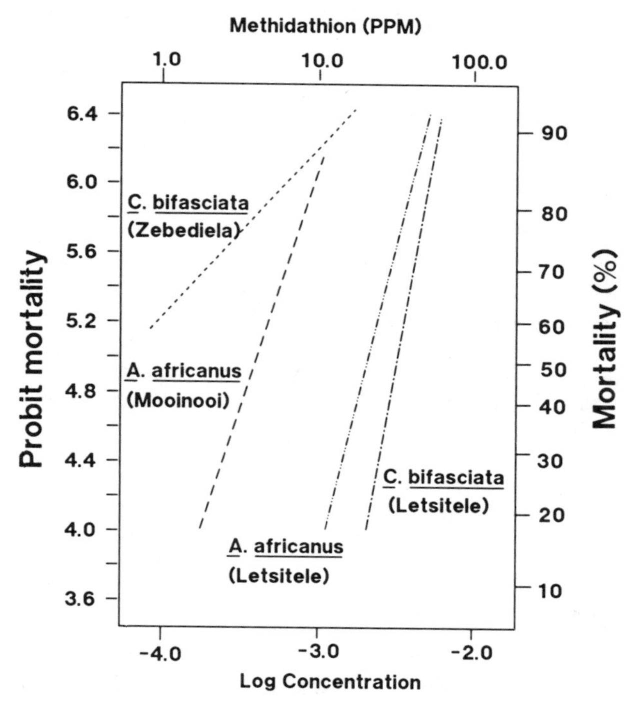

Figure 14.2. Log-concentration versus probit mortality lines for two strains of *Aphytis africanus* and *Comperiella bifasciata* exposed to methidathion (after Schoones and Giliomee 1982).

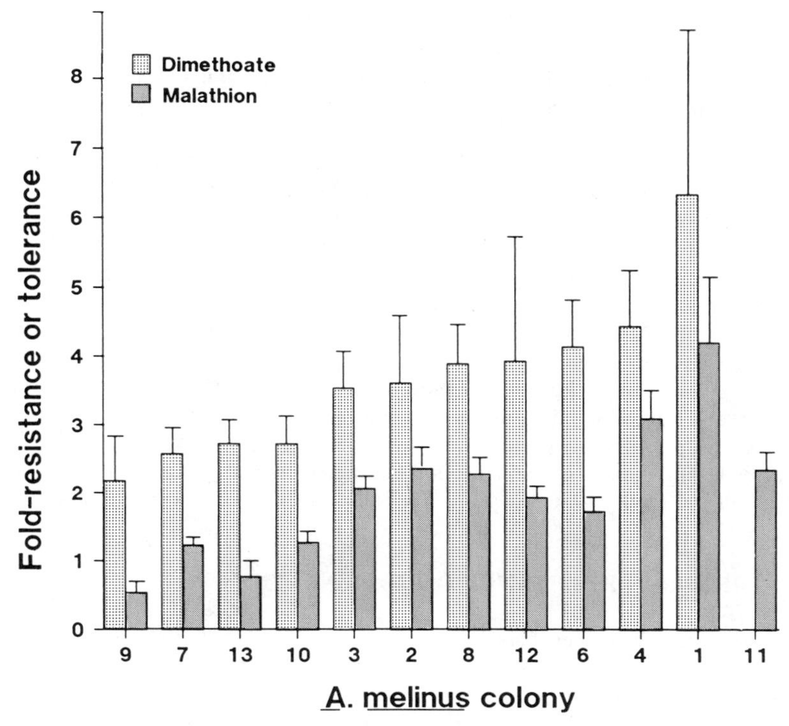

Figure 14.3. Variation in LC_{50} for *Aphytis melinus* colonies collected from California citrus for two insecticides (after Rosenheim and Hoy 1986).

the citrus-growing regions of California. For each population, the history of pesticide exposure both within orchards and regionally was compared to observed patterns of resistance. Low levels of resistance ranging from 2- to 8-fold were found to carbaryl, chlorpyrifos, dimethoate, malathion, and methidathion (Table 14.1). The variable levels of resistance to dimethoate and malathion are depicted in Fig. 14.3. Rosenheim and Hoy reported that LC_{50} for each strain correlated with both in-grove and regional pesticide use histories. Patterns of variability in the response of each strain were best explained in multiple regression analysis by the inclusion of both pesticide-use histories.

After surveying for resistance in field populations of *A. melinus*, genetic improvement studies were conducted using carbaryl, chlorpyrifos, dimethoate, malathion, and methidathion (Table 14.1). High levels of resistance were not obtained through artificial selection; however, carbaryl selection resulted in 5-fold resistance as compared to an *S* field strain and 20-fold resistance compared to an *S* laboratory strain over a 2-year selection period.

The status of resistance was also evaluated in apple populations of the braconid parasitoid *Pholetesor ornigis*, which attacks the spotted tentiform leafminer (Trimble and Pree 1987). Toxicity tests for permethrin, fenvalerate, azinphosmethyl, and methomyl conducted with parasitoid populations in 1984 were compared with similar tests run 6 years earlier with populations from another area of Ontario, Canada, before SPs were widely used in the field. Results indicated that parasitoids had developed low levels of resistance to permethrin, fenvalerate, and methomyl (Table 14.1), but apparently not to azinphos-methyl, which had been used for over 30 years in the area.

Other genetic improvement studies of beneficial parasitoids have focused on several hymenopteran species, including the citrus scale parasitoid *Aphytis holoxanthus* (Rosen 1967, Havron and Rosen 1988) and *Trioxys pallidus*, a braconid parasitoid of the walnut aphid (Hoy and Cave 1988). Both studies demonstrated considerable genetic variability in the responses of populations to pesticides, and low levels of resistance to several OP insecticides were found (Table 14.1). Further discussions of the development and exploitation of resistance in these species are presented in Chapters 19 and 20.

It was predicted several decades ago that the Aphelinidae would be one of the first parasitoid families to develop pesticide-resistant species (Croft 1972). This prediction was based on aspects of aphelinid life history, ecology, and management in the crops on which they occur. Factors taken into consideration included the number of natural enemy generations per season, the rates of movement or dispersal, the intensity of pesticide treatment, and the perennial nature of the apple and citrus orchard systems where species from this family commonly occur. These factors are important parameters for predicting resistance development in other arthropod natural enemies as well as in their prey or hosts (Georghiou and Taylor 1976, 1977a, 1977b, Tabashnik and Croft 1982, 1985; Chapter 16).

14.5. TRENDS IN RESISTANCE DOCUMENTATION

Resistance research on beneficial species has increased since 1980 in both scope and precision. The potential for genetic improvement (Chapter 19) of resistance in predators and parasitoids has been a major stimulus of this interest. Resistance research on natural enemies is being undertaken in many laboratories around the world with encouraging results (e.g., Hoy 1985). In addition, more basic studies of the genetics, toxicology, and factors influencing population selection for resistance in natural enemies are being published. A more detailed understanding of the ecological differences between pests and beneficials is particularly important for maintaining and managing stable levels of resistance in field populations of natural enemies.

Improved methods for monitoring pesticide resistance in natural enemies are also of critical importance. In this regard, interest is growing in the development of standardized methods for resistance monitoring (FAO 1977, 1979, Croft 1977). As compared to susceptibility tests (Chapter 5), resistance tests can be very simple. In many cases these tests are identical to or only slightly modified from those developed previously for pest species (Busvine 1980). An example of such an adaptation has been demonstrated by research scientists working with phytoseiid mites. The standard slide-dip method accepted internationally by FAO (1977) for assessing resistance in spider mites has, with minor modifications, been proposed for resistance monitoring in predaceous mites. Regional monitoring for resistance in other important natural enemies, such as that carried out for the parasitoid *A. melinus* (Rosenheim and Hoy 1986) and the predaceous mites *A. fallacis* (Brown 1976) and *M. occidentalis* (Hoy 1985), will probably be important in managing resistance in these species (Chapter 17).

14.6. SUMMARY

Relatively few cases of resistance have been documented in natural enemy populations. High, stable levels of resistance have been far more commonly reported for pedators than for parasitoids. The propensity for resistance seems to vary greatly among natural enemy groups and is distinctly greater for some entomophagous families or species (e.g., phytoseiid mites) than for others. Predicting the propensity for resistance in natural enemy species to individual pesticides is very tenuous, but some general patterns can be seen across families of natural enemies and in relation to different crop production systems. Resistance appears to be determined principally by the life history characteristic and the ecological, genetic, and environmental setting of exposure in the field.

In spite of the meager data available on resistance, existing patterns of resistance suggest some common principles governing resistance evolution in natural enemies. Perhaps these principles can be exploited for beneficial purposes in practicing resistance management in IPM systems. Principles governing resistance selection and the management of resistant natural enemies are discussed in Chapters 16–19. The application of these principles to the design of IPM systems using insecticide-resistant natural enemies is discussed in Chapter 20.

15

TOXICOLOGY AND GENETICS

15.1. INTRODUCTION

Most cases of insecticide resistance in arthropods are based on several well-known toxicological and behavioral mechanisms. Georghiou and Saito (1983) listed as examples of such mechanisms: 1) altered sites of action, such as modified acetylcholinesterases (AChE) or the knock-down resistance factor (kdr), which involves changes in ionic balance in the nervous system; 2) increased activity of detoxification enzymes such as general esterases, cytochrome P-450 monooxygenases (also called mixed-function oxidases), or MFOs, glutathione-*S*-transferases, and epoxide hydrolases; 3) reduced penetration, which affects the rate of pesticide movement through the integument or tissues of the arthropod to the site of action. Other factors such as increased sequestration and excretion of toxicant may also confer resistance in arthropod populations (Chapter 6).

Behavioral resistance usually involves changes in arthropod activity resulting in different rates of pesticide uptake between susceptible (S) and resistant (R) strains (Georghiou 1972, Pluthero and Singh 1981). One common behavioral mechanism of resistance is avoidance of pesticide residues. Generally, less information is available for behavioral resistance than for biochemical and physiological mechanisms of resistance (Gould 1984, Lockwood et al. 1984).

Most mechanisms of pesticide resistance known for pest arthropods have been found among predatory and parasitic species as well (Croft and Strickler 1983). Only reduced

Table 15.1. Known Mechanism of Insecticide Resistance in Predaceous Mites and Insects

	Mechanism of Resistance[1]						
Species/Compound	AChE	kdr	DDTase	GSH	Esterase	MFO	Pen
Amblyseius fallacis							
DDT		Yes	Yes			No	
Azinphosmethyl	No			Yes	Yes	No	No
Carbaryl						Yes	
Permethrin		Yes			Yes	No	
Amblyseius potentillae (= *andersoni*)							
Parathion/propoxur	Yes						
Phytoseiulus persimilis							
Methidathion	No			Yes			No
Metaseiulus occidentalis							
Carbaryl						Yes	
Propoxur						Yes	
Typhlodromus pyri							
Paraoxon/propoxur	Yes			No			
Chrysoperla carnea							
Carbaryl					Yes	Yes	

[1] AChE = modified acetylcholinesterase; kdr = knock down resistance factor; DDTase = DDTase detoxification enzyme; GSH = glutathione-*S*-transferase detoxification enzyme; esterase = esterase detoxification enzymes; MFO = mixed-function oxidase detoxification enzymes; Pen = penetration factor.

penetration (see discussion on penetration and its relevance to physiological selectivity in Chapters 6 and 9) and behavioral resistance have yet to be identified for this group (Table 15.1). Very likely these mechanisms will soon be found among natural enemies.

The genetic basis of pesticide resistance is also reasonably well known among pests (Georghiou and Saito 1983), but has been less well studied among entomophagous species. Almost all known genetic mechanisms of resistance among natural enemies are for phytoseiid mites. Hoy (1985) published an extensive review of the genetics of the Phytoseiidae, including aspects of pesticide resistance. Limited research on the genetics of resistance has been conducted with *Chrysoperla carnea*. *Bracon hebetor* is another natural enemy for which extensive genetic research has been done in relation to pesticide exposure. While yet to be studied for resistance genetics, this braconid parasitoid has been used extensively as a model for screening mutagenic properties of pesticides (Grosch and Valcovic 1967, Grosch 1970, 1972, 1975, Grosch and Hoffman 1973).

In this chapter, the underlying biochemical, toxicological, and genetic mechanisms of resistance in arthropod natural enemies are reviewed. Because these mechanisms tend to be specific to individual species and compounds, considerable detail is given for the limited examples available. The predatory phytoseiid mite *Amblyseius fallacis* has probably been most extensively studied in this regard and illustrates the diversity of approaches used to identify toxicological and genetic mechanisms of pesticide resistance in natural enemies.

While information on the toxicology and genetics of resistance among natural enemies is presently limited, this type of study may soon be increasing. Research in artificial selection and genetic engineering is being considered for many beneficial species, including a number of predators and parasitoids (Beckendorf and Hoy 1985, Hoy 1987). Studies of the mechanisms and genetics of resistance will undoubtedly become better defined and more extensive as biotechnology is applied to natural enemies (see Chapter 19).

15.2. STUDY OF RESISTANCE MECHANISMS

A variety of diagnostic tests have been employed to evaluate mechanisms of pesticide resistance (Oppenoorth and Welling 1979, Georghiou and Saito 1983). Comparative bioassay tests involve the use of diagnostic compounds that are known to be detoxified in particular ways in insects (e.g., carbaryl is generally detoxified by oxidative detoxification enzymes). Synergist studies employ compounds known to block particular detoxification systems. Examples of synergists include piperonyl butoxide (PBO), a known inhibitor of MFO systems, and DEF (s,s,s,-tributyl phosphorotrithioate), which inhibits esterase systems. Biochemical analysis of detoxification enzymes is conducted by using key indicator reactions (e.g., aldrin epoxidation) to quantify the level of differential detoxification occurring between R and S strains. Detoxification routes of a compound may also be elucidated using a radiolabeled pesticide or related substrate. In this type of test, the actual fate of the metabolized toxicant can be followed comparatively in R and S strains. Other methods of analysis have also been used to follow the metabolic breakdown of pesticides and by-products in R versus S strains (Oppenoorth and Welling 1979, Georghiou and Saito 1983).

A combination of the above diagnostic tests is frequently used to identify mechanisms contributing to single or multiple resistance factors. Even when several types of analysis are employed, the relative contributions of different resistance factors can only be estimated qualitatively.

15.2.1. Mechanisms of Resistance in Phytoseiid Mites

15.2.1.1. ***Amblyseius fallacis.*** The earliest studies of resistance mechanisms in the Phytoseiidae were reported by Motoyama et al. (1972) who observed that an organophosphate (OP) resistance strain of *A. fallacis* degraded radiolabeled isotopes of azinphosmethyl faster than an S strain. They found identifical sensitivity to AChE in R and S strains to P = O analogues of azinphosmethyl, thus indicating that altered AChE was not a significant factor in resistance. No differences were observed in the penetration of radiolabeled compounds between the two strains. However, significantly different glutathione-*S*-transferase activity between strains was observed, which was 3–6 times as active (depending on the method of evaluation) in the R strain in degrading a thioester (P = S) analogue of azinphosmethyl. Moreover, several R strains were shown to have extra protein bands by gel electrophoresis, which indicated that unique esterases were probably involved in the observed resistance. When general esterase assays were conducted by *in vivo* and *in vitro* inhibition using alpha naphthyl acetate, the R strain had greater nonspecific esterase activity than did the S strain. Specific hydrolytic esterases conferring resistance in this species were later identified by Chang and Whalon (1986).

Motoyama et al. (1977) further studied the role of glutathione-*S*-transferase in OP-resistant *A. fallacis* by comparing the toxicity of *O*-alkyl analogues of azinphosmethyl to S and R strains. They found that resistance was greatest with methyl analogues, but decreased with ethyl and *n*-propyl analogues. These data indicated that glutathione-*S*-transferases were probably involved in the observed azinphosmethyl resistance as well. Motoyama et al. (1972, 1977) predicted that resistant *A. fallacis* would exhibit strong cross resistance to other OP insecticides with *O*-methyl groups.

Croft et al. (1976a) tested the hypothesis of Motoyama et al. (1972, 1977) by evaluating cross resistance in a Michigan strain of *A. fallacis* to 16 OP analogues of compounds to which this species was resistant. Previously the tested strain had been exposed primarily to azinphosmethyl and had limited exposure to diazinon. Significantly high resistance ratios between R and S strains were observed to the methyl phosphorodithioates (dimethoate, malathion, and phosmet). Low resistance ratios were found to most ethyl phosphorodithioates (carbophenothion and ethion); however, cross resistance to the closely related analogue of azinphosmethyl, azinphosethyl, was relatively high. Resistance was appreciable to methyl- and ethyl-substituted phosphorothioates, although particularly high resistance ratios were associated with the ethyl derivatives (parathion and diazinon). Only two straight-chain phosphates, phosphamidon (methyl) and TEPP (ethyl), were evaluated; cross resistance was moderate to the former and extremely low to the latter. Thus, resistance was not consistently higher for methyl OP derivatives as predicted by Motoyama et al. (1972). However, because the strain studied by Croft et al. (1976a) had been exposed to some diazinon, selected resistance mechanisms may have been different than those in the strain studied by Motoyama. Multiple detoxification mechanisms or predominant detoxification enzymes other than glutathione-*S*-transferase may have been selected for.

Further insight into the multiple-resistance mechanisms present in Michigan strains of *A. fallacis* was gained by Croft and coworkers (Croft and Wagner 1981, Croft et al. 1982, Mullin et al. 1982, Strickler and Croft 1981, 1982, Scott et al. 1983) using bioassays, synergists, and detoxification tests. In comparative bioassays, Strickler and Croft (1981) reported little cross resistance between the synthetic pyrethroid (SP) permethrin and azinphosmethyl; however, they observed a slow increase in azinphosmethyl resistance in predators selected with permethrin in the laboratory. A slight association between permethrin and azinphosmethyl resistance was obtained, indicating at least in part a common resistance mechanism (possibly esterases; Chang and Whalon 1986).

Resistance levels to permethrin, azinphosmethyl, DDT, and carbaryl are presented in Table 15.2 for strains of *A. fallacis* exhibiting multiple resistance (Croft et al. 1982). In a greenhouse-adapted (GH-1) strain, which had been selected with permethrin only (Strickler and Croft 1982), relatively high levels of resistance to permethrin, azinphosmethyl, and DDT were present, but virtually no cross resistance to carbaryl was observed. A Geneva, N.Y., strain was even more resistant to azinphosmethyl and DDT than the GH-1 strain and was moderately resistant to permethrin. A Monroe strain, which was not resistant to azinphosmethyl, had low levels of DDT and permethrin resistance. A Collins strain showd high azinphosmethyl and relatively high DDT resistance, but virtually no permethrin resistance (Table 15.2). A Fenville strain had moderate to high azinphosmethyl and DDT resistance, low permethrin resistance, and a slightly elevated LC_{50} value for carbaryl as compared to other strains.

The data of Croft et al. (1982) indicated that multiple resistance was present in most field strains of *A. fallacis* from Michigan apple orchards. (Fortunately, a susceptible strain found earlier was available for comparative study; Croft et al. 1976a.) None of the field populations sampled appeared to have only a single predominant mechanism of resistance. They concluded that characterizing the individual mechanisms present in these

Table 15.2. Levels of Permethrin, Azinphosmethyl, DDT, and Carbaryl Resistance among Nine Strains of the Predaceous Mite *Amblyseius fallacis* (after Croft et al. 1982)

	Permethrin 3.4EC[2]		Azinphosmethyl 50WP		DDT 50WP		Carbaryl 50WP	
Strain	LC_{50}	Fold-R	LC_{50}	Fold-R	LC_{50}	Fold-R	LC_{50}	Fold-R
Rose Lake[1]	0.00018[3]	1	0.007[3]	1	0.032[3]	1	—	—
Monroe	0.00083	5	0.010	1	0.208	7	—	—
Composite	0.0043	3	0.013	2	0.088	3	—	—
Collins	0.00017	1	0.18	26	>1.308	41[4]	0.0125[3]	2
Graham	0.00075	4	0.24	34	0.840	26	—	—
Kleins	0.0013	8	0.09	13	0.131	4	—	—
Geneva	0.0025	15	0.21	30	>1.380	43[4]	0.0142	2
GH-1	0.022	129	0.11	16	>1.440	45[4]	0.0072	1
Fenville	0.00063	4	0.13	19	0.930	29	0.0196	3

[1] See treatment history of strains in Strickler and Croft (1981, 1982).
[2] Data taken from Strickler and Croft (1981, 1982).
[3] Percent active ingredient of material in water.
[4] A conservative estimate based on mortality at 1.2% solution and assuming a maximum slope of 4.0 probit units of change per 10-fold increase in concentration.

strains with multiple resistance would be difficult. However, several further attempts were made to qualitatively assess mechanisms of resistance in these predatory mites. Croft et al. (1982) concluded that MFOs were probably not a major factor in these *A. fallacis* strains with multiple resistance. This conclusion was based on the difficulty in selecting for carbamate resistance in the field (Croft and Meyer 1973, Croft and Hoying 1975) and the lack of carbaryl resistance present in individual field strains (with the possible exception of the Fenville strain; Table 15.2). Carbaryl resistance is most commonly associated with MFO detoxification enzymes in arthropods (Kuhr and Dorough 1976), although other mechanisms such as modified AChE can also be involved (van de Baan et al. 1985, Anber and Oppenoorth 1986, Anber and Overmeer 1988). As noted, LC_{50} values for carbaryl were similar for all predatory mite strains tested and were near the values observed for S strains as reported by Croft and Meyer (1973) and Croft and Hoying (1975).

Croft et al. (1982) concluded that several different mechanisms of DDT and SP resistance were present in field populations of *A. fallacis*. While some strains were both DDT and SP resistant (e.g., GH-1), others were DDT resistant, but SP susceptible (e.g., Collins). Thus data indicated that DDT and permethrin resistances were due to separate mechanisms, which could be present alone or together in individual strains. Both kdr and MFO can confer resistance to both compounds. Other resistance mechanisms common to OPs and SPs such as esterases could have contributed to permethrin resistance. Such diversity in resistance mechanisms might be expected in orchard populations that had been exposed to a wide range of insecticides for 30 years or more.

To further evaluate the mechanisms of resistance in Michigan strains of *A. fallacis*, Scott et al. (1983) studied their responses to several diagnostic compounds with and without synergists (Table 15.3). Studies of methoxychlor toxicity with the synergists piperonyl butoxide and DEF indicated that most strains had low oxidative activity, further indicating that MFOs were not a major resistance mechanism in this case. (See confirmation of the lack of MFO activity in aldrin epoxidation studies by Mullin et al. 1982.) Scott et al. (1983) concluded that resistance in certain strains (e.g., GH-1, Fenville) to

Table 15.3. Toxicity and Synergism Studies of Resistance Mechanisms in Six Strains of the Predaceous Mite *Amblyseius fallacis* (after Scott et al. 1983)

	LC_{50} Tests, Compounds					LT_{50} Synergist Tests, Compounds + Synergist				
Strain	DDT	Methoxy-chlor	Permethrin	Azinphos-methyl	Carbaryl	DDT + PBB	Methox + PBB	Azin + PBB[2]	Azin + DEF	Perm + DEF
Rose Lake	1	1	1	1	1	NS[3]	NS	NS	NS	NS
Collins	41[1]	6	1	26	—	NS	NS	*[4]	*	NS
Kleins	4	2	8	13	—	NS	NS	NS	*	NS
Geneva	43	3	15	30	2	NS	NS	+[5]	*	+
GH-1	45	18	50	16	1	NS	NS	NS	*	*
Fenville	29	13	4	19	3	*	*	NS	NS	*

[1] Fold resistance level over the susceptible strain Rose Lake (1.0 level for Rose Lake: DDT = 0.032, methoxychlor = 0.068, permethrin = 0.0002, azinphosmethyl = 0.018, carbaryl = 0.009% a.i.).
[2] LT_{25} values instead of LT_{50}.
[3] Not significantly different from 1:1 ratio.
[4] Significantly greater than 1:1 ratio at $P \leq 0.05$ level.
[5] Significantly greater than 1:1 ratio at $P \leq 0.01$ level.

Table 15.4. Detoxification Capability of the Predaceous Mite *Amblyseius fallacis* and Its Prey *Tetranychus urticae* (Activity is expressed in pmole/mg protein for a combined microsomal plus cytosolic fraction; mean ± SD from 3–5 separate enzyme preparations; after Mullin et al. 1982; also see Fig. 6.4)

	Enzyme Activity				
		Epoxide Hydrolase			
Strain	Aldrin Epoxidase	trans	cis	Glutathione-S-Transferase	Esterase $\times 10^{-3}$
Amblyseius fallacis					
S strain	0.27	278	431*	1095**	318
R strain	0.23	600###	834##	1683#	1788###
Tetranychus urticae					
S Strain	1.44**	1587***	117	102	389#
R Strain	1.60	1613	124	219##	263

Interspecific differences between susceptible strains of predator or prey: $*P < 0.01$; $**P < 0.005$; $***P < 0.001$. Intraspecific differences between susceptible or resistant strains of predator and prey: $\#P < 0.10$; $\#\#P < 0.05$; $\#\#\#P < 0.005$

both DDT and permethrin was due primarily to kdr, while in other strains DDT resistance was due to DDT dehydrochlorinase (e.g., Geneva, Collins). Hydrolytic esterases played a significant role in both OP and SP resistance in several strains with multiple resistance.

In biochemical studies of intrinsic tolerance and resistance factors, Mullin et al. (1982) examined differences in whole-body detoxification enzyme levels in S and R (GH-1) strains of *A. fallacis* (OP/SP resistant) and its prey *T. urticae* (OP resistant) (Table 15.4; also Fig. 6.4). As measured by the conversion of aldrin to dieldrin by aldrin epoxidase, MFO activity was 5 times lower in the S strain of the predator than in the S strain of the prey mite; however, there was no difference in aldrin degradation between R and S strains. This was consistent with the data from synergist studies with respect to MFO activity (Table 15.4). Levels of cytosolic *cis*-epoxide hydrolase, the fraction usually associated with food substrate detoxification, were significantly higher in the prey than in the predator. Significant differences in the levels of this enzyme were also observed between SP-resistant and SP-susceptible strains of the predator (Table 15.4). *A. fallacis* had higher glutathione-*S*-transferase activity than *T. urticae*, and activity was significantly higher in R versus S strains of the predator. Susceptible prey and predators had similar esterase levels, but that of the resistant predator was much higher, again confirming that either SP or OP resistance, or both, was due to higher levels of these enzymes.

In a further examination of the hydrolytic esterase detoxification systems in *A. fallacis*, Whalon et al. (1982a) examined the electrophoretic banding patterns of S and R strains. Their goal was to develop genetic markers to verify the establishment of two genetically improved permethrin-R strains after field releases. The esterase banding patterns for several S (i.e., composite, susceptible, indigenous) and R (i.e., GH-1, Geneva, field-collected) predator strains are presented in Fig. 15.1. Seven distinct esterase bands were observed among the different mite strains tested. Certain bands were associated with enzymes taken up by predators when feeding on prey mites, while other bands were associated with esterase resistance mechanisms. Generally, susceptible strains were very similar in banding characteristics (Fig. 15.1), with little or no esterases present at certain

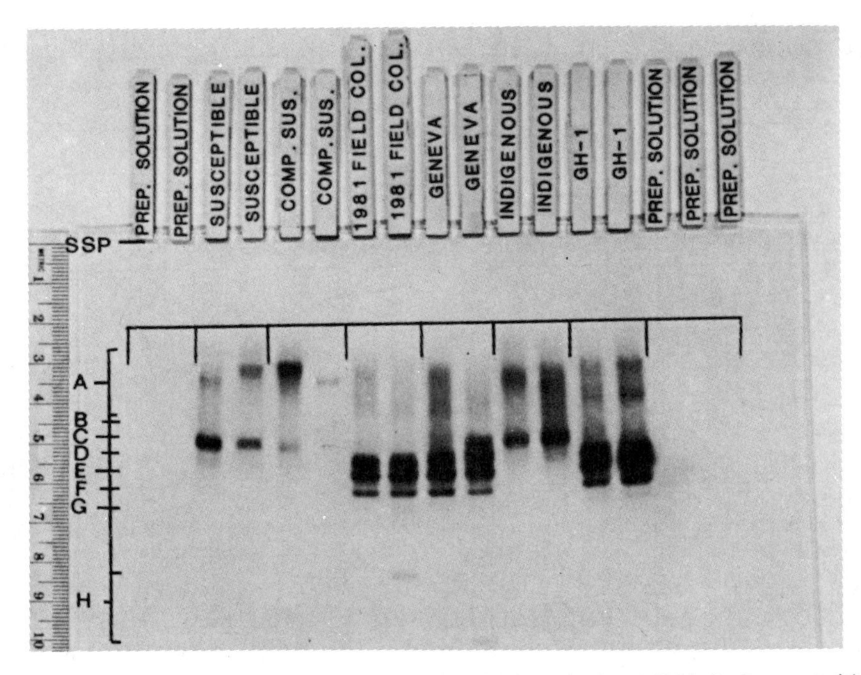

Figure 15.1. Esterase band formation in polyacrylamide gel for six strains (two individual mites per strain) of *Amblyseius fallacis*. Strains = susceptible, composite susceptible, 1981 field-collected, Geneva, indigenous, and greenhouse 1; two preparations were included as controls. (After Whalon et al. 1982a.)

locations on the gel. The two resistant released strains (Geneva and GH-1) as well as the overwintered strain (1981 field collected) also produced very similar patterns. They had several additional bands that were not found or were much less distinguishable in the susceptible strains. Electrophoretic banding patterns indicated that either greater enzyme quantities or several enzymes of the same approximate size were present in the released R strains and the established field-collected strain. Their data also conclusively showed that genetic improvement had been successful: the resistant trait had been introduced into field populations of this mite and had survived the first season in the field. (See further discussion of these data in Chapter 19.)

Chang and Whalon (1986) examined the hydrolytic esterase factors involved in SP resistance (and also OP resistance) in the *A. fallacis* strains mentioned above. The highest levels of pyrethroid-hydrolyzing activity in whole-body homogenates were found in the GH-1 strain that was highly resistant to both permethrin and azinphosmethyl. Three different pyrethroid-R strains had esterases that hydrolyzed *trans*-permethrin at least two times faster than *cis*-permethrin, while isomeric specificity was not observed in a susceptible strain. Fifteen esterase bands were resolved from whole-body homogenates by gel electrophoresis in this strain and were identified as carboxylesterases. The major pyrethroid-hydrolyzing activity was located in the E5–E12 bands (Fig. 15.2); six esterase bands exhibiting low pyrethroid-hydrolyzing activity were exhibited by the predators. It was determined that these enzymes were obtained by feeding on the two-spotted spider mite. Their data further confirmed the results of Motoyama et al. (1972), Mullin et al.

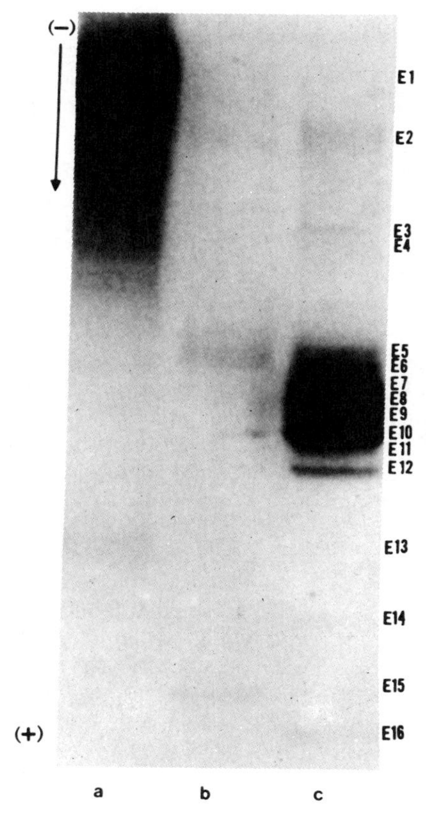

Figure 15.2. Gradient gel electrophoresis of soluble esterases from *Tetranychus urticae* (prey, a) and the composite susceptible (predator, b) and greenhouse 1 (predator, c) strains of *Amblyseius fallacis*. Esterase activity was visualized with alpha-naphthyl acetate as substrate. (After Chang and Whalon 1986.)

(1982), Whalon et al. (1982a) and Scott et al. (1983), implicating esterases as a major resistance mechanism in *A. fallacis*, conferring resistance to permethrin and azinphosmethyl (and probably several other compounds as well).

After reviewing resistance mechanism studies conducted with *A. fallacis*, it is clear that characterizing the individual factors responsible for resistance in field strains of natural enemies can be a formidable task. Researchers often obtain some indication of the collective range of mechanisms, but specific associations between compounds and resistance mechanisms are tenuous at best. The identification of individual resistance factors in *A. fallacis* and their genetic basis might better be accomplished by starting with a field-collected susceptible strain and selecting only single-line populations with individual compounds. This type of laboratory study is common with pest arthropods (e.g., house fly, cockroach, *Heliothis*) as an essential step in modern toxicological investigation. However, selection in the laboratory may not produce the same patterns and mechanisms of resistance that would develop in the field (Roush and McKenzie 1987; see further discussion, Section 15.3.4). Despite this limitation, it seems that such laboratory studies would give additional insight into the basic factors of resistance associated with individual predator populations. Both approaches—working with laboratory strains selected with individual compounds and with field strains selected with several different compounds simultaneously—present different types of problems to researchers trying to elucidate causal factors and interrelationships between resistance mechanisms. (See later discussions of differences associated with laboratory- and field-selected resistance in genetic improvement research; Chapter 19.)

15.2.1.2. Other Phytoseiid Species. Fournier et al. (1987a), working with strains of *Phytoseiulus persimilis* selected for resistance to methidathion in the laboratory, observed no differences in penetration between R and S strains or in inhibition rates of oxon analogues of this compounds by AChE (Table 15.1). However, analysis of detoxification enzymes using C^{14}-labeled substrate implicated increased pesticide breakdown by glutathione-*S*-transferases in the R strain. This resistance mechanism was very similar to that reported by Motoyama et al. (1972) for *A. fallacis.*

The only study of this kind with *M. occidentalis* has been for carbaryl and benomyl resistance mechanisms in laboratory-selected populations during the course of genetic improvement experiments (Roush and Plapp 1982; Table 15.1). The mechanism of resistance in this case was suspected to be oxidative, and two techniques were used to investigate this possibility. The first involved the use of piperonyl butoxide (PBO), a synergist of oxidatively metabolized xenobiotics. In the second test, *in vivo* conversion of heptachlor to heptachlor epoxide was measured in S and R strains.

In synergist studies with *M. occidentalis*, carbaryl and benomyl toxicity were synergized in both adults and eggs of the predator. Mortality in resistant strains increased over controls from 17 to 98% for carbaryl and from 0 to 90–100% for benomyl, respectively, when the synergist was added in each test. Results of *in vivo* MFO activity showed that the R strain converted over 50% of absorbed heptachlor to heptachlor epoxide, whereas the S strain converted only about 40%. These differences were statistically significant. The authors felt that both synergist and heptachlor metabolism studies confirmed that carbaryl and benomyl resistance in this species was primarily due to oxidative detoxification. Furthermore, the absence of cross resistance to methomyl, dimethoate, acephate, azinphosmethyl, or permethrin (Table 15.1) implied that other resistance mechanisms were probably not involved in carbamate resistance.

OP and carbamate resistance mechanisms were explored in a Dutch strain of *Typhlodromus pyri* by studying the activity of glutathione-*S*-transferase and the reaction rate of AChE with paraoxon and propoxur in R and S strains (van de Baan et al. 1985; Table 15.1). A 36-fold reduction in inhibition of AChE by paraoxon and a 14-fold reduction by propoxur were found in the R strain. There was no difference between R and S strains in glutathione-*S*-transferase activity. The authors concluded that an insensitive AChE was primarily responsible for observed levels of OP and carbamate resistance in this mite, considering the monogenic nature of the resistance (see Section 15.3.1) and their experimental results.

Insensitive AChE has also been implicated as the main cause of OP and carbamate resistance in a strain of *Amblyseius potentillae* (Anber and Oppenoorth 1986, Anber and Overmeer 1988). Whole-body homogenates of adult female mites incubated with ACh and various concentrations of paraoxon or propoxur showed inhibition rate reductions of 311- and 781-fold, respectively, in strains showing greater than 1000- and 14-fold levels of resistance. Less dramatic reductions of inhibition rates were observed for tetrachlorvinphos, omethoate, and azinphosmethyl.

In genetic improvement studies of dimethoate resistance in *Amblyseius pseudolongispinosus*, Ke (1987) reported that electrophoretic banding patterns of nonspecific esterases were different between S and R strains, similar to those found among *A. fallacis* populations resistant to OPs and SPs (Chang and Whalon 1986).

*15.2.1.3. **Summary of Resistance Mechanisms in the Phytoseiidae.*** As summarized earlier (Table 15.1), resistance mechanisms of phytoseiids have been identified to the major insecticide classes, including DDT and related compounds, OPs, carbamates, and SPs. Most work has been done with *A. fallacis*; however, this work suffers from a lack of study of individual resistance mechanisms in strains having clearly defined histories of selection. All field populations of this species, which were sampled, exhibited multiple-resistance mechanisms. Research with *M. occidentalis* and *T. pyri* has largely been limited to the study of individual mechanisms; however, it is well known that multiple-resistance mechanisms are present in most field populations of *M. occidentalis* and *T. pyri*, as is the case with *A. fallacis* (see Tables 14.1 and 15.1).

With regard to mechanisms of resistance within the Phytoseiidae as a whole, those discovered to date include most of the important mechanisms identified for pest arthropods (Table 15.1; Georghiou and Saito 1983). A limited study of penetration as a resistance mechanism in natural enemies has ben attempted, and results have so far proven negative (Motoyama et al. 1970, Fournier et al. 1987a). The infrequent study of penetration is undoubtedly due to the small size of these arthropods (Motoyama et al. 1970, Fournier et al. 1987a). Pesticide penetration has been more commonly examined in larger natural enemies, both in examining selectivity relationships with pests (Bull and Ridgway 1969, Hagstrum 1970, Zhuravskaya et al. 1976, Bull 1980; Chapter 6) and in a limited way for resistance factors (Atallah and Nettles 1966).

There are several interesting questions relating to the patterns of detoxification, tolerance, and resistance to pesticides observed in phytoseiid species and natural enemies in general that have yet to be addressed. It is well known that detoxification enzymes are commonly involved in pesticide tolerance among natural enemies (Chapter 6). Are these detoxification enzymes the same as those conferring pesticide resistance in these species (Chapter 6)? What is the relationship between intrinsic detoxification enzyme activity and induced activity in conferring resistance in natural enemy species (Yu et al. 1979, Yu 1983; Chapter 16)? Comparative studies of tolerance and resistance mechanisms are almost nonexistent among natural enemies, or between these species and their prey or hosts (Tabashnik and Johnson 1988).

One study that examined detoxification enzyme activity in R and S strains of a natural enemy and its prey was that of Mullin et al. (1982). They compared detoxification enzymes of susceptible *A. fallacis* with those of *T. urticae* and observed that the predator had significantly lower activity of aldrin epoxidase, *trans*-epoxide hydrolase, and glutathione-*S*-transferase, higher levels of *cis*-epoxide hydrolase, and similar levels of general esterase (Table 15.4). However, in comparing susceptible and multiple-R strains of the predator, aldrin epoxidase activities were the same, but activities of *cis*- and *trans*-epoxide hydrolase, glutathione-*S*-transferase, and general esterase were higher in the R strain (Table 15.4).

Among the natural enemies in which resistance mechanisms have been studied, oxidative detoxification mechanisms are much less common than hydrolytic mechanisms such as general esterases (Table 15.1). Furthermore, pesticide tolerance in natural enemies seems to be more commonly due to hydrolytic rather than oxidative mechanisms (Mullin et al. 1982, Mullin and Croft 1985; see Chapter 6). The oxidative mechanism of carbaryl resistance in *M. occidentalis* reported by Roush and Plapp (1982) is an exception to this generalization. However, van de Baan et al. (1985) felt that the slightly higher (1.3-fold) *in vivo* heptachlor epoxide formation observed in R versus S mites and the level of synergism by piperonyl butoxide of carbaryl observed by Roush and Plapp (1982) were not sufficient evidence to conclude that increased oxidative metabolism was the primary mechanism of resistance observed. They suggested that altered AChE should also have been examined.

In general, data on pesticide resistance and tolerance mechanisms in the Phytoseiidae support the general trends of resistance development (and selectivity) proposed by Plapp (1980). Oxidative mechanisms appear to be less common than hydrolytic mechanisms in conferring insecticide resistance to phytoseiid mites. A further discussion of the evolutionary basis and significance of these patterns can be found in Chapter 16.

15.2.2. Mechanisms of Resistance in Predaceous Insects

The amount of toxicological study that has been conducted on resistant natural enemies in the class Insecta is very limited (Tables 14.1 and 15.1). In fact, more research has been focused on tolerance mechanisms than on resistance mechanisms in this group of species (Ishaaya and Casida 1981, Bashir and Crowder 1983; Chapter 6). While resistance mechanisms in predatory mites and one predatory insect have received limited study, no such work has been conducted with insect parasitoids.

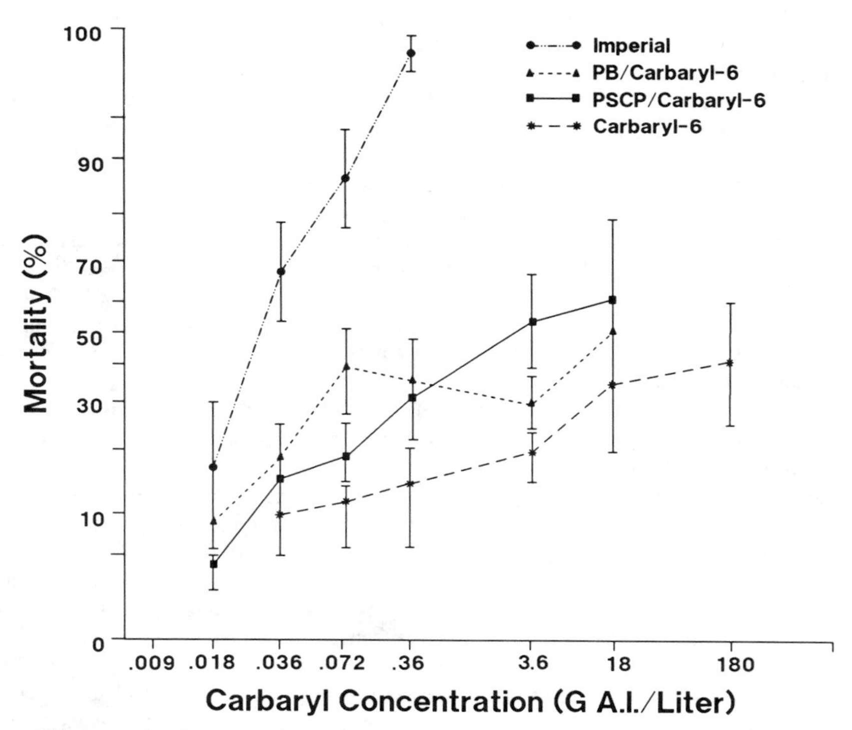

Figure 15.3. Log-carbaryl concentration–mortality relationships for first instar *Chrysoperla carnea* from the Imperial Valley, California (S strain) and the C-6 colony (Resistant) pretreated with either the esterase inhibitor PSCP or the oxidase inhibitor piperonyl butoxide (after Grafton-Cardwell and Hoy 1986b).

One of the first studies of resistance mechanisms in insect predators was conducted by Atallah and Nettles (1966). They studied DDT metabolism and excretion in R and S strains of the coccinellid *Coleomegilla maculata* in an attempt to explain the observed 6-fold resistance to this pesticide. Their findings indicated that resistance was partially due to an increased rate of dehydrochlorination of DDT to DDE and excretion in the feces, rather than to differential rates of penetration in the two strains.

In other studies, *C. carnea* populations were selected in genetic improvement trials in the laboratory (Grafton-Cardwell and Hoy 1986b). Larvae showing low levels of resistance to phosmet, diazinon, methomyl, and carbaryl were tested with several synergists, including piperonyl butoxide and phenyl saligenin cyclic phosphate (PSCP). Both synergists increased the toxicity of carbaryl, indicating that both oxidative (MFOs) and hydrolytic esterases were involved in resistance (Fig. 15.3).

Resistance mechanism studies on predaceous or parasitic insects are most certainly limited by the number of documented cases of resistant species (Table 14.1). The neuropteran *Chrysoperla carnea* and the coccinellids *Stethorus punctum* and *Coleomegilla maculata* represent good candidates for future investigations of resistance mechanisms. *C. carnea* has probably been the subject of more toxicity and toxicological investigations than any other insect predator (Chapters 2, 5, and 9). It is remarkably tolerant to a wide range of pesticides. Both *S. punctum* and *C. maculata* exhibit high levels of resistance to a number of pesticides on a regional scale (Table 14.1). Also, biochemical and toxicological data from R and S strains of these predaceous beetles could be compared with those of the closely related phytophagous coccinellid *Epilachna varivestris*, the Mexican bean beetle. Basic studies on the penetration, detoxification, and synergism of pesticides have been conducted on this pest (Sternburg and Kearns 1952, Brattsten and Metcalf 1970).

15.3. GENETICS OF RESISTANCE IN NATURAL ENEMIES

As with toxicological research, genetic studies of resistance in natural enemies have focused on the Phytoseiidae (Table 15.5). Similar work has been done on *C. carnea*, but it consists only of a single study (Grafton-Cardwell and Hoy 1986b). In all genetic studies of resistant natural enemies, experiments have been confined to standard crossing procedures designed to test whether monogenic or polygenic resistance factors were involved. These tests usually involved basic parental, F_1, and backcross tests between S and R individuals. Determination of the basis of more complex polygenic mechanisms of resistance in these species has not been attempted.

15.3.1. Genetics of Resistance in Phytoseiid Mites

As Hoy (1985) pointed out in her review, knowledge of the fundamental genetics of the Phytoseiidae has accumulated rapidly since 1972. Research has been stimulated by the need to understand the underlying basis for pesticide resistance as well as an interest in genetic improvement of these species. Genetic studies have involved aspects of cytogenetics, sex determination, sex ratio, interspecific variability, reproductive incompatibility, response to inbreeding, and artificial selection.

Studies of the genetics of insecticide resistance in the Phytoseiidae have been confined to six species (Table 15.5), the most extensive work having been done with *M. occidentalis*, followed by *A. fallacis* and *P. persimilis*. One of the first studies on phytoseiid resistance genetics was conducted by Croft et al. (1976a) with *A. fallacis*. Populations from Michigan

Table 15.5. Studies of the Genetics of Resistance to Insecticides in Arthropod Natural Enemies

Species	Compound	Strain Type	Resistance Type	Reference
Amblyseius fallacis	Azinphosmethyl	Fld.	Monogenic/ semidominant	Croft et al. 1976a
	Azinphosmethyl/ carbaryl	Lab.	OP-monogenic/ carbaryl?	Croft & Meyer 1973
	Permethrin	Lab.	Polygenic/ recessive	Croft & Whalon 1982
	Azinphosmethyl permethrin	Lab.	mixed	
Amblyseius pseudolongi-spinosus	Dimethoate	Lab.	Polygenic	Ke 1987
Amblyseius potentillae (= *andersoni*)	Parathion carbaryl	Lab.	Polygenic	Anber & Overmeer 1988
Phytoseiulus persimilis	Parathion	Lab.	Monogenic/ semidominant	Schulten and van de Klashorst 1974
	Oxydemeton-methyl	Lab.	Monogenic/ semidominant	Schulten et al. 1976
	Synthetic pyrethroid?	Lab.	Polygenic/ semidominant	Fournier et al. 1987a
	Organophosphate	Lab.	Monogenic/ semidominant	Zilbermints & Petrushov 1980
Metaseiulus occidentalis	Carbaryl	Lab.	Monogenic/ semidominant	Roush & Hoy 1981a
	Propoxur	Lab.	Monogenic/ semidominant	Roush & Plapp 1982
	Sulfur	Fld.	Monogenic/ semidominant	Hoy & Standow 1982
	Azinphosmethyl	Fld.	Monogenic/ semidominant	Hoy 1985
	Dimethoate	Fld.	Polygenic	Roush & Hoy 1981a
	Permethrin	Lab.	Polygenic/ recessive	Hoy et al. 1980 Hoy & Knop 1981
	Bioresmethrin	Lab.	Polygenic	van de Klashorst unpubl., in Hoy 1984a
	Azinphosmethyl/ carbaryl/sulfur/ permethrin	Lab.	Mixed	Hoy 1984a
Typhlodromus pyri	Parathion/ carbaryl	Lab.	Monogenic/ semidominant	Overmeer & van Zon 1983 van de Baan et al. 1985
Chrysoperla carnea	Carbaryl	Lab.	Polygenic/ recessive	Grafton-Cardwell & Hoy 1986b

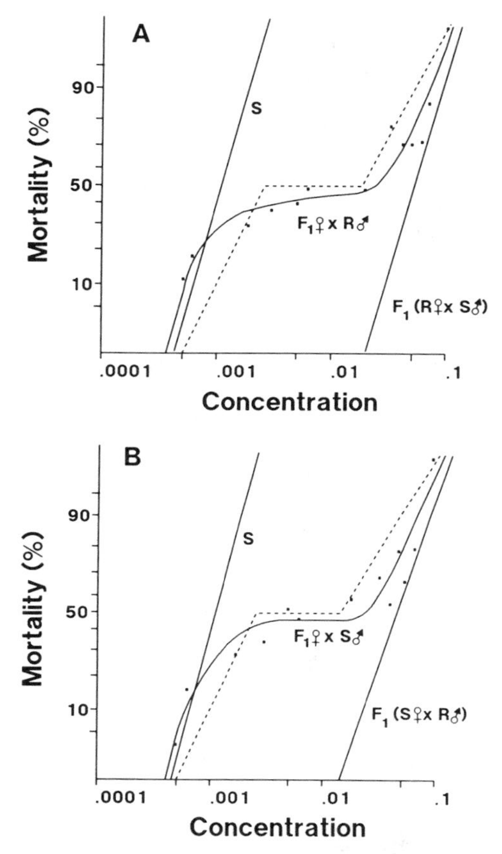

Figure 15.4. Concentration–mortality lines for the backcross of azinphosmethyl-resistant and -susceptible strains of *Amblyseius fallacis* which demonstrate semidominant, monogenic resistance (after Croft et al. 1976a).

were tested using standard crossing and backcrossing methods with both azinphosmethyl-R and -S strains. As shown in Fig. 15.4, reciprocal parental crosses showed almost identical concentration–mortality lines, indicating that the heritability of the resistance factor was semidominant. When F_1 females were backcrossed with S males, the concentration–mortality line had a broad flat inflection at 40–50% mortality. This line was reasonably congruent with the expected theoretical line calculated for a 1:1 proportion of hybrid R and S genotypes. Data reflected a clear segregation, indicating that the resistance factor was principally due to a single gene allele.

In contrast, Croft and Whalon (1982) observed recessive, polygenic resistance to permethrin in a genetically improved permethrin/OP-R strain of *A. fallacis*. Log concentration–probit mortality lines for the permethrin-R backcross (combined) matings of R and S parental strains are presented in Fig. 15.5 (LC_{50} values were 0.017 and 0.000072% a.i., respectively). The LC_{50} for the parental cross hybrid was intermediate in susceptibility, but closer to the S parent line than to the R parent line. The concentration–mortality line of the backcross had an LC_{50} value very near that of the S strain, indicating

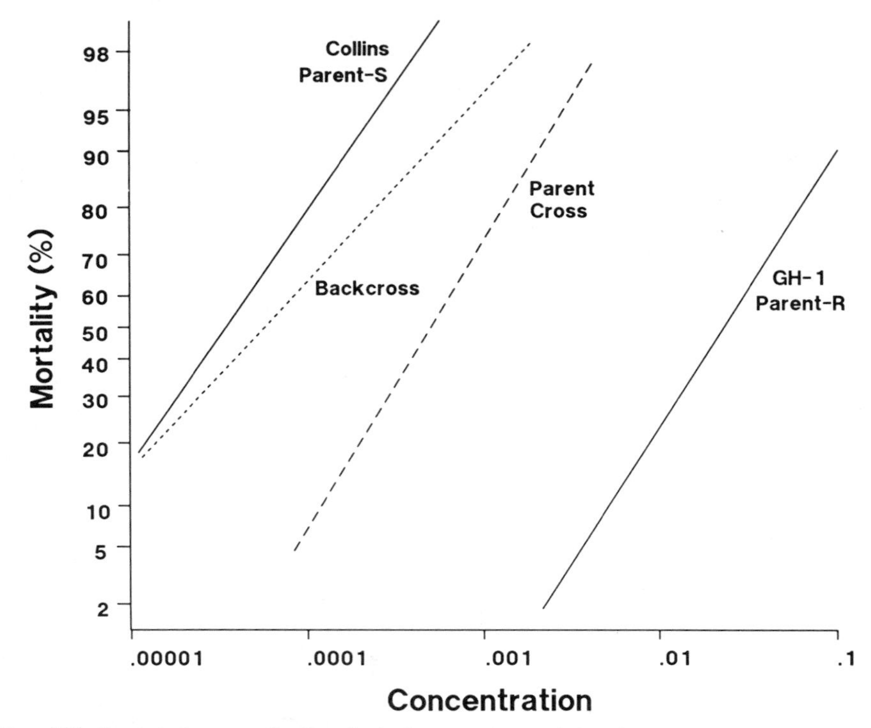

Figure 15.5. Concentration–mortality lines for backcrosses of permethrin-resistant and -susceptible strains of *Amblyseius fallacis* demonstrating polygenic, recessive resistance (after Croft and Whalon 1983).

relatively recessive heritability, and the backcross line did not have a broad inflection in the midconcentration range characteristic of a monogenic model of resistance. Croft and Whalon (1982) predicted from these genetic studies that SP-resistant *A. fallacis*, if released into the field, would be easily diluted by hybridization with susceptible strains. (See Chapter 17 for a verification of this conclusion when actual field releases were attempted; Whalon et al. 1982a, Croft 1983, Croft and Whalon 1983.)

California researchers have determined the resistance genetics of *M. occidentalis* for a wide range of field-collected and genetically improved populations in the laboratory (Table 15.5). Semidominant, monogenic characteristics have been identified in field-collected populations resistant to azinphosmethyl (Hoy 1985) and to sulfur (Hoy and Standow 1982). Field-selected resistance to dimethoate has been shown to be due to a polygenic mechanism (Roush and Hoy 1981a). Conversely, with the exception of carbaryl resistance (Roush and Hoy 1981a), most laboratory-selection attempts with this species (i.e., to permethrin by Hoy and Knop 1981; to bioresmethrin by van de Klashorst 1984) have resulted in polygenic, relatively recessive resistance factors.

It has been a goal of researchers to identify or breed strains of natural enemies having multiple resistance in a single strain. The most success has been achieved with *M. occidentalis.* Hoy (1984a) reported strains of this species resistant to azinphosmethyl/carbaryl/sulfur and azinphosmethyl/carbaryl/permethrin. To date, the genetic properties of the multiple-R strains mentioned above have not been studied extensively. It has been assumed that the genetic basis of individual resistance factors in multiple-R strains are not different from those in source populations with a single resistance factor. However, interactions among resistance factors in multiple-R strains may alter their genetic determination compared to other R strains.

Croft and Hoying (1975) reported genetic incompatibilities between resistance mechanisms when they released an OP/carbaryl-R strain of *A. fallacis* into an apple orchard. They observed expression of OP resistance, but carbamate resistance was lost rapidly. In the laboratory, both resistance factors had been stable when chemical selection pressures were similarly maintained. As further exploitation of multiple-R strains is attempted, more attention to the interaction of resistance factors will be necessary.

Parathion and demeton-*S*-methyl resistance in *P. persimilis* (Schulten and van de Klashorst 1974) is also due to a major, almost completely dominant gene. When resistance levels in this laboratory-selected strain were compared to resistance that developed in a field-selected strain were compared to resistance that developed in a field-selected strain, higher levels of resistance were observed in field populations. This suggested that modifier genes might be augmenting the basic mechanism, or that an entirely different genetic mechanism was operative in the field strain. In USSR strains of *P. persimilis*, OP resistance was shown to be based on a major, dominant gene (Zilbermints and Petrushov 1980). However, Fournier et al. (1987a) observed a polygenic genetic mechanism in this species after selection with the OP methidathion for genetic improvement in the laboratory. Similarly, polygenic resistance to dimethoate was observed in *Amblyseius pseudolongispinosus* by Ke (1987) after populations were selected by three different methods in the laboratory.

Monogenic, semidominant resistance to parathion was documented in *T. pyri* by Overmeer and van Zon (1983) in a field-collected strain (Table 15.5). During genetic experiments, backcrossed females were exposed to discriminating concentrations of both carbaryl and bromophos. No mortality was observed to either compound, indicating that resistance to parathion conferred cross resistance to carbaryl and bromophos at near recommended field rates.

In summary, the genetic mechanisms for resistance are best understood for the Phytoseiidae, primarily for the species *M. occidentalis, A fallacis*, and *P. persimilis* (Table 15.1). Monogenic, semidominant resistance factors have been most commonly observed in field-selected populations. Several cases of polygenic resistance resulting from laboratory selection studies have been reported. To date, the detailed genetic analyses of individual components of polygenic resistance mechanisms have not been attempted.

15.3.2. Genetics of Resistance and Sex Determination

Study of the genetics of pesticide resistance has shed considerable light on the mechanisms of sex determination in the Phytoseiidae (Hoy 1985). Conversely, studies of sex determination are relevant to understanding rates of resistance development of predatory mites and other natural enemies (Havron 1983, Croft and van de Baan 1988). Male phytoseiids are normally haploid; however, evidence from irradiation and cytological studies suggests that in some phytoseiids (e.g., *M. occidentalis, P. persimilis*) males are of biparental origin and lose one haploid chromosome set during embryogenesis. This system of sex determination has been referred to as parahaploidy (Hoy 1979). Helle et al. (1978), in genetically analyzing OP-resistant *P. persimilis* males, concluded that the paternal chromosome set is normally lost during embryogenesis. Thus even though phytoseiid males may be biparental, the inheritance pattern is similar to that which would result from arrhenotokous reproduction.

Detailed studies of sulfur resistance in *M. occidentalis* also supported the existence of parahaploid sex determination in phytoseiid mites. In genetic studies with R and S populations, Hoy and Standow (1981) observed that F_1 progeny had resistance levels similar to those expected in a species with parahaploid males. Thus, the F_1 progeny with R mothers was more resistant than the F_1 progeny with S mothers. These studies shed light on the unique influences that sexual reproduction mechanisms can have on the genetics of insecticide resistance. In the future, a better understanding of these mechanisms in the Phytoseiidae will be essential for understanding and managing pesticide resistance in these species. (For a further discussion of the cytogenetics and mechanisms of sex determination in the Phytoseiidae see Hoy 1985.)

Effects of arrhenotokous reproduction (particularly in phytoseiid mites and hymenopteran parasitoids) on pesticide resistance have been reported in reviews by Havron (1983), Crozier (1985), Schulten (1985), Cranham and Helle (1985), and Croft and van de Baan (1988). Arrhenotoky seems to confer an evolutionary advantage over more normal forms of arthropod reproduction (e.g., diploidy) because in each generation, the whole genetic complement is exposed to selective forces through hemizygous males. Advantages of sexual reproduction are maintained in biparental females by meiotic segregation, crossing over, and fertilization. When genetic recombination occurs in the diploid females, the new characteristics are directly exposed to selection in male progeny. Male haploidy reduces the effective population size and thus genetic variation, although this depends on the sex ratio (Crozier 1985). It is estimated that the additive genetic variance in arrhenotokous species is 33% larger than that of similar but fully diploid species (Hartl 1971). Presumably, the rate of resistance evolution of species with this type of reproduction would be accelerated by a factor similar to this ratio.

Rapid resistance development in predatory mites as compared to other arthropods might be partly explained by parahaploid and arrhenotokous reproductive mechanisms; however, many hymenopteran parasitoids also show similar types of reproductive mechanisms. Yet among natural enemies, parasitoids have shown the least potential for developing resistant strains (Chapter 14). A wide range of other genetic, biotic, and ecological factors determine the rates of pesticide resistance evolution in arthropods (Georghiou and Taylor 1977a, 1977b, Tabashnik and Croft 1982). In contrasting the frequency of resistance development between phytoseiids and hymenopteran parasitoids, the latter factors must confer appreciable differences in this regard. (See further discussion of factors influencing rates of pesticide resistance evolution in Chapter 16.)

15.3.3. Genetics of Resistance in Predaceous Insects

The only study of the genetics of resistance in a predaceous insect is with genetically improved strains of *Chrysoperla carnea* (Grafton-Cardwell and Hoy 1986b). After four rounds of selection with carbaryl in the laboratory, significant decreases in mortality to standard rates of this insecticide were observed in populations collected from treated alfalfa fields in southern and central California. When standard crossing and backcrossing tests were performed on these strains, data indicated that the mode of inheritance was not due to a single dominant or recessive gene, but was probably polygenic.

15.3.4. Selection of Resistant Natural Enemies in the Laboratory and Field

With the exception of carbaryl and dimethoate resistance in *M. occidentalis* and parathion/demeton-*S*-methyl resistance in *P. persimilis* (Table 15.5), all cases of field-developed resistance have been monogenic and semidominant. On the other hand, laboratory selection trials have tended to produce polygenic resistance (Table 15.5). (However, it is not always certain that genetic improvement stocks used initially in selection studies are susceptible and have not previously developed some level of resistance in the field.) Low gene frequencies or low incipient levels of resistance could have been present in both the laboratory-selected carbaryl-R strain of *M. occidentalis* and the parathion/demeton-*S*-methyl-R strain of *P. persimilis* due to previous field selection. For example, Hoyt (unpublished) noted that carbaryl resistance had long been present in field populations of *M. occidentalis* before it could be detected by standard slide-dip bioassay procedures. Similarly, the susceptible strain employed by Schulten et al. (1976) had previously been exposed to greenhouse applications of parathion, thus raising the possibility of the presence of resistance genes in low frequency. The case of polygenic dimethoate resistance in California field populations of *M. occidentalis* (Hoy 1985) is not problematic, since it appears to be a definitive case of polygenic resistance that developed from field selections.

The point of the above discussion is to raise several questions about pesticide-resistance selection in natural enemy species. Do laboratory selection methods in which challenge dosages or concentrations are incrementally increased tend to produce polygenic resistance; whereas field selection methods, which usually select with much higher dosages, tend to produce monogenic, semidominant resistance traits? Is polygenic resistance generally less stable than the monogenically determined resistance? Would it be more valuable to genetically improve populations by using semifield methods, or should selection for monogenic, dominant factors in the laboratory be attempted? Further discussion of these questions in relation to genetic improvement methods is presented in Chapter 20.

15.4. SUMMARY

Most early studies of resistance mechanisms in natural enemies were undertaken using techniques similar to those used to study resistance in pest arthropods. This type of research was usually conducted by toxicologists studying resistance mechanisms in pests, who then decided to look for a similar mechanism in predaceous species (e.g., Motoyama et al. 1970). In other cases, early studies of resistance mechanisms were undertaken by field entomologists, who were less well versed in basic toxicological research methods (e.g., Croft et al. 1976a). While their research was somewhat limited from a toxicological perspective, they focused on the practical value of resistance in the field. Thus, they sought to improve pesticide selectivity for natural enemies and to employ pesticide-R strains of natural enemies for management or release in the field (Chapters 17–21). As a result, the practical benefits of pesticide resistance were quickly incorporated into field IPM programs, probably faster than they would have been otherwise.

Research by both toxicologists and field entomologists tended to be limited in scope. Most studies examined a subset of possible mechanisms, often in isolation from one another. Integrated evaluations of a wide range of mechanisms conferring resistance were virtually nil. The early work of Motoyama et al. (1972, 1977) came closest to approaching this level of synthesis, especially when one considers their association with Rock and coworkers who studied insecticide-resistant strains of *A. fallacis* in the field (Rock and Yeargan 1970, 1971, 1972, Rock 1972).

A research team composed of members ranging from the toxicologist to the field entomologist should be encouraged to conduct integrated studies of resistance mechanisms in natural enemies in the future. The same points made for toxicological research apply to studies of the genetics of pesticide resistance. The current status of research in the toxicology and genetics of resistance in natural enemies reflects a generalization made earlier in this volume: seldom has basic study of natural enemy–pesticide relationships been considered a high research priority by either chemical or biological control specialists familiar with toxicological and genetic methods. However, in the near future, as interest in genetically engineering pesticide resistance into natural enemies increases, levels of support and emphasis in research may change dramatically (Chapter 21).

16

FACTORS AFFECTING RESISTANCE

16.1. INTRODUCTION

The development of pesticide resistance in arthropod natural enemies is an evolutionary phenomenon that is influenced by factors operating at the cellular, organismal, and population levels of biological organization (Croft and Morse 1979, Tabashnik and Croft 1985, Croft and van de Baan 1988). These factors include aspects of the genetics (Chapters 6 and 12), biochemistry and toxicology (Chapters 6 and 15), life history and bionomics (Chapters 3 and 4), ecology (Chapter 8), and management of resistant natural enemies and their hosts or prey (Chapters 17–20). Factors influencing resistance

This chapter was coauthored by B. E. Tabashnik, Department of Entomology, University of Hawaii, Honolulu, Hawaii.

Table 16.1. A Summary of Hypotheses Proposed to Explain the Different Frequencies of Pesticide Resistance among Arthropod Natural Enemies and Their Prey or Hosts

Level of Effect	Hypothesis	References
Genetic	Genetic variability	Huffaker 1971, Georghiou 1972, Croft 1972, Tabashnik & Croft 1985
Organismal	Preadaptation or intrinsic detoxification	Croft & Morse 1979, Croft & Strickler 1983, Tabashnik 1986a
Population/ ecological	Immigration/ residency ratio	Croft 1982, Tabashnik & Croft 1985, Whalon & Croft 1985
Interspecific/ ecological	Food limitation	Croft & Morse 1979, Hoy 1979, Tabashnik & Croft 1985, Tabashnik 1986a

development in arthropod natural enemies are in most instances similar to those affecting resistance evolution in arthropod pests (Georghiou and Taylor 1976, Tabashnik and Croft 1982, Roush and Croft 1986). In addition, there are factors influencing resistance that are unique to predators and parasitoids (Croft and Morse 1979, Tabashnik and Croft 1985, Tabashnik 1986a).

In this chapter, factors influencing pesticide resistance development in entomophagous insects and mites are examined. Comparisons with pests provide insights into processes occurring in natural enemies. The chapter focuses first at suborganismal and organismal levels, and then on interacting populations of species. Relevant empirical and theoretical studies are reviewed to suggest ways that different factors may impact on resistance evolution.

Several hypotheses (Table 16.1) have been proposed to explain the greater than 100-fold difference in the incidence of resistance among arthropod pests compared to natural enemies (Chapter 14). These hypotheses include 1) the lower resistance gene frequency hypothesis, 2) the lesser genetic-variability hypothesis, 3) the preadaptation or intrinsic detoxification-difference hypothesis, 4) the differential immigration/residency hypothesis, and 5) the food limitation hypothesis. Evidence for each of these hypotheses and their integrated effects are discussed.

16.2. GENETIC FACTORS

Genetic factors influencing resistance development in any population include the number of loci conferring resistance, the frequency of resistant alleles at each locus, and the fitness associated with resistance (Georghiou and Taylor 1976, Georghiou 1983). Genetic factors influencing resistance development have been studied on a modest scale for arthropod predators and parasitoids (Croft and Strickler 1983, Hoy 1985).

16.2.1. Number and Dominance of Resistance Alleles

For natural enemy species, fewer than 10 studies have examined the dominance and number of alleles conferring resistance (Table 15.2). Both monogenic and polygenic resistance has been documented. Dominance has ranged from partially dominant or recessive to almost completely recessive (Table 15.2).

Roush and McKenzie (1987) concluded that monogenic resistance is more likely to spread through a population than polygenic resistance, particularly when there is spatial or temporal variation in pesticide exposure. Theoretical studies using single-gene locus models suggest that the functional dominance of resistance, which depends on whether or not heterozygotes are killed by insecticides, can influence the rate of resistance development (Comins 1977b, Curtis et al. 1978, Taylor and Georghiou 1979). Functional dominance is particularly influential when there is a gene influx from susceptible populations (Tabashnik and Croft 1982).

Results from genetic improvement projects with phytoseiid mites have shown that monogenic, dominant resistance is easier to manage than recessive, polygenic resistance (Hoy 1985; Chapters 15–21). It is thought that the difficulty in maintaining polygenic, recessive resistance in field populations of natural enemies is due to the immigration of susceptible (S) genotypes and their subsequent hybridization with resistant (R) genotypes (Whalon et al. 1982a, Hoy 1985; Chapter 20). For example, where genetically improved, synthetic pyrethroid (SP) resistant strains of *Metaseiulus occidentalis* (Hoy et al. 1980, 1983) and *Amblyseius fallacis* (Whalon et al. 1982a, Croft and Whalon 1983, Croft 1983) were released in the field, polygenic SP resistance was not maintained after selection pressure was reduced.

16.2.2. Fitness Costs of Resistance

One factor that can be important in determining the propensity and stability of resistance in field populations of arthropods is the extent to which alleles for resistance affect fitness in the absence of pesticide selection (Georghiou 1983). Often it is assumed that R individuals are at a reproductive disadvantage in the absence of pesticides, otherwise they would be more common in natural populations. Fitness includes characteristics such as developmental rate, survival rate, fecundity, fertility, sex ratio, and mating ability.

Based on a review of available literature, Roush and Croft (1986) concluded that for pest arthropods reduced fitness in the absence of pesticides was not always associated with resistance. Furthermore, laboratory-selected R strains of pests usually exhibited greater reproductive disadvantages than R strains that developed in the field (Roush and Croft 1986). Laboratory rearing alone can greatly reduce the fitness of many different groups of arthropods (Mackauer 1976). However, reduced fitness in pesticide-R populations selected in the laboratory might also indicate that artificial selection is a very different process than that which occurs in the field (see Chapters 14, 19, and 20). Only a few studies comparing the fitness of R and S strains of natural enemy species have been made. Almost all studies have involved artificially selected populations rather than field-selected, resistant natural enemy populations.

Schulten and van de Klashorst (1974) compared OP-R and -S strains of the predaceous mite *Phytoseiulus persimilis* selected in a greenhouse environment. At 25 °C and 70–80% relative humidity, the developmental period, preovipositional period, total egg production over a 10-day period, egg consumption over a 5-day period, and sex ratio were not significantly different between the two strains. They concluded that the resistance factor was either neutral for fitness or that any deleterious effects were readily balanced by the remaining gene pool characteristics.

Roush and Hoy (1981a, 1981b) compared carbaryl-S and laboratory-selected, carbaryl-R strains of the predatory mite *M. occidentalis*. Mites were reared on their native prey *T. urticae* under laboratory conditions, and complete generational life table estimates were made. In the absence of carbaryl, the R strain of the predator did not differ significantly from the S strain in developmental time, sex ratio, fecundity, mating ability, diapause, or ability to control spider mites. In fact, due to small difference in the sex ratio between the two strains, the R mites had a higher intrinsic rate of increase (r_m) than did the S strain.

Mueller-Beilschmidt and Hoy (1987) reported on the activity levels of males and females of field-selected and laboratory-selected resistant strains of *M. occidentalis* as an index of fitness. Using a computerized video tracking system, they observed that adult females of a strain resistant to SPs and OPs had a significantly lower average activity level than either the wild strain or three other genetically improved resistant strains. They also found that the movement patterns of individual strains differed. Their studies indicated that the activity level is heritable and might be used to evaluate fitness in genetically improved strains in the field.

Koenig and Hassan (1986), in studies of genetically improved strains of *Phytoseiulus persimilis*, observed that resistance to pirimiphos-methyl and pyrazophos was associated with a higher intrinsic mortality (i.e., non-pesticide-induced) and a reproductive disadvantage. They surmised that these differences could result in a gradual loss of resistance to both compounds; however, in the laboratory, resistance to pyrazophos was stable while resistance to pirimiphos-methyl decreased rapidly after selection was discontinued.

Grafton-Cardwell and Hoy (1986b) found that a carbaryl-R strain of *Chrysoperla carnea* that was reared in the absence of carbaryl exhibited lower larval and pupal survival and produçed fewer females than the carbaryl-*S* strain from which it was derived. They also noted that the fecundity and adult longevity of the R strain were significantly greater, which compensated in part for the reduced survival of the immature stages. They calculated the total progeny from 100 females of the R and S strains and estimated that over a single generation the S strain produced 2000 more progeny than the R strain.

In light of these limited studies, it is difficult to generalize the influence of genetic fitness on the stability of resistance in field populations. Several studies have associated fitness disadvantages with resistance, but it is unclear how these results apply to field populations. Deleterious effects present in a less fit organism that might be seen under variable field conditions, might be masked under the more uniform, ideal conditions maintained in the laboratory. Follow-up studies are needed to assess the fitness of field-established, pesticide-resistant natural enemies. Methods for such studies are not well developed, but estimates of life table parameters, including rates of development and survivorship, would be a good place to start. Tracking the relative abundance of R and S strains established in the field in the absence of sprays would also indicate fitness, but such a study would be difficult to carry out.

16.2.3. Intrinsic Resistance-Allele Frequencies

A largely untested hypothesis, suggested to explain the observed differences in resistance development between pests and natural enemies, is that natural enemies may have a lower initial resistance-allele frequency than pests (Tabashnik and Croft 1985). Such resistant alleles might govern enhanced detoxification, reduced penetration, nerve insensitivity, or other resistance mechanisms.

Mutation rates and allele frequencies for certain traits are known to differ significantly among very different groups of organisms (e.g., between microorganisms and higher organisms; Falconer 1981), but such differences may be less significant among arthropods. Intrinsic resistance-allele frequencies and mutation rates have not been measured in either field or laboratory populations of natural enemies. Even for pest species, intrinsic resistance-allele frequencies for susceptible populations are not well known (Roush and Croft 1986).

There has been considerable scientific discussion on how different allele frequencies for a resistance trait might arise, be selected for, and be maintained in a population (e.g., Bishop 1982). In addition, theories and studies of factors influencing mutation rates and genetic variability—including environmental stresses as inducers—pose interesting questions regarding the origin and maintenance of resistance alleles (McDonald 1983). Many of these ideas are controversial and their relevance to resistance is not known. For example, resistance to drugs in mammalian cell lines can be induced by compounds unrelated to the drugs for which selection pressure is being applied (Schimk et al. 1978). In deleterious microbes, higher frequencies of resistance alleles can be induced by exposure to either radiation or chemical inducers before selection with a pesticide. Could pesticide exposure influence mutation rates leading to resistance development in pests and natural enemies? Several authors have proposed that exposing natural enemies to inducers might increase the mutation rate, leading to a higher incidence of pesticide resistance; however, little success with these methods has been reported (e.g., Roush and Plapp 1982; Chapter 19).

Although no information is currently available on resistance-allele frequencies or mutation rates in field population of natural enemies, one can explore the potential consequences of lower initial resistance-allele frequencies in natural enemy and pest populations by using models that assume that such differences exist. General analytical models (e.g., May and Dobson 1986) have demonstrated that in a single-gene locus, two-allele system the rate of resistance development increases approximately linearly as the logarithm of the initial resistance-allele frequency. Due to this logarithmic relationship, large differences in initial resistance-allele frequency result in relatively small changes in evolutionary rates of resistance development. For example, if there is a millionfold difference between the initial resistance gene frequency in a pest (10^{-3}) and a natural enemy (10^{-9}), the expected difference in the rate of resistance development is only 3-fold.

Computer simulations for 12 species of natural enemies of apple pests showed that 10- and 100-fold reductions in the assumed initial resistance-allele frequency (reduced from 10^{-3} to 10^{-4} and 10^{-5}) had little impact on predicted times to resistance development for most species (Tabashnik and Croft 1985). Assumptions of this model were that resistance is controlled primarily at one locus with two alleles and that there is a fixed dose-mortality line associated with each genotype (SS, SR, RR). Thus, both analytical and numerical modeling approaches have indicated that rates of resistance development are relatively insensitive to changes in initial resistance-allele frequency.

16.2.4. Genetic Variation in Resistance Alleles

A related hypothesis which accounts for differences in resistance potential between pests and natural enemies states that phytophagous arthropod pests have greater intrinsic genetic variation in response to pesticides than do natural enemies. This variation then provides a higher incidence and greater variety of resistance alleles for selection.

The genetic variation hypothesis may explain the lack of resistance development particularly in parasitoids, which in evolving their highly specialized parasitic habit, may have sacrificed some genetic diversity. This in turn could limit their ability to adapt to stresses such as pesticides (Huffaker 1971, Georghiou 1972). Indeed, members of the order Hymenoptera have been found to display the lowest electrophoretically detectable genetic variation of any of the major insect orders (Graur 1985). Although solitary hymenopteran species are closer to the average heterozygosity than social hymenopteran species, they still exhibit less variation than nonhymenopteran orders.

Natural enemies—and parasitoids in particular—differ from pests in genetic variation as manifested in their responses to pesticides. This was discussed in Chapter 2 where analysis focused on interspecific variation between predators and parasitoids. When extensive susceptibility response data were compared, greater variation was observed among predators than parasitoids (Table 2.7 and Fig. 2.9). Similar trends between species from these two groups were also apparent (Theiling and Croft unpubl.). Although these comparisons were made at a very broad level of aggregation and many factors could be influencing these indices (see Chapter 2), the observed patterns may be significant. If present in field populations, these trends lend support to the genetic variation hypothesis: natural enemies are less variable in response to pesticides than pests and parasitoids are less variable compared to predators and phytophagous arthropods. Explanations of these trends in susceptibility response variation are not known, but they raise questions for further research.

According to quantitative genetics theory, the rate of increase in pesticide resistance should be directly proportional to the additive genetic variance associated with the species in question (Via 1986). Therefore, large differences in additive genetic variance could affect resistance development and may contribute to observed differences between pests and natural enemies. In this regard, most economically significant cases of pesticide resistance in pests are thought to be under monogenic control; however, there are also many examples of polygenic resistance among pest species (Roush and McKenzie 1987). As noted in Chapter 15, a few cases of polygenic resistance are known among biological control agents (Table 15.5). No quantitative comparison of the relative incidence and impact of polygenic resistance on resistance evolution in pests and natural enemies has been made (see a qualitative comparison between spider mites and predatory mites in Croft and van de Baan 1988). The sample for natural enemies is probably too small to make such comparisons meaningful.

No paired-species studies of a pest–natural enemy system have been conducted to experimentally assess how genetic variability contributes to resistance development. Some indirect evidence that may shed light on this relationship comes from studies of susceptible strains of a spider mite and its closely associated phytoseiid predator.

For a wide variety of pesticides tested from several different classes (e.g., organophosphates, carbamates, synthetic pyrethroids, botanicals), the slope of the log concentration–probit mortality (LCP) line for the herbivorous mite *T. urticae* is invariably less than that for the predaceous mite *A. fallacis* (e.g., see Croft and Wagner 1981, Strickler and Croft 1985, B.A.C. unpubl. data). Because the slope of the LCP line is inversely proportional to the standard deviation of log tolerances (Finney 1978), these results may indicate that the pest has more phenotypic variation in response to pesticides than the predator. Other cases of more homogeneous responses to pesticides in predaceous forms compared to their prey can be found in the literature (e.g., between the reduviid predator *Pristhesancus papercaulus* of the soybean looper *Pseudoplusia includens* to acephate; Martin and Brown 1984), but the reverse appears to be less common. It is likely that some portion of observed phenotypic variation in response to pesticides is genetically based, but as pointed out by Tabashnik and Cushing (1989), genetic variation cannot be inferred from the slope of LCP lines without explicit estimates of the genetic component of phenotypic variation.

Rosenheim and Hoy (1986) found considerable variation in pesticide resistance in field populations of the citrus scale parasitoid *Aphytis melinus* (see Chapter 14, Section 14.4.4). This species was introduced to California in the late 1950s from a single culture collected in Pakistan and India. Rosenheim and Hoy (1986) concluded that their results demonstrated considerable evolutionary divergence in a parasitic wasp that might have been expected to have low genetic variability.

These findings for arthropod pests and natural enemies at large, predators and parasitoids, herbivorous and predatory mites, and the citrus scale parasitoid suggest that the magnitude of genetic variation between arthropod pests and natural enemies may differ. However, direct evidence requires estimates of genetic variation in pesticide tolerance or susceptibility within populations of pests and natural enemies (Tabashnik and Cushing 1989). Furthermore, this information must be linked to the propensities for resistance development in these groups. No such data are presently available.

16.3. BIOCHEMICAL, PHYSIOLOGICAL, AND TOXICOLOGICAL FACTORS

16.3.1. Preadaptation Hypothesis

One of the early hypotheses proposed to explain observed differences in resistance between arthropod pests and natural enemies was called the preadaptation or intrinsic detoxification-difference hypothesis (Tabashnik 1986a). Gordon (1961) first proposed that herbivores are preadapted to detoxify pesticides because they have evolved systems to detoxify such plant defensive compounds as alkaloids, terpenoids, phytosteroids, and other allelochemicals. Subsequent research has generally confirmed this hypothesis among plant-feeding species, although a number of exceptions have also been reported (Rosenthal and Jansen 1979, Berenbaum 1985, Berenbaum and Neal 1987). Later it was reasoned that entomophagous species may never have evolved extensive detoxification systems or that they may have lost some of their capacity to detoxify due to limited exposure to plant defensive compounds (Croft and Morse 1979, Plapp 1981).

In terms of actual impact, whether a natural enemy never had a high detoxification capacity or had it and subsequently lost it is irrelevant. The question is an interesting one, however, to those studying the evolution of detoxification systems in arthropods (e.g., Wilkinson 1979). The relevance of the preadaptation hypothesis to the observed lack of pesticide resistance among insect natural enemies is difficult to determine since the evolutionary chronology of detoxification mechanisms is poorly understood for any insect group (Wilkinson 1979).

Intrinsic and inducible detoxification in a natural enemy may determine its ability to evolve detoxification-based resistance. For example, esterase enzymes that are used to detoxify allelochemicals could also be used to detoxify pesticides. One would expect to find greater esterase activity in an R population than in an S population of predators or parasitoids.

No studies comparing the detoxification of secondary plant compounds and pesticide resistance potential in natural enemies are available in the literature. However, the ability to detoxify plant-defensive compounds and its relation to resistance development might be reflected by the diverse feeding habits of different natural enemies. For example, facultative entomophages should exhibit higher rates of resistance development than obligate, monophagous entomophages. Entomophagy is manifested in many different forms. Specialized, host-specific parasitoids are probably the most strictly entomophagous species with progeny that feed only on insect tissues and adults which host feed or do not feed at all (Thompson 1981, 1986). Some parasitoids, however, feed on plant nectar or indirectly on plant compounds in the form of aphid honeydews (Bartlett 1964). Many obligatory predators also feed on plant nectar or pollen, or both (Clausen 1962). Several families of predominantly predaceous natural enemies are not taxonomically distant from phytophagy. For example, in the order Hemiptera, certain species of Miridae, Lygaeidae, and Pentatomidae are facultative herbivores, feeding on both plant tissues and insect prey. Several predaceous beetle families contain both entomophagous and phytophagous forms (e.g., Coccinellidae). Also, some families of insects (e.g., Cynipidae, Eurytomidae) include facultative herbivores as well as insect parasitoids (Clausen 1962). Species from these groups would seem to be ideal models for studying the relationships between feeding strategies and pesticide resistance potential. At present no general surveys of detoxification enzymes across such diverse groups of natural enemies have been conducted. This research would be one way to critically examine the preadaptation hypothesis.

The preadaptation hypothesis has been refined by discriminating among different groups of detoxification enzymes. For example, preliminary research indicates that several relatively specialized entomophagous species may have retained some of their basic detoxification capacities in the form of certain specific enzymes (e.g., epoxide hydrolases; Mullin et al. 1982, Mullin and Croft 1984, Croft and Mullin 1984). This is especially true for detoxification enzymes involved in basic physiological functions such as esterases used in hydrolytic metabolism (Chapter 6). Natural enemy species may be less well adapted than pests in terms of detoxification enzymes less involved in the basic aspects of physiology and development (e.g., see oxidative enzymes; Mullin et al. 1982, Croft and Mullin 1984, Yu 1987, 1988). Furthermore, since most pesticides developed to date can be rendered innocuous principally by rapid detoxification (Chapter 6), it is not surprising that detoxification-deficient natural enemies are more severely affected by these agents than are the target pest herbivores (Chapter 2).

Based on patterns of pesticide susceptibility and detoxification observed in comparative studies of cotton pests and their predators and parasitoids, Plapp (1981) proposed that there are major differences between oxidative and hydrolytic detoxification enzyme systems in herbivores and natural enemies. Plapp's hypothesis stated that whereas pests have well-developed oxidative detoxification systems (e.g., mixed-function oxidases) which function primarily to detoxify xenobiotics, natural enemies generally lack such systems. However, natural enemies and pests have similar capacities for hydrolytic detoxification using esterases or glutathione-*S*-transferases, which also function in a number of basic metabolic pathways. Mullin and Croft (1985) postulated that entomophagous species may have retained (or acquired) epoxide hydrolase detoxification enzymes to break down products of fatty-acid metabolism from host or prey tissues as a function of their entomophagous feeding strategy, perhaps to a greater extent than phytophagous species (Chapter 6). The importance of these enzymes in pesticide detoxification is not well understood.

16.3.2. Experimentation on Preadaptation

Published studies which test the preadaptation hypothesis for pests and natural enemies are limited. Several studies have measured the impact of plant compounds on the survivorship and reproduction of natural enemies where compounds were taken up directly or via the food chain by feeding on pests or on plants. The few comparative detoxification–pesticide susceptibility studies available indicate that entomophagous predators and parasitoids differ quantitatively in their detoxification enzyme activities compared to their plant-feeding hosts or prey. Only a single study has been conducted which compares detoxification and resistance development in pests, and no research on detoxification enzyme induction as it relates to susceptibility or resistance has been conducted on natural enemies, as has been done with pest species (Yu and Terriere 1973, Yu et al. 1979, Terriere 1984). Research needs in each of these areas are discussed in greater detail hereafter.

Toxins produced by plants can affect natural enemy development, metabolism, and susceptibility to pesticides either through their hosts or prey or by occasional feeding on plant substrates (Chapter 6). Jones et al. (1962) and Jones (1981) reported differences in rhodanase production, an enzyme that detoxifies cyanogenic compounds, between two hymenopteran parasitoids. *Apanteles zygaeranum*, an internal parasitoid of the lepidopteran *Zygaena* spp., was able to detoxify HCN; however, *Mesochorus temporalis*, which parasitizes principally *Z. filipendulae*, was not. *Zenillia adamsonii*, a tachinid parasitoid of the monarch butterfly, contains toxic cardenolides in its body. It must therefore be able to deal with these poisons, which originate in the milkweed plant and are mediated through its host (Reichstein et al. 1968). A more convincing example of the interaction of secondary plant compounds and detoxification systems of natural enemies was reported by Campbell and Duffey (1979), who measured the effects of tomatine on the ichneumonid *Hyposoter exiguae* by adding it to artificial diet fed to its host *Heliothis zea*. *H. exiguae* is normally exposed to high levels of this secondary plant compound via its lepidopteran host. Parasitoid larval development was slowed; percent pupal eclosion, adult weight, and adult longevity were reduced; and the extracts of larvae contained tomatine. Initially, Gilmore (1938) found that larval survivorship of *Apanteles congregatus* was drastically lowered when its host, the tobacco hornworm, was fed a high-nicotine strain of tobacco as opposed to tomato or jimson weed. This was later confirmed in studies by Thurston and Fox (1972). More recently, Barbosa et al. (1982, 1986) investigated this tritrophic relationship in greater detail. They compared rates of development and survival of the two parasitoids *A. congregatus* and *Hyposoter annilipes* when their host *Spodoptera frugiperda* fed on nicotine-free or nicotine-containing diets. The effects of nicotine were more severe on the generalist parasitoid *H. annilipes* than on the specialist species *A. congregatus*. These data further confirmed the evolutionary response of parasitoids to natural biochemical defense systems of plants and herbivorous insects, although differences in the responses of generalist and specialist parasitoids contradicted patterns observed in some other natural enemy groups, which appeared to be favored in their detoxification capabilities by their previous exposure to many host or prey types.

Some of the first detoxification studies that shed light on differences between natural enemies and pests were conducted by Brattsten and Metcalf (1970, 1973). They examined mixed function oxidase (MFO) activity in 54 species of arthropods, including 32 phytophages and 17 entomophages, using carbaryl and the MFO inhibitor piperonyl butoxide (PBO). Their initial conclusions were that there was considerable variability among the species tested and that no apparent pattern of detoxification capability was present. However, their conclusions were based on comparisons of species classified by feeding specificity such as degree of omnivory or herbivory rather than between pests and natural enemies or predators and parasitoids.

Mullin and Croft (1985) reanalyzed the data of Brattsten and Metcalf (1970, 1973), focusing more specifically on comparisons between phytophages and entomophages (data combined for both predators and parasitoids). Variation about the mean LD_{50} of each group to the carbamate insecticide carbaryl was examined. Synergistic ratios in the presence and absence of PBO were also evaluated. Considerable variation in the responses of species within each ecological group was observed: herbivores had a 10-fold higher

Table 16.2. Toxicity of Carbaryl to Phytophagous Pests Relative to Predaceous and Parasitic Arthropod Natural Enemies (Adapted from Brattsten and Metcalf 1970, 1973, Croft and Morse 1979, and Mullin and Croft 1985)

		Entomophages		
	Phytophages ($n = 32$)	Predators ($n = 7$)	Parasitoids ($n = 8$)	Predators + Parasitoids ($n = 17$)
Lethal dose 50%	378 ± 198[1,3]	70.3 ± 43[2]	26 ± 8[2]	37 ± 18[1,3]
Synergistic ratio	15.4 ± 7.3[1]	10.6 ± 2.4[2]	8.2 ± 1.7[2]	7.2 ± 1.3[1]

[1]Data from summary by Mullin and Croft (1985).
[2]Data from summary by Croft and Morse (1979).
[3]Data significantly different at $P \leqslant 0.001$ between arthropod groups.

mean LD_{50} and a 2-fold higher mean MFO activity than did natural enemies (Table 16.2). While mean LD_{50} values were significantly different between pests and natural enemies, mean synergistic ratios were not. Earlier, Croft and Morse (1979) had similarly compared the pest, predator, and parasitoid response data of Brattsten and Metcalf (1970, 1973) (Table 16.2). Again, differences were observed between predators and parasitoids and also between either natural enemy group and pests, with differences being greatest for mean LD_{50} values and less so for synergistic ratio means. Results were generally consistent with the degree of feeding specialization that occurs in these groups as discussed above: the more polyphagous predators were less susceptible and more variable in their responses than parasitoids.

More specific studies of detoxification enzymes in individual natural enemies and their hosts or prey were made by Mullin et al. (1982), Mullin and Croft (1984), and Croft and Mullin (1984). Since these comparisons are related to intrinsic susceptibility as well as to resistance potential, these experiments were discussed in Chapter 6; however, several points relevant to detoxification and resistance can be added here. In a comparative study with its susceptible prey *Tetranychus urticae*, S strains of the predatory mite *Amblyseius fallacis* had 5-fold lower MFO activity as measured by aldrin epoxidation, no difference in general esterase levels, 6-fold lower *trans*-epoxide hydrolase activity, but 11-fold and 4-fold higher glutathione-*S*-transferase and *cis*-epoxide hydrolase activities, respectively (Table 15.4). In comparing R and S strains of *A. fallacis*, the multiple-R strain had similar MFO activity as the S strain, a 6-fold increase in general esterase, a 1.6-fold increase in glutathione-*S*-transferase, and a 2-fold increase in *cis*- and *trans*-epoxide hydrolase (Mullin et al. 1982; Table 15.4). While the interspecific comparisons (S strains of predator versus pest) presumably reflect long term evolutionary adaptations in detoxification systems, intraspecific comparisons (R versus S strains of each species) reflect short-term evolutionary responses to selection primarily by insecticides. The relationship between individual detoxification enzymes and specific resistance factors in each species was not examined.

Other comparisons of phytophagous pests and entomophagous species, primarily with S strains, have been made for 20 pests and 7 beneficial species (Chapter 6, see partial presentation in Table 6.3; Mullin and Croft 1984, Croft and Mullin 1984, Mullin 1985). Included in the comparisons were species from the insect families Anthomyiidae,

Arctiidae, Braconidae, Chrysomelidae, Chrysopidae, Coccinellidae, Lasiocampidae, Meloidae, Noctuidae, Pieridae, Tortricidae, and the acarine families Tetranychidae and Phytoseiidae. In general, oxidative detoxification systems appeared to be more active among phytophagous species and no consistent differences in glutathione-*S*-transferase or esterase activity were apparent. In all studies made to date, the ratio of *cis*- to *trans*-epoxide hydrolase has been consistently different between the two groups. Entomophages tend to have a lower ratio while that of phytophages is higher.

The pattern in epoxide hydrolase ratios between pests and natural enemies has been interpreted to represent a greater reliance by natural enemies on *trans*-epoxide hydrolase to metabolize fatty acids associated with insect as compared to plant foods (Croft and Mullin 1984, Mullin and Croft 1985; Chapter 6). Selective pesticide development may be possible by exploiting differences in this particular detoxification system. A toxicant that acts primarily on biochemical pathways involved in *cis*-epoxide hydrolase catabolism might selectively interfere with the capacity of herbivores to detoxify materials associated with their plant diet. Exploitation of epoxide hydrolase enzyme systems for resistance development in natural enemies is even more speculative at this point, since the role of this enzyme in pesticide resistance among arthropod species is not well understood (Mullin et al. 1982).

The relative importance of induction versus constitutive production of detoxification enzymes in natural enemies is not known. One might expect that natural enemies would not maintain high constitutive levels of detoxification enzymes due to high physiological costs; rather they would depend more on enzyme induction for detoxification. The relationship between induction and constitutive maintenance of detoxification enzymes is being explored in pest species (Yu et al. 1979, Yu 1983). Similar comparisons on natural enemies could be informative.

The impact of differences in the intrinsic susceptibility of pests and natural enemies on resistance development was studied in a simulation model for 12 species of natural enemies and 12 species of pests of apple (Tabashnik and Croft 1985; see earlier discussion of intrinsic gene frequencies and resistance in Section 16.2.3). Assumptions of the model were that resistance is controlled primarily at a single-gene locus with two alleles (S, R) and that there is a fixed dose–mortality line associated with each genotype (SS, SR, RR). To examine the potential impact of greater intrinsic susceptibility to pesticides in natural enemies on their rate of resistance development, the LD_{50} values for susceptible (SS) natural enemies were reduced 10- and 100-fold relative to the LD_{50} values for susceptible (SS) pests. Selectivity studies (Chapter 2) indicated that these assumptions were not unreasonable in light of the empirically observed differences in the intrinsic susceptibility between pests and natural enemies (Chapter 2). Assuming higher intrinsic susceptibility for (SS) natural enemies did not substantially alter the model's predictions for the relative difference in resistance rates in both the pest and the natural enemy species studied.

More extreme interpretations of the preadaptation hypothesis, however, greatly reduced the simulated rates of resistance development of the natural enemy species studied (Tabashnik and Croft 1985). Under the assumption that the LD_{50} values of all genotypes (SS, SR, RR) for natural enemies were 10- or 100-fold lower than the corresponding values for pests, resistance development was predicted to take more than 25 years for nearly all natural enemies studied. (For pests, rates of resistance development remained the same since assumptions were unchanged for these species.) This interpretation of the preadaptation hypothesis assumes, in effect, that natural enemies are more intrinsically susceptible and further, that the resistance attained by natural enemies will be lower (in absolute terms) than the resistance attained by pests.[1] The fact that some natural enemies of pests have evolved resistance in less that 25 years in the field (Croft 1982, Tabashnik and Croft 1985) suggests that these more extreme interpretations are not applicable to all natural enemy species. However, most levels of resistance in natural enemies evolved to date are generally lower than those observed in pest species (Croft and Strickler 1983; Chapter 14), and this trend may lend support to the importance of the preadaptation hypothesis.

Sensitivity studies of these representations of the preadaptation hypothesis (Tabashnik and Croft 1985) generally indicated that biochemical or toxicological factors were probably *not* the primary determinants accounting for different rates of resistance development observed in the field between apple pests and natural enemies (see also Sections 16.4.2 and 16.4.3). While factors operating at the suborganismal level certainly influence resistance development in entomophages, ecological factors operating at the population level appear to be more important (see Section 16.4 and conclusions of Tabashnik and Croft 1985, Tabashnik 1986b).

16.4. LIFE HISTORY AND ECOLOGICAL TRAITS

Natural enemy interactions with their prey or hosts can affect resistance development (Section 16.5), but in this section, we consider the ecology of natural enemies as single species (autecology). For now, interspecific interactions are ignored and resistance development is examined as it is influenced by life history traits such as survivorship, fecundity, dispersal, and the number of generations per year. The influence of life history traits on resistance development should be similar in natural enemies and pests.

16.4.1. General Models

General models show that resistance development is influenced by many life history traits and ecological factors, including developmental time, fecundity, sex ratio, and initial population size. One consistent generalization that can be made from these types of studies is that resistance develops faster as fecundity and the number of generations per year increase. Conversely, increased natural (nonpesticide) mortality and immigration of susceptible individuals slow resistance development (e.g., Comins 1977a, 1977b, Taylor and Georghiou 1979, Tabashnik and Croft 1982, May and Dobson 1986). The influence of some life history traits depends on the insecticide dose and the presence or absence of immigration by susceptibles (e.g., fecundity, survivorship; Tabashnik and Croft 1982). Nonetheless, models suggest that the number of generations per year and immigration by susceptible individuals can have a strong influence on resistance development under most conditions (Tabashnik and Croft 1982, May and Dobson 1986).

16.4.2. Correlation Studies

Tabashnik and Croft (1985) found statistical correlations between selected life history traits and observed rates of development of resistance to the OP insecticide azinphos-

[1] The assumption of lower LD_{50} values for all three genotypes implies that the resistance allele increases the LD_{50} by a fixed multiple. Thus, as SS individuals are less tolerant, so are RS and RR individuals. Lowering the LD_{50} of SS individuals without altering the LD_{50} of RS or RR individuals assumes that a resistance allele provides a fixed level of tolerance, regardless of the tolerance of SS individuals. It is unknown if either assumption is widely applicable.

methyl over a 30-year period for 12 natural enemies of apple pests and 12 species of their prey or hosts. The number of generations per year was strongly correlated with observed rates of resistance development. Of the apple pests and natural enemies examined, few species with less than three generations per year evolved resistance to azinphosmethyl; however, those with three or more generations per year evolved resistance more rapidly. Similar relationships exist between generation turnover and observed rates of resistance development for many other arthropods (Georghiou 1980, May and Dobson 1986). These empirical observations support theoretical findings that the number of generations per year has a strong influence on the rate of resistance development (Tabashnik and Croft 1982, May 1985, May and Dobson 1986).

Although predictions of the Tabashnik and Croft (1985) model corresponded well with the order of observed times to resistance development for natural enemies and their principal prey or hosts, predicted rates tended to be faster than observed rates, especially for natural enemies (Fig. 16.1). The model predicted that 6 natural enemies would not evolve resistance in 25 years; 4 of these species are not yet resistant, including the coccinellid *Coleomegilla maculata*, the mirid *Hyaliodes harti*, the anthocorid *Orius*

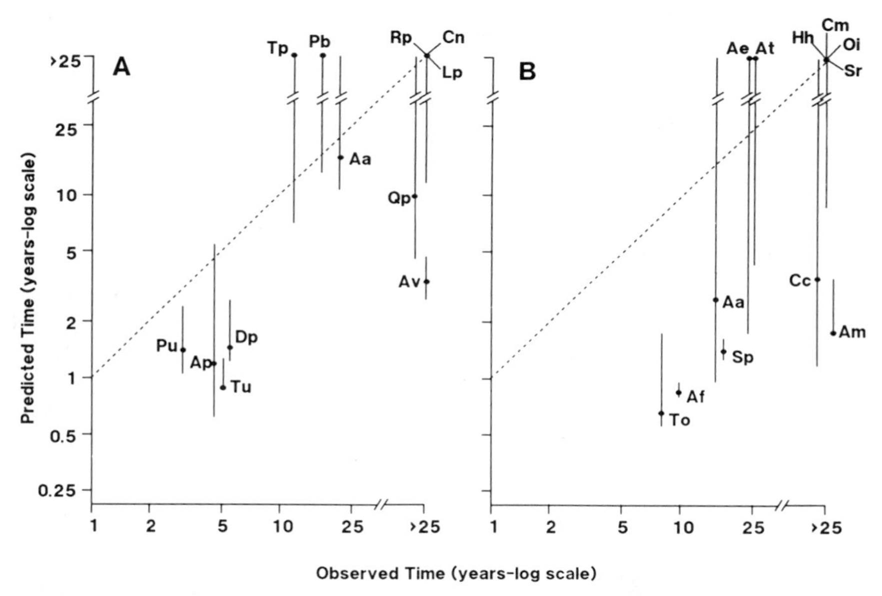

Figure 16.1. Predicted versus observed times to evolve resistance to azinphosmethyl for apple arthropods. Predicted time ● = simulated time to evolve resistance using means of estimates of population ecology parameters. Observed time = years after 1955 (first widespread use of azinphosmethyl) to first report of resistance. Vertical bars show range of predicted times from sensitivity analysis. Predicted = observed along dashed lines. (a) Pest $n = 12$, Spearman's rank correlation coefficient rs = 0.652; $P < 0.05$. Aa = *Archips argyrospilus*; Ap = *Aphis pomi*; Av = *Agyrotaenia velutinana*; Cn = *Conotrachelus nenuphar*; Dp = *Dysaphis plantaginea*; Lp = *Laspeyresia pomonella*; Pb = *Phyllonorycter blancardella*; Pu = *Panonychus ulmi*; Qp = *Quadraspidiotus perniciosus*; Rp = *Rhagoletis pomonella*; Tp = *Typhlocyba pomaria*; Tu = *Tetranychus urticae*. (b) Natural enemies (same assumptions as pests). $n = 12$, rs = 0.692; $P < 0.05$. Aa = *Aphidoletes aphidimyza*; Ae = *Anagrus epos*; Af = *Amblyseius fallacis*; Am = *Aphelinus mali*; At = *Aphelopus typhlocyba*; Cc = *Chrysoperla carnea*; Cm = *Coleomegilla maculata*; Hh = *Hyaliodes harti*; Oi = *Orius insidiosus*; Sp = *Stethorus punctum*; Sr = *Syrphus ribesii*; Mo = *Metaseiulus occidentalis*. (Adapted from Tabashnik and Croft 1985.)

insidiosus, and the syrphid *Syrphus ribesii*. The mymarid *Anagrus epos* and the dryinid *Aphelopus typhlocyba*, both of which parasitize leafhoppers of apple (Fig. 16.1b), were reported to be resistant after 24 years.

Two natural enemy species that showed the greatest difference between observed and predicted rates of resistance development were the chrysopid *Chrysoperla carnea* and the aphelinid *Aphelinus mali* (Fig. 16.1b). The observed time for *Chrysoperla* was more than 20 years less than the predicted time. Sensitivity analysis showed that the predictions were extremely sensitive to variations in fecundity, initial population size, and immigration levels, indicating that small changes in estimates for any of these parameters might change simulation results appreciably. *A. mali* was also predicted to develop resistance 20 years faster than was observed. This species previously developed strains resistant to DDT (Johansen 1957) after years of intensive use. *A. mali* is suspected to be resistant to azinphosmethyl because it is commonly found in treated orchards (Croft 1982), although no verification of this situation has been attempted.

As noted in the study of Tabashnik and Croft (1985), significant rank correlations between observed and simulated times to resistance were found for both pest and natural enemy species of apple. These simulations incorporated variation in life history traits among species, but assumed that the genetics and initial levels of resistance were alike for all species. These results suggest therefore that life history and ecological differences among species incorporated in the simulation are sufficient to explain variation in the rates of resistance evolution among the natural enemies of apple pests. Further modification of the model based on another resistance hypothesis even more closely simulated the resistance differences between pest and natural enemy species. (See later discussion of the food limitation hypothesis and interspecific effects on resistance development; Section 16.5.)

16.4.3. Immigration

Theoretical work suggests that resistance evolves more slowly as the ratio of susceptible immigrants to residents in the treated population increases (Comins 1977a, 1977b, Taylor and Georghiou 1979, Tabashnik and Croft 1982). Tabashnik and Croft (1985) noted that while the size of any treated population fluctuates, a rough index of resident population size for a given species can be approximated by two ecological parameters: initial overwintering population size and reproductive potential, where

$$[P_r] = [\mathrm{S}][F_a][P]^g[\mathrm{OW}]$$

P_r = resident population size
S = survival from egg through adults
F_a = % of adults that are female
P = progeny per female
g = generations per year
OW = % overwintering survival

They observed that the rate of resistance development was negatively correlated both with the ratio of immigrants to their overwintering population sizes and with the ratio of immigrants to their reproductive potential. These negative correlations are consistent with the idea that immigration by susceptibles can retard resistance development. However, Tabashnik and Croft added that a more thorough evaluation of the impact of immigration in these models requires better data on immigration for almost all species studied.

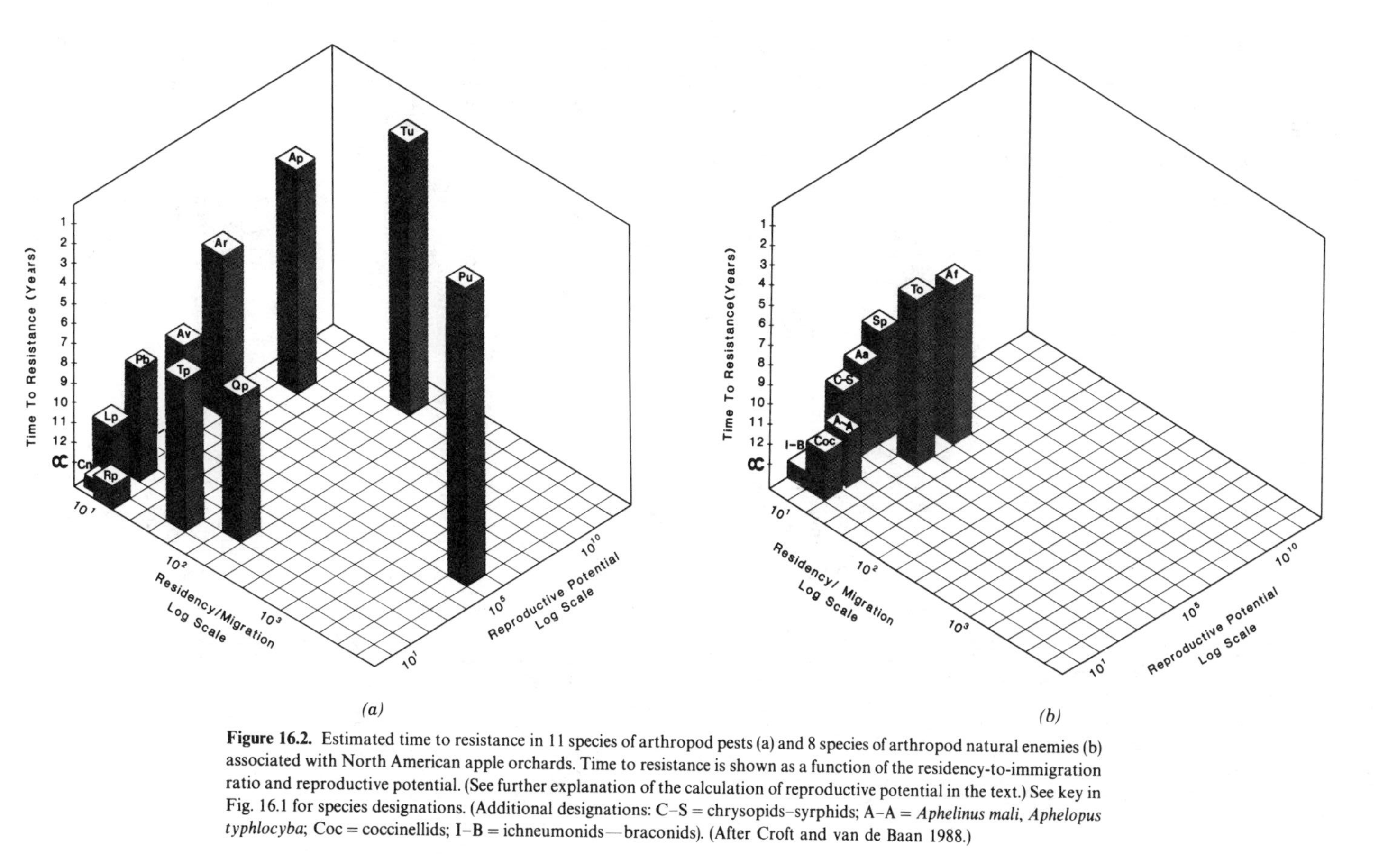

Figure 16.2. Estimated time to resistance in 11 species of arthropod pests (a) and 8 species of arthropod natural enemies (b) associated with North American apple orchards. Time to resistance is shown as a function of the residency-to-immigration ratio and reproductive potential. (See further explanation of the calculation of reproductive potential in the text.) See key in Fig. 16.1 for species designations. (Additional designations: C–S = chrysopids–syrphids; A–A = *Aphelinus mali, Aphelopus typhlocyba*; Coc = coccinellids; I–B = ichneumonids—braconids). (After Croft and van de Baan 1988.)

Estimates of the influence of immigration on the dilution of resistance in the field for both pests and natural enemies were sought using two different methods. Initially, a survey of expert opinion was conducted to gather information on seasonal reproductive rates and immigration/residency levels (see Croft 1982, Tabashnik and Croft 1985 for further details on this survey). Results of this survey are presented in Fig. 16.2 as an index for both pests (a) and natural enemies (b). Observed time to resistance development for individual species was plotted against estimates for reproduction and the immigration-to-residency ratio. The reproductive index was calculated as discussed above.

As shown in Fig. 16.2, there was a strong positive correlation between resistance development and both reproduction rates and immigration/residency estimates. As reproductive rates and the ratio of residents to immigrants increased, the probability of resistance increased, or the predicted time to resistance development decreased. The most rapid rates of resistance development for pests and natural enemies were computed for the spider mites *Tetranychus urticae* and *Panonychus ulmi* and their mite predators *Metaseiulus occidentalis* and *Amblyseius fallacis*. Next in probability of rapidly developing resistant strains (Fig. 16.2b) were arthropods with high reproductive potentials and low immigration rates, such as aphids, leafhoppers, leafminers, and the natural enemies *Stethorus punctum* and *Aphidoletes aphidimyza*.

Interestingly, whereas both arthropods pest and natural enemies showed similar ranges in reproductive estimates, major differences were observed in immigrant-to-resident ratios (Fig. 16.2). Natural enemies, in general, were far more likely than pests to have higher immigrant-to-resident ratios. These data suggest that the more host- or prey-specific natural enemies, having high residency levels in treated environments and minimum immigration tendencies (e.g., phytoseiid mites, aphelinid parasitoids), would be most likely to develop resistance in the field. For the same reasons, resistance would also be the most stable in these species in the field. Highly mobile natural enemies such as *C. carnea* would be less likely to develop stable resistance in the field, and resistance would be expected to persist only when large R populations occurred in the surrounding environment. As noted in Chapter 14, the first case of reported resistance in *C. carnea* was reported from the San Joaquin valley in California, where treated orchards and fields are contiguous over much of the region (Grafton-Cardwell and Hoy 1986b).

A second estimate of immigration was made using small apple trees located in proximity to species pools of apples pests and their natural enemies (Fig. 16.3; taken from Whalon and Croft 1985, 1986). Portable apple trees from which pests and natural enemies had been eradicated were placed in two regional species-pool environments (i.e., in areas of dense and less dense apple production) and under one of four treatments of potential immigration: 1) within 20 m of an unsprayed apple orchard, 2) within 20 m of a sprayed apple orchard, 3) within 185 m of an unsprayed apple orchard, and 4) within 4800 m of an unsprayed apple orchard.

For pests (Fig. 16.3a and b), trees placed near abandoned orchards had the largest number of species and the fastest rates of colonization in either regional species-pool environment; however, there were no significant differences among immigration rates for natural enemies across the four species-pool treatments in either year's experiments (Fig. 16.3a and b). These results indicated that specialist apple pests tend to move from apple tree to apple tree or, more likely, from one treated orchard to another. Therefore, in areas of intense apple production, resistant pests would move from one site to another and resistance would probably not be diluted by these immigrants. Conversely, generalist predators would tend to come from sources other than orchards and be almost ubiquitous in any of the species-pool treatments. Generalist predators would move from unsprayed

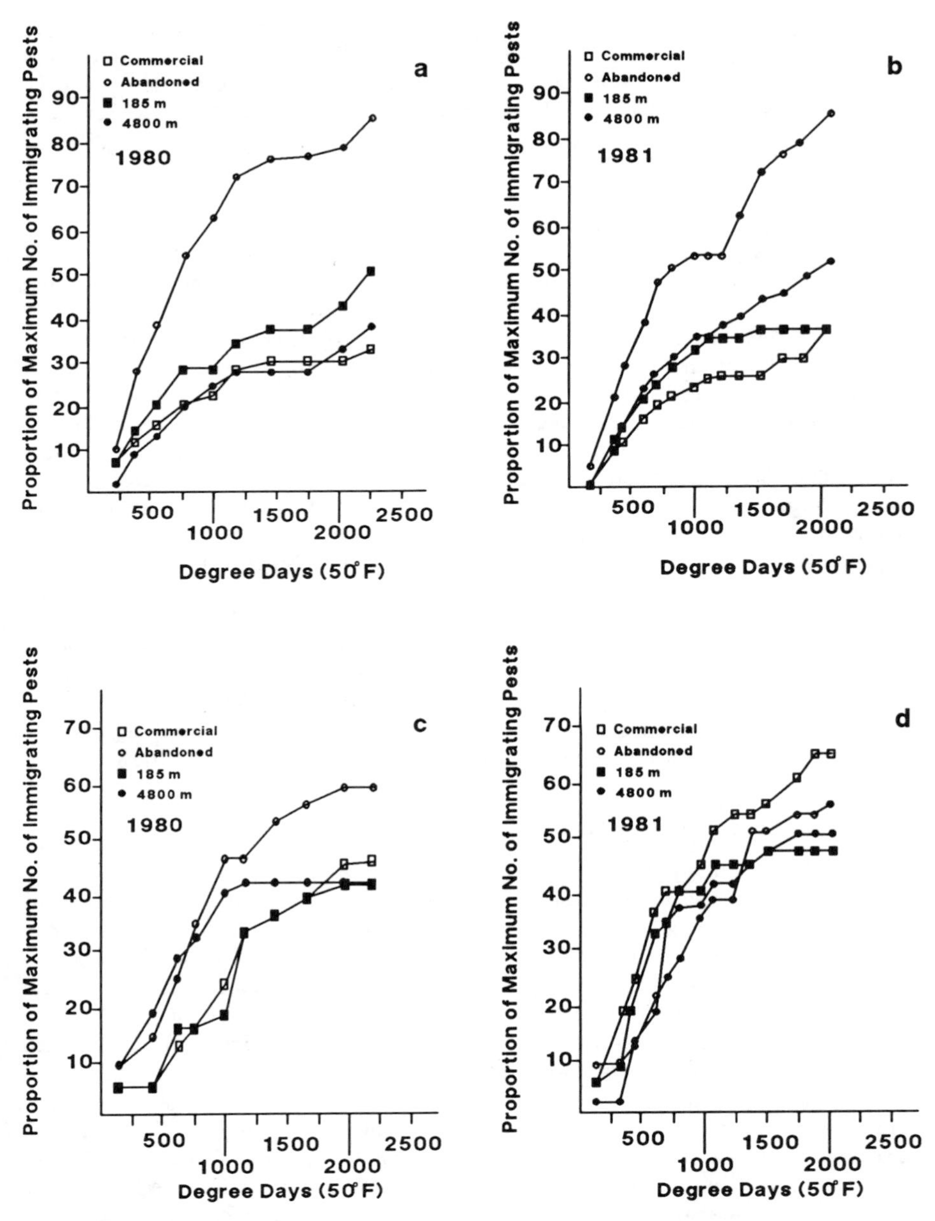

Figure 16.3. The proportion of maximum number of pests (a, b) and natural enemies (c, d) immigrating into portable orchards located in proximity to four species pools in intense (1980) and less intense (1981) apple production areas in Michigan. (Adapted from Whalon and Croft 1985.)

habitats into treated habitats, diluting pesticide resistance in those few resident natural enemies in the treated habitat. Also, because the treated area is only a small portion of the generalist predator's habitat, a relatively small percentage of predators would be directly exposed to pesticides, thus further reducing the likelihood that these beneficial populations would maintain resistance.

16.5. INTERSPECIFIC INTERACTIONS

Many aspects of interspecific interactions between pests and natural enemies influence resistance development in natural enemies. Croft and Brown (1975) discussed specific examples of such factors in their review of pesticide side effects on arthropod natural enemies. Predators are generally more polyphagous than parasitoids. Thus, while predators surviving pesticide selection may more readily turn to alternate prey to sustain them following treatment, parasitoids may be less able to locate suitable hosts after the elimination of their specific hosts. Predators often occur in the same habitat with their prey for many life stages where they may be exposed to pesticides. Parasitoids may more often be protected internally within their hosts. Therefore, pesticide selection may be greater for predators than for parasitoids. Survival of a resistant endo- or ectoparasitoid (depending on the stage of parasitoid development) usually depends on the survivorship of its host. Therefore, the frequency of resistant alleles for both parasitoid and host determines parasitoid survivorship. (See, for example, Abu and Ellis 1977 for a specific study of how host survivorship affects parasitoid survivorship.) Thus, if the frequency of the resistance allele was 1 in 100,000 in the parasitoid and 1 in 10,000 in the host, the chance of the resistant parasitoid surviving would be 1 in 10 million.

Since the documentation of these specific factors, several more general hypotheses have been proposed to explain how interspecific interactions may affect resistance development.

16.5.1. The Food Limitation Hypothesis

The food limitation hypothesis is based on the dynamics of interacting populations of pests and natural enemies (Huffaker 1971, Croft and Morse 1979). It contends that while resistant pests surviving pesticide exposure have unlimited food (the crop) to exploit thereafter, surviving natural enemies usually find their prey or hosts severely reduced to a level below their needs for survival and reproduction. In these instances, resistant natural enemies may die or leave the treated habitat. Thus, resistance evolves more slowly in natural enemies than in pests because natural enemies starve, emigrate, or have limited reproduction following spraying, while pests do not.

The food limitation hypothesis is more applicable to monophagous than to generalist natural enemies or facultative entomophages. If pesticide sprays reduce the abundance of a specific prey, generalist predators may be able to exploit other prey. However, if broad-spectrum insecticides severely reduce all prey of a generalist, the net effect is essentially the same as for monophagous natural enemies. The lack of resistance among generalist predators and facultative entomophages may be influenced by their residency/immigration characteristics as well (Whalon and Croft 1985, 1986, Croft and van de Baan 1988). Consideration of the combined effects of reduced food supply following selection and the residency/immigration characteristics of natural enemies highlights the important conclusion that there is probably no single explanation for observed differences in resistance development between pests and natural enemies. More likely, the limited incidence of resistance in natural enemies may be due to a combination of the factors discussed above.

Tests of the food limitation hypothesis have involved correlation studies using population models of resistance development in natural enemies and empirical patterns of resistance in field populations (Croft 1972, 1982, Tabashnik and Croft 1985). Laboratory selection experiments (Morse and Croft 1979, 1981, Hoy 1985) and modeling studies (Tabashnik and Croft 1985, Tabashnik 1986a, 1986b) have also been employed. Theoretical ecologists have added additional support to this hypothesis (May 1985, May and Dobson 1986).

Some empirical evidence to support the food limitation hypothesis can be found in the literature. Morse and Croft (1979, 1981) selected predatory mites in the presence of high and low population densities of prey spider mites (Fig. 16.4). When prey were abundant, *Amblyseius fallacis* rapidly acquired resistance to a level comparable to *Tetranychus urticae*. This contrasts with field situations where prey are greatly reduced by selection, and resistance development by the predator lags severely behind that of the spider mite (Morse and Croft 1981; Fig. 16.4). In related experiments, Hoy (1985) observed rapid rates of resistance development in genetic improvement studies in which unlimited spider mite prey were available to the predator *Metaseiulus occidentalis*.

The food limitation hypothesis implies that a natural enemy may begin to evolve resistance in the field only after substantial resistance or tolerance is present in its host or prey. Croft (1972) noted that in all cases of resistance in natural enemies developed up to that time, their respective principal prey or hosts either had developed resistance or were intrinsically tolerant to the compound in question. These patterns are also supported by resistance trends in apple pests and natural enemies after 30 years of exposure to azinphosmethyl (Croft 1982).

In modeling studies with the 24-species complex of apple pests and natural enemies, Tabashnik and Croft (1985) observed that for natural enemies in general, rates of resistance evolution were significantly slower in the field compared to model predictions (Fig. 16.1). In a modified run of the model, natural enemies were assumed to evolve resistance only after their prey or hosts became resistant. Predicted times for natural enemies were obtained by adding the observed times to resistance for prey or host species to the original predicted times for each natural enemy species (Fig 16.5). Correspondence between predicted and observed times to resistance (Fig. 16.5) improved significantly for all 6 natural enemies that were initially predicted to evolve resistance too fast. With the food limitation hypothesis incorporated in this general way, there was also a better rank correlation between prediction and observation for all species.

In a study of the variability in levels of pesticide resistance observed in field populations, Rosenheim and Hoy (1986) concluded that patterns observed in *Aphytis melinus* did not support the contention that food limitation was a key factor in determining the extent to which resistance had developed. Their conclusion was based on comparing the levels of resistance developed by the aphelinid to two compounds applied for general control of citrus pests with resistance levels developed to three more specific scalicides. In this case it was expected that resistance would be higher to the two broad-spectrum compounds, but such was not the case.

To summarize, there have been few rigorous tests of the food limitation hypothesis. Available evidence is generally correlative and inconclusive, and virtually no test has examined the hypothesis directly. However, several studies do indicate that food limitation following sprays is a major factor retarding resistance in natural enemies. Further and more direct study of this hypothesis is warranted. In one sense the food

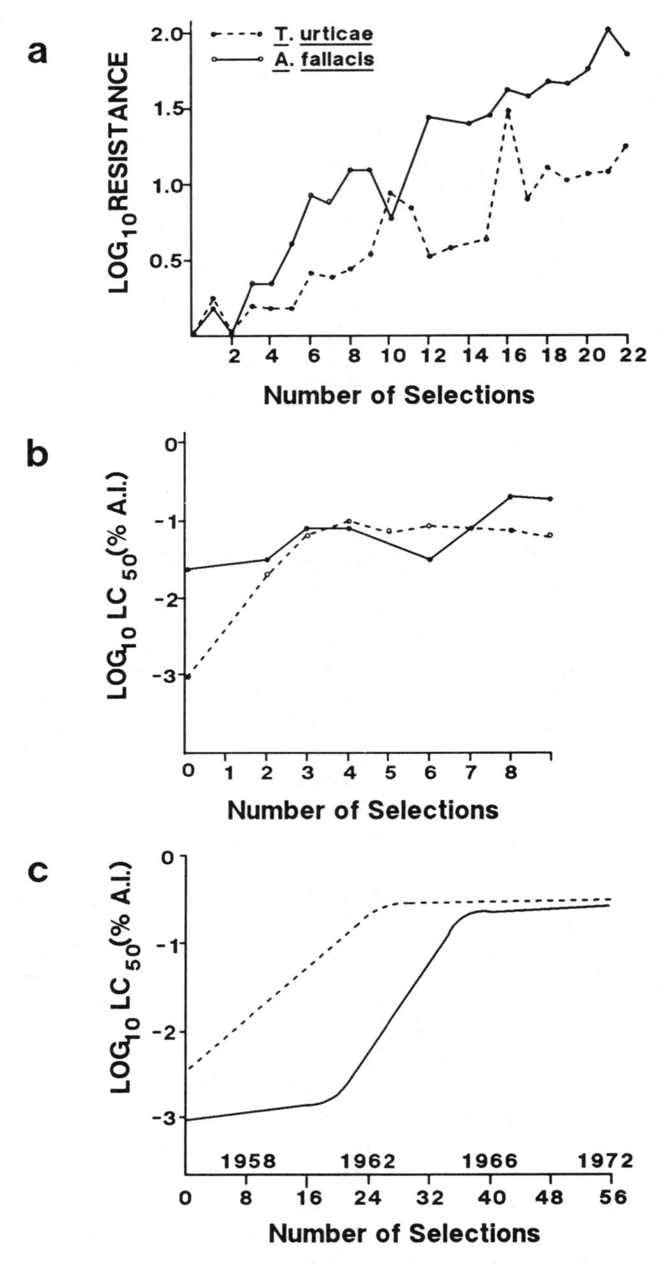

Figure 16.4. Development of resistance to azinphosmethyl in *Amblyseius fallacis* (predator) and *Tetranychus urticae* (prey) populations under different experimental and field conditions. (a) Log_{10} proportional change in LC_{50} for predator and prey: susceptible strains, independent prey and predator selection, unlimited food for predator and prey, 75% adult selection for both predator and prey. (b) Log_{10} proportional change in LC_{50} for predator and prey: susceptible strains, coupled selection, unlimited food for pest, limited food for predator, 75% selection for pest. (c) Log_{10} of LC_{50} for *Panonychus ulmi* and *A. fallacis* in commercial apple orchards in midwestern United States: susceptible strains, unlimited food for pest, limited food for predator, 99.9% selection of prey initially, use of constant application rate. (After Morse and Croft 1981.)

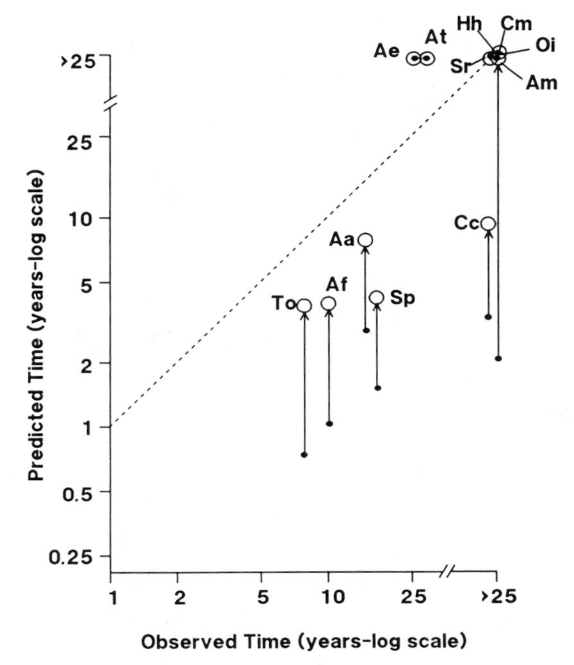

Figure 16.5. Effects of food limitation hypothesis on predicted times for apple arthropod natural enemies to develop resistance to azinphosmethyl. $n = 12$; Spearman's rank correlation coefficient rs = 0.806; P < 0.005. (See Fig. 16.1 for key to species names.) ○ = predictions with food limitation hypothesis incorporated; ● = predictions under initial assumptions of no food limitation (Fig. 16.1). Arrows show changes in predictions due to food limitation hypothesis. (Adapted from Tabashnik and Croft 1985.)

limitation hypothesis essentially predicts the same outcome as the preadaptation hypothesis—facultative herbivores should develop resistance faster than obligate carnivores. The food limitation and preadaptation hypotheses are complementary; their relative importance may depend on the species and pesticide.

16.5.2. Density Dependence

A mathematical evaluation of the interspecific dynamics affecting resistance development in natural enemies was reported by May and Dobson (1986). They discounted the preadaptation hypothesis, but supported both food limitation and immigration–residency aspects of predator–prey population dynamics. Based on analytical models of density dependence in pests and natural enemies, they predicted that subsequent to pesticide applications, which cause equivalent mortality to pest and predators, prey populations will exhibit an overcompensating density-dependent response and will return to above-normal levels in the next generation. Conversely, natural enemies, due principally to food limitation, will exhibit an undercompensating density-dependent response. Natural enemies will tend to recover steadily following the disturbance, but at a slower rate and initially at a level far below that present before spraying. Immigration of susceptibles has a greater impact in diluting small, resistant populations of natural enemies

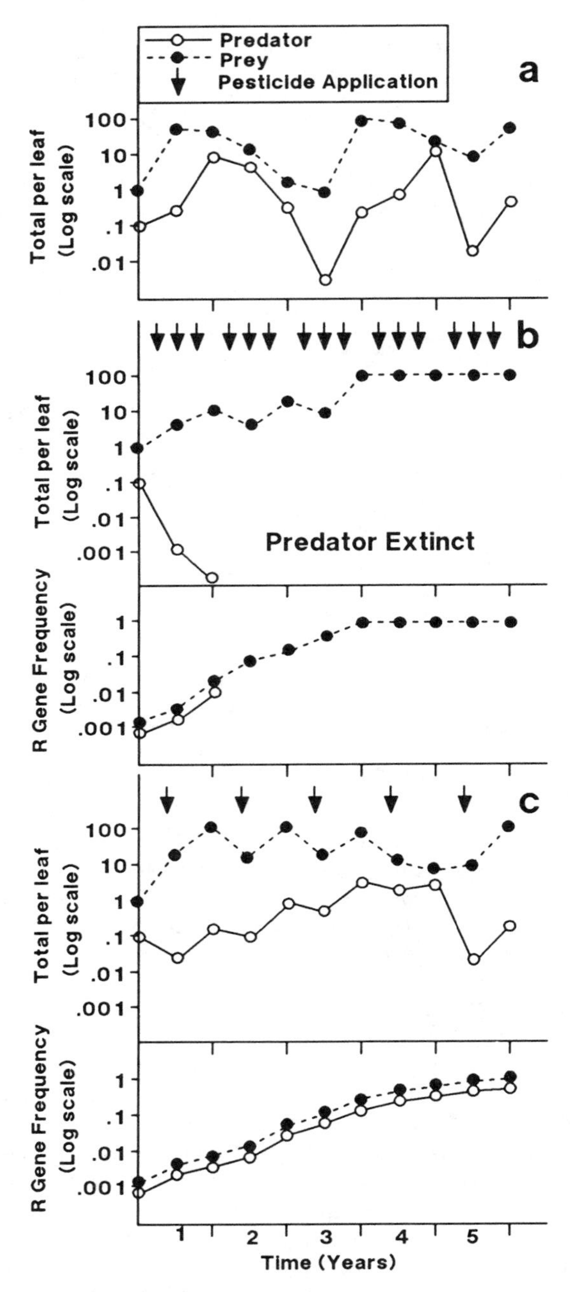

Figure 16.6. Patterns of pesticide resistance development in a predator–prey system. (a) No pesticide use. (b) High pesticide use, biweekly sprays, dose = 0.002% a.i., immigration = 0. (c) Low pesticide use, monthly sprays, dose = 0.002% a.i., immigration = 0 (After Tabashnik 1986b.)

than similar effects on larger pest populations. These points agree with the conclusions reached by Tabashnik and Croft (1982, 1985) and Tabashnik (1986b) in models simulating the more specific orchard environment.

16.5.3. Coupled Predator–Prey Systems

Initial theoretical tests of the food limitation hypothesis assumed that a natural enemy did not begin to develop resistance until after its prey or host had become fully resistant (Tabashnik and Croft 1985); however, evolution of resistance is not an all-or-none process. It is more likely than soon after a pest develops limited resistance, its natural enemies also begin to develop resistance. To better understand resistance development in natural enemies, Tabashnik (1986b) simulated a coupled predator–prey system in which both species had the potential to evolve resistance. Basic assumptions about resistance were the same as in earlier models (Tabashnik and Croft 1985), except that predator and prey population dynamics were linked. The predator's survival and fecundity were a function of prey density and, conversely, prey density was affected by predation. This simulation was used to examine the effects of immigration, dose, application frequency, pesticide selectivity, and food availability on resistance evolution in predatory–prey systems.

Tabashnik (1986b) concluded that intensive pesticide use could cause rapid resistance development in the past (prey or host), while suppressing resistance development or causing local extinction in the natural enemy (predator or parasitoid) (Fig. 16.6). This

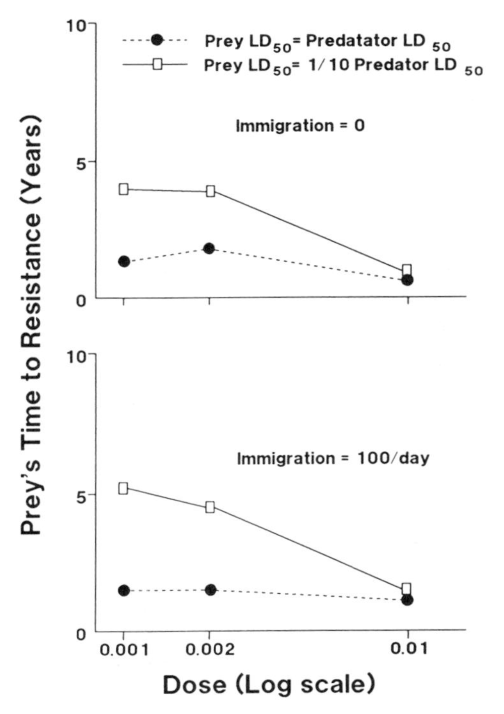

Figure 16.7. Effects of threshold sprays and pesticide selectivity on time to develop resistance to pesticides for prey. Sprays made only when the number of active prey exceeded 20 per leaf. (After Tabashnik 1986b.)

occurred because high insecticide doses and frequent applications killed most of the prey, thereby severely limiting the predator's food supply. Simulations showed that reduced pesticide use and lower doses could promote resistance development in the natural enemy and retard it in the pest. Lower doses and reduced application frequency resulted in weaker selection for resistance in the pest and preserved the predator's food supply (see Chapter 20.)

Immigration had little impact on the prey's time to develop resistance because the resident prey population size was large (about 1–100 per leaf) compared to the number of immigrants (10^{-4} per leaf daily). Although the immigration rate of predators was equal to that of prey, immigration had a greater influence on resistance in the predator because the resident predator population was relatively small (about 0.01–1 per leaf). These results are consistent with previous findings that the immigrant-to-resident ratio determines the impact of immigration on resistance development (Comins 1977a, 1977b, Taylor and Georghiou 1979, Tabashnik and Croft 1982, 1985).

Tabashnik (1986b) also examined the potential impact of using selective pesticides

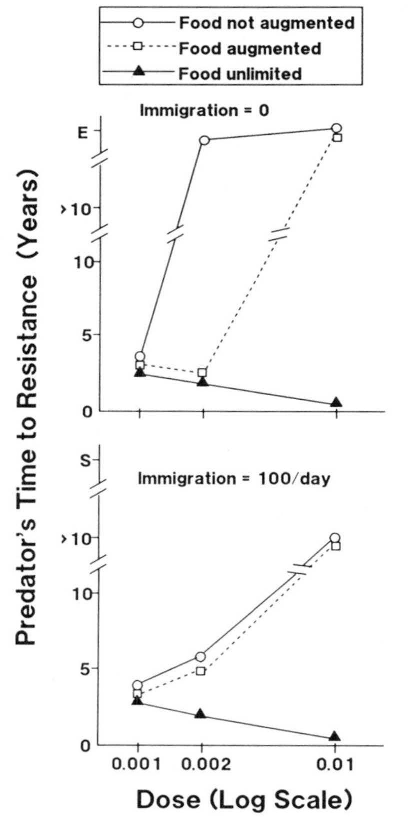

Figure 16.8. Effects of predator's food supply on predator's rate of pesticide resistance development. Biweekly sprays; E = local extinction of the predator; S = permanent suppression of predator resistance due to immigration of susceptibles (After Tabashnik 1986b.)

which are more toxic to pests than natural enemies. Simulations showed that if a selective pesticide was used on a regularly scheduled basis, pest resistance developed just as rapidly as when a nonselective pesticide was used. If, however, sprays were made at moderate rates only when an economic threshold was exceeded, then pest resistance developed more slowly with a selective pesticide than with a nonselective pesticide (Fig. 16.7). In the latter case, the benefits of the selective pesticide were realized because the predators reduced prey density and therefore reduced the number of sprays needed. (Verification of these results in an actual case history of cyhexatin resistance in spider mites and predaceous phytoseiid mites in the field is discussed in Chapter 20, Section 20.6.2.)

The simulations of Tabashnik (1986b) also showed that augmentation of the predator's food supply could promote resistance development (Fig. 16.8). The predator developed resistance more rapidly when it had an alternate food source that reduced starvation and emigration due to low prey density. When the predator had an unlimited food supply, it developed resistance as rapidly as its prey.

The results of Tabashnik (1986b) affirmed the idea that reduced pesticide use can retard resistance development in pests while promoting resistance development in natural enemies (Croft 1982). Reduced pesticide use slows the evolution of resistance in pests by reducing selection intensity. Increased pest survival results in a greater food supply for natural enemies, which reduces natural enemy starvation and emigration following sprays. Thus, decreased pesticide use enables natural enemies to maintain populations in treated habitats and increases their potential to develop pesticide resistance.

Other studies also demonstrate that intensive pesticide use can suppress resistance in natural enemies while promoting it in pests (Morse and Croft 1981, Tabashnik 1986a, 1986b). These findings are consistent with previous work, showing that a high dose strategy for managing resistance cannot be recommended for general use due to its many serious shortcomings (Tabashnik and Croft 1982).

The results of Tabashnik (1986b) further support the idea that an adequate food supply is a prerequisite for resistance development in predators and for maintenance of resistant predators in the field. Empirical research shows that when adequate prey are available, predators develop resistance (Morse and Croft 1981) and artificially selected resistant predators can be established in the field (Hoy 1985). Tactics to ensure adequate food for predators in the field include allowing greater survival of the pest by raising economic thresholds, providing alternative prey, and augmenting the food supply with food sprays (Rabb et al. 1976). Such methods should promote resistance development in natural enemies, enhancing their effectiveness as biological control agents in treated environments.

16.6. INTEGRATION AND SUMMARY

Resistance development is similar in natural enemies and pests in several respects. Physiological, biochemical, toxicological, and genetic factors determine whether or not resistance can develop, but ecological factors such as immigration–residency relationships, life history traits, and interspecific population dynamics determine the rates of resistance development and the stability of resistance in the field.

Natural enemies do have some unique characteristics that make them less likely to develop and maintain high levels of resistance in the field compared to their prey or hosts. Natural enemies depend on the target of pesticide applications (pests) for food, whereas the pest's food is intentionally conserved by the use of pesticides. Natural enemies also seem to have higher rates of immigration relative to their densities in treated habitats than pests.

As alluded to here and discussed in greater detail in the remainder of the book, management actions can be taken to minimize resistance in pests and maximize resistance in natural enemies. Emerging principles may alter current practices which tend to produce highly resistant pests and leave natural enemies in a highly susceptible state.

PART SIX

PESTICIDE RESISTANCE MANAGEMENT

17

ENDEMIC SPECIES

17.1. INTRODUCTION

Since insecticide-resistant (R) natural enemies were first discovered in the late 1960s (Chapter 14), their use in IPM has expanded greatly. New ways to increase or diversify pesticide resistance and better means to manage these biological control agents have been developed (Croft and Strickler 1983, Hoy 1985). Management of endemic, resistant natural enemies, which began in the 1960s, was followed by introductions of these agents beyond their native ranges. Since the early 1970s, resistant natural enemies have been introduced into many areas of the world (Chapter 18). Deliberate genetic improvement of pesticide resistance in entomophagous species using hybridization and artificial selection has been conducted primarily in the 1980s (Chapters 19 and 20).

The most extensive research on insecticide-resistant natural enemies, particularly phytoseiid mites, has been conducted in deciduous tree fruit and nut crops. Discovery, transfers, and genetic improvement of resistant natural enemies on other crops is increasing rapidly (Hoy 1985; Chapters 18–21). Resistant natural enemies have been most widely exploited on deciduous tree fruits and nuts because 1) resistant natural enemies were first discovered on these crops, 2) pesticide selection intensity is high, 3) few selective pesticides are available, and 4) the potential for biological control of spider mites on these crops is favorable. Chapter 20 is devoted exclusively to resistance management of pests and natural enemies on deciduous tree fruits.

This chapter focuses on the use of endemic, resistant natural enemies in IPM. Case histories involving several phytoseiid mites and a predaceous insect are reviewed. Successful implementation and attendant environmental and economic benefits of these programs are discussed. Examples are presented in which resistant natural enemies have been introduced into other areas within their native ranges where resistance was absent or had only minimally developed.

It should be kept in mind that a specific pest–natural enemy pair (which is the subject of resistance management) is usually part of a larger, more complex arthropod community. The resistance status or potential of all pests and natural enemies in a given crop must be considered before pesticides are applied. Other aspects of pest management and crop management establish the environment in which resistance management operates. The need to structure resistance management around basic IPM and agronomic practices is addressed in Chapter 20.

17.2. ENDEMIC, RESISTANT PHYTOSEIIDS IN NORTH AMERICA

Phytoseiid mites on deciduous tree fruits were the first insecticide-resistant natural enemies to be discovered (Chapter 14) and exploited in IPM. The initial discovery of resistant *Metaseiulus occidentalis* in the western United States was serendipitous (Hoyt 1969a, Whalon and Croft 1984). In the early 1960s, Hoyt (1969a) observed that *M. occidentalis* continued to control spider mites in Washington apple orchards despite regular sprays of azinphosmethyl. Previously, *M. occidentalis* had been locally eradicated by these applications. Resistance was first detected in orchards after azinphosmethyl had been used for 5–8 years. Hoyt's observation of resistance was based on the survivorship of predators in the field. More complete documentation of organophosphate (OP) resistance with R and S strains came later (Croft and Jeppson 1970).

In 1961, an investigation was initiated on the ecology of the entire mite complex on apple (Hoyt 1969b). After relationships between chemical use patterns, spider mite levels, and predation by *M. occidentalis* were elucidated, an IPM program was developed based on selective chemical control of the codling moth and other pest insects and biological control of mites. In 1965, when a heavy frost in one area greatly reduced the potential crop, many growers decided not to spray for mites and to adopt integrated mite management. IPM orchards relying on OP-resistant predaceous mites remained green, while orchards receiving acaricides were bronzed due to poor control of resistant spider mites. Thereafter, IPM programs based on the biological control of spider mites were widely implemented in Washington State (Whalon and Croft 1984).

The success of the Washington program influenced research in nearby areas where *M. occidentalis* was present. In British Columbia, California, Oregon, Utah, Idaho, and Colorado, integrated mite management was widely implemented over the next few years (Table 17.1). Resistance to OPs in *M. occidentalis* and other predators was the *major* factor that allowed these programs to succeed, since chemical control of other orchard pests was

Table 17.1. Major Integrated Mite Control Programs in North America Based on Insecticide-Resistant Natural Enemies

State or Province	Principal Pest Controlled[1]	Principal Natural Enemies[2]	Research Reference	Implementation Reference
WESTERN PROGRAMS				
Washington	Tm, Pu, As	To[3], Spp, Zm[4]	Hoyt 1969a, 1969b	Hoyt & Retan 1967, Hudson et al. 1974
British Columbia	Tm, Pu, As	To[3], Tp[3], Zm[4]	Downing & Moilliet 1971	Downing & Arrand 1968, B.C. Minist. Agric. 1982
Oregon	Pu, Tm	To[3], Tp[3]	Zwick 1972, AliNiazee 1974, Croft & AliNiazee 1983, Hadam et al. 1986	Whalon & Croft 1984
California	Tp, Tu, Pu	To[3], Ss	Croft & Barnes 1972, Rice et al. 1976	Davis et al. 1979
Idaho	Tm, Tu, Pu	To[3]	Larson 1970	Larson 1970
Utah	Tu, Tm, Pu, As	To[3]	Davis 1970, Croft 1972	Whalon & Croft 1984
Colorado	Tu, Pu	To[3]	Quist 1974	Quist 1974, Whalon & Croft 1984
MIDWESTERN AND EASTERN PROGRAMS				
Missouri	Pu, Tu	Af[3]	Poe & Enns 1969	—
Illinois	Pu, Tu	Af[3]	Meyer 1974, 1975	Meyer 1981
Michigan	Pu, Tu, As	Af[3], Sp[3], Ag[4], Zm	Croft & McGroarty 1977, Croft 1983, Whalon et al. 1982b	Croft 1975
Ohio	Pu, Tu, As	Af[3], Zm[4], Ag[4]	Holdsworth 1968	Holdsworth 1974
Ontario	Pu, Tu, As	Af[3]	—	Hagley et al. 1978
Quebec	Pu, Tu	Af[3]	Parent 1961, 1967, Bostanian & Coulombe 1986	Paradis 1981

Table 17.1. (*Continued*)

State or Province	Principal Pest Controlled[1]	Principal Natural Enemies[2]	Research Reference	Implementation Reference
Pennsylvania	Pu, Tu	Sp[3], Af[3]	Asquith 1971, Hull et al. 1978	Asquith 1972, Hull 1979b
New York	Pu, Tu	Af[3], Tp[3]	Watve & Lienk 1976	Tette 1981, Tette et al. 1982
		Af[3]	Weires et al. 1979	
Massachusetts	Pu, Tu, As	Af[3]	Prokopy et al. 1981, Hislop & Prokopy 1979, 1981	Coli et al. 1979, Prokopy et al. 1981
New Jersey	Pu, Tu, As	Af[3]	Swift 1970	Christ 1971
Nova Scotia	Pu, As	Several	Sanford & Herbert 1970	Anon. 1970, Nova Scotia Minist. Agric. 1982
Virginia, West Virginia	Pu, Tu	Af[3]	Clancy & Mcalister 1958 Parrella et al. 1980, 1981a, 1981b, 1982	W.V.U. Extension Service 1982
North Carolina	Pu, Tu, As	Af[3], Sp[3]	Rock & Yeargan 1971	Rock 1972, Rock & Apple 1983

[1] Pests of importance in each area are listed sequentially. Tm = *Tetranychus mcdanieli*; Pu = *Panonychus ulmi*; As = *Aculus schlechtendali*; Tu = *Tetranychus urticae*; Tp = *Tetranychus pacificus*.

[2] Natural enemies of importance in each area are listed sequentially. To = *Typhlodromus occidentalis*; Sp = *Stethorus punctum*; Af = *Amblyseius fallacis*; Ag = *Agistemus fleschneri*; Zm = *Zetzellia mali*; Tp = *Typhlodromus pyri*; Ss = *Scolothrips sexmaculata*; Spp = *Stethorus punctillum*.

[3] Presence of an OP-resistant natural enemy.

[4] Presence of an OP-tolerant natural enemy.

essential. Eventually, resistance to azinphosmethyl in *M. occidentalis* was documented in most apple-growing areas in western North America (Croft and Brown 1975; see exception in southern California, Croft and Jeppson 1970, Croft and Barnes 1972).

Almost simultaneously, *Amblyseius fallacis* and the coccinellid *Stethorus punctum* began to survive OP sprays in midwestern and eastern fruit-growing areas of the United States and Canada (Table 17.1). In New Jersey (Swift 1970), North Carolina (Rock and Yeargan 1971), Pennsylvania (Asquith 1971), and Michigan (Croft and Nelson 1972) early integrated mite management programs were based on OP-resistant predators. Appreciable biological control of spider mites was achieved by controlling other tree fruit pests with chemicals to which predators had developed resistance. When unfavorable predator to prey ratios allowed pest mites to build up, selective acaricides such as cyhexatin or propargite were used to reduce prey populations (e.g., see Croft 1975, Croft and McGroarty 1977).

It is reiterated here that resistance in natural enemies was the critical development that made integrated mite management possible. Resistance allowed the use of certain broad-spectrum insecticides without eliminating these pesticide-adapted natural enemies. In

Table 17.2. Toxicity of Orchard Pesticides to Plant-Feeding Mites and Their Principal Predators Occurring in Michigan Apple Orchards (Adapted from Croft 1975)

		Toxicity[2]				
Pesticide	Rate per Acre[1]	European Red Mite/ Two-Spotted Spider Mite	Apple Rust Mite	*Amblyseius fallacis*	*Agistemus fleschneri*/ *Zetzellia mali*	*Stethorus punctum*
Benlate 50WP	1–1.5 lb	M	M	H	H	NA
Captan 50WP	8 lb	0	0	0	0	0
Cyprex 65WP	1.5–2 lb	0	0	0	0	0
Dikar 80WP	8 lb	H	HR	M	HR	0
Karathane 25WP	2 lb	H	HR	M	HR	0
Lime sulfur	8 gal	0	HR	H	HR	H
Polyram 80WP	8 lb	0	NA	0	0	0
Dimethoate 50WP	8 lb	M	M	M*	0+	M*
Diazinon 50WP	4 lb	M	H	M*	0+	NA
Guthion 50WP	2 lb	0	M	0*	0+	M*
Imidan 50WP	4 lb	0	M	0*	0+	H
Parathion 15WP	8 lb	0	NA	M*	0+	H
Phosphamidon 8EC	1 pt	0	M	H	M	H
Sevin 50WP	8 lb	0	HR	HR	HR	HR
Systox 6EC	1.3 pt	H	M	H	0+	H
Thiodan 50WP	4 lb	0	HR	M	HR	M
Zolone 3EC	4 pt	M	H	HR	H	H
Omite 30WP	5 lb	H	HR	0	HR	0
Plictran 50WP	1–1.5 lb	HR	HR	M	HR	M

[1] Highest rates recommended for commercial fruit production.
[2] 0 = non- to slightly toxic; M = moderately toxic; H = highly toxic, short residue; HR = highly toxic, long residue; NA = no information available.
* Lack of toxicity is due to a resistant natural enemy.
+ Lack of toxicity is due to a tolerant natural enemy.

essence, certain nonselective insecticides became selective ones because of resistance in natural enemies. The details of the integrated mite management programs that evolved from the discovery of resistant natural enemies are reviewed here. Throughout this discussion, the role that resistance plays in integrated mite management is emphasized.

17.2.1. *Amblyseius fallacis* in the Midwest and East

Management of *Amblyseius fallacis* demonstrates many of the elements of integrated mite management used with other insecticide-resistant predatory mites. While species may differ from area to area, the concepts of using predator-to-prey ratios for decision-making, applying selective acaricides, and managing alternate prey are applicable. Case histories for other natural enemies focus on unique aspects of integrated mite management which differ from those discussed here for *A. fallacis*.

Management scenarios featuring resistant *A. fallacis* and the pest mites *Panonychus ulmi, Tetranychus urticae*, and *Aculus schlechtendali* during a growing season are depicted in Fig. 17.1. These predator-to-prey density data were taken from field studies in Michigan apple orchards (Croft 1975, Croft and McGroarty 1977). Pesticides used to control orchard pests and their side effects on predators are indicated in Table 17.2. Basic components of integrated mite management programs utilizing resistant *A. fallacis* are discussed below in reference to Fig. 17.1.

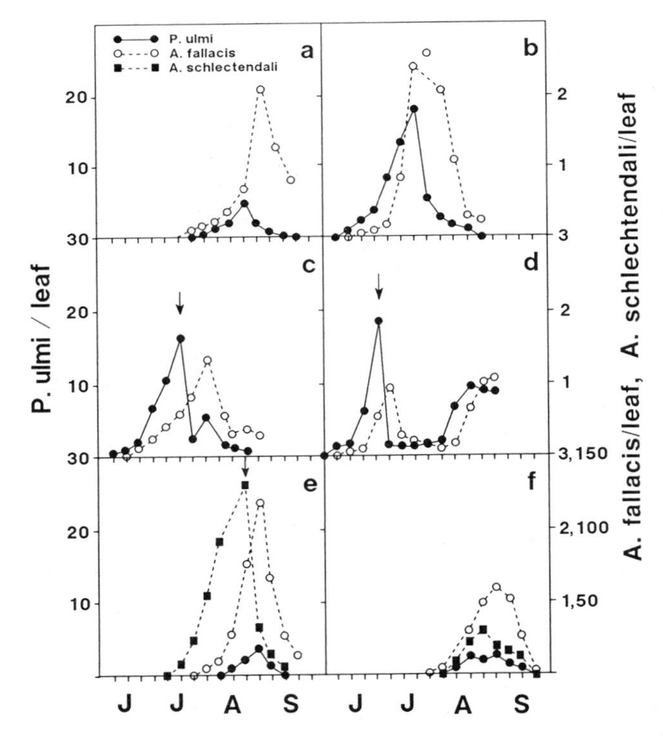

Figure 17.1. Population abundance of the predatory mite *Amblyseius fallacis* and the plant-feeding mites *Panonychus ulmi* and *Aculus schlechtendali* in Michigan apple orchards (adapted from Croft and McGroarty 1977).

1. *Predator/prey ratios.* Biological control of pest mites can be predicted using numerical dynamics of pest and predatory mites. More specifically, management decisions and biological control by *A. fallacis* are determined by the number of predators and pest mites present in an orchard. Predator-to-prey ratios which favor biological control of pest mites have been defined in a number of different situations (e.g., Croft and Nelson 1972, Croft 1975, Croft and McGroarty 1977, Dover et al. 1979, Bower and Murison 1984, Clark and Buckley 1984).

Figure 17.1 shows how predator-to-prey ratios are used in decision making. In Fig. 17.1a, an effective ratio of *A. fallacis* to spider mites (*P. ulmi.*, *T. urticae*) was present early in the season and predators provided complete biological control before pest mites exceeded 5 per leaf (Croft and McGroarty 1977). A less effective ratio of predators to prey early in the growing season is shown in Fig. 17.1b. In this case, pest mites reached a density of 20 per leaf before biological control was attained (economic threshold is 15 mites per leaf). In the former case, no intervention was necessary to augment biological control; in the latter, a selective acaricide should have been used to create a more favorable balance between pest and predator.

2. *Physiologically selective acaricides.* Figure 17.1c illustrates how an acaricide such as cyhexatin or propargite can be used at a reduced rate to decrease pest mite density, thus changing the predator-to-prey ratio to favor biological control. Figure 17.1d shows the effect of completely eliminating pest mites by using a high rate of one of these acaricides. Predators subsequently starved and a secondary outbreak of spider mite occurred. (Note the absence of late-season control by predators as compared with Fig. 17.1c.)

3. *Alternate prey.* Figures 17.1e and f demonstrate how the apple rust mite *A. schlechtendali* serves as alternate prey for pedators and influences the biological control of spider mites. In Fig. 17.1e, apple rust mites entered apple trees before spider mites developed. *A. fallacis* fed on rust mites and became well distributed throughout the tree. Endosulfan was applied to selectively control rust mites as spider mites began to increase. Rust mites declined and predators, which were unaffected by the spray, switched to spider mite prey and quickly controlled that population. In Fig. 17.1f, rust mites were present only at moderate levels and predators provided complete control of both pest mites without sprays for rust mites.

4. *Selectivity and resistance monitoring.* The density relationships presented in Fig. 17.1 show the potential for *A. fallacis* to control pest mites. Orchards depicted in Fig. 17.1 employed selective pesticides for other pests of apple, thus no major reduction in *A. fallacis* was observed seasonally. In each orchard example (Fig. 17.1), 3–10 applications of insecticides and fungicides were applied for the control of disease and insect pests other than mites. These applications were required for the suppression of such key pests as codling moth *Cydia pomonella*, plum curculio *Conotrachelus nenuphar*, apple maggot *Rhagoletis pomonella*, and apple scab *Venturia inaequalis* (Croft 1982).

Pesticides used to control orchard pests and their selectivity to *A. fallacis* and other natural enemies of phytophagous mites are indicated in Table 17.2. Again, resistance in *A. fallacis* has allowed more flexible integration of chemicals and predators (Motoyama et al. 1970, Croft and Nelson 1972, Croft and Meyer 1973, Croft et al. 1976c, Croft 1982; see also Chapter 4, Fig. 4.3). Most fungicides, insecticides such as endosulfan, and acaricides such as propargite and cyhexatin are physiologically selective to predatory mites, while others which are normally toxic to predators can be used with proper timing or placement to ensure ecological selectivity (Chapter 10). (In Table 17.2,* denotes compounds where R strains make field use possible and $^{+}$ where intrinsic tolerances are involved.) Research measuring resistance, or tolerance or susceptibility, of *A. fallacis* to most pesticides (Croft

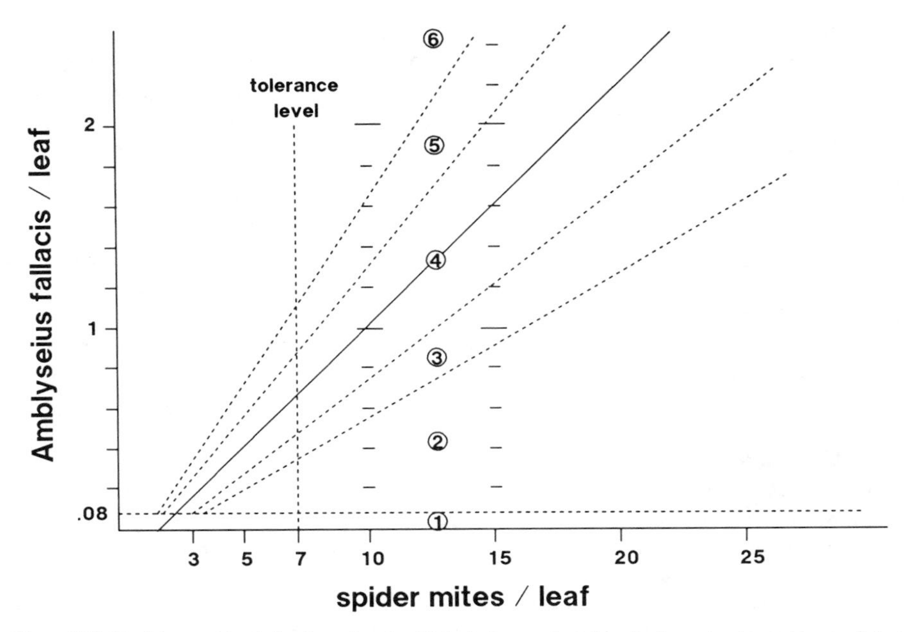

Figure 17.2. Decision-making index for estimating biological control of plant-feeding mites by *Amblyseius fallacis* or the need to use supplemental physiologically selective acaricide sprays to adjust predator-to-prey ratios (adapted from Croft 1975, Croft and McGroarty 1977).

and Nelson 1972, Croft and Meyer 1973, Strickler and Croft 1982) provided the basis for the pesticide recommendations in Table 17.2 (Croft 1975).

5. *Predicting biological control success.* As shown in Figs. 17.1a–c, whether to expect *A. fallacis* to provide biological control before spider mites exceed an economic threshold or whether to spray a selective acaricide, reducing prey density and establishing a more favorable predator-to-prey ratio depends on assessments made before pest mites reach damaging levels. A management index was developed to equate mite sampling data with control decisions (Fig. 17.2 and Table 17.3). It is based on a 5-year field study of biological control of spider mites by *A. fallacis* (Croft and McGroarty 1977). In the field, it is not necessary to make a mite-control decision until spider mites begin to increase (Fig. 17.2). If prey mite density exceeds 7 per leaf and only a few *A. fallacis* are present, then a full-strength acaricide spray is recommended (see region 1 in Fig. 17.2 and Table 17.3). If the *A. fallacis*-to-spider mite ratio falls in regions 2 or 3, application of a selective acaricide at lower rates will reduce spider mites. (A complete elimination of spider mites would cause predators to starve or to leave the apple tree; compare Figs. 17.1c and d.) Thereafter, predators will regulate spider mites at low densities. If a predator-to-prey ratio falls in regions 4, 5, or 6, biological control of spider mites is almost certain (Table 17.3).

Although this index is useful in decision-making, population dynamics are affected by a number of variables. For example, development of the European red mite and of *A. fallacis* is affected differently by temperature. At low temperatures the predator's development is slower than that of the prey, reducing its effectiveness during long, cool periods. Non-selective pesticides to which predators are not resistant can substantially alter pre-

Table 17.3. Spray Recommendations for Using Physiologically Selective Acaricides to Control Spider Mites and to Conserve Insecticide-Resistant *Amblyseius fallacis* (Adapted from Croft and McGroarty 1977)

Region of Decision Index (Fig. 17.2)	Suggested Recommendation	Probability for Biological Control
1	As bronzing appears, spray recommended acaricide at full rate (see Fig. 17.2).	Very low
2	If bronzing appears, spray Plictran 50 WP[1] at 8 oz/acre.	Equal to or less than 10%
3	If bronzing appears, spray Omite 30WP[2] at 1.25 lb or Plictran 50WP at 6 oz/acre.	Greater than 10%, but less than 50%
4	Wait one week, biological control should occur soon; if not, spray Omite at 1.25 lb or Plictran at 6 oz/acre.	Approximately 50%
5	Same as 4.	Greater than 50%, but less than 90%
6	Wait one week; biological control is almost certain.	Greater than 90%

[1] Plictran 50WP is a proprietary product of Dow Chemical Corp. (common name = cyhexatin); this compound has been withdrawn from commercial use in the United States. Fenbutatin oxide (= Vendex) is another organotin with similar selective properties to predatory and spider mite species.
[2] Omite 30WP is a proprietary product of Uniroyal Corp. (common name = propargite).

dator-to-prey ratios, invalidating predictions of biological control effectiveness. To correct for temperature and pesticide effects, these variables were incorporated into a population model of biological control by *A. fallacis* (Dover et al. 1979, Croft and Hoyt 1983). Use of this model further improves spider mite control predictions based on samples of insecticide-resistant predators, spider mites, and current weather data taken daily in the field (Dover et al. 1979).

17.2.2. *Metaseiulus occidentalis* in the West

Some of the earliest and most effective uses of insecticide-resistant *M. occidentalis* have been in the Pacific Northwest of the United States. The success of these programs depends on the broad tolerance of *M. occidentalis* to pesticides and the apparent ease with which it develops pesticide resistance (Chapter 14). Also of value is its effectiveness in regulating the pest mites *T. urticae, T. mcdanieli, T. pacificus, A. schlechtendali*, and even *P. ulmi* under certain conditions (Hoyt 1969a, 1969b, Croft and Barnes 1971, Hoy 1985).

Management programs that depend on insecticide-resistant *M. occidentalis* are ongoing in most fruit-producing areas in western North America (Table 17.1). Individual programs in each area are similar to those for *A. fallacis*, with some exceptions. *M. occidentalis* tends to be more effective in controlling aggregated prey mites such as *T. urticae* and *T. mcdanieli* (Hoyt 1969a, Croft and Nelson 1972). *M. occidentalis* can develop exclusively on apple rust mites, whereas *A. fallacis* requires some spider mites (Croft and

Hoying 1977, Croft and McGroarty 1977). *M. occidentalis* does not control spider mites well if populations are composed only of *P. ulmi* (Hoyt 1969a, unpublished data). Other aspects of management of *M. occidentalis* can be reviewed in the literature listed in Table 17.1 and in the many studies of the introduction of this mite into areas where it did not occur naturally (Chapter 18).

17.3. ENDEMIC, RESISTANT PHYTOSEIID MITES, WORLDWIDE

European scientists conducted some of the earliest studies of phytoseiid biology to ascertain their potential as biological control agents (Table 17.4; see Chant 1959, van de Vrie 1962, Collyer 1964a, 1964b). Phytoseiids eventually developed resistance to insecticides in Europe, but it was less dramatic than in North America (Hoyt 1969a; see literature in Table 17.4). Initially, levels of azinphosmethyl resistance that developed in European populations of *Typhlodromus pyri, Amblyseius chilenensis*, and *A. andersoni* were lower than those of *A. fallacis* and *M. occidentalis* in North America (Table 14.1). To some extent, the lower propensity for resistance slowed the development of IPM programs; however, the major stumbling block to integrated mite management in European orchards was the lack of selective fungicides that would control orchard diseases such as apple scab without harming predatory mites.

Because of problems with selective control of disease and insect pests and the lower levels of resistance in the principal phytoseiid mites species, European programs focused more on the identification of physiologically selective pesticides to conserve predators. More recently, integrated mite management programs have been recommended in several areas of Europe (Cranham and Solomon 1981, Gruys 1980c, 1980a; Table 17.4).

Table 17.4. Examples of Major Integrated Mite Management Programs (Excluding North America) Using Endemic, Insecticide-Resistant Phytoseiid Mite Predators

Country	Principal Pest Controlled[1]	Principal Natural Enemies[2]	Research Reference	Implementation Reference
United Kingdom	Pu, As	Tp	Cranham & Solomon 1981	Cranham & Solomon 1981
The Netherlands	Pu, As	Tp	Gruys 1980a, 1980b, 1982	Gruys 1980a, 1980b, 1982
Italy	Pu, As	Ap	Gambaro 1975, 1983, 1986	Gambaro 1975, 1983, 1986
USSR	Pu, As	To	Chubinnishvili et al. 1982	—
New Zealand	Pu, Tu, As	Tp	Penman et al. 1976, 1979	Cook 1980a, 1980b
Uruguay	Pu, Tu, As	Ac	Croft et al. 1976b	Carbonell & Briozzo 1981
Chile	Pu, Tu, As	Ac	Gonzalez 1981	Gonzalez 1981

[1] Pu = *Panonychus ulmi*; Tu = *Tetranychus urticae*; As = *Aculus schlechtendali*.
[2] Ac = *Amblyseius chilenensis*; Af = *Amblyseius fallacis*; Ap = *Amblyseius potentillae*; To = *Typhlodromus occidentalis*; Tp = *Typhlodromus pyri*.

17.3.1. *Typhlodromus pyri* in The Netherlands

An integrated mite management program was developed in Dutch orchards near the Schalmburg region using endemic, insecticide-resistant *T. pyri* (Gruys 1980a, 1980b, 1980c, Trapman and Blommers 1985, Blommers and Overmeer 1986). This program illustrates the extent to which physiological selectivity in pesticides can extend to the entire pest–natural enemy complex in apple orchards. Emphasis was placed on the conservation of *T. pyri* and several other key natural enemies. Major pests and natural control agents, selective pesticides, and alternative controls used in this program are listed in Table 17.5. In some cases, physiological selectivity is achieved because a natural enemy has developed resistance to a pesticide, while in others, physiological selectivity is due to innate tolerance.

Partial biological control and selective chemical control can be achieved for several aphid, leafroller, and leafminer pests (Table 17.5). Control of the European red mite and apple rust mite is achieved by pesticide-resistant *T. pyri*. The fungicides, insecticides, and acaricides used in integrated mite management and their selectivity to *T. pyri* are listed in Table 17.5. Tolerance to pirimicarb, carbaryl, and some OPs used to control lepidopteran and other insect pests are due to acquired resistance in this predatory mite (Table 14.1).

One interesting tactic used in the Dutch program, which is employed in several other areas of apple and grape production in Europe (e.g., in Switzerland, Baillod 1984; in England, Cranham and Solomon 1981), involves the intentional distribution of *T. pyri* into orchards where resistance has not developed or where predators have been eliminated (usually by nonselective pesticides). Apparently *T. pyri* disperses slowly from tree to tree and orchard to orchard (Gruys 1980c, Trapman and Blommers 1985, Blommers and Overmeer 1986). Growers overcome this problem by transferring predators to new sites. Within the first or second growing season after reintroducing *T. pyri*, biological control of spider mites can be reestablished and acaricide use reduced (Cranham and Solomon 1981, Baillod 1984, Blommers and Overmeer 1986, Genini and Baillod 1987).

Some problems were encountered during the development of this program due to the degree of physiological selectivity required in pesticides. Some growers were reluctant to implement the program because it required careful timing and application of 4–8 relatively specialized pesticides, including some insect growth regulators (Chapter 12). In some cases, growers preferred using one or two applications of less selective insecticides, which were toxic to key pests and nontoxic to some secondary pests and their natural enemies (Chapter 9; Croft 1981a, 1982). The complexity of the Dutch integrated mite management program has been overcome through experience in implementation and its proven effectiveness (Gruys 1980a, 1980b, 1980c, Trapman and Blommers 1985, Blommers and Overmeer 1986).

17.3.2. *Typhlodromus pyri* in the United Kingdom

In the United Kingdom, integrated mite management programs are based on resistant *T. pyri*. They emphasize more traditional selective pesticides (compared to the Dutch program) to conserve this natural enemy (Cranham and Solomon 1981). Several broad-spectrum pesticides are used to ecologically selective ways. Table 17.6 summarizes the effects of various pesticides on *T. pyri* populations in the United Kingdom. Again, resistance plays a major role in the use of several compounds (Kapetanakis and Cranham 1983; Chapter 14).

Changes in the fungicides used to control powdery mildew have significantly improved conservation of *T. pyri* in the United Kingdom (Cranham and Solomon 1981). Dinocap, binapacryl, and sulfur suppress spider mites and nearly eliminate predatory mites;

Table 17.5. Summary of Selected Major IPM Tactics Used in Northwestern European Apple Orchards, Including Insecticide-Resistant Predatory Mites, Other Natural Enemies, Other Alternative Controls, and Selective Pesticides (after Gruys 1980a, 1980b, 1982)

Pest Species	Natural Enemies	Selective Pesticides	Alternative Control Tactics
Panonychus ulmi (European red mite)	*Typhlodromus pyri* (insecticide-R) ***	White oil*, benzoximate**, fenbutatin oxide**	—
Aculus schlechtendali (apple rust mite)	*Typhlodromus pyri* (insecticide-R) ***	Endosulfan**	—
Eriosoma lanigerum (woolly aphid)	Earwigs***, *Aphelinus mali**	Endosulfan**	Resistant varieties ***
Rhopalosiphum insertum (apple grain aphid)	Predators* *Monoctonus cerasi**	Pirimicarb***	Late-leafing varieties ***
Dysaphis plantaginea (rosy apple aphid)	Predators**, *Ephedrus persicae***	Pirimicarb***	Resistant or late leafing variables**
Aphis pomi (apple aphid)	Predators*, *Trioxys angelicae**	Pirimicarb***	—
Lygus pabulinus (lygus bug)	—	Bromophos***	Control of dicot weeds nearby***
Adoxophyes orana (summer fruit tortrix moth)	Parasitoids**	IGRs***, *B.t.*,** NP-virus**	Mating disruption with pheromones*
Archips podana (fruit tree tortrix)	—	Diflubenzuron,* IGRs***	*B.t.***
Cydia pomonella (codling moth)	Predators and parasioids**	Diflubenzuron,*** Granulosis virus**	Sanitation**
Stigmella malella (apple pygmy moth)	Parasitoid: *Chrysocharis prodice****	Diflubenzuron***	—
Hopocampa testudinea (apple sawfly)	—	Thiophanate-methyl,** diflubenzuron*	Resistant varieties ***
Anthonomus pomorum (apple blossom weevil)	Parasitoid: *Syrrhizus delusorius*	Diflubenzuron**	—
Lepidosaphes ulmi (mussel scale)	Predators, earwigs **	White oil**, IGRs***	Sanitation**

Asterisks indicate control efficacy:*** effective for one year of longer,** good temporary control;* useful additional control; no asterisk; insufficient experience to evaluate.

Table 17.6. Effects of Pesticides on Insecticide-Resistant Strains of *Typhlodromus pyri* Used for Biological Control of Spider Mites in the Unites Kingdom (after Cranham and Solomon 1981)

Pest or Disease Controlled	Harmless or of Low Toxicity to *T. pyri*	Harmful or of High Toxicity to *T. pyri*
Aphids	Pirimicarb, oxydemeton-methyl	Vamidothion
Apple sucker	HCH, azinphosmethyl + oxydemetonmethyl, sulphone	
Apple sawfly	HCH	
Winter moth, codling moth, fruit tree tortrix, summer fruit tortrix	Diflubenzuron carbaryl, phosalone, azinphosmethyl	Permethrin, cypermethrin, deltamethrin
Spider mites	Tetradifon, propargite, cyhexatin	Dicofol, amitraz
Apple rust mite	Carbaryl	Pirimiphos-methyl
Mildew	Bupirimate, triadimefon, fenarimol, thiophanate-methyl	Dinocap, binapacryl, sulfur, benomyl
Scab	Captan, dodine, dithianon, urea	Mancozeb, propineb, benomyl, carbendazim
Storage rots	Captan	
Fruit thinners	Carbaryl, ethiophon	

however, newer compounds such as bupirimate, triadimefon, and fenarimol control disease without harming spider mites, phytoseiid predators, or rust mites. These fungicides allow biological control to operate in an environment less structured by pesticides. Most other fungicides used for the control of apple scab are harmless to *T. pyri*. Benzimidazole fungicides (benomyl, carbendazim, thiabendazole, thiophanate-methyl) cause some sterility in *T. pyri*, but occasional sprays of thiophanate-methyl can be used without harm to this mite.

Studies with insecticides have established that diflubenzuron (Dimilin) and pirimicarb are harmless to *T. pyri*, but reliance on these two insecticides allows too much fruit damage from tortricid larvae and other pests. In addition, minor pests such as scale and apple sucker may become a problem when using these broadly selective insecticides (Cranham and Solomon 1981; Chapter 9). Research has shown that carbaryl and certain OPs cause little or no harm to resistant strains of *T. pyri* in some orchards (Kapetanakis and Cranham 1983). The geographical extent of resistance in *T. pyri* is not yet known. Considering that regional OP resistance is present in this species in other areas of the world (Chapter 14), it seems unlikely that it should not be present in the United Kingdom.

The physiologically selective acaricides tetradifon and propargite are harmless to *T. pyri* (Cranham and Solomon 1981). Cyhexatin is more toxic to spider mites than to predatory mites, and applications at half the recommended rate prevent overkill of spider mites. Researchers have reported devastating effects of pyrethroid (SP) insecticides on *T. pyri* (AliNiazee and Cranham 1980). A single application of permethrin, cypermethrin, or deltamethrin causes mortality of *T. pyri* and resurgence of spider mites or rust mites. SPs have been only cautiously recommended in United Kingdom tree fruit production, and even then growers expect increased mite problems due to the disruption of *T. pyri*.

Insecticide-resistant *T. pyri* usually become well established and regulate pest mites within the first two seasons of transition from broad-spectrum to selective pesticide programs. The transition process can be hastened by introducing carbaryl- and OP-resistant *T. pyri* from seed orchards, transferring a single branch for every 1–3 trees (Cranham and Solomon 1981). During establishment it is usually necessary to use one or two selective acaricide sprays to prevent damage by spider mites until predators are well distributed. Maintenance of moderate spider mite populations helps *T. pyri* become established. Thereafter, acaricides are seldom required and if left undisturbed, predators will regulate prey at low levels. Integrated mite management in the United Kingdom has been promising enough to warrant testing under a wide variety of commercial conditions (Cranham and Solomon 1981).

17.3.3. *Typhlodromus pyri, Amblyseius andersoni*, and *Amblyseius aberrans* in Southern Europe

Management of phytophagous mites using insecticide-resistant phytoseiid mites in southern Europe involves three species: *Typhlodromus pyri, Amblyseius andersoni* (= *potentillae*), and *Amblyseius aberrans* (Baillod et al. 1982, Baillod 1986, Gambaro 1986, Corino et al. 1986). These mites have biological control characteristics that complement each other. The simultaneous occurrence of all three species may provide better biological control of pest mites on apple than any one alone (Baillod 1986). The effectiveness of *T. pyri* as a predator was discussed for Dutch and English mite management programs; this species is similarly effective in southern Europe. The predatory potential of insecticide-resistant *Amblyseius aberrans* is less well known (Baillod 1986). *A. andersoni* has the remarkable ability to regulate spider mite populations in southern Europe.

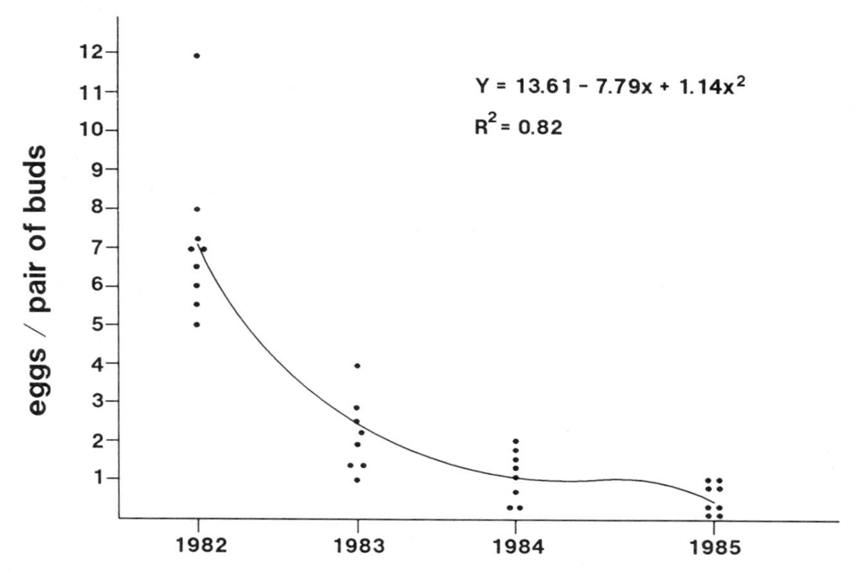

Figure 17.3. Annual trends in the number of *Panonychus ulmi* overwintering eggs in an apple orchard practicing integrated mite control using the predaceous mite *Amblyseius andersoni* (= *A. potentillae*) (After Gambaro 1986).

Gambaro (1986) conducted a long-term ecological study of *A. andersoni* in the Po valley of northern Italy. This species is an excellent regulator of *P. ulmi* due to its rapid numerical response and ability to reproduce on alternate foods such as pollen and tree sap (Gambaro 1986). In some Po valley orchards, biological control of pest mites has been maintained continuously for 10–12 years. During this time, no outbreaks of spider mites have been observed. Gambaro (1986) followed counts of *P. ulmi* winter eggs at the end of five consecutive growing seasons and observed a continuous decline in the average of eggs/bud (Fig. 17.3). Insecticide-resistant *A. andersoni* is indeed an effective natural enemy, especially when complemented with resistant *T. pyri* and *A. aberrans*. It appears that an ideal mix of resistant beneficials is available for management in the Po valley (Baillod 1986).

In this example, resistant strains of predatory mites have been transferred beyond their native ranges. For this reason, a further description this complex of predatory mites is presented in Chapter 18. Keep in mind that the combined use of endemic, resistant natural enemies as well as introduced R strains provides for the integration of pesticides and biological control of plant-feeding mites in this example (e.g., Gambaro 1975, 1983, 1986, Baillod and Guignard 1984, Corino 1985, Corino et al. 1986).

17.3.4. *Typhlodromus pyri* in New Zealand

In New Zealand, long-established (and therefore considered native) strains of OP-resistant *T. pyri* are the mainstay of integrated mite management programs. These predaceous mites, which were probably introduced with apple culture from Europe, have been studied to determine their range and degree of resistance, population dynamics, and compatibility with pesticides used for the control of arthropod, disease and weed pests (Penman et al. 1976, Wearing et al. 1978a, 1978b, Collyer 1980).

In addition to *T. pyri*, several other insecticide-resistant phytoseiid mites have been successfully imported into New Zealand for the control of *T. urticae* (Chapter 18). The predators used in seven different fruit-growing districts are listed in Table 17.7. Generally, *T. pyri* occurs throughout the country and is widely managed. *T. pyri* from New Zealand exhibit the highest levels of azinphosmethyl resistance found anywhere in the world (Table 14.1). This is probably due to the extensive pesticide use required for pest-free export standards in New Zealand (Penman et al. 1976).

In contrast to Dutch populations of *T. pyri* (Gruys 1980a, 1980b, 1980c; Table 17.5), azinphosmethyl resistance in New Zealand populations allows this compound to be compatible with integrated mite management (Penman et al. 1976). Other broad-spectrum pesticides are carefully used to suppress key insect and disease pests while conserving predatory mites (Cook 1980b). A list of pesticides recommended for key pest control is given in Table 17.7 with notes for appropriate use. Compounds recommended for integrated mite management as well as those to be avoided or used with care are identified. Selectivity of these compounds to *P. persimilis, A. fallacis, T. occidentalis* (introduced species), as well as *T. pyri* is presented, although the program in New Zealand focuses mostly on *T. pyri* and *P. persimilis* (Chapter 18, Section 18.3).

17.3.5. *Amblyseius chilenensis* in South America

In South America, *Amblyseius chilenensis* is widespread in apple-growing areas of Uruguay, Argentina, and Chile (Table 17.4). Resistance to OP insecticides such as azinphosmethyl, phosmet, and tetrachlorvinphos as well as cross resistance or tolerance to carbaryl are present in several areas (Croft et al. 1976b, Gonzalez 1981). In Uruguay,

Table 17.7. Toxicity of Pesticides Used in New Zealand to Endemic and Introduced Insecticide-Resistant Phytoseiid Mites Including *Typhlodromus pyri, Amblyseius fallacis, Phytoseiulus persimilis*, and *Typhlodromus occidentalis* (after Cook 1980b)

Pest Controlled	Compound	Toxicity to Predaceous Mites	Comments[1]
Leafrollers, codling moth	Azinphosmethyl	Mixed	May be somewhat toxic to Tp and Pp
	Phosmet	Low	—
	Chlorpyrifos	Low	May eliminate prey mites and cause predators to starve
Mealybug	Chlorpyrifos	Low	May eliminate prey mites and cause predators to starve
	Parathion	Mixed	May be somewhat toxic to Pp
Scale	Azinphosmethyl	Low	May be somewhat toxic to Tp and Pp
	Chlorpyrifos	Low	May eliminate prey mites and cause predators to starve
	Parathion	Mixed	May be somewhat toxic to Pp
Woolly aphid	Chlorpyrifos	Low	May eliminate prey mites and cause predators to starve
	Lindane prebloom	Low	—
	Vamidothion	Mixed	May be toxic to some predators and not to others
Spider mites	Propargite	Low	—
	Cyhexatin	Low	High dosages may cause mortality to predators
	Fenbutatin oxide	Low	—
	Azocyclotin	Low	—
	Dicofol	Mixed	Nontoxic to Tp only;
	Bromopropylate	Low	can cause starvation of predators by overkill of prey mites

[1] Tp = *Typhlodromus pyri; Pp = Phytoseiulus persimilis.*

Carbonell and Briozzo (1981) reported predator–prey relationships between *A. chilenensis* and *P. ulmi*, focusing on overwintering and dispersal into apple trees. Azinphosmethyl and tetrachlorvinphos are used to control codling moth and the leafroller *Argyrotaenia sphaleropa*, while resistant *A. chilenensis* provide biological control of the *P. ulmi*. If acaricides are needed to augment predators, physiologically selective acaricides such as cyhexatin or propargite are recommended. Fungicides such as dicofol and benomyl can be used for apple scab control as well as for selective control of *P. ulmi* without substantially harming. *A. chilenensis.*

In Chile, *A. chilenensis* is widely resistant to many orchard pesticides (Gonzalez 1981). Along with the stigmaeid mite *Agistemus longisetus*, which is tolerant to OPs, *A. chilenensis* can control *P. ulmi* on apple and pear, *T. urticae* on apple, and rust mites on several deciduous tree fruit crops. Selective pesticide programs have been developed around *A.*

chilenensis strains which are resistant to azinphosmethyl and diazinon. This mite readily tolerates field sprays of the OPs phosmet and phosalone, which can be used to control codling moth and several other minor pests. Acaricides used in Chile that are selective to *A. chilenensis* include cyhexatin, propargite, and fenbutatin oxide.

17.4. MANAGEMENT OF ENDEMIC, RESISTANT PREDACEOUS INSECTS

Of the insect predators that have developed resistance (Table 14.1), the only case in which a species has been extensively managed within its native range involves *Stethorus punctum*. This coccinellid feeds exclusively on plant-feeding mites. Other resistant predatory insects such as *Coleomegilla maculata* or the cecidomyiid *Aphidoletes aphidimyza* have not been as extensively exploited. This may be because resistance was not sufficiently high to justify manipulation of these natural enemies in the field, or perhaps these species occur only sporadically in the treated environment. If a resistant species is a minor part of a predatory complex associated with the pest, its importance as a biological control agent may not warrant management on crops where resistance has developed.

17.4.1. *Stethorus punctum* in the Eastern United States

IPM programs in which insecticide-resistant *Stethorus punctum* are managed differ from those developed for predatory mites (Table 17.1). Resistance or tolerance attributes of predatory mites and *S. punctum* are quite different. Also, alate *S. punctum* recolonize treated plots much more rapidly than do wingless predatory mites. Extensive research on pesticide selectivity and management of endemic, OP-resistant *S. punctum* has been conducted by Asquith, Hull, and coworkers in Pennsylvania (Asquith 1971, 1972, Asquith and Colburn 1971, Mowrey et al. 1977).

Because of the mobility of *S. punctum*, pesticides can be used in ecologically selective ways to further increase its survival. Selectivity is widely achieved through the use of an alternate-row method of pesticide application (Chapter 10). Because one side of each tree row is sprayed at any one treatment period, mobile predators are conserved or move to untreated refugia in trees (Fig. 10.2; Asquith 1971). In Pennsylvania, the insecticide and acaricide efficiency guide for apple IPM focuses on major pests and the important mite predators *S. punctum* (adult and larval forms) and *A. fallacis* (all life stages) (Table 17.8). The relatively low toxicity of azinphosmethyl, dimethoate, parathion, permethrin, phosalone, phosmet, and several other compounds to *S. punctum* is due to a wide spectrum of resistance development to many compounds used on apple (Hull and Starner 1983; Chapter 14). In addition, OP-resistant *A. fallacis* complement the predatory coccinellid when *S. punctum* populations are low due to poor survivorship or use of particularly toxic compounds (e.g., carbaryl; Table 17.2).

A detailed simulation model has been developed to estimate the need for acaricides and the potential for biological control of the European red mite *P. ulmi* by insecticide-resistant *S. punctum* in Pennsylvania (Mowrey et al. 1977). The model predicts pest and predator density and the need to apply selective acaricides using current weather information and initial orchard counts of adult and larval predators and spider mites. Tabular output of the model has been used successfully by growers for on-site management of pest mites and predators for several years.

Table 17.8. Effects of Insecticides and Acaricides on Insecticide-Resistant Populations of *Stethorus punctum* in Pennsylvania Apple Orchards (after Hull and Beers 1985)

Compound	Rate/Hectare	Toxicity Rating to *S. punctum*[1]	
		Adults	Larvae
Azinphosmethyl 50WP	1.12 kg	+	+
Bacillus thuringiensis	1.68 kg	0	0
Carbaryl 50WP	2.24 kg	+ + +	+ + +
Chlorpyrifos 50WP	3.36 kg	+	+
Cyhexatin 50WP	0.87 kg	+	+
Demeton 6F	0.73 l	+	+ +
Diazinon 50WP	2.24 kg	+ +	+ +
Dicofol 35WP	4.48 kg	+	+
Dimecron 8EC	0.07 l	+	+ +
Dimethoate 25 WP	2.80 kg	+	+
Endosulfan 50WP	2.24 kg	+ +	+ +
Fenvalerate 2.4EC	0.44 l	+ + +	+ + +
Formetanate 92SP	0.73 l	+	+ +
Methomyl 2E	4.68 l	+ +	+ +
Oxamyl 2L	3.50 l	+ +	+ +
Parathion (Encap.)	2.91 kg	+	+
Parathion 15WP	5.60 kg	+	+
Permethrin 3.2EC	0.73 l	+ + +	+ + +
Phosalone 50WP	4.48 kg	+	+
Phosmet 50WP	2.24 kg	+	+
Propargite 30WP	3.36 kg	+	+
Vendex 50WP	0.87 kg	+	+

[1]Natural enemy toxicity: 0 = nontoxic; [+]slightly toxic; [++]moderately toxic; [+++]highly toxic.

17.5. TRANSFER OF RESISTANT PHYTOSEIIDS WITHIN NATIVE RANGES

Early detection and redistribution of resistant phytoseiid strains within their native ranges are commonly practiced in many areas of the world where these mites occur. In many cases growers and extension personnel are heavily involved in these efforts (Thomas and Chapman 1978, Cook 1980a, 1980b), but such efforts often are not published.

Documentation of this tactic was first reported by Croft and Barnes (1971, 1972), who established R strains of *M. occidentalis* from Washington and Utah in southern California orchards, where this species was not yet resistant to azinphosmethyl. In the presence of azinphosmethyl, resistant populations from Washington (Fig. 17.4a) and Utah (Fig.17.4b) provided control of McDaniel spider mites *Tetranychus mcdanieli* early in the season before populations reached damaging levels. Similar releases of S predators in azinphosmethyl-treated plots did not provide control until spider mites attained moderate densities (Fig. 17.4c). In trees where no releases were made and the only source of predators was immigration (Fig. 17.4d), pest density attained more than 100 per leaf.

Subsequently, it was shown that survival of indigenous, susceptible *M. occidentalis* was not sufficient to control the pest in the treated environment. Introduced, resistant strains interbred with native predatory mites and eventually established a population similar in

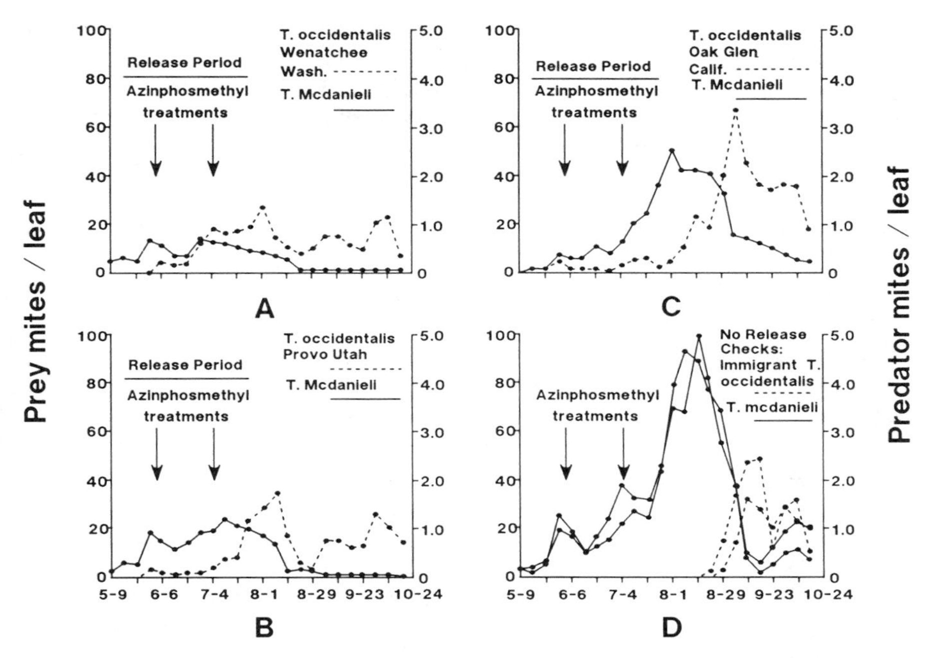

Figure 17.4. Densities of *Metaseiulus occidentalis* and *Tetranychus mcdanieli* following release of OP-resistant strains (a and b) and a susceptible strain (c) of the predatory mite for comparison with no-release controls (d) (adapted from Croft and Barnes 1972).

resistance to the introduced strains, but with morphological characteristics of the native susceptible strain (Croft and Barnes 1971, 1972). This hybrid strain persisted thereafter and provided good control of the pest. While this experiment initially demonstrated the practical benefits of transferring resistant natural enemies within their native ranges, it also ultimately proved to be one of the first practical successes in genetic improvement of an insecticide-resistant strain of a natural enemy by hybridization methods (Chapter 19).

While subsequent studies of genetic improvement would be conducted under more controlled laboratory conditions using artificial selection methods, resistance was effectively incorporated into the native population in these field tests. Since these early experiments, additional transfers of OP-resistant *M. occidentalis* have been made in the western United States (e.g., Larson 1970, Croft and Brown 1975); however, this operation has become routine and is seldom reported in the literature.

In New Zealand, OP-resistant *T. pyri* were transferred from orchards near Nelson on the South Island to Waikato, a small, relatively isolated apple-producing area on the North Island in 1977–1978 (Wearing and Proffitt 1982). Previously, *T. pyri* had been collected infrequently in Waikato, probably due to heavy fungicide use and the susceptibility of indigenous strains. Initially, resistant predators from the Nelson area were aspirated from leaves into containers, and about 600 individuals were released onto two European red-mite infested Red Delicious trees. Establishment and spread were monitored by regular sampling and counting of both prey and predators. Although initial buildup of predators was hampered somewhat by fungicide sprays and heavy OP use, predators were common throughout the release orchard at the beginning of the 1981–1982 growing season. In a survey of 11 Waikato apple orchards, OP-resistant *T. pyri* had spread to 3 orchards by the end of the 1982 growing season. Predators from the original release orchard were distributed to other Waikato and Bay of Plenty orchards. Soon thereafter, growers in these areas were advised of means to conserve and manage these mites for improved pest control (Cook 1980a, 1980b).

In New Zealand, transfer of OP-resistant *T. pyri* is supervised by the Division of Scientific Industrial Research and local horticulture officers (Cook 1980a, 1980b). Detailed monitoring of resistance in *T. pyri* is conducted periodically and is the basis for interarea introductions. Where azinphosmethyl-resistant *T. pyri* are not present, a regular program of introduction and distribution is practiced and orchard management techniques are recommended. Key points of the program include the following:

1. The development of local OP-resistant *T. pyri* is encouraged by avoiding use of other chemicals which are harmful to this species.
2. If a local strain of OP-resistant *T. pyri* is not present, introductions are made from other areas and conservation measures include selective pesticides.
3. Growers are taught to identify major mite species and to understand the basic life cycles of *T. pyri* and its prey, the European red mite. Careful monitoring of predator/prey ratios is encouraged from November through February, with special attention given to the critical period in early January.
4. Recommended pesticides are compatible with integrated mite management.
5. Populations of *T. pyri* are transferred under supervision of horticulture advisory officers.

The New Zealand program includes careful monitoring and managemnt of OP resistance in *T. pyri*, as well as transfer of highly resistant strains from region to region within the country to obtain maximum levels of OP resistance and biological control (Table 17.7; Cook 1980b). Procedures for the transfer of insecticide-resistant strains of this species into European orchards and grape vineyards were referred to earlier (Sections 17.3.1 and 17.3.2). Transfer appears to be a primary component of integrated mite management programs in areas where this species predominates (Trapman and Blommers 1985, Blommers and Overmeer 1986, Genini and Baillod 1987). Apparently, most other phytoseiids have greater powers of dispersal than *T. pyri*, probably accounting for less emphasis on transfer with other species.

The transfer and manipulation of endemic, insecticide-resistant predatory mites raises important questions regarding intraspecific competition, genetic adaptation to new environments (e.g., diapause requirements), and management considerations. What type of management program best favors the establishment of insecticide-resistant predators in a given habitat? How many mites must be released to ensure establishment? How can the success of transfer programs best be evaluated? Answers to these questions are applicable to introductions of exotic (Chapter 18) and genetically improved biological control agents (Chapter 19) as well. Some of these factors are discussed more extensively in Chapter 19.

Croft (1976) presented some general recommendations on factors to consider in transfering insecticide-resistant predatory mites into new areas. He discussed methods of culture and distribution of mites in the field, release strategies, management needs, and evaluation procedures. Critical factors identified were climatic conditions, prey and alternate prey, aspects of habitat (e.g., ground cover beneath trees aids in the establishment of *A. fallacis*), and maintenance of a favorable chemical environment so that released resistant predators could outcompete native susceptible predators.

17.6. IMPACT OF ENDEMIC, RESISTANT NATURAL ENEMIES IN IPM

Integrated mite management programs that are based on the management of endemic, insecticide-resistant predators have been in place longer than those programs based on introductions of exotic or genetically improved insecticide-resistant beneficials. Preliminary estimates of ecological, environmental, and economic impacts as reflected in control costs, pesticide use levels, and pesticide resistance trends have been made. Impact data are most abundant from North America, where insecticide-resistant predatory mites were first discovered and exploited. The following discussion of benefits involves the successful exploitation of these organisms in IPM programs within their native ranges (Whalon and Croft 1984).

17.6.1. Economic Benefits

Croft et al. (1985) reported data from S. C. Hoyt, who monitored the impact of an integrated mite control program based on management of OP-resistant *M. occidentalis* in a 1000-ha apple orchard in central Washington State from 1957 to 1975 (Fig. 17.5). Data are presented in terms of control costs reflecting reduced pesticide usage. Using the period of 1957–1966 to represent preintegrated mite management and 1967–1975 to reflect the impact of integrated mite management, Hoyt found that the total cost for mite control declined from nearly $60 per acre to $20–30 per acre despite inflation. The dotted line from 1969 to 1975 represents the general trend for the 5-year period (S. C. Hoyt pers. comm.).

On a regional scale in Washington State, Hoyt (1982) estimated that integrated mite

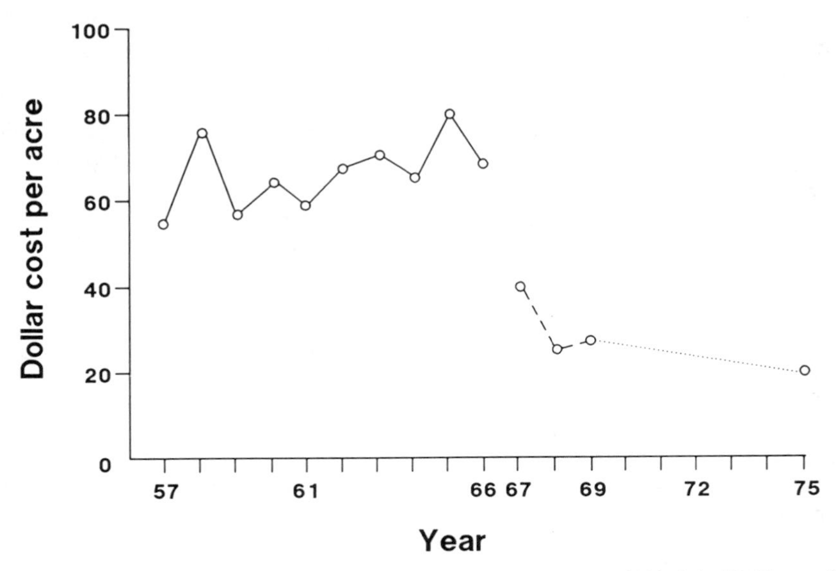

Figure 17.5. Cost per acre for control of plant-feeding mites in a 1000-acre apple block in Washington State representing pre-IPM (1957–1966) versus post-IPM (1967–1975) practices (adapted from Hoyt and Caltagirone 1971).

management based on OP-resistant *M. occidentalis* reduced the amount spent on acaricides and their application by approximately $5 million per year on apples from 1967 to 1981, for a total of $70 million. He projected that similar programs based on the use of resistant predatory mites implemented worldwide probably provided a savings of $20–25 million per year. Beyond direct cost benefits, Hoyt (1982) noted that the use of resistant predaceous mites for the management of pest mites had averted disastrous losses in crop yield and quality as well as reducing losses of effective pesticides due to resistance (Section 17.6.4). Seldom are the more indirect benefits of IPM and resistance management programs quantified.

Similar successful implementation of integrated mite management has been realized in Pennsylvania and Michigan, although pesticide use reductions are more in the range of 30–40%. Larger complexes of insect pests and diseases must be managed in Pennsylvania and Michigan relative to mite control costs, thus savings in mite control has resulted in a less dramatic reduction in the overall pesticide bill. Based on a management program of resistant *S. punctum*, acaricide use for *P. ulmi* in Pennsylvania was reduced by over 75% during its 12-year history (Croft et al. 1985). Over a 6-year period (1970–1976) in Michigan, OP-resistant *A. fallacis* reduced mite control costs in an experimental plot by 90% from $20.00–30.00 per acre (2–3 applications per year) to $2.50 per acre. Over the same time period, a 60% decline in costs to about $8.00 per acre for mite control (in 1976) was attained in grower-implemented integrated mite control programs in commercial orchards (Croft et al. 1985).

17.6.2. Adoption Levels

Whalon and Croft (1984) estimated the implementation of IPM programs on deciduous tree fruit crops, including integrated mite management programs which are principally based on endemic, insecticide-resistant natural enemies in North America. They divided their estimates into those acres in North America under IPM and those indirectly influenced by IPM. Since the late 1960s, adoption of mite IPM has increased exponentially to approximately 25% of deciduous tree fruit acreage in 1982 (120,800 acres). They defined implementation as involving monitoring, multiple management tactics, action thresholds, IPM education, and IPM communication systems. They estimated that the actual acreage influenced by IPM probably exceeded 95% of the industry. This was mostly represented by integrated "mite programs where selective use of acaricides and concern for predator survival (if not actual monitoring) was practiced by nearly 75% of commercial growers."

17.6.3. Environmental Benefits

Large-scale reductions in pesticide use made possible by the management of resistant natural enemies undoubtedly has considerable environmental benefits. Ecological benefits are difficult to quantify beyond immediate acaricide reductions, because many of these benefits are expressed in complex species associations. These benefits often involve processes that operate with long time delays. For example, reductions in toxic effects on honeybees and other pollinators, on birds, fish, and other wildlife, and even on human applicators have been brought about to some degree by resistance management and IPM using pesticide-adapted natural enemies (Brown 1978a; Chapters 20 and 21).

17.6.4. Resistance Development

Croft (1985) used field data and modeling studies (Tabashnik and Croft 1985) to estimate the reduction in the rate of acaricide resistance development in spider mites as a result of

Table 17.9. Simulation Model Results of Time to Resistance Development in the European Red Mite *Panonychus ulmi* under Various Selection Levels and Intensities with Acaricides (Unpublished Studies of Tabashnik and Croft; see Methods of Investigation in Tabashnik and Croft 1985)

Selection Level (Mortality)	Applications per Year		
	6	3	1
93%	1.5 years[1]	1.3 years	2.6 years
75%	1.6 years	1.9 years	6.5 years
50%	1.5 years	2.2 years	13.6 years

[1] Simulated time to resistance.

IPM programs using OP-resistant phytoseiid mites. Standard acaricide programs (without biological controls) led to acaricide resistance in 1–2 years, while phytophagous mites in IPM programs took 13 years to become resistant (Table 17.9). When OP-resistant predaceous mites were employed, acaricide use was reduced 8-fold. This reduction in selection intensity greatly extended the effective use of acaricides in the field. Economic benefits to the public were not estimated because cost benefits from IPM programs to such diverse groups as industry, marketing organizations, and pesticide users have not been studied in detail (Mirianowski and Carlson 1986). More recent experience with acaricide resistance management in the field has verified predictions of resistance and attendent costs for pest control (Croft et al. 1984, Miller et al. 1985, Flexner 1988; Chapter 21).

17.7. CONCLUDING REMARKS

Insecticide resistance in endemic natural enemies is known for a number of species in addition to those that have been exploited in IPM programs (compare Tables 14.1 and 17.1). The utility of a resistant natural enemy in pest control strategies depends on the level of resistance developed and whether or not the compound can still be used for pest control. Also important is the effectiveness of the resistant natural enemy as a biological control agent. If the natural enemy is not sufficiently effective to provide economic control of a particular pest, then the benefit of a resistance factor may not be sufficient to justify further study. The management of resistant natural enemies which effectively regulate key pest species will be a major focus of future research. While most research on the management of endemic, resistant natural enemies has been conducted on species attacking mite pests, resistant predators and parasitoids of scale and aphids will probably soon be exploited for similar purposes.

Pesticide resistance in an endemic natural enemy population is of greatest use when detected and exploited as early as possible in the useful life of a pesticide. The feasibility of managing a resistant natural enemy is critically dependent on the overall performance of the pesticide. For example, resistance development in other pests can make resistance in a natural enemy a moot point. These points emphasize the importance of monitoring and managing resistance in both pests and natural enemies, and that resistance must be recognized as a dynamic evolutionary phenomenon.

18

INTRODUCED SPECIES

18.1. INTRODUCTION

The first introduction of an insecticide-resistant arthropod natural enemy to an area outside of its native range involved an organophosphate (OP) resistant (R) strain of *Metaseiulus occidentalis* (Table 18.1). A population from apple in the western United States was introduced into eastern Australia in the early 1970s, and successful establishment was reported shortly thereafter (Waterhouse 1973, Readshaw 1975). This importation occurred soon after OP resistance was first documented in *M. occidentalis*, and introductions of OP-resistant strains from Washington and Utah were made into a southern California apple orchard where susceptible (S) strains had previously been present (Hoyt 1969a, Croft and Jeppson 1970, Croft and Barnes 1971, 1972, Croft and McMurtry 1972a). This successful establishment set a precedent for research on the importation of resistant natural enemies, and importations have continued on a wide geographic scale to the present.

Other introductions of exotic insecticide-resistant phytoseiids have been attempted, some producing outstanding results (Table 18.1). Introductions of pesticide-resistant natural enemy species from other families have been limited. This is due in part to low levels or the absence of resistance among other families of predators and parasitoids (Table 14.1). Other limitations may involve mass production and maintenance of resistant natural enemies in the field (Section 18.8).

A number of case histories of exotic introductions of insecticide-resistant phytoseiid mites are reviewed in this chapter. In several case histories, the technology for transferring and establishing pesticide-resistant strains is briefly discussed. A final section reviews methods of collection, colonization, introduction, and establishment of resistant natural enemies as well as evaluations of these programs.

Table 18.1. Examples of Successful Introductions of Insecticide-Resistant Natural Enemies beyond Their Native Ranges

Species	Site of Origin	Site of Establishment	Reference
Amblyseius andersoni	Italy	Switzerland	Caccia et al. 1985
Amblyseius chilenensis	South America	France	Fournier et al. 1985b
Amblyseius fallacis	Michigan, U.S.	New Zealand	Penman et al. 1979
	Indiana, U.S.	Michigan, U.S.	Croft & Meyer 1973
	Michigan, U.S.	Sungmore, Taiwan	Lo et al. 1986
	Michigan, U.S.	China	Zhang & Kong 1985
Phytoseiulus persimilis	The Netherlands, England	Germany	Hassan 1982a, 1982b
Metaseiulus occidentalis	Washington, U.S.	California, U.S.	Croft & Barnes 1972
	Washington, U.S.	Canberra, Australia	Readshaw 1975
	Washington, U.S.	New Zealand	Penman et al. 1979, Butcher & Penman 1983
	Washington, U.S.	Georgia, USSR	Chubinnishvili et al. 1982
	California, U.S.	Oregon, Washington, U.S.	Hoy et al. 1983
	California, U.S.	Sungmore, Taiwan	Lo et al. 1986
Typhlodromus pyri	New Zealand	United Kingdom	AliNiazee & Cranham 1980, Cranham & Solomon 1981, Kapetanakis & Cranham 1983
	New Zealand	Australia	Seymour 1982a
	New Zealand	The Netherlands	van de Baan et al. 1985

Introductions of field-selected or genetically improved strains of insecticide-resistant natural enemies are specialized cases of classical biological control. Research on introductions of exotic biological control agents is well developed, with an extensive theoretical and methodological foundation for foreign exploration, introduction, colonization, culture, and evaluation. For a review of the general topic of exotic natural enemy species introductions, see DeBach (1964) and Huffaker and Messenger (1976).

18.2. INTRODUCTIONS OF *METASEIULUS OCCIDENTALIS* INTO AUSTRALIA

It was fortuitous that the first introduction of an exotic, pesticide-resistant natural enemy was so successful, for it clearly documented the value of this type of research. In some ways, however, success led to some similar attitudes as those which arose following the successful importation of the vedalia beetle *Rodolia cardinalis* from Australia to the United States for biological control of the cottony cushion scale *Icerya purchasi* on citrus in the late 1800s (Doutt 1964). Both made this approach to biological control look easy. The successful introduction of *M. occidentalis* into Australia allowed researchers to avoid many of the problems of genetic improvement and selective pesticide development that have been encountered more recently (Chapters 20 and 21). Conversely, the progress achieved in this area of research to date is probably due in part to the success of this initial effort.

Because of the extensive documentation and large number of personnel involved in the release, distribution, and management of *M. occidentalis* in Australia, this program is reviewed here in considerable detail. For more information on Australian importations of insecticide-resistant predatory mites, see Readshaw (1979), Field et al. (1979), Marsden et al. (1980), Seymour (1982a, 1982b), Readshaw et al. (1982), Bower and Thwaite (1982), Bower (1984), and James (1988).

The Australian *M. occidentalis* introduction project began in 1972 when releases of OP-resistant *M. occidentalis* from Washington, United States, were made in the fall (February) and spring (November and December) into a single commercial apple orchard near Canberra where two-spotted spider mites *Tetranychus urticae* were the principal pest mites. Results were spectacular (Fig. 18.1; Readshaw 1975). Despite six sprays of

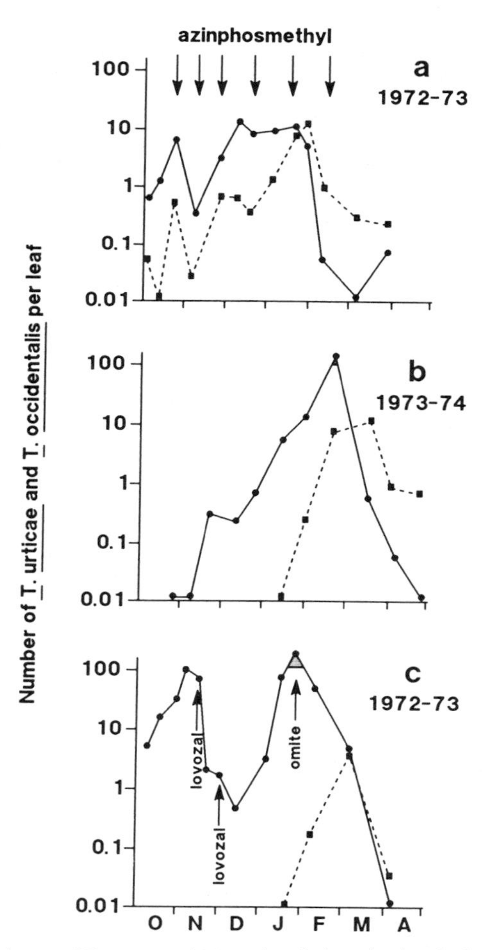

Figure 18.1. Population estimates of *T. urticae* and *M. occidentalis* in orchard at Canberra (1972–1974). All trees sprayed with 0.05% a.i. azinphosmethyl as indicated by arrows in (a). Predators released in autumn (a) and spring (b), 1972. Acaricide control (c) sprayed twice. Darkened areas on graphs indicate periods in which leaf damage by *T. urticae* occurred (after Readshaw 1975).

Table 18.2. Summary of Trials Conducted to Evaluate the Effectiveness of OP-Resistant *M. occidentalis* in Controlling *T. urticae* in Australian Apple Orchards (Adapted from Readshaw 1975)

	Site[1]						
Status	Tatura, Vic.	Lenswood, S.A.	Batlow, N.S.W.	Bathurst, N.S.W.	Stoneville, W.A.	Leeton, N.S.W.	Stanthorpe, Qld.
Established	Y[2]	Y	Y	Y	Y	Y	Y
Overwintered	Y	Y	Y	Y	Y	NA	NA
Survival, 1973–1974	Y	Y	N[3]	Y	N	Y	Y
Effective as a predator 1973–1974	Y[4]	Y	N	Y	N	Y	Y

[1]Vic. = Victoria; S.A. = South Australia; N.S.W. = New South Wales; W.A. = western Australia; Qld. = Queensland.
[2]Y = effective biological control attained; N = not attained; NA = not applicable, predators released in the spring.
[3]Poor or no survival as a result of the application of chemicals that were toxic to predators.
[4]Some predators survived and eventually controlled *T. urticae* with the help of the ovicide propargite.

Figure 18.2. Distribution and dates of release of insecticide-resistant phytoseiid mites introduced into tree fruit producing areas of Australia (1972–1978) (taken from Readshaw 1975, Field et al. 1979, Seymour 1982a).

azinphosmethyl for codling moth control, which previously had eliminated all native mite predators (e.g., *Stethorus* spp.), OP-resistant *M. occidentalis* successfully overwintered and maintained *T. urticae* well below damaging levels throughout the 1972–1973 and 1973–1974 growing seasons (Fig. 18.1). Fall releases exhibited earlier biological control than did spring releases of predators (Fig. 18.1). Released resistant predators appeared in acaricide check trees only late in the growing season. Densities of *M. occidentalis* and spider mites were monitored at the original release site for the next 12 years. Over this period, the grower sprayed for spider mites only twice, and then at only half of the recommended field rate to improve predator-to-prey ratios. It was estimated that without the introduction of the predators, this grower would have applied acaricides at least 36 times at the full recommended rate over the same period (Seymour 1982a, 1982b).

The next phase of the introduction project began in November 1973, when this same strain of *M. occidentalis* was released in seven different apple plots located throughout mainland Australia (Readshaw 1975). These experiments were conducted to evaluate the optimal climatic, cultural, and host–prey conditions under which the predator would become established and be effective. A summary of these introductions is presented in Table 18.2. (See Fig. 18.2 for specific locations of field trials.) Under intensive schedules of azinphosmethyl and other insecticides, *M. occidentalis* controlled *T. urticae* in four out of seven trials. In the cases where introduced predators were not effective (Table 18.2), other pesticides may have reduced predator populations (Readshaw 1975).

Following the initial research by Readshaw (1973, 1975), other researchers and field-advisory personnel became involved in the project as it expanded into various fruit-growing regions of the country. R. P. Field and colleagues, working in Victorian peach and apple orchards, conducted extensive trials during 1974–1977 with R strains of *M. occidentalis*. Initially, this group worked with a parathion-tolerant, azinphosmethyl-susceptible strain which was probably accidentally introduced at an earlier time (Field 1976, 1978, 1982, Field et al. 1979, Schicha 1978). Researchers observed that while parathion sprays permitted survival of this strain of *M. occidentalis*, azinphosmethyl treatments did not, and spider mites reached outbreak proportions. Field (1974) recommended the prompt introduction of North American azinphosmethyl-R *M. occidentalis* into Victoria.

In 1974, the initial introduction of the azinphosmethyl-R strain was made into a commercial peach orchard in the Goulburn valley, Victoria (Field 1978; Fig. 18.2). Its establishment and effectiveness over three successive field seasons were followed closely. The predator quickly established itself on release trees and subsequently controlled *T. urticae* each season under spray programs that included azinphosmethyl or phosmet for codling moth control. Predators did not become established in plots treated with the fungicide benomyl, probably because of direct mortality. Predators spread rapidly throughout the remainder of the orchard during the first season, and only one application of the acaricide propargite was required on nonrelease trees over a 3-year period.

With the success achieved in trial orchards, Field et al. (1979) developed mass-rearing and release methods for *M. occidentalis* reared on *T. urticae*. Soybean plants grown under flood-irrigation methods were seeded with spider mites and predators by transferring them from infested apple shoots. The density of predators in mass-rearing plots was monitored regularly by collecting soybean leaflets. Predators were then distributed to all apple- and peach-growing regions in Victoria within 150 km of the mass-rearing field. Half of a soybean plant was placed in every fifth tree of every fifth row (release density of about 315 motiles and 160 eggs/tree). A total of 2990 ha of apples and peaches in 216 orchards were seeded in the Goulburn valley in these tests (Table 18.3, Fig. 18.2). The following year,

Table 18.3. Release and Establishment of OP-Resistant *M. occidentalis* in Victoria, Australia, Apple and Peach Orchards (1978) (after Field et al. 1979)

District	Hectares Seeded with *M. occidentalis*	No. of Orchards Seeded	Orchards Sampled for *M. occidentalis*	Orchards Confirming Establishment of *M. occidentalis*
Goulburn valley:				
Cobram/Invergordon	910	55	6	6
Shepparton East	415	42	3	3
Ardmona	450	27	3	3
Kyabram	170	16	5	5
North East	150	14	1	1
Harcourt	120	8	5	5
Baccus Marsh	200	14	8	7
West Gippsland	85	7	5	4
Metro. and Hills	245	19	6	5
Mornington Peninsula	245	14	7	7
Total	2990	216	49	46

predators were recovered from overwintering sites in 46 of 49 areas where releases were made (Table 18.3).

As indicated in early nationwide trials conducted by Readshaw (1975) in 1973–1974, establishment and effective control of spider mites (including both *T. urticae* and *P. ulmi*) by *M. occidentalis* were not equally successful in all fruit-growing areas of Australia. Williams (1978) reported difficulties in achieving successful control of spider mites on Tasmania, where initial releases were made in 1977 and 1978. Although *M. occidentalis* became established and in some cases prevented serious damage to apple foliage by *T. urticae* and *P. ulmi*, it did not control mites in orchards where *P. ulmi* alone was present (*P. ulmi* is not a preferred prey of *M. occidentalis*; Hoyt 1969a). Similarly, at Batlow, New South Wales, the efficacy of this predator against *P. ulmi* was generally mixed and applications of acaricides were required more regularly than in other release areas (Marsden et al. 1980). In western Australia near Stoneville, establishment of the predator was not documented in surveys made in 1981, even though spray schedules were modified to conserve predators.

In summary, *M.* occidentalis was eventually released in over 30 areas throughout the country (Ford 1976), with only a few instances of poor establishment and control. Significant control of *T. urticae* was achieved in almost every release area within a few years after establishment. Overall, the program was deemed a successful introduction of a pesticide-resistant natural enemy into a new country (Readshaw 1975). The few failures in the Australian program offer insights into factors limiting the establishment of pesticide-resistant natural enemies in new areas. The presence of nonpreferred prey and the use of nonselective acaricides or other pesticides during the establishment period were mentioned earlier. Other limiting factors might include a failure to adapt to new climates (*M. occidentalis* did not prosper in cooler, more humid climates such as in Tasmania), poorly adapted photoperiod-induced diapause that would be needed for survival in the new environment, and limited release numbers.

More recent studies have indicated that introduced strains of *M. occidentalis* may be adapting to the winter environment of certain regions of Australia (e.g., New South Wales).

This adaptation will likely enable the resistant predatory mite to better survive and provide biological control of spider mites (James 1988). Whereas the original California progenitor strains of this species showed definite short-day responses to photoperiod (Hoy and Flaherty 1970, Croft 1971, Hoy 1975), recently collected field strains did not exhibit such responses, indicating a substantial change in the established population (James 1988).

For the control of spider mites in orchards where *P. ulmi* predominated, Australian researchers imported azinphosmethyl-R *Amblyseius fallacis* from Michigan, United States, and azinphosmethyl/carbaryl-R *Typhlodromus pyri* from New Zealand and the United States (Readshaw 1979, Williams 1978, Seymour 1982a, 1982b, Thwaite and Bower 1980). Both of these species readily feed on *P. ulmi* and are commonly associated with this pest in the field (e.g., see Croft and McGroarty 1977). *A. fallacis* was released into Tasmanian orchards in 1976 and predators became well established (Williams 1978); however, after overwintering in the groundcover and at the bases of trees, they failed to reestablish on foliage early enough during the following spring to prevent severe leaf damage by *P. ulmi*. Others (e.g., Seymour 1982a, 1982b) noted poor establishment of *A. fallacis* at experimental release sites. In 1978, *T. pyri* was released in this same region and good biological control of *P. ulmi* was attained (Williams 1978).

In the highlands of New South Wales and Queensland, a New Zealand strain of OP/carbamate-resistant *T. pyri* was released and established, and performed well from the very first field season after release (Seymour 1982a, 1982b). In study orchards in both regions, this predator successfully controlled European red mites for five consecutive years. As reported by Seymour (1982a), further trials to expand the distribution of this beneficial were being made by CSIRO. Similarly, Thwaite and Bower (1980) reported promising results in controlling *P. ulmi* in preliminary trials with *T. pyri* near Bathurst, New South Wales. In New South Wales, control of *P. ulmi* by *M. occidentalis* has not been effective, but *T. pyri* provided good control of this pest, especially in the coastal highlands, which have high seasonal rainfall and humidity and the lowest temperatures in the region (Bower 1984).

Another interesting turn in the Australian predatory mite importation story is the introduction (possibly unintentional) of *Phytoseiulus persimilis* (Ridland et al. 1986). This mite was first found near Sydney in 1978 on strawberries, but more recently it has been reported in high numbers in commercial apple and nectarine orchards in Victoria. Populations were found overwintering on weed hosts beneath orchard trees, and later in the season they were found in trees responding to *Tetranychus urticae* populations. The importance of *P. persimilis* in integrated mite management programs in Australia is not yet known, but there is concern about its impact on programs where OP-resistant *M. occidentalis* are used. *P. persimilis* has a very high reproductive rate and is known to overexploit spider mite populations. Other mite predators might then be unable to persist following the decline of prey to very low levels.

In 1979, Readshaw (1979) gave a status report on the biological control project on spider mites in Australian orchards and noted that there appeared to be little potential for improved biological control using native predatory species, which included several predaceous mites and *Stethorus* spp. (He apparently had ruled out importation of OP-resistant strains of *Stethorus punctum* from Pennsylvania, United States; Chapter 14). He stated "there is good reason to expect that orchard mites will be successfully controlled throughout eastern Australia with introduced insecticide-resistant predatory mites (including *M. occidentalis, T. pyri*, and *A. fallacis*)." He cautioned that using pyrethroid (SP) insecticides against other orchard pests would likely cause major mite problems to return to Australia unless SP-resistant predators were found and released. (See Chapters

13 and 20 on problems associated with pyrethroid use and possibilities for developing SP-resistant predatory mites.)

In 1979, CSIRO commissioned an independent economic study of agricultural research projects of the Division of Entomology (Marsden et al. 1980). One aspect of this cost–benefit analysis of applied research covered orchard mite control projects, focusing principally on biological control and importations of insecticide-resistant phytoseiid mites conducted from 1968 to 1975. Major findings were that *M. occidentalis* survived, overwintered, and effectively controlled *T. urticae* under standard commercial spray programs and that this mite had been distributed to all major pome-fruit areas of the country with good success. Marsden et al. (1980) listed some constraints of the integrated mite management programs using OP-resistant predatory mites:

1. Maximum benefit from predatory mites was achieved when acaricides were totally abandoned. This would depend on the continued susceptibility of codling moth to OPs, thus preventing the need for other nonselective insecticides (see problems with SPs discussed earlier; also Chapters 13 and 20).
2. Predatory mites did not always provide complete control of spider mites and acaricidal sprays were occasionally needed to augment predators (once every 3 years on the average).
3. It would take several years for predatory mites to become completely established in all areas of the country. Marsden et al. (1980) anticipated that establishment would near completion by 1981, given moderate efforts by growers and state extension services.
4. Even when establishment was complete, they thought it unlikely that all growers would adopt a reduced acaricide program.
5. Introduced predatory mites in 1979–1980 did not provide similar high levels of biological control of *P. ulmi* in several areas of the country. (See later successes achieved in this regard; Williams 1978, Seymour 1982a.)

In spite of these difficulties, Marsden et al. (1980) estimated the potential saving to growers (and ultimately to the public through lower prices) for the period from 1981 to 2000. Their calculations were based on the difference between conventional chemical control programs requiring an average of two acaricide sprays for mite control per year, and integrated programs projected to require only one spray every 3 years. The net savings at 1975 prices was $86.00 (Australian) per hectare per year. Marsden et al. (1980) estimated the overall area in the country where predatory mites could be used successfully to be 52,000 ha. Although in 1981 only 8667 ha were under integrated mite management, it was projected that in 1986, full implementation in the 52,000 ha would be achieved. Using a 10% inflation rate, the value of the expected benefits from 1981 to the year 2000 was expected to be about $17 million (Australian). (This total referred only to savings on deciduous tree fruits; additional savings would be realized in cotton, strawberries, and ornamentals.) The overall cost–benefit ratio for this research program was estimated at 1:2–4.

For Australian integrated mite management programs in the 1980s, guidelines have been developed for managing spider mites, including both *T. urticae* and *P. ulmi* based on detailed survey procedures, predator–prey thresholds, and selective acaricides (Readshaw et al. 1982, Bower and Thwaite 1982, Field 1982). An extensive program of cooperative implementation by the State Department of Agriculture (in collaboration with CSIRO scientists) has been undertaken to educate growers so that maximum utilization of insecticide-resistant natural enemies can be achieved.

A final note regarding the introduction of predatory mites into Australia concerns a highly publicized and politically astute demonstration project that was carried out in the late 1970s to early 1980s (Seymour 1982a, 1982b). CSIRO released OP-resistant *M. occidentalis* to control *T. urticae* in the National Rose Garden on the Australian Capitol grounds in Canberra. Results were spectacular, with visible differences between prerelease and postrelease plots. The use of these plots for public relations and education of the benefits of biological control research demonstrates how positive attitudes towards this type of research can be fostered (see similar policy-related topics in Chapter 21).

In the early to mid 1980s, city parks administrators (Carmody et al. 1981, Clark and Buckley 1984) reported their experiences from Canberra and Sidney on the biological control of spider mites using *M. occidentalis* in the National Rose Gardens. Predators successfully released in 1981 survived winters until monitoring ceased in 1983. They found that low rates of selective acaricide (cyhexatin) were occasionally necessary to prevent excessive mite damage to these highly visible ornamentals. Their recommendation was that when the pest incidence reached approximately 70% of leaves, the selective spray was necessary and predators would control populations thereafter.

Orchid producers have also benefited from the introduction of predatory mites into trial nurseries in New South Wales. CSIRO has investigated possibilities of using OP-resistant *M. occidentalis* to control spider mites on soybeans, corn, and several greenhouse crops with encouraging results (Seymour 1982a, 1982b).

18.3. INTRODUCTIONS OF MULTIPLE SPECIES INTO NEW ZEALAND

Another successful introduction of insecticide-resistant natural enemies into a foreign country occurred in New Zealand. (See a series of papers entitled, "Integrated Control of Apple Pests in New Zealand," referred to by Penman et al. 1979.) This research was begun after extensive studies indicated that previously established phytoseiids (particularly *T. pyri*) provided only limited biological control under commercial orchard spray regimes in certain areas of the country (Collyer 1964a, Hoyt 1972, Collyer and Geldermalsen 1975; Chapter 17).

To date, insecticide-resistant strains of *A. fallacis*, *M. occidentalis*, and *P. persimilis* have been imported into New Zealand. The introduction, establishment, and successful biological control achieved with each of these species has been augmented by research programs aimed at the management of endemic R strains of *T. pyri* as well (see Section 17.3.4).

Amblyseius fallacis was first introduced into New Zealand using quarantine, mass-rearing, and release procedures described by Thomas and Chapman (1978). Populations were initially obtained from Michigan, United States, in shipments of 100–200 active stages in 1973 and 1975 (Penman et al. 1979). After evaluation of releases for several seasons in different areas of the country, several general conclusions regarding the establishment and effectiveness were reported by Penman et al. (1976).

In single season releases and evaluations when azinphosmethyl was intensively applied for the control of codling moth, OP-resistant *A. fallacis* provided effective control of European red mites when directly placed on foliage (Penman and Chapman 1980). Following an early application of cyhexatin, 300 *A. fallacis* per tree were released into 10-

year old apple trees at Lincoln College near Canterbury, New Zealand, in mid-December of 1976. Predator and prey populations were carefully monitored. European red mites did not exceed 12 per leaf where *A. fallacis* was released. In contrast, nearby trees where no predators were released had over 300 *P. ulmi* per leaf late in the season (Fig. 18.3; Penman and Chapman 1980).

In the season following the release of *A. fallacis*, control of *P. ulmi* was delayed until late in the summer and was generally inadequate (Fig. 18.3). Whereas the pest mite population was virtually unchecked early in the season and achieved very high levels by midsummer, the predator did not appear until midseason and always lagged behind the prey numerically. The reason for the delayed numerical response of predators and the outbreak of spider mites was attributed to maladapted diapause that caused delayed spring activity

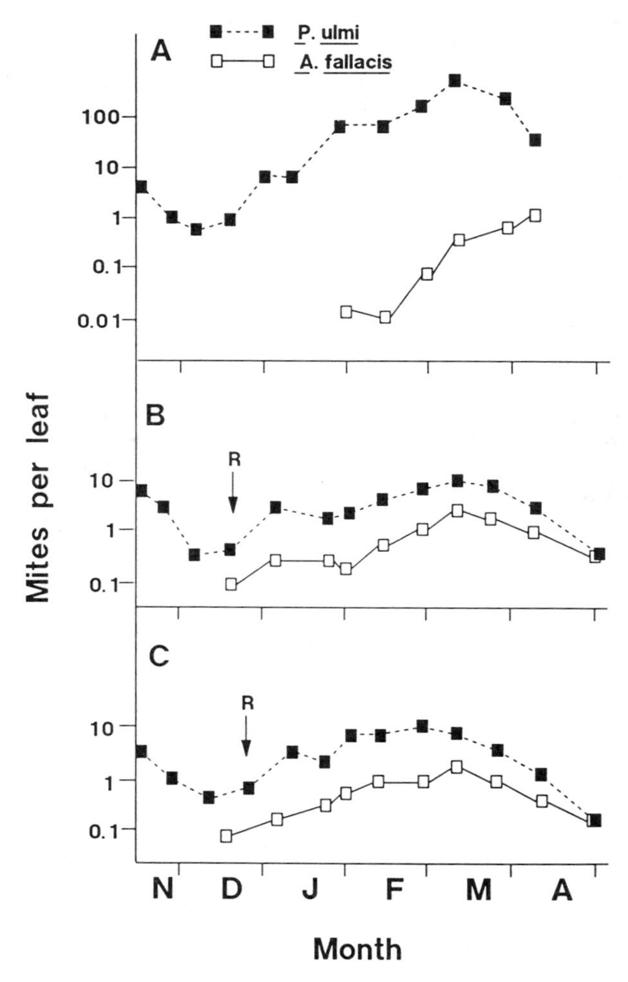

Figure 18.3. Population estimates of *Amblyseius fallacis* following field release into New Zealand apple orchards (after Penman and Chapman 1980).

of predators or perhaps led them to overexploit prey mites earlier in the spring.

A. fallacis failed to establish in plots where native *T. pyri* were already present, and in most cases *P. ulmi* was the principal prey mite. Thomas and Chapman (1978) felt this may have been due to the absence of *T. urticae* as prey for *A. fallacis* rather than to direct competition between the two phytoseiids. In food preference studies in release plots, *A. fallacis* seemed to prefer *T. urticae* over *P. ulmi*. On occasion, selective predation by *A. fallacis* changed the composition of spider mite populations, although *P. ulmi* was eventually controlled provided predators were abundant early in the predator–prey interaction cycle (Thomas and Chapman 1978). The fact that New Zealand orchards do not support high populations of the alternate prey *Aculus schlechtendali* (apple rust mite) may have contributed to the poor performance of *A. fallacis* after releases were made (Penman et al. 1979).

Maintenance of some ground cover beneath apple trees is important in providing food and shelter for *A. fallacis*, since this mite overwinters and develops throughout the growing season in this habitat. In Michigan, *A. fallacis* develops high populations on orchard foliage only in the summer, after moving into the tree from the groundcover (Croft and McGroarty 1977). In contrast, both *T. pyri* and *M. occidentalis* tend to occur primarily in the orchard trees throughout the year, as do spider mites. In tests made in New Zealand, clover was maintained beneath trees where releases of *A. fallacis* were made; however, periodic sampling revealed that few predators, *T. urticae*, or alternate prey were present in the groundcover habitat. Early season releases of *T. urticae* into orchard ground cover resulted in significantly greater populations of *A. fallacis* in the ground cover and their earlier movement into apple trees (Penman and Chapman 1980).

High overwintering mortality of *A. fallacis* was also implicated as contributing to the late buildup of this mite in orchard trees. New Zealand winters are much milder than are winters in the native range of *A. fallacis* (i.e., Michigan, United States). At milder temperatures predators become active, begin feeding, and may even oviposit earlier in the season (Thomas unpubl. data). Predators may then overexploit their prey to such an extent that subsequent predator mortality may be high. Later in the spring, predator densities may be insufficient to control increasing populations of *P. ulmi*. Penman et al. (1979) also suggested that the Michigan strain of *A. fallacis* may have been maladapted to New Zealand conditions due to factors influencing diapause regulation such as photoperiod and temperature.

In New Zealand orchards where *T. urticae* was the primary pest mite, OP-resistant *Metaseiulus occidentalis* from Washington, United States, were initially released in 1976. They provided a similar solution to the spider mite problem as in Australia (Readshaw 1975; see Section 18.2); however, *T. urticae* is a major apple pest only in central Oetago. In this area, *M. occidentalis* has been established and has provided effective control over several seasons (Penman et al. 1979).

Near Canterbury, New Zealand, *M. occidentalis* was released into strawberry plots in 1976–1977. In the absence of predators, two-spotted mites reached levels of 21.4 per leaflet, whereas in release plots they reached only 9.1 per leaflet before declining due to predation. Subsequent reports (e.g., Butcher and Penman 1983) indicated that both *M. occidentalis* and *A. fallacis* had established and were being managed in IPM programs.

Introductions of *Phytoseiulus persimilis* into New Zealand had somewhat different results than those of *A. fallacis* and *M. occidentalis*. *P. persimilis* was initially introduced to control *T. urticae* on small fruits where it has had considerable impact (Walker et al. 1981, Charles et al. 1985). For some time after the initial introduction, its spread and distribution were not apparent. More recently, *P. persimilis* has appeared in mid to late season in apple trees of northern, subtropical parts of the North Island, such as Waikato and Bay of Plenty. While most often associated with *T. urticae*, *P. persimilis* also feeds readily on European red mite (Wearing and Proffitt 1982). The insecticide resistance status of these mites has not been extensively documented to date, although considering their occurrence in orchard trees during periods of intensive spraying, some degree of resistance to pesticides must be present in this species. Genetic improvement of SP resistance in this species has been reported from New Zealand by Markwick (1984, 1986). SP resistance would expand the use of this biological control agent in green houses as well as on a number of field crops. (See further discussion of genetic improvement of this species in Chapter 19.)

Finally, because of their extensive efforts in establishment, genetic improvement, and the management of insecticide-resistant phytoseiid mites, New Zealand has become a major exporter of these populations to other areas of the world. The transfer of strains from New Zealand to Australia and the United Kingdom is cited elsewhere in this chapter. As noted in a recent DSIR report (DSIR 1985), considerable demand has been made by researchers to obtain OP-resistant and more recently SP-resistant strains of *T. pyri* for importation into many countries (Chapters 19 and 20).

18.4. *PHYTOSEIULUS PERSIMILIS* IN GLASSHOUSES

Serendipitous use of *P. persimilis* in New Zealand orchards and strawberries represents only a small portion of the many uses of this mite, which has been transferred from its native Chile throughout the world for the management of glasshouse spider mites. The use of insecticide-resistant strains of *P. persimilis* is only part of its biological control success story, since relatively susceptible strains have been used repeatedly in inundative-release programs for the control of spider mites after broad-spectrum pesticides decimated predators. International transfer, introduction, and utilization are common with multiple-resistant strains (Table 14.1) which have acquired their pesticide tolerances since leaving their native home, especially in western Europe (Anon. 1977, Hassan 1982a, 1982b, Fournier et al. 1987a) and eastern Europe (Storozhkov et al. 1977). In these cases, establishment is less critical, since extensive mass-rearing and release programs have been developed for *P. persimilis* in closed glasshouse environments (Gould et al. 1969, Markkula and Tiittanen 1976b, 1980).

18.5. THE UNITED KINGDOM IMPORTS A BETTER NATIVE *TYPHLODROMUS PYRI*

Generally, most scientists consider *T. pyri* to be native to western Europe, and it was probably introduced into North America and New Zealand sometime in the last century in connection with apple culture (Croft and Strickler 1983). In British orchards and many other areas of western Europe, *T. pyri* is the principal species associated with *P. ulmi*; however, in Europe, due to the cool temperatures and other factors, diseases are generally a greater problem than insects (Gruys 1982; Chapter 13). Codling moth and various leafrollers are the principal species requiring limited OP applications. Because low levels of OPs are used in Europe and *T. pyri* acquires azinphosmethyl resistance relatively slowly (Penman et al. 1976; Chapter 11), *T. pyri* populations typically have exhibited low levels of resistance in Europe compared to other more intensively sprayed fruit-growing areas of the world (e.g., New York, New Zealand).

The introduction of New Zealand strains of *T. pyri* into the United Kingdom where this species is native does not fit the classification proposed initially for this chapter; however, it is considered more typical of a new species introduction than an endemic transfer of a resistant strain within its native range (Chapter 17). Regardless of classification, this example of a managed, insecticide-resistant beneficial is unique from other cases previously discussed.

Cranham and Solomon (1978) first reported plans made in 1977 in the United Kingdom to: 1) study the toxicity of individual pesticides to native phytoseiids including *T. pyri*, 2) assess the value of introducing OP-resistant *T. pyri* from New Zealand in a trial orchard, and 3) conduct more thorough searches for OP resistance in native populations of *T. pyri*. Results from both literature surveys and field trials to assess the effects of fungicides, insecticides, and acaricides on native *T. pyri* were reported in Cranham and Solomon (1981). They concluded that native strains were less resistant than New Zealand strains of *T. pyri* by a factor of about 5- to 10-fold, and that the introduction and management of the exotic strain should be attempted.

AliNiazee and Cranham (1980) reported on initial introductions of an OP-resistant strain of this species from New Zealand into a research orchard at the East Malling Station in 1977. Field observations of survival after azinphosmethyl sprays indicated that it eventually dispersed throughout the whole block in the summer of 1978. The authors reasoned that it was likely that hybridization with the susceptible indigenous strain had occurred since no seasonal asynchronies with spider mite populations were observed. (Additional study of the fate of the New Zealand strain using classical genetic or electrophoretic techniques was not attempted.) In 1979, azinphosmethyl sprays reduced predator populations in release plots somewhat, but not to previously observed levels. This may have reflected that both S and R predators were still present in treatment plots.

More recently, native *T. pyri* as resistant as New Zealand strains to azinphosmethyl, parathion, and carbaryl were found in English orchards (Kapetanakis and Cranham 1983, Cranham et al. 1983). Populations were obtained from areas quite distant from those where New Zealand populations were released. Since the early 1980s, exploitation of these endemic strains has been pursued in the United Kingdom rather than further distributing the New Zealand strain.

Observations in southeastern England suggest that resistance has developed in native *T. pyri* and these predators have colonized many orchards where pesticide programs are based on selective products (Solomon and Fitzgerald 1984). In areas where colonization has not occurred naturally, insecticide-resistant predators have been artificially introduced using hedgerows or small seedling plantings as nurseries for producing *P. ulmi* and the beneficial. Cuttings taken from these nurseries are distributed to orchards where they are placed in individual trees (Solomon and Easterbrook 1983). Predators disperse and rapidly colonize trees where biological control can be obtained within the season after release. A number of research papers detailing selective insecticide programs for these mites have been published (e.g., Easterbrook 1984). Five-year evaluations of programs based on selective pesticides and insecticide-resistant predators compared to standard spray programs using broad-spectrum pesticides have shown that no acaricide applications were required in the IPM plots in the last 3 years of the trial, whereas multiple applications were necessary on an annual basis in the standard plots (Easterbrook et al. 1985).

18.6. IMPORTS AND NATIVES IN SOUTH CENTRAL EUROPE

As noted in Chapter 17 (Section 17.3.3.), this example presently involves three phytoseiid species, which together have complementary characteristics as biological control agents of *P. ulmi*, *T. urticae*, *Eutetranychus carpini*, and *A. schlechtendali* on apple and grape (Baillod 1982, Baillod et al. 1986, Genini and Baillod 1987). OP resistance in *A. andersoni* was first reported from northern Italy by Gambaro (1975, 1986). Populations of this species along with endemic and exotic strains of OP/carbamate-resistant *T. pyri* have been employed in Switzerland, Italy, Spain, France, Austria, and Germany (Genini et al. 1983, Baillod and Guignard 1984, Caccia et al. 1985, Corino and Ruaro 1986, Baillod 1986). Reports from a number of areas indicate that not all populations of this species are resistant, even after many years of OP and carbamate use (e.g., Boller 1985). The more recent discovery of resistance to parathion in populations of *A. aberrans* from vineyards (Corino et al. 1986) provides an additional source of improved predators that can be transferred throughout the region and utilized in mite management programs, both on apple and grape.

A summary of the resistance status of these mites is indicated in Table 18.4. A wide range of resistance to several classes of insecticides and fungicides has been documented or is suspected due to the survival of predators following applications of these compounds in certain fields. Transfers of predators among countries in the region is widely practiced (Gambaro 1986). Field techniques for the establishment, colonization, and transfer have been tested extensively (e.g., Caccia et al. 1985). In many cases, displacement of

Table 18.4. Insecticide-Resistant Strains of Three Species of Phytoseiid Mites Used in Integrated Mite Management Programs in European Apple and Grape Production (from Baillod 1986)

Phytoseiid Species	Country					
	Germany	Austria	France	Spain	Italy	Switzerland
			Insecticides (Organophosphates + Carbamates)			
A. abberans					o P[7] o R[7]	
A. andersoni				+P	o+P[6] o+R[4]	o+P[6] o+R[4]
T. pyri	o P o R[7]	+P	o P		o+P[6]	+R[5] o+P[8]
			Fungicides (Sulfur + Thiocarbamates)			
A. abberans			+P[2]		o+P[3]	
A. andersoni					o+P[3]	
T. pyri	o P[1]	+P			o+P[3]	

Key: + = in apple; o = in grape; R = resistance proven; P = resistance probable.
[1] = manzeb
[2] = sulfur
[3] = thiocarbamates (in general)
[4] = azinphosmethyl, tetrachlorvinphos
[5] = azinphosmethyl, phosmet
[6] = azinphosmethyl, parathion, omethoate, dimethoate, phosalone, acephate
[7] = parathion
[8] = parathion, tetrachlorvinphos, phosalone.

susceptible native strains of these species and other phytoseiids occurs. Species introduction strategies vary greatly by species and by location. For example, a species with a more limited dispersal capacity like *T. pyri* must be placed in 1 out of every 3 trees, but a more mobile species like *A. andersoni* can be released more sparingly.

Conditions under which different predatory mite species are most effective have been characterized so that predators can be introduced to complement one another. For example, *T. pyri* is particularly effective in regulating low-density populations, but is less effective in suppressing high-density outbreak populations of spider mites (Baillod 1986). In contrast, *A. andersoni* is more efficient at higher prey levels, often completely controlling spider mites due to its rapid numerical response (Baillod 1986). Gambaro (1986) reports that *T. pyri* can provide long-term regulation of phytophagous mites at endemic densities of 1 per 300–400 leaves for periods of 10–12 years without the use of supplemental acaricides.

18.7. OTHER CASES OF EXOTIC INTRODUCTIONS

In addition to introductions cited thus far, a number of other insecticide-resistant natural enemies have been transferred beyond their native ranges. In many cases, it is too early to evaluate their impact. For example, only casual notes on the importation of *Amblyseius chilenensis* from South America into France have been reported (Fournier et al. 1985b). This species has many biological and resistance characteristics that are similar to *A. fallacis* (Croft unpubl.), and it would appear to be an equally valuable species to consider for importation into areas where insecticide-resistant predatory mites do not already exist.

Other introductions of insecticide-resistant natural enemies have been less well documented. For example, OP-resistant *A. fallacis* and *M. occidentalis* have been sent to almost every continent for experimental evaluation and release. Many scientists have provided source colonies of resistant phytoseiids for importation on a regular basis. As noted earlier, New Zealand scientists are involved in shipments of *T. pyri* and *P. persimilis* to many other areas of the world (DSIR 1985). Some lesser publicized introductions include those in the USSR (*M. occidentalis*, Chubinnishvili et al. 1982) and China (*M. occidentalis*, Naixin 1983; *A. fallacis*, Zhang and Kong 1985, 1986).

In the Soviet Union, *M. occidentalis* obtained from Michigan, United States, was initially released in 1978 into vineyards after laboratory mass-rearing on soybeans infested with *T. urticae* (Chubinnishvili et al. 1982). Mites were released at levels of 100 per plant, and subsequent observations indicated that predators survived fungicide and OP sprays. Control of pest mites was excellent with no noticeable damage, whereas mite damage was appreciable in nonrelease control plots by the end of the season. In 1979, mites successfully overwintered, resumed feeding, and oviposited. Chubinnishvili et al. (1982) concluded that resistant *M. occidentalis* could be successfully released and established in Georgia and other regions of the USSR.

The successful importation of insecticide-resistant strains of *M. occidentalis* and *A. fallacis* and susceptible *A. californicus* into temperate-zone Taiwanese pear orchards to control populations of *P. ulmi* and *T. urticae* has been reported by Lo (1986) and Lo et al. (1986). Among the three species, *A. fallacis* was the most effective predator. It was established in 1985 and recovered from several weedy hosts of *T. urticae* in 1986. *A. fallacis* effectively controlled spider mites and is now selectively conserved in an integrated mite management program.

Only a few apple production regions probably remain where the potential for importing insecticide-resistant phytoseiid mites has not been explored (Croft and Strickler

1983; e.g., South Africa; Giliomee pers. comm.). Resistance may already be present in endemic strains of phytoseiids in some of these areas. Pesticide-resistant phytoseiid mites are now used for integrated mite management on tree fruit crops, worldwide. Their value on strawberries, almonds, grapes, and hops is increasing, and their exploitation will very likely expand in the future to include other crops on which spider mites are pests.

18.8. SUMMARY OF EXOTIC INTRODUCTIONS

Exploitation of resistance in natural enemy species is a specialized type of classical biological control. Classical introduction of biological control agents is a well-developed discipline within the field of biological control. Principles of exploration, culture, mass-rearing, release, and evaluation of exotic biological control agents are relevant to the importation of resistant natural enemies. Most of the principles relating to the management of newly introduced insecticide-resistant phytoseiid mites are similar to these discussed for endemic strains of these mites in Chapter 17.

Introductions of exotic insecticide-resistant natural enemies are a short-term alternative to the use of physiologically selective pesticides (Chapter 9). This generalization applies to the management of endemic or genetically improved predators and parasitoids as well (Chapters 17 and 19–21). Exploitation of insecticide-resistant natural enemies will probably continue to provide considerable benefits on perennial crops such as deciduous tree fruits where heavy pesticide use is the norm. The management of pesticide-resistant natural enemies on these crops is widely accepted and is not likely to be replaced by a complete reliance on selective pesticides in the near future for several reasons. First, it is unlikely that selective pesticides will be developed quickly and second, biological control of many secondary pests of tree fruits is a superior alternative to selective chemical control.

Establishing and managing pesticide-resistant natural enemies places a focus on chemical management practices. As discussed in Chapter 17, a chemical environment must be created in which the released, resistant predators would be favored over an endemic, susceptible strain. Croft (1976) emphasized the need to maintain some selection pressure on an introduced, pesticide-resistant natural enemy. This is necessary to ensure that competition by native species does not diminish the opportunity for the new species to establish itself in the new environment.

How much selection pressure is needed to maintain an adequate level of resistance in released populations so that the survival of pesticide applications used to control other pests is certain? Little research has been conducted on this topic, although some empirical information is available. Generally, it is thought that once resistance develops in a population, lower selection intensity will maintain the resistance factor. With exotic natural enemies, hybridization with susceptible individuals is less of a problem than with endemic species; however, genetic instability within the resistance genome can cause reversion of resistance if selection is not maintained above some threshold.

The maintenance and management of resistance in introduced natural enemies requires additional study. To a considerable extent this research can be done empirically; however, a better understanding of the genetic stability of resistance in the absence of hybridization would probably lead to improved management of these factors in the field.

A final point relates more to laboratory culture and maintenance of insecticide-resistant beneficials as source populations for scientific groups interested in exotic importations. In the same way that biological control insectaries are commonly maintained in a number of countries, such facilities should be developed for resistant strains of natural enemies whether resistance is developed in the field or through artificial, laboratory selection. These unique strains have similar requirements for quarantine and colonization as other organisms used in biological control, but with an added degree of complexity—the resistance factor. Quality control of the resistance and maintenance of genetic variability have not been addressed adequately. Problems such as inbreeding, strain vigor, and genetic bottlenecks that may plague the colonization of biological control agents in general are also detrimental to resistant strains of natural enemies. Methods of rearing and maintenance will receive increasing attention as genetic improvement and genetic engineering (Chapter 19) of beneficial natural enemies become more common.

19

HISTORY, METHODS, AND CONSTRAINTS OF GENETIC IMPROVEMENT

This chapter was coauthored by M. E. Whalon, Department of Entomology, Michigan State University, East Lansing, Michigan, and R. T. Roush, Department of Entomology, Cornell University, Ithaca, New York.

19.1. INTRODUCTION

Genetic improvement is the alteration of genetic characteristics of a species through hybridization, artificial selection, or genetic engineering to suit human needs (Hoy 1985). Genetic improvement has long been recognized as a means to increase the usefulness of beneficial arthropods, including biological control agents (Malley 1916, DeBach 1958, Hoy 1976). Only recently, however, has research been devoted specifically to manipulating the attributes of arthropod natural enemies. Attempts have been made to select for increased fertility (Wilkes 1947, Voroshilov and Kolmakova 1977), arrhenotoky (Legner 1987a), thelytoky (Hoy and Cave 1986), increased oviposition (Legner 1987b), changes in diapause (Hoy 1984b, Gilkeson and Hill 1986a, 1986b), heat tolerance (White et al. 1970, Voroshilov 1979), and pesticide resistance.

Hoy (1982, 1984a, 1985) and Legner and Warkentin (1985) have published broad reviews on the genetic improvement of biological control agents. Scientists in the USSR have long advocated genetic improvement to enhance the effectiveness of natural enemies (Gusev and Voroshilov 1977, Voroshilov and Kolmakova 1977, Voroshilov 1979). Recent reviews on the genetic improvement of natural enemies have emphasized gene manipulation techniques (Beckendorf and Hoy 1985, Hoy 1987, Berge et al. 1986).

In this chapter, the history of genetic improvement using hybridization and artificial selection to enhance insecticide resistance in natural enemies is reviewed. Improvements in biological control are briefly discussed. We focus mostly on artificial selection used over many generations. A substantial amount of research using this approach has been reported. Constraints encountered in genetic improvement are addressed.

Some general concepts and methods of genetic improvement are outlined later in the chapter. Prospects for using recombinant DNA technology are considered. Practical results using recombinant DNA may be years away (Beckendorf and Hoy 1985, Berge et al. 1986, Hoy 1987), but the potential benefits of this technology merit discussion.

19.2. GENETIC IMPROVEMENT OF PESTICIDE RESISTANCE IN NATURAL ENEMIES

Few studies have documented the use of hybridization alone to genetically improve pesticide resistance in arthropods, and almost all have involved phytoseiid mites (Section 19.2.1). Artificial selection has been much more commonly employed with or without hybridization. *Successful* and *unsuccessful* cases of genetic improvement are listed in Table 19.1.[1] Each project has been rated as to whether high levels of resistance were achieved and whether releases of the improved natural enemy were of value in the field. Where available, information is included on the number of selections used in each trial.

Of the 34 cases cited in Table 19.1, nearly equal numbers involve predators and parasitoids; however, in only a few instances has the level of resistance achieved with parasitoids been sufficient to be of use in the field. No field releases of genetically improved parasitoids have been reported, but several attempts are expected soon (Section 19.2.3). Levels of resistance in predators have been of practical benefit in two-thirds of the cases, and about 50% of the improved strains have been released in the field. In their survey

[1] There have been a number of unpublished, unsuccessful attempts made to genetically improve insecticide resistance in natural enemies. Unfortunately, these results are often considered unpublishable. Unsuccessful results should be documented to guide future genetic improvement research.

of resistance in natural enemies, Tabashnik and Johnson (1988) also noted differences in the resistance levels reported for predators versus parasitoids: LC_{50} or LD_{50} values were considerably higher for predaceous forms. Selected trials from Table 19.1 are discussed below.

19.2.1. Genetic Improvement of Phytoseiid Mites

The earliest artificial selection experiments were conducted on parasitoids (Table 19.1; also Section 19.2.3), but neither high nor stable resistance was obtained in these experiments (Pielou and Glasser 1952, Robertson 1957, Adams and Cross 1967, Kot et al. 1971, 1975, Abdelrahman 1973; see also Chapter 14). The first *successful* case of genetic improvement in which a high level of pesticide resistance was obtained involved a predatory phytoseiid mite. However, the early record of success with phytoseiids was mixed. Genetic improvement of pesticide resistance has been most successful with *Amblyseius fallacis, Metaseiulus occidentalis, Phytoseiulus persimilis*, and *Typhlodromus pyri*.

19.2.1.1. Amblyseius fallacis. An early genetic improvement test with *A. fallacis* was reported by Croft and Meyer (1973), Meyer (1975), and Croft and Hoying (1975) using hybridization methods. Starting with organophosphate (OP)-resistant (100-fold) and carbaryl-resistant (25- to 77-fold) strains which had developed independently in the field, they hybridized an OP/carbamate-resistant strain and attempted to maintain it in the field. Alternating selections of azinphosmethyl and carbaryl were used in the laboratory and field after predators were released. Relatively stable resistance to both carbaryl and azinphosmethyl was maintained for 10–25 generations in the laboratory. In the field, OP resistance persisted, while carbaryl resistance declined rapidly. Azinphosmethyl resistance in the hybrid *A. fallacis* was monogenic and semidominant (Croft and Meyer 1973, Meyer 1975, Croft and Hoying 1975). The genetic basis of carbaryl resistance in *A. fallacis* is not known for any strain. Thus, a definitive explanation for the failure to maintain the azinphosmethyl/carbaryl-resistant strain in the field was not possible. Carbaryl resistance in *A. fallacis* may be polygenic, considering the time required for resistance development (Chapter 14) and the selection intensity needed to maintain moderate levels in the field (Croft and Meyer 1973). If carbaryl resistance is monogenic, it must be highly unstable (Meyer 1975). Establishment of the OP/carbamate-resistant strain of *A. fallacis* was never conclusively documented.

Genetic improvement of pyrethroid (SP) resistance in *A. fallacis* was achieved by artificial selection using a DDT/OP/SP-resistant strain (see resistance attributes in Fig. 4.2). This strain was created by hybridizing resistant strains which had been collected from apple orchards in the midwestern and eastern United States (Strickler and Croft 1981, 1982, Whalon et al. 1982a, Croft 1983). One source strain from Geneva, N.Y., had already developed a 14.7-fold level of resistance to permethrin (Strickler and Croft 1981; Table 19.1).

Hybrid and field-collected strains of *A. fallacis* were selected in a greenhouse to obtain higher SP resistance. Both strains were selected with permethrin, and a third strain received alternating selections of permethrin and azinphosmethyl. The permethrin-selected hybridized strain (GH-1) attained 64-fold resistance after 12 selections over 36 months (Fig. 19.1), but no further increase in resistance was obtained over the next 10 selections. Hybrid mites exposed to alternating SP/OP selection (GH-2) achieved a similar level of SP resistance in 18–20 selections (10 permethrin selections) as the GH-1

Table 19.1. Attempts at Genetic Improvement of Arthropod Natural Enemies Using Artificial Selection Techniques

Group/Species	Family	Pesticide	Fold R	Selec-tions	Success Indices[1]	References
PARASITOIDS						
Aphytis melinus	Aphelinidae	Malathion	4	8	N, N	Abdelrahman 1973,
		Carbaryl	5–20	10	Y, N	Rosenheim & Hoy 1988
		Chlorpyrifos	<2	8	N, N	
		Dimethoate	2	8	N, N	
		Malathion	<2	10	N, N	
		Methidathion	2	8	N, N	
Aphytis holoxanthus	Aphelinidae	Malathion	Nil	30	N, N	Havron 1983, Havron et al.
		Malathion	Nil	25	N, N	1987a, 1987b
		Azinphosmethyl	15–30	15	Y, N	Havron & Rosen 1988
Trioxys pallidus	Aphelinidae	Azinphosmethyl	8	14	N, N	Hoy & Cave 1988
Encarsia formosa	Aphelinidae	Ethyl-parathion	3	21	N, N	Delorme et al. 1984, 1985
		Deltamethrin	4	21	N, N	
Encarsia formosa	Aphelinidae	Lindane	3	9	N, N	Walker & Thurling 1984
Trichogramma evanescens	Trichogrammatidae	Demeton-methyl	22	34	N, N	Kot et al. 1971
Trichogramma japonicum	Trichogrammatidae	Several compounds	Nil	5 yr	N, N	Li-ying et al. 1988
Trichogramma confusum	Trichogrammatidae	Several compounds	Nil	5 yr	N, N	Li-ying et al. 1988
Bracon mellitor	Braconidae	DDT	4	5–7	N, N	Adams & Cross 1967
		Carbaryl	4		N, N	
		Parathion	4		N, N	
		DDT/toxaphene	9		N, N	
Macrocentrus ancylivorus	Braconidae	DDT	12	71	N, N	Pielou & Glasser 1951, Robertson 1957
PREDATORS						
Amblyseius fallacis	Phytoseiidae	Permethrin	66	12	Y, Y	Strickler & Croft 1982
Amblyseius longispinosus	Phytoseiidae	Carbaryl	—	—	?, ?	Lo et al. 1984
		Dimethoate	—	—	?, ?	
		Carbaryl/dimethoate	—	—	?, ?	

Amblyseius nicholsi	Phytoseiidae	Phosmet	19	33	?, N	Huang et al. 1987
Amblyseius	Phytoseiidae	Dimethoate	36	7	?, ?	Ke 1987
pseudolongispinosus			19	10	?, ?	
			15	13	?, ?	
Metaseiulus occidentalis	Phytoseiidae	Carbaryl	—[2]	18	Y, Y	Roush & Hoy 1981a
		Methomyl	12	12	N, N	Roush 1979
		Dimethoate	11	—	N, N	Roush 1979
		Permethrin	10	18	Y, Y	Hoy & Knop 1981
		Permethrin	13	18	Y, N	Ouyang & Hoy 1988
		Bioresmethrin	—	—	Y, Y	van de Klashorst 1984
		Abamectin	4	20	Y, N	Ouyang & Hoy 1988
Phytoseiulus persimilis	Phytoseiidae	DDVP	4	5	N, N	Avella et al. 1985
		Rotenone	0	5	N, N	
		Methidathion	120	50	Y, Y	Fournier 1981
		Deltamethrin	10	7	N, N	Fournier et al. 1985a
		Methidathion	200	40	Y, Y	
		Parathion	100	8	Y, Y	Schulten & van de Klashorst 1974
						Schulten et al. 1976
		Ethion	4	9	?, ?	Beglyarov et al. 1978
		Pirimicarb	> 10	?	Y, Y	Hassan 1980
		Pyrazophos	?	?	?, ?	Koenig & Hassan 1986
		Pirimiphos-methyl	?	?	?, ?	
		Bioresmethrin	2	50	N, N	Manttari 1980
		Fenvalerate	Nil	14	N, N	Markwick 1986
Typhlodromus pyri	Phytoseiidae	Cypermethrin	10	6	Y, Y	Markwick 1986
		Cypermethrin	36	35	Y, Y	Markwick 1988
Chrysoperla carnea	Chrysopidae	Carbaryl	3	4	Y, N	Grafton-Cardwell & Hoy 1986a, 1986b

[1] Resistance level sufficient to provide for natural enemy survival in the field at recommended rate use for pest control (Yes, No); field releases of genetically improved strain made (Yes, No).
[2] Although resistance levels to carbaryl were undefined, there was more than 100-fold cross resistance to propoxur (Roush and Plapp 1982).

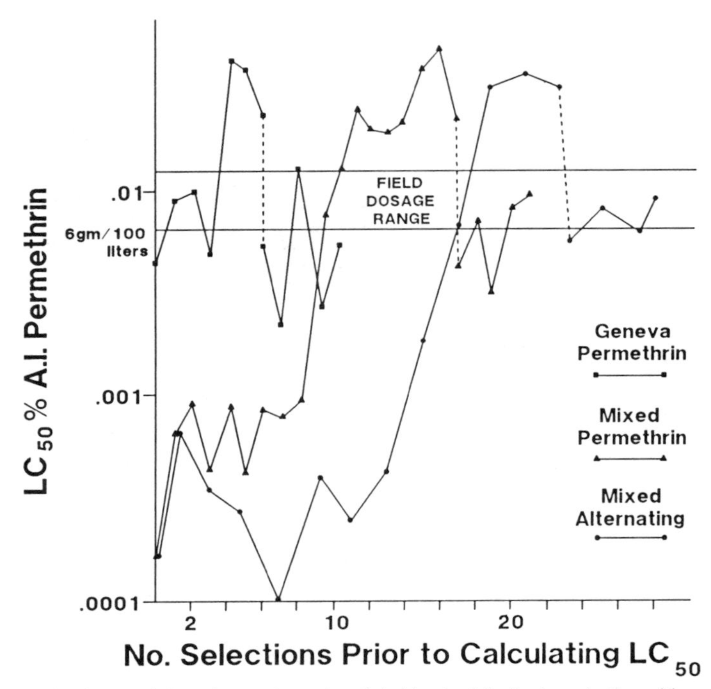

Figure 19.1. Levels of permethrin resistance in strains of *Amblyseius fallacis* after selection with permethrin and alternating permethrin/azinphosmethyl in greenhouse experiments (after Strickler and Croft 1982).

strain. The moderate level of permethrin resistance which was already present in the field-collected Geneva strain did not increase after 10 selections (Fig. 19.1; Strickler and Croft 1982). In all three treatments, previously established azinphosmethyl resistance was maintained (see Fig. 19.1; Strickler and Croft 1982).

The GH-1 hybrid, Geneva field-collected, and susceptible strains of *A. fallacis* were released in an apple orchard at rates of 2500/tree (Table 19.2). Predator and prey mites were sampled after sprays of permethrin and fenvalerate were applied (Whalon et al. 1982a). Prior to releases, no predatory mites were present in the orchard. The first application of permethrin at half the field rate greatly reduced the susceptible predatory mite population (Table 19.2). Resistant predatory mites were generally unaffected by this spray. A second permethrin application at the full field rate virtually eliminated all susceptible predatory mites. This application reduced GH-1 and Geneva strains almost equally to low levels. When fenvalerate was applied at the recommended field rate in September (Table 19.2), the GH-1 strain survived, but the Geneva strain was all but eliminated.

As illustrated in Fig. 15.1, electrophoretic carboxylesterase banding provided another means to monitor populations of released SP-resistant *A. fallacis* (Whalon et al. 1982a, Croft and Whalon 1983, Croft 1983). Tests confirmed establishment following applications of permethrin. Indigenous predators were not detected in release trees after late June, but were readily collected from surrounding trees. Predators found in the check and susceptible release trees (Table 19.2) in July and August had carboxylesterase bands

Table 19.2. Densities of *Amblyseius fallacis* Present in Plots Following Release of Genetically Improved, Pyrethroid-Resistant Strains into a Commercial Apple Orchard in Michigan (after Whalon et al. 1982a)

Date	Action Taken	Release Plots			
		GH-1	Geneva	Susceptible	Indigenous[1]
5/29	Permethrin sprayed (1780 ml/ha)				
6/15	Fenvalerate sprayed (2500 ml/ha)				
6/21	Permethrin sprayed (1780 ml/ha)				
6/21	Sample	0.00	0.00	0.00	0.00
6/28	Sample	0.00	0.00	0.00	0.00
6/28	Release of GH-1 strain into plots				
7/3	Release of Geneva and susceptible strains into plots				
7/10	Sample	0.85	0.82	0.42	0.0
7/16	Sample	1.43	1.76	2.60	0.0
7/23	Permethrin sprayed (200 ml/ha)				
7/25	Sample	1.85	0.59	0.22	0.01
8/4	Sample	1.33	1.73	0.16	0.00
8/7	Permethrin sprayed (780 ml/ha)				
8/11	Sample	0.86	0.63	0.00	0.41
8/27	Sample	0.49	0.43	0.21	0.17
9/3	Fenvalerate sprayed (1482 ml/ha)				
9/12	Sample	0.59	0.16	0.11	0.11
9/22	Sample	0.63	0.05	0.01	0.00

[1] No releases were made into indigenous plots; however, samplings for survival of indigenous or migrant mites were made on the dates specified.

that were characteristic of the GH-1 or Geneva strains. These mites probably dispersed from nearby release trees. Several predatory mites exhibited uncharacteristic banding patterns and were thought to be either long-range dispersers or hybrids.

In the second season after release, predators were monitored by leaf sampling, carboxylesterase evaluations, and dosage–mortality assessments (Croft and Whalon 1983). One pyrethroid application was made to the entire orchard early in the season. In the spring, the LC_{50} of predators in the orchard was lower than that of the original released population, but was still 10- to 20-fold higher than the LC_{50} of a susceptible strain (Table 19.3). Data suggested that SP resistance had been affected by hybridization with susceptible mites. By midseason, predators exhibited a higher LC_{50} to permethrin than the resistant strains that had been released (Table 19.3). Carboxylesterase banding patterns of field-collected *A. fallacis* again confirmed the presence of the released resistant mites in the orchard.

In companion laboratory and greenhouse studies, the stability of SP resistance in *A.*

Table 19.3. LC_{50} Values for *Amblyseius fallacis* Released in 1980 and Subsequently Collected in 1981 from a Commercial Apple Orchard Treated with Pyrethroid Insecticides (after Croft and Whalon 1983)

Strain (Date of Collection)	LC_{50}[1]	CI (95%)	Slope	Fold Resistance Level
RELEASED STRAINS (1980)				
Resistant	0.0090	0.0070–0.011	1.94	53
Susceptible	0.00017	0.00009–0.00025	1.31	1
Indigenous	0.0003	0.00011–0.00057	1.73	2
FIELD-COLLECTED STRAINS (1981)				
5/8	0.0034	0.0024–0.0043	1.96	20
6/16	0.16	0.013–0.024	1.25	94
9/5	0.00077	0.001–0.0005	1.08	4

[1] Compared to the susceptible released strain.
[2] Percent active ingredient permethrin in water.

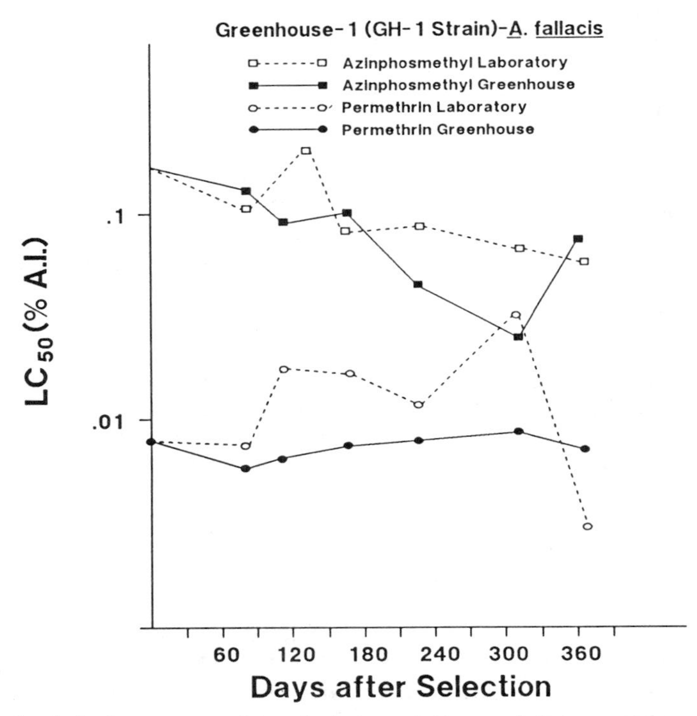

Figure 19.2. Levels of resistance to permethrin and azinphosmethyl in a genetically improved, permethrin-resistant strain of *Amblyseius fallacis* left unselected for one year in a greenhouse or laboratory environment (after Croft 1983).

fallacis was determined under conditions of selection versus no selection and immigration of susceptible mites versus no immigration (Croft and Whalon 1983). In the absence of selection, resistance to both permethrin and azinphosmethyl remained high for 25 generations or about one year (Fig. 19.2). Data confirmed that SP resistance in *A. fallacis* would be reasonably stable provided that hybridization with susceptible strains did not occur.

It was concluded that at least two selections per season would be necessary to maintain the SP resistance in field populations of *A. fallacis* (Croft and Hoyt 1978, Croft 1982; see Chapter 17). In order to prevent the development of resistance in the codling moth *Cydia pomonella*, more extensive use of pyrethroids was not recommended (Croft and Hoyt 1978). One pyrethroid spray was recommended to control early season apple pests and another was recommended after predator–prey mite interactions occurred in August. The timing of these sprays would bracket the influx of susceptible predators (Johnson and Croft 1981), providing maximum selection of resistant predatory mites without inducing SP resistance in orchard pests. (See discussion of simultaneous resistance management in apple natural enemies and pests in Chapter 20.)

More recently, Strickler et al. (1987) reported on the status of *A. fallacis* at the original Michigan release site. Since 1980, SP resistance has stabilized in predators as pyrethroids have become more widely used in the area. Whether this stability originated with the initial releases of resistant predators or whether it has arisen independently due to selection in the field is unknown (Strickler et al. 1987). Electrophoretic, toxicological, and genetic studies are needed to determine the origins of SP resistance in this population.

19.2.1.2. Metaseiulus occidentalis. More genetic improvement research has been conducted on *M. occidentalis* than on any other natural enemy. Most attempts have involved artificial selection (Hoy 1982, 1985). However, an early project involved field hybridization and transfer of OP resistance from a Washington strain into a susceptible population in California apple orchards (Croft and Barnes 1971, 1972; see also Section 17.5 and Fig. 17.4). This case of genetic improvement was fortuitous, since it was not clear whether the resistant strain from Washington would be established (Chapter 17) or whether resistance would be transferred into the local population by hybridization. Following hybridization studies by Croft and Jeppson (1970), releases of both native-susceptible and OP-resistant *M. occidentalis* from Washington were made into individual orchard trees (Croft and Barnes 1971). After 2 years, mites persisting in plots that had been treated with OPs were morphologically similar to susceptible native mites, thus indicating that resistance had been transferred into the native strain (Croft and Barnes 1972).

Genetic improvement of *M. occidentalis* using artificial selection began in the late 1970s. Mixed hybrid colonies collected from 18 California tree fruit, nut, and vineyard sites were selected for resistance to carbaryl (Roush and Hoy 1981a). A high level of resistance was obtained after 15–19 selections on treated pinto bean foliage and filter paper disks (Roush and Hoy 1981a). Cross resistance to benomyl and propoxur was present after selection (Table 14.2). Although simultaneous selection with OPs was not maintained, prior levels of OP resistance persisted (Roush and Plapp 1982). Because OP resistance is monogenic, such stability is not surprising (Table 15.5). Roush and Hoy (1981b) reported high vigor and no reproductive incompatibilities in the carbaryl/OP-resistant strain (Chapter 16, Section 16.2.2).

The carbaryl/OP-resistant strain was initially released into California almonds in 1979 for control of *T. urticae* (Roush and Hoy 1981b). Almonds are treated with carbaryl to control lepidopteran pests. As determined by population sampling, predators successfully established, controlled *T. urticae*, and survived field rates of carbaryl. Overwintering success was verified (Roush and Hoy 1981b), and predators dispersed into surrounding habitats. Although dispersing predators interbred with native susceptible mites, resistance to OPs and carbaryl persisted (Hoy 1982). Since the initial trials, the carbaryl/OP-resistant strain has been established in many almond orchards in the San Joaquin valley (Hoy 1985).

The dispersal and persistence of carbaryl/OP-resistant *M. occidentalis* were evaluated by Hoy (1982). Late in the summer of 1979, this strain was released into a few trees of an almond orchard. By the next season, predators collected from seven widely separated orchards had moderate to high levels of resistance to carbaryl, indicating that the resistant strain had overwintered and dispersed from the release site. These predators had a substantial impact on the spider mite population, even during the 1980 season. Seasonal samples in 1981 indicated that carbaryl-resistant *M. occidentalis* had survived the second year.

Further study of the dispersal and persistence of the OP/carbaryl-resistant strain of *M. occidentalis* was reported by Hoy et al. (1985). They noted peak dispersal late in the season each year and that predators left the orchard on prevailing winds. Predators were trapped on sticky panels 200 m from the orchard. A survey of carbaryl susceptibility in predators collected from almond trees surrounding the release site indicated that carbaryl-resistant predators had dispersed at least 800 m between 1981 and 1983.

Roush et al. (1980) and Roush and Hoy (1981a) selected *M. occidentalis* for resistance to dimethoate, starting with a composite colony of mites collected from pears, apples, blackberries, and vineyards. In each selection, 500–1000 individuals were treated with dimethoate concentrations which caused 70–90% mortality. After eight selections, resistance had increased beyond that found in the field-collected strains, but it leveled off thereafter. After the dimethoate-resistant strain was released into a vineyard, the laboratory-selected strain proved to be more resistant than native *M. occidentalis*; however, resistant mites still suffered mortality at recommended field rates. Applications of dimethoate caused higher mortality to predators in subsequent generations. These results indicated that released mites interbred with native susceptible predators. Genetic studies demonstrated that resistance to dimethoate was polygenic (Roush and Hoy 1981a), which further explained the difficulty encountered in establishing and maintaining dimethoate resistance in the field.

Similar techniques were used to select for resistance to methomyl in another *M. occidentalis* strain (Roush and Hoy 1981a). A nearly 12-fold increase in resistance was obtained after 12 selections. The genetic basis of methomyl resistance was never determined (Hoy 1985), nor were field releases made, since resistance levels were insufficient for survival at field rates used in California vineyards.

Strains of *M. occidentalis* selected with permethrin achieved near 10-fold levels of resistance after 18 rounds of selection; however, further increases were not observed in the laboratory (Hoy et al. 1980, Hoy and Knop 1981). Genetic studies demonstrated that resistance to permethrin was polygenic, and bioassays indicated cross resistance to other pyrethroids (Hoy and Knop 1981, Hoy et al. 1980). OP resistance was maintained in this strain even in the absence of selection with OPs. When releases were made into Washington apple, Oregon pear, and California almond orchards, predators successfully overwintered and survived low rates of permethrin (Hoy et al. 1983). Hoy (1985) concluded that the polygenic basis of permethrin resistance required that native susceptible strains be eliminated to prevent hybridization with resistant mites. The permethrin-resistant strain persisted for several seasons at the release sites, but continuous applications of permethrin were necessary to maintain resistance (Hoy 1982).

Another mechanism of pyrethroid resistance may have developed in strains of *M. occidentalis* from Washington. Ouyang and Hoy (1988) reported a slightly higher level of resistance in this field-collected population, which they further selected for pyrethroid resistance in the laboratory (Table 14.1). At present this strain better survives field rates of pyrethroids than any of the earlier genetically improved strains (Ouyang and Hoy 1988). The genetics of the newer SP resistance factor has not been reported yet, but it will be interesting to compare results from such studies with similar data for laboratory-selected populations (Hoy et al. 1980, Hoy and Knop 1981).

In Australia, van de Klashorst (1984) selected *M. occidentalis* with bioresmethrin, and preliminary indications of a dominant, polygenic mechanism were reported (Hoy 1985). Cross resistance to fenvalerate and permethrin was present, and OP resistance persisted throughout selection. The biological control performance of the selected strain was tested

Table 19.4. Cost–Benefit Analysis for Integrated Mite Management in California Almonds Using Genetically Improved Strains of Pesticide-Resistant *Metaseiulus occidentalis* (adapted from Headley and Hoy 1986)

Plan	Item	Value
NO RELEASE OF PREDATORS NECESSARY		
	Itemization of costs	
	Cost of conventional treatment per acre (includes materials and application costs)	$75.00
	Minus reduced-rate acaricide treatment (materials and application costs per acre)	(21.00)
	Minus cost of mite monitoring per acre	(10.00)
Savings per acre		$44.00
Savings per acre over 5-year period		$158.62[1]
RELEASE OF PREDATORS NECESSARY		
	Itemization of costs	
	Cost of conventional treatment per acre (includes materials and application costs)	$75.00
	Minus reduced-rate acaricide treatment (materials and application costs per acre)	(21.00)
	Minus cost of mite monitoring per acre	(10.00)
	Minus cost of first-year predator releases	(20.00)
First year savings per acre		$24.00
Second and following years' savings per acre		44.00
Savings per acre over 5-year period		$140.76[1]

[1]Present value at 12% discount rate; value = savings year $1/1.12$ + savings year $2/(1.12)^2$ + savings year $3/(1.12)^3$ + savings year $4/(1.12)^4$ + savings year $5/(1.12)^5$ where savings are the same each year. This is the same as the present value of an annuity.

in "mini-orchards" maintained in greenhouses. Spider mite populations were regulated, and continued low-level tolerance to bioresmethrin applications was observed.

Hoy and coworkers used hybridization and artificial selection to obtain carbamate/OP/permethrin-resistant and carbaryl/OP/sulfur-resistant strains of *M. occidentalis* (Hoy 1984b; Chapter 14). Commercial production and the establishment of the latter strain in almonds and vineyards are widespread in central California (Hoy 1984b, Hoy et al. 1984). Headley and Hoy (1986, 1987) evaluated the economic benefits that California almond growers obtained through the use of *M. occidentalis* with multiple resistance. Studies were based on standard costs for 1) recommended chemical controls, 2) the release of insecticide-resistant predators, 3) monitoring predators and spider mites, and 4) costs of reduced rates of acaricides which were figured at 0.1% of the standard acaricide program. Savings were estimated for plans with and without releases of resistant predators (Table 19.4).

Growers utilizing the nonrelease programs saved $44.00 per acre per year. Since predators persist indefinitely without supplementation unless unusual environmental conditions occur or nonselective pesticides are used, the savings over a 5-year period were $158.62 per acre, computed with a discount rate of 12%. Growers releasing predators saved only $24.00 per acre the first year, but savings increased in subsequent years to $44.00. Savings over a 5-year period were $140.76 per acre (Table 19.4).

Headley and Hoy (1986) estimated the net value of the integrated mite management program at 25, 50, and 75% adoption by growers to range from $11.6 to $18.3 million in California. These were net returns above the initial research costs of $823,000. The cost–benefit ratio based on net returns ranged from 1:15 to 1:35, depending on the rate of adoption. The annual return on investment ranged from 280 to 370%. An informal follow-up survey by California extension personnel in 1985 found that nearly 25% of growers had adopted the program. By 1984–1985, strains of *M. occidentalis* with multiple resistance had been released on at least 12,000 acres of almonds. By 1987, it was anticipated that 60–70% of growers with pest mites would adopt the program.

19.2.1.3. Phytoseiulus persimilis. Genetic improvement of resistance in *Phytoseiulus persimilis* has been attempted in several laboratories in Europe (Schulten and van de Klashorst 1974, Gusev and Voroshilov 1977, Manttari 1980, Hassan 1982a, Fournier et al. 1985a, Avella et al. 1985, Koenig and Hassan 1986; see also Chapter 20). This research with *P. persimilis* may have begun even earlier than that with *A. fallacis* and *M. occidentalis*. Initially, artificial selection was used to improve resistance which had been detected in greenhouses (Gusev and Voroshilov 1977, Schulten and van de Klashorst 1974). *P. persimilis* is resistant to most families of insecticides, including OPs, carbamates, and SPs (see Table 14.1), largely as a result of genetic improvement. Levels of resistance in this species are generally lower than those of *A. fallacis* and *M. occidentalis* (Table 14.1), although resistance to methidathion approaches the high levels achieved with other phytoseiids (Fournier et al. 1987b; Table 19.1).

While artificial selection of *P. persimilis* has been encouraging, obtaining levels of resistance sufficient for field use has been a problem. In one case, gamma radiation was used to increase genetic variation in laboratory populations in an effort to obtain higher levels of resistance. Manttari (1980) selected *P. persimilis* for 50 successive generations while increasing the rates of bioresmethrin, but the highest concentration used was 20 times lower than field rates. The maximum resistance ratio obtained was only 2-fold. Manttari (1980) concluded that resistance in this strain could not be developed further.

Avella et al. (1985) and Fournier et al. (1985a) sought to increase genetic variability in

P. persimilis to improve the probability of artificially selecting for resistance to DDVP and deltamethrin, respectively. Hybridization with other strains and selection with other pesticides did not produce the desired results. Gamma radiation was also used, but limited increases in resistance were observed. After each experiment, the initial concentration used for selection was still lethal to adults of selected strains.

The effectiveness of genetically improved *P. persimilis* has not been widely documented, perhaps because inundative releases of this mite in greenhouses are common. Less time is required to establish predatory mites in greenhouses than on perennial crops. Inundative releases may be so frequent in greenhouses that the benefits of resistant strains are of less importance.

19.2.1.4. Typhlodromus pyri. In studies begun in the late 1970s, Markwick (1984, 1986, 1988) selected *Typhlodromus pyri* for cypermethrin resistance. After five selections, survival of adult females was over 70% at one-fifth of the field rate. Subsequent laboratory selections increased the level of resistance to above 10-fold, but resistance was insufficient for practical use in the field. Higher levels of resistance to pyrethroids have been obtained by selecting *T. pyri* in the field over a 9-year period (Markwick 1988, Walker pers. comm.). This field-selected strain readily survives pyrethroid sprays. Transfer of this strain to additional sites within New Zealand has been widely implemented. The genetic basis and the stability of cypermethrin resistance in the absence of selection have yet to be determined (Walker pers. comm.).

19.2.1.5. Other Studies with Phytoseiid Mites. Lo et al. (1984) reported that 17 acaricides and 6 insecticides used in Taiwan have been evaluated on *Amblyseius longispinosus* to determine resistance potential. Strains resistant to dimethoate, dimethoate/carbaryl, and carbaryl have been successfully selected. The genetic basis of resistance and field release data have not been reported for these strains.

Similarly in China, Ke (1987) has selected *Amblyseius pseudolongispinosus*, an important predator of spider mites of cotton, for resistance to dimethoate. Levels of 6.7-, 10.0-, and 12.5-fold were achieved after 36, 19, and 15 selections, using three different methods. Resistance was polygenic in all three cases. Tests made under greenhouse conditions with limited or no immigration showed that resistant predators could survive recommended field rates of dimethoate and provide biological control of spider mites on cotton plants. The status of resistance in the field has not been reported.

In other Chinese experiments, Huang et al. (1987) collected *Amblyseius nicholsi* from citrus and selected them with the OP phosmet. After 33 generations, 19-fold resistance was obtained. Genetic analysis showed that phosmet resistance is semidominant and monogenic. Cross-resistance to phoxim was observed as well as to optunal, dimethoate, vapona, and trichlorfon to a lesser degree.

19.2.1.6. Trends in Genetic Improvement of Phytoseiids. Successful artificial selection of pesticide resistance in phytoseiid mites has been encouraging (Table 19.1). However, there are trends in these studies which raise questions regarding the types and levels of resistance that are most desirable for use in the field.

Insight can be gained by examining experiments conducted to select phytoseiid mites for resistance to pyrethroids in the laboratory (Table 19.1). Such studies have been conducted on at least four different species with remarkably similar results: low levels of resistance were selected in previously unexposed populations, but in each case, continued selection was unproductive (Hoy and Knop 1981, Strickler and Croft 1982, van de Klashorst 1984, Avella et al. 1985, Fournier et al. 1985a, Markwick 1986, 1988). Pyrethroid resistance was found to be polygenic in *M. occidentalis* (Hoy and Knop 1981, van de Klashorst 1984), *A. fallacis* (Strickler and Croft 1982), *A. pseudolongispinosus* (Ke 1987), and *P. persimilis* (Fournier et al. 1985a).

While initial surveys for carbaryl resistance in California populations of *M. occidentalis* were unsuccessful, selection of a composite colony resulted in rapid resistance development (Roush 1979, Roush and Hoy 1981a). Analysis of cross resistance to propoxur gave more definitive discrimination of the carbaryl-resistance factor (Roush and Plapp 1982). Following carbaryl selection, Roush (1979) estimated the original gene frequency in the source colony to be about 4×10^{-3}, considerably higher than normal estimates of intrinsic resistance gene frequencies of 10^{-6} to 10^{-8} (Roush and Croft 1986). Data suggest that the resistance gene had already been selected to some degree in the source populations. Thus resistance might have been more easily detected had a bioassay been available that was more sensitive to differences in carbaryl toxicity.

Two additional studies emphasize the value of selecting strains that have already developed a degree of resistance in the field or which have been deliberately selected in the field (Ouyang and Hoy 1988, Markwick 1988, Walker pers. comm.). As noted, a laboratory-selected strain of *M. occidentalis* with polygenic resistance to permethrin was successfully maintained in tree fruit and nut orchards using multiple SP applications to prevent hybridization with susceptible mites (Section 19.2.1.2). However, resistance was not high enough to prevent appreciable reduction of resistance in *M. occidentalis* by field rates of permethrin (Hoy et al. 1983, Mueller-Beilschmidt and Hoy 1987). More recently, a field-selected population from Washington apple orchards has been found which exhibits higher and apparently more stable SP resistance. This strain was collected some distance from the earlier release site and is most likely a result of exposure to SPs or cross-resistant compounds.

Laboratory and field-selection experiments to obtain cypermethrin resistance were begun in 1978 with New Zealand populations of *T. pyri* (Markwick 1986, 1988, Walker pers. comm.). Laboratory selections conducted on a composite strain produced only a 14-fold resistance after 17 selections. A field colony selected four times per season demonstrated a significantly improved level of resistance in 1984. After nine seasons of continuous selection, resistant predators readily controlled spider mites in orchards where multiple applications of cypermethrin were used.

To summarize, the highest levels of resistance among genetically improved phytoseiids have arisen where populations have been deliberately selected in the field, or where resistant field populations have been further selected in the laboratory (Table 19.1.; Schulten and van de Klashorst 1974, Roush and Hoy 1981a, Koenig and Hassan 1986, Ouyang and Hoy 1988, Markwick 1988). In populations where no known field-derived resistance was already present, artificial selection has produced only modest levels of resistance (i.e., $\leq$ 10-fold). Resistance selected in the laboratory has primarily been polygenic (see reviews of Croft and Strickler 1983, Hoy 1985), which is consistent with patterns in the genetics of resistance in pests (Roush and McKenzie 1987).

19.2.2. Genetic Improvement of Predaceous Insects

Resistance to carbaryl in the lacewing *Chrysoperla carnea* was the first record of genetic improvement among predaceous insects (Grafton-Cardwell 1985, Grafton-Cardwell and Hoy 1986a, 1986b). First instar *C. carnea* collected in the Imperial Valley, California, were selected with carbaryl for four generations. Larval mortality was reduced from 98% to 0–10% when assayed at concentrations well above field rates of carbaryl. The survival of adults of the resistant strain was significantly greater than that of an unselected strain, but common field rates of carbaryl caused some mortality. Carbaryl resistance in *C. carnea* was found to be polygenic (Grafton-Cardwell and Hoy 1986b).

Pree et al. (1989) have identified resistant and susceptible populations of *C. carnea* from Canadian apple orchards. Their data suggest that source colonies used for artificial selection in California probably had appreciable resistance to carbaryl. While studies with *C. carnea* have demonstrated rapid, successful selection for resistance in the laboratory, field release of this genetically improved strain has not been reported (Grafton-Cardwell 1985, Grafton-Cardwell and Hoy 1986a, 1986b). Possibly, further selection and exploitation of the field-developed resistant *C. carnea* reported by Pree et al. (1989) will be even more successful.

Preliminary trials of artificial selection of insecticide resistance in *Anthocoris* were mentioned by Pralavorio et al. (1988), but substantial levels of resistance have not yet been reported.

Genetic improvement of predaceous insects might be more successful if candidates are chosen on the basis of their potential for resistance or tolerance exhibited in the field, as was done with *C. carnea* (Grafton-Cardwell 1985, Grafton-Cardwell and Hoy 1986a, 1986b). These species might have higher resistance gene frequencies or a greater variety of resistance mechanisms. *Stethorus punctum* and *Coleomegilla maculata* appear to be good candidates for genetic improvement based on these criteria (Table 14.1).

19.2.3. Genetic Improvement of Parasitoids

Genetic improvement of parasitoids was first attempted just after resistance to DDT became well documented among arthropod pests (Table 19.1). Pielou (1950), Pielou and Glasser (1951), and Robertson (1957) selected the braconid *Macrocentrus ancylivorus* for over 70 generations with DDT and observed 12-fold resistance; however, total reversion occurred in only 13 generations. Adams and Cross (1967) noted even more modest levels of resistance after selecting *Bracon mellitor* with several different pesticides for 5–7 generations (Table 19.1). The stability of this resistance was not studied. Kot et al. (1971) selected *Trichogramma evanescens* with metasystox 34 times over 70 generations and achieved a 22-fold resistance, but resistance was unstable and reverted when selection was discontinued. After selecting *A. melinus* with malathion for 8 generations, a resistance factor of less than 4-fold was obtained by Abdelrahman (1973). Although further increase in resistance was probable, the study was discontinued because of the 200-fold disparity between the LD_{50} of the resistant natural enemy and that of its scale hosts. Apparently, further selection was not expected to produce a high enough level of resistance that the natural enemy would survive field rates of malathion. These early studies emphasize the difficulty in selecting parasitoids to resistance levels that are of practical value.

From their analysis of the SELCTV database, Theiling and Croft (1988) reported that on the average, parasitoids are about 10 times more susceptible to pesticides than pests and 5 times more susceptible than predators. They considered these estimates to be conservative of differences in the susceptibility of parasitoids and their hosts. Many specific studies have documented the high susceptibility of parasitoids relative to their hosts (Brattsten and Metcalf 1970, 1973, Croft and Brown 1975, Croft and Morse 1979). These differences in intrinsic susceptibility can pose a major impediment to genetic improvement by determining the levels of resistance that must be achieved. Hence, while a low level of resistance in a pest can cause field failure of a pesticide, a parasitoid (or predator) must develop a level of resistance greater than the susceptibility of its host to achieve favorable selectivity. This means that 10-fold to 1000-fold resistance is needed in many parasitoids before genetic improvement can be of any practical value. In many cases, this is unreasonable.

While more moderate levels of resistance may allow natural enemies to reinvade a treated habitat earlier than could a susceptible strain (Fournier et al. 1987a, Hoy and Cave 1988, Rosenheim and Hoy 1988), higher levels of resistance might afford continuous or more consistent biological control. Even minor discontinuities in spatial or temporal synchrony of natural enemies and their host or prey may greatly diminish biological control (Chapters 7 and 8). High levels of pesticide resistance are desirable—and in many cases necessary—to substantially improve biological control where pesticides are used.

In more recent attempts to artificially select pesticide resistance in parasitoids, similar limitations as those identified for predators have been encountered, but in a few cases, higher and more stable levels of resistance have been attained (Table 19.1). Researchers have sought to overcome the perceived limits of earlier studies by using large and highly outbred source colonies, by maintaining maximum population sizes during selection, and by other novel methods.

Delorme et al. (1984, 1985) selected the aphelinid *Encarsia formosa* for 21 successive generations with either ethyl parathion or deltamethrin and observed a maximum difference of only 3- to 4-fold between selected and unselected strains (Table 19.1). Resistance to each compound was unstable. They concluded that the genetic heterogeneity of source colonies was probably insufficient for substantial resistance to occur.

Similar results were obtained with *E. formosa* from The Netherlands when selected with lindane for nine generations (Walker and Thurling 1984; Table 19.1). A mere 3-fold resistance at the LC_{50} was detected, although levels of resistance observed at the LC_{25} and LC_{90} levels were higher due to a nonlinear response curve. Walker and Thurling (1984) considered that the thelytokous, parthenogenetic reproduction of this species may have limited its genetic plasticity, thereby hindering the development of resistance. Thelytoky could be a benefit to genetic improvement, should a resistance mechanism be present, by minimizing gene loss, easily fixing new gene combinations, and rapidly passing the trait to all progeny (see Section 19.3.1).

Havron and Rosen (1984) reported no significant differences in resistance to malathion in seven stocks of *Aphytis holoxanthus* from Israeli citrus. These parasitoids were subsequently selected 25–30 times by exposure to pesticide-laden foods. Survivors of field populations which had been treated with malathion were introduced into selection lines at generations 11, 12, 22, and 27. Selection of diploid females and haploid males together as well as haploid males alone were tried. No method produced detectable resistance. In subsequent studies with *A. holoxanthus* (Havron and Rosen 1988), 8-fold resistance to azinphosmethyl was achieved after 19 generations of selection. Further selection produced a resistance in the range of 16- to 24-fold at the LC_{80}, but resistance leveled off thereafter.

Rosenheim and Hoy (1988) selected several field-collected colonies of *Aphytis melinus* which had shown considerable variability in response to pesticides used in citrus orchards (Rosenheim and Hoy 1986). Parasitoids were exposed to pesticide residues applied to emergence cages. No substantial increases in resistance to dimethoate, methidathion, chlorpyrifos, or malathion were observed, but 5- to 20-fold resistance to carbaryl was obtained after 10 selections in the laboratory. Survival of carbaryl-resistant adults was 50% or more when caged on 18–75-day-old residues. Survival of a susceptible strain was only 0–10% under these experimental conditions (Rosenheim and Hoy 1988).

Hoy and Cave (1988) selected *Trioxys pallidus*, a parasitoid of the walnut aphid, 14 times with azinphosmethyl and observed 7.5-fold resistance. When caged on treated walnut foliage, 26% of the resistant strain survived field rates of pesticide for at least 72 hr on either 3- or 17-day-old residues (Fig. 19.3). In contrast, susceptible *T. pallidus* experienced close to 100% mortality after 72 hr under the same experimental conditions (Table 19.1). In this case, even a modest level of resistance increased survival, suggesting that resistant *T. pallidus* may survive after immigrating into treated plots or emerging from hosts soon after azinphosmethyl has been applied.

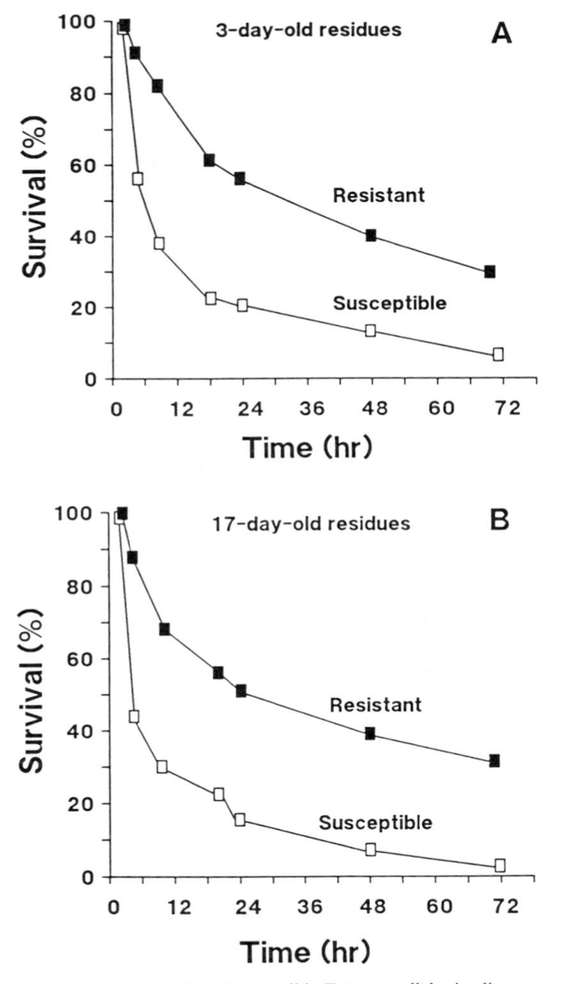

Figure 19.3. Survival of genetically improved and susceptible *Trioxys pallidus* in clip-cage tests on treated walnut foliage bearing azinphosmethyl residues (after Hoy and Cave 1988).

Walker and Thurling (1984) reported a near 15-fold level of resistance to permethrin in Dutch strains of *Encarsia formosa* (Table 19.1). Resistance did not appear to be entirely due to their selection experiments with permethrin, but probably resulted from previous exposure to other pesticides used in greenhouses.

One question raised by the limited resistance in parasitoids arising from genetic improvement is whether parasitoids have comparable ability to develop resistance as pests and predaceous arthropods, either in the laboratory or in the field. In a study of genetic variability across many arthropod orders, parasitic Hymenoptera were shown to have less genetic variation as measured by electrophoresis (Graur 1985; see also Chapter 16, Section 16.2.4). Lower genetic variation may limit the potential for resistance development among these species. On the other hand, pesticide resistance often appears to be under the control of a single, initially rare gene (Roush and McKenzie 1987). A rare, monogenic mechanism of resistance resulting from a single amino acid substitution may not be detected with electrophoresis or any other general quantitative measure.

Whatever the reason, very few cases of appreciable resistance among parasitoids have been reported in the literature, whether by field or laboratory selection (Croft and Strickler 1983). The only exception involves a South African strain of *Comperiella bifasciata* which attained 66-fold resistance to methidathion after many years of continuous selection in the field (Schoones and Giliomee 1982; Chapter 14). One wonders whether the method of testing accurately reflects the level of resistance actually present in the field. Even at the levels reported, Schoones and Giliomee (1982) concluded that resistance was insufficient for adult parasitoids to survive in the field, since recommended rates of methidathion were 12–18 times higher than the LC_{50} of the resistant parasitoid. On the other hand, Schoones and Giliomee did not test the parasitoid under field conditions. Lower levels of resistance as measured in the laboratory allowed some survival of *Trioxys pallidus* on treated walnut foliage (Hoy and Cave 1988).

The mere fact that Schoones and Giliomee (1982) could find a 66-fold difference between resistant and susceptible strains suggests that the field-selected strain must have been surviving pesticide exposure under at least some circumstances, either at the fringes of the treated area or after residues had decayed somewhat. A significant difference in LC_{50} values, as long as it is repeatable and outside the range of variation found in susceptible populations of the species, can almost certainly be due only to adaptation. If the field-selected strain did not survive pesticide pressure under at least some circumstances, no selection pressure would have been operating to favor the resistant genotype. Unfortunately, the genetics of resistance in this population was never determined.

Genetic analysis of pesticide resistance has not been reported for any parasitoid species, as it has for resistant predators (Table 15.5). Several factors probably contribute to this disparity. First, resistance in parasitoids has not reached high enough levels to use classical genetic techniques to elucidate mechanisms. Since resistance has seldom developed to a level sufficient to be of practical benefit in the field, the need to understand the genetic basis of resistance for management of these species has been less critical.

19.3. AMELIORATION OF GENETIC IMPROVEMENT

Constraints to the genetic improvement of predators and parasitoids target basic research needed to better manipulate pesticide resistance in arthropod natural enemies. For example, the genetic basis for resistance appears to be fundamentally different depending on whether selection was carried out in the laboratory or in the field. Field selection tends to produce monogenic resistance, while laboratory selection tends to produce polygenic resistance (Whitten and McKenzie 1982, Oppenoorth 1984, Roush and McKenzie 1987).

Several factors may contribute to the different outcomes of field and laboratory selection. First and foremost, major resistance genes are usually rare prior to selection, and are believed to be carried by fewer than one in 10,000 individuals and perhaps by fewer than one in 10,000,000 (Whitten and McKenzie 1982). Almost all laboratory populations are far too small or were founded by too few individuals to have a high probability of carrying such rare genes. Consequently, laboratory populations tend to accumulate more common genes, each contributing to the total level of resistance. Field selection, in contrast, may operate on a nearly infinite gene pool. During the course of genetic improvement trials, resistance occasionally develops in the field before laboratory selection studies are complete (e.g., Markwick 1986). Although artificial selection in the laboratory occasionally produces monogenic resistance, this usually occurs when genetic material is incorporated from field populations that have had prior exposure to the selecting pesticide or one that is cross resistant to it (e.g., Roush and Plapp 1982).

Several additional factors contribute to the dichotomy between laboratory and field selection, including the precision with which the selecting dose is applied and dispersal characteristics of the selected species (Roush and McKenzie 1987). In the field, a wide range of residue levels may be present, whereas dose can be controlled in the laboratory. Controlled dose allows for more precise discrimination between, and selection for, minor genes. Monogenic resistance is favored in the field because it is spread more readily throughout populations, whereas polygenic resistance is diluted through dispersal and hybridization.

If, as suggested, one major problem with current genetic improvement programs is that they have not had access to major resistance genes, then selection procedures must be modified to provide them. There appear to be two ways to accomplish this. First, one might focus on generating resistance via mutagenesis in the laboratory, as was attempted on a limited scale by Fournier et al. (1985a) and Manttari (1980). Second, one could attempt selection in the field, as has been tried in a preliminary way with predatory mites in apple orchards by Markwick (1986).

19.3.1. Generating Resistance in the Laboratory

The use of mutagenesis to generate pesticide resistance was pioneered by Kikkawa (1964), who used X rays to induce a parathion-resistance gene in *Drosophila melanogaster*. The resistant gene induced in *Drosophila* appeared to be the same as the one found in a field-collected strain; however, this method is not without limitations. Although Fournier et al. (1985a) and Manttari (1980) also used radiation to induce mutation in natural enemies prior to selecting for pesticide resistance, neither were successful. Neither of these studies included estimates of the number of irradiated animals screened for the resistant phenotype. Mutation rates are such that, even with enhancement by mutagenesis, considerable numbers of offspring must be screened. For example, Kikkawa (1964) tested 28,500 flies to isolate the resistant mutant in *Drosophila*.

More importantly, radiation may not be the most efficient way to produce pesticide-resistant mutants. Radiation generates many kinds of mutations, but many resistance mutations may be simple base-pair substitutions (Roush and McKenzie 1987). The DNA sequences for several insecticide resistance genes in pests have been identified, and in several cases have been shown to be due to single amino acid substitutions (e.g., Mouches et al. 1986, Berge and Fournier 1988, Feyereisen et al. 1989). Geneticists in the 1960s found that chemical mutagen, ethyl methane sulfonate (EMS), generates single base-pair substitutions much more efficiently than X rays. EMS produces fewer deleterious alterations and chromosomal rearrangements that can seriously affect the viability of offspring.

Resistance has been induced using EMS in at least two insect species, *D. melanogaster* (Pluthero and Threlkeld 1984, Wilson and Fabian 1986) and the green peach aphid (Buchi 1981). The study by Wilson and Fabian (1986) is particularly interesting because the mutation conferred nearly 100-fold resistance to methoprene, apparently by an altered target site. This resistance gene was isolated from one of 3490 family lines. EMS was found to induce new methoprene resistance alleles much more frequently than X rays (T. G. Wilson unpubl.). Although EMS is commonly administered in sugar solutions to adult males, vacuum injection is a more efficient method of delivery (Saga and Lee 1970) and can be used for any arthropod. Because F_1 mutants are often mosaics, it may be necessary to screen the F_2 offspring.

These studies show that mutagenesis can overcome at least one of the problems encountered in traditional laboratory selection: the absence of suitable genetic variation. Mutagenesis is under investigation in several laboratories, although it may be premature to define specific details for its use. Before field testing, each resistant mutant should be crossed and backcrossed to a wild-type strain for several generations and then reselected for resistance. This procedure, repeated backcrossing, is outlined by Roush and McKenzie (1987) and aids in the elimination of deleterious mutations that may have been induced by the mutagen.

19.3.2. Selecting for Resistance in the Field

Field selection, by definition, closely resembles actual field exposure. There are several advantages to screening for major resistance genes in the field. A large number of natural enemies can be screened easily. Field selection experiments only differ from natural selection in the more limited size of the treatment area and the number of total natural enemies being selected. In field selections, rough estimates can be placed on the size of populations being selected and the plot size required to screen for resistance at a given gene frequency. For example, based on general mutation rates, resistance gene frequencies may be on the order of 10^{-6} to 10^{-9}. Sampling theory tells us [eq. (1) in Roush and Miller 1986] that there would be a 95% probability of a resistance gene being present in a sample of 3 million individuals, assuming a frequency of 10^{-6}. Realistically, a single individual may be lost or go undetected, so a safety factor of 10 × is recommended. After sampling a natural enemy population to estimate density, one can easily calculate the approximate size of the plot required for field selection (e.g., approximately 20–200 apple trees would span this range for predatory mites occurring at a density of 3/leaf on 50,000 leaves/tree).

During field selection, it may be necessary to continually reinfest the plot with hosts or prey to assure that surviving natural enemies find food, since this may limit the development of resistance in the field (Huffaker 1971, Croft and Morse 1979, Morse and Croft 1981). In addition, selected natural enemies must mate and produce viable progeny.

Early detection of resistant variants, which are present in small numbers, may be ameliorated by improving bioassays and biochemical monitoring tools (Brown and

Brogdon 1987). Small populations of selected beneficials may be concentrated to assay for resistance using pheromone traps or by other means, as has been done for pests (Riedl et al. 1985, Haynes et al. 1986). Concentrated sources of hosts or prey might also be highly effective. In a variation of the field plot method, one might devise a system for collecting large numbers of natural enemies in self-dosing traps that could be checked regularly for survivors; however, care must be taken to avoid selection for resistance in pests. In many agroecosystems, key pests have far greater dispersal tendencies that natural enemies (e.g., Whalon and Croft 1985, 1986). Such dispersal tendencies would tend to suppress the development of resistance in pests and hasten it in natural enemies (Tabashnik and Croft 1982). Alternatively, field selection could be attempted in nontarget crops. For example, predatory mites to be used on apple could be selected on beans or roses to avoid the selection of insect pests of apple. Selection could also be conducted in areas isolated from major cropping regions.

19.3.3. Improving Selection for Resistance

Selection for resistance in natural enemies can sometimes be improved by selecting with a compound that is cross resistant to the compound to which resistance is sought. Examples were cited earlier in which these methods have proven useful in increasing resistance levels in predatory mites. A cross-resistant compound may allow more sensitive discrimination between slightly resistant and susceptible individuals because it discriminates between genotypes (Roush and McKenzie 1987). This was illustrated with the predatory mite *M. occidentalis* in detecting carbamate resistance (Roush and Hoy 1981a, 1981b). Although carbaryl was used in the initial selection program, cross resistance was

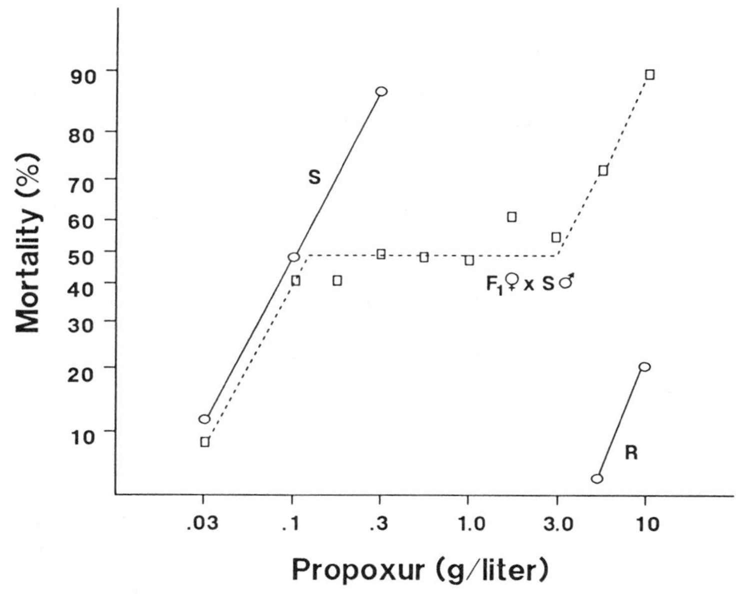

Figure 19.4. A discriminating dose used to identify benomyl-resistant genotypes in a genetically improved strain of *Metaseiulus occidentalis* (after Roush and Plapp 1982).

later identified to propoxur (Roush and Plapp 1982), which more clearly discriminated between resistance and susceptibility (see Fig. 19.4). The exact reason for the differential discrimination by propoxur and carbaryl is not well understood, but since discrimination between genotypes is the key to resistance selection (Roush and McKenzie 1987), selection with propoxur would have been more direct.

Any selection procedure that discriminates between resistant and susceptible genotypes will increase the efficiency of selection. Although traditional methods have emphasized topical assays, recent research on spider mites (Dennehy et al. 1983) and the tobacco budworm *Heliothis virescens* (Roush and Luttrell 1987) suggest that residual exposure more closely correlates with field mortality and discriminates more precisely between resistant and susceptible genotypes.

19.3.4. Overcoming Problems Associated with Polygenic Resistance

Several ways to overcome problems associated with polygenic resistance have been proposed and tested. For example, polygenically determined traits can be better preserved if they occur in a thelytokous strain. Thelytoky is not very common among natural enemies, although it does occur within several important natural enemy groups including the Acari, Hymenoptera, and Diptera (Hoy and Cave 1986). Recently, Legner (1987a) demonstrated that thelytoky could be selected and fixed in the arrhenotokous parasitoid *Muscidifurax raptor* by mating hybrid females. A change in ovipositional behavior was observed in hybrid females, which subsequently produced thelytokous offspring for 10 generations in the laboratory. Legner considered that this technique might be useful in biological control to reduce outbreeding in a field-released, genetically improved strain of parasitoid.

To circumvent the outbreeding of polygenically determined resistance in the predaceous mite *M. occidentalis*, Hoy and Cave (1986) screened virgin females collected from 59 laboratory and field-collected colonies to determine whether thelytoky occurred or might be selected. Although their results were negative for 13,000 females, they felt that this type of research could be successful for other species which showed variability in reproductive genetics. Recent success fixing thelytoky in *Muscidifurax raptor* by mating hybrid females (Legner 1987a) demonstrates the feasibility of these methods for natural enemy species.

A better understanding of the pleiotrophic effects of resistance selection and their impact on fitness might lead to improved artificial selection. The impact of fitness and pleiotrophy on resistance development and stability is virtually unknown, except that failure to establish resistant natural enemies in the field is often attributed to such effects (Chapter 15). Mueller-Beilschmidt and Hoy (1987) described the use of activity parameters to estimate the fitness of genetically improved strains. They have shown that activity level is heritable. Detailed video monitoring of other aspects of behavior may also provide indices of fitness. Further development of such assays may allow fitness to be monitored during genetic improvement trials and perhaps also following field release of the organism.

A better understanding of the ecological factors influencing resistance and its stability may enable more efficient use of genetically improved natural enemies. The release environment, immigration/residency, reproductive capacity, and density-dependent food relationships of species are extremely important in determining the fate of a resistant natural enemy (Chapters 16 and 20).

19.4. IMPLEMENTATION OF A GENETIC IMPROVEMENT PROGRAM

Steps involved in the implementation of a genetic improvement program can be generalized from the case histories presented in Table 19.1. An outline proposed by Hoy and coworkers (1976, 1979, 1982, 1985; Fig. 19.5) provides a framework to examine some general concepts that are relevant to the implementation phase of a program designed to genetically improve insecticide resistance in natural enemies.

19.4.1. Choice of Organism and Resistance Trait

Selecting an appropriate test organism or effective natural enemy for genetic improvement will depend on the objectives of a researcher (Fig. 19.5). If interested in

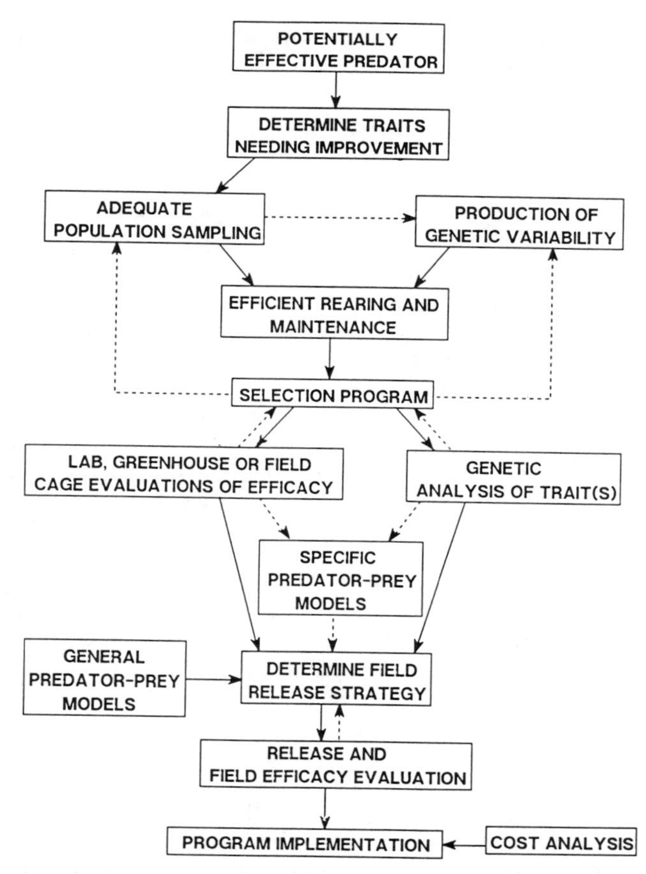

Figure 19.5. Steps involved in the design of a project to genetically improve pesticide resistance in an arthropod natural enemy (after Hoy 1982).

studying basic principles of genetic improvement, one might choose any appropriate arthropod model for testing, but preferably one that has been well-studied genetically, is easily reared in the laboratory, and is easily sampled in the field, such as *Drosophila melanogaster*.

If one's interest is in developing an improved natural enemy for use in a specific, chemically structured agroecosystem, then one's primary concerns are the natural enemy's biological control effectiveness, its capacity to adapt to new environments, and its propensity for resistance. A primary requisite is to obtain a representative sample of initial stock with sufficient genetic variability upon which selection can operate.

Compared with pests, it is more difficult to maintain large populations of natural enemies in culture because several trophic levels of organisms must be maintained. Considering that intrinsic resistance gene frequencies in arthropods most often are at levels of 10^{-4} to 10^{-6} (Roush and Croft 1986), ensuring that sufficient genetic variability is present for the selection of a resistant strain seems improbable. In these cases, the use of artificial diets or other natural or surrogate hosts is necessary to maintain large, diverse natural enemy populations. It is important to minimize unwanted genetic changes in laboratory populations that may result in reduced fitness upon release into natural environments (Mackauer 1976). Recent advances in our knowledge of nutrition and artificial rearing of predators and parasitoids are encouraging (Thompson 1986).

19.4.2. Methods and Monitoring of Selection

When undertaking the selection of insecticide resistance in natural enemies, one must decide which life stages should be selected, the appropriate level of selection, and the best monitoring methods for early detection of resistance. Most genetic improvement projects proceed on a trial and error basis, attempting to mimic field exposure and to maintain other natural processes by selection of natural enemies on treated plants in greenhouses (Strickler and Croft 1981, 1982, Roush and Hoy 1981b) or on treated leaf disks (Roush and Hoy 1981b). Because of the limited size of laboratory colonies, most selecting concentrations are conservative to prevent colony extinction. For example, most selections are conducted to cause 50–90% mortality rather than >95%, as often occurs in the field. Incrementally increasing selection may favor polygenic rather than monogenic resistance (Table 15.5; Hoy 1985, Roush and Croft 1986; see also Chapters 14 and 20).

In general, the operational factors (Georghiou and Taylor 1977b) which limit resistance in pests apply in reverse to maximize resistance in natural enemies. Selection will generally proceed faster in natural enemies when multiple life stages are exposed to selection, when natural enemies with rapid rates of increase are used, when persistent compounds are applied, and so forth (Chapter 20).

In monitoring for resistance during selection trials, it is difficult to detect the very early stages of resistance. Initial or early selection frequencies of resistance genes may be very low, and incipient resistance may be difficult to detect (Roush and Miller 1986). Generally, differences in susceptibility below 3- to 5-fold at the LC_{50} or LD_{50} are unlikely to be statistically significant, and yet early resistance is often in this range (Chapter 14). More sensitive biochemical and immunological techniques for the identification of resistant genotypes (Brown and Brogdon 1987) are needed to monitor the development of resistance in natural enemies.

19.4.3. Evaluation of Selected Traits

While preliminary studies of genetically improved natural enemies are not essential to a successful program, they may allow scientists to anticipate problems and to reduce possibilities for failure during implementation. One such evaluation might involve tests of fitness, although these types of studies can be difficult to conduct. Roush and Hoy (1981b) evaluated fitness of a carbaryl/OP-resistant strain of *M. occidentalis* by examining mating competitiveness, diapause, persistence of resistance in hybrids of resistant and susceptible strains, and ability to control spider mites in the greenhouse. They found no evidence that problems would be encountered in introducing the resistant strain. In contrast, Croft and Whalon (1983) studied a pyrethroid-resistant strain of *A. fallacis* in the laboratory and observed that in the absence of hybridization with susceptibles, resistance did not decline appreciably over a 25-generation period. However, when released in the field, resistance was flooded out of the population very rapidly (see also Chapter 16, Section 16.2.2).

It is possible to detect factors that may limit successful field releases in laboratory, semifield, or greenhouse studies. Of course, some limitations may arise in field tests that could not be otherwise foreseen or detected (Chapter 5). Immigration and hybridization with native susceptible strains are difficult to control or mimic in laboratory tests.

Several researchers have discussed (Mackauer 1976, Hoy 1985) how preliminary releases of resistant natural enemies may identify problems associated with laboratory-selected strains, such as the expression of deleterious traits acquired during the selection process. These types of problems have not been encountered in the field when insecticide-resistant phytoseiid mites have been released. Roush and McKenzie (1987) questioned how often and to what extent fitness is reduced by laboratory rearing and by the selection of pest arthropods with pesticides in the field.

To generalize, laboratory tests that attempt to evaluate the efficiency of natural enemies once they are released in the field are usually of questionable value. While major problems may be identified, these tests are usually poor indicators of subtle factors that may influence a successful establishment of genetically improved natural enemies in the field.

19.4.4. Field Release Strategy

The techniques used in field releases of insecticide-resistant natural enemies have largely been empirically derived, beginning with inoculative and inundative releases in the early 1970s (e.g., Croft and Barnes 1972, Croft and McMurtry 1972a). Establishment usually occurs if adequate hosts or prey are present and if deleterious sprays and competitive native natural enemies are excluded. Orchard and vineyard releases typically consist of 350–1000 females per plant, released into every plant or every few plants. In one California orchard, release of 350 predators into every third tree in every third row resulted in the establishment and spread to adjacent trees within 2 to 3 weeks.

Beyond these empirical approaches, models may be useful in developing field release strategies. Acreage, plant density, natural enemy mobility, prey or host density, and chemical treatment schedules should be inputs to such models. Minimum release rates for establishment might be ascertained if variables such as genetic mechanisms, fitness, the fate of resistance in the presence of hybridization with susceptible mites, population levels of susceptible strains, population growth rates, and dispersal rates were known. Models might also be used to predict the outcome of different release and establishment strategies.

One predator–prey simulation model by Rabbinge and Hoy (1980) was developed to evaluate releases of *M. occidentalis* in greenhouse roses and other crops. Numerical dynamics of predators and prey mites could be predicted based on initial predator–prey ratios and intervening environmental conditions. Several more general models of the population dynamics of tetranychid and phytoseiid mite species have been developed (reviewed in Welch 1979 and Logan 1982). With the incorporation of genetic attributes, these models might be adapted for use in genetic improvement projects.

19.4.5. Field Release and Impact

Field release methods for insecticide-resistant phytoseiid mites range from the transfer of mite-infested foliage (Roush and Hoy 1981b, Whalon et al. 1982a) to direct transfer of mites (Croft and Barnes 1971) to attempted aerial application on inert carriers (G.T. Scriven unpubl.). Very few studies, however, have compared different types and levels of releases and their success rates. (See Croft and McMurtry 1972a and Boscheri et al. 1986 for studies of release levels of insecticide-resistant natural enemies.) Although tchniques for releasing genetically improved natural enemies are still being developed, many of the principles of inundative releases of biological control agents are applicable (Ridgway and Vinson 1977). Releasing genetically improved organisms may require some additional considerations. For example, maintenance of certain genes may require a different release level than would standard inundative or inoculative releases for biological control.

The spread of genetically improved beneficials, once released, raises some interesting questions of regional impact and benefit. With the release of predatory mites whose mobility is limited, benefits can be measured at the individual farm level. However, mobile arthropods will disperse, benefiting growers regionally. Biological control districts or regional organizations similar to those used by California citrus growers for parasitoid releases (DeBach 1964) may be needed for more mobile natural enemies.

19.4.6. Program Implementation and Economic Evaluation

Few genetic improvement programs have been in place long enough to allow economic evaluation for a large segment of growers. However, it is anticipated that costs and benefits would be similar to those realized for introductions of insecticide-resistant natural enemies that developed naturally in the field (except for basic research selection costs; Chapters 17 and 18). The cost–benefit ratios of biological control projects are generally positive, ranging from 1:2 to 1:10 (See Marsden et al. 1980, Hoyt 1982, Croft et al. 1985). The best documentation of economic benefits of a genetic improvement program is that of carbaryl/OP-resistant *M. occidentalis* in California (Hoy 1985, Headley and Hoy 1986, 1987). Rearing, release, and management of *M. occidentalis* have been taken on by the private sector for further promotion and economic development. Because of technology transfer, the scientific community may have less access to economic data, but the performance of these companies over time should indicate the economic status of these programs.

19.5. GENETIC ENGINEERING OF PESTICIDE RESISTANCE

Hybridization and artificial selection for genetic improvement of arthropod natural enemies are currently being widely researched. In the future, however, the use of recombinant DNA technology may produce natural enemies with higher and more stable levels of pesticide resistance (Beckendorf and Hoy 1985, Berge et al. 1986, Mouches et al.

1986). The utilization of genetically engineered natural enemies in the field, however, faces formidable challenges beyond those of a technical nature. Of primary concern are environmental hazards and regulatory constraints. Other problems will undoubtedly arise as implementation progresses. Biological control via genetically engineered natural enemies should be implemented within the framework of current ecological and IPM theory. Scientists will be challenged to integrate genetically engineered organisms into established IPM systems (Croft 1986).

19.5.1. Methods

Beckendorf and Hoy (1985) and Hoy (1987) reviewed the application of recombinant DNA technology to biological control agents, stressing constraints to development and use. Their discussion was speculative in nature rather than a treatise of experimental research, because very little actual research has been conducted on the subject.

Beckendorf and Hoy (1985) discussed some of the concepts of genetic engineering as applied to insects. They outlined the steps required to construct a genomic library and to isolate clones of specific resistance genes including 1) genomic DNA isolation, 2) fragmentation of DNA with restriction enzymes, 3) insertion of DNA into phage, cosmid or plasmid vectors to produce a library, 4) screening of the library with a specific DNA or RNA probe, and 5) isolating and cloning of individual fragments carrying target DNA.

The application of these techniques to arthropod biological control agents faces formidable limitations. There is a lack of basic knowledge of the natural enemy genome; this is true for any arthropod species (with the possible exception of *Drosophila*), let alone for any specific biological control agent. Reliable vectors to transfer genetic material are currently being investigated. Genetic markers for the identification of transformed arthropod genomes have not been identified. Finally, few scientists interested in biological control have mastered molecular biology. Still, one might reasonably assume that these limitations will be overcome.

Essential for the genetic improvement of pesticide resistance in natural enemies is the identification of specific genes conferring pesticide resistance or other desired traits. Only recently have genes conferring resistance to an insecticide been cloned. These include the OP-resistance gene for mosquitoes (Mouches et al. 1986), the *ace* gene of *Drosophila* (Berge et al. 1986, Berge and Fournier 1988), and the P-450 gene from the house fly (Feyereisen et al. 1989). While suitable vectors and phenotypic markers may be found, their utilization depends on research on mapping and cloning of pesticide resistance genes.

Following the cloning of a resistance gene, this genetic information must be inserted into the natural enemy genome. A possible strategy is the use of transposable elements from *Drosophila* as vectors. However, there are many questions that must be answered before transposable elements can be used for gene insertion in natural enemies. For example, will such elements produce natural enemy strains with functional progeny? Some transposable elements like the P element in *D. melanogaster* only act as a vector in populations that do not contain repressor genes (Spradling and Rubin 1982). Are repressor genes present in the natural enemy being considered for gene manipulation?

In the event that transposable elements are unsuitable vectors, another approach would be to analyze the natural enemy's genomic DNA for indigenous transposable elements. It must be recognized, however, that a vast amount of research would be required to bring our understanding of natural enemy genetics to the current state of knowledge of the *Drosophila* genome. Insect viruses such as the Baculoviruses are also being examined as possible vectors and their use is growing rapidly (Miller 1983). Other, more efficient techniques for achieving genetic transformation may emerge in the future. These procedures may involve the analysis of DNA sequences, isolation, and cloning, construction of vectors, and microinjection. The scope of such research lies beyond any current genetic improvement programs for natural enemies.

Provided a vector is identified, research on genetic improvement can proceed. The vector should be designed to contain a phenotypic marker. The insecticide resistance gene is then cloned into the vector (e.g., Mouches et al. 1986). Thus constructed, the vector is microinjected (Spradling and Rubin 1982) into embryos of the natural enemy. Injected embryos are reared and mated with wild-type individuals. If a phenotypic marker was incorporated into the genome and expressed, it should be apparent in F_1 offspring. Dosage mortality or discriminating dose tests can then be used to identify insecticide resistance expression in subsequent generations of the organism. If the phenotypic marker and insecticide resistance are not expressed, a genomic DNA blot can be prepared of the F_1 generation. It should be probed to detect the presence of the transferred gene(s). If the resistance gene was incorporated into the target genome, but not expressed, it will be necessary to try different promoters for the pesticide resistance gene (several promoters are currently under investigation for this purpose) or it may be necessary to reinsert the gene for incorporation in a different location on the genome.

In summary, the practical application of genetic engineering techniques to generate pesticide-resistant natural enemies depends on further research to identify suitable vectors, phenotypic markers, and, most importantly, the resistance genes themselves. If genetic transformation of a natural enemy can be accomplished, then the fitness and performance of the organism must be evaluated in the field.

19.5.2. Environmental Considerations

Beyond the technical aspects of genetically engineering pesticide-resistant natural enemies, environmental and regulatory constraints governing their use in the field must be satisfied. Problems likely to be encountered may be similar to those experienced by groups releasing genetically improved microbes such as ice-nucleating bacteria and crops containing *Bacillus thuringiensis* toxins (Adang and Binns 1988). A National Academy of Science panel (NAS/NRC 1987) pointed out that genetically improved organisms are not substantially different from those organisms produced in the past through traditional cross breeding and artificial selection methods. Whitten (1988) emphasized that attention should be paid to the products of genetic engineering rather than the processes used to obtain them. Balanced regulatory controls are needed to govern release and use of genetically improved natural enemies.

Presumably, genetically engineered microbes and plants will pave the way for the introduction of other such organisms. Release of genetically altered natural enemies may face fewer public, policy, and legal challenges. Hopefully, these arthropods will be perceived as less of a threat than released microbes. However, unforseen problems associated with the release of genetically improved natural enemies may arise as these agents become candidates for use.

19.6. SUMMARY

Of the case of pesticide resistance involving deliberate genetic improvement (Table 19.1), only 20% involve levels of resistance that allow natural enemies to survive recommended field rates of pesticides. Where resistance has developed in the field through natural selection (Table 14.1), approximately 68% of the cases have achieved levels of resistance that allow similar survival of natural enemies. Genetically improved strains have been most successfully exploited where resistance levels of 10-fold or higher have been obtained. In a few cases, strains were already tolerant to a pesticide before they were selected for resistance. With polygenic resistances, field implementation has been more limited.

Successful utilization of genetically improved natural enemies is hampered by our poor understanding of their population genetics. A better foundation in ecological genetic research would aid in the development of resistant strains of natural enemies (Roush and McKenzie 1987). Other areas of research which merit attention include means to increase or maintain genetic heterogeneity, artificial selection to mimic field selection, and development of better monitoring and collection techniques. The use of genetic markers and population genetic models would further facilitate the use of population genetics theory for genetic improvement research.

Genetic improvement through artificial selection has been highly successful in horticulture, crop breeding, and livestock production. However, artificial selection of beneficial arthropods, with the exception of species such as the silk worm and the honey bee, is still at a primitive level. Support for this area of scientific inquiry is minimal. Few scientists interested in the genetic improvement of natural enemies have adequate training in population genetics and population ecology (Croft and Roush 1987). This expertise and support are necessary before the principles of genetic improvement can be more fully applied to the development of pesticide-resistant natural enemies.

20

DECIDUOUS TREE FRUIT SPECIES

20.1. INTRODUCTION

The development of pesticide resistance in a single key pest can threaten an IPM program (Croft et al. 1985). There are cases where pesticide resistance in a single key pest has curtailed crop production for several years (e.g., cotton in Mexico, pear in northern Italy; Dover and Croft 1984). The history of resistance for *Heliothis* spp. on cotton, diamondback moth (*Plutella xylostella*) on vegetables, and spider mites (e.g., *Tetranychus urticae*) on greenhouse crops attests to the impact that resistance can have on crop production (Georghiou and Saito 1983, Dover and Croft 1984). These case histories emphasize the need for pest control that minimizes pesticide resistance and exploits biological control via selective pesticides.

In this chapter, tactics used in resistance management are discussed. Following a discussion of principles, examples of their application are given for a pest–natural enemy complex on apple in North America. The biology of this apple pest–natural enemy complex is reviewed in Croft and Hoyt (1983).

20.2. GOALS OF RESISTANCE MANAGEMENT IN PEST–NATURAL ENEMY COMPLEXES

Pesticide resistance is often not limited to a single species, although a key pest may be the focus of resistance management in a given crop ecosystem. Other pests and natural enemies are continually undergoing selection by pesticides. Resistance among species other than primary pests may be discerned when control failures or pest outbreaks occur. Sometimes, however, they are only detected with broad resistance surveys, but these are rarely conducted. Some resistant secondary pests do not become problems because they are still partially controlled by pesticides. Sometimes the damage caused by resistant secondary pests can be tolerated until effective pesticides are employed.

Pesticide resistance among secondary pests and natural enemies may go unnoticed due to limited monitoring (Chapter 14). Sometimes it is difficult to tell whether a pest control failure is the result of poor pesticide application, changing environmental conditions, or pesticide resistance. Efficient monitoring for resistance among pests and natural enemies is critical to resistance management (Brown and Brogdon 1987). For a complex of species, resistance monitoring on a timely basis can be challenging.

Many cases of resistance in pests are due to mismanagement. Under ideal pesticide management, resistance would not occur in pests; however, this goal is probably unattainable on many crops. A more realistic goal is to avoid creating highly resistant pests and to exploit resistant biological control agents. Resistance may be manageable through more precise and selective pesticide use. With a greater reliance on physiologically selective pesticides, selection pressure on pests may be lessened through more effective biological control. Prevention of highly resistant pests and encouragement of highly resistant natural enemies should be sought rather than crisis intervention.

20.3. OPERATIONAL FACTORS

Minimizing resistance in pests and *maximizing* resistance in natural enemies with a single pest control tactic is difficult. Understanding how operational factors influence resistance evolution is the key to achieving this end. Georghiou and Taylor (1977a) defined operational factors as those under the control of the field manager or IPM practitioner, and presented a list of such factors. One operational factor is the choice of an appropriate pesticide, taking into account the toxicant's chemical nature, its relationship to previously used chemicals, its persistence, formulation, and so forth. Also under control of the pest manager is application of the pesticide, including the economic threshold, life stage selected, and mode of application. Recently, the concept of operational measures has been broadened to include ecological factors, such as manipulation of the immigration/residency of a species (Croft 1982, Suckling et al. 1984, 1985).

Operational factors for resistance management of natural enemies are similar to those for pests (Croft and Strickler 1983, Tabashnik and Croft 1985), but apply in the reverse direction. For example, dilution of resistance in pests may be maximized by reducing resistant pest populations prior to immigration of susceptibles. With natural enemies, minimal immigration of susceptibles is desirable so that resistance is not diluted.

Table 20.1. Operational Factors Influencing the Development of Pesticide Resistance in Arthropod Natural Enemies and Pests (adapted from Georghiou and Taylor 1977a, Georghiou 1983, Tabashnik and Croft 1982, Tabashnik 1986a)

Natural Enemy Species	Pest Species
OBJECTIVE: MAXIMIZE RESISTANCE	OBJECTIVE: MINIMIZE RESISTANCE
Minimize impact of S immigrants	Maximize impact of S immigrants
Maintain high levels of R residents	R Residents undesirable
High dose, eliminating S genotype	Low dose, sparing S genotype
Selection of all life stages	Selection of one or few life stages, preferably adults
Frequent application	Infrequent application
Persistent pesticides	Nonpersistent pesticides
Areawide application	Local application
Selection of entire population	Preservation of individuals in refugia
Use of cross-resistant compounds	Avoid cross-resistant compounds
Avoid use of mixtures, rotations, and synergists	Use mixtures, rotations, and synergists
Provide a food supply following selection	—

A list of operational measures for natural enemies and pests is given in Table 20.1. These operational factors apply to resistance management of individual species. When an operational measure affects both pests and natural enemies, it is more difficult to apply them effectively (Section 20.4).

20.4. SIMULTANEOUS RESISTANCE MANAGEMENT IN PESTS AND NATURAL ENEMIES

Simultaneous resistance management of a pest and natural enemy using a reduced pesticide dosage is illustrated in Fig. 20.1. Pesticides are often more toxic to natural enemies than to pests by a factor of 10- to 100-fold (Croft and Morse 1979, Theiling and Croft 1988; Chapter 2). A high dose applied to control a pest often exerts greater selection pressure on a natural enemy (Fig. 20.1a). However, greater selection pressure does not always translate into higher resistance in a natural enemy due to food limitation following selection (Chapter 16, Section 16.5.1).

When a high dose of pesticide is applied, resistance may be rapidly selected in a pest population (Fig. 20.1a). Resistant pests have abundant food—the crop—on which to reproduce. In contrast, the natural enemy that survives a spray usually find its food—the pest—severely reduced, and often emigrates or starves. Therefore, resistance in the natural enemy is delayed or does not occur (Croft and Morse 1979, Tabashnik 1986a).

The potential for resistance in a natural enemy can often only be realized after the pest has developed resistance. Resistant pests remain as a source of food for the natural enemy following selection. Unfortunately, once a pest develops resistance, use of the pesticide is usually discontinued. This stops selection of the natural enemy just as resistance begins to develop.

More moderate dosages of pesticide can have very different effects on resistance

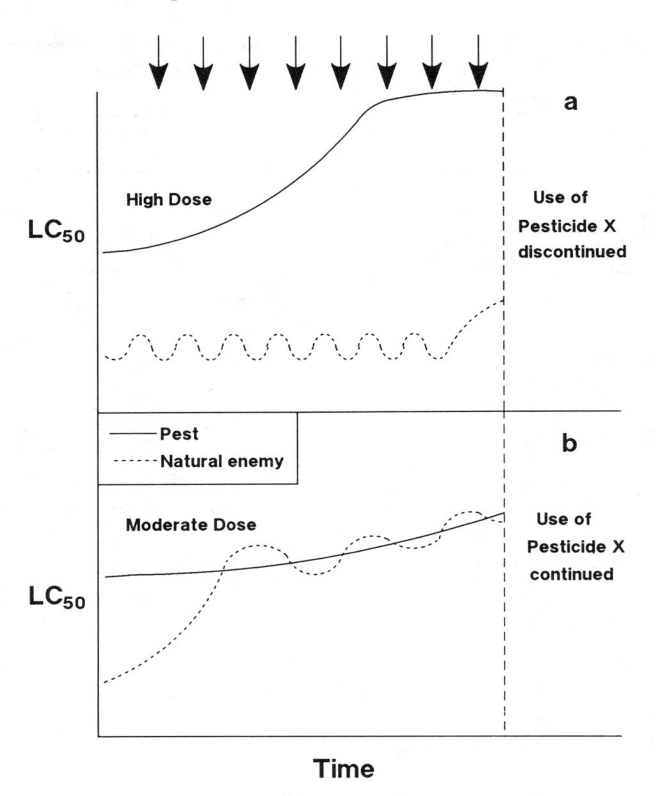

Figure 20.1. Generalized effects of high dose versus moderate dose on resistance development in an arthropod natural enemy and its host or prey (after Morse and Croft 1981).

development in a pest and natural enemy from that shown in Fig. 20.1a. Again, since the natural enemy is more susceptible, a dose that is moderately toxic to a pest and nonpersistent will leave food for the natural enemies that survive selection (Fig. 20.1b.; see Morse and Croft 1981 for a specific example of this effect on plant-feeding and predatory mites). A moderate dose reduces selection for resistance in the pest (Fig. 20.1b). Thereafter, resistance in the natural enemy increases at about the same rate as in the pest (Croft and Morse 1979, Morse and Croft 1981, Tabashnik 1986a). While this case oversimplifies the process of selection in the field, it demonstrates in principle how resistance management in pests and natural enemies may be achieved simultaneously.

Differences in pest and natural enemy biology allow other resistance management tactics to be applied (Table 20.1). Implementation of these tactics is not always as clear as is suggested in Fig. 20.1, but pesticide resistance management can be improved over past history. Specific examples of resistance management tactics for pests and natural enemies of apple are presented in Section 20.6.

Resistance management requires more extensive monitoring and knowledge about resistance in both pest and natural enemies as well as the ecological environment in which a resistance episode occurs. Maintenance of a more extensive information base may be justified only for high-value crops. Conversely, some systems of resistance management developed on high-value crops may apply to crops of lesser value.

Tabashnik and Croft (1982) developed a decision pathway to choose resistance management tactics for pests based on pesticide selection, application, and immigration of susceptible pests. A similar decision tool for natural enemies is presented in Fig. 20.2. This flowchart includes the most sensitive factors that influence resistance development in natural enemies: immigration/residency levels and food availability. While both factors do not directly involve pesticides, they can be affected by pesticides. In Fig. 20.2, it is assumed that there is a genetic mechanism for resistance in a natural enemy population and that it can be selected for in the field.

Initially, decisions are based on immigration of susceptibles and residency levels of resistant predators or parasitoids in a treated environment (Fig. 20.2). Immigration/resi-

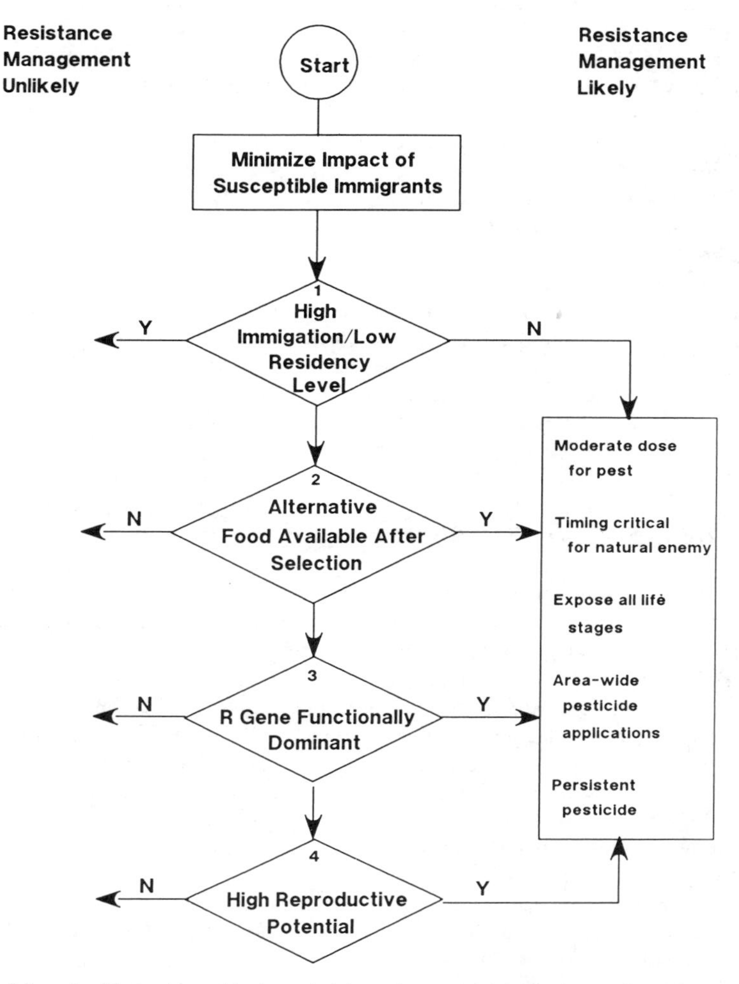

Figure 20.2. Schematic of factors to consider in maximizing resistance development in an arthropod natural enemy (modified from Tabashnik and Croft 1982).

dency attributes indicate the likelihood of maintaining rsistance. Natural enemies with low immigration rates and high residency levels should be easiest to manage for resistance (e.g., phytoseiid mites and aphelinid parasitoids; Chapters 13–17). Treated areas isolated from susceptible immigrants or areas surrounded by resistant populations are optimal for managing resistance in natural enemies.

Resistance in natural enemies can be altered by minimizing the immigration of susceptibles through crop location, control of surrounding vegetation, killing immigrants before they reproduce, or maintaining high levels of resistant residents just before immigration occurs. The latter can be accomplished by supplemental releases, supplemental feeding, selective pesticide use, or other augmentations (Ridgway and Vinson 1977). It is desirable to maximize selection of a natural enemy without greatly limiting its food supply (Fig. 20.2). Conserving primary hosts or prey (pests) may be difficult, since they often are the targets of pesticide applications and may be intolerable at levels needed to sustain natural enemies.

In managing resistance in both pests and natural enemies, timely decisions must be made as to when selection occurs or when other management tactics are employed. Windows of opportunity change as phenology, migration, and refugia change. Practicing resistance management requires considerable knowledge of the population dynamics of a pest–natural enemy complex.

While a moderate pesticide-use strategy can effectively minimize resistance in a pest and simultaneously encourage resistance in a natural enemy (Fig. 20.1b), the interim pest damage sustained may be intolerable. A low-dose strategy works best for pests that can be tolerated somewhat. The most effective pesticide-resistant natural enemies in agriculture tend to be those attacking indirect pests such as aphids, mites, scales, and whiteflies. Also, physiologically selective pesticides or ecologically selective uses of pesticides are not always available for pest control. Therefore, one may seek to detect or field select a pesticide-resistant natural enemy. This is only possible if its host or prey does not become resistant first and a food supply is available to sustain the resistant natural enemy in the field.

20.5. THE PEST–NATURAL ENEMY COMPLEX ON APPLE

Having discussed some general principles of resistance management for a pest–natural enemy complex, some of the operational measures used in North American apple orchards are discussed below. Resistance management for this arthropod complex is possible because of previous problems with pesticide resistance, continued susceptibility in certain pests, and the successful use of resistant natural enemies (Croft 1983).

While similar resistance patterns may have evolved in other intensively sprayed crops such as cotton and citrus, these phenomena have been documented more extensively on apple (see reviews in Glass 1960, Croft 1983). Also, apple in North America has been sprayed for three or four decades with the same organophosphate (OP) insecticide, azinphosmethyl. Because of widespread continuous use of azinphosmethyl, several arthropods have adapted to this pesticide. The unique features of resistance in arthropod pests and natural enemies associated with apple provide insight into the factors influencing resistance and resistance management in both trophic groups (Chapter 16).

Glass (1960) and Brader (1977) previously reviewed resistance to insecticides among deciduous tree fruit insect pests. Similar reviews for acarines are numerous (Helle 1965, Helle and van de Vrie 1975, Dittrich 1975, Herne et al. 1979, Cranham and Helle 1985);

one deals specifically with spider mites of deciduous tree fruit (Herne et al. 1979). Most reviews have focused on the disadvantages of pest resistance and have offered alternative chemical solutions. Limited emphasis has been given to resistance in natural enemies.

As concluded in earlier reviews, the published literature seldom fully indicates the status of resistance on any crop. In many cases, precise documentation using resistant and susceptible strains is not available. Pest control failures are sometimes the only evidence that resistance has developed in the field, and this information is seldom published. A survey of field researchers may better reflect the history and status of resistance on apple. Such a survey was conducted in 1981 for pests and natural enemies of apple in North America (Croft 1982, Tabashnik and Croft 1985). Estimates of when resistance first developed were obtained. These data, with a recent update, are presented here as background for a discussion of resistance management.

20.5.1. Pesticide Resistance in Apple Pests

Cases of pesticide resistance in arthropod pests of apple are listed in Table 20.2. They begin in 1908 with resistance to lime sulfur in San Jose scale in Washington State (Melander 1914) and extend to pyrethroid resistance in the spotted tentiform leafminer *Phyllonorycter blancardella* in Ontario, Canada (Pree et al. 1987).

The distribution of resistance in pests and natural enemies over time is illustrated in Fig. 20.3. Most cases have involved spider mites and predaceous phytoseiid mites. The greatest frequency of resistant pests (species-chemical combinations) occurred just after the introduction of OP insecticides. During the 1960s and 1970s, the frequency of pest resistance declined somewhat. However, resistance is still considered the single biggest threat to the stability of IPM in North America (Brunner and Cameron 1985). Cases of resistance among natural enemies have increased greatly since the mid 1960s (Fig. 20.3; see also Table 14.1).

In the pre-IPM era of apple pest control (1920–1950s), resistance in key pests—most commonly the codling moth—was geographically widespread (Table 20.2) and led to the discontinued use of lead arsenate in North America in the early 1940s and DDT in the 1950s. The OPs parathion and azinphosmethyl were introduced in the mid 1950s. For some reason, OP resistance has never developed in codling moth or other key pests of apple, even though OPs have been used for more than 30 years at 3–10 applications per season. Resistance has never evolved in the plum curculio *Conotrachelus nenuphar* or the apple maggot *Rhagoletes pomonella* (Table 20.2). These species have only a single generation per season in most fruit-growing areas, compared to 2–5 for the codling moth and the redbanded leafroller *Argyrotaenia velutinana.* The latter pest developed resistance to TDE in eastern and midwestern states in the 1950s (Table 20.2).

More recently, OP resistance has appeared in several moderately destructive leafrollers, including the fruit tree leafroller *Archips argyrospilus* in British Columbia (Madsen and Carty 1977), the oblique-banded leafroller *Choristoneura rosaceana* in New York (Reissig 1978), and the variegated leafroller *Platynota flavedona* and the fruitworm *Spargonathis sulfureana* in New York (Weires 1985). In general, these pests are more sporadic than the codling moth or redbanded leafroller. Resistance in tortricid pests of apple in North America reflects a general trend of increasing OP resistance in this group, worldwide (Croft and Hull 1988).

Pesticide resistance in secondary pests of apple, including leafhoppers, leafminers, aphids, and mites, make up the majority of cases in Table 20.2. Resistance to DDT in the leafhopper *Erythroneura lawsoniana* occurred in localized orchards in Kentucky in

Table 20.2. Cases of Resistance in Arthropod Pests of Apple in North America, 1908–1985 (adapted from Croft 1982)

Species	Date Reported	Compound	Area	Reference
PRIMARY AND SPORADIC PESTS				
Codling moth (*Cydia pomonella*)	1928–1930	Lead arsenate	Most areas	Hough 1928, Glass 1960
	1951	DDT	WA, VA (11 other states)	Glass 1960
	1985	Diflubenzuron	OR	Moffitt et al. 1988
Redbanded leafroller (*Argyrotaenia velutinana*)	1954	TDE	NY, VA (7 other states)	Glass 1960
Fruit tree leafroller (*Archips argyrospilus*)	1977	Diazinon	B. C. Canada	Madsen & Carty 1977
Oblique-banded leafroller (*Choristoneura rosaceana*)	1978	Phosalone	NY	Reissig 1978
Variegated leafroller (*Platynota flavedona*)	1984	Azinphosmethyl	NY	Weires 1985
Tufted apple budmoth (*Platynota idaeusalis*)	1985	Azinphosmethyl	PA	Meaghers & Hull 1986
Spargonathis fruitworm (*Spargonathis sulfureana*)	1984	Azinphosmethyl	NY	Weires 1985
SECONDARY PESTS				
Erythroneura lawsoniana	1953	DDT	KY	Glass 1960
White apple leafhopper (*Typhlocyba pomaria*)	1959	DDT, TDE	WA	Glass 1960
	1970	Azinphosmethyl	NY, WA	Trammel 1974, Hoyt unpubl.

San Jose scale (*Quadraspidiotus perniciosus*)	1908	Lime sulfur	WA	Melander 1914
Spotted tentiform leafminer (*Phyllonorycter blancardella*)	1976	Azinphosmethyl, phosmet	Eastern, midwestern U.S. & Canada	Weires 1977
	1979	Diazinon	Ont., Canada	Pree et al. 1980
	198?	Permethrin, fenvalerate	Ont., Canada	Pree et al. 1987
Green apple (*Aphis pomi*)	1958	Parathion	WA, NC	Hoyt, Rock unpubl.
	1959	Azinphosmethyl	WA, most other areas	FAO 1967
	1965	Diazinon	WA	FAO 1967
Rosy apple aphid (*Dysaphis plantaginea*)	1959	Azinphosmethyl	WA	FAO 1967
European red mite (*Panonychus ulmi*)	1950–51	Parathion, malathion, TEPP, EPN	Most areas	Glass 1960
	1954–59	Azinphosmethyl, demeton, ethion, ovex, genite, fenson, mitox, chlorbenside, chlorobenzilate	CA, WA & other areas	Glass 1960
	1958	Dicofol	OR, WA, NC, IL, B.C.	Herne et al. 1979
	1961	Tetradifon	Pacific Northwest	Hoyt unpubl.
	1966	Oxythioquinox	WA, OR, B.C.	Herne et al. 1979
	1972	Chlordimeform	MI, WA, OR, B.C.	Croft & McGroarty 1977, Herne et al. 1979
	1986	Cyhexatin	NY, WA	Welty et al. 1987, Tanigoshi unpubl.
Two-spotted spider mite (*Tetranychus urticae*)	1955	Parathion, malathion	Most areas	Glass 1960
	1979	Cyhexatin	OR, WA, CA	Croft et al. 1982

Table 20.2. (*Continued*)

Species	Date Reported	Compound	Area	Reference
McDaniel mite (*Tetranychus mcdanieli*)	1951	Parathion	WA	O'Neil & Hantsbarger 1952, Glass 1960
	1954–59	Azinphosmethyl, demeton, ethion, ovex, genite, fenson, mitox, dicofol, chlorobenzilate	WA, B.C.	Hoyt, Madsen unpubl.
	1961	Tetradifon	WA	Hoyt & Kinney 1964, Herne et al. 1979
	1964	Binapacryl	WA	Hoyt & Kinney 1964, Herne et al. 1979
	1966	Oxythioquinox	WA, OR	Herne et al. 1979
	1972	Chlordimeform	WA, OR	Hoyt, Westigard unpubl.
Apple rust mite (*Aculus schlechtendali*)	1952	Parathion	WA	Herne et al. 1979
	1965	Endosulfan	WA	Herne et al. 1979
	1972	Chlordimeform	WA	Herne et al. 1979
	1972	Carbaryl	WA	Hoyt unpubl.

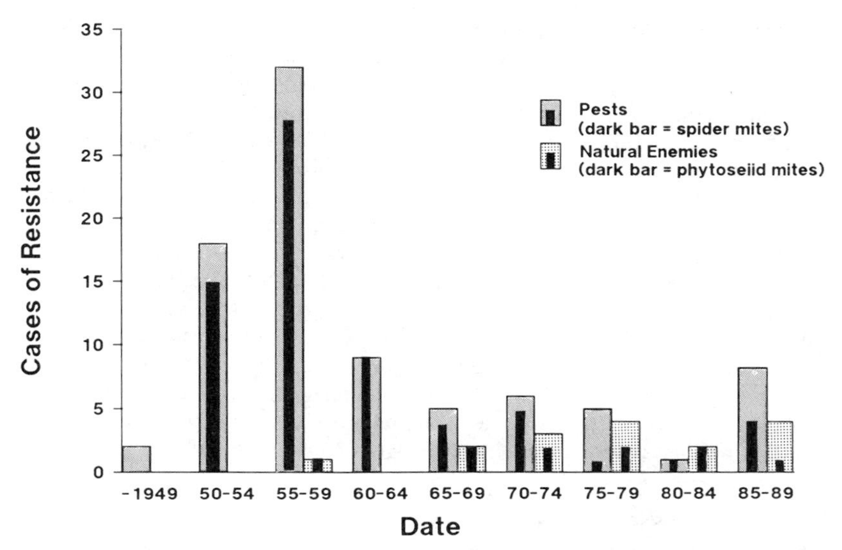

Figure 20.3. Cases of insecticide resistance in arthropod pests and natural enemies of apple from 1908 to 1980. Bars within pest and natural enemy classes represent cases of resistance in spider mites (pests) and phytoseiid mites (natural enemies), respectively. Each unit represents a unique species–compound combination (see Table 20.2, adapted from Croft 1982).

1953. Similarly, the white apple leafhopper *Typhlocyba pomaria* developed DDT and TDE resistance in the 1950s in New York (Glass 1960). In the 1970s and 1980s, the white apple leafhopper and spotted tentiform leafminer *Phyllonorycter blancardella* developed resistance to azinphosmethyl throughout eastern and midwestern North America (Trammel 1974, Weires 1977, Pree et al. 1980). The most recent case of resistance in this group involves permethrin resistance in *P. blancardella* in Ontario, Canada (Pree et al. 1987). Multiple summer applications of pyrethroids have been applied in Ontario for 3–6 years (this use has not been common in most North American apple regions). White apple leafhopper and leafminers such as *Phyllonorycter elmaella* exhibit widespread OP resistance in the western United States (E. H. Beers unpubl., H. W. Riedl unpubl.).

Resistant secondary pests have made it difficult to maintain IPM systems on apple due to the frequent use of nonselective pesticides for their control. Those apple pests with the greatest frequency of resistant species and cases include aphids and spider mites (aphids = 2 species, 4 cases; mites = 7 species, 52 cases; Table 20.2). Resistance to OPs was first observed in aphids in the 1950s (Table 20.2, Fig. 20.3). Both the apple aphid *Aphis pomi* and the rosy apple aphid *Dysaphis plantaginea* are resistant to parathion and azinphosmethyl in most areas of North America.

OP resistance in *T. urticae*, *Panonychus ulmi*, and *Tetranychus mcdanieli* occurred throughout North American apple production areas and caused severe problems in the late 1950s and early 1960s (Table 20.2, Fig. 20.3). Thereafter, spider mites developed resistance to almost every new acaricide developed, including the sulfones, dicofol, tetradifon, binapacryl, and oxythioquinox. Resistance occurred almost predictably on a 3–5-year cycle. Replacement compounds frequently were introduced to combat resistance and reinstate effective chemical control (Table 20.1; Hoyt and Caltagirone 1971).

Since IPM was introduced in 1965 (Chapter 17), acaricide resistance in spider mites has been limited to chlordimeform, propargite, and the organotins (Table 20.2, Fig. 20.3). These resistance cases have not been industry wide, nor have they occurred in orchards practicing IPM (Croft et al. 1987). The organotins and propargite are still effective when used with caution in most production areas. Extensive efforts have been made to develop resistance management programs for spider mites, especially with physiologically selective acaricides, such as the organotins and several newer products (Section 20.7).

Resistance to acaricides in the apple rust mite *Aculus schlechtendali* has been most pronounced in the Pacific Northwest, even though this pest, an alternative prey for predatory mites, is common throughout North America (Table 20.2). In the early 1950s, *A. schlechtendali* developed resistance to OPs, as did spider mites (Table 20.2). More recently, it has developed resistance to endosulfan, chlordimeform, and carbaryl in Washington State (S. C. Hoyt unpubl.). The development of resistance to endosulfan was a disappointment, since this compound is physiologically selective to many natural enemies (Chapter 9, Section 9.5.3.2). Previously, endosulfan was toxic to the apple rust mite, but not to spider mites or most mite predators. It was a useful compound for manipulating biological control systems (e.g., Hoyt 1969a, Croft 1975).

20.5.2. Pesticide Resistance in Apple Natural Enemies

Cases of pesticide resistance among arthropod natural enemies on all crops were reviewed in Chapter 14 (Table 14.1). Almost 50% of cases cited involve predators of apple pests. The predatory mites *Amblyseius fallacis* (Oatman 1966, Croft et al. 1982) and *Metaseiulus occidentalis* (Hoyt 1969a) first developed DDT resistance in the 1950s, although documentation came later. The first documented cases of resistance in natural enemies of apple were parathion and azinphosmethyl resistance in *A. fallacis* (Motoyama et al. 1970) and azinphosmethyl resistance in *M. occidentalis* (Croft and Jeppson 1970) (Chapter 14, Section 14.1). OP resistance in *Stethorus punctum* was first reported in the early 1960s (Dean Asquith pers. comm., Asquith and Hull 1973). Since these early cases, OP resistance has been reported in several other natural enemies of apple pests (Table 14.1).

Aphidoletes aphidimyza, a cecidomyiid predator of aphids, has developed resistance to azinphosmethyl in Massachusetts and Michigan apple orchards (Table 20.2). Resistance may extend throughout the midwestern and eastern United States, but monitoring has been limited. Resistance ranges from 10- to 20-fold in eggs; however, higher resistance levels may be present in other life stages (Warner 1981). Resistance is sufficient to allow survival at recommended field rates. Greater use of this resistant predator for apple aphid management is being emphasized (Prokopy et al. 1981, Warner 1981).

Pree et al. (1989) documented resistance in populations of *Chrysoperla carnea* to many pesticides including DDT, azinphosmethyl, parathion, phosmet, carbaryl, methomyl, and fenvalerate (Table 14.1). The presence of DDT and parathion resistance suggests that resistant strains of this species have long been present in apple orchards. In the modeling studies of Tabashnik and Croft (1985), resistance in *C. carnea* was predicted long before it was observed in the field. The data of Pree et al. (1989) also suggest that resistance was present long before it was observed in the field. Exploitation of these resistant strains of *C. carnea* has not yet been reported.

While the above cases constitute the clearly documented resistance in natural enemies of apple pests, other species may have developed resistant strains or may be in the process of doing so (see Table 20.2). Sayedoleslami (1979) noted that the mymarid *Anagrus epos* and the dryinid *Aphelopus typhlocyba*, which parasitize the white apple leafhopper, readily tolerate sprays of azinphosmethyl in the field. Similarly the aphelinid parasitoids of the woolly apple aphid *Aphelinus mali* and the San Jose scale *Prospaltella perniciosus* are commonly found in apple orchards which are heavily sprayed with OPs. Previously, Johansen (1957) noted that *A. mali* developed resistance to DDT in Washington apples in the 1950s. While high levels of resistance have been less common among parasitoids than predators (Croft 1977, Croft and Strickler 1983), OP resistance found in parasitoids of citrus pests may indicate the potential for resistance in this group (Schoones and Giliomee 1982, Rosenheim and Hoy 1986).

One instance where azinphosmethyl resistance appears not to have developed is in the leafminer parasitoid *Pholetesor ornigis.* Trimble and Pree (1987) compared populations from the same orchard in Ontario, Canada, in samples taken 6 years apart. Adult parasitoids showed low-level resistance to permethrin, fenvalerate, and methomyl, but not to azinphosmethyl (Table 14.1). These data may confirm that some parasitoids have ecological attributes, such as high immigration/residency ratios, that do not favor the persistence of resistance. However, these parasitoids may lack the genetic mechanisms to evolve high levels of resistance (Chapter 19).

20.5.3. The Impact of OP Resistance on Species Composition and Management

Cases of resistance to azinphosmethyl and other OPs are summarized graphically for key pests, secondary pests, and natural enemies of apple in Fig. 20.4. Over time, the development of OP resistance has greatly influenced species composition and has presented many unique problems and opportunities for IPM in apple orchards (Croft 1982). For example, when first employed on apple (early transition period, Fig. 20.4), azinphosmethyl was a broad-spectrum insecticide which had good activity on most primary and secondary pests, but was highly toxic to most natural enemies (Croft and Whalon 1982). Because most mites and aphids quickly developed azinphosmethyl-resistant strains while their natural enemies did not, there was a difficult transition period in the early 1960s when one specialized pesticide after another was required to control resistant mites and aphids.

Because of its effectiveness on the key pests of apple, azinphosmethyl continued to be used. A number of natural enemies have developed azinphosmethyl resistance since the mid 1960s (IPM era, Fig. 20.4). In essence, Fig. 20.4 shows how many elements of the apple ecosystem have adapted or are adapting to this long-term chemical perturbation. Over time, azinphosmethyl has become more selective due to the development of resistance among natural enemies (Croft and Whalon 1982).

20.5.4. Monitoring OP Resistance in Key Pests

Continued susceptibility of key pests of apple to the OPs is the cornerstone of resistance management for the entire pest–natural enemy complex of this crop. Through the use of OPs, selective chemical control and biological control of many secondary pests is possible. The absence of OP resistance in certain key pests is puzzling. It seems highly improbable that such a large group of species as the codling moth, plum curculio, apple maggot, and redbanded leafroller would all lack genetic mechanisms for OP resistance. Evidently, it took 6–20 years for DDT and lead arsenate resistance to develop in the codling moth (Glass 1960, Morgan and Madsen 1976). If selection with OPs is similar to that of lead

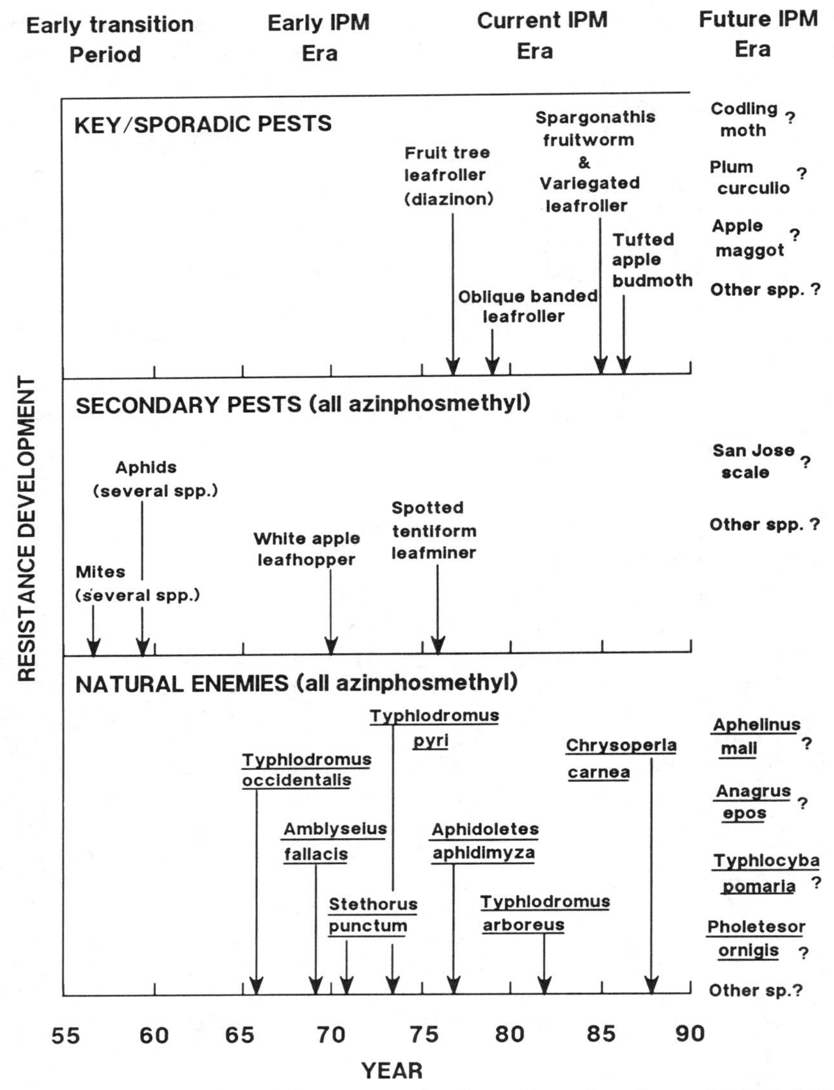

Figure 20.4. Patterns of OP resistance in key pests, secondary pests, and natural enemies associated with deciduous tree fruit crops in North America (after Croft 1982).

arsenate and DDT, then key pests with multiple generations per season should have developed resistance to OPs by now.

Monitoring for OP resistance in key pests associated with apple has been a major component of pesticide resistance management in western North America. Resistance in the codling moth has been of primary concern. Data for azinphosmethyl resistance in *C. pomonella* (LD_{50} values) are summarized in Table 20.3. Barnes and Moffitt (1963) had the insight to develop standard techniques and to establish baselines for both sexes of this

Table 20.3. LD_{50} Values for Strains of Codling Moth to Azinphosmethyl Taken in North American and Australian Apple Orchards (1959–1986) (Adapted from Croft 1982)

Strain	Collection Site	LD_{50} (g/moth)	Reference
R_2	California	0.07	Barnes & Moffitt 1963
Barnes (SS)	California	0.18	
Christie	California	0.10	Bailey & Madsen 1964
Simoni	California	0.10	
Burnley	Australia	0.09 male 0.09 female	D. S. Morris unpubl.
Burnley	Australia	0.19 male 0.23 female	Rose & Hooper 1969
Scarlett	California (Solano Co.)	0.06 male 0.08 female	Riedl et al. 1985
San Jose	California (Santa Clara Co.)	0.12 male 0.12 female	
Wheeler	California (Sacramento Co.)	0.11 male 0.14 female	
Maxwell	California (Eldorado Co.)	0.09 male 0.13 female	
Irving	California (Eldorado Co.)	0.08 male 0.11 female	
Geneva	New York (Ontario Co.)	0.16 male 0.21 female	

species in the late 1950s. Several research groups subsequently tested field strains from western and midwestern North America without finding evidence of significant changes in susceptibility (Bailey and Madsen 1964). Similar data were published in Australia (Rose and Hopper 1969).

Riedl et al. (1985) improved techniques for monitoring resistance in low-density populations of codling moth while conducting extensive monitoring and selection studies with azinphosmethyl. Using pheromone traps to collect moths and topical doses to treat them while entrapped in the sticky surface, they were able to compare results from trap bioassays to standard diagnostic dosage assays. They demonstrated that these tools could be used to detect the development of resistance in moths beyond such variation associated with vigor tolerance (i.e. 3- to 4-fold, Table 20.3). This new tool for monitoring resistance requires no rearing or experimental handling. A regional monitoring program for codling moth is now in place in the western United States using this procedure. Recent surveys of commercial and abandoned apple orchards in Washington, Oregon, Utah, and California have detected a possible resistance factor in the range of 10- to 20-fold using the pheromone trapping system (Riedl unpubl.). The significance of these results is not clear, since field failures have not been noted at the sites surveyed. In selection studies, Riedl et al. (1985) obtained only 3- to 6-fold resistance to azinphosmethyl in larval populations after 9 generations of exposure in the laboratory.

Table 20.4. Estimated Values for Bionomic and Management Characteristics that May Influence the Development of Insecticide Resistance in Key and Secondary Pests and Natural Enemies of Apple (from an expert survey[1] by C. D. Jorgensen and B. A. Croft, unpublished; see Croft 1982)

Species	Generations /Season	Progeny /Female	Carrying Capacity /unit	EIL	Dispersal /Tree /Season	Overwinter /Tree /Season	Adult Mobility Index	% Time in Orchard	% not in Refugia
PEST SPECIES									
Codling moth	2.0	64	1.3/fruit	0.02/fruit	4	6	7.2	100	80
Redbanded leafroller	2.8	180	0.8/fruit	0.01/fruit	16	<1	8.9	60	95
Plum curculio	1.1	114	2.9/fruit	0.03/fruit	8	<1	7.0	90	80
Apple maggot	1.1	185	11.0/fruit	0.02/fruit	7	4	9.7	98	80
San Jose scale	2.6	227	73/cm bark	7.5/cm bark	5	800	2.3	100	80
Green apple aphid	11.1	67	1500/terminal	33/terminal	80	1,200	6.3	97	79
Rosy apple aphid	7.9	50	219/terminal	3.2/terminal	10	115	5.2	45	72
Woolly apple aphid	5.0	40	122/terminal	19/terminal	10	800	2.4	100	75
Spotted tentiform leafminer	3.2	41	25/leaf	2.2/leaf	60*	100	8.2	100	60
European red mite	7.4	49	134/leaf	17/leaf	78	100,000	3.5	100	95
Two-spotted spider mite	12.2	79	80/leaf	10/leaf	50	10,000	2.4	100	99

NATURAL ENEMY SPECIES									
Amblyseius fallacis	7.6	50	8/leaf	—	18	200	4.0	100	92
Metaseialus occidentalis	7.2	27	5/leaf	—	5	85	3.3	100	97
Stethorus punctum	2.8	307	4/leaf	—	55	12	9.0	76	95
Aphidoletes aphidimyza	3.5	62	200/terminal	—	20	10	5.0	97	74
Chrysoperla spp.	2.9	166	1.1/leaf	—	12	3	10.0	94	79
Coccinellids	2.0	183	1/leaf	—	50	1	8.3	8	98
Syrphids	3.3	210	30/leaf	—	40	10	9.5	95	79
Anthocorids	3.7	38	0.05/leaf	—	35	15	8.0	88	79
Aphelinus mali	3.0	75	100/terminal	—	1	10	7.0	99	80
Anagrus epos	3.0		10/cm bark	—	10	100	5.7	98	77
Aphelopus typhlocyba	2.0	50	1/leaf	—	7	50	7.0	95	68

[1]Data listed here are estimates based on expert opinion rather than experimentally derived estimates. They are useful in making comparative assessments of differences between species, which may influence their propensities to develop resistant strains, but they should not be cited as experimental estimates.

*Whalon and Croft (1986) have reported considerably higher levels of immigration of this species into orchards in Michigan under outbreak conditions of this pest.

20.5.5. Species Attributes that Influence Resistance

Considering the patterns of OP resistance that have evolved in the apple arthropod complex, it is useful to consider certain bionomic and management factors that influence resistance. In this regard, apple pest and natural enemy species represent a range of life history and ecological attributes (Table 20.4; see also Chapter 16).

For example, organisms with many generations per season and high reproductive rates would be expected to develop resistance faster than those with fewer generations and progeny. Correlations between reproductive rates and resistance potential among apple species is evident in the data for mites, aphids, leafhoppers, and leafminers. This also applies to generations per season (Table 20.4). The codling moth and redbanded leafroller, key pests that have several generations per season, have not demonstrated as much potential for resistance as have many secondary pests (Table 20.2). The plum curculio and apple maggot are univoltine and apparently have never developed resistance. Among natural enemy species, resistance in phytoseiid mites also supports the association of many generations per season and many progeny with a high propensity for resistance.

The relationship between the number of susceptible organisms colonizing or dispersing into an orchard each season and the number of resistant organisms overwintering within the orchard—the immigration/residency ratio—is undoubtedly one of the most important factors determining the propensity of a species to develop resistance (Georghiou and Taylor 1977a, Comins 1977a, 1981). In this regard, most key pests of apple have relatively small overwintering populations in orchards, but also relatively low colonization rates (almost equal proportions, Table 20.4). This would result in a significant dilution of selected resistant genotypes on a seasonal basis.

Among most secondary pests of apple (Table 20.4), the number of potentially resistant individuals overwintering in an orchard is usually much greater than the number of susceptible organisms colonizing each season (Whalon and Croft 1986). Mites and apple aphids represent the most extreme examples. Within these groups of pests there are two interesting exceptions to this generalization. The San Jose scale has a high overwintering population in orchards (Table 20.4) and a low colonization rate, yet there is no evidence of resistance to azinphosmethyl or other OPs in this species. It may not be highly selected with pesticides due to its tendency to colonize gnarled bark and other refugia that would tend to reduce its rate of resistance development (see Table 20.4). The fact that the OPs used to control this pest are usually combined with oils may also be a factor contributing to its lack of resistance.

The frequency of resistance development in the tentiform leafminer seems to contradict some of the generalizations about resistance. Even though this species has a relatively high ratio of immigration to residency, it has developed high levels of resistance to OPs (Weires 1977, Pree et al. 1980) and more recently to the pyrethroids (Pree et al. 1987). Large population size within a region and a high potential for movement may contribute to the observed propensity to develop resistance rapidly in the field (Whalon and Croft 1986).

Among natural enemy species (Table 20.4), the number of susceptible colonists is almost always equal to or greater than the number of potentially resistant residents, except for predatory phytoseiid mites and possibly certain host-specific microhymenopterous parasitoids (Table 20.4; Whalon and Croft 1986). As noted earlier, the phytoseiid mites have shown the greatest propensity for resistance of all of the natural enemies of apple pests, and there is some evidence that aphelinid species may have adapted also. *S. punctum* is an interesting case of a species that tends to emigrate rather than overwinter in the orchard, yet it has readily developed resistance to azinphosmethyl and other pesticides (Hull and Starner 1983; see also Chapter 14).

In summary, the combined effects of many genetic and ecological factors determine a species' potential to develop resistance in the field. Basd on OP resistance observed to date, ecological factors appear to play the most important role in the development of resistance in species associated with North American apple production over the past 30 years (Fig. 20.4). Those characteristics that appear to be most closely related with the development of resistance are generations per season and immigration to residency ratios. Thus, considerably more attention should be paid to insect population dynamics relative to the development of resistance. Managing resistance in these species may be more effective as these relationships are better understood.

20.5.6. Future Use of Pesticides

Because of the pattern of OP resistance that has developed in the pest–natural enemy complex of apple, researchers are cautious about the use of new broad-spectrum insecticides that might disturb existing IPM systems. They are particularly concerned about compounds such as the pyrethroids, which lack the selectivity provided by azinphosmethyl (Croft and Hoyt 1978, Croft and Whalon 1982, Hull and Starner 1983). Even though azinphosmethyl is not as effective as it once was on some leafrollers, leafminers, and leafhoppers, its selectivity to natural enemies justifies its continued use until resistance develops in a key pest or until a compound with superior selectivity is registered.

In contrast, the SPs permethrin and fenvalerate show more favorable mammalian toxicity and are highly effective on most key pests. They also are more toxic to most sporadic and some secondary pests than azinphosmethyl. Researchers anticipate that the propensity for key pests to develop resistance to the pyrethroids may be high based on cross resistance to DDT and other insecticides used in the past (Table 20.1).

Most tree fruit researchers feel that compounds such as the pyrethroids have a place in apple IPM, but not before their impact within the context of current IPM programs is carefully examined. Pyrethroids could be used to control severe outbreaks of OP-resistant pests. The incorporation of physiological selectivity into these chemicals would greatly enhance their usefulness (Chapter 13), as will manipulation of the entomophagous fauna to achieve the same ends. Selection of SP-resistant predaceous mites (e.g., Hoy et al. 1980, Strickler and Croft 1981, 1982; Chapter 19) is an example of research that will facilitate use of these products where IPM is practiced.

Resistance and the use of pyrethroids might best be managed in the future by carefully phasing these compounds into current, OP-dominated IPM systems. At present pyrethroids are recommended only on a limited basis. For example, a single prebloom application before spider mites are a problem suppresses OP-resistant leafminers or OP-tolerant *Lygus* spp. without serious disruption of biological control. Research will probably identify other specific uses for pyrethroids which similarly preserve biological control.

One final note on the development of broad-spectrum, selective pesticides for orchard pest management is in order. Integration of pesticides and biological control agents in orchard systems should improve as highly selective compounds such as diflubenzuron, other growth regulators, and microbial pesticides are developed and registered (Gruys 1982, Croft and Hull 1988, Croft and Riedl 1988). These products have relatively good activity against key pest of pear and apple, but are more selective to natural enemies than most conventional insecticides used in the past. While these properties may stimulate widespread adoption, these pesticides should be used with caution as the potential for resistance in key pests has already been demonstrated. In experimental plots where diflubenzuron has been intensively applied, resistance in the codling moth has been documented in southern Oregon, United States (Moffitt et al. 1988).

20.6. RESISTANCE MANAGEMENT OF SPIDER MITES AND PREDACEOUS MITES

A discussion of resistance management in spider mites and predaceous mites encompasses both pesticide selectivity and resistance management. The goal is to minimize resistance in spider mites and to encourage resistance in predaceous mites. In mite predators, resistance is needed to the insecticides used to control insect pests. Most current acaricides are physiologically selective and nontoxic to predaceous mites. However, if spider mites develop resistance to these acaricides, field rates may be increased. Higher doses may be toxic to natural enemies and at this point, acaricide resistance is of significance in the entomophage. Of greater concern is the continued effectiveness of acaricides on spider mites. Biological control is the single most important alternative used to limit resistance in spider mites. Where biological control is effective, resistance problems are least acute.

In reviewing the history of resistance management in spider mites and predatory mites, a more general perspective of the types of problems encountered can be discussed. Here, a specific resistance management problem is embedded within a larger resistance management sphere or management problem. Consideration of both predatory and pest mites as well as the entire pest–natural enemy complex demonstrates the scope of resistance management that is often necessary in most crop production systems.

Not only does resistance management require technological improvement in pest control methods, but enlightened policies involving pesticide production, marketing, regulation, and usage must be implemented. Later in this section, the human dimension of scientific development and technology transfer of resistance management research is discussed. This provides an introduction to policy issues and selective pesticide development, which are treated in Chapter 21.

20.6.1. Use and Misuse of the Organotin Acaricides

Studies in the late 1960s and early 1970s with the organotin cyhexatin established its range of selectivity to spider mites and the predatory Phytoseiidae. Rock and Yeargan (1970) compared the toxicity of cyhexatin to adult female *Panonychus ulmi* and *Tetranychus urticae* and to OP-resistant strains of the predatory mite *Amblyseius fallacis*. LC_{50} values demonstrated a 5- to 100-fold greater tolerance in the predaceous mite. Mowrey et al. (1977) reported similar selectivity to the coccinellid *Stethorus punctum* in Pennsylvania apples. Croft and McGroarty (1977) developed a relatively precise concentration–mortality model for selectively reducing *P. ulmi* and *T. urticae* with cyhexatin without affecting populations of *A. fallacis*. Reduced rates of organotins (OTs) could be used to manipulate predator–prey ratios so that predators would subsequently regulate prey at low population densities, as long as prey were susceptible. The OTs were found to be similarly selective to other predatory mite species (e.g., Lienk et al. 1976). The closely related fenbutatin-oxide had selective properties similar to cyhexatin, although it was slightly less active on spider mites. It also was slightly less toxic to predators at equivalent or higher concentrations (e.g., Lienk et al. 1976, Croft et al. 1987).

Relatively high toxicity to spider mites and selectivity to predators allowed OTs to compete in the marketplace with earlier nonselective acaricides. OTs were not only as good or better than older acaricides in killing spider mites, they were also nontoxic to predatory mites at normal field rates. The high activity of OTs led to their misuse and the eventual development of resistance to them.

Because cyhexatin and fenbutatin-oxide were such good acaricides, they were marketed by industry first as acaricides and second as selective products useful in the conservation of natural enemies. The newness of the concept of selectivity, concern over efficacy problems due to reduced kill of pests, and uncertainty as to supplemental biological control provided by predators made companies unwilling to put less than full rate information on labels during the early stages of product sales. Some research and extension scientists promoted conservative use of these products by recommending reduced rates in combination with biological control for spider mites. As noted, many growers used these products as conventional nonselective acaricides. Biological control sometimes followed these applications. Postspray control by predators resulted in fewer subsequent mite control problems compared to nonselective acaricides. The utilization of high doses of these products was undoubtedly unnecessary.

In spite of their misuse, spider mites did not become resistant to the OTs in the normal 5–10 years as they did to earlier nonselective acaricides (Hoyt 1969a). In retrospect, either these products had a unique mode of action or resistance was hard to fix genetically in populations (Croft et al. 1984, Hoyt et al. 1985, Flexner 1988). The continued effectiveness of the OTs only added to their unilateral use. In the 1970s, these products virtually dominated the acaricide markets for deciduous tree fruits. Some wondered if OT resistance would ever develop. Others thought that the lack of OT resistance was due to successful integrated mite management. At a local level there was good evidence for the latter conclusion (Dover and Croft 1984, S. C. Hoyt unpubl.). Where integrated mite management was implemented in the early 1970s, mite populations in many orchards have never shown any signs of OT resistance (Croft et al. 1987). However, many apple growers in the same area who did not implement integrated mite management eventually encountered resistance problems where OTs were used (Croft et al. 1987).

20.6.2. Organotin Resistance

The earliest reports of OT resistance in treated populations of *T. urticae* were from pear in the Rogue River valley of Oregon (Hoyt et al. 1985). Biological control of spider mites on pear is limited by inadequate predator populations and low economic thresholds for mites. Also the use of nonselective pesticides to control pear psylla and codling moth often increases mite problems due to predator mortality. Initially, only 1–2 applications of OTs at low rates were necessary to control spider mites without predators. Due to resistance, application rates doubled, as did the number of applications. Mite control costs increased from about \$25–30/ha in 1972 to \$125/ha resistance-affected orchards by 1983. In some cases, costs for mite control approach 10% of crop value (Hoyt et al. 1985).

In contrast to the situation on pears, the McDaniel spider mite was a serious problem on apples in the mid 1960s in several areas of western North America. During this period, this species developed resistance to virtually all registered acaricides (Hoyt 1969a). Following the establishment of integrated mite management in the mid 1960s, McDaniel spider mite problems were alleviated and this pest was replaced by a complex of apple rust mites, European red mites, and predators. Petroleum oils, an occasional acaricide application, and predators prevented European red mites from reaching damaging levels in most orchards. OTs were registered in the early 1970s and within a few years, growers and consultants realized that a single, low-rate application could control the European red mite. This worked because of the effectiveness of the acaricide and the survival of adequate predaceous mites to prevent resurgence. One component of an integrated program was lacking, however, and that was population monitoring and applications on the basis of need only. A specific timing of the OT application had been established, and these sprays were applied to some orchards regardless of the mite populations. Petroleum oils were eliminated from the program, so selection pressure was exerted solely by OTs.

Many new orchardists and consultants during the 1970s had not experienced the severe mite problems of the 1960s and were quite satisfied with this mite control program. They did not understand the need for acaricide conservation and the strategy of integrated mite management. By the early 1980s, the single OT application was no longer adequate in these orchards. A low level of resistance to OTs had developed in spider mites, and rates were subsequently increased. This reduced the survival of predators, and resurgences of European red mite occurred. It became necessary to use multiple applications of OTs at high rates to control the mites, and predators were virtually eliminated. Resistance levels climbed quickly in these orchards. At present it is again difficult to establish an effective IPM program in these areas. In addition to inadequate control, growers are experiencing high costs, and in some orchards McDaniel spider mites are becoming prevalent again. In contrast, growers who used integrated mite management, applying acaricides only when needed, have not experienced resistance problems or high costs.

Laboratory research with *T. urticae* in the United States (Croft et al. 1984, Hoyt et al. 1985, Flexner et al. 1987) and Australia (Edge and James 1982, 1986) initially indicated that cyhexatin resistance reached a plateau at 5- to 15-fold. Cyhexatin-resistant populations are cross resistant to fenbutatin oxide. At these resistance levels, OTs continue to provide at least partial kill of field populations of spider mites (Hoyt et al. 1985). OT resistance in *T. urticae* is not monogenic nor due to a dominant set of traits (Croft et al. 1984). It is unstable and reverts toward susceptibility in the field when applications are discontinued (Croft et al. 1984, Edge and James 1986). After a moderate break from OT selection, resistance rebounds slowly when selection is resumed (Flexner 1988).

Because OT resistance is unstable and local in occurrence, considerable dilution of resistance by hybridization with OT-susceptible mites occurs in the field (Flexner et al. 1987). Immature stages are less resistant than adults, allowing for improved control, even of resistant mites, at certain times such as early spring. Also, different OT formulations seem to have a differential effect on OT-resistant and susceptible strains of *T. urticae* (Edge and James 1986, Flexner et al. 1987). All of these factors provide a range of features around which resistance management programs have been developed.

20.6.3. New Physiologically Selective Acaricides

There has been a shortage of registered, selective acaricides. As constraints have developed to the unilateral use of OTs, several new compounds have appeared on the market that have equal or greater selectivity (Bouron 1985, Aveyard et al. 1986). Clofentezine, hexythiazox, and abamectin are new acaricides that are chemically unrelated to any currently or previously registered materials. Both clofentezine and hexythiazox are ovicides. They kill oviposited eggs and cause adult female mites to lay sterile eggs. Clofentezine and hexythiazox show good activity on the European red mite and other tetranychids (e.g., Hoy and Ouyang 1986), but are less effective against active stages of these mites.

Because clofentezine and hexythiazox are primarily effective as ovicides, researchers and industry personnel are cautious about using them alone as acaricides. In the past, spider mites have rapidly developed resistance to ovicidal compounds (Hoyt 1969a). Already, resistance to these products has been observed in field and laboratory selection studies when selected for only 15–20 generations (unpublished data). Therefore, research is being conducted on the use of these products at reduced rates or in rotations along with compounds more toxic to active mite stages. As will become more evident, the activity of these compounds on a limited number of life stages may be fortuitous insofar as resistance management is concerned.

Both clofentezine and hexythiazox are at least as compatible with biological control as the OTs. Generally, they are more selective than the OTs to predatory mites (Hoy and Ouyang 1986), for high rates of OTs will kill many phytoseiid species (e.g., Rock and Yeargan 1970). Second, even though the newer compounds kill spider mite eggs, their action is not immediate. Rather, the eggs may accumulate and predatory mites may continue to feed on them (Croft et al. 1987). Finally, these compounds are not highly toxic to the apple rust mite *A. schlechtendali*, an alternate host for predatory mites. The OTs and propargite would almost eliminate these relatively innocuous mites (Croft and McGroarty 1977, Lienk et al. 1976).

Abamectin, a by-product of ascomycete fungi, is a broad-spectrum toxin that is active on many veterinary and agricultural arthropod pests. It is moderately selective to phytoseiid mites (Grafton-Cardwell and Hoy 1983, Croft et al. 1987). Abamectin is similar to the OTs in that it is effective on active life stages of *T. urticae*, although trials with *P. ulmi* have been less promising (S. C. Hoyt unpubl.). Abamectin is active at lower concentrations than earlier acaricides. With regard to resistance, it is speculated that there is the same need for caution with abamectin as with the ovicides discussed earlier. High activity and intense selection on active life stages of mites may lead to rapid resistance development.

There is considerable impetus to conserve these new acaricides by implementing a preventative program of resistance management. Of course, integrated mite management relying on maximum biological control by predators is a major focus of these efforts. It is the overall goal of tree fruit personnel not only to prevent resistance to the new products, but to limit resistance to the OTs and prolong their use.

20.6.4. Combinations and Alternations of Acaricides

With the problems of cyhexatin resistance and the availability of the newer selective acaricides, a multipoint program of resistance management focusing on both groups of products has been developed. The particulars of these programs in different fruit-growing areas depend on the species of phytophagous mites and predators, the levels of OT resistance in spider mites, and the status of IPM programs.

Selective acaricides are best used in combination with biological control by predaceous mites. Predation usually increases during the growing season as predator populations expand. Any early season (or late season in the previous year) mite management tactic that does not select for acaricide resistance improves chances for optimal mite control compatible with resistance management. For example, high levels of biological control in a previous season which decimate overwintering spider mites or careful use of early season oils or fungicides with acaricidal properties are useful means to suppress spider mites. They generally ensure that, early in the growing season, full-rate acaricide sprays are not required for mite control. Where OT-resistant strains of spider mites are present, using OTs or newer selective acaricides during early season will not usually be supplemented by adequate predation because of the small size of predator populations. Therefore, it is best to minimize chemical control of mites at these times by using either the OTs or the newer selective ovicides at full rates.

If *T. urticae* becomes a problem shortly after bloom, then the following uses of OTs or the newer acaricides may be necessary. Mixtures of ovicides with compounds providing good kill of active mite forms may be required to achieve adequate control. Abamectin provides excellent control of two-spotted mite at normal field rates, but higher rates may not be conducive to minimizing the selection of resistance, even though in the short term, excellent spider mite control can be achieved. Therefore, this compound should be used with caution.

Studies of cyhexatin efficacy in Australia (Edge and James 1986) and the Pacific Northwest of the United States (Flexner et al. 1987) have shown that small, early season populations consisting primarily of larvae and overwintering adults of OT-resistant *T. urticae* can be more readily controlled than this same population later in the season. The reason for this difference may be greater susceptibility in larval and early nymphal stages (Edge and James 1986, Flexner et al. 1987). Reasonable control of early season populations of *T. urticae*, even highly resistant strains, may often be obtained with recommended field rates. Sprays must be carefully timed to control populations when eggs are approaching eclosion or have recently eclosed.

When summer applications of physiologically selective acaricides are used, a more heterogeneous distribution of life stages is usually present. Summer use of OTs to control OT-resistant *T. urticae* may not provide adequate control. Therefore, combinations with the newer ovicidal acaricides, normally at reduced rates, will often be effective (Edge and James 1986). As noted, these compounds have different modes of action than the OTs and are active against different life stages. Use of combinations and rotations of acaricides will reduce the chance of resistance development and allow for stability or possibly reversion of OT resistance.

During the summer, alternative uses of both the OTs and newer experimental acaricides are possible because of increased predation. When predators are present at high densities, no acaricides are necessary to achieve spider mite control. Obviously, this is the best tactic for resistance management. When predators are present at insufficient levels to provide complete biological control of *T. urticae*, then reduced rates of OTs, combinations of low rates of OTs and a selective, non-cross-resistant acaricide, or low rates of the new acaricides alone will adjust predator–prey ratios to favor predation.

Because *P. ulmi* differs in its overwintering stage and susceptibility to chemicals, the management of resistance in *P. ulmi* differs appreciably from that for *T. urticae*. (See Croft et al. 1987 for resistance management of this species using acaricides.)

20.6.5. Policy Needs

Considering the past problems encountered in trying to implement selective use of acaricides in integrated mite management programs in North America, recent research has focused on ways to preserve OT acaricides and do a better job of minimizing resistance to the new acaricides. Accomplishing these objectives has required new approaches to resistance problems on the part of the scientific community, extension personnel, pesticide industry personnel, consultants, and growers.

First, the scientific community has done a better job of maintaining integrated mite management programs as crop production, chemical, processing, and marketing factors changed over time. Extension personnel have developed new and better educational aids for integrated mite management for many of the major production regions in North

America. In most cases, in the 1960s and 1970s, implementation of integrated mite management programs was developed primarily by researchers, and delivery to the public was not carefully planned (Whalon and Croft 1984). Appropriate bulletins and educational programs have done much to restimulate interest in integrated mite management and resistance management.

The chemical industry has made resistance management and integrated mite management high priorities in promoting acaricides. This difference in philosophy is not only held more commonly by researchers, but also by upper management, marketing, and sales personnel. In several instances, resistance management research programs for the new selective ovicides have been put in place even before these products have been widely marketed. Label information proposed by industry and supported by regulatory agencies would restrict ovicide use to once per season. Industry-sponsored detection and monitoring systems are being put into place to evaluate the status of resistance in the field.

Private consultants and growers have become better informed and more committed to practicing integrated mite management. The development of extension materials and of chemical products for their use has facilitated this transition. Lastly, there has been increased coordination and cooperation among all of the groups discussed above as well as among cooperatives and marketing groups. State and federal regulation also has improved.

In retrospect, research and management tools were in place to implement integrated mite management and resistance management of spider mites on apple in several areas of North America nearly two decades ago. Successful implementation was reported from several fruit-growing areas, and it appeared that these programs would replace unilateral chemical control systems on a widespread basis. However, because of problems at several levels of research and implementation, the agricultural community failed to obtain or maintain adoption on as widespread a basis as seemed possible. In other cases, certain crop conditions or ecological situations were not amenable to these methods. In the early 1980s, this community was again faced with increasing resistance problems, in part because of the inappropriate use of the OTs and other acaricides. The agricultural community now has available to it a new round of selective acaricides with which to practice integrated mite management and acaricide resistance management. Hopefully, with both the technological and the policy developments mentioned above, these valuable products will not be misused and the supply of chemicals needed for integrated mite management will not be depleted.

20.7. SUMMARY

Pesticide resistance among arthropod pests and natural enemies of apple in North America has evolved and is being managed to the extent that there are many desirable attributes of selectivity and stability as far as IPM is concerned. While control of a few secondary pests is presently a problem due largely to the development of resistant populations, this is a minor detriment compared to the difficulties that resistance in a key pest species such as the codling moth could cause. In the event of OP resistance among key pests, growers will undoubtedly move to a new class of chemical pesticides, most likely the pyrethroids. In such a case, IPM on apple would be set back somewhat, since there is evidence that the insect and mite fauna would change appreciably. In apple ecosystems there is a large pesticide-adapted fauna of pests and natural enemies. Preliminary research has shown that pests such as spider mites and woolly apple aphid are likely to increase with pyrethroid use (e.g., Hoyt et al. 1978, Croft and Hoyt 1978). Also, natural enemy complexes may be altered substantially by pyrethroid-based programs. In Pennsylvania, the predatory mirid *Hyaliodes vitripennis*, largely a scarce species, became the predominant natural enemy associated with mites and aphids under the common use of pyrethroids (Hull and Asquith 1978). Whalon (unpubl.) observed low toxicity to hemipteran species in studies of arthropod pests and natural enemies on newly established apple orchards receiving regular pyrethroid treatments. Not only did the predator *Deraeocoris nebulosis* become more common in permethrin-sprayed plots, but there were marked increases in two plant-feeding mirids, *Campylomma verbasci* and *Plagiognathus politus flaveolus*.

Since these studies with pyrethroid insecticides, researchers have continued to explore ways to selectively integrate new pesticides into apple IPM systems so as to minimize the disruption of OP-resistant natural enemies (e.g., Hull and Starner 1983, Hull et al. 1985a, 1985b). New compounds with selective toxicity to different pests have been discovered (Croft and Hull 1988), but only modest improvements in selectivity to natural enemies have been made. The greatest range of selectivity has been obtained with insect growth regulators and chitin inhibitors (Chapter 12). As discussed by Croft and Whalon (1983), either selective chemicals must be developed or resistant natural enemies must be obtained so that nonselective compounds can be used in current IPM systems.

Rather than making an abrupt change from one insecticide class to another, a gradual transition would be more desirable. This will be possible only if resistance in one or more key pests does not develop soon. Certainly the steps taken to develop predatory mite strains resistant to SPs are preparation for such a transition (Chapter 19). Although these projects have often resulted in resistance that is relatively unstable in the field (Chapter 19), more long-term research on field and artificial selection should overcome these problems. Furthermore, screening and testing to identify more selective pyrethroids for use in orchards are continuing with modest success. Several research activities are essential to further the transition of IPM and resistance management. Effective monitoring systems are necessary to determine the status and potential for resistance among the key pests and natural enemies associated with tree fruits. A variety of operational tactics such as chemical rotations, alternative application methods, nonpersistent compounds, reduced selection levels of pests, and others are needed to reduce the frequency and extent of resistance in pests while exploiting the benefits of resistance in natural enemies. In more specific pest–natural enemy systems, resistance management programs are making good progress in conserving such important chemical tools as physiologically selective acaricides. With their continued development and greater reliance on biological control, continued use of these products should be possible for years to come.

Within the scope of a future resistance management scenario for tree fruits, it is not unreasonable to speculate that instead of experiencing a 5- to 15-year cycle before resistance evolves to a particular chemical class, as was typical with earlier pesticides, it may be possible to use more diversified systems of chemical use combined with biological control to prolong the period of insecticide efficacy considerably. Resistance could be of much less importance in future systems of pest control on tree fruits. In these cases, long-term resources and the attendant costs associated with new insecticide production, registration, and use would be reduced appreciably. Resistance management is viewed as a desirable and not unrealistic goal for future IPM systems. Similar or improved systems of resistance management should be sought for other high-value crops harboring diverse groups of arthropod pests and natural enemies.

PART SEVEN

CONCLUSIONS

21

TECHNOLOGY AND POLICY ISSUES OF SELECTIVE PESTICIDE DEVELOPMENT

21.1. INTRODUCTION

The efforts made to develop pesticide-resistant natural enemies and to use pesticides in ecologically selective ways (Chapters 10–20) are symptoms of the limited availability of physiologically selective pesticides. This final chapter addresses some of the constraints to having more broadly selective pesticides for use in agriculture and natural resource management.

In this regard, I remember a day in 1969 when, as a graduate student, I went to Dr. B. L. Bartlett at the University of California, Riverside, and inquired as to the number of cases in which selective pesticides had been used in commercial agriculture. Dr. Bartlett was rather pessimistic about the possibility that we would soon see widespread development of selective pesticides. He pointed out that many complex factors were involved in the development, production, and registration of these products.

It is disappointing that we can ask this question today, some 20 years later, and the answer is the same. Few physiologically selective insecticides are available to practice IPM for a broad range of species (Chapter 9), and the reasons why involve a complex set of factors. Some of these are technological, but human and institutional factors or policies also limit their development and use.

I view the limited availability of selective pesticides to be a problem of policy more than of technology. The lack of physiologically selective pesticides is a case in which technology could provide for reduced inputs and increased sustainability of food production. Many environmental benefits could be achieved through such developments. Instead, only minor improvements have been made to conserve natural enemies through selective pesticide use in the past three decades. In most cases, improvements have involved the use of nonselective pesticides in ecologically selective ways rather than the development of physiologically selective compounds (Chapters 9–13).

21.2. TECHNICAL PROGRESS IN SELECTIVE PESTICIDE DEVELOPMENT

It is feasible that the scarcity of physiologically selective pesticides will be overcome in the next decade through emerging technologies of pesticide design. However, there are some formidable problems that must be overcome before these tools can be widely employed in the field (Jackson 1984, Brown 1988). These constraints may come from the potential side effects and means of delivering such agents in the field (Chapter 19). There are several possible sources from whence such selective pesticides could arise.

New selective pesticides may include novel insect pathogens or their by-products. Strains of *Bacillus thuringiensis*, which show high activity to lepidopteran (variety Kurstaki), dipteran (variety *israelensis*), and coleopteran (variety Tenebrionidae) pests, may be manipulated using genetic engineering techniques to increase their range, their specificity, or their activity (Kirschbaum 1985, Comai and Stalker 1984, Barnes and Lavrik 1986). Already, genes that produce beta endotoxin of *B.t.* have been transferred into tobacco, cotton, soybean, cabbage, potato, sunflower, tomato, and maize (Barnes and Lavrik 1986, Vaeck et al. 1987, Adang and Binns 1988). These genes undoubtedly will be transformed into many other commercial crops (Comai and Stalker 1984).

Genetically engineered microbial by-products such as the *B.t.* beta endotoxin can be highly toxic to both pests and natural enemies when placed in direct contact with these organisms (e.g., Herbert and Harper 1986, Hoy and Ouyang 1987). Their incorporation into plants should confer ecological selectivity to predators and parasitoids; however, any facultative natural enemies which feed on both pests and plants may be impacted by these agents through food chain effects.

Undoubtedly, the presence of *B.t.* endotocin genes in all plants of a population will put extreme selection pressure on pests, which may lead to rapid resistance development. Steps to combat the rapid development of resistance to genetically transformed crops have been proposed (Beegle 1988). Most involve monitoring for changes in successful pest attacks on these plants, and replacing failing varieties with new genetically improved variants of the crop. Some have proposed that crops containing *B.t.* endotoxin genes should be less frequently employed and used in combination with alternative controls such as predators and parasitoids. Based on experience with pesticide resistance in pests, these more diversified and ecologically based strategies may be more effective in stemming the loss of genetically improved crop plants. One encouraging sign is that the biotechnology industry seems to be concerned about the problems of resistance to these products. Hopefully, preventive resistance management programs will be used, similar to those being developed for acaricides (Chapter 20).

Another possibility for developing improved selective pesticides may come from the design of inhibitors of neurohormonal proteins that are specific for various groups of pests (Greenhalgh and Roberts 1987). Recently, five neuropeptides from *Periplanata americana, Schistocera gregaria, Locusta migratoria, Manduca sexta,* and *Heliothis virescens* have been identified (Schooley et al. 1987). Once inhibitors are discovered which degrade key neuropeptides at a single site, then appropriate insect control agents may be developed. Another application of this class of agents may come from the insertion of neuropeptide genes into viral or bacterial pathogens of insects, or even gut symbionts. There is some justified concern whether neuropeptide-associated control mechanisms will be too species-specific to be economically feasible. Can a broader range of selectivity such as between groups of phytophages and entomophages be incorporated into such metabolic systems?

New selective pesticides may be developed as proinsecticides, which are inactive in their original form but transformed to active states by plants or pests, but ideally not by natural enemies (Fukuto 1984, Brattsten unpubl.). Certain selective systemic pesticides currently have broad physiological selectivity in addition to having appreciable ecological selectivity attributes (e.g., pirimicarb; see also Chapters 9 and 10). Further emphasis on the development of physiologically selective proinsecticides is warranted.

More successful identification of physiologically selective pesticides may be attained with traditional screening studies, if additional resources are made available for basic research. Structure–activity studies for pesticide selectivity were reported by Stevenson and Walters (1984) and Elliott et al. (1983). Rather than randomly screening different products against candidate pests, they tested the impact of various pyrethroids against *Venturia canescens*, the ichneumonid parasite of *Ephestia kuehniella*. Pyrethroids that are 8–100 times more toxic to the pest than to the parasite were identified. More studies of this type are needed in the future.

21.3. POLICY ISSUES OF SELECTIVE PESTICIDE DEVELOPMENT

There are policies that constrain the development of physiologically selective pesticides. They involve public and private institutions, laws, and attitudes associated with pesticide research, development, and regulation. Policy aspects of selective pesticide use are more controversial than technological issues. Policy and selective pesticide development are seldom treated in the scientific or pesticide industry literature (Jackson 1984, Brown 1988).

Why has it been so difficult to develop and implement physiologically selective pesticides for field use? Considering differences in herbivorous pests and entomophagous natural enemies, there are many avenues to selectivity mechanisms (selectrophores) that might be exploited (Mullin and Croft 1985; Chapter 6). The research on physiological selectivity mechanisms per se is meager. For example, there are only a handful of comparative studies of detoxification in pests and natural enemies (Chapters 6 and 16). The limited research available on the uptake of pesticides (Chapter 3) and on factors affecting susceptibility to pesticides in natural enemies (Chapter 4) is indicative of this paucity of study. Both fields of study reflect a lack of effort by insect toxicologists and biological control specialists to pursue basic research on selectivity. Funding limitations for work in these fields are a major constraint. Perhaps industry believes that there are too few leads on how to go about developing selective pesticides or that such research is cost prohibitive. In general, conventional screening has most commonly been employed to detect selective compounds in the past. Often screening for selectivity occurs after a pesticide has been registered or at best, after it is in the field development stages of testing. Very few companies screen for selectivity to natural enemies in experimental work to identify candidate materials for commercial development.

Evidence that the most important impediments to developing selective pesticides are not technological is indicated by the development of selective herbicides and acaricides. Breakthrough in the development of selective herbicides began in the late 1950s with the discovery of phenoxy herbicides (Hatzios and Penner 1982, Duke 1985). Research has continued with discoveries of a vast array of products. Some of these pesticides operate on a specific basis, but many are broadly selective (i.e., toxic to several weeds as well as not affecting several crops). As with arthropods, differential uptake, penetration, translocation, and metabolism of these herbicides are the basic mechanisms that account for the selectivity observed among herbicides to different plant species (Hatzios and Penner 1982). In general, plants metabolize xenobiotic toxins by oxidation, hydrolysis, and glutathion conjugation. In addition to these basic mechanism, plants have a number of photochemically reductive mechanisms that can be involved in the metabolism of these chemicals.

To date, both quantitative and qualitative differences in metabolism have been implicated as the basis for herbicide selectivity. In terms of qualitative metabolic differences, the phenoxy herbicides (e.g., 2-4-D) are selective primarily against dicotyledonous weeds, but are not toxic to the monocotyledonous crop plants (Hatzios and Penner 1982). Even among dicotyledonous weeds and crops, different metabolic pathways have been identified and exploited to achieve broad selectivity with these compounds. Examples in which quantitative differences in metabolism have been exploited to achieve a broad level to selectivity between weeds and crops include: 1) the benzoic acid herbicides chloramben and dicamba used on squash, snap bean, soybean, corn, and cucumber against velvetleaf, giant foxtail, barnyard grass, and wild mustard weeds; 2) the trazinone herbicide metribuzin used on sugar beet against pigweed, nightshade, and barnyard grass; and 3) bentazon, to which rice, soybean, corn, and navy bean are tolerant while flatsedge, dwarf arrowhead, kuroguawai, black nightshade, cocklebur, and thistle are susceptible (see Hatzios and Penner 1982).

The point is that crops and weeds are no more distantly related and biochemically or toxicologically different than natural enemies and their prey or hosts. In fact, a number of arguments can be made to support the contention that arthropods that feed on animal versus plant tissues are more likely to have different physiological systems than weed and crop plants (Mullin and Croft 1985; Chapter 6).

Another example in which high priority has been given to physiological selectivity involves acaricides used in integrated mite management (Chapters 17 and 20). This class of pesticides contains the greatest number of physiologically selective compounds surveyed to date (Chapter 9). While selectivity was initially considered a bonus or a secondary requirement to overall pest mite activity, selectivity to predatory mites has become almost an equal prerequisite for a new acaricide. Most companies develop acaricide selectivity data very early in the registration process. Selectivity attributes are widely heralded in marketing and advertising programs for acaricidal products.

A plausible explanation for why selectivity has been less commonly researched in pest insect–entomophage systems as compared to weed–crop and spider mite–predatory mite systems is because of the perceived economics of the research. The value of conserving crops and predatory mites is well established, but similar values for many insect natural enemies may be less well documented. The value of selective pesticides in different crop ecosystems would be variable, depending on the effectiveness of biological control. One of the important tasks for the scientific community is to better document the cost /benefits of selective pesticides in the field.

Another area of policy that requires attention relates to the regulatory laws that influence the registration and use of selective pesticides. The Environmental Protection Agency (EPA) could implement regulatory provisions that favor physiologically selective compounds to a greater degree. Using selectivity or natural enemy conservation as part of the benefit–risk and efficacy equation of pesticides would be useful. Examples of regulatory mechanisms that might support the development and use of selective pesticides would be incorporating selectivity data into pesticide registration requirements, establishing selective pesticide use by prescription in sensitive situations, and developing criteria for using selectivity as a basis for emergency use.

An example in which economics, the policies of industry, and the regulation of EPA have limited the registration of a selective pesticide involves pirimicarb in the United States. Early studies demonstrated that this compound is selective and can be used effectively on many crops to conserve biological control agents while providing control of plant-sucking insect pests (e.g., Helgesen and Tauber 1974, Brown et al. 1983). In analyses from the SELCTV database, pirimicarb had the highest favorable selectivity ratio for a wide range of natural enemy species compared to their prey or hosts (Chapters 2 and 9). In many countries, this compound has been registered for use for decades (e.g., Gruys 1980a, Kalushkov 1982). In the United States, petitions to the EPA for registration in the early 1980s were stopped by questions regarding the high mammalian toxicity of pirimicarb. Studies to clarify these effects were estimated to cost up to $1 million. Because of these added requirements for registration, the parent company of the product chose to rescind its petition for registration (Federal Register 1981). They anticipated that the market would be insufficient to justify the added cost for gathering this information. Also, the patent life of this product was about to end, so the company decided to spend its resources working with the product outside of the United States, where registration was more readily obtainable and not subject to such testing.

A decade later, insecticide resistance in aphids on tree fruit crops in the United States precipitated a request to reconsider the registration of pirimicarb in the western United States. Resistance to several nonsystemic OPs, such as diazinon, azinphosmethyl, and phosalone, is widespread in several aphid pests of tree fruits. Scientists asked the parent pesticide company and EPA to reconsider the registration of this potentially valuable compound (S. C. Hoyt pers. corresp. to EPA). In justifying this action it was stated that:

> Results of tests conducted in several apple growing areas in the late 1970s indicated that Pirimor (pirimicarb) provided excellent control of the apple aphid (*Aphis pomi*), including organophosphate-resistant strains. Research on the selectivity of Pirimor to natural enemies, including a wide variety occurring on deciduous tree fruits, has shown this compound to be one of the least harmful aphicides. This selectivity extends to almost all families of natural enemies tested. Therefore, the compound had high potential for use in IPM ... We feel the current difficulties in apple aphid control warrant a reexamination by the EPA ... of the potential for registration of Pirimor.

As yet no response to this request has been received.

While the example cited above implies that regulatory and industry groups are doing little to promote the development and registration of physiologically selective pesticides, this is not entirely true. In some cases, the value of physiological selectivity has been recognized. One example of an industry group putting together packages of pesticides to emphasize a selective approach was reported by Collins et al. (1984). Scientists from different parent companies identified different ways to use the physiologically selective chlorfluazuron, a benzoylphenyl urea growth regulator; buprofezin, another novel insect growth regulator; a proprietary sex pheromone product of the pink bollworm; and pirimicarb. They suggest that in this case, "the correct tools are now available for IPM of major pest species. It only remains for applied entomologists to devise strategies for the optimal use of these compounds in particular crops."

Davies (1982) cited a number of examples of ways that industry is trying to assist in the development and use of selective pesticides in IPM. Aspects of screening, field assessment, monitoring, and cooperation among pest management personnel were emphasized. In summarizing the discussion of industry representatives who addressed selective pesticide development, the bottom line was the marketing and economic side of the question. In the past, the value of selective pesticides has not been adequately addressed by segments of the pest control research community. Further research to document the value of biological control agents is needed to justify expenditures for research on selective pesticide development in the future.

21.4. CONCLUSIONS

Our inability to develop physiologically selective products represents somewhat of a failure by public and private institutions to act with vision. The problem is akin to pesticide resistance problems which have plagued agriculturists for years without significant progress having been made to solve the problem. Only recently have the complex policy constraints surrounding pesticide resistance come under scrutiny (Georghiou and Saito 1983, Dover and Croft 1984, NAS/NRC 1986). One outcome is a greater commitment to lessen the problems of pesticide resistance and to better conserve these products. Similar efforts are needed to develop physiologically selective pesticides which conserve effective biological control agents. The policies of selective pesticide development are even more basic to effective IPM than those of pesticide resistance management.

APPENDIX

CASES OF PESTICIDE SELECTIVITY TO ARTHROPOD NATURAL ENEMIES OVER THEIR PREY OR HOSTS

Table A.1. Cases of Pesticide Selectivity to Arthropod Natural Enemies over Their Prey or Hosts: Early Inorganics and Botanicals

Crop	Natural Enemy	Prey/Host	Test Method[1]	Selectivity Class[2]	Selectivity Type[3]	Selectivity Ratio	Reference
BORDEAUX							
??	*Chrysopa formosa*	Several	C	n-p	P	—	Babrikova 1980
Apple	*Anystis agilis, Zetzellia mali, Typhlodromus pyri, Deraeocoris fasciolus, Hyaliodes harti, Plagiognathus obscurus, Haplothrips faurei, Aphytis mytilapidis*	Several	F	nn-pp	?	—	MacPhee & Sanford 1961
LEAD ARSENATE							
Apple	*Stethorus punctum*	*Panonychus ulmi*	F	n-pp	E	—	Asquith & Hull 1973
	Several phytoseiid spp.	Several	F	nn-pp	E	—	Dabrowski 1970a
	Amblyseius fallacis	Several	LC	n-pp	E	—	Daugherty 1953
	Stethorus nigripes, Tetranychus urticae	*Panonychus ulmi*	C	n-pp	E	—	Edwards & Hodgson 1973
	Psallus ambiguus, Blepharidopterus angulatus, Anthocorus nemorum, Propylea 14-*punctata, Adalia punctata*	Aphids, mites, psyllids	C	nn-pp	E	—	Gratwick 1965

Table A.1. (*Continued*)

Crop	Natural Enemy	Prey/Host	Test Method[1]	Selectivity Class[2]	Selectivity Type[3]	Selectivity Ratio	Reference
Bean	*Hippodamia convergens*	Aphids	C	n-p	E	—	Haug & Peterson 1938
Pear	*Coccinella septempunctata, Harmonia axyridus Propylaea japonica, Theridion octomaculatum, Misumena tricuspidata*	*Toxoptera pyricola, Myzus malisuctus*	F	nn-pp	E	—	Hukusima & Kondo 1962
Apple	*Zetzellia mali, Typhlodromus pyri, Atractotomus mali, Deraeocoris fasciolus, Hyaliodes harti, Anthocoris musculus, Haplothrips faurei, Leptothrips mali, Aphytis mytilaspidis*	Several	F	nn-pp	?	—	MacPhee & Sanford 1961
	Anthocoris nemorum, Chrysoperla carnea, Coccinella septempunctata, Psallus ambiguus, Deraeocoris lutescens, Himacerus apterus, Malacoris chlorizans, Orthotylus marginalis, Phytocoris reuteri	Many	C	nn-pp	B	—	Niemczyk et al. 1979
Citrus	*Leptomastidea abnormis, Scynmus includens*	*Planococcus citri*	C	nn-p	P	—	Viggiani et al. 1972

NICOTINE SULFATE							
Greenhouse	*Encarsia formosa*	*Trialeurodes vaporariorum*	C	n-p	B	—	Ledieu 1979
Citrus	*Leptomastidea abnormis, Scymnus includens*	*Planococcus citri*	C	nn-p	P	—	Viggiani et al. 1972
Beans	*Adalia bipunctata, Coccinella septempunctata*	*Aphis fabae*	F	n-p	?	—	Way et al. 1954
Alfalfa	*Hippodamia* spp.	*Therioaphis maculata*	LC	nn-p	P	45 >240	Bartlett 1958
Apple	Several phytoseiid spp.	Several	F	nn-pp	E	—	Dabrowski 1970a
	Zetzellia mali, Phytoseius macropilis, Typhlodromus spp., *Atractotomus mali, Campylomma verbasci, Deraeocoris* spp.	Several	F	nn-pp	?	—	MacPhee & Sanford 1961
	Aphelinus mali	*Eriosoma lanigerum*	F	n-p	?	—	Noble 1935
Vegetables	*Coccinella septempunctata*	*Lipaphis erysimi*	LC	n-p	P	—	Pradhan et al. 1959
Greenhouse	*Encarsia formosa*	*Trialeurodes vaporariorum*	C	n-p	P	—	Wiackowski & Herman 1971
Several	*Chrysoperla carnea*	Several aphids	C	n-pp	P	—	Wiackowski 1968
Apple	*Holocothorax testaceipes,* other parasites	*Phyllonorycter ringoneiella*	F	nn-p	?	—	Yamada & Kawashima 1983

Table A.1. *(Continued)*

Crop	Natural Enemy	Prey/Host	Test Method[1]	Selectivity Class[2]	Selectivity Type[3]	Selectivity Ratio	Reference
PYRETHRUM							
Apple	*Typhlodromus tilarium*	*Panonychus ulmi*	F	n-p	?	—	Karg 1978
Greenhouse	*Encarsia formosa*	*Trialeurodes vaporariorum*	C	n-p	B	—	Ledieu 1979
Vegetables	*Phygaedeuon trichops*	*Delia* spp.	C	n-p	P	—	Naton 1978
Strawberries	Carabids, staphylinids, spiders	?	F	nn-p	?	—	Nikusch & Gernot 1986b
Citrus	*Leptomastidea abnormis, Scymnus includens*	*Planococcus citri*	C	nn-p	P	—	Viggiani et al. 1972
RYANIA							
Apple	*Psallus ambiguus, Blepharidopterus angulatus, Anthocorus nemorum, Propylea* 14-*punctata, Adalia punctata*	Mites, aphids, psyllids	C	nn-pp	?	—	Gratwick 1965
	Typhlodromus tilarium, Amblyseius finlandicus, Zetzellia mali	*Panonychus ulmi, Tetranychus urticae*	C	nn-pp	?	—	Karg 1970
	Anystis agilis, Zetzellia mali, Phytoseius macropilis,	Several	F	nn-pp	?	—	MacPhee & Sanford 1961

	Typhlodromus spp., *Campylomma verbasci*, *Deraeocoris fasciolus*, *Haplothrips faurei*, *Leptothrips mali*, *Aphytis mytilaspidis*						
	Trichogramma embryophagum	*Galleria mellonella*	C	n-p	?	—	Stein 1961
Fruit	*Ooencyrtus kuwanae*, *Telenomus terebrans*	*Malacosoma neustria*	C	nn-p	P	—	Tadic 1979
Apple	*Stethorus loxtoni*, *S. nigripes*, *S. vagans*	*Tetranychus urticae*	C	n-p	P	—	Walters 1976b
Pear	*Deraeocorus brevis piceatus*, *D. fasciolus*, *Chrysoperla carnea*, *C. oculata*	*Psylla pyricola*	F	n-pp	?	—	McMullen & Jong 1967
TARTAR EMETIC							
Citrus	*Comperiella bifasciata*	*Aonidiella aurantii*	F	n-p	P	—	Brandt 1982

[1] F = field; C = contact; LC = lethal concentration; LD = lethal dosage; S = systemic; O = oral; R = residue.
[2] n = single natural enemy species; nn = multiple; p = single prey or host species; pp = multiple.
[3] P = physiological selectivity; E = ecological; B = both P & E; ? = uncertain.

Table A.2. Cases of Insecticide Selectivity to Arthropod Natural Enemies over Their Prey or Hosts: DDT and Organochlorines

Crop	Natural Enemy	Prey/Host	Test Method[1]	Selectivity Class[2]	Selectivity Type[3]	Selectivity Ratio	Reference
ALDRIN							
Mustard	*Coccinella septempunctata*	*Liaphis erysimi*	LC	n-p	P	—	Singh & Malhotra 1975a
BHC							
Apple	*Aphelinus mali*	*Eriosoma lanigerum*	F	n-p	E	—	Evenhuis 1959
CHLORDANE							
Poultry	*Macrocheles muscadomestica*	*Musca domestica*	LC	n-p	P	9	Axtell 1966
DDT							
Apple	*Aphelinus mali*	*Eriosoma lanigerum*	F	n-p	E	—	Evenhuis 1959
DIELDRIN							
Structural pests	*Loxosceles spinulosa, Oxopes longispinosa, Salticus* sp.	*Hodotermes nossambicus*	F	nn-p	?	—	Dippenaar et al. 1978
Cereal	*Coleomegilla maculata*	*Oulema melanopa*	C	n-p	P	—	Yun & Ruppel 1964
ENDOSULFAN							
Sunflower	*Coccinella repanda*	*Heliothis* spp.	C	n-p	P	—	Broadly 1983
Many	*Trichogramma* spp.	Several	C, R, F	nn-pp	P	—	Bull & Coleman 1985
Rice	*Cyrtorhinus lividipennis*	Rice hoppers	C	n-pp	P	—	Chelliah & Rajendran 1984
Apple	*Amblyseius fallacis*	*Panonychus ulmi, Tetranychus urticae*	C	n-pp	P	—	Croft & Nelson 1972
	Neoseiulus caudiglans	*Panonychus ulmi*	C	n-p	?	—	Downing 1966

Citrus	*Aphytis melinus*	*Aonidiella aurantii*	C	n-p	P	—	Davies & McLaren 1977
Rape	*Philonthus fuscipennis, Tachyporus hypnorum*	Several	C	nn-pp	P	—	Eghtedar 1969
ENDOSULFAN							
Apple	*Aphelinus mali*	*Eriosoma lanigerum*	F	n-p	E	—	Evenhuis 1959
Rice	*Cyrtorhinus lividipennis, Microvelia atrolineata, Lycosa pseudoannulata*	*Nilaparvata lugens*	C	nn-p	P	—	Fabellar & Heinrichs 1984
Several	*Trichogramma cacoeciae, Pales pavida, Pimpla turionellae, Chrysoperla carnea, Leptomastix dactylopii*	Several	C	nn-pp	P	—	Franz et al. 1980
Alfalfa	*Trioxys complanatus*	*Therioaphis trifolii, Therioaphis maculata, Acyrthosiphon kondoi*	C, R	n-pp	B	—	Franzmann & Rossiter 1981
Sorghum	*Tetrastichus* spp.	*Contarinia sorghicola*	F	n-p	?	—	Garg & Taley 1977
Citrus	*Cales noacki*	*Aleurothrixus floccosus*	C	n-p	E	—	Garrido et al. 1982
Apple	*Typhlodromus pyri, Chrysocharis prodice, Aphelinus mali*	Several aphids	C, F	nn-pp	B	–	Gruys 1980b
?	*Trichogramma japonicum*	*Corcyra cephalonica*	C	n-p	?	—	Gupta et al. 1984
Apple	*Pterostichus melanarius, Harpalus affinis, Amara* spp.	Several	LD, F	nn-pp	P	—	Hagley et al. 1980
	Amblyseius fallacis	*Panonychus ulmi, Tetranychus urticae, Aculus schlechtendali*	C	n-pp	P	—	Hislop et al. 1978

Table A.2. (*Continued*)

Crop	Natural Enemy	Prey/Host	Test Method[1]	Selectivity Class[2]	Selectivity Type[3]	Selectivity Ratio	Reference
Greenhouse	*Scolothrips longicornis, Encarsia formosa*	Several	LC	nn-pp	B	—	Iacob & Posoiu 1981 Iacob et al. 1981
Cotton	*Trichogramma pretiosum*	*Heliothis zea*	R	n-p	P	—	Jacobs et al. 1984
Alfalfa	*Coccinella repanda*	*Therioaphis trifolii, Acyrthosiphon kondoi*	LD	n-pp	P	—	Kay 1979
Vegetables	*Amblyseius tetranychivorus*	Spider mites	R	n-pp	P	—	Krishnamoorthy 1983
Apple	*Amblyseius fallacis, Typhlodromus pyri*	*Panonychus ulmi, Aculus schlechtendali*	C	n-p	P	—	Lienk et al. 1976, 1978
Potato	*Copidosoma kowhleri*	*Phthorimaea operculella*	C	n-p	P	—	Makar & Kadhav 1980
?	*Menochilus sexmaculatus*	Several aphids	C	n-pp	P	—	Makar & Kadhav 1981
Cabbage	*Coccinella repanda*	*Brevicoryne brassicae*	C	n-p	P	—	Mishra & Satpathy 1985
Apple	*Hippodamia convergens*	Several aphids	LD	n-pp	P	—	Moffitt et al. 1972
Rose	*Coccinella septempunctata, Adonia variegata, Chrysoperla carnea*	*Microsiphum rosae*	C	nn-p	P	—	Natskova 1974
Sugarcane	*Trichogramma australicum, T. japonicum*	Sugarcane borers	?	nn-pp	E	—	Paul et al. 1976

Cotton	*Chrysoperla carnea*	*Heliothis virescens*	LC	n-p	P	—	Plapp & Bull 1978
?	*Chrysoperla oculata*	Many	C	n-pp	P	—	Pree & Hagley 1985
Sorghum	*Aphidencyrtus aphidivorus*	*Rhopalosiphum maidis*	?	n-p	?	—	Radke & Barwad 1978
Hazelnut	Several phytoseiids	*Phytoptus avellanae*	F	n-p	?	—	Ragusa 1977
Several	*Bracon brevicornis*	?	LD	n-?	?	—	Sarup et al. 1972
Cabbage	*Coccinella septempunctata*	*Brevicoryne brassicae*	LC	n-p	P	—	Sharma & Adlakha 1981
Sugarcane	Several ants	*Chilo infucsatella*	F	nn-p	?	—	Sithanantham 1980a
Cowpea, bean, pea	*Coccinella septempunctata*	*Aphis craccivora*	LC	n-p	P	—	Singh et al. 1978
?	*Coccinella septempunctata*	?	O	n-pp	P	—	Singh & Malhotra 1975b
?	*Coccinella septempunctata*	*Aphis craccivora, Lipaphis erysimi, Uroleucon carthami, Aphis gossypii*	C	n-pp	P	—	Singh et al. 1979
Greenhouse	*Phytoseiulus persimilis*	Spider mites	C	n-p	P	—	Stamenkovic & Peric 1984a
Citrus	*Amblyseius swirskii*	*Prays citri, Phyllocoptruta oleivora*	C	n-pp	P	—	Swirski et al. 1967
Mustard	*Coccinella septempunctata*	*Lipaphis erysimi*	C	n-p	P	—	Teotia & Tiwari 1972
Aubergine	*Menochilus sexmaculatus*	*Aphis gossypii*	F	n-p	?	—	Tiwari & Moorthy 1985
Apple	*Typhlodromus pyri*	Hemipteran spp.	F	n-pp	P	—	Trapman & Blommers 1985
	Coccinella novemnotata	*Aphis pomi*	LD	n-p	P	—	Travis et al. 1978
Citrus	*Leptomastidea abnormis, Scymnus includens*	*Planococcus citri*	C	nn-p	P	—	Viggiani et al. 1972

Table A.2. (*Continued*)

Crop	Natural Enemy	Prey/Host	Test Method[1]	Selectivity Class[2]	Selectivity Type[3]	Selectivity Ratio	Reference
Greenhouse	*Encarsia formosa*	*Trialeurodes vaporariorum*	C	n-p	P	—	Wiackowski & Herman 1971
ENDRIN							
Cotton	*Coccinella undecimpunctata*	*Lipaphis erysimi*	LD	n-p	P	5	Abdel-Aal et al. 1979
Apple	*Amblyseius fallacis*	*Panonychus ulmi, Aculus schlechtendali*	C	n-p	P	—	Lienk et al. 1976, 1978
Mustard	*Coccinella septempunctata*	*Lipaphis erysimi*	LC	n-p	P	—	Singh & Malhotra 1975a
	Coccinella septempunctata	*Lipaphis erysimi*	C	n-p	P	—	Teatia & Tiwari 1972
Cereal	*Coleomegilla maculata*	*Oulema melanopa*	C	n-p	P	—	Yun & Ruppel 1964
LINDANE							
Greenhouse	*Phytoseiulus persimilis, Encarsia formosa*	Thrips, springtails	?	nn-p	B	—	Costello & Elliott 1981
Vegetables	*Coccinella septempunctata*	*Lipaphis erysimi*	LC	n-p	P	—	Pradhan et al. 1959
Cowpea, bean, pea	*Coccinella septempunctata*	*Aphis craccivora*	LC	n-p	P	—	Singh et al. 1978
?	*Coccinella septempunctata*	*Aphis craccivora, Lipaphis erysimi, Uroleucon carthami, Aphis gossypii*	C	n-pp	P	—	Singh et al. 1979
Citrus	*Leptomastidea abnormis, Scymnus includens*	*Planococcus citri*	C	nn-p	P	—	Viggiani et al. 1972

Greenhouse	*Phytoseiulus persimilis*	Several	C	n-pp	P	—	Stenseth 1979
Cereal	*Coleomegilla maculata*	*Oulema melanopa*	C	n-p	P	—	Yun & Ruppel 1964
METHOXYCHLOR							
Alfalfa	*Bathyplectes curculionis*	*Hypera postica*	F	n-p	?	—	Davis 1970
Apple	*Amblyseius fallacis*	*Panonychus ulmi, Tetranychus urticae Aculus schlechtendali*	C	n-pp	P	—	Hislop et al. 1978
Alfalfa	*Microctonus aethiopoides*	*Empoasca fabae, Hypera postica*	F	n-pp	?	—	Hower & Davis 1984
Apple	*Typhlodromus tillarium*	*Panonychus ulmi*	F	n-p	?	—	Karg 1978
PERTHANE							
Rice	*Lycosa pseudoannulata, Micraspis crosea*	*Nilaparvata lugens*	F	nn-p	?	—	Dyck & Orlido 1977
TOXAPHENE							
Rape	*Philonthus fuscipennis, Tachyporus hypnorum*	Several	C	nn-pp	P	—	Eghtedar 1969
Cotton	*Coccinella undecimpunctata, Chrysoperla vulgaris, Paederus alferi, Scymnus* sp.	Cotton leafworm	F	nn-p	?	—	Hassanein & Khalil 1968
Soybean	*Geocoris punctipes, Nabis* spp., *Orius insidiosus*	*Sericothrips variablilis*	F	nn-p	E	—	Huckaba et al. 1983
Cotton	*Micropletis croceipes*	*Heliothis* spp.	LD	n-p	P	—	Powell et al. 1986a
Several	*Bracon brevicornis*	?	LD	n-?	?	—	Sarup et al. 1972
Potato	*Pseudophonus rufipes*	Many	LD	n-pp	?	—	Van Dinther 1963

[1] F = field; C = contact; LC = lethal concentration; LD = lethal dosage; S = systemic; O = oral; R = residue.
[2] n = single natural enemy species; nn = multiple; p = single prey or host species; pp = multiple.
[3] P = physiological selectivity; E = ecological; B = both P & E; ? = uncertain.

Table A.3. Cases of Insecticide Selectivity to Arthropod Natural Enemies over Their Prey or Hosts: Organophosphates

Crop	Natural Enemy	Prey/Host	Test Method[1]	Selectivity[2] Class	Selectivity[3] Type	Selectivity Ratio	Reference
		Early Organophosphates					
SCHRADAN*							
Apple	*Stethorus punctillum*	*Tetranychus urticae*	LC	n-p	P	20	Bravenboer 1959
Cotton	*Coccinella undecimpunctata, Scymnus scyriacus, Chrysopa vulgaris*	*Aphis gossypii*	F	nn-p	B	—	Ahmed 1955
Alfalfa	*Hippodamia* spp.	*Therioaphis maculata*	LC	nn-p	P	9– > 20	Bartlett 1958
Cotton	Syrphidae, Coccinellidae, Chrysopidae	Several	F	nn-pp	E	—	Ripper et al. 1951
Cabbage	*Aphidius brassicae, Coccinella septempunctata*	*Brevicoryne brassicae*	F	nn-p	E	—	Ripper et al. 1951
		Aliphatic Derivatives of Phosphorus Compounds					
ACEPHATE*							
Rice	*Lycosa pseudoannulata, Oedothorax insecticeps*	*Nilaparvata lugens*	F	nn-p	?	—	Chen & Chiu 1979
Rice	*Lycosa pseudoannulata, Oedothorax insecticeps*	*Nilaparvata lugens*	S	nn-p	?	—	Chiu & Cheng 1976

Forest	*Coeliodes pissodis, Medetera bistriata, Roptrocerus xylophagorum, Thanasimus dubius*	*Dendroctonus frontalis*	S	nn-p	?	—	Coster & Ragenovich 1976
Tobacco	*Coccinella septempunctata, Macrolophus rubi, Dicyphus eckerleini*	Several aphids	C	nn-pp	?	—	Dirimanov et al. 1974
Rice	*Cyrtorhinus lividipennis, Microvelia atrolineata, Lycosa pseudoannulata*	*Nilaparvata lugens*	C	nn-p	P	—	Fabellar & Heinrichs 1984
Crucifers	*Apanteles plutella*	*Plutella xylostella*	LC	n-p	P	—	Feng & Wang 1984
Citrus	*Amitus hesperidum, Encarsia opulenta*	*Aleurocanthus woglumi*	C	nn-p	?	—	Fitzpatrick & Dowell 1981
Bean	*Pediobius foveolatus*	*Epilachna varivestris*	F	n-p	E	—	Flanders et al. 1984
Alfalfa	*Trioxys complanatus*	*Therioaphis trifolii, T. maculata, Acyrthosiphon kondoi*	C, R	n-pp	P	—	Franzmann & Rossiter 1981
Alfalfa	*Microctonus aethiopoides*	*Empoasca fabae, Hypera postica*	F	n-pp	?	—	Hower & Davis 1984
Cabbage	*Diaeretiella rapae*	*Myzus persicae*	C, F	n-p	P	—	Hsieh 1984
Artichoke	*Chrysocharis ainsliei*	*Platyptilia carduldactyla, Phytomyza syngenesiae*	F	n-pp	?	—	Lange et al. 1980

Table A.3. (*Continued*)

Crop	Natural Enemy	Prey/Host	Test Method[1]	Selectivity[2] Class	Selectivity[3] Type	Selectivity Ratio	Reference
Forest	*Apanteles fumiferanae, Glypta fumiferanae*	*Choristoneura fumiferana*	F	nn-p	?	—	Otvos & Raske 1980
Forest	*Compsilura concinnata*	*Lymantria dispar*	LD	n-p	P	2–11	Respicio & Forgash 1984
DEMETON-METHYL*							
Cotton	*Xanthogramma aegyptium*	*Aphis gossypii*	O	n-p	P	—	Azab et al. 1971
Cotton	*Chrysopa vulgaris*	*Aphis gossypii*	F	nn-p	B	—	Ahmed 1955
Cotton	*Coleomegilla maculata, Chrysopa rufilabris, C. oculata*	*Aphis gossypii*	F	nn-p	B	—	Ahmed et al. 1954
Alfalfa	*Hippodamia* spp.	*Therioaphis maculata*	LC	nn-p	P	10–> 21	Bartlett 1958
Bean	*Phytoseiulus persimilis*	*Tetranychus urticae*	S	n-p	E	—	Binns 1971
Apple	*Metaseiulus occidentalis*	*Tetranychus urticae*	LC	nn-p	P	5–20	Bravenboer 1959
Safflower	*Menochilus sexmaculatus*	*Acanthiophilus helianthi, Uroleucon compositae*	F	nn-pp	?	—	Chaudhary et al. 1983
Cereal	Several carabids, staphylinids	Several aphids	F	nn-pp	?	—	Feeney 1983

Several	*Trichogramma cacoeciae, Pales pavida, Pimpla turionellae, Chrysoperla carnea, Leptomastix dactylopii*	Several	C	nn-pp	P	—	Franz et al. 1980
Alfalfa	*Trioxys complanatus*	*Therioaphis trifolii, T. maculata, Acyrthosiphon kondoi*	C, R	n-pp	E	—	Franzmann & Rossiter 1981
Alfalfa	*Hippodamia convergens*	*Acyrthosiphon pisum*	F	n-p	?	—	Garcia 1975
Cotton	*Coccinella septempunctata, C. undecimpunctata, Adonia variegata*	*Aphis gossypii, A. craccivora*	LC	nn-pp	P	5–200	Gargov 1968
Vegetables	*Cycloneda sanguinea*	*Brevicoryne brassicae*	LC	n-p	P	15	Gravena & Batiste 1979a, b
Conifer	*Copidosoma deceptor*	*Exoteleia nepheos*	F	n-p	?	—	Hain & Wallner 1974
Ornamental	*Encyrtus infelix*	*Saissetia coffeae*	F	n-p	?	—	Hamlen 1975
?	*Chrysoperla carnea*	?	C	n-?	P	—	Kowalska & Pruszynski 1969
?	*Chrysoperla carnea*	?	C	n-?	P	—	Miszczak 1975
Cabbage	*Coccinella repanda*	*Brevicoryne brassicae*	C	n-p	P	—	Mishra & Satpathy 1985
Alfalfa	*Aphidius ervi*	*Acyrthosiphon onobrychuis*	F, C	n-p	E	—	Orbtel 1961

Table A.3. (*Continued*)

Crop	Natural Enemy	Prey/Host	Test Method[1]	Selectivity[2] Class	Selectivity[3] Type	Selectivity Ratio	Reference
Vegetables	*Coccinella septempunctata*	*Lipaphis erysimi*	LC	n-p	P	—	Pradhan et al. 1959
Cotton	*Orius tristicolor*	Mites, aphids	F	n-pp	?	—	Reynolds et al. 1953
Cotton	*Menochilus sexmaculatus*	*Aphis gossypii*	O	n-p	P	—	Satpathy et al. 1968
Rose	Chrysopids, coccinellids	*Anuraphis devecta*	F	nn-p	?	—	Selivanova 1967
Bean	*Coccinella septempunctata*, *aphelinids, braconids*	*Aphis fabae*	F	nn-p	B	—	Stacherska 1963
Alfalfa	*Austromicromus tasmaniae, Coccinella undecimpunctata*	*Acyrthosiphon kondoi, A. pisum*	LD	nn-pp	P	16–400	Syrett & Penman 1980
Mustard	*Coccinella septempunctata,*	*Lipaphis erysimi*	C	n-p	P	—	Teotia & Tiwari 1972
Aubergine	*Menochilus sexmaculatus*	*Aphis gossypii*	F	n-p	?	—	Tiwari & Moorthy 1985
Alfalfa	*Orius* spp., *Geocoris* spp., *Nabis* spp., *Chrysoperla* spp., *Hippodamia* sp.	*Therioaphis maculata*	F	nn-p	?	—	van den Bosch et al. 1956
Wheat	*Lysiphlebus testaceipes, Orius tristicolor*	*Schizaphis graminum*	F	nn-p	P	—	Vazquez & Carrillo 1972

Citrus	*Leptomastidea abnormis*, *Scymnus includens*	*Planococcus citri*	C	nn-p	P	—	Viggiani et al. 1972
Several	*Coccinella septempunctata*, *C. quinquepunctata*, syrphids, *Diaretiella rapae*	Several aphids	C	nn-pp	?	—	Wiackowski & Dronka 1968
Vegetables	*Derostenus variipes*	*Liriomyza munda*	C	n-p	P	—	Wolfenbarger & Getzin 1963
DICHLORVOS							
Cotton	*Chrysoperla carnea*	*Heliothis virescens*	LD	n-pp	P	3	Bull & Ridgeway 1969
Greenhouse	*Encarsia formosa*	*Trialeurodes vaporariorum*	C	n-p	P	—	Delorme & Angot 1983
Coconut	*Microbracon brevicornis*	*Nephantis serinopa*	LD	n-p	P	—	Saradamma & Nair 1968
Crucifers	*Apanteles plutella*	*Plutella xylostella*	LC	n-p	P	—	Feng & Wang 1984
DICROTOPHOS*							
Rice	*Lycosa pseudoannulata*, *Oedothorax insecticeps*	*Chilo suppressalis*, *Nilaparvata lugens*, *Nephotettix cincticeps*	C	nn-pp	?	—	Chu et al. 1977
Cotton	*Micropletis croceipes*	*Heliothis* spp.	LD	n-p	P	—	Powell et al. 1986a

Table A.3. (*Continued*)

Crop	Natural Enemy	Prey/Host	Test Method[1]	Selectivity[2] Class	Selectivity[3] Type	Selectivity Ratio	Reference
DIMETHOATE*							
?	*Chrysopa formosa*	Several	C	n-p	P	—	Babrikova 1980
Safflower	*Menochilus sexmaculatus*	*Acanthiophilus helianthi, Uroleucon compositae*	F	nn-pp	?	—	Chaudhary et al. 1983
Bean	*Anthocoris nemorum, A. confusus*	*Acyrthosiphum pisum*	S	nn-p	P	—	Elliott 1970
Cotton	*Chrysoperla carnea*	*Spodoptera exigua, Lygus hesperus*	F	n-pp	?	—	Eveleens et al. 1973
Cotton	*Agistemus exertus*	*Tetranychus cinnabarinus*	F	n-p	?	—	Farghaly et al. 1974
Citrus	*Amitus hesperidum, Encarsia opulenta*	*Aleurocanthus woglumi*	C	nn-p	?	—	Fitzpatrick & Dowell 1981
Bean	*Pediobius foveolatus*	*Epilachna varivestris*	F	n-p	E	—	Flanders et al. 1984
Cotton	*Coccinella septempunctata, C. undecimpunctata, Adonia variegata*	*Aphis gossypii, A. craccivora*	LC	nn-pp	P	1–3	Gargov 1968
Vegetables	*Derostenus variipes*	*Liriomyza munda*	F	n-p	?	—	Getzin 1960
Alfalfa	*Microctonus aethiopoides*	*Empoasca fabae, Hypera postica*	F	n-pp	?	—	Hower & Davis 1984
Citrus	*Cryptolaemus montrouzieri*	*Aonidiella aurantii*	C	n-p	P	—	Morse & Bellows 1986

Rose	*Coccinella septempunctata, Adonia variegata, Chrysoperla carnea*	*Macrosiphum rosae*	C	nn-p	P	—	Natskova 1974
Many	*Chrysoperla oculata*	Many	C	n-pp	P	—	Pree & Hagley 1985
Chili	*Aphelinus mali, Aphidius platensis, Menochilus sexmaculatus*	*Myzus persicae*	F	nn-p	?	—	Rajagopal & Kareem 1983
Mustard	*Coccinella septempunctata*	*Liaphis erysimi*	LC	n-p	P	—	Singh & Malhotra 1975a
Cereal	*Aphidius ervi*	*Sitobion avenae*	C	n-p	B	—	Suss 1983
Aubergine	*Menochilus sexmaculatus*	*Aphis gossypii*	F	n-p	?	—	Tiwari & Moorthy 1985
Citrus	*Cycloneda sanguinea, Chrysoperla* spp.	*Toxoptera citricida*	F	nn-p	E	—	Trevizoli & Gravena 1979
Sugarcane	*Epipyrops melaneoleuca*	*Pyrilla perpusillae*	F	nn-p	B	—	Varma & Bindra 1980
Wheat	*Lysiphlebus testaceipes, Orius tristicolor*	*Schizaphis graminum*	F	nn-p	P	—	Vazquez & Carrillo 1972
Citrus	*Leptomastidea abnormis, Scymnus includens*	*Planococcus citri*	C	nn-p	P	—	Viggiani et al. 1972
Vegetables	*Derostenus variipes*	*Liriomyza munda*	C	n-p	P	—	Wolfenbarger & Getzin 1963

Table A.3. (*Continued*)

Crop	Natural Enemy	Prey/Host	Test Method[1]	Selectivity[2] Class	Selectivity[3] Type	Selectivity Ratio	Reference
Greenhouse	*Encarsia formosa*	*Trialeurodes vaporariorum*	C	n-p	P	—	Zseller & Budai 1982
DISULFOTON*							
Cotton	*Campoletis perdistinctus*	*Heliothis* spp.	LD, O, S	n-p	P	—	Cate et al. 1972
Tobacco	*Apanteles congregatus*	*Manduca sexta*	F	n-p	?	—	Cherry & Pless 1969
Rice	*Lycosa pseudoannulata*, *Oedothorax insecticeps*	*Nilaparvata lugens*	S	nn-p	?	—	Chiu & Cheng 1976
Cotton	Several coccinellids (*C. sanguinea*)	*Eutinobothrus brasiliensis*, *Aphis gossypii*	F	nn-p	E	—	Gravena et al. 1976
Sorghum	*Hippodamia convergens*, *Chrysoperla* spp.	*Schizaphis graminae*	F	nn-p	?	—	Teetes 1972
ETHION							
Grape	*Metaseiulus occidentalis*	*Tetranychus pacificus*	F	n-p	?	—	AliNiazee et al. 1974
Citrus	*Cales noacki*	*Aleurothrixus floccosus*	C	n-p	E	—	Garrido et al. 1982
Apple	*Metaseiulus occidentalis*	*Tetranychus mcdanieli*	F	n-p	?	—	Hoyt 1969a
Citrus	*Chrysoperla rufilabris*	Several	C	n-pp	P	—	Lawrence 1974

MALATHION							
Cereal	*Hippodamia convergens, Nabis alternatus*	*Macrosiphum avenae*	LC	nn-p	P	700	Hamilton & Kieckhefer 1968, 1969
Rice	*Lycosa pseudoannulata*	*Chilo suppressalis, Nephotettix cincticeps*	LD	nn-pp	P	2–16	Miyata & Saito 1982
Alfalfa	*Aphidius ervi*	*Acyrthosiphon onobrychuis*	F, C	n-p	E	—	Orbtel 1961
Greenhouse	*Phytoseiulus persimilis*	Spider mites	C	n-p	P	—	Stamenkovic & Peric 1984b
METHAMIDOPHOS							
Potato	*Coleomegilla maculata, Chrysoperla oculata*	*Myzus persicae*	LD	nn-p	P	5–12	Lecrone & Smilowitz 1980
Mustard*	*Diaeretus rapae*	*Lipaphis erysimi*	F	n-p	?	—	Singh & Rawat 1981
MONOCROTOPHOS*							
Rice	*Lycosa pseudoannulata, Oedothorax insecticeps*	*Nilaparvata lugens*	S	nn-p	?	—	Chiu & Cheng 1976
Rice	*Lycosa pseudoannulata, Oedothorax insecticeps*	*Chilo suppressalis, Nilaparavata lugens, Nephotettix cincticeps*	C	nn-pp	?	—	Chu et al. 1977
Alfalfa	*Trioxys complanatus*	*Therioaphis trifolii, T. maculata, Acyrthosiphon kondoi*	C, R	n-pp	E	—	Franzmann & Rossiter 1981

Table A.3. (*Continued*)

Crop	Natural Enemy	Prey/Host	Test Method[1]	Selectivity[2] Class	Selectivity[3] Type	Selectivity Ratio	Reference
?	*Trichogramma japonicum*	*Corcyra cephalonica*	C	n-p	?	—	Gupta et al. 1984
?	*Menochilus sexmaculatus*	Several aphids	C	n-pp	P	—	Makar & Kadhav 1981
?	*Trichogramma brasiliensis*	*Corcyra cephalonica*	C	n-p	P	—	Navarajan et al. 1979
Rice	*Lycosa pseudoannulata, Cyrtorhinus lividipennis*	*Sogatella furcifera*	C	n-pp	P	—	Salim & Heinricks 1985
Cotton	*Trichogramma chelonis*	*Earias vitella*	C	n-p	P	—	Santharam & Kumara-swamy 1985
OMETHOATE							
Wheat	*Lysiphlebus testaceipes, Orius tristicolor*	*Schizaphis graminum*	F	nn-p	P	—	Vazquez & Carrillo 1972
PHORATE*							
Bean	*Anthocoris nemorum, A. confusus*	*Acyrthosiphon pisum*	S	nn-p	P	—	Elliott 1970
Alfalfa	*Bathyplectus curculionis*	*Hypera postica*	F	n-p	?	—	Miller et al. 1973

PHOSPHAMIDON*							
Cotton	*Xanthogramma aegyptium*	*Aphis gossypii*	O	n-p	P	—	Azab et al. 1971
Cotton	*Apanteles angaleti*	*Pectinophora gossypiella*	C	n-p	P	—	Mani & Nagarakatti 1983
Apple	*Anthocoris nemoralis*, mirids, chrysopids, hemerobiids, empidids, spiders, *Allothombium* sp., Hymenoptera, tachinids	*Psylla pyri*	F	nn-pp	?	—	Staubli et al. 1984
Aubergine	*Menochilus sexmaculatus*	*Aphis gossypii*	F	n-p	?	—	Tiwari & Moorthy 1985
Apple	*Aphidoletes aphidimyza*	*Aphis pomi*	C	n-p	P	—	Warner & Croft 1982
THIOMETON*							
Cotton	*Xanthogramma aegyptium*	*Aphis gossypii*	O	n-p	P	—	Azab et al. 1971
Apple	Several phytoseiids	Several	F	nn-pp	?	—	Dabrowski 1970b
Apple	*Amblyseius finlandicus*, *Phytoseiulus macropilis*	?	C	nn-pp	P	—	Dabrowski 1969 a
Alfalfa	*Trioxys complanatus*	*Therioaphis trifolii*, *T. maculata*, *Acyrthosiphon kondoi*	C, R	n-pp	B	—	Franzmann & Rossiter 1981
Alfalfa	*Coccinella repanda*	*Therioaphis trifolii*, *Acyrthosiphon kondoi*	LD	n-pp	P	—	Kay 1979

Table A.3. (*Continued*)

Crop	Natural Enemy	Prey/Host	Test Method[1]	Selectivity[2] Class	Selectivity[3] Type	Selectivity Ratio	Reference
?	*Coccinella septempunctata*	*Myzus persicae*	C	n-p	P	—	Kirknel 1975
?	*Menochilus sexmaculatus*	Several aphids	C	n-pp	P	—	Makar & Kadhav 1981
Rose	*Coccinella septempunctata, Adonia variegata, Chrysoperla carnea*	*Macrosiphum rosae*	C	nn-p	P	—	Natskova 1974
Cotton	*Menochilus sexmaculatus*	*Aphis gossypii*	O	n-p	P	—	Satpathy et al. 1968
Cowpea, bean, pea	*Coccinella septempunctata*	*Aphis craccivora*	LC	n-p	P	—	Singh et al. 1978
?	*Coccinella septempunctata*	*Aphis craccivora, Lipaphis erysimi, Uroleucon carthami, Aphis gossypii*	C	n-pp	P	—	Singh et al. 1979
Bean	*Coccinella septempunctata,* aphelinids, braconids	*Aphis fabae*	F	nn-p	B	—	Stacherska 1963
TRICHLORFON							
Cotton	*Agistemus exsertus*	Several spider mites	C	n-pp	P	—	Abo Elghar et al. 1971
Cabbage	Ichneumonids	*Plutella maculipennis*	F	nn-p	E	1	Adaskevich 1966
Poplar	*Euderus caudatus*	*Saperda carcharias*	F	n-p	?	—	Arru 1973

Turf	Staphylinids, eriogonids, parasitoids, eupodids	Many spp.	F	nn-pp	?	—	Cockfield & Potter 1983
Cotton	*Chrysoperla carnea*	*Heliothis virescens*	LD	n-pp	P	260	Bull & Ridgeway 1969
Alfalfa	*Bathyplectes curculionis*	*Hypera postica*	F	n-p	?	—	Davis 1970
Alfalfa	*Trioxys complanatus*	*Therioaphis trifolii, T. maculata, Acyrthosiphon kondi*	C, R	n-pp	E	—	Franzmann & Rossiter 1981
Sugar beet	*Chrysoperla carnea*	*Myzus persicae*	C	n-p	P	—	Hassan et al. 1985
Alfalfa	*Geocoris pallens, G. bullatus, Nabis alternata, Aphidius smithi, A. ervi*	*Lygus hesperus, L. elisus, Acyrthosiphon pisum*	F	nn-pp	?	—	Johansen & Eves 1973
Cereal	*Telenomus* sp.	*Eurygaster integriceps*	F	n-p	B	—	Kamenkova 1971
Apple	*Trichogramma cacoeciae, T. embryophagum*	?	C	nn-?	?	—	Kapustina 1975
Several	*Chrysoperla carnea, C. formosa, C. perla, Coccinella septempunctata*	Many	L	nn-pp	?	—	Kharizanov & Babrikova 1978
Cotton	*Chrysoperla carnea, Collops balteatus, Hippodamia convergens*	*Heliothis* spp.	C	nn-pp	P	—	Lingren & Ridgway 1967
Apple	*Trichogramma cacoeciae*	*Orgyia antiqua*	F	n-p	B	—	Niemczyk et al. 1979

Table A.3. (*Continued*)

Crop	Natural Enemy	Prey/Host	Test Method[1]	Selectivity[2] Class	Selectivity[3] Type	Selectivity Ratio	Reference
Forest	*Brachymeria intermedia*	*Lymantria dispar*	LD	n-p	P	3–8	Respicio & Forgash 1984
Olive	*Habrobracon hebetor*, *Chelonus eleaphilus*	*Prays citri*	LD	nn-p	P	—	Soultanopoulos & Broumas 1979
Legumes	*Micromus tasmaniae*	*Acrythosiphon pisum*, *A. kondoi*	F	n-pp	?	—	Wightman & Whitford 1982
Beans	Several spiders	*Pseudoplusia includens*	F	nn-p	?	—	Yabar 1980
?	*Telenomus* spp. *Trichogramma* spp.	Several lepidopterans	L	nn-p	P	—	Xie et al. 1984
Ornamentals	*Cycloneda sanguinea limbifer*	*Rhopalosiphon maidis*, *Toxoptera aurantii*	C	n-pp	P	—	Zeleny 1969
VAMIDOTHION*							
Citrus	*Cales noacki*	*Aleurothrixus floccosus*	F	n-p	?	—	Carrero 1979
Apple	Coccinellidae, Syrphidae	Several aphids	F	nn-pp	?	—	Burgaud & Cessac 1962
Apple	*Stethorus nigripes*	*Panonychus ulmi*, *Tetranychus urticae*	C	n-pp	P	—	Edwards & Hodgson 1973
Cotton	*Coccinella*, *septempuntata* *C. undecimpunctata*, *Adonia variegata*	*Aphis gossypii*, *A. craccivora*	LC	nn-pp	P	28–428	Gargov 1968

Apple	*Stethorus loxtoni, S. nigripes, S. vagans*	*Tetranychus urticae*	C	n-p	P	—	Walters 1976b
Carbon Cyclic Derivatives of Phosphorus Compounds							
BROMOPHOS							
Sugar beet	*Chrysoperla carnea*	*Myzus persicae*	C	n-p	P	—	Hassan et al. 1985
Greenhouse	*Aphidoletes aphidimyza*	Several aphids	C	n-pp	P	—	Sell 1985
Olive	*Eupelmus urozonus*	*Dacus olea*	C	n-p	E	—	Soultano-poulos & Broumas 1977
Apple	*Typhlodromus pyri*	Hemipteran spp.	F	n-pp	P	—	Trapman & Blommers 1985
Several	*Chrysoperla carnea*	Several aphids	C	n-pp	P	—	Wiackowski 1968
CARBOPHENOTHION							
Grape	*Metaseiulus occidentalis*	*Tetranychus pacificus*	F	n-p	?	—	AliNiazee et al. 1974
Alfalfa	*Hippodamia* spp.	*Therioaphis maculata*	LC	nn-p	P	6–25	Bartlett 1958
Apple	*Metaseiulus occidentalis, Stethorus punctillum*	*Tetranychus urticae*	LC	nn-p	P	1–5	Bravenboer 1959
Cereal, Sugar beet	*Pterostichus melanarius, P. maldidus, Harpalus rufipes, Bembidion lampros*	Several	F	nn-pp	?	—	Edwards et al. 1984b
Rice	*Cyrtorhinus lividipennis, Microvelia atrolineata, Lycosa pseudoannulata*	*Nilaparvata lugens*	C	nn-p	P	—	Fabellar & Heinrichs 1984

Table A.3. (*Continued*)

Crop	Natural Enemy	Prey/Host	Test Method[1]	Selectivity[2] Class	Selectivity[3] Type	Selectivity Ratio	Reference
Alfalfa	*Microctonus aethiopoides*	*Empoasca fabae, Hypera postica*	F	n-pp	?	—	Hower & Davis 1984
Citrus	*Chrysoperla rufilabris*	Several	C	n-pp	P	—	Lawrence 1974
Castor	*Stethorus pauperculus*	*Tetranychus neocaledonicus, T. cucurbitae*	C	n-p	P	—	Puttaswamy & Channabasa-vanna 1976
Apple	*Aphidoletes aphidimyza*	*Aphis pomi*	C	n-p	P	—	Warner & Croft 1982
CHLORFENVINPHOS							
Cabbage	Several carabids	Several	C, F	nn-pp	B	—	Edwards & Thompson 1975
Cereal, sugar beet	*Pterostichus melanarius, P. maldidus, Harpalus rufipes, Bembidion lampros*	Several	F	nn-pp	?	—	Edwards et al. 1984b
Cauliflower	*Aleochara bilineata, Bembidion lampros*	*Delia* sp.	F	nn-p	?	—	Finlayson et al. 1980
Cabbage	*Trybliographa rapae*	*Hylemya brassicae*	C	n-p	P	—	Hassan 1973
?	*Menochilus sexmaculatus*	Several aphids	C	n-pp	P	—	Makar & Kadhav 1981

Vegetables	*Bembidion lampros*	*Delia brassicae*	F	n-p	B	—	Obadofin & Finlayson 1977
?	*Trichogramma cacoeciae*	*Spodoptera littoralis*	C	n-p	P	—	Shires et al. 1984
?	*Coccinella septempunctata*	?	O	n-pp	P	—	Singh & Malhotra 1975b
Cabbage	*Pterostichus melanarius, Senolophus comma*	*Hylemya brassicae*	C	nn-p	?	—	Tomlin 1975
Potato	*Perillus bioculatus*	*Leptinotarsa decemlineata*	F	n-p	?	—	Wegorek & Pruszynski 1979
FENITROTHION							
Cereals	*Pterostichus melanarius*	*Sitobion avenae, Metopolopium dirhodum, Rhopalosiphum padi*	F	n-pp	?	—	Chiverton 1984
Rice	*Lycosa pseuannulata, Oedothorax insecticeps*	*Chilo suppressalis, Nilaparvata lugens*	LD	nn-pp	P	—	Chu et al. 1976a
Forest	*Apanteles fumiferanae, Glypta fumiferanae*	*Choristoneura fumiferana*	F	nn-p	?	—	Otvos & Raske 1980
Apple	*Trichogramma cacoeciae*	*Orgyia antiqua*	F	n-p	B	—	Niemczyk et al. 1979
FENTHION							
Apple	Several phytoseiids	Several	F	nn-pp	?	—	Dabrowski 1970b

Table A.3. (*Continued*)

Crop	Natural Enemy	Prey/Host	Test Method[1]	Selectivity[2] Class	Selectivity[3] Type	Selectivity Ratio	Reference
HEPTENOPHOS*							
Sugar beet	*Chrysoperla carnea*	*Myzus persicae*	C	n-p	P	—	Hassan et al. 1985
?	*Pimpla turionellae*	*Galleria mellonella*	C, F	n-p	P	—	Bogenschutz 1984
PARATHION-METHYL							
Cereal	*Agonum dorsale*, *Nebria brevicollis*, *Pterostichus melanarius*, *Erigone* spp.	*Metopolophium dirhodum*	ED	nn-p	P	33 27 196 42	Brown et al. 1983
PARATHION							
Cereal	*Hippodamia convergens*, *Nabis alternatus*	*Macrosiphum avenae*	LC	nn-p	P	600 22	Hamilton & Kieckhefer 1968, 1969
PROFENFOS							
Crucifers	*Apanteles plutella*	*Plutella xylostella*	LC	n-p	P	—	Feng & Wang 1984
STIROPHOS							
?	*Chrysopa formosa*	Several	C	n-pp	P	—	Babrikova 1980
?	*Chrysopa carnea*	Several	C	n-pp	P	—	Babrikova 1979

Corn	*Scarites substriatus, Pterostichus chalcites, Bembidion rapidum, Bembidion quadrmaculatum*	*Agrotis ipsilon*	C, O	nn-p	E	—	Gholson et al. 1978
Several	*Chrysopa carnea, C. formosa, C. perla, Coccinella septempunctata*	Many	L	nn-pp	?	—	Kharizanov & Babrikova 1978
Cabbage	*Pterostichus melanarius, Senolophus comma*	*Hylemya brassicae*	C	nn-p	?	—	Tomlin 1975
Citrus	*Leptomastidea abnormis, Scymnus includens*	*Planococcus citri*	C	nn-p	P	—	Viggiani et al. 1972
TETRACHLORVINPHOS							
Several	*Chrysoperla carnea*	Many	C	n-pp	?	—	Suter 1978
		Heterocyclic Derivatives of Phosphorus Compounds					
AZINPHOSMETHYL							
Apple	*Stethorus punctum*	*Panonychus ulmi*	F	n-pp	E	—	Asquith & Hull 1973
CHLORPYRIFOS							
Cotton	*Coccinella undecimpunctata*	*Spodoptera littoralis*	LD	n-p	P	17	Abdel-Aal et al. 1979
Citrus	*Cryptolaemus montrouzieri*	*Aonidiella aurantii*	C	n-p	P	—	Morse & Bellows 1986
Cotton	Several phytoseiids, *Agistemus exsertus*	*Tetranychus* spp.	F	nn-p	?	—	Osman & Keie 1975

Table A.3. (*Continued*)

Crop	Natural Enemy	Prey/Host	Test Method[1]	Selectivity[2] Class	Selectivity[3] Type	Selectivity Ratio	Reference
DIALFOR							
Grape	*Metaseiulus occidentalis*	*Tetranychus pacificus*	F	n-p	?	—	AliNiazee et al. 1974
DIAZINON							
Rice	*Gonatonarium dentatum*	*Nilaparvata lugens*	LD	nn-p	P	14	Chang 1981
Rice	*Pirata subpiraticus*	*Nilaparvata lugens*	LD	nn-p	P	2	Chang 1981
Apple	Several phytoseiids	Several	F	nn-pp	?	—	Dabrowski 1970b
Rice	*Lycosa pseudoannulata, Micraspis crosea*	*Nilaparvata lugens*	F	nn-p	?	—	Dyck & Orlido 1977
Apple	*Aphelinus mali*	*Eriosoma lanigerum*	F	n-p	E	—	Evenhuis 1959
Cabbage	*Coccinella repanda*	*Brevicoryne brassicae*	C	n-p	P	—	Mishra & Satpathy 1985
Rice	*Lycosa pseudoannulata, Cyrtorhinus lividipennis, Harmonia ocomaculata, Paederus fascipes*	*Sogatella furcifera*	C	pp-n	P	—	Salim & Heinricks 1985
Olive	*Eupelmus urozonus*	*Dacus olea*	C	n-p	E	—	Soultano-poulos & Broumas 1977

DIOXATHION							
Vegetables	*Derostenus variipes*	*Liriomyza munda*	F	n-p	?	—	Getzin 1960
Citrus	*Amblyseius stipulatus, Iphisieus degenerans, Phytoseiulus persimilis*	*Panonychus citri*	C	nn-p	P	—	Jeppson et al. 1975
MENAZON*							
?	*Chrysoperla formosa*	Several	C	n-p	P	—	Barbrikova 1980
?	*Chrysoperla carnea*	Several	C	n-p	P	—	Barbrikova 1979
Citrus	*Cales noacki*	*Aleurothrixus floccosus*	F	n-p	?	—	Carrero 1979
?	Several carabid spp.	?	LD, F	nn-p	B	—	Critchley 1972b
Tobacco	*Coccinella septempunctata, Macrolophus rubi, Dicyphus eckerleini*	Several aphids	C	nn-pp	?	—	Dirimanov et al. 1974
Apple	*Stethorus nigripes*	*Panonychus ulmi, Tetranychus urticae*	C	n-pp	P	—	Edwards & Hodgson 1973
Apple	*Synharmonia* spp.	Several aphids	F	n-pp	?	—	Mangutova 1971
Tobacco	Many species	*Myzus persicae*	F	nn-p	?	—	Mardzhanyan & Ust' yan 1966
Rose	*Coccinella septempunctata, Adonia variegata, Chrysoperla carnea*	*Microsiphum rosae*	C	nn-p	P	—	Natskova 1974

Table A.3. (*Continued*)

Crop	Natural Enemy	Prey/Host	Test Method	Selectivity[2] Class	Selectivity[3] Type	Selectivity Ratio	Reference
Mustard	*Coccinella septempunctata*	*Liaphis erysimi*	C	n-p	P	288	Raj & Rama-krishnan 1983
Cabbage	*Coccinella septempunctata*	*Brevicoryne brassicae*	LC	n-p	P	—	Sharma & Adlakha 1981
Mustard	*Coccinella septempunctata*	*Liaphis erysimi*	LC	n-p	P	—	Singh & Malhotra 1975a
?	*Coccinella septempunctata*	?	O	n-pp	P	—	Singh & Malhotra 1975b
Several	*Chrysoperla carnea*	Several aphids	C	n-pp	P	—	Wiackowski 1968
MEPHOSFOLAN*							
Cotton	Several phytoseiids, *Agistemus exsertus*	*Tetranychus* spp.	F	nn-p	?	—	Osman & Keie 1975
METHIDATHION							
Cotton	*Geocoris pallens*	Several	C	n-pp	?	—	Yokoyama & Pritchard 1984

PHOSALONE*							
Cabbage	Hymenoptera, Coccinellidae, Syrphidae, Cecidomyiidae	*Brevicoryne brassicae*	?	nn-p	B	—	Alexandrescu & Hondru 1981
Apple	Hymenoptera, Syrphidae, Carabidae, Coccinellidae	Aphididae, Lepidoptera	O, F	nn-pp	B	—	Akhmedov 1981
Apple	*Stethorus punctum*	*Panonychus ulmi*	F	n-pp	E	—	Asquith & Hull 1973
Several	*Coccinella septempunctata*	Several	?	n-pp	?	—	Babrikova 1982
?	*Chrysopa formosa*	Several	C	n-pp	P	—	Babrikova 1980
?	*Chrysopa carnea*	Several	C	n-pp	P	—	Babrikova 1979
Apple	*Coccinella septempunctata*, Syrphidae	Several	C	n-pp	?	—	Cessac & Burgaud 1964
Vegetables	*Diaretiella rapae*	*Brevicoryne brassicae*, *Myzus persicae*	C	n-pp	?	—	Delorme 1976
Sorghum	*Tetrastichus* spp.	*Contarinia sorghicola*	F	n-p	?	—	Garg & Taley 1977
Apple	*Pterostichus melanarius*, *Harpalus affinis*, *Amara* spp.	Several	LD, F	nn-pp	P	—	Hagley et al. 1980
Greenhouse	*Encarsia formosa*, *Phytoseiulus persimilis*	Several	LC	nn-pp	B	—	Iacob & Posoiu 1981, Iacob et al. 1981

Table A.3. (*Continued*)

Crop	Natural Enemy	Prey/Host	Test Method[1]	Selectivity[2] Class	Selectivity[3] Type	Selectivity Ratio	Reference
?	*Pterostichus melanarius, Harpalus rufripes*	?	C	nn-pp	?	—	Isaichev 1978
Apple	*Typhlodromus tilarium*	*Panonychus ulmi*	F	n-p	?	—	Karg 1978
Apple	*Trichogramma cacoeciae, T. embryophagum*	?	C	nn-?	?	—	Kapustina 1975
Peas	*Aphidius ervi*	*Acyrthosiphon pisum*	L	n-p	?	—	Khukhrii 1976
Several	*Chrysopa carnea, C. formosa, C. perla, Coccinella septempunctata*	Many	L	nn-pp	?	—	Kharizanov & Babrikova 1978
Artichoke	*Chrysocharis ainsliei*	*Platyptilia, carduldactyla, Phytomyza syngenesiae*	F	n-pp	?	—	Lange et al. 1980
Cabbage	*Apanteles plutellae*	*Plutella xylostella*	C	n-p	P	—	Mani & Krishna-moorthy 1984
Cotton	*Apanteles angaleti, A. kirkpatricki*	*Pectinophora gossypiella*	C	nn-p	P	—	Mani & Nagarkatti 1983
?	*Trichogramma brasiliensis*	*Corcyra cephalonica*	C	n-p	P	—	Navarajan et al. 1979
Apple	Hemiptera, Chrysopidae, Coccinellidae, Syrphidae, Forficulidae, Araneae, Hymenoptera	Many	F	nn-pp	P	—	Nikusch & Gernot 1986a

Apple	*Leptothrips mali, Haplothrips subtilissimus, Orius insidiosus Stethorus punctum, Deraeocoris nebulosus, Chrysoperla* spp.	*Panonychus ulmi, Aculus schlechtendali, Aphis citricola, Dysaphis plantaginea*	F	nn-pp	?	—	Parrella et al. 1981b
Apple	*Stethorus punctillum*	*Panonychus ulmi*	F	n-p	?	—	Pasqualini & Malavolta 1985
Wheat	Carabids, staphylinids	Cereal aphids	F	nn-pp	P	—	Poehling et al. 1985a
Many	*Chrysoperla oculata*	Many	C	n-pp	P	—	Pree & Hagley 1985
Hop	*Metaseiulus occidentalis*	*Tetranychus urticae*	F	n-p	?	—	Pruszynski & Cone 1973
Chili	*Aphelinus mali, Aphidius platensis, Menochilus sexmaculatus*	*Myzus persicae*	F	nn-p	?	—	Rajogopal & Kareem 1983
Apple	Miridae, Coccinellidae, Chrysopidae	*Panonychus ulmi*	F	nn-p	?	—	Reboulet 1986
Cotton	*Trichogramma chilonus*	*Earias vitella*	C	n-p	P	—	Santharam & Kumaraswami 1985
Cabbage	*Coccinella septempunctata*	*Brevicoryne brassicae*	LC	n-p	P	—	Sharma & Adlakha 1981
Mustard	*Diaeretus rapae*	*Lipaphis erysimi*	F	n-p	?	—	Singh & Rawat 1981

Table A.3. (*Continued*)

Crop	Natural Enemy	Prey/Host	Test Method[1]	Selectivity[2] Class	Selectivity[3] Type	Selectivity Ratio	Reference
Cotton	*Chrysoperla carnea*, *Trichogramma brasiliensis*	*Corycra cephalonica*	R	nn-p	P	—	Singh & Varma 1986
Apple	*Anthocoris nemoralis*, mirids, chrysopids, hemerobiids, empidids, spiders, *Allothrombium* sp., Hymenoptera, tachinids	*Psylla pyri*	F	nn-pp	?	—	Staubli et al. 1984
Apple	*Coccinella novemnotata*	*Aphis pomi*	LD	n-p	P	—	Travis et al. 1978
Apple	*Aphidoletes aphidimyza*	*Aphis pomi*	C	n-p	P	—	Warner & Croft 1982
Potato	*Perillus bioculatus*	*Leptinotarsa decemlineata*	F	n-p	?	—	Wegorek & Pruszynski 1979
PHOSMET							
Apple	*Stethorus punctum*	*Panonychus ulmi*	F	n-pp	E	—	Asquith & Hull 1973
Alfalfa	*Bathyplectes curculionis*	*Hypera postica*	F	n-p	?	—	Davis 1970
Greenhouse	*Phytoseiulus persimilis*	Spider mites	C	n-p	P	—	Stamenkovic & Peric 1984b
Apple	*Anthocoris nemoralis*, mirids, Chrysopids, hemerobiids, empidids, spiders, *Allothrombium* spp., Hymenoptera, tachinids	*Psylla pyri*	F	nn-pp	?	—	Staubli et al. 1984

Cotton	*Chrysoperla carnea*	*Aphis gossypii*	LD	n-p	P	—	Zhuravskaya et al. 1976
PHOSPHOLAN							
Cotton	*Microgaster rufiventris*	*Spodoptera littoralis*	O, LD	n-p	P	—	Hegazi et al. 1982
PYRIDAFENTHION							
Rice	*Pirata subpiraticus*	*Nilaparvata lugens*	LD	n-p	P	36	Chang et al. 1979
Rice	*Lycosa pseudoannulata, Oedothorax insecticeps*	*Chilo suppressalis, Nilaparvata lugens*	LD	nn-pp	P	—	Chu et al. 1976a
Rice	*Lycosa pseudoannulata*	*Chilo suppressalis, Nephotettix cincticeps*	LD	nn-pp	P	4–12	Miyata & Saito 1982
QUINALPHOS							
Safflower	*Menochilus sexmaculatus*	*Acanthiophilus helianthi, Uroleucon compositae*	F	nn-pp	?	—	Chaudhary et al. 1983
THIONAZIN							
Bean	*Phytoseiulus persimilis*	*Tetranychus urticae*	S	n-p	P	—	Binns 1971

[1] F = field; C = contact; EC = effective concentration; LC = lethal concentration; LD = lethal dosage; S = systemic; O = oral; R = residue; ED = effective dosage.
[2] n = single natural enemy species; nn = multiple; p = single prey or host species; pp = multiple.
[3] P = physiological selectivity; E = ecological; B = both P & E; ? = uncertain.
*Astenote indicates compounds with systemic activity.

Table A.4. Cases of Insecticide Selectivity to Arthropod Natural Enemies over Their Prey or Hosts: Carbamates

Crop	Natural Enemy	Prey/Host	Test Method[1]	Selectivity Class[2]	Selectivity Type[3]	Selectivity Ratio	Reference
ALDICARB*							
Soybean	Many spp.	Many spp.	F	nn-pp	?	—	Farlow & Pitre 1983
Sorghum	*Cycloneda sanguinea*, *Scymnus sp.*	*Schizaphis graminae*	F	nn-p	?	—	Gravena & Batiste 1979b
Cotton	Several coccinellids (*Cycloneda sanguinea*)	*Eutinobothrus brasiliensis*, *Aphis gossypii*	F	nn-p	?	—	Gravena et al. 1976
Artichoke	*Chrysocharis ainsliei*	*Platyptilia carduldactyla*, *Phytomyza syngenesiae*	F	n-pp	?	—	Lange et al. 1980
Corn	*Peterostichus chalcites*, *Harpalus pensylvanicaus*, *Calosoma sayi*	?	F	nn-p	?	—	Lesiewicz et al. 1984
Greenhouse	*Phytoseiulus persimilis*	*Tetranychus urticae*	S	n-p	B	—	McClanahan 1967
Alfalfa	*Aphidius smithi*	*Acyrothosiphon pisum*	S	n-p	E	—	Tamaki et al. 1969
AMINOCARB							
Forest	*Apanteles fumiferanae*, *Glypta fumiferanae*	*Choristoneura fumiferana*	F	nn-p	?	—	Otvos & Raske 1980

BENDIOCARB							
Turf	Staphylinids, eriogonids, parasitic mites, eupodids	Many	F	nn-pp	?	—	Cockfield & Potter 1983
CARBARYL							
Cotton	*Agistemus exsertus*	Several spider mites	C	n-pp	P	—	Abo Elghar et al. 1971
Alfalfa	*Microctonus aethiopoides*, *Bathyplectus curculionis*	*Hypera postica*	LD	nn-p	P	—	Abu & Ellis 1977
Cotton	*Chrysopa boninensis*, *Diparopsis castanea*	*Heliothis armigera*	LD	n-pp	P	—	Brettell 1979
Rice	*Lycosa pseudoannulata*, *Oedothorax insecticeps*	*Nilaparvata lugens*	S	nn-p	?	—	Chiu & Cheng 1976
Greenhouse	*Phytoseiulus persimilis*	?	C, F	n-p	?	—	Coulon et al. 1979
Rape	*Philonthus fuscipennis*, *Tachyporus hypnorum*	Several	C	nn-pp	P	—	Eghtedar 1969
Tobacco	*Jalysus spinosus*	Several	F	n-pp	?	—	Elsey 1973
Alfalfa	*Trioxys complanatus*	*Therioaphis trifolii*, *T. maculata*, *Acyrthosiphon kondoi*	C, R	n-pp	E	—	Franzmann & Rossiter 1981
Rice	*Brumoides saturalis*	*Sogatella furcifera*	F	n-p	?	—	Garg & Sethi 1984

Table A.4. (*Continued*)

Crop	Natural Enemy	Prey/Host	Test Method[1]	Selectivity Class[2]	Selectivity Type[3]	Selectivity Ratio	Reference
Corn	*Scarites substriatus, Pterostichus chalcites, Bembidion rapidum, B. quadrmaculatum*	*Agrotis ipsilon*	C, O	nn-p	E	—	Gholson et al. 1978
Potato	*Coleomegilla maculata, Chrysoperla oculata*	*Myzus persicae*	LD	nn-p	P	5–7	Lecrone & Smilowitz 1980
Cotton	*Chrysoperla carnea*	*Heliothis virsecens*	LC	n-p	P	—	Plapp & Bull 1978
Forest	*Compsilura concinnata*	*Lymantria dispar*	LD	n-p	P	70–800	Respicio & Forgash 1984
Apple	*Typhlodromus pyri*	?	C	n-?	P	—	Sekita 1986
Forest	*Enoclerus lecontei*	*Dendroctonus brevicomis*	LD	n-p	P	50	Swezey et al. 1982
Rice	*Lycosa pseudoannulata, Conocephalus maculatus*	*Nephotettix cincticeps*	LD	nn-p	?	2–90	Takahashi & Kiritani 1973
Soybean	*Geocoris punctipes, Nabis roseipennis,* spiders	Many	F	nn-pp	?	—	Turnipseed et al. 1975
Beans	Several spiders	*Pseudoplusia includens*	F	nn-p	?	—	Yabar 1980

Cotton	*Geocoris pallens*	Several	C	n-pp	?	—	Yokoyama et al. 1984
Cotton	*Clubiona japonicona*	Several	C	n-pp	P	—	Zhao et al. 1981
CARBOFURAN*							
Rice	*Gonatonarium dentatum*	*Nilaparvata lugens*	LD	nn-p	P	227	Chang 1981
Rice	*Pirata subpiraticus*	*Nilaparvata lugens*	LD	nn-p	P	8	Chang 1981
Tobacco	*Apanteles congregatus*	*Manduca sexta*	F	n-p	?	—	Cherry & Pless 1969
Rice	*Platygaster* spp.	*Nilaparvata lugens*	S	n-pp	?	—	Chiu et al. 1980
Cauliflower	*Aleochara bilineata, Bembidion lampros*	*Delia* sp.	F	nn-p	?	—	Finlayson 1979
Corn	*Scarites substriatus, Pterostichus chalcites, Bembidion rapidum, B. quadrmaculatum*	*Agrotis ipsilon*	C, O	nn-p	E	—	Gholson et al. 1978
Cotton	Several coccinellids (*Cycloneda sanguinea*)	*Eutinobothrus brasiliensis, Aphis gossypii*	F	nn-p	E	—	Gravena et al. 1976
Rice	*Anagrus* spp., spiders	Several plant hoppers	F	n-pp	B	—	Kim et al. 1984
Corn	*Pterostichus chalcites, Harpalus pensylvanicaus, Calosoma sayi*	?	F	nn-p	?	—	Lesiewicz et al. 1984
Rice	Spiders	*Nilaparvata lugens, Spogatella furcifera*	F	n-pp	?	—	Vorley 1985

Table A.4. (*Continued*)

Crop	Natural Enemy	Prey/Host	Test Method[1]	Selectivity Class[2]	Selectivity Type[3]	Selectivity Ratio	Reference
CARTAP							
Rice	*Gonatonarium dentatum*	*Nilaparvata lugens*	LD	nn-p	P	132	Chang 1981
Rice	*Pirata subpiraticus*	*Nilaparvata lugens*	LD	nn-p	P	126	Chang 1981
Cotton	*Trichogramma pretiosum*	*Heliothis* spp.	F	n-p	?	—	Garcia 1980
Apple	*Holocothorax testaceipes*	*Phyllonorycter ringoneiella*	F	nn-p	?	—	Yamada & Kawashima 1983
ETHIOFENCARB							
?	*Chrysoperla formosa*	Several	C	n-p	P	—	Babrikova 1980
?	*Chrysoperla carnea*	Several	C	n-p	P	—	Babrikova 1979
?	*Cales noacki*	*Aleurothrixus floccosus*	C	n-p	P	—	Beitia & Garrido 1985
?	*Pimpla turionellae*	*Galleria mellonella*	C, F	n-p	P	—	Bogenschutz 1984
Tobacco	*Coccinella septempunctata, Macrolophus rubi, Dicyphus eckerleini*	Several aphids	C	nn-pp	?	—	Dirimanov et al. 1974
Sugar beet	*Pterostichus melanarius*	*Aphis fabae, Myzus persicae*	C	n-pp	P	—	Dunning et al. 1982
Citrus	*Cales noacki*	*Aleurothrixus floccosus*	C	n-p	E	—	Garrido et al. 1982
Cereal	*Aphidius ervi*	*Sitobion avenae*	C	n-p	B	—	Suss 1983

Several	*Chrysoperla carnea*	Many	C	n-pp	?	—	Suter 1978
FENOBUCARB							
Rice	Several spiders	*Nilaparvata lugens*	F	nn-p	?	—	Gopinath et al. 1981
Rice	*Lycosa pseudoannulata, Oedothorax insecticeps*	*Nephotettix cincticeps*	LC, F	nn-p	P	—	Kawahara et al. 1971
Rice	Spiders	*Nilaparvata lugens, Sogatella furcifera*	F	n-pp	?	—	Vorley 1985
Rice	*Pirata subpiraticus*	*Nilaparvata lugens*	C, F	n-p	P	—	Yoo et al. 1984
FORMETANATE							
Citrus	*Cryptolaemus montrouzieri*	*Aonidiella aurantii*	C	n-p	P	—	Morse & Bellows 1986
ISOCARB							
Rice	*Lycosa pseudoannulata, Micraspis crosea*	*Nilaparvata lugens*	F	nn-p	?	—	Dyck & Orlido 1977
ISOPROCARB							
Rice	*Pirata subpiraticus*	*Nilaparvata lugens*	LD	n-p	P	66	Chang et al. 1979
Rice	*Lycosa pseudoannulata, Micraspis crosea*	*Nilaparvata lugens*	F	nn-p	?	—	Dyck & Orlido 1977
Rice	*Pirata subpiraticus*	*Nilaparvata lugens*	C, F	n-p	P	—	Yoo et al. 1984
METHOMYL*							
Many	*Trichogramma* spp.	Several	C, R, F	nn-pp	P	—	Bull & Coleman 1985
Apple	*Chrysoperla* spp.	Several	C	n-pp	?	—	David & Horsburgh 1985

Table A.4. (*Continued*)

Crop	Natural Enemy	Prey/Host	Test Method[1]	Selectivity Class[2]	Selectivity Type[3]	Selectivity Ratio	Reference
Bean	*Pediobius foveolatus*	*Epilachna varivestris*	F	n-p	E	—	Flanders et al. 1984
Alfalfa	*Trioxys complanatus*	*Therioaphis trifolii, Therioaphis maculata, Acyrthosiphon kondoi*	C, R	n-pp	B	—	Franzmann & Rossiter 1981
Cabbage	*Diaeretiella rapae*	*Myzus persicae*	C, F	n-p	P	—	Hsieh 1984
Cabbage	*Pteromalus puparum*	*Pieris rapae*	C	n-p	P	—	Lasota & Kok 1986
Cotton	*Microplitis croceipes*	*Heliothis* spp.	LD	n-p	P	—	Powell et al. 1986a
Cotton	*Geocoris pallens*	Several	C	n-pp	?	—	Yokoyama et al. 1984
MTMC							
Rice	*Cyrtorhinus lividipennis, Lycosa pseudoannnulata*	*Nilaparvata lugens, Nephotettix cincticeps*	C	nn-pp	P	—	Ku & Wang 1981
OXAMYL*							
Ornamental	*Encyrtus infelix*	*Saissetia coffeae*	F	n-p	?	—	Hamlen 1975
Greenhouse	*Encarsia formosa*	*Trialeurodes vaporariorum*	C	n-p	P	—	Hatalane & Budai 1982
Apple	*Sympiensis marylandensis*	*Phyllonorycter crataegella*	LC	n-p	P	—	van Driesche et al. 1985
Apple	*Apanteles ornigis*	*Phyllonorycter blancardella*	C	n-p	B	—	Weires et al. 1982

Greenhouse	*Encarsia formosa*	*Trialeurodes vaporariorum*	C	n-p	P	—	Zseller & Budai 1983
PIRIMICARB*							
?	*Aphidoletes aphidimyza*	Several aphids	C	n-pp	P	—	Adashkevich Nurmukhamedov 1985
Cabbage	Coccinellidae, Syrphidae, Cecidomyiidae, Hymenoptera	*Brevicoryne brassicae*	F	nn-pp	?	—	Alexandrescu & Hondru 1981
?	*Chrysoperla formosa*	Several	C	n-pp	P	—	Babrikova 1980
?	*Chrysoperla carnea*	Several	C	n-pp	P	—	Babrikova 1979
Cereals	Coccinellidae, Chrysopidae, *Aphidius* spp.	Several	F	nn-pp	?	—	Ba-Angood & Stewart 1980
?	*Cales noacki*	*Aleurothrixus floccosus*	C	n-p	P	—	Beitia & Garrido 1985
Bean	*Phytoseiulus persimilis*	*Tetranychus urticae*	S	n-p	P	—	Binns 1971
Cotton	*Chrysoperla boninensis, C. congrua, C. pudica*	Spider mites, *Aphis gossypii*	LC	nn-pp	P	—	Brettell 1984
Wheat	*Hippodamia variegata*	*Sitobion avenae, Schizaphis graminium*	F	n-pp	B	—	Broodryk 1980
Cereal	*Syrphus* spp., *Coccinella septempunctata, Nebria brevicollis, Pterostichus melanarius*	*Metopalophium dirhodum*	ED	nn-p	P	4 158 76 >1000	Brown et al. 1983

Table A.4. (*Continued*)

Crop	Natural Enemy	Prey/Host	Test Method[1]	Selectivity Class[2]	Selectivity Type[3]	Selectivity Ratio	Reference
Citrus	*Cales noacki*	*Aleurothrixus floccosus*	F	n-p	?	—	Carrero 1979
Cereal	Aphidiids, carabids, staphylinids	Several aphids	F	nn-pp	?	—	Cole & Wilkinson 1984
Citrus	*Aphytis melinus*	*Aonidiella aurantii*	C	n-p	P	—	Davies & McLaren 1977
Vegetables	*Diaretiella rapae*	*Brevicoryne brassicae, Myzus persicae*	C	n-pp	?	—	Delorme 1976
Greenhouse	*Encarsia formosa*	*Trialeurodes vaporariorum*	C	n-p	P	—	Delorme & Angot 1983
Tobacco	*Coccinella septempunctata, Macrolophus rubi, Dicyphus eckerleini*	Several aphids	C	nn-pp	?	—	Dirimanov et al. 1974
Cereal	Several carabids, staphylinids	Several aphids	F	nn-pp	?	—	Feeney 1983
Cereal	*Pterostichus melanarius, Trechus quadristriatus, Philonthus* spp., *Tachyrorus hypnorum*	Several	F	nn-pp	?	—	Feeny 1983

Cereal	*Forficula ayrucykarua*	Cereal aphids	R	n-pp	?	—	Ffrench-Constant & Vickerman 1985
Sugar beet	*Pterostichus melanarius*	*Aphis fabae, Myzus persicae*	C	n-pp	P	—	Dunning et al. 1982
Cereal	*Cycloneda sanguinea*	*Macrosiphum avenae*	C, F	n-p	P	—	Fagundes et al. 1985
Peach	*Metaseiulus occidentalis*	*Myzus persicae*	F	n-p	P	—	Field 1978
Several	*Trichogramma cacoeciae, Pales pavida, Pimpla turionellae, Chrysoperla carnea, Leptomastix dactylopii*	Several	C	nn-pp	P	—	Franz et al. 1980
Alfalfa	*Trioxys complanatus*	*Therioaphis trifolii, T. maculata, Acyrthosiphon kondoi*	C, R	n-pp	B	—	Franzmann & Rossiter 1981
Cereal	*Forficula auricularia*	Several aphids	LD	n-pp	P	—	Ffrench-Constant & Vickerman 1985
Citrus	*Cales noacki*	*Aleurothrixus floccosus*	C	n-p	E	—	Garrido et al. 1982
Bean	*Coccinella septempunctata, Chrysoperla carnea*	*Acyrthosiphon pisum*	C	nn-p	B	—	Grapel 1981, 1982
Vegetables	*Cycloneda sanguinea*	*Brevicoryne brassicae*	LC	n-p	P	15	Gravena & Batiste 1979a

Table A.4. (*Continued*)

Crop	Natural Enemy	Prey/Host	Test Method[1]	Selectivity Class[2]	Selectivity Type[3]	Selectivity Ratio	Reference
Greenhouse	*Phytoseiulus persimilis, Amblyseius longispinosus*	Several aphids	C, R	nn-pp	P	—	Goodwin & Bowden 1984
Apple	*Typhlodromus pyri, Chrysocharis prodice, Aphelinus mali*	Several aphids	C, F	nn-pp	B	—	Gruys 1980b
Apple	*Aphelinus mali*	*Eriosoma lanigerum*	C	n-p	?	—	Haidari & Georgis 1978
Alfalfa	*Coccinella transversoguttata, Hippodamia parenthesis, H. quinquesignata, H. tredecimpunctata, H. sinuata, Orius tristicolor, Nabis alternatus, Aeolothrips fasciatus*	*Acyrothosiphon pisum*	F	nn-p	?	—	Harper 1978
Greenhouse	*Phytoseiulus persimilis*	*Tetranychus urticae*	C	n-p	P	—	Hassan 1982a
Sugar beet	*Chrysoperla carnea*	*Myzus persicae*	C	n-p	P	—	Hassan et al. 1985
Greenhouse	*Encarsia formosa*	*Trialeurodes vaporariorum*	C	n-p	P	—	Hatalane & Budai 1982

Greenhouse	*Chrysoperla carnea, Encarsia formosa, Phytoseiulus persimilis*	*Myzus persicae*	C, F	nn-p	P	—	Helgesen & Tauber 1974
Cereal	*Coccinella septempunctata, Chrysoperla carnea, Syrphus corollae*	*Sitobion avenae, Metopolophium dirhodim, Rhopalosiphum padi*	F	nn-pp	B	—	Hellpap 1982
?	*Coccinella septempunctata, Chrysoperla carnea, Syrphus vitripennis*	*Acyrthosiphon pisum*	F	nn-p	B	—	Hellpap & Schmutterer 1982
Many	Many spp.	*Myzus persicae, Aphis fabae, Acyrthosiphon pisum*	F	nn-pp	E	—	Hofer 1979
Hops	*Coccinella quinque-punctata, Adalia bipunctata, Cycloneda sanguinea, Propylea quattuordecimpunc-tata*	*Phorodon humuli, Aphis fabae*	C	nn-pp	P	—	Kalushkov 1982
Apple	*Typhlodromus tilarium*	*Panonychus ulmi*	F	n-p	?	—	Karg 1978
Alfalfa	*Coccinella repanda*	*Therioaphis trifolii, Acyrthosiphon kondoi*	LD	n-pp	P	—	Kay 1979

Table A.4. (*Continued*)

Crop	Natural Enemy	Prey/Host	Test Method[1]	Selectivity Class[2]	Selectivity Type[3]	Selectivity Ratio	Reference
Strawberry	*Phytoseiulus persimilis*	*Tetranychus urticae*, several aphids	F	n-p	P	—	Kennedy et al. 1976
Broccoli	Several parasitoids	*Brevicoryne brassicae*, *Myzus persicae*	F	nn-pp	P	—	Kennedy & Oatman 1976
?	*Coccinella septempunctata*	*Myzus persicae*	C	n-p	P	—	Kirknel 1975
Alfalfa	*Trioxys complanatus*	*Therioaphis trifolii*, *Acyrthosiphon kondoi*	F	n-pp	?	—	Lambert 1981
Pepper	*Coccinella septempunctata*	*Myzus persicae*, *Aphis fabae*, *Brevicoryne brassicae*	C, F	n-pp	?	—	Laska 1972
Greenhouse	*Encarsia formosa*	*Trialeurodes vaporariorum*	C	n-p	B	—	Laska et al. 1980
Potato	*Coleomegilla maculata*, *Chrysoperla oculata*	*Myzus persicae*	LD	nn-p	P	4030–4099	Lecrone & Smilowitz 1980
Alfalfa	Many spp.	Aphids, thrips, *Contarinia medicaginis*, *Subcoccinella vigintiquattuorpunctata*	F	nn-pp	P	—	Margarit & Hondru 1981

Alfalfa	*Chrysoperla oculata*	*Acyrthosiphon pisum*	C	n-p	P	—	McDonald & Harper 1978
Vegetables	*Phygaedeuon trichops*	*Delia* spp.	C	n-p	P	—	Naton 1978
Rose	*Coccinella septempunctata, Adonia variegata, Chrysoperla carnea, Erisyrphus balteatus*	*Microsiphum rosae*	C	nn-p	P	—	Natskova 1974
Apple	*Stethorus punctillum*	*Panonychus ulmi*	F	n-p	?	—	Pasqualini & Malavolta 1985
Sugar beet	*Thaumatomyia glabra, T. rufa*	*Pemphigus fuscicornis*	C	nn-p	P	—	Petrukka & Gres 1974
Citrus	*Cycloneda sanguinea*	*Toxoptera citricida*	C	n-p	P	—	Portillo 1977
Cereal	Carabids, Staphylinids, spiders	*Metopolophium dirhodum, Sitobion avenae, M. festucae*	F	nn-pp	?	—	Powell et al. 1985a
Many	*Chrysoperla oculata*	Many	C	n-pp	P	—	Pree & Hagley 1985
Greenhouse	*Phytoseiulus persimilis*	Spider mites	F	n-p	P	—	Pruszynski et al. 1983
Many	*Macrolophus nubilus*	Many	F	n-pp	?	—	Prutenskaya et al. 1984
Citrus	*Aphytis melinus*	*Aonidiella aurantii, Aspidiotus nerii*	F	n-pp	P	—	Saba 1978
Nursery	*Phytoseiulus persimilis*	*Tetranychus urticae*	F	n-p	P	—	Simmonds 1973
Alfalfa	*Nabis* spp., *Orius* spp., *Hippodamia convergens, Chrysoperla* spp., *Aphidius smithi*	*Acyrthosiphon pisum*	F	nn-p	P	—	Summers et al. 1975

Table A.4. *(Continued)*

Crop	Natural Enemy	Prey/Host	Test Method[1]	Selectivity Class[2]	Selectivity Type[3]	Selectivity Ratio	Reference
Cereal	*Aphidius ervi*	*Sitobion avenae*	C	n-p	B	—	Suss 1983
Greenhouse	*Phytoseiulus persimilis*	Spider mites	C	n-p	P	—	Stamenkovic & Peric 1984a
Several	*Chrysoperla carnea*	Many	C	n-pp	?	—	Suter 1978
Alfalfa	*Austromicromus tasmaniae, Coccinella undecimpunctata*	*Acyrthosiphon kondoi, A. pisum*	LD	nn-pp	P	1,000–10,000	Syrett & Penman 1980
Apple	*Typhlodromus pyri*	Hemipteran spp.	F	n-pp	B	—	Trapman & Blommers 1980
Citrus	*Cycloneda sanguinea, Chrysoperla* spp.	*Toxoptera citricida*	F	nn-p	B	—	Trevizoli & Gravena 1979
Citrus	*Leptomastidea abnormis, Scymnus includens*	*Planococcus citri*	C	nn-p	P	—	Viggiani et al. 1972
Apple	*Typhlodromus pyri, Amblyseius andersoni*	Several aphids	C	nn-pp	P	—	Vigi et al. 1985
Apple	*Aphidoletes aphidimyza*	*Aphis pomi*	C	n-p	P	—	Warner & Croft 1982
Cereal	*Erythraeus* spp., lycosid spiders, *Tachyporus* spp., *Bambidion lampros, Hapalus rufipes, Pterostichus melanarius Coccinella septempunctata*	Cereal aphids	C	nn-pp	P	—	Wiktelius 1986

Sugar beets	Coccinellids, syrphids	Several aphids	C	nn-pp	P	—	Wilhelm 1979
Greenhouse	*Encarsia formosa*	*Trialeurodes vaporariorum*	C	n-p	P	—	Zseller & Budai 1982
PROPOXUR							
Rice	*Lycosa pseudannulata*	*Nephotettix cincticeps, Nilaparvata lugens*	LD	n-pp	P	—	Chu et al. 1975
Forest	*Coeliodes pissodis, Medetera bistriata, Roptrocerus xylophagorum Thanasimus dubius*	*Dendroctonus frontalis*	S	nn-p	?	—	Coster & Ragenovich 1976
Rice	*Lycosa pseudoannulata, Tetragnatha* sp., *Araneus* sp., *Microvelia atrolineata*	*Nilaparvata lugens*	F	nn-p	?	—	Reissig et al. 1982
THIODICARB							
Cotton	*Chrysoperla boninensis*	*Heliothis armigera, Diparopsis castanea*	LC	nn-pp	P	—	Brettell 1984
Many	*Trichogramma* spp.	Several	C,R,F	nn-pp	P	—	Bull & Coleman 1985
Cotton	*Trichogramma pretiosum*	*Heliothis virescens*	C	n-p	B	—	Bull & House 1983
Cotton	*Micropletis croceipes*	*Heliothis* spp.	R	n-p	P	—	Powell & Scott 1985

[1] F = field; C = contact; EC = effective concentration; ED = effective dosage; LC = lethal concentration; LD = lethal dosage; S = systemic; O = oral; R = residue.
[2] n = single natural enemy species; nn = multiple; p = single prey or host species; pp = multiple.
[3] P = physiological selectivity; E = ecological; B = both P & E; ? = uncertain.

Table A.5. Cases of Insecticide Selectivity to Arthropod Natural Enemies over Their Prey or Hosts: Pyrethroids

Crop	Natural Enemy	Prey/Host	Test Method[1]	Selectivity Class[2]	Selectivity Type[3]	Selectivity Ratio	Reference
BIORESMETHRIN							
Greenhouse	*Encarsia formosa*	*Trialeurodes vaporariorum*	C	n-p	P	—	Delorme & Angot 1983
CYPERMETHRIN							
Cabbage	Hymenoptera, Coccinellidae, Syrphidae, Cecidomyiidae	*Brevicoryne brassicae*	?	nn-p	B	—	Alexandrescu & Hondru 1981
Cotton	*Coccinella undecimpunctata*	*Spodoptera littoralis*	LD	n-p	P	156	Abdel-Aal et al. 1979
Cereal	*Nebria brevicollis, Pterostichus melanarius*	*Metopalopium dirhodum*	ED	nn-p	P	5	Brown et al. 1983
Cereal	*Forficula auricularia*	Several aphids	LD	n-pp	P	4	Ffrench-Constant & Vickerman 1985
Cabbage	*Apanteles plutellae*	*Plutella xylostella*	R	n-p	P	—	Mani & Krishnamoorthy 1984
Strawberry	Carabids, staphylinids, spiders	?	F	nn-p	?	—	Nikusch & Gernot 1986b
DECAMETHRIN							
Cotton	*Chrysoperla carnea*	*Heliothis virescens*	LC	n-p	P	—	Plapp & Bull 1978
DELTAMETHRIN							
?	*Pimpla turionellae*	*Galleria mellonella*	C, F	n-p	P	—	Bogenschutz 1984

Cotton	*Chrysoperla boninensis*	*Heliothis armigera, armigera, Diparopsis castanea*	LC	n-pp	P	—	Brettell 1984
Cereal	*Forficula auricularia*	Several aphids	LD	n-pp	P	—	Ffrench-Constant & Vickerman 1985
Strawberry	Carabids, staphylinids, spiders	?	F	nn-p	?	—	Nikusch & Gernot 1986b
Cabbage	*Apanteles plutellae*	*Plutella xylostella*	R	n-p	P	—	Mani & Krishnamoorthy 1984
FENVALERATE							
Cotton	*Coccinella undecimpunctata*	*Spodoptera littoralis*	LD	n-p	P	37	Abdel-Aal et al. 1979
Cotton	*Chrysoperla boninensis*	*Heliothis armigera, Diparopsis castanea*	LC	nn-pp	P	—	Brettell 1984
Cereal	*Pterostichus melanarius*	*Sitobion avenae, Metopolopium dirhodum, Rhopalosiphum padi*	F	n-pp	?	—	Chiverton 1984
Cereal	*Coccinella transversoguttata, C. trifasciata, Hippodamia convergens, H. glacialis, H. parenthesis*	*Diabrotica longicornis, Oulema melanopus*	LD	nn-pp	P	—	Coats et al. 1979

Table A.5. (*Continued*)

Crop	Natural Enemy	Prey/Host	Test Method[1]	Selectivity Class[2]	Selectivity Type[3]	Selectivity Ratio	Reference
Crucifers	*Apanteles plutella*	*Plutella xylostella*	LC	n-p	P	—	Feng & Wang 1984
Apple	*Apanteles ornigis*	*Phyllonorycter blancardella*	LC	n-p	P	2	Hagley et al. 1981
Cotton	*Coccinella septempunctata*, *Chrysoperla sinica*	*Heliothis armigera*	F	nn-p	?	—	Hu 1981
Cotton	*Chrysoperla carnea*	?	C	n-p	P	—	Kismir & Sengonca 1980
Cabbage	*Pteromalus puparum*	*Pieris rapae*	C	n-p	P	—	Lasota & Kok 1986
Cabbage	*Apanteles plutellae*	*Plutella xylostella*	R	n-p	P	—	Mani & Krishnamoorthy 1984
Cotton	*Chrysoperla carnea*	*Heliothis virescens*	LC	n-p	P	—	Plapp & Bull 1978
Cereal	Carabids, staphylinids	Cereal aphids	F	nn-pp	?	—	Poehling et al. 1985b
Cotton	*Micropletis croceipes*	*Heliothis* spp.	LD	n-p	P	—	Powell et al. 1986a
Cotton	*Micropletis croceipes*	*Heliothis* spp.	R	n-p	P	—	Powell & Scott 1985
Cotton	*Chrysoperla carnea*, *Trichogramma brasiliensis*	*Corycra cephalonica*	R	nn-p	P	—	Singh & Varma 1986
Alfalfa	*Austromicromus tasmaniae*, *Coccinella undecimpunctata*	*Acyrthosiphon kondoi*, *A. pisum*	LD	nn-pp	P	5–1000	Syrett & Penman 1980
Vegetables	*Apanteles* sp., *Diglyphus intermedius*, *Opius bruneipes*	*Liriomyza sativae*, *Keiferia lycopersicella*	C	nn-pp	P	—	Waddill 1978

FLUCYTHRINATE							
Cotton	*Chrysoperla boninensis*	*Heliothis armigera, Diparopsis castanea*	LC	nn-pp	P	—	Brettell 1984
Cotton	*Micropletis croceipes*	*Heliothis* spp.	LD	n-p	P	—	Powell et al. 1986a
Cotton	*Micropletis croceipes*	*Heliothis* spp.	R	n-p	P	—	Powell & Scott 1985
Cotton	*Chrysoperla carnea*	*Heliothis virescens*	LC	n-p	P	—	Rajakulendran & Plapp 1982b
FLUVALINATE							
Cotton	*Chrysoperla carnea*	*Heliothis virescens*	LC	n-p	P	—	Rajakulendran & Plapp 1982b
PERMETHRIN							
Cotton	*Coccinella undecimpunctata*	*Spodoptera littoralis*	LD	n-p	P	2	Abdel-Aal et al. 1979
Cabbage	Hymenoptera, Coccinellidae, Syrphidae, Cecidomyiidae	*Brevicoryne brassicae*	?	nn-p	B	—	Alexandrescu & Hondru 1981
Sunflower	*Nealiolus curculionis*	*Cylindrocopturus adspersus*	F	nn-p	?	—	Charlet & Oseto 1983
Cereal	*Coccinella transversoguttata, C. trifasciata, Hippodamia convergens, H. glacialis, H. parenthesis, H. tredecimpunctata tibialis*	*Diabrotica longicornis, Oulema melanopus*	LD	nn-pp	P	—	Coats et al. 1979
Apple	*Apanteles ornigis*	*Phyllonorycter blancardella*	LC	n-p	P	19	Hagley et al. 1981

Table A.5. (*Continued*)

Crop	Natural Enemy	Prey/Host	Test Method[1]	Selectivity Class[2]	Selectivity Type[3]	Selectivity Ratio	Reference
Cabbage	*Diaeretiella rapae*	*Myzus persicae*	C, F	n-p	P	—	Hsieh 1984
Cabbage	*Pteromalus puparum*	*Pieris rapae*	C	n-p	P	—	Lasota & Kok 1986
Cabbage	*Apanteles plutellae*	*Plutella xylostella*	R	n-p	P	—	Mani & Krishnamoorthy 1984
Cotton	*Chrysoperla carnea*	*Heliothis virescens*	LC	n-p	P	—	Plapp & Bull 1978
Cotton	*Micropletis croceipes*	*Heliothis* spp.	LD	n-p	P	—	Powell et al. 1986a
?	*Bracon hebetor*	*Ephestia cautella*	C	n-p	P	—	Press et al. 1981
Forest	*Brachymeria intermedia, Compsilura concinnata*	*Lymantria dispar*	LD	nn-p	P	4–31	Respicio & Forgash 1984
Cereal	*Aphidius ervi*	*Sitobion avenae*	C	n-p	B	—	Suss 1983
RESMETHRIN							
Greenhouse	*Encarsia formosa*	*Trialeurodes vaporariorum*	C	n-p	E	—	Osborne 1981
TRALOMETHRIN							
Cotton	*Chrysoperla carnea*	*Heliothis virescens*	LC	n-p	P	—	Rajakulendran & Plapp 1982b

[1] F = field; C = contact; EC = effective concentration; ED = effective dosage; LC = lethal concentration; LD = lethal dosage; S = systemic; O = oral; R = residue.
[2] n = single natural enemy species; nn = multiple; p = single prey or host species; pp = multiple.
[3] P = physiological selectivity; E = ecological; B = both P & E; ? = uncertain.

Table A.6. Cases of Pesticide Selectivity to Arthropod Natural Enemies over Their Prey or Hosts: Miscellaneous Pesticides and Acaricides

Crop	Natural Enemy	Prey/Host	Method[1]	Selectivity Class[2]	Selectivity Type[3]	Selectivity Ratio	Reference
AMITRAZ							
Cotton	*Chrysoperla boninensis, C. congrua, C. pudica*	Spider mites	LC	nn-pp	P	—	Brettell 1984
AZOCYCLOTIN							
Grape	*Amblyseius chilinensis*	*Brevipalpus chilensis, Oligonychus vitis, Colomerus vitus*	F	n-pp	P	—	Gonzalez 1983
Apple	*Typhlodromus pyri, Amblyseius fallacis, M. occidentalis, Phytoseiulus persimilis*	*Panonychus ulmi, Tetranychus urticae*	C	nn-pp	P	—	Wearing & Ashley 1982
BENZOMATE							
Apple	*Metaseiulus occidentalis, Typhlodromus columbiensis, Zetzellia mali*	*Panonychus ulmi*	C	nn-p	P	—	Downing & Moilliet 1975
Greenhouse	*Phytoseiulus persimilis*	*Tetranychus urticae*	LD	n-p	P	—	Everson & Tonks 1981
Apple	*Typhlodromus pyri, Chrysocharis prodice, Aphelinus mali*	*Panonychus ulmi*	C, F	nn-p	B	—	Gruys 1980b
Many	*Trichogramma cacoeciae, Pales pavida, Pimpla turionellae, Chrysoperla carnea, Leptomastix dactylopii, Amblyseius potentillae, Anthocoris nemorum, Crytolaemus montrouzieri, Drino inconspicua, Encarsia formosa, Opius* spp., *Phygadeuon trichops, Phytoseiulus persimilis, Syrphus vitripennis*	Many	C	nn-pp	P	—	Franz et al. 1980, Hassan et al. 1983

Table A.6. (*Continued*)

Crop	Natural Enemy	Prey/Host	Method[1]	Selectivity Class[2]	Selectivity Type[3]	Selectivity Ratio	Reference
Citrus	*Amblyseius eharai*	*Panonychus citri*	R	n-p	P	—	Kashio & Tanaka 1979
?	*Stethorus loi, Amblyseius longispinosus*	*Tetranychus truncatus*	C	nn-p	P	—	Lo & Chao 1975
Greenhouse	*Phygadeuon trichops*	*Tetranychus urticae*	C	n-p	P	—	Naton 1983
Apple	*Amblyseius potentillae*	*Panonychus ulmi*	C	n-p	P	—	Overmeer & van Zon 1982
Apple	*Amblyseius fallacis*	*Tetranychus urticae*	C	n-p	P	—	Streibert 1981
Apple	*Stethorus punctillum, Orius* spp.	*Panonychus ulmi*	F	nn-p	P	—	Waldner 1985
BINAPACRYL							
Cotton	*Chrysoperla boninensis, C. congrua, C. pudica*	Spider mites	LC	nn-pp	P	—	Brettell 1984
Apple	*Stethorus nigripes*	*Panonychus ulmi, Tetranychus urticae*	C	n-pp	P	—	Edwards & Hodgson 1973
Apple	*Hippodamia convergens*	Several aphids	LD	n-pp	P	—	Moffitt et al. 1972
?	*Amblyseius californicus*	*Tetranychus urticae*	C	n-p	P	—	Wu et al. 1985
BROMOPROPYLATE							
?	*Amblyseius californicus*	*Tetranychus urticae*	C	n-p	P	—	Wu et al. 1985
BUPROFEZIN							
Greenhouse	*Encarsia formosa, Cales noacki*	*Trialeurodes vaporariorum, Aleurothrixus floccosus*	C	nn-pp	P	—	Garrido et al. 1984

Greenhouse	*Encarsia formosa*	*Trialeurodes vaporariorum*	C	n-p	P	—	Martin & Workman 1986
CHLOROBENZILATE							
Apple	*Stethorus punctillum*	*Tetranychus urticae*	LC	n-p	P	20	Bravenboer 1959
Greenhouse	*Phytoseiulus persimilis*	*Tetranychus urticae*	LD	n-p	P	—	Everson & Tonks 1981
Citrus	*Cales noacki*	*Aleurothrixus floccosus*	C	n-p	P	—	Garrido et al. 1982
Citrus	*Amblyseius stipulatus*	*Panonychus citri*	F	n-p	P	—	Jones & Parrella 1983
Greenhouse	*Aphidoletes aphidimyza*	*Microsiphum rosae*, spider mites	C	n-p	P	—	Markkula & Tiittanen 1976a
Castor	*Stethorus pauperculus*	*Eutetranychus banksi*	C	n-p	P	—	Rai et al. 1964
Citrus	*Aphytis holoxanthus*	*Chrysomphalus aonidum*	C	n-p	P	—	Rosen 1967
Citrus	*Sympherobius sanctus*, *Cryptolaemus montrouzieri*,	Several	C	nn-pp	?	—	Sekeroglu & Uygun 1980
Apple	*Amblyseius fallacis*	*Tetranychus urticae*	C	n-p	P	—	Streibert 1981
Citrus	*Amblyseius swirskii*, *A. limonicus*, *Typhlodromus rhenanus*	*Phyllocoptruta oleivora*, *Brevipalpus* spp.	C, F	nn-p	P	—	Swirski et al. 1967
CHLORDIMEFORM							
Apple	*Stethorus* spp.	*Tetranychus urticae*	F	n-p	P	—	Bower & Kaldor 1980
Many	*Trichogramma* spp.	Several	C, R, F	nn-pp	P	—	Bull & Coleman 1985
Cotton	*Hippodamia convergens*, *Chrysoperla carnea*	*Heliothis virescens*, *Heliothis zea*	C, F	nn-pp	B	—	Bull & House 1978
Cotton	*Trichogramma pretiosum*	*Heliothis virescens*	C	n-p	B	—	Bull & House 1983
Rice	*Lycosa pseudoannulata*, *Micraspis crosea*	*Nilaparvata lugens*	F	nn-p	?	—	Dyck & Orlido 1977

Table A.6. (*Continued*).

Crop	Natural Enemy	Prey/Host	Method[1]	Selectivity Class[2]	Selectivity Type[3]	Selectivity Ratio	Reference
Apple	*Stethorus nigripes*	*Panonychus ulmi, Tetranychus urticae*	C	n-pp	P	—	Edwards & Hodgson 1973
Cotton	*Trichogramma pretiosum*	*Heliothis* spp.	F	n-p	?	—	Garcia 1980
Cabbage	*Apanteles glomeratus*	*Pieris rapae*	LC	n-p	P	—	Hamilton & Attia 1976
Rice	*Enoplognatha japonica, Oedothorax insecticeps, Erigone prominens, Gnathonarium dentatum, Pachygnatha clercki, Dolomedes pallitarsis, Pardosa laura, Pirata subpiraticus, P. clercki, Xysticus* sp., *Clubiona japonicola*	Rice stem borer	F	nn-p	?	—	Koyama 1975
Tobacco	Many generalist predators	*Spodoptera litura*	F	nn-p	?	—	Nakasuji et al. 1973
Cotton	*Chrysoperla carnea*	*Heliothis virescens*	LC	n-p	P	—	Plapp & Bull 1978
Cotton	*Micropletis croceipes*	*Heliothis* spp.	LD	n-p	P	—	Powell et al. 1986a
Apple	*Stethorus* spp.	*Tetranychus urticae*	C, F	n-p	P	—	Walters 1976a
Apple	*Stethorus loxtoni, S. nigripes, S. vagans*	*Tetranychus urticae*	C	n-p	P	—	Walters 1976b, 1976c
Pear	*Deraeocoris brevis piceatus*	*Psylla pyricola*	F	n-p	P	—	Westigard 1973
Cotton	*Clubiona japonicola*	Several	C	n-pp	P	—	Zhao et al. 1981

CLOFENTEZINE							
Apple	Phytoseiid species	*Panonychus ulmi, Tetranychus urticae*	F	nn-pp	P	—	Baillod et al. 1986
Apple	*Typhlodromus* spp.	*Panonychus ulmi*	F	n-pp	P	—	Bouron 1985
Apple	*Stethorus punctillum, Typhlodromus* sp.	*Panonychus ulmi*	F	n-pp	?	—	Comai 1985
Apple	*Typhlodromus pyri*	*Aculus schlechtendali*	F	n-p	P	—	Easterbrook 1984
Several	*Metaseiulus occidentalis*	*Tetranychus urticae*	C	n-pp	P	—	Hoy & Ouyang 1986
Greenhouse	*Phytoseiulus persimilis, Encarsia formosa*	Spider mites	C	n-p	P	—	Stamenkovic & Peric 1984
CYHEXATIN							
?	*Cales noacki*	*Aleurothrixus floccosus*	C	n-p	P	—	Beitia & Garrido 1985
Apple	*Amblyseius chilenensis*	*Panonychus ulmi*	F	n-p	P	—	Carbonell & Briozzo 1981
Greenhouse	*Phytoseiulus persimilis*	*Tetranychus urticae*	C	n-p	P	—	Caprioli et al. 1983
Apple	*Typhlodromus pyri*	*Panonychus ulmi*	F	n-p	P	—	Collyer 1980
Apple	*Stethorus punctum*	*Panonychus ulmi*	C	n-p	P	—	Colburn & Asquith 1971
Greenhouse	*Phytoseiulus persimilis, Encarsia formosa*	*Tetranychus urticae*	?	nn-p	B	—	Costello & Elliott 1981
Apple	*Amblyseius fallacis*	*Panonychus ulmi, Tetranychus urticae*	C	n-pp	P	—	Croft & Nelson 1972
Apple	*Metaseiulus occidentalis, Typhlodromus columbiensis*	*Panonychus ulmi*	C	nn-p	P	—	Downing & Moilliet 1975
Several	*Trichogramma cacoeciae, Pales pavida, Pimpla turionellae, Chrysoperla carnea, Leptomastix dactylopii*	Several	C	nn-pp	P	—	Franz et al. 1980

Table A.6. (*Continued*)

Crop	Natural Enemy	Prey/Host	Method[1]	Selectivity Class[2]	Selectivity Type[3]	Selectivity Ratio	Reference
Greenhouse	*Phytoseiulus persimilis*	*Tetranychus urticae*	LD	n-p	P	—	Everson & Tonks 1981
Grape	*Amblyseius chilinensis*	*Brevipalpus chilensis, Oligonychus vitis, Colomerus vitus*	F	n-pp	P	—	Gonzalez 1983
Apple	*Typhlodromus pyri*	*Panonychus ulmi*	F	n-p	P	—	Hoyt 1973
Citrus	*Amblyseius hibisci, A. stipulatus, Iphiseius degenerans, Phytoseiulus persimilis*	*Panonychus citri, Tetranychus pacificus*	C	nn-pp	P	—	Jeppson et al. 1975
Strawberry	*Phytoseiulus persimilis*	*Tetranychus urticae*	F	n-p	P	—	Kennedy et al. 1976
Greenhouse	*Encarsia formosa*	*Tetranychus urticae*	C	n-p	B	—	Ledieu 1979
Apple	*Amblyseius fallacis, Typhlodromus pyri*	*Panonychus ulmi, Aculus schlechtendali*	C	n-p	P	—	Lienk et al. 1976, 1978
Apple	*Chiracanthium mildei*	*Spodoptera littoralis,* spider mites	C	n-pp	P	—	Mansour et al. 1981
Greenhouse	*Phygadeuon trichops*	*Tetranychus urticae*	C	n-p	P	—	Naton 1983
Strawberry	*Scolothrips sexmaculatus, Feltiella acarivora, Stethorus picipes*	*Tetranychus urticae*	F	n-pp	P	—	Oatman et al. 1985
Cotton	*Amblyseius gossipi*	*Tetranychus arabicus, T. cucurbitacearum*	C	n-pp	P	—	Osman & Zohdy 1976
Several	*Amblyseius bibens, A. potentillae, Typhlodromus pyri*	*Tetranychus urticae*	LC	nn-p	P	—	Overmeer & van Zon 1981

Apple	*Stethorus punctillum*	*Panonychus ulmi*	F	n-p	?	—	Pasqualini & Malavolta 1985
Apple	Miridae, Coccinellidae, Chrysopidae, Anthocoridae	*Panonychus ulmi*	F	nn-p	?	—	Reboulet 1986
Grape	*Agistemus exsertus, Amblyseius gossipi, Tydeus californicus*	*Tenuipalpus ganati*	F	n-pp	P	—	Rizk et al. 1983a
Citrus	*Agistemus exsertus, Cheletogenes ornatus*	*Eutetranychus orientalis*	F	n-pp	?	—	Rizk et al. 1983b
Apple	*Amblyseius fallacis*	*Panonychus ulmi, Tetranychus urticae*	LC	n-pp	P	—	Rock & Yeargan 1970
Greenhouse	*Phytoseiulus persimilis*	*Tetranychus urticae*	C	n-p	P	—	Stenseth 1975
Apple	*Amblyseius fallacis*	*Tetranychus urticae*	C	n-p	P	—	Streibert 1981
Citrus	*Leptomastidea abnormis, Scymnus includens*	*Planococcus citri*	C	nn-p	P	—	Viggiani et al. 1972
Apple	*Typhlodromus pyri*	*Panonychus ulmi*	F	n-p	P	—	Wearing et al. 1978b
Pear	*Deraeocoris brevis, Phytocoris* sp., *Pilophorus* sp., *Orius tristicolor, Chrysoperla* sp., *Hemerobius,* spiders, *Amblyseius schusteri, Typhlodromus arboreus, T. caudiglans*	*Epitrimerus pyri, Tetranychus urticae, Eotetranychus carpini*	F	nn-pp	P	—	Westigard et al. 1986
Apple	*Typhlodromus pyri, Amblyseius fallacis, M. occidentalis, Phytoseiulus persimilis*	*Panonychus ulmi, Tetranychus urticae,*	C	nn-pp	P	—	Wearing & Ashley 1982
Greenhouse	*Phytoseiulus persimilis*	Spider mites	C	n-p	P	—	Stamenkovic & Peric 1984b
DICOFOL							
Greenhouse	*Phytoseiulus persimilis*	*Tetranychus atlanticus*	F	n-p	P	—	Atanasov 1974

Table A.6. (*Continued*)

Crop	Natural Enemy	Prey/Host	Method[1]	Selectivity Class[2]	Selectivity Type[3]	Selectivity Ratio	Reference
Greenhouse	*Phytoseiulus persimilis*	*Tetranychus turkestani*	?	n-p	P	—	Balevski 1976
?	*Chrysoperla carnea*	Several	C	n-p	P	—	Babrikova 1979
?	*Cales noacki*	*Aleurothrixus floccosus*	C	n-p	P	—	Beitia & Garrido 1985
Apple	*Metaseiulus occidentalis,* *Stethorus punctillum*	*Tetranychus urticae*	LC	nn-p	P	4 80	Bravenboer 1959
Vegetables	*Diaeretiella rapae*	*Brevicoryne brassicae, Myzus persicae*	C	n-pp	?	—	Delorme 1976
Greenhouse	*Phytoseiulus persimilis*	*Tetranychus urticae*	LD	n-p	P	—	Everson & Tonks 1981
Several	*Trichogramma cacoeciae, Pales pavida, Pimpla turionellae, Chrysoperla carnea, Leptomastix dactylopii*	Several	C	nn-pp	P	—	Franz et al. 1980
Citrus	*Cales noacki*	*Aleurothrixus floccosus*	C	n-p	E	—	Garrido et al. 1982
Cotton	*Agistemus exertus*	*Cenopaplus pulcher*	C	n-p	P	—	Hassen et al. 1970
Greenhouse	*Encarsia formosa, Phytoseiulus persimilis*	Several	C	nn-pp	P	—	Helyer 1982
Greenhouse	*Phytoseiulus persimilis*	*Tetranychus urticae*	LC	n-p	P	6	Herne & Chant 1965

Cotton	*Coccinella undecimpunctata Paederus alfierii, Orius* spp., *Chrysoperla carnea, Agistemus exsertus, Typhlodromus zaheri*	Spider mites	F	nn-p	?	—	Hoda & Hafez 1981
Greenhouse	*Encarsia formosa*	*Tetranychus urticae*	C	n-p	B	—	Ledieu 1979
?	*Telenomus remus*	*Spodoptera litura*	C	n-p	P	—	Mani & Krishnamoorthy 1986
Cotton	*Apanteles angaleti, Apanteles kirkpatricki*	*Pectinophora gossypiella*	C	nn-p	P	—	Mani & Nagarkatti 1983
Greenhouse	*Aphidoletes aphidimyza*	*Microsiphum rosae*, spider mites	C	n-p	P	—	Markkula & Tiittanen 1976a
Okra	*Amblyseius tetranychivorus*	*Tetranychus ludeni*	F	n-p	P	—	Nangia & Channabasavanna 1983
Castor	*Stethorus pauperculus*	*Tetranychus neocaledonicus, T. cucurbitae*	C	n-p	P	—	Puttaswamy & Channabasavanna 1976
Citrus	*Aphytis holoxanthus*	*Chrysomphalus aonidum*	C	n-p	P	—	Rosen 1967
Citrus	*Aphytis melinus*	*Aonidiella aurantii, Aspidiotus nerii*	F	n-pp	P	—	Saba 1978
Citrus	*Pauridia peregrina*	*Planococcus citri*	LC	n-p	P	—	Searle 1965
Citrus	*Sympherobius sanctus, Cryptolaemus montrouzieri*	Several	C	nn-pp	?	—	Sekeroglu & Uygun 1980
Apple	*Amblyseius fallacis*	*Tetranychus urticae*	C	n-p	P	—	Streibert 1981
Cotton	*Chrysoperla carnea, Coccinella septempunctata, C. undecimpunctata, Hippodamia variegata, Orius niger, Nabis palifer, Deraeocoris punctalatus, Geocoris arenarius, Campylomma verbasci, Scolothrips acariphagus*	*Tetranychus urticae, Heliothis armigera*	F, C	nn-pp	P	—	Sukhoruchenko et al. 1977

Table A.6. (*Continued*)

Crop	Natural Enemy	Prey/Host	Method[1]	Selectivity Class[2]	Selectivity Type[3]	Selectivity Ratio	Reference
Several	*Chrysoperla carnea*	Many	C	n-pp	?	—	Suter 1978
Citrus	*Amblyseius swirskii, A. limonicus, Typhlodromus rhenanus*	*Phyllocoptruta oleivora, Brevipalpus* spp.	C, F	nn-p	P	—	Swirski et al. 1967
Citrus	*Leptomastidea abnormis, Scymnus includens*	*Planococcus citri*	C	nn-p	P	—	Viggiani et al. 1972
Cotton	*Geocoris pallens*	Several	C	n-pp	?	—	Yokoyama & Pritchard 1984
Cotton	*Amblyseius gossipi*	*Tetranychus arabicus*	LC	n-p	P	—	Zhody et al. 1979
DIENCHLOR							
Greenhouse	*Encarsia formosa, Phytoseiulus persimilis*	*Tetranychus urticae*	C	n-p	B	—	Ledieu 1979
HEXYTHIAZOX							
Apple	Phytoseiid sp.	*Panonychus ulmi, Tetranychus urticae*	F	nn-pp	P	—	Baillod et al. 1986
Apple	*Stethorus punctillum, Typhlodromus* sp.	*Panonychus ulmi*	F	n-pp	?	—	Comai 1985
Several	*Metaseiulus occidentalis*	*Tetranychus urticae*	C	n-pp	P	—	Hoy & Ouyang 1986
OVEX							
Hazelnut	Phytoseiids	*Phytoptus avellanae*	F	nn-p	?	—	Ragusa 1983

OXYTHIOQUINOX							
?	*Chrysoperla formosa*	Many	C	n-p	P	—	Babrikova 1980
Greenhouse	*Phytoseiulus persimilis*	*Tetranychus turkestani*	?	n-p	P	—	Balevski 1976
Greenhouse	*Encarsia formosa*	*Trialeurodes vaporariorum*	C	n-p	P	—	Elenkov et al. 1975
Greenhouse	*Aphidoletes aphidimyza*	*Microsiphum rosae*, spider mites	C	n-p	P	—	Markkula & Tiittanen 1976a
Castor	*Stethorus pauperculus*	*Tetranychus neocaledonicus, T. cucurbit*	C	n-p	P	—	Puttaswamy & Channabasavanna 1976
Citrus	*Aphytis melinus*	*Aonidiella aurantii, Aspidiotus nerii*	F	n-pp	P	—	Saba 1978
Citrus	*Sympherobius sanctus, Cryptolaemus montrouzieri*	Several	C	nn-pp	?	—	Sekeroglu & Uygun 1980
Plum	Several phytoseiids	*Panonychus ulmi*	F	nn-p	?	—	Simova 1976
?	*Amblyseius californicus*	*Tetranychus urticae*	C	n-p	P	—	Wu et al. 1985
PROPARGITE							
Apple	*Typhlodromus pyri*	*Panonychus ulmi*	F	n-p	P	—	Collyer 1980
DINOCHLOR							
Greenhouse	*Aphidoletes aphidimyza*	*Microsiphum rosae*, spider mites	C	n-p	P	—	Markkula & Tiittanen 1976a
Roses	*Phytoseiulus persimilis*	*Tetranychus urticae*	F	n-p	?	—	Samways 1979
DINOBUTON							
?	*Chrysoperla* carnea	Several	C	n-p	P	—	Babrikova 1979
Greenhouse	*Phytoseiulus persimilis*	*Tetranychus urticae*	F	n-p	P	—	Berendt 1974

Table A.6. (*Continued*)

Crop	Natural Enemy	Prey/Host	Method[1]	Selectivity Class[2]	Selectivity Type[3]	Selectivity Ratio	Reference
Greenhouse	*Aphidoletes aphidimyza*	*Microsiphum rosae*, spider mites	C	n-p	P	—	Markkula & Tiittanen 1976a
Cotton	*Chrysoperla carnea*, *Coccinella septempunctata*, *C. undecimpunctata*, *Hippodamia variegata*, *Orius niger*, *Nabis palifer*, *Deraeocoris punctalatus*, *Geocoris arenarius*, *Camplylomma verbasci*, *Scolothrips acariphagus*	*Tetranychus urticae*, *Heliothis armigera*	F, C	nn-pp	P	—	Sukhoruchenko et al. 1977
?	*Amblyseius californicus*	*Tetranychus urticae*	C	n-p	P	—	Wu et al. 1985
FENBUTATIN OXIDE							
Apple	*Amblyseius chilinensis*	*Panonychus ulmi*	F	n-p	P	—	Carbonell & Briozzo 1981
Apple	*Typhlodromus pyri*	*Panonychus ulmi*	F	n-p	P	—	Collyer 1980
Greenhouse	*Phytoseiulus persimilis*, *Encarsia formosa*	*Tetranychus urticae*	?	nn-p	B	—	Costello & Elliott 1981
Apple	*Metaseiulus occidentalis*, *Typhldromus columbiensis*	*Panonychus ulmi*	C	nn-p	P	—	Downing & Moilliet 1975
Many	*Trichogramma cacoeciae*, *Pales pavida*, *Pimpla turionellae*, *Chrysoperla carnea*, *Leptomastix dactylopii*, *Amblyseius potentillae*, *Anthocoris nemorum*, *Cryptolaemus montrouzieri*, *Drino inconspicua*, *Encarsia formosa*, *Opius* spp., *Phygadeuon trichops*, *Phytoseiulus persimilis*, *Syrphus vitripennis*	Many	C	nn-pp	P	—	Franz et al. 1980, Hassan et al. 1983

Grape	*Amblyseius chilinensis*	*Brevipalpus chilensis, Oligonychus vitis, Colomerus vitus*	F	n-pp	P	—	Gonzalez 1983
Apple	*Typhlodromus pyri, Chrysocharis prodice, Aphelinus mali*	*Panonychus ulmi*	C, F	nn-p	B	—	Gruys 1980b
Apple	*Amblyseius fallacis*	*Panonychus ulmi, Tetranychus urticae, Aculus schlechtendali*	C	n-pp	P	—	Hislop et al. 1978
Greenhouse	*Encarsia formosa*	*Trialeurodes vaporariorum*	C	n-p	P	—	Hoogcarspel & Jobsen 1984
Citrus	*Amblyseius hibisci*	*Panonychus citri, Tetranychus pacificus*	C	n-pp	P	—	Jeppson et al. 1975
Citrus	*Amblyseius eharai*	*Panonychus citri*	R	n-p	P	—	Kashio & Tanaka 1979
Greenhouse	*Encarsia formosa*	*Tetranychus urticae*	C	n-p	B	—	Ledieu 1979
Apple	*Amblyseius fallacis, Typhlodromus pyri*	*Panonychus ulmi, Aculus schlechtendali*	C	n-p	P	—	Lienk et al. 1976, 1978
Greenhouse	*Phytoseiulus persimilis*	*Tetranychus urticae*	C	n-p	P	—	Lindquist et al. 1979
Greenhouse	*Phygadeuon trichops*	*Tetranychus urticae*	C	n-p	P	—	Naton 1983
Several	*Amblyseius bibens, A. potentillae, Typhlodromus pyri*	*Tetranychus urticae*	LC	nn-p	P	—	Overmeer & van Zon 1981
Apple	*Stethorus punctillum*	*Panonychus ulmi*	F	n-p	?	—	Pasqualini & Malavolta 1985
Greenhouse	*Phytoseiulus persimilis*	*Tetranychus urticae*	F	n-p	P	—	Pruszynski 1984
Hazelnut	Phytoseiids	*Phytoptus avellanae*	F	nn-p	?	—	Ragusa 1983
Greenhouse	*Phytoseiulus persimilis*	spider mites	C	n-p	P	—	Stamenkovic & Peric 1984b
Apple	*Amblyseius fallacis*	*Tetranychus urticae*	C	n-p	P	—	Streibert 1981

Table A.6. (*Continued*)

Crop	Natural Enemy	Prey/Host	Method[1]	Selectivity Class[2]	Selectivity Type[3]	Selectivity Ratio	Reference
Greenhouse	*Amblyseius bibens*	*Tetranychus urticae*	C	n-p	P	—	van Zon & van der Geest 1980
Citrus	*Leptomastix dactylopii*	*Planococcus citri*	C	n-p	P	—	Viggiani & Tranfaglia 1978
Apple	*Stethorus loxtoni, S. nigripes, S. vagans*	*Tetranychus urticae*	C	n-p	P	—	Walters 1976b
Apple	*Typhlodromus pyri, Amblyseius fallacis, M. occidentalis,*	*Panonychus ulmi, Tetranychus urticae*	C	nn-pp	P	—	Wearing & Ashley 1982
Apple	*Amblyseius fallacis*	*Panonychus ulmi, Tetranychus urticae*	C	n-pp	P	—	Croft & Nelson 1972
Citrus	*Aphytis melinus*	*Aonidiella aurantii*	C	n-p	P	—	Davies & McLaren 1977
Grape	*Amblyseius chilinensis*	*Brevipalpus chilensis, Oligonychus vitis, Colomerus vitus*	F	n-pp	P	—	Gonzalez 1983
Citrus	*Cales noacki*	*Aleurothrixus floccosus*	C	n-p	E	—	Garrido et al. 1982
Apple	*Amblyseius fallacis*	*Panonychus ulmi, Tetranychus urticae, Aculus schlechtendali*	C	n-pp	P	—	Hislop et al. 1978
Almond	*Metaseiulus occidentalis*	*Panonychus ulmi, Tetranychus urticae*	F, C	n-pp	P	—	Hoy et al. 1984

Apple	*Amblyseius fallacis, Typhlodromus pyri*	*Panonychus ulmi, Aculus schlechtendali*	C	n-p	P	—	Lienk et al. 1976, 1978
?	*Stethorus loi, Amblyseius longispinosus*	*Tetranychus truncatus*	C	nn-p	P	—	Lo & Chao 1975
Greenhouse	*Phytoseiulus persimilis*	Spider mites	C	n-p	P	—	Stamenkovic & Peric 1984a
Apple	*Amblyseius fallacis*	*Tetranychus urticae*	C	n-p	P	—	Streibert 1981
Apple	*Typhlodromus pyri, Amblyseius fallacis, M. occidentalis, Phytoseiulus persimilis*	*Panonychus ulmi, Tetranychus urticae*	C	nn-pp	P	—	Wearing & Ashley 1982
?	*Amblyseius californicus*	*Tetranychus urticae*	C	n-p	P	—	Wu et al. 1985
SOAP							
Greenhouse	*Phytoseiulus persimilis*	*Tetranychus urticae*	C	n-p	P	—	Osborne & Petitt 1985
Greenhouse	*Encarsia formosa*	*Trialeurodes vaporariorum*	C	n-p	P	—	Puritch et al. 1982
TETRADIFON							
Cotton	*Agistemus exsertus*	Several spider mites	C	n-pp	P	—	Abo Elghar et al. 1971
?	*Cales noacki*	*Aleurothrixus floccosus*	C	n-p	P	—	Beitia & Garrido 1985
?	*Pimpla turionellae*	*Galleria mellonella*	C, F	n-p	P	—	Bogenschutz 1984
Apple	*Metaseiulus occidentalis, Stethorus punctillum*	*Tetranychus urticae*	LC	nn-p	P	40 80	Bravenboer 1959
Cotton	*Chrysoperla boninensis, C. congrua, C. pudica*	Spider mites	LC	nn-pp	P	—	Brettell 1984
Greenhouse	*Phytoseiullus persimilis*	*Tetranychus urticae*	C	n-p	P	—	Caprioli et al. 1983
Citrus	*Cales noacki*	*Aleurothrixus floccosus*	F	n-p	?	—	Carrero 1979

Table A.6. (*Continued*)

Crop	Natural Enemy	Prey/Host	Method[1]	Selectivity Class[2]	Selectivity Type[3]	Selectivity Ratio	Reference
Vegetables	*Diaeretiella rapae*	*Brevicoryne brassicae, Myzus persicae*	C	n-pp	?	—	Delorme 1976
Apple	*Stethorus nigripes*	*Panonychus ulmi, Tetranychus urticae*	C	n-pp	P	—	Edwards & Hodgson 1973
Citrus	*Cales noacki*	*Aleurothrixus floccosus*	C	n-p	E	—	Garrido et al. 1982
Greenhouse	*Encarsia formosa, Phytoseiulus persimilis*	Several	C	nn-pp	P	—	Helyer 1982
Apple	*Typhlodromus tilarium, Amblyseius finlandicus, Zetzellia mali*	*Panonychus ulmi, Tetranychus urticae*	R	nn-pp	?	—	Karg 1970
Citrus	*Amblyseius eharai*	*Panonychus citri*	R	n-p	P	—	Kashio & Tanaka 1979
Greenhouse	*Encarsia formosa*	*Tetranychus urticae*	C	n-p	B	—	Ledieu 1979
?	*Stethorus loi, Amblyseius longispinosus*	*Tetranychus truncatus*	C	nn-p	P	—	Lo & Chao 1975
Apple	*Typhlodromus tiliae*	*Panonychus ulmi*	F	n-p	?	—	Muller 1960
Greenhouse	*Phygadeuon trichops*	*Tetranychus urticae*	C	n-p	P	—	Naton 1983

Greenhouse	*Phytoseiulus persimilis*	Spider mites	C	n-p	P	—	Stamenkovic & Peric 1984a
Greenhouse	*Phytoseiulus persimilis*	*Tetranychus urticae*	C	n-p	P	—	Stenseth 1975
TETRASUL							
Greenhouse	*Phytoseiulus persimilis*	*Tetranychus urticae*	C, F	n-p	?	—	Coulon et al. 1979
TRIFORINE							
Greenhouse	*Phytoseiulus persimilis*	*Tetranychus urticae*	F	n-p	P	—	Babikir 1978

[1]F = field; C = contact; EC = effective concentration; LC = lethal concentration; LD = lethal dosage; S = systemic; O = oral; R = residue.
[2]n = single natural enemy species; nn = multiple; p = single prey or host species; pp = multiple.
[3]P = physiological selectivity; E = ecological; B = both P & E; ? = uncertain.

Table A.7. Cases of Pesticide Selectivity to Arthropod Natural Enemies over Their Prey or Hosts: Microbial Insecticides and Insect Growth Regulators

Crop	Natural Enemy	Prey/Host	Method[1]	Selectivity Class[2]	Selectivity Type[3]	Selectivity Ratio	Reference
ABAMECTIN							
Pluchea discoridis	*Phytoseius finitimus*	*Eriophyes dioscoridis*	C	n-p	P	400 1985	El-Banhawy & El-Bagoury 1985
Apple	*Metaseiulus occidentalis*	*Tetranychus urticae*, *Panonychus ulmi*	C	n-pp	P	—	Grafton-Cardwell & Hoy 1983
Apple	*Amblyseius fallacis*	*Quadraspidiotus perniciosus*	F	n-p	P	—	Pfeiffer 1985
Celery	Several eulophids, pteromalids	*Liriomyza trifolii*	F	nn-p	B	—	Trumble 1985
BACILLUS THURINGIENSIS							
Cotton	*Doru lineare*	*Heliothis* spp.	F	n-p	P	—	Compos & Gravena 1984
Sweet potato	*Pheidole megacephala*	*Cylas formicarinus*	C	n-p	P	—	Castineiras & Calderon 1982
Citrus	*Aphytis melinus*	*Aonidiella aurantii*	C	n-p	P	—	Davies & McLaren 1977
Tobacco	*Jalysus spinosus*	Several	F	n-pp	?	—	Elsey 1973
Grape	*Typhlodromus pyri*	*Panonychus ulmi*, *Tetranychus urticae*, *Eriophyes vitis*	F	n-pp	?	—	Englert & Kettner 1983
Corn	*Meterous* sp., *Chelonus* sp., *Apanteles* sp., *Eiposoma*	*Spodoptera fugiperda*	F	nn-p	?	—	Fernadez & Clavijo 1984
Cauliflower	*Aleochara bilineata*, *Bembidion lampros*	*Delia* sp.	F	nn-p	?	—	Finlayson 1979

Many	*Trichogramma cacoeciae*, *Pales pavida*, *Pimpla turionellae*, *Chrysoperla carnea*, *Leptomastix dactylopii*, *Amblyseius potentillae*, *Anthocoris nemorum*, *Crytolaemus montrouzieri*, *Drino inconspicua*, *Encarsia formosa*, *Opius* spp., *Phygadeuon trichops*, *Phytoseiulus persimilis*, *Syrphus vitripennis*	Many	C	nn-pp	P	—	Franz et al. 1980, Hassan et al. 1983
Coffee	*Solenopsis* sp.	*Perileucoptera coffeella*	F	n-p	?	—	Gravena 1984
Soybean	*Macrocharops bimaculata*, *Calosoma granulatum*	*Anticarsia gemmatalis*	F	nn-p	?	—	Habib & Amaral 1985
Several	*Chrysoperla carnea*, *Hippodamia convergens*, *Aphytis melinus*	Several	C	nn-pp	P	—	Haverty 1982
Collards	*Diaeretiella rapae*, syrphids, *Asaphes lucens*, *Aphidencyrtus, aphidivorus*	*Myzus persicae*	C, F	nn-pp	P	—	Horn 1983
Cabbage	*Diadegma eucerophaga*	*Plutella xylostella*	F, C	n-p	P	—	Iman et al. 1986
Apple	*Diaphnocoris* spp., *Pilophorous plexsus*, *Campylomma verbasci*, *Anystis agilis*, *Atomus* spp., spiders	Several	F	nn-pp	P	—	Jaques 1965

Table A.7. (*Continued*)

Crop	Natural Enemy	Prey/Host	Method[1]	Selectivity Class[2]	Selectivity Type[3]	Selectivity Ratio	Reference
Apple	*Typhlodromus tilarium, Amblyseius finlandicus, Zetzellia mali*	*Panonychus ulmi, Tetranychus urticae*	C	nn-pp	?	—	Karg 1970
Apple	*Trichogramma cacoeciae, Trichogramma embryophagum*	?	C	nn-?	?	—	Kapustina 1975
Broccoli	Several parasitoids	*Trichoplusia ni, Pieris rapae, Plutella xylostella*	F	nn-pp	P	—	Kennedy & Oatman 1976
Apple	*Trichogramma embryophagum*	*Cydia pomonella*	C	n-p	P	—	Kostadinov 1979
Cabbage	*Apanteles plutellae*	*Plutella xylostella*	F	n-p	P	—	Lim et al. 1986
Vegetables	*Phygadeuon trichops*	*Delia* spp.	C	n-p	P	—	Naton 1978
Vegetables	*Bembidion lampros*	*Delia brassicae*	F	n-p	B	—	Obadofin & Finlayson 1977
Forest	*Apanteles fumiferanae, Glypta fumiferanae*	*Choristoneura fumiferana*	F	nn-p	?	—	Otvos & Raske 1980
Greenhouse	*Phytoseiulus persimilis*	Spider mites	F	n-p	P	—	Pruszynski et al. 1983
?	*Trichogramma evanescens*	*Spodoptera littoralis, Anagasta kuehniella*	O	n-pp	P	—	Salama & Zaki 1985
Several	*Chrysoperla carnea*	Many	C	n-pp	?	—	Suter 1978
Forest	*Apanteles melanoscelus*	*Lymantria dispar*	F	n-p	?	—	Ticehurst et al. 1982
Corn	*Trichogramma nubilale*	*Ostrinia nubilalis*	C	n-p	P	—	Tipping & Burbutis 1983

Citrus	*Leptomastidea abnormis, Scymnus includens*	*Planococcus citri*	C	nn-p	P	—	Viggiani et al. 1972
Apple	*Stethorus loxtoni, S. nigripes, S. vagans*	*Tetranychus urticae*	C	n-p	P	—	Walters 1976b
Pear	*Deraeocoris brevis, Phytocoris* sp., *Pilophorus* sp., *Orius tristicolor, Chrysoperla* sp., *Hemerobius,* Spiders, *Amblyseius schusteri, Typhlodromus arboreus, T. caudiglans*	*Choristoneura rosaceana*	F	nn-pp	P	—	Westigard et al. 1986
Apple	*Stethorus loxtoni, S. nigripes, S. vagans*	*Tetranychus urticae*	C	n-p	P	—	Walters 1976b
BEAUVERIA BASSIANA							
Sweet potato	*Pheidole megacephala*	*Cylas formicarinus*	C	n-p	P	—	Castineiras & Calderon 1982
Apple	*Trichogramma cacoeciae, T. embryophagum*	?	C	nn-?	?	—	Kapustina 1975
BUPROFEZIN							
?	*Cales noacki, Encarsia formosa*	*Aleurothrixus floccosus, Trialeurodes vaporariorum*	C	nn-pp	P	—	Garrido et al. 1985
Rice	*Lycosa pseudoannulata, Cyrtorhinus lividipennis Microvelia atrolineata*	*Nilaparvata lugens, Sogatella furcufera, Nephotettix cincticeps*	?	nn-pp	?	—	Heinrichs et al. 1984

Table A.7. (*Continued*)

Crop	Natural Enemy	Prey/Host	Method[1]	Selectivity Class[2]	Selectivity Type[3]	Selectivity Ratio	Reference
DIFLUBENZURON							
Cotton	*Trichogramma pretiosum*	*Heliothis* spp.	F	n-p	?	—	Ables et al. 1977
Apple	*Metaseiulus occidentalis, Zetzellia mali*	*Cydia pomonella*	F	nn-p	P	—	Anderson & Elliott 1982
Forest	*Ooencyrtus kuwanae*	*Lymantria dispar*	C	n-p	P	—	Brown & Respicio 1981
Pear	*Trechnites insidiosus, Anthocoris* spp., *Deraeocoris brevis, Campylomma verbasci*	*Psylla pyricola, Cydia pomonella*	F	nn-pp	?	—	Burts 1983
Greenhouse	*Phytoseiulus persimilis*	?	C, F	n-p	?	—	Coulon et al. 1979
Apple	*Typhlodromus pyri*	*Panonychus ulmi*	F	n-pp	?	—	Cranham 1978b
Cotton	*Coleomegilla maculata, Hippodamia convergens, Orius insidiosus, Geocoris punctipes, Chrysoperla* spp.	*Heliothis* sp.	F	nn-p	?	—	Deakle & Bradley 1982
Pine	Several chalcidoids	*Thaumetopoea pityocampa*	F	nn-p	?	—	Ferrari & Tiberi 1979
Many	*Trichogramma cacoeciae Pales pavida, Pimpla turionellae, Chrysoperla carnea, Leptomastix dactylopii, Amblyseius potentillae, Anthocoris nemorum,*	Many	C	nn-pp	P	—	Franz et al. 1980, Hassan et al. 1983

	Cryptolaemus montrouzieri, *Drino inconspicua*, *Encarsia formosa*, *Opius* spp., *Phygadeuon trichops*, *Phytoseiulus persimilis*, *Syrphus vitripennis*						
Cabbage	*Aleochara bilineata*	*Delia radicum*	C	n-p	P	—	Gordon & Cornect 1986
Apple	*Typhlodromus pyri*, *Chrysocharis prodice*, *Aphelinus mali*	Many	C, F	nn-pp	B	—	Gruys 1980b
Greenhouse	*Scolothrips longicornis*, *Phytoseiulus persimilis*	Several	LC	nn-pp	B	—	Iacob & Posoiu 1981, Iacob et al. 1981
Celery	*Hippodamia convergens*	*Trichoplusia ni*	F, C	n-p	P	—	Jones et al. 1983
Cotton	*Nabis* spp., *Hippodamia convergens*, *Coleomegilla maculata*, *Orius insidiosus*, *Chrysoperla* spp., spiders	Several	F	nn-pp	B	—	Keever et al. 1977
Greenhouse	*Encarsia formosa*	*Trialeurodes vaporariorum*	C	n-p	B	—	Ledieu 1979
Vegetables	*Phygadeuon trichops*	*Delia* spp.	C	n-p	P	—	Naton 1978
Larch	*Chrysocharis laricellae*, *Dicladocerus nearcticus*	*Coleophora laricella*	LD	nn-p	P	—	Page et al. 1982
Apple	*Stethorus punctillum*	*Panonychus ulmi*	F	n-p	?	—	Pasqualini & Malavolta 1985
Cotton	*Microplitis croceipes*	*Heliothis* spp.	LD	n-p	P	—	Powell et al. 1986a
Pear	*Amblyseius hibisci*	*Tetranychus urticae*, *Panonychus ulmi*	F	n-pp	?	—	Riedl & Hoying 1980

Table A.7. (*Continued*)

Crop	Natural Enemy	Prey/Host	Method[1]	Selectivity Class[2]	Selectivity Type[3]	Selectivity Ratio	Reference
Greenhouse	*Phytoseiulus persimilis*	Spider mites	C	n-p	P	—	Stamenkovic & Peric 1984a
Apple	*Amblyseius fallacis*	*Tetranychus urticae*	C	n-p	P	—	Streibert 1981
Potato	*Doryphorophaga doryphorae*	*Leptinotarsus decimlineata*	C, R	n-p	P	—	Tamaki et al. 1984
Forest	*Ooencyrtus pityocampae*, *Tetrastichus servadeii*, *Anastatus bifaciatus*, *Trichogramma embryophagum*	*Thaumetapoea pityocampa*	F	nn-p	?	—	Tsankov & Mirchev 1983
Forest	*Apanteles melanoscelus*	*Lymantria dispar*	F	n-p	?	—	Ticehurst et al. 1982
Pear	*Trichnites insidiosus*, mirids, chrysopids, coccinellids	*Psylla pyricola*	F	nn-p	?	—	Westigard 1979
Pear	*Deraeocoris brevis*, *Phytocoris* sp., *Pilophorus* sp., *Orius tristicolor*, *Chrysoperla* sp., *Hemerobius*, spiders, *Amblyseius schusteri*, *Typhlodromus arboreus*, *T. caudiglans*	*Cydia pomonella*,	F	nn-pp	P	—	Westigard et al. 1986
Apple	*Holocothorax testaceipes*, other parasites	*Phyllonorycter ringoneiella*	F	nn-p	?	—	Yamada & Kawashima 1983

EPOFENONANE							
Apple	*Typhlodromus pyri*, *Chrysocharis prodice*, *Aphelinus mali*	Many	C, F	nn-pp	B	—	Gruys 1980b
Apple	*Apanteles ater*, *Colpoclypeus florus*	*Archips rosanus*, *Adoxophyes orana*	F	nn-pp	P	—	Reede et al. 1984
FENOXYCARB							
Apple	*Apanteles ater*, *Colpoclypeus florus*	*Archips rosanus*, *Adoxophyes orana*	F	nn-pp	P	—	Reede et al. 1984
Citrus	*Mapaphycus bartletti*, *Aphytis holoxanthus*, *Comperiella bifasciatus*, *Aphytis chrysomphali*, *A. hispanicus*, *Encarsia inbquirenda*	*Saissetia oleae*, *Chrysomphalus aonidum*	C, F	nn-pp	B	—	Peleg 1983
HYDROPRENE							
?	*Pseudogonia rufifrons*	*Galleria mellonella*	C	n-p	P	—	Mellini & Boninsegni 1983
KINOPRENE							
?	*Coccinella* spp.	?	C	n-?	?	—	Kismali & Erkin 1984a, b
Several	*Chrysoperla carnea*	Many	C	n-pp	?	—	Suter 1978
Greenhouse	*Encarsia formosa*	*Trialeurodes vaporariorum*	C	n-p	P	—	Zseller & Budai 1982
METHOPRENE							
Greenhouse	*Phytoseiulus persimilis*	?	C, F	n-p	?	—	Coulon et al. 1979

Table A.7. (*Continued*)

Crop	Natural Enemy	Prey/Host	Method[1]	Selectivity Class[2]	Selectivity Type[3]	Selectivity Ratio	Reference
NOSEMA PYRAUSTA							
Corn	*Trichogramma evanescens*	*Ostrinia nubilalis*	F	n-p	?	—	Huger 1984
PENTAFLURON							
Celery	*Hippodamia convergens*	*Trichoplusia ni*	F, C	n-p	P	—	Jones et al. 1983
TRICHOTHECIN							
Greenhouse	*Propylea quattuor-decimpunctata, Coccinella quatrodecimpunctata*	Aphids	O	nn-pp	P	—	Lyashova & Evtushenko 1983
TRIFLUBENZERON							
Greenhouse	*Encarsia formosa*	*Trialeurodes vaporariorum*	C	n-p	P	—	Oomen & Wiegers 1984

[1] F = field; C = contact; EC = effective concentration; LC = lethal concentration; LD = lethal dosage; S = systemic; O = oral; R = residue.
[2] n = single natural enemy species; nn = multiple; p = single prey or host species; pp = multiple.
[3] P = physiological selectivity; E = ecological; B = both P & E; ? = uncertain.

REFERENCES

Abdel-Aal, Y. A. I., El-sayed, A. M. K., Negm, A. A., Hussein, M. H., and El-sebae, A. H. 1979. The relative toxicity of certain insecticides to *Spodoptera littoralis* (Boisd.) and *Coccinella undecimpunctata* L. Int. Pest Control 21(4): 79–80, 82.

Abdel-Salam, F. 1967. Über die Wirkung von Phosphorsäureestern auf einige Arthropoden innerhalb der Apfelbaum-Biozonose in Abhängigkeit von ihrer Dichte. Z. Angew. Zool. 54: 233–283.

Abdelrahman, I. 1973. Toxicity of malathion to the natural enemies of California red scale, *Aonidiella aurantii* (Mask.) (Hemiptera: Diaspididae). Aust. J. Agric. Res. 24: 119–133.

Abdul Kareem, A. A., Thangavel, P., and Balasubramaniam 1977. Studies on the predatism of the lady beetle, *Menochiles sexmaculatus* (F.), on bean aphid, *Aphis craccivora* Koch., treated with antifeeding compounds, Z. Angew. Entomol. 83: 406–409

Ables, J. R., Jones, S. L., and Bee, M. J. 1977. Effect of diflubenzuron on beneficial arthropods associated with cotton. Southwest. Entomol. 2: 66–72.

Ables, J. R., Jones, S. L., House, V. S., and Bull, D. L. 1980. Effect of diflubenzuron on entomophagous arthropods associated with cotton. Southwest. Entomol. Supp. (1): 31–35.

Ables, J. R., West, R. P., and Shepard, M. 1975. Response of the house fly and its parasitoids to dimilin. J. Econ. Entomol. 68: 622–624.

Abo Elghar, M. R., Elbadry, E. A., Hassan, S. M., and Kilany, J. M. 1971. Some effects of pesticides on the predatory mite *Agistemus exsertus*. J. Econ Entomol. 61: 26–27.

Abou-awad, B. A., and El-Banhawy, E. M. 1985. Comparison between the toxicity of synthetic pyrethroids and other compounds to the predacious mite *Amblyseius gossipi* (*Mesostigmata: Phytoseiidae*). Exp. Appl. Acarol. 1: 185–191.

Abu, J. F., and Ellis, C. R. 1977. Toxicity of five insecticides to the alfalfa weevil, *Hypera postica*, and its parasites, *Bathyplectes curculionis* and *Microctonus aethiopoides*. Environ. Entomol. 6: 385–389.

Adams, C. H., and Cross, W. H. 1967. Insecticide resistance in *Bracon mellitor*, a parasite of the boll weevil. J. Econ. Entomol. 60: 1016–1020.

Adams, J. B. 1960. Effects of spraying 2, 4-d amine on coccinellid larvae. Can. J. Zool. 38: 285–288.

Adams, R. G., and Prokopy, R. J. 1977. Apple aphid control through natural enemies. Mass. Fruit Notes 42(6): 6–10.

Adang, M. J., and Binns, A. N. 1988. A new era of systemic pesticides genetically engineered in plants, p. 249c. *In* Proc. 18th Intern. Cong. Entomol., Vancouver, B.C., Canada, July 3–9, 1988.

Adashkevich, B. P. 1966. Effect of chemical and microbiological treatments upon parasites of *Plutella maculipennis* Curt. Zool. Zh. 45: 1040–1046.

Adashkevich, V. P., and Nurmukhamedov, D. 1985. *Aphidoletes* and pesticides. Zash. Rest. 5: 36.

Adjei-Maafo, I. K., and Wilson, L. T. 1983. Factors affecting the relative abundance of arthropods on nectaried and nectariless cotton. Environ. Entomol. 12: 349–352.

Afifi, S. E. D., and Knutson, H. 1956. Reproductive potential, longevity, and weight of house flies which survived one insecticidal treatment. J. Econ. Entomol. 49: 310–313.

Aguayo, M. I. and Villaneueva, F. R. 1985. Susceptibility of *Meteorus hyphantriae* Riley to methyl parathion. Southwest. Entomol. 10(2): 107–109.

Ahlstrom, K. P., and Rock, G. C. 1973. Comparative studies on *Neoseiulus fallacis* and *Metaseiulus occidentalis* for azinphosmethyl toxicity and effects of prey and pollen on growth. Ann. Entomol. Soc. Amer. 66:1109–1113.

Ahmad, S. 1982. Roles of mixed-function oxidases in insect herbivory, pp. 41–47. Proc. 5th Intern. Symp. Insect–Plant Relationships, Wageningen, The Netherlands, Mar. 1–4, 1982.

Ahmad, S. and Knowles, C. O. 1972. Biochemical mode of action of tricyclohexylhydroxytin. Comp. Gen. Pharmac. 3: 125–133.

Ahmad, S., O'Neill, J. R., Mague, D. L., and Nowalk, R. K. 1978. Toxicity of *Bacillus thuringiensis* to gypsy moth larvae parasitized by *Apanteles melanoscelus*. Environ. Entomol. 7: 73–76.

Ahmed, M. K. 1955. Comparative effect of systox and schradan on some predators of aphids in Egypt. J. Econ. Entomol. 48: 530–532.

Ahmed, M. K., Newsom, L. D., Emerson, R. B., and Roussel, J. S. 1954. The effect of systox on some common predators of the cotton aphid. J. Econ. Entomol. 47: 445–449.

Ajami, A. M., and Riddiford, L. M. 1973. Comparative metabolism of the *Cecropia* juvenile hormone. J. Insect. Physiol. 19: 635–645.

Akhmedov, M. A. 1981. Pesticides and aphid enemies. Zash. Rast. 111: 30.

Albert, R., and Bogenschutz, H. 1984. Testing the effect of pesticides on the beneficial arthropod *Coelotes terrestris* (Wider) (Araneida, Agelenidae) by means of a glass-pane test. Anz. Schädl., Pflschutz., Umweltschutz 57(6): 111–117.

Alexandrescu, S., and Hondru, N. 1981. Selectivity of some insecticides used for the control of the cabbage aphid (*Brevicoryne brassicae* L.) on cabbage crops. Analele Institutului de Cercetari pentru Protectia Plantelor 16: 375–383.

Ali, A. U. D. D., Abdellatif, M. A., Bakry, N. M., and El-Sawaf, S. K. 1973. Studies on biological controls of the greater wax moth, *Galleria mellonella*. 1. Susceptibility of wax moth larvae and adult honeybee workers to *Bacillus thuringiensis*. J. Agric. Res. 12: 117–123.

AliNiazee, M. T. 1974. Role of a predatory mite, *Typhlodromus arboreus*, in biological control of spider mites on apple in western Oregon, *In* Proc. 4th Intern. Cong. Acarol. Vol.4. pp.637–642.

AliNiazee, M. T. 1982. Effect of two synthetic pyrethroids on a predatory mite *Typhlodromus arboreus* in apple orchards of western Oregon, pp. 655–658. *In* Proc. VI Intern. Cong. Acarol. Wiley, New York.

AliNiazee, M. T., and Cranham, J. E. 1980. Effects of four synthetic pyrethroids on a predatory mite, *Typhlodromus pyri* and its prey, *Panonychus ulmi*, on apples in Southeast England. Environ. Entomol. 9: 436–439.

AliNiazee, M. T., Stafford, E. M., and Kido, H. 1974. Management of grape pests in central California vineyards: toxicity of some commonly used chemicals to *Tetranychus pacificus* and its predator, *Metaseiulus occidentalis*. J. Econ. Entomol. 67: 543–547.

Allen, H. W. 1958. Orchard studies on the effect of organic insecticides on parasitism of the oriental fruit moth. J. Econ. Entomol. 51: 82–87.

Amaya, N. M. 1982. Effect of some insecticides on the parasitic activity of *Trichogramma pretiosum* (Riley) (Hymenoptera: Trichogrammatidae) released after applications, pp. 195–199. *In* Proc. Trichogrammes Symp. Intern. Antibes, France. Apr. 20–23, 1982.

Anber, H. A. I., and Oppenoorth, F. J. 1986. Insensitive acetylcholinesterase as the main cause of organophosphorus and carbamate resistance in a strain of *Amblyseius potentillae* from Switzerland. Abst. Intern. Cong. Pesticide Chem., Toronto, Ont., Canada, 1986.

Anber, H. A. I. and Overmeer, W. P. J. 1988. Resistance to organophosphates and carbamates in the predaceous mite, *Amblyseius potentillae* (Garman) due to insensitive acetylcholinesterase. Pestic. Biochem. Physiol. 30: 91–97.

Andersen, A. 1982. The effect of different dosages of isofenphos on Carabidae and Staphylinidae. Z. Angew. Entomol. 94: 61–65.

Anderson, D. W., and Elliott, R. H. 1982. Efficacy of diflubenzuron against the codling moth, *Laspeyresia pomonella* (Lepidoptera: Olethreutidae), and impact on orchard mtes. Can. Entomol. 114: 733–737.

Andreadis, T. G. 1980. *Nosema pyrcusta* infection in *Macrocentrus grandii*, a braconid parasite of the European corn borer. *Ostrinia nubilalis*. J. Invert. Pathol. 35: 229–233.

Andreadis, T. G. 1982. Impact of *Nosema pyrausta* on field populations of *Macrocentrus grandii*, an introduced parasite of the European corn borer, *Ostrinia nubilalis*. J. Invert. Pathol. 39: 298–302.

Andreadis, T. G., and Hall, D. W. 1979. Significance of transovarial infections of *Amblyospora* sp. (Microspora: Thelohaniidae) in relation to parasite maintenance in the mosquito *Culex salinarius*. J. Invert. Pathol. 34: 152–157.

Anonymous 1970. Tree Fruit Spray Schedules and Pesticide Guide for Nova Scotia. Nova Scotia Dept. Agric. and Market. Bull. 134 pp.

Anonymous 1977. Annu. Rept. of Dept. Agric. Res., Rept. Royal Trop. Inst., The Netherlands, 72 pp.

Apperson, C. S., Leidy, R. B., and Powell, E. E. 1984. Effects of Amdro on the red imported fire ant (Hymenoptera: Formicidae) and some nontarget ant species and persistence of Amdro on a pasture habitat in North Carolina. J. Econ. Entomol. 77: 1012–1018.

Armstrong, K. F., and Bonner, A. B. 1985. Investigation of a permethrin-induced antifeedant effect in *Drosophila melanogaster*: an ethological approach. Pestic. Sci. 16: 641–650.

Arru, G. M. 1973. Survival of *Euderus caudatus* Thom. (Chalcidoidea: Eulophidae), internal egg-parasite of *Saperda chacharias* L. (Coleoptera: Cerambycidae) after treatments against the newly hatched larvae of its host. Boll. Zool. Agraria Bachicoltura 11: 1–10.

Ascerno, M. E. 1975. Effects of the insect growth regulator Altozar on the parasitoid, *Microctonus aethiops*, and its host *Hypera postica*. N.Y. Entomol. Soc. 83: 135.

Ascerno, M. E., Hower, A. A., and Smilowitz, Z. 1983. Effects of the insect growth regulator hydroprene on nondiapausing *Microctonus aethiopiodes* (Hymenoptera: Braconidae), a parasite of the alfalfa weevil (Coleoptera: Curculionidae). Environ. Entomol. 12: 158–160.

Ascerno, M. E., Smilowitz, Z., and Hower, A. A. 1980. Effects of the insect growth regulator hydroprene on diapausing *Microctonus aethiopoides*, a parasite of the alfalfa weevil. Environ. Entomol. 9: 262–264.

Askari, A., Abivardi, C., and Alishah, A. 1984. Effect of camphor on cabbage aphid and its primary parasitoid. Ann. Appl. Biol. 104(Suppl.): 24–25.

Askew, R. R. 1971. Parasitic Insects. Elsevier, New York, 316 pp.

Asquith, D. 1971. The Pennsylvania integrated control program for apple pests—1971. Penn. Fruit News 50: 43–47.

Asquith, D. 1972. The economics of integrated pest management. Penn. Fruit News 31: 27–31.

Asquith, D., and Colburn, R. 1971. Integrated pest management in Pennsylvania apple orchards. Bull. Entomol. Soc. Amer. 17: 89–91.

Asquith, D. and Hull, L. A. 1973. *Stethorus punctum* and pest-population responses to pesticide treatments on apple trees. J. Econ. Entomol. 66: 1197–1203.

Asquith, D., Hull, L. A., Travis, J. W., and Mowry, P. D. 1976. Apple, tests of insecticides 1975. Insecticide & Acaricide Tests. Entomol. Soc. Amer. Publ. 1: 17–19.

Asyakin, B. D. 1973. Use of *Aphidoletes aphidimyza* Rond. (Diptera: Cecidomyiidae) against aphids on greenhouse cucumbers. Lapiski SLKh. 212: 10–14.

Atallah, Y. H., and Nettles, C. W. 1966. DDT-metabolism and excretion in *Coleomegilla maculata* De Geer. J. Econ. Entomol. 59: 560–564.

Atallah, Y. H., and Newsom, L. D. 1966. Ecological and nutritional studies on *Coleomegilla maculata* De Geer (Coleoptera: Coccinellidae). III. The effect of DDT, toxaphene, and endrin on the reproductive and survival potentials. J. Econ. Entomol. 59: 1181–1187.

Atanasov, N. 1974. Possibilities of using the predatory mite (*Phytoseiulus persimilis* A-H.) for biological control of *Tetranychus atlanticus* McGerg. on glasshouse cucumbers. Gradinarska i Lozarska Nauka 11(5): 113–117.

Aveling, C. 1981. Action of mephosfolan on anthocorid predators of *Phorodon humuli*. Ann. Appl. Biol. 97: 155–164.

Avella, M., Fournier, D., Pralavorio, M., and Berge, J. B. 1985. Selection of a strain of *Phytoseiulus persimilis* Athias-Henriot for resistance to deltamethrin. Agronomie 5(2): 177–180.

Aveyard, C. S., Peregrine, D. J., and Bryan, K. M. G. 1986. Biological activity of clofentezine against egg and motile stages of tetranychid mites. Exp. Appl. Acarol. 2: 223–229.

Axtell, R. C. 1966. Comparative toxicities of insecticides to house fly larvae and *Macrocheles muscaedomestica*, a mite predator of the house fly. J. Econ. Entomol. 59: 1128–1130.

Azab, A. K., Tawfik, M. F. S., Fahmy, H. S. M., and Awadallah, K. T. 1971. Effect of some insecticides on the larvae of the aphidophagous syrphid, *Xanthogramma aegyptium* Wied. (Diptera: Syrphidae). Bull. Entomol. Soc. Egypt 5: 37–45.

Ba-Angood, S. A., and Stewart, R. K. 1980. Effect of granular and foliar insecticides on cereal aphids (Hemiptera) and their natural enemies on field barley in Southwestern Quebec. Can. Entomol. 112: 1309–1313.

Babenko, V. A. 1980. A test on the rational control of aphids. Zash. Rast. 6: 14–15.

Babikir, E. T. A. 1978. Factors affecting biological control of the red spider mite in glasshouses. The effect of fungicides and light intensity on the population dynamics of *Tetranychus urticae* and its predator *Phytoseiulus persimilis* with reference to the efficiency of biological control in glasshouses. Thesis, Bradford University, UK.

Babrikova, T. 1979. The effect of pesticides on the individual stages of the common lacewing (*Chrysopa carnea* Steph.). Rasteniev'dni Nauki 16(8): 105–115.

Babrikova, T. 1980. Studies on the effect of some pesticides on various stages of the lacewing—*Chrysopa formosa* Br. Nanchni Trudove, Entomologiya, Mikrobiologiya, Fitopatologiya 25(3): 31–40.

Babrikova, T. 1982. Toxicity of some pesticides to individual stages of the seven-spotted lacewing. Rastitelna Zashchita 30(10): 23–26.

Babrikova, T., and Kuzmanova, I. 1984. The toxicity of biological preparations based on *Bacillus thuringiensis* to some stages of *Chrysopa septempunctata* Wesm., *Chrysopa formosa* Br. and *Chrysopa perla* L. Grad. Lozarska Nauka 21(7): 55–59.

Babrikova, T., Kuzmanova, I., and Lai, N. T. 1982. The effect of biological preparations based on *Bacillus thuringiensis* on some stages of the lacewing, *Chrysopa carnea* Steph. Grad. Lozarska Nauka 19: 40–45.

Bai Shiang, T., Lin, H., and Hsu, T. S. 1981. Studies on the cholinesterases from the Tussah silkworm *Anthereae pernyi* Guer and its tachinid parasite *Blepharipa tibialis* Chao. Acta Entomol. Sinica 24: 1–8.

Bailey, J. B., and Madsen, H. F. 1964. A laboratory study of three strains of codling moth, *Carpocapos pomonella* (L.), exhibiting tolerance to DDT in the field. Hilgardia 35: 185–210.

Baillod, M. 1982. Work of the sub-group on mites. Boll. Zool. Agraria Bachicoltura 16: 35–48.

Baillod, M. 1984. Lutte biologique contre les acariens phytophages. Rev. Suisse Vitic., Arboric., Hortic. 16: 137–142.

Baillod, M. 1986. Régulation naturelle des teranyques en vergers de pommiers et perspectives actuelles de lutte biologique à l'aide d'acariens prédateurs phytoseiides. Bull. OILB/SROP IX/3: 5–16.

Baillod, M., and Guignard, E. 1984. Resistance of *Typhlodromus pyri* Scheuten to azinphos and biological control of phytophagous mites in fruit-tree cultures. Rev. Suisse Vitic., Arboric., Hortic. 16: 155–160.

Baillod, M., Schmid, A., Guignard, E., Antonin, P., and Caccia, R. 1982. Lutte biologique contre l'acarien rouge en viticulture. II. Equilibres naturels dynamiques des populations et expériences de lachers de typhlodromes. Rev. Suisse Vitic., Arboric., Hortic. 14: 345–352.

Baillod, M., Guignard, E., Genini, M., and Antonin, P. 1985. Biological control trials in 1984 against phytophagous mites in apple orchards, susceptibility and resistance to insecticides in *Typhlodromus pyri* Scheutan. Rev. Suisse Vitic., Arboric., Hortic. 17(2): 129–135.

Baillod, M., Guignard, E., and Antonin, P. 1986. A new generation of growth-inhibiting specific acaricides. Rev. Suisse Vitic., Arboric., Hortic. 18(4): 213–219.

Baker, R. S., Laster, M. L., and Kitten, W. F. 1985. Effects of the herbicide monosodium methanearsonate on insect and spider populations in cotton fields. J. Econ. Entomol. 78: 1481–1484.

Balevski, N. 1976. The ovicidal effect of certain pesticides on the eggs of *Tetranychus atlanticus* and *Phytoseiulus persimilis*. Rastitelna Zashchita 24: 25–27.

Ban, J. N. 1979. Note on the susceptibility of *Trichogrammatoidea lutea* (Gir)–Hymenoptera: Trichogrammatidae to pesticides pp. 259–265. *In* Proc. Cong. Control Insects in Tropical Envir., Marseilles, France, Mar., 1979, Pt. 1.

Bandong, J. P., and Litsinger, J. A. 1986. Egg predators of rice leaffolder (LF) and thier susceptibility to insecticides. IRRI Newsl. 11(3): 21

Barbosa, P., Saunders, J. A., and Waldvogel, M. 1982. Plant-mediated variation in herbivore suitability and parasitoid fitness, pp. 63–71. *In* Proc. 5th Intern. Symp. Insect–Plant Relationships, Wageningen, The Netherlands, Mar. 1–4, 1982.

Barbosa, P., Saunders, J. A., Kemper, J., Trumble, R., Oleghno, J., and Martinat, P. 1986. Plant allelochemicals and insect parasitoids: effects of nicotine on *Cotesia congregata* (Say) (Hymenoptera: Braconidae) and *Hyposoter annulipes* (Cresson) (Hymenoptera: Ichneumonidae). J. Chem. Ecol. 12: 1319–1328.

Barclay, H. J. 1982. Models for pest control using predator release, habitat management, and pesticides in combination. J. Appl. Ecol. 19: 337–348.

Barclay, H., and van den Driessche, P. 1977. Predator–prey models with added mortality. Can. Entomol. 109: 763–768.

Barker, P. S. 1968. Effectiveness of malathion against four species of mites that inhabit stored grain. J. Econ. Entomol. 61: 944–946.

Barlow, F., and Hadaway, A. B. 1952. Studies on aqueous suspensions of insecticides. Part II: Quantitative determinations of weights of DDT picked up and retained. Bull. Entomol. Res. 42: 769–776.

Barlow, J. S. 1972. Some host–parasite relationships in fatty acid metabolism, pp. 438–451. *In* Insect and Mite Nutrition. Ed. J. G. Rodriguez, North-Holland, Amsterdam.

Barnes, G., and Lavrik, P. 1986. Improvement in the efficacy of microbial control agents, pp. 669–676. *In* 1986 Brit. Crop Prot. Conf., Pests & Diseases. Vol. II. Proc. Conf. Brighton Metropole, England, Nov. 17–20, 1986.

Barnes, M. M., and Moffitt, H. R. 1963. Resistance to DDT in the adult codling moth and reference curves for guthion and carbaryl. J. Econ. Entomol. 56: 722–725.

Baronio, P., and Sehnal, F. 1980. Dependence of the parasitoid *Gonia cinerascens* on the hormones of its lepidopterous hosts. J. Insect Physiol. 26: 619–626.

Barras, K. G., Kisner, R. L., Lewis, W. J., and Jones, R. L. 1982. Effects of the parasitoid, *Microplitis croceipes*, on the haemolymph proteins of the corn earworm, *Heliothis zea*. Comp. Biochem. Physiol. 43B: 941–947.

Barrett, G. W. 1968. The effects of an acute insecticide stress on a semi-enclosed grassland ecosystem. Ecology 49: 1019–1035.

Bartell, D. P., Sanborn, J. R., and Wood, K. A. 1976. Insecticide penetration of cocoons containing diapausing and nondiapausing *Bathyplectes curculionis*, an endoparasite of the alfalfa weevil. Environ. Entomol. 5: 659–661.

Bartlett, B. R. 1951. The action of certain "inert" dust materials on parasitic hymenoptera. J. Econ. Entomol. 44: 891–896.

Bartlett, B. R. 1953. Retentive toxicity of field-weathered insecticide residues to entomophagous insects associated with citrus pests in California. J. Econ. Entomol. 46: 565–569.

Bartlett, B. R. 1956. Natural predators—can selective insecticides help to preserve biotic control? *In* Agricultural Chemicals, pp. 42–44, 107.

Bartlett, B. R. 1958. Laboratory studies of selective aphicides favoring natural enemies of spotted alfalfa aphid. J. Econ. Entomol. 51: 374–378.

Bartlett, B. R. 1963. The contact toxicity of some pesticide residues to hymenopterous parasites and coccinellid predators. J. Econ. Entomol. 56: 694–698.

Bartlett, B. R. 1964. Integration of chemical and biological control, pp. 489–514. *In* Biological Control of Insect Pests and Weeds. Ed. P. DeBach, Reinhold Press, New York, 844 pp.

Bartlett, B. R. 1966. Toxicity and acceptance of some pesticides fed to parasitic Hymenoptera and predatory coccinellids. J. Econ. Entomol. 59: 1142–1149.

Basedow, T. 1985. The effects of pesticides on surface-living predatory beetles and spiders in agriculture. Ber. über Landwirtsch. 198: 189–200.

Basedow, T., Borg, A., and Scherney, F. 1976. Effects of insecticides upon the terrestrial predaceous arthropods in cereal fields, especially the ground beetles (Coleoptera, Carabidae). Entomol. Exp. Appl. 19: 37–51.

Basedow, T., Rzehak, H., and Voss, K. 1985. Studies on the effect of deltamethrin sprays on the numbers of epigeal predatory arthropods occurring in arable fields. Pestic. Sci. 16: 325–331.

Bashir, N. H., and Crowder, L. A. 1983. Mechanisms of permethrin tolerance in the common green lacewing (Neuroptera: Chrysopidae). J. Econ. Entomol. 76: 407–409.

Battu, S. 1977. Occurrence of *Parasarcophaga misera* (Walker) and *Campoletis* sp. as parasites of *Spodoptera litura* (Fabricius) from India. Curr. Sci. 46 (16): 568–569.

Battu, S., and Dilawari, V. K. 1978. Preliminary investigations on the safety evaluation of *Spodoptera litura* (Fabricius) nuclear polyhedrosis virus (SLNPV) against a parasitoid, *Parasarcophaga misera* (Walker). Entomol. Newsl., 8: 6.

Beckage, N. E. 1985. Endocrine interactions between endoparasitic insects and their hosts. Annu. Rev. Entomol. 30: 371–413.

Beckage, N. E., and Riddiford, L. M. 1982. Effects of parasitism by *Apanteles congregatus* on the endocrine physiology of the tobacco hornworm *Manduca sexta*. Gen. Comp. Endocrinol. 47: 308–322.

Beckage, N. E., and Riddiford, L. M. 1983. Lepidopteran anti-juvenile hormones: effects on development of *Apanteles congregatus* in Manduca sexta. J. Insect Physiol. 29: 633–638.

Beckage, N. E., Templeton, T. J., Stirling, B. A., and Nielsen, B. D. 1987. Disruptive developmental effects of benzyl-1-3-benzodiooxole derivatives on unparasitized and parasitized *Manduca sexta* larvae. J. Insect Physiol. 33: 603–611.

Beckendorf, S. K., and Hoy, M. A. 1985. Genetic improvement of arthropod natural enemies through selection, hybridization or genetic engineering techniques, pp. 167–187. *In* Biological Control of Agricultural Integrated Pest Management Systems. Eds. M. A. Hoy and D. C. Herzog, Academic Press, New York, 589 pp.

Beegle, C. C. 1988. The *Bacillus thuringiensis* story: flagship of microbial control, p. 249a. *In* Proc. 18th Intern. Cong. Entomol., Vancouver, B.C., Canada, July 3–9, 1988.

Beegle, C. C., and Oatman, E. R. 1975. Effect of a nuclear polyhedrosis virus on the relationship between *Trichoplusia ni* (Lepidoptera: Noctuidae) and the parasite, *Hyposoter exiguae* (Hymenoptera: Ichneumonidae). J. Invert. Pathol. 25: 59–71.

Beeman, R. W. 1982. Recent advances in mode of action of insecticides. Annu. Rev. Entomol. 27: 253–281.

Beeman, R. W., and Matsumura, F. 1973. Chlordimeform: a pesticide acting upon amine regulatory mechanisms. Nature 242: 273–274.

Beesley, S. G., Compton, S. G., and Jones, D. A. 1985. Rhodanese in insects. J. Chem. Ecol. 11: 45–50.

Beglyarov, G. A., and Maslienko, L. V. 1978. The toxicity of certain pesticides to *Encarsia*. Zash. Rast. 11: 36–37.

Beglyarov, G. A., Zil'bermints, I. L., and Petrushov, A. A. 1978. A *Phytoseiulus persimilis* strain resistant to insecticides. Means to use it in an integrated control system against glasshouse pests, pp. 51–63. *In* Biologicheskij Method Bor'by s Meditelyami i Boleznyami Rastenij v Zakrytom Grunte. Lolos, Moscow (in Russ.).

Beitia, F., and Garrido, A. 1985. Study of the sterilization of *Cales noacki* How (Hym. Aphelinidae) by the use of pesticides. Anales Inst. Nac. Investigaciones Agarias, Agricola 28: 147–155.

Bell, M. R., and McGovern, W. L. 1975. Susceptibility of the ectoparasite, *Bracon mellitor*, to infection by *Microsporidan pathogens* in its host, *Anthonomous grandis*, J. Invert. Pathol. 25: 133–134.

Bellows, T. S., Morse, J. G., Hadjidemetriou, D. G., and Iwata, Y. 1985. Residual toxicity of four insecticides used for control of citrus thrips (Thysanoptera: Thripidae) on three beneficial species in a citrus agroecosystem. J. Econ. Entomol. 78: 681–688.

Beraldo, M. J. A. H., Rocha, E. A., and Machado, V. L. L. 1981. Toxicidade de inseticidas (em laboratorio) para *Polybia* (*Myrapetra*) *paulista* (Ihering, 1896) (Hymenoptera–Vespidae). Anais da Soc. Entomol. do Brasil 10: 261–267.

Berenbaum, M. 1985. Brementown revisited: allelochemical interactions in plants. Rec. Advan. Phytochem. 19: 139–169.

Berenbaum, M., and Neal, J. J. 1987. Allelochemical interactions in crop plants. Wash. ACS Ser. 333, pp. 416–430.

Berendt, O. 1973. Influence of prey density on acaricidal effect on the predacious mite, *Phytoseiulus persimilis*. Statens Plantepathologiske Forsog, pp. 36–40.

Berendt, O. 1974. The effect of the acaricide dinobuton on population density of the predator *Phytoseiulus persimilis* A.-H. and its prey *Tetranychus urticae* Koch (Acarina; Phytoseiidae and Tetranychidae). Tidsskrift Planteav. 78: 103–115.

Berge, J. B., and Fournier, D. 1988. Advances in molecular genetics of acetylcholinesterase insensitivity in insecticide-resistant insects, p. 461a. *In* Proc. 18th Intern. Cong. Entomol., Vancouver, B.C., Canada, July 3–9, 1988.

Berge, J. B., Mouches, C., and Fournier, D. 1986. Molecular biology of some insecticide resistance genes suitable to improve resistance in beneficial arthropods. Abst. 3E-15, VI Cong. Pestic. Chem. Ottawa, Ont., Canada, Aug. 10–15, 1986.

Berkett, L. P., and Forsythe, H. Y. 1980. Predaceous mites (Acari) associated with apple foliage in Maine. Can. Entomol. 112: 497–502.

Bernays, E. A., and Simpson, S. J. 1982. Control of food intake. Adv. Insect Physiol. 16: 59–118.

Berry, R. E., Yu, S. J., and Terriere, L. C. 1980. Influence of host plants on insecticide metabolism and management of variegated cutworm. J. Econ. Entomol. 73: 771–774.

Bess, H. A. 1964. Populations of the leaf-miner *Leucoptera meyricki* Ghesa. and its parasites in sprayed and unsprayed coffee in Kenya. Bull. Entomol. Res. 55: 59–81.

Biache, G. 1975. Effects of *Bacillus thuringiensis* on *Pimpla instigator* (Ichneumonidae-Pimplinae). Ann. Soc. Entomol. France 11: 609–617.

Biernbaum, M. R. 1987. Plant synergism of allelochemicals and pesticides. Symp. Pres. Ann. Mtg. E. S. A. Boston, MA, Nov. 30, 1987.

Bigger, M. 1973. An investigation by Fourier analysis into the interaction between coffee leafminers and their larval parasites. J. Animal Ecol. 42: 417–434.

Bigler, F., Shanin, F., and Hassan, S. A. 1984. A procedure for testing side-effects of pesticides on the predator *Chrysopa carnea*, p. 730. Abst. XVII Intern. Cong. Entomol., 960 pp.

Binns, E. S. 1971. The toxicity of some soil-applied systemic insecticides to *Aphis gossypii* (Hom.: Aphididae) and *Phytoseiulus persimilis* (Acarina: Phytoseiidae) on cucumbers. Ann. Appl. Biol. 67: 211–222.

Bishop, A. L., and Blood, P. R. B. 1980. Arthropod ground strata composition of the cotton ecosystem in south-eastern Queensland, and the effect of some control strategies. Aust. J. Zool. 28: 693–697.

Bishop, J. A. 1982. The Neo-Darwinian theory and pesticide resistance. Pestic. Sci. 13: 97–103.

Blackman, R. L. 1967. The effects of different aphid foods on *Adalia bipunctata* L. and *Coccinella 7-punctata* L. Ann. Appl. Biol. 59: 207–291.

Blais, J. R. 1977. Effects of aerial application of chemical insecticides on spruce budworm parasites. Bimonth. Res. Notes 3: 41–42.

Blaisinger, P. 1979. Standardised methods for the evaluation in the laboratory and in orchards of the effect of pesticides on auxilliary arthropods. Proc. Intern. Symp. IOBC/WPRS, pp. 103–106.

Blaisinger, P. 1986. Research on a method for estimating the impact of insect growth regulators on the orchard fauna. Bull. SROP 9(3): 75–84.

Blanc, M. 1986. Short-term effects of insecticides and acaricides on the beneficial fauna of orchards: results of investigations and comments on the methodology. Bull. SROP 9(3): 4–11.

Blommers, L., and Helsen, H. 1986. Host plant influence on the effect of pesticides on the predacious mite *Typhlodromus pyri*. Bull. SROP 9(3): 55–59.

Blommers, L. H. M., and Overmeer, W. P. J. 1986. On the fringes of natural spider mite control. Bull. SROP IX/4: 48–61.

Blum, M. S. 1981. Chemical Defenses of Arthropods. Academic Press, New York, 562 pp.

Boethel, D. J., and Eikenbary, R. D., Eds. 1986. Interactions of Plant Resistance and Parasitoids and Predators of Insects. Wiley, New York 213 pp.

Bogenschutz, H. 1975. Prüfung des Einflusses von Pflanzenshutzmitten auf Nutzinsekten. Z. Ang. Entomol. 77: 438–444.

Bogenschutz, H. 1979. Standardized evaluation of side effects of pesticides on *Coccygomimus turionellae* (L.) (Hymenoptera: Ichneumonidae), pp. 440–441. Abt. Waldschutz, Forstliche Versuchs- und Forschungsanstalt, D-7801 Stegen-Wittental, German Federal Republic.

Bogenschutz, H. 1984. The effect of plant protection compounds on the capacity for parasitism in the hymenopteran *Coccygomimus turionellae*. Nachr. Pflschutz. 36: (5) 65–67.

Boller, E. 1985. An outdoor study on the side effects of pesticides on predatory mites in vine growing in eastern Switzerland. Schweiz. Z. Obst- und Weinbau 121(12): 322–325.

Boller, E. F., Janser, E., and Potter, C. 1984. Testing of the side-effects of herbicides used in viticulture on the common spider mite *Tetranychus urticae* and the predacious mite *Typhlodromus pyri* under laboratory and semi-field conditions. Z. Pflkrankh. Pflschutz. 91: 561–568.

Boness, M. 1983. Peropal, alsystin and cropotex: evaluations of their effect on beneficial arthropods. Nachr. Pflschutz. 36: 38–53.

Bonnemaison, L. 1962. Toxicity of various contact and systemic insecticides to the predators and parasites of aphids. Phytiatrie-Phytopharmacie 11: 67–84.

Boscheri, S., Rellich, C., Obrist, J., and Dissertori, A. 1986. Praktische Erfahrungen mit der Übertragung von Raubmilben. Obstbau Weinbau 23(4): 93–95.

Bostanian, N. J., and Belanger, A. 1985. The toxicity of three pyrethroids to *Amblyseius fallacis* (Garman) Acari. Phytoseiidae and their residues on apple foliage. Agric., Ecosys., and Environ. 14: 243–250.

Bostanian, N. J., and Coulombe, L. J. 1986. An integrated pest management program for apple orchards in southwestern Quebec. Can. Entomol. 118: 1131–1142.

Bostanian, N. J., Belanger, A., and Rivard, I. 1985. Residues of four synthetic pyrethroids and azinphos-methyl on apple foliage and their toxicity to *Amblyseius fallacis* (Acari: Phytoseiidae). Can. Entomol. 117: 143–152.

Bouron, H. 1985. A new conception of the control of the fruit-tree red spider mite. Phytoma 364: 10–11.

Bower, C. C. 1984. Integrated control of European red mite, *Panonychus ulmi* (Koch), on apples in N. S. W., pp. 61–67. *In* Proc. 4th Aust. Appl. Entomol. Res. Conf., Adelaide, Australia, Sept. 24–28, 1984.

Bower, C. C., and Kaldor, J. 1980. Selectivity of five insecticides for codling moth control: effects on the twospotted spider mite and its predators. Environ. Entomol. 9: 128–132.

Bower, C. C., and Murison, R. D. 1984. Deciding when to spray in integrated mite control on apples, pp. 188–196. *In* Proc. 4th Aust. Appl. Entomol. Res. Conf., Adelaide, Australia, Sept. 24–28, 1984.

Bower, C. C., and Thwaite, W. G. 1982. Development and implementation of integrated control or orchard mites in New South Wales, pp. 177–190. *In* Proc. Australasian Workshop on Devel. and Implemen. of IPM, Auckland, New Zealand.

Bowers, W. S. 1981. How anti-juvenile hormones work. Amer. Zool. 21: 737–742.

Bowers, W. S. 1982. Endocrine strategies for insect control. Entomol. Exp. Appl. 31: 3–14.

Boyce, H. R., and Dustan, G. G. 1955. Parasitism of twig-infesting larvae of the oriental fruit moth., *Grapholitha molesta* (Bussck) (Lepidoptera: Olethreutidae), in Ontario, 1939–1953. Annu. Rept. Entomol. Soc. Ont. 84: 48–55.

Bracken, G. K., and Barlow, J. S. 1967. Fatty acid composition of *Exeristes comstockii* (Cress) reared on different hosts. Can. J. Zool. 45: 57–61.

Bracken, G. K., and Bucher, B. E. 1967. Mortality of hymenopterous parasite caused by *Serratia marcescens*. J. Invert. Pathol. 9: 130–132.

Brader, L. 1977. Resistance in mites and insects affecting orchard crops, pp. 353–377. *In* Pesticide Management and Insecticide Resistance, Eds. D. L. Watson and A. W. A. Brown, Acadmic Press, New York 638 pp.

Bragg, D. E. 1974. Influence of methyl and ethyl parathion on parasitoids of *Phytomyza syngenesiae* (Diptera: Agromyzidae) in artichokes. Environ. Entomol. 3: 576–577.

Brandenburg, R. L. 1985. Interaction of alfalfa plant height and insecticide carrier application rates on beneficial arthropod populations. Agric. Ecosys, Environ. 13: 159–166.

Brandt. J. M. 1982. Residual toxicity of field-weathered insecticide residues on citrus leaves during spring to a parasite of red scale. Citrus Subtrop. Fruit J. 587: 16–19, 21.

Brattsten, L. B. 1979a. Ecological significance of mixed-function oxidations. Drug Metab. Rev. 10: 35–58.

Brattsten, L. B. 1979b. Biochemical defense mechanisms in herbivores against plant allelochemicals, pp. 199–270. *In* Herbivores, Their Interactions with Secondary Plant Metabolites. Eds. G. A. Rosenthal and D. H. Janzen, Academic Press, New York.

Brattsten, L. B., and Metcalf, R. L. 1970. The synergistic ratio of carbaryl with piperonyl butoxide as an indicator of the distribution of multifunction oxidases in the Insecta. J. Econ. Entomol. 36: 101–104.

Brattsten, L. B., and Metcalf, R. L. 1973. Synergism of carbaryl toxicity in natural insect populations. J. Econ. Entomol. 66: 1347–1348.

Brattsten, L. B., Wilkinson, C. F., and Eisner, T. 1977. Herbivore–plant interaction: mixed function oxidases and secondary plant substances. Science 196: 1349.

Braun, A. R., Guerrero, J. M., Bellotti, A. C., and Wilson, L. T. 1987a. Relative toxicity of permethrin to *Monoychellus progresivus* Doreste and *Tetranychus urticae* Koch (Acari: Tetranychidae) and their predators *Amblyseius limonicus* Garman and McGregor (Acari: Phytoseiidae) and *Oligota minuta* Cameron (Coleoptera: Staphylinidae): bioassays and field validation. Environ. Entomol. 16: 545–550.

Braun, A. R., Guerrero, J. M., Bellotti, A. C., and Wilson, L. T. 1987b. Evaluation of possible non-lethal side effects of permethrin used in predator exclusion experiments to evaluate *Amblyseius limonicus* (Acari: Phytoseiidae) in biological control of cassava mites (Acari: Tetranaychidae). Environ. Entomol. 16: 1012–1018.

Bravenboer, L. 1959. Die Empfindlichkeit von *Tetranychus urticae* und ihren naturlichen Feinden *Typhlodromus longipilus* und *Stethorus punctillum* gegen Insektizide, Akarizide und Fungizide, pp. 937–938. Proc. 4th Intern. Cong. Plant Prot., 1957.

Brettell, J. H. 1979. Green lacewings (Neuroptera: Chrysopidae) of cotton fields in central Rhodesia 1. Biology of *Chrysopa boninensis* Okamoto and toxicity of certain insecticides to the larva. Rhodes. J. Agric. Res. 17: 141–150.

Brettell, J. H. 1982. Green lacewings (Neuroptera: Chrysopidae) of cotton fields in central Zimbabwe. 2. Biology of *Chrysopa congrua* Walker and *C. pudica* Navas and toxicity of certain insecticides to their larvae. Zimbabwe J. Agric. Res. 20: 77–84.

Brettell, J. H. 1984. Green lacewings (Neuroptera: Chrysopidae) of cotton fields in central Zimbabwe. 3. Toxicity of certain acaricides, aphicides and pyrethroids to larvae of *Chrysopa boninensis* Okamoto, *Chrysopa congrua* Walker and *Chrysopa pudica* Navas. Zimbabwe J. Agric. Res. 22: 133–139.

Brettell, J. H., and Burgess, W. M. 1973. A preliminary assessment of the effect of some insecticides on predators of cotton pests. Rhod. Agric. J. 70: 103–104.

British Columbia Ministry of Agriculture 1982. Production guide for interior districts—1982. Tree Fruit, 74 pp.

Broadbent, A. B., and Pree, D. J. 1984. Effects of diflubenzuron and BAY SIR 8514 on beneficial insects associated with peach. Environ. Entomol. 13: 133–136.

Broadly, R. H. 1983. Toxicity of insecticides to *Coccinella repanda* Thunberg and *Harmonia octomaculata* (Fabricius) (Coleoptera: Coccinellidae). Queensland J. Agric. Anim. Sci. 40: 125–127.

Broodryk, S. W. 1980. Some aspects of pest control in the Loskop irrigation scheme. J. Entomol. Soc. S. Africa 43: 1–5.

Brooks, G. T. 1978. The metabolism of xenobiotics in insects, pp. 151–214. *In* Progress in Drug Metabolism, vol 3. Eds. J. W. Bridges and L. F. Chasseaud, Wiley, New York.

Brooks, G. T. 1980. Biochemical targets and insecticide action. pp. 41–55. *In* Insect Neurobiology and Pesticide Action. Soc. Chemical Industry Pesticides Group, London.

Brooks, W. M., and Cranford, J. D. 1972. Microsporidoses of the hymenopterous parasites, *Campoletis sonorensis* and *Cardiochiles nigriceps*, larval parasites of *Heliothis* species. J. Invert. Pathol. 20: 77–94.

Brown, A. W. A. 1958. The spread of insecticide resistance in pest species, pp. 351–414. *In* Advances in Pest Control Research, Ed. R. L. Metcalf, Wiley-Interscience, New York.

Brown, A. W. A. 1971. Pest resistance to pesticides, pp. 457–552. *In* Pesticides in the Environment, vol. 1. Ed. R. White-Stevens, Marcel Dekker, New York.

Brown, A. W. A. 1976. Epilogue: resistance as a factor in pesticide management, pp. 816–822. *In* Proc. 15th Intern. Cong. Entomol., Washington, D.C., 1976.

Brown, A. W. A. 1977. Considerations of natural enemy susceptibility and developed resistance in light of the general resistance problem. Z. Pflkrankh. Pflschutz. 84: 132–139.

Brown, A. W. A. 1978a. Ecology of Pesticides. Wiley Interscience, New York, 525 pp.

Brown, A. W. A. 1978b. Insecticides and the arthropod fauna of plant communities, pp. 28–62. *In* Ecology of Pesticides. Ed. A. W. A. Brown, Wiley-Interscience, New York, 525 pp.

Brown, G. C. 1987. A modeling approach to epizootiological dynamics of insect host–pathogen interactions and an implementation example. *In* Insect Pest Management Modeling, Eds. C. A. Shoemaker and G. C. Brown, Wiley, New York.

Brown, G. C., and Shanks, C. H. 1976. Mortality of two-spotted spider mite predators caused by the systemic insecticide, carbofuran. Environ. Entomol. 5: 1155–1159.

Brown, K. C. 1988. The design of experiments to assess the effects of pesticides on beneficial arthropods in orchards: replication versus plot size, p. 463d. *In* Proc. 18th Intern. Cong. Entomol., Vancouver, B.C., Canada, July 3–9, 1988.

Brown, K. C., Lawton, J. H., and Shires, S. W. 1983. Effects of insecticides on invertebrate predators and their cereal aphid (Hemiptera: Aphididae) prey: laboratory experiments. Environ. Entomol. 12: 1747–1750.

Brown, M. A., and Casida, J. E. 1983. Oxime ether analogs of pyrethroids help determine the contribution of pyrethroid esterases toward selective toxicity. 185th Nat. Mtg. Amer. Chem. Soc., Seattle, Wash.

Brown, M. A., and Casida, J. E. 1984. Influence of pyrethroid ester, oxime ether, and other central linkages on insecticidal activity, hydrolytic detoxification, and physicochemical parameters. Pestic. Biochem. Physiol. 22: 78–85.

Brown, M. W., and Respicio, N. C. 1981. Effect of diflubenzuron on the gypsy moth egg parasite *Ooencyrtus kuwanae* (Hymenoptera: Encyrtidae). Melsheimer Entomol. Ser. 31: 1–7.

Brown, T. M., and Brogdon, W. G. 1987. Improved detection of insecticide resistance through conventional and molecular techniques. Ann. Rev. Entomol. 32: 145–162.

Brun, L. O., Chazeau, J., and Edge, V. E. 1983. Toxicity of four insecticides to *Phytoseiulus macropilis* (Banks) and *P. persimilis* Athias-Henriot (Acarina: Phytoseiidae). J. Aust. Entomol. Soc. 22: 303–305.

Brunner, J. F., and Cameron, R. 1985. White paper: pest management research. CSRS Rev. Paper, Washington State University, Pullman, WA, 16 pp.

Brust, G. E., Stinner, B. R., and McCartney, D. A. 1986. Predator activity and predation in corn agroecosystems. Environ. Entomol. 15: 1017–1021.

Buchi, R. 1981. Evidence that resistance against pyrethroids in aphids *Myzus persicae* and *Phorodon humuli* is not correlated with high carboxylesterase activity. J. Plant. Dis. Prot. 88: 631–634.

Buehler, A., Hanzlik, T. N., and Hammock, B. D. 1985. Effects of parasitization of *Trichoplusia ni* by *Chelonus* spp. Physiol. Entomol. 10: 383–394.

Bull, D. L. 1980. Fate and efficacy of sulprofos against certain insects associated with cotton. J. Econ. Entomol. 73: 262–264.

Bull, D. L., and Coleman, R. J. 1985. Effects of pesticides on *Trichogramma* spp. Southwestern Entomol. (Suppl) 8: 156–168.

Bull, D. L., and House, V. S. 1978. Effects of chlordimeform on insects associated with cotton. Southwest. Entomol. 3: 284–291.

Bull, D. L., and House, V. S. 1983. Effects of different insecticides on parasitism of host eggs by *Trichogramma pretiosum* Riley. Southwest. Entomol. 8: 46–53.

Bull, D. L., and Ridgway, R. L. 1969. Metabolism of trichlorfon in animals and plants. J. Agric. Food Chem. 17: 837–841.

Bull, D. L., Ridgway, R. L., Buxkemper, W. E., Schwarz, M., McGovern, T. P., and Sarmiento, R. 1973. Effects of synthetic juvenile hormone analogues on certain injurious and beneficial arthropods associated with cotton. J. Econ. Entomol. 66: 623–626.

Burgaud, L., and Cessac, M. 1962. Efficacité pratique d'un insecticide acaricide endothérapique nouveau le vamidothion. Phytiatrie-Phytopharmacie 11: 117–128.

Burges, H. D., and Hussey, H. W. 1971. Microbial Control of Insects and Mites. Academic Press, London, 861 pp.

Burts, E. C. 1983. Effectiveness of a soft-pesticide program on pear pests. J. Econ. Entomol. 76: 936–941.

Bushchik, T. N., and Lazurina, M. V. 1982. Benlate and *Phytoseiulus*. Zash. Rast. 4: 22.

Busvine, J. R. 1980. Recommended methods for measurement of pest resistance to pesticides. FAO Plant Prot. Prod. Papers 21, 51 pp.

Butcher, M. R., and Penman, D. R. 1983. The effects of chemicals used on strawberry crops on introduced phytoseiid mites, pp. 67–70. *In* Proc. 36th New Zealand Weed and Pest Contr. Conf.

Butler, G. D., and Las, A. S. 1983. Predaceous insects: effect of adding permethrin to the sticker used in gossyplure applications. J. Econ. Entomol. 76: 1148–1451.

Butt, B. A. 1975. Bibliography of the codling moth. Bull. ARS/W-31 USDA Publ., 221 pp.

Caccia, R., Baillod, M., Guignard, E., and Kreiter, S. 1985. Introduction d'une souche de *Amblyseius andersoni* Chant (Acarina: Phytoseiidae) résistant à' l'azinphos dans la lutte contre les acariens phytophages en viticulture. Rev. Suisse Vitic., Arboric. Hortic. 17(5): 285–290.

Cadogan, B. L., and Laing, J. E. 1982. A study of *Balaustium putmani* (Acarina: Erythraeidae) in apple orchards in southern Ontario. Proc. Entomol. Soc. Ont. 112: 13–22.

Campbell, A., Frazer, B. D., Gilbert, N., Gutierrez, A. P., and Mackauer, M. 1974. Temperature requirements of some aphids and their parasites. J. Appl. Ecol. 11: 431–438.

Campbell, B. C., and Duffey, S. S. 1979. Tomatine and parasitic wasps: potential incompatibility of plant antibiosis with biological control. Science 205: 700–702.

Campbell, B. C., and Duffey, S. S. 1981. Alleviation of a tomatine-induced toxicity to the parasitoid, *Hyposoter exiquae*, by phytosterols in the diet of the host, *Heliothis zea*. J. Chem. Ecol. 7: 927–946.

Campbell, M. M. 1975. Duration of toxicity of residues of malathion and spray oil on citrus foliage in south Australia to adults of a California red scale parasite *Aphytis melinus* Debach (Hymenoptera: Aphelinidae). J. Aust. Entomol. Soc. 14: 161–164.

Cantwell, G. E., and Lehnert, T. 1979. Lack of effect of certain microbial insecticides on the honeybee. J. Invert. Pathol. 33: 381–382.

Cantwell, G. E., Knox, D. A., Lehnert, T., and Michael, A. S. 1966. Mortality of the honey bee, *Apis mellifera*, in colonies treated with certain biological insecticides. J. Invert. Pathol. 8: 228–233.

Caprioli, V., Trematerra, D., and Piccardi, P. 1983. Toxicity of organic acaricides to *Tetranychus urticae* Koch and *Phytoseiulus persimilis* Athias-Henriot and method of evaluation pp. 615–620. *In* Atti XIII Cong. Naz. Italiano Entomol., 1983.

Carbonell, B. J., and Briozzo, B. J. 1981. Notes on the management of the red spider mite *Panonychus ulmi* (Koch) and its predator, *Amblyseius chilenensis* (Dosse), in apple orchards in Uruguay. Cent. Invest. Agric. Alberto Boerge. 2: 3–8.

Carmody, M. A., Clark, J. D., Lee, S. B., Nazer, C. J., Nicholls, J. C., and Readshaw, J. L. 1981. Biological control of two-spotted mite in the National Rose Garden. Austr. Parks Recreation, May: 47–51.

Carolin, V. M., and Coulter, W. K. 1971. Trends of western spruce budworm and associated insects in Pacific Northwest forests sprayed with DDT. J. Econ. Entomol. 64: 291–297.

Carrero, J. M. 1979. Toxicity in the field to *Cales noacki* How., a parasite of the citrus whitefly *Aleurothrixus floccusus* Mask., of several insecticides. Anales Inst. Nac. Invest. Agrarias. Protec. Vegetal. 9: 75–91.

Carruthers, R. I., Whitfield, G. H., and Haynes, D. L. 1985. Pesticide-induced mortality of natural enemies of the onion maggot, *Delia antiqua* (Diptera: Anthomyiidae). Entomophaga 30: 151–161.

Casegrande, R., and Haynes, D. L. 1976. An analysis of strip spraying for the cereal leaf beetle. Environ. Entomol. 5: 612–620.

Casida, J. E., Gammon, D. W., Glickman, A. H., and Lawrence, L. F. 1983. Mechanisms of selective action of pyrethroid insecticides. Annu. Rev. Pharmacol. Toxicol. 23: 413–438.

Castineiras, A., and Calderon, A. 1982. Susceptibility of *Pheidole megacephala* to three microbial insecticides: Dipel, Bitoxobacillin 202 and *Beauveria bassiana* under laboratory conditions. Ciencia Tecnica Agric., Protec. Plantas (Suppl): 61–66.

Cate, J. R., Ridgway, R. L., and Lingren, P. D. 1972. Effects of systemic insecticides applied to cotton on adults of an ichneumonid parasite, *Campoletis perdistinctus*. J. Econ. Entomol. 65: 484–488.

Cessac, M., and Burgaud, L. 1964. Efficacite pratique d'un nouvel insecticide-acaricide de contact: la phosalone (11 974 RP). Phytiatrie-Phytopharmacie 13: 45–54.

Chambers, H. W. 1973. Comparative tolerance of selected beneficial insects to methyl parathion. Commun. to Annu. Mtg. Entomol. Soc. Amer. Nov. 28, p. 68.

Chang, C. K., and Whalon, M. E. 1986. Hydrolysis of permethrin by pyrethroid esterases from resistant and susceptible strains of *Amblyseius fallacis* (Acari: Phytoseiidae). Pestic. Biochem. Physiol. 25: 446–452.

Chang, C. P., and Plapp, F. W. 1983. DDT and synthetic pyrethroids: mode of action, selectivity, and mechanisms of synergism in the tobacco budworm (Lepidoptera: Noctuidae) and a predator, *Chrysopa carnea* Stephens (Neuroptera: Chrysopidae). J. Econ. Entomol. 76: 1206–1210.

Chang, K., and Huang, L. 1963. Preliminary study on the control of cottony cushion scale by the Australian ladybeetle. Acta Entomol. Sinica 12: 688–700.

Chang, Y. D. 1981. Feasibility studies on the biological control by augmentation and conservation of natural enemies of rice paddy. Res. Rep. Agri. Sci. Tech. 8(1): 18–28.

Chang, Y. D., Song, Y. H., and Choi, S. Y. 1979. Effects of insecticide application on the populations of the paddy rice insect pests and their natural enemies (1). Selective toxicity of insecticides fro brown planthopper, *Niiaparvata lugens*, and predaceous paddy spiders. Korean J. Plant Pathol. 18(4): 149–152.

Chant, D. A. 1959. Phytoseiid mites (Acarina: Phytoseiidae). Pt. I. Bionomics of seven species in south-eastern England. Pt. II. A taxonomic review of the family Phytoseiidae, with descriptions of thirty-eight new species. Can. Entomol. 91(Suppl. 12), 166 pp.

Chapman, R. K., and Allen, T. C. 1948. Stimulation and suppression of some vegetable plants by DDT. J. Econ. Entomol. 41: 616–623.

Charles, J. G., Collyer, E., and White, V. 1985. Integrated control of *Tetranychus urticae* with *Phytoseiulus persimilis* and *Stethorus bifidus* in commercial raspberry gardens. New Zealand J. Exp. Agric. 13: 385–393.

Charlet, L. D., and Oseto, C. Y. 1983. Toxicity of insecticides on a stem weevil, *Cylindrocopturns adspersus* (Coleoptera: Curculionidae), and its parasitoids in sunflower. Environ. Entomol. 12: 959–960.

Chaudhary, B. S., Singh, O. P., and Rawat, R. R. 1983. Field evaluation of some insecticides against the safflower aphid, the capsule fly and the predator. Pesticides 17: 30–32.

Chelliah, S., and Rajendran, R. 1984. Toxicity of insecticides to the predatory mirid bug *Cyrtorhinus lividipennis* Reuter. IRRI Newsl. 9: 15–16.

Chen, B. H., and Chiu, S. C. 1979. The annual occurrence of *Lycosa* and *Oedothorax* spiders and their response to insecticides. J. Agric. Res. China 28: 285–290.

Cherry, E. T., and Pless, C. D. 1969. Effect of carbofuran and disulfoton on parasitism of tobacco budworms and hornworms on burley tobacco. J. Econ. Entomol. 64: 187–190.

Cherwonogrodzky, J. W. 1980. Microbial agents as insecticides. Residue Rev. 76: 73–96.

Chesnut, T. L., and Cross, W. H. 1971. Arthropod parasites of the boll weevil, *Anthonomus grandis*: 2. Comparisons of their importance in the United States over a period of thirty-eight years. Ann. Entomol. Soc. Amer. 64: 549–556.

Chiu, S. C., and Cheng, C. H. 1976. Toxicity of some insecticides commonly used for rice insect control to the predators of rice-hoppers. Plant Prot. Bull. Taiwan 18: 254–260.

Chiu, S. F., Huang, Z. X., Haung, B. Q., and Xu, M. 1980. Root zone application of systemic insecticides for pest control in China. IRRI Newsl. 5: 21–22.

Chiverton, P. A. 1984. Pitfall-trap catches of the carabid beetle *Pterostichus melanarius*, in relation to gut contents and prey densities, in insecticide treated and untreated spring barley. Entomol. Exp. Appl. 36: 23–30.

Choi, S. Y., Lee, H. R., and Ryu, J. K. 1978. Effects of carbofuran root-zone placement on the spider populations in the paddy fields. Korean J. Plant Protect. 17: 99–103.

Chou, H. C., Chung, H. C., Wang, C. S., Wei, D. Y., Hu, C. Y., Quo, F., and Chen, S. J. 1981. Effects of juvenoids on the reproduction of the adult lady beetle, *Coccinella septempunctata* L. Sinozoologica 1: 185–192.

Christ, E. G. 1971. Coordinator, Tree Fruit Recommendations for New Jersey. New Jersey Extension Service leaflet 466-A.

Chu, Y. I., Ho, C. C., and Chen, B. J. 1975. Relative toxicity of some insecticides to green rice leafhopper, brown planthopper and their predator *Lycosa pseudoannulata*. Plant Prot. Bull. Taiwan 17: 424–430.

Chu, Y. I., Lin, D. S., and Mu, T. 1976a. Relative toxicity of 9 insecticides against rice insect pests and their predators. Plant Prot. Bull. Taiwan 18: 369–376.

Chu, Y. I., Lin, D. S., and Mu, T. 1976b. The effect of Padan, Ofunack and Sumithion on the feeding amount of *Lycosa pseudoannulata* (B&S) and *Oedothorax insecticeps* (B&S) (Lycosidae and Micryphantidae). Plant Prot. Bull. Taiwan 18: 377–390.

Chu, Y. I., Ho, C. C., and Chen, B. J. 1976c. The effect of BPMC and urburn on the predation of *Lycosa* spider (*Lycosa pseudoannulata*). Plant Prot. Bull. Taiwan 18: 42–57.

Chu, Y. I., Lin, D. S., and Mu, T. 1977. Relative toxicity of 5 insecticides against insect pests of rice and their predators, with effect of bidrin on the extent of feeding by *Lycosa pseudoannulata* and *Oedothorax insecticeps*. Plant Prot. Bull. Taiwan 19: 1–12.

Chubinnishvili, T. I., Koblianidze, Y. V., Petrushov, A. Z., and Zil'bermints, I. V. 1982. Introduction of resistant *Metaseiulus*. Zash. Rast. 1: 30–31.

Clancy, D. W., and McAlister, H. J. 1958. Effects of spray practices on apple mites and their predators in West Virginia, pp. 597–601. *In* Proc. 10th Intern. Cong. Entomol., Montreal, P. Q., Canada, 1956, vol. 4.

Clark, J., and Buckley, P. 1984. Control of twospotted mite in the National Rose Garden. Aust. Horticult. 82(1): 42–47.

Clark, J. M., and Matsumura, F. 1982. Two different types of inhibitory effects of pyrethroids on nerve Ca^- and Ca^+ Mg-ATPase activity in the squid, *Loligo pealei*. Pestic. Biochem. Physiol. 18: 180–190.

Clausen, C. P. 1962. Entomophagous Insects. Hafner, New York 688 pp.

Coaker, T. H. 1966. The effect of soil insecticides on the predators and parasites of the cabbage root fly (*Erioshcia brassicae* (Bouche)) and on the subsequent damage caused by the pest. Ann. Appl. Biol. 57: 397–407.

Coats, S. A., Coats, J. R., and Ellis, C. R. 1979. Selective toxicity of three synthetic pyrethroids to eight coccinellids, a eulophid parasitoid, and two pest chrysomelids. Environ. Entomol. 8: 720–722.

Cockfield, S. D., and Potter, D. A. 1983. Short-term effects of insecticidal applications on predaceous arthropods and oribatid mites in Kentucky bluegrass Turf. Environ. Entomol. 12: 1260–1264.

Colburn, R., and Asquith, D. 1971. Tolerance of the stages of *Stethorus punctum* to selected insecticides and miticides. J. Econ. Entomol. 64: 1072–1074.

Cole, J. F. H., and Wilkinson, W. 1984. Selectivity of pirimicarb in cereal crops, pp. 311–316. *In* Proc. Brit. Crop Protection Conf. Pests and Diseases, Brighton Metropole, England, Nov. 19–22, 1984, vol. 1.

Cole, J. F. H., Everett, C. J., Wilkinson, W., and Brown, R. A. 1986. Cereal arthropods and broad-spectrum insecticides, pp. 181–188. *In* Proc. Brit. Crop Protection Conf. Pests and Diseases. Brighton Metropole, England, Nov. 17–20, 1986, vol. 1.

Coli, W. R., Prokopy, R. J., and Hislop, R. G. 1979. Integrated management of apple pests in Massachusetts commercial orchards—1979 results. Insects and nuts. Mass. Fruit News 44: 9–16.

Collins, M. D., Perrin, R. M., Jutsum, A. R., and Jackson, G. J. 1984. Insecticides for the future: a package of selective compounds for the control of major crop pests, pp. 299–304. *In* Proc. Brit. Crop Prot. Conf., Pests and Diseases, Brighton Metropole, England, Nov. 19–22, 1984, vol. 1.

Collyer, E. 1964a. Phytophagous mites and their predators in New Zealand orchards. New Zealand J. Agric. Res. 7: 551–568.

Collyer, E. 1964b. A summary of experiments to demonstrate the role of *Typhlodromus pyri* Scheut. in the control of *Panonychus ulmi* (Koch) in England. C.R. 1st Intern. Cong. Acarol., Fort Collins, Colo., 1963. Acarologia, pp. 363–371.

Collyer, E. 1976. Integrated control of apple pests in New Zealand. 6. Incidence of European red mite, *Panonynchus ulmi*, and its predators. New Zealand J. Zool. 3: 39–50.

Collyer, E. 1980. Integrated control of apple pests in New Zealand. 16. Progress with integrated control of European red mite. New Zealand J. Zool. 7: 271–279.

Collyer, E., and Geldermalsen, M. van 1975. Integrated control of apple pests in New Zealand. 1. Outline of experiment and general results. New Zealand J. Zool. 2: 101–134.

Comai, M. 1985. Degree of control of the red spider exerted by new acaricide products. Informatore Agrario 41: 65–68.

Comai, M., and Stalker, D. M. 1984. Impact of genetic engineering on crop protection. Crop Protec. 3: 399–408.

Comins, H. N. 1977a. The development of insecticide resistance in the presence of migration. J. Theoret. Biol. 64: 177.

Comins, H. N. 1977b. The management of pesticide resistance. J. Theoret. Biol. 64: 117.

Comins, H. N. 1981. The mathematical evaluation of options for managing pesticide resistance, pp. 1–36. *In* Pest and Path. Management Network. Working Paper IIASA, vol. 9, 54 pp.

Compos, A. R., and Gravena, S. 1984. Insecticides, *Bacillus thuringiensis* and predaceous arthropods for the control of budworms on cotton. Anais Soc. Entomol. do Brasil 13: 95–105.

Comstock, J. H. 1880. Introduction of the 1880 report of the USDA. USDA Bull. pp. 289–290.

Congdon, B. D., and Tanigoshi, L. K. 1983. Indirect toxicity of dimethoate to the predaceous mite *Euseius hibisci* (Chant) (Acari: Phytoseiidae). Environ. Entomol. 12: 933–935.

Cook, C. 1980a. Mites identification pests and predator species. Horticultural Produce and Practice, New Zealand Advisory Bull., 2 pp.

Cook, C. 1980b. Mites, integrated control on apples. Horticultural Produce and Practice HPP 197, New Zealand Advisory Bull., 4 pp.

Corino, L. 1985. The species of phytoseiids (Acarina: Phytoseiidae) in Piedmont vineyards. Vignevini 12(6): 53–58.

Corino, L., and Ruaro, P. 1986. Inroduzione di fitosidi (Acarina: Phytoseiida) nel vigneto per la lotta biologica contro gli acari fitofagi *Panonychus ulmi* Koch e *Tetranychus urticae* Koch. Atti Giomate Fitopath. 1: 365–374.

Corino, L., Baillod, M., and Duvernez, C. 1986. Resistancia *Kampimodromus aberrans* (Oudeman) al parathione lotta biologica contro gli acari fitofagi in viticoltura. Ricerca Vit. Enol. 4: 39–42.

Cossentine, J. E., and Lewis, L. C. 1986. Impact of *Vairimorpha necatrix* and *Vairimorpha* sp. (Microspora: Microsporida) on *Bonnetia comta* (Diptera: Tachinidae) within *Agrostis ipsilon* (Lepidoptera: Noctuidae) hosts. J. Invert. Pathol. 47: 303–309.

Costello, R. A., and Elliott, D. P. 1981. Integrated control of mites and whiteflies in greenhouses. Victoria, B.C., Ministry of Agriculture and Food, 16 pp.

Coster, J. E., and Ragenovich, I. R. 1976. Effects of six insecticides on emergence of some parasites and predators from southern pine beetle infested trees. Environ. Entomol. 5: 1017–1025.

Coudron, T. A., Law, J. H., and Koeppe, J. K. 1981. Insect hormones. Trends Biochem. Soc. 6: 248–251.

Coulon, J., and Delorme, R. 1981. Effects secondaires des fongicides sur les entomophages utilises en culture protégée dans la technique de la lutte integrée, pp. 69–90. Troisieme Colloque sur les Effets Non-Intentionnels des Fongicides, Paris, 1981.

Coulon, J., Barres, P., and Daurade, M. H. 1979. Laboratory studies cncerning the action of different plant protection chemicals on *Phytoseiulus persimilis*, a predatory mite utilised against phytophagous mites on crops under cover. Phytiatrie-Phytopharmacie 28: 145–156.

Crabtree, B., and Newsholme, E. A. 1975. Comparative aspects of fuel utilization and metabolism by muscle, pp. 405–500. *In* Insect Muscle. Ed. P. N. R. Usherwood, Academic Press, New York.

Cranham, J. E. 1978a. Control of codling moth with diflubenzuron, pp. 108–110. *In* Mitteilungen aus der Biologischen Bundesanstalt für Land- und Forstwirtschaft Berlin-Dahlen no. 180, 120 pp.

Cranham, J. E. 1978b. Apples. East Malling Res. Rept. 1978, pp. 110–113.

Cranham, J. E., and Helle, W. 1985. Pesticide restance in Tetranychidae, pp. 405–419. *In* Spider Mites: Their Biology, Natural Enemies and Control. Eds. W. Helle and M. W. Sabelis, Elsevier, Amsterdam, 458 pp.

Cranham, J. E., and Solomon, M. G. 1978. Establishment of predaceous mites in orchards. East Malling Res. Rept. for 1977, 110 pp.

Cranham, J. E., and Solomon, M. G. 1981. Mite management in commercial apple orchards. Rept. E. Malling Res. Sta. for 1980, pp. 171–72.

Cranham, J. E., Kapetanakis, E. G., and Fisher, A. J. 1983. Resistance to insecticides in the predatory mite *Typhlodromus pyri* and its spider mite prey. p. 638. *In* Proc. 12th Intern. Cong. Plant Prot., Brighton, 1983, vol. 2.

Crawford, M. A. 1970. The progression of long-chain fatty acids from herbivore to carnivore and the evolution of the nervous system. Biochem. J. 119: 47P.

Critchley, B. R. 1972a. Field investigations on the effects of an organophosphorus pesticide, thionazin, on predacious Carabidae (Coleoptera). Bull. Entomol. Res. 62: 327–342.

Critchley, B. R. 1972b. A laboratory study of the effects of some soil-applied organophosphorus pesticides on Carabidae (Coleoptera). Bull. Entomol. Res. 62: 229–241.

Critchley, B. R. 1972c. Effects of three soil fumigants on Carabidae. Plant Pathol. 21: 188–194.

Croft, B. A. 1971. Comparative studies on four strains of *Typhlodromus occidentalis* (Acarina: Phytoseiidae) V. Photoperiodic induction of diapause. Ann. Entomol. Soc. Amer. 64: 962–964.

Croft, B. A. 1972. Resistant natural enemies in pest management systems. SPAN 15(1): 19–22.

Croft, B. A. 1975. Integrated control of apple mites. Extension Bull. 3–825. Mich. State Univ. Ext. Ser., 12 pp.

Croft, B. A. 1976. Establishing insecticide-resistant phytoseiid mite predators in deciduous tree fruit orchards. Entomophaga 21: 383–399.

Croft, B. A. 1977. Susceptibility surveillance to pesticides among arthropod natural enemies: modes of uptake and basic responses. Z. Pflkrankh. Pflschutz. 84: 140–157.

Croft, B. A. 1981a. Use of crop protection chemicals for integrated pest control. Phil. Trans. R. Soc. Lond. 295: 125–141.

Croft, B. A. 1981b. Development, use and management of insecticide-resistant natural enemies of orchard pests in North America. Proc. Joint USA/USSR Conf. Use of Beneficial Organisms in Control of Crop Pests. ESA Spec. Bull. 62: 54–59.

Croft, B. A. 1982. Arthropod resistance to insecticides: a key to pest control failures and successes in North American apple orchards. Entomol. Exp. Appl. 31: 88–110.

Croft, B. A. 1983. Status and management of pyrethroid resistance in the predatory mite, *Amblyseius fallacis*. Great Lakes Entomol. 16: 17–32.

Croft, B. A. 1986. Integrated pest management: the agricultural–environmental rationale, pp. 712–728. *In* CIPM. Integrated Pest Management of Major Agricultural Systems. Eds. R. F. Frisbie and P. L. Adkisson, Texas A&M Exp. Sta. Publ. MP-1616, 743 pp.

Croft, B. A., and AliNiazee, M. T. 1983. Differential resistance to insecticides in *Typhlodromus arboreus* Chant and associate Phytoseiid mites of apple in the Willamette Valley, Oregon. J. Econ. Entomol. 12: 1420–1422.

Croft, B. A., and AliNiazee, M. T. 1989. Biological control in deciduous tree fruit crops. *In* Principles and Applications of Biological Control. Ed. T. W. Fisher, Univ. Calif. Riverside (in press).

Croft, B. A., and Barnes, M. M. 1971. Comparative studies on four strains of *Typhlodromus occidentalis*. III. Evaluations of releases of insecticide resistant strains into an apple orchard ecosystem. J. Econ. Entomol. 64: 845–850.

Croft, B. A., and Barnes, M. M. 1972. Comparative studies on four strains of *Typhlodromus occidentalis*. VI. Persistence of insecticide resistant strains in an apple orchard ecosystem. J. Econ. Entomol. 65: 211–216.

Croft, B. A., and Blythe, E. J. 1980. Aspects of the functional, numerical and starvation responses of *Amblyseius fallacis* to prey density. Rec. Adr. Acarol. 1:41–47. Academic Press, New York.

Croft, B. A., and Brown, A. W. A. 1975. Responses of arthropod natural enemies to insecticides. Annu. Rev. Entomol. 20:285–335.

Croft, B. A., and Hendriks, J. 1988. Data integration for risk assessment: a case study of microbial pathogen impact on terrestrial arthropods. Proc. EPA Workshop. Integration of Research and Model Development in Biotechnology Risk Assessment (in press).

Croft, B. A., and Hoying, S. A. 1975. Carbaryl resistance in native and released populations of *Amblyseius fallacis*. Environ. Entomol. 4:895–898.

Croft, B. A., and Hoying, S. A. 1977. Competetive displacement of *Panonychus ulmi* Koch by *Aculus schlechtendali* Nolepa in apple orchards. Can. Entomol. 109:1025–1034.

Croft, B. A., and Hoyt, S. C. 1978. Considerations for the use of pyrethroid insecticides for deciduous fruit pest control in the U.S.A. Environ. Entomol. 7:627–630.

Croft, B. A., and Hoyt, S. C., Eds. 1983. Integrated Management of Insect Pests of Pome and Stone Fruits. Wiley Interscience, New York, 456pp.

Croft, B. A., and Hull, L. A. 1983. The orchard as an ecosystem, pp. 19–42. *In* Integrated Management of Insect Pests of Pome and Stone Fruits. Eds. B. A. Croft and S. C. Hoyt, Wiley Interscience, New York.

Croft, B. A., and Hull, L. A. 1987. Chemical control and resistance in tortricoid pests of pome and stone fruits. *In* Tortricoid Pests. Eds. L. P. S. van der Geest and H. H. Evenhius, Elsevier, Amsterdam.

Croft, B. A., and Hull, L. A. 1988. Chemical control and resistance in tortricoid pests of pome and stone fruits, ch. 2.3.5. *In* Tortricoid Pests. Eds. L. P. S. van der Geest and H. H. Evenhius, Elsevier, Amsterdam.

Croft, B. A., and Jeppson, L. R. 1970. Comparative studies on four strains of *Typhlodromus occidentalis*. II. Laboratory toxicity to ten compounds common to apple pest control. J. Econ. Entomol. 63:1528–1531.

Croft, B. A., and McGroarty, D. L. 1977. The role of *Amblyseius fallacis* in Michigan apple orchards. Res. Rept. 33, Mich. Agric. Exp. Sta., 48 pp.

Croft, B. A., and McMurtry, J. A. 1972a. Comparative studies on four strains of *Typhlodromus occidentalis* Nesbitt (Acarina: Phytoseiidae): life history studies. Acarologia 13:460–470.

Croft, B. A., and Meyer, R. H. 1973. Carbamate and organophosphorus resistance patterns in populations of *Amblyseius fallacis*. Environ. Entomol. 2:691–695.

Croft, B. A., and Morse, J. G. 1979. Recent advances in natural enemy-pesticide research. Entomophaga 24:3–11.

Croft, B. A., and Mullin, C. A. 1984. Comparison of detoxification enzyme systems in *Argyrotaenia citrana* (Leptidoptera: Tortricidae) and the ectoparasite, *Oncophanes americanus* (Hymenoptera: Braconidae). Environ. Entomol. 13:1330–1335.

Croft, B. A., and Nelson, E. E. 1972. Toxicity of apple orchard pesticides to Michigan populations of *Amblyseius fallacis*. Environ. Entomol. 1:576–579.

Croft, B. A., and Riedl, H. W. 1988. Chemical control and resistance to pesticides in codling moth *Cydia pomonella* L., ch. 2.1.6. *In* Tortricid Pests, Eds. L. P. S. van der Geest and H. H. Evenhuis, Elsevier, Amsterdam.

Croft, B. A., and Roush, R. T. 1987. Technical and policy issues in management of pesticide resistance in arthropod pests of agriculture. Proc. AAAS Symp., Washington, D.C.

Croft, B. A., and Strickler, K. 1983. Natural enemy resistance to pesticides: documentation, characterization, theory and applications, pp. 669–702. *In* Pest Resistance to Pesticides. Eds. G. P. Georghiou and T. Saito, Plenum Press, New York, 809 pp.

Croft, B. A., and van de Baan, H. E. 1988. Ecological and genetic factors influencing evaluation of pesticide resistance in tetranychid and phytoseiid mites. Exper. Appl. Acarol. 4:277–300.

Croft, B. A., and Wagner, S. W. 1981. Selectivity of acaricidal pyrethroids to permethrin-resistant strains of *Amblyseius fallacis*. J. Econ. Entomol. 74:703–706.

Croft, B. A., and Whalon, M. E. 1982. Selective toxicity of pyrethroid insecticides to arthropod natural enemies and pests of agricultural crops. Entomophaga 27:3–21.

Croft, B. A., and Whalon, M. E. 1983. The inheritance and persistence of permethrin resistance in the predatory mite, *Amblyseius fallacis*. Environ. Entomol. 12:215–218.

Croft, B. A., Brown, A. W. A., and Hoying, S. A. 1976a. Organophosphorus-resistance and its inheritance in the predaceous mite *Amblyseius fallacis*. J. Econ. Entomol. 69:64–68.

Croft, B. A., Briozzo, J., and Carbonell, J. B. 1976b. Resistance to organophosphorus insecticides in a predaceous mite, *Amblyseius chilenensis*. J. Econ. Entomol. 69:563–565.

Croft, B. A., Welch, S. M., and Dover, M. J. 1976c. Dispersion statistics and sample size estimates for populations of the mite species, *Panonychus ulmi* (Koch) and *Amblyseius fallacis* Garman on apple. Environ. Entomol. 5:227–234.

Croft, B. A., Wagner, S. W., and Scott, J. G. 1982. Multiple and cross resistances to insecticides in pyrethroid resistant strains of the predatory mite, *Amblyseius fallacis*. Environ. Entomol. 11:161–164.

Croft, B. A., Miller, R. W., Nelson, R. D., and Westigard, P. H. 1984. Inheritance of early-stage resistance to formetanate and cyhexatin in *Tetrancychus urticae* Koch (Acarina: Tetranychidae). J. Econ. Entomol. 77:574–578.

Croft, B. A., Adkisson, P. L., Sutherst, R. W., and Simmons, G. A. 1985. Applications of ecology for better pest control, pp. 763–745. *In* Ecological Entomology. Eds. C. B. Huffaker and R. L. Rabb, Wiley, New York, 844 pp.

Croft, B. A., Hoyt, S. C., and Westigard, P. H. 1987. Spider mite management on pome fruits revisited: organotin and acaricide resistance management. J. Econ. Entomol. 80:304–311.

Crozier, R. H. 1985. Adaptive consequences of male-haploidy, pp. 201–219. *In* Spider Mites: Their Biology, Natural Enemies and Control. Eds. W. Helle and M. W. Sabelis, Elsevier, Amsterdam, 458 pp.

Culin, J. D., and Yeargan, K. V. 1983. The effects of selected insecticides on spiders in alfalfa. J. Kansas Entomol. Soc. 56:151–158.

Curtis, C. F., Cook, L. M., and Wood, R. J. 1978. Selection for and against insecticide resistance and possible methods of inhibiting the evolution of resistance in mosquitoes. Ecol. Entomol. 3:273–287.

Dabrowski, Z. T. 1968. Studies on the toxicity of pesticides used in orchards in Poland to predatory mites. Roczniki Nauk Rolniczych 93:655–670.

Dabrowski, Z. T. 1969a. Laboratory studies on the toxicity of pesticides for *Typhlodromus finlandicus* (Oud.) and *Phytoseius macropilis* (Banks) (Phytoseiidae, Acarina). Roczniki Nauk Rolniczych 95:337–369.

Dabrowski, Z. T. 1969b. Changes in associations of predatory mites (Phytoseiidae, Acarina) in apple orchards brought about by the application of pesticides. Zeszyty Naukowe Szkoly Glowenej Gospodarstwa Wiejskiego Ogrodnictow 5:101–139.

Dabrowski, Z. T. 1970a. Studies of the subsequent action of pesticides on spider mites (Tetranychidae) and predacious mites (Phytoseiidae) in apple orchards. Roczniki Nauk Rolniczych 1:7–26.

Dabrowski, Z. T. 1970b. Effect of pesticides on the associations of predatory mites in apple orchards. Ekologia Polska 18:817–836.

Dahl, G. H., and Lowell, J. R. 1984. Microencapsulated pesticides and their effects on non-target insects, pp. 141–150. *In* Advances in Pesticide Formulation Technology. Ed. H. B. Scher, Amer. Chem. Soc., Washington, D.C.

Dahlman, D. L., and Vinson, S. B. 1980. Glycogen content in *Heliothis virescens* parasitized by *Campoletis sonorensis*. Ann. Entomol. Soc. Amer. 69:523–524.

Daily, J. C., and McKenzie, J. A. 1987. Resistance management strategies in Australia: the "*Heliothis*" and "wormkill" programmes. Proc. Brit. Crop. Prot., Bristol, UK.

Daneshvar, H., and Rodriquez, J. G. 1975. Toxicity of organophosphorus systemic pesticides to predator mites and prey. Entomol. Exp. Appl. 18: 297–301.

Daugherty, D. M. 1953. A practical survey of the insecticidal tolerance of the predaceous mite, *Typhlodromus fallacis* (Garman). M.S. thesis, Ohio State Univ., Columbus.

David, B. V., and Somasundaram, L. 1985. Indirect benefits and risks of pyrethroid usage in crop protection. Pesticides 19(5): 13–17.

David, P. J., and Horsburgh, R. L. 1985. Ovicidal activity of methomyl on eggs of pest and beneficial insects and mites associated with apples in Virginia. J. Econ. Entomol. 78: 432–436.

Davies, R. A. H. 1982. The role of the agrochemical industry in the development and implementation of IPM, pp. 77–84. *In* Proc. Aust. Workshop on Development and Implementation of IPM.

Davies, R. A. H., and McLaren, I. W. 1977. Tolerance of *Aphytis melinus* DeBach (Hymenoptera: Aphelinidae) to 20 orchard chemical treatments in relation to integrated control of red scale, *Aonidiella aurantii* (Maskell) (Homoptera: Diaspididae). Aust. J. Exp. Agric. Animal Husbandry 17: 323–328.

Davis, D. W. 1970. Insecticidal control of the alfalfa weevil in northern Utah and some resulting effects on the weevil parasite *Bathyplectes curculionis*. J. Econ. Entomol. 63: 119–125.

Davis, D. W., Hoyt, S. C., McMurtry, J. A., AliNiazee, M. T. (Eds.) 1978. Biological Control and Insect Pest Management. Univ. of Calif. priced publ. 4096, 102 pp.

de Kort, C. A. D., and Granger, N. A. 1981. Regulation of the juvenile hormone titer. Annu. Rev. Entomol. 26: 1–28.

de Kort, C. A. D., Wieten, M., and Kramer, S. J. 1979. The occurrence of juvenile hormone specific esterases in insects. A comparative study. Proc. Konk. Ned. Akad. Wet., Ser. C 82: 325–331.

de Kort, C. A. D., Peter, M. G., and Koopmanschap, A. B. 1983. Binding and degradation of juvenile hormone III by haemolymph proteins of the Colorado potato beetle: a re-examination. Insect Biochem. 13: 481–487.

de Kort, C. A. D., Koopmanschap, A. B., and Ermens, A. A. M. 1984. A new class of juvenile hormone binding proteins in insect haemolymph. Insect Biochem. 14: 619–623.

de Loof, A., van Loon, J., and Vanderroost, C. 1979a. Influence of ecdysterone, precocene and compounds with juvenile hormone activity on induction, termination and maintenance of diapause in the parasitoid wasp, *Nasonia vitripennis*. Physiol. Entomol. 4: 319–328.

de Loof, A., van Loon, J., and Hadermann, F. 1979b. Effects of juvenile hormone I, methoprene and kinoprene on development of the hymenopteran parasitoid *Nasonia vitripennis*. Entomol. Exp. Appl. 26: 301–313.

Deakle, J. P., and Bradley, J. R. 1982. Effects of early season applications of diflubenzuron and azinphosmethyl on population levels of certain arthropods in cottonfields. J. Georgia Entomol. Soc. 17: 200–204.

DeBach, P. 1958. Selective breeding to improve adaptations of parasitic insects, pp. 759–768. *In* Proc. 10th Intern. Cong. Entomol., vol. 4.

DeBach, P., Ed. 1964. Biological Control of Insect Pests and Weeds. Reinhold, New York, 844 pp.

DeBach, P., and Bartlett, B. 1951. Effects of insecticides on biological controls of insect pests of citrus. J. Econ. Entomol. 44: 372–383.

DeBach, P., and Bartlett, B. 1964. Methods of colonization, recovery and evaluation, pp. 702–726. *In* Biological Control of Insect Pests and Weeds. Ed. P. DeBach, Reinhold, New York, 844 pp.

DeBach, P., and Sundby, R. A. 1963. Competetive displacement between ecological homologues. Hilgardia 34: 105–166.

Delorme, R. 1975. Toxicty for *Diaeretiella rapae* (Hymenoptera Aphididae) and for its host *Myzus persicae* of twelve pesticides used in sprays. Phytiatrie-Phytopharmacie 24: 265–278.

Delorme, R. 1976. Laboratory evaluation of the toxicity of pesticides used in the treatment of aerial parts of plants to *Diaeretiella rapae* (Hym.: Aphidiidae). Entomophaga 21: 19–29.

Delorme, R., and Angot, A. 1983. Toxicités relatives à divers pesticides pour *Encarsia formosa* Gahan (Hym., Aphelinidae) et pour son hôte, *Trialeurodes vaporariorum* Westw. (Hom., Aleyrodidae). Agronomie 3: 577–584.

Delorme, R., Angot, A., and Augé, D. 1984. Variations de sensibilité d'*Encarsia formosa* Gahan (Hymenoptera: Aphelinidae) soumis à des pressions de sélection insecticide: approaches biologique et biochimique. Agronomie 4: 305–309.

Delorme, R., Berthier, A., and Augé, D. 1985. The toxicity of two pyrethroids to *Encarsia formosa* and its host *Trialeurodes vaporariorum*: prospecting for a resistant strain of the parasite. Pesticide Sci. 16: 332–336.

Demolin, G. 1978. Action of Dimilin on the caterpillars of *Lymantria dispar* L.: incidence on their tachinid endoparasites. Ann. Sci. Forestières 35: 229–234.

Dempster, J. P. 1968. The sublethal effect of DDT on the rate of feeding by the ground-beetle *Harpalus rufipes*. Entomol. Exp. Appl. 11: 51–54.

Dennehy, T. J., Granett, J., and Leigh, T. F. 1983. Relevance of slide-dip and residual bioassay comparisons to detection of resistance in spider mites. J. Econ. Entomol. 76: 1225–1230.

Desaiah, D., Cutkomp, L. K., and Koch, R. B. 1973. Inhibition of spider mite ATPases by Plictran and three organochlorine acaricides. Life Sci. 13: 1693–1703.

Ding, Y., Xiong, J. J., and Huang, M. D. 1983. Resistance of *Amblyseius nicholsi* Ehara et Lee (Acar.: Phytoseiidae) to some pyrethroids. Natural Enemies of Insects 5(3): 124–128.

Dippenaar, A. S., Genis, N. L., van Rak, H., and Viljoen, J. H. 1978. The effect of dieldrin coverspraying on some South African spiders and scorpions. Phytophylactica 10(4): 115–122.

Dirimanov, M., Stefanov, D., and Dimitrov, A. 1974. The effect of certain insecticides on the more important species of useful insects on tobacco. Rastitelna Zashchita 22: 18–20.

Dittrich, V. 1975. Acaricide resistance in mites. Z. angew. Entomol. 78: 28–45.

Divakar, B. J. 1980. The effect of a juvenile hormone analogue on *Eucelatoria* sp. (Diptera: Tachinidae) through its host, *Heliothis armigera* (Hubn.) (Lepidoptera: Noctuidae). Experientia 36: 1332–1333.

Dmitrienko, V. K. 1979. The effect of pesticides containing chlorine and phosphorus on ants. Ekologiya 1: 53–60.

Dorn, S., Frischknecht, M. L., Martinez, V., Zurflüh, R., and Fischer, U. 1981. A novel non-neurotoxic insecticide with a broad activity spectrum. Z. Pflkrankh. Pflschutz. 88: 269–275.

Doutt, R. L. 1964. The historical development of biological control, ch. 3, pp. 21–42. *In* Biological Control of Insect Pests and Weeds. Ed. P. DeBach, Reinhold, New York, 844 pp.

Doutt, R. L., and Hagan, K. S. 1950. Biological control measures against *Pseudococcus maritimus* on pears. J. Econ. Entomol. 43: 94–96.

Doutt, R. L., and Nakata, J. 1965. Parasites for control of grape leafhopper. Calif. Agric. 19: 3.

Dover, M. J., and Croft, B. A. 1984. Getting tough: policy issues in management of pesticide resistance. Study 1. World Res. Inst. Policy Paper, Nov. 1984, 80 pp.

Dover, M. J., Croft, B. J., Welch, S. M., and Tummala, R. L. 1979. Biological control of *Panonychus ulmi* (Acarina: Tetranychidae) by *Amblyseius fallacis* (Acarina: Phytoseiidae): a predator-prey model. Environ. Entomol. 8: 282–292.

Dowd, P. F., Smith, C. N., and Sparks, T. C. 1983. Influence of soybean leaf extracts on ester cleavage in cabbage and soybean loopers (Lepidoptera: Noctuidae). J. Econ. Entomol. 76: 700–703.

Downing, R. S. 1966. The effect of certain miticides on the predacious mite *Neoseiulus caudiglans* (Acarina: Phytoseiidae). Can. J. Plant Sci. 46: 521–524.

Downing, R. S., and Arrand, J. C. 1968. Integrated control of orchard mites in British Columbia. Brit. Columbia Dept. Agric., Entomol. Branch Publ., 77 pp.

Downing, R. S., and Moilliet, T. K. 1971. Occurrence of phytoseiid mites in apple orchards in south central British Columbia. Entomol. Soc. Brit. Columbia 68: 33–36.

Downing, R. S., and Moilliet, T. K. 1975. Preliminary trials with citrazon—a selective acaricide. Entomol. Soc. Brit. Columbia 72: 3–5.

Dreyer, D. L., and Jones, K. C. 1981. Feeding deterrency of flavonoids and related phenolics towards *Schizaphis graminum* and *Myzus persicae*: aphid feeding deterrents in wheat. Phytochemistry 20: 2489–2493.

Drummond, F. A., van Driesche, R. G., and Logan, P. A. 1985. Model for the temperature dependent emergence of overwintering *Phyllonorycter crataegella* (Clements) (Lepidoptera: Gracillariidae), and its parasite, *Sympiesis marylandensis* Girault (Hymenoptera: Eulophidae). Environ. Entomol. 14: 305–311.

DSIR 1985. Integrated mite control. Dept. Rept. Sci. & Indust. Res. New Zealand, pp. 11–12.

Duffey, S. S. 1980. Sequestration of plant natural products by insects. Annu. Rev. Entomol. 25: 447–477.

Duke, S. O. 1985. Weed Physiology, vol. II. Herbicide Physiology. CRC Press, Boca Raton, Fla.

Dulmadge, H. T., Martinez, J. T., and Pena, T. A. 1976. Bioassay of *Bacillus thuringiensis* (Berliner) endotoxin using the tobacco budworm. U.S. Dept. Agri., Tech. Bull. 1528: 1–15.

Dumbre, R. B., and Hower, A. A. 1976a. Sublethal effects of insecticides on the alfalfa weevil parasites *Microctonus aethiopoides*. Environ. Entomol. 5: 683–687.

Dumbre, R. B., and Hower, A. A. 1976b. Relative toxicities of insecticides to the alfalfa weevil parasite *Microctonus aethiops* and the influence of parasitism on host susceptibility. Environ. Entomol. 5: 311–315.

Dumbre, R. B., and Hower, A. A. 1977. Contact mortality of the alfalfa weevil parasite *Microctonus aethiopiodes* from insecticide residues on alfalfa. Environ. Entomol. 6: 893–894.

Dunbar, J. P., and Johnson, S. W. 1975. *Bacillus thuringiensis*: effect on the survival of a tobacco budworm parasitoid and predator in the laboratory. Environ. Entomol. 4: 352–354.

Dunning, R. A., Cooper, J. M., Wardman, J. M., and Winder, G. H. 1982. Susceptibility of the carabid *Pterostichus melanarius* (Illiger) to aphicide sprays applied to the sugar-beet crop. Ann. Appl. Biol. 100: 32–33.

Dustan, G. G., and Boyce, H. R. 1966. Parasitism of the oriental fruit moth, *Grapholitha molesta* (Busck) (Lepidoptera: Tortricidae) in Ontario, 1956–1965. Proc. Entomol. Soc. Ontario 96: 100–102.

Dyck, V. A., and Orlido, G. C. 1977. Control of the brown planthopper (*Nilaparvata lugens*) by natural enemies and timely application of narrow-spectrum insecticides. IRRI Ann. Rept. 1976, pp. 28–72.

Easterbrook, M. A. 1984. Chemical and integrated control of apple rust mite, pp. 1107–1111. *In* Proc. 1984 Brit. Crop Protection Conf., Pests and Diseases. Brighton Metropole, England, Nov. 19–22, 1984, vol. 3.

Easterbrook, M. A., Solomon, M. G., Cranham, J. E., and Souter, E. F. 1985. Trials of an integrated pest management programme based on selective pesticides in English apple orchards. Crop Protec. 4(2): 215–230.

Edelson, J. V. (Ed.) 1977–1986. Insecticide and acaricide tests: Vol. 2–10. Entomol. Soc. Amer. Publ., College Park, MD.

Edge, V. E., and James, D. G. 1982. Detection of cyhexatin resistance in twospotted mite, *Tetranychus urticae* Koch (Acarina: Tetranychidae) in Australia. J. Aust. Entomol. Soc. 21: 198.

Edge, V. E., and James, D. G. 1986. Organotin resistance in *Tetranychus urticae* Koch (Acarina: Tetranychidae) in Australia. J. Econ. Entomol. 79: 1477–1483.

Edwards, B. A. B., and Hodgson, P. J. 1973. The toxicity of commonly used orchard chemicals to *Stethorus nigripes* (Coleoptera: Coccinellidae). J. Aust. Entomol. Soc. 12: 222–224.

Edwards, C. A. 1966. Insecticide residues in soils. Residue Rev. 13: 83–132.

Edwards, C. A., and Thompson, A. R. 1975. Some effects of insecticides on predatory beetles. Ann. Appl. Biol. 80: 132–135.

Edwards, C. A., Lofty, J. R., and Stafford, C. J. 1970. Effect of pesticides on predatory beetles. Rep. Rothamsted Exper. Sta. 1969 (part 1), 246 pp.

Edwards, C. A., Sunderland, K. D., and George, K. S. 1979. Studies on polyphagous predators of cereal aphids. J. Appl. Ecol. 16: 811–823.

Edwards, C. A., Thornhill, W. A., Jones, B. A., Bater, J. E., and Lofty, J. R. 1984a. The influence of pesticides on polyphagous predators of pests, pp. 317–323. *In* Proc. 1984 Brit. Crop Protection Conf. Pests and Diseases, Brighton Metropole, England, Nov. 19–22, 1984, vol. 1.

Edwards, P. J., Wilkinson, W., and Coulson, M. 1984b. A laboratory toxicity test for carabid beetles, pp. 359–362. *In* Proc. 1984 Brit. Crop Protection Conf. Pests and Diseases, Brighton Metropole, England, Nov. 19–22, 1984, vol. 1.

Eggers-Schumacher, H. A. 1983. A comparison of the reproductive performance of insecticide-resistant and susceptible clones of *Myzus persicae*. Entomol. Exp. Appl. 34: 301–307.

Eghtedar, E. 1969. The susceptibility of *Philonthus fuscipennis* Mannh. and *Tachyporus hypnorum* L. (Col., Staphylinidae) to insecticides. Nachr. Pflschutz. 21: 182–185.

Ehler, L. E., Endicott, P. C., Hertlein, M. B., and Alvarado-Rodriquez, B. 1984. Medfly eradication in California: impact of malathion-bait sprays on an endemic gall midge and its parasitoids. Entomol. Exp. Appl. 36: 201–208.

Eijsackers, H. 1978. Side effects of the herbicide 2, 4, 5-T affecting the carabid *Notiophilus biguttatus*, a predator of springtails. Z. Angew. Entomol. 86: 113–128.

Eisner, T. 1970. Chemical defenses against predation in arthropods, pp. 157–217. *In* Chemical Ecology, Eds. E. Sondheimer and J. B. Simeone, Academic Press, New York.

El-Banhawy, E. M. 1980. Comparison between the response of the predacious mite *Amblyseius brazilli* and its prey *Tetranychus desertorum* to the different IGRs methoprene and Dimilin (Acari: Phytoseiidae, Tetranychidae). Acarologia 21: 221–227.

El-Banhawy, E. M., and Abou-Awad, B. A. 1985. Effects of synthetic pyrethroids and other compounds on the susceptibility and development of the egg stage of the predaceous mite *Amblyseius gossipi* (Mesostigmata: Phytoseiidae). Entomophaga 30: 265–270.

El-Banhawy, E. M., and El-Bagoury, M. E. 1985. Toxicity of avermectin and fenvalerate to the eriophyid gall mite *Eriophyes dioscoridis* and the predacious mite *Phytoseius finitimus* (Acari: Eriophyidae; Phytoseiidae). Intern. J. Acarol. 11: 237–240.

El-Sufty, R., and Fuhrer, E. 1981. Interrelationships between *Pieris brassicae* L. (Lep., Pieridae), *Apanteles glomeratus* L. (Hym., Braconidae) and the fungus *Beauveria bassiana* (Bals.) Vuill. Z. Angew. Entomol. 92: 321–329.

El-Sufty, R., and Fuhrer, E. 1985. Interrelations between *Cydia pomonella* L. (Lep., Tortricidae), *Ascogaster quadridentatus* Wesm. (Hym., Braconidae), and the fungus *Beauvarian bassiana* (Bals.) Vuill. Z. Angew. Entomol. 99: 504–511.

Elenkov, E., Kristova, E., Shanab, L. M., and Spasova, P. 1975. Toxicity of some fungicides against the greenhouse whitefly (*Trialeurodes vaporariorum* West.) and its parasite (*Encarsia formosa* Gah.). II. Toxicity of Morestan against the greenhouse whitefly and its parasite. Acta Phytopathol., Acad. Sci. Hungaricae 10: 171–176.

Elenkov, E. S., Khristova, E., Loginova, E., Spasova, P., and Popova, R. 1980. Toxicity of metalaxyl to the greenhouse whitefly (*Trialeurodes vaporariorum* Westw.) and its parasite *Encarsia formosa* Gah. Gradinarska i Lozarska Nauka 17(7/8): 67–73.

Elenkov, E. S., Khristova, E., Loginova, E., and Spasova, P. 1984. Toxicity of some fungicides to the glasshouse whitefly (*Trialeurodes vaporariorum* Westw.) and its parasite *Encarsia formosa* Gah. Gradinarska i Lozarska Nauka 21: 70–77.

Elliott, M., Ed. 1977. Synthetic Pyrethroids. Amer. Chem. Soc. Symp. Ser., 229 pp.

Elliott, M., Janes, N. F., Stevenson, J. H., and Walters, J. H. H. 1983. Insecticidal activity of the pyrethrins and related compounds. Part XIV: Selectivity of pyrethroid insecticides between

Ephestia kuhniella and its parasite *Venturia canescens*. Pestic. Sci. 14: 423–426.

Elliott, W. M. 1970. The action of some systemic aphicides on the nymphs of *Anthocoris nemorum* (L.) and *A. confusus* Reut. Ann. Appl. Biol. 66: 313–321.

Elliott, W. M., and Way, M. J. 1968. The action of some systemic aphicides on the eggs of *Anthocoris neumorum* (L.) and *A. confusus* Reut. Ann. Appl. Biol. 62: 215–226.

Elsey, K. D. 1973. *Jalysus spinosus*: effect of insecticide treatments on this predator of tobacco pests. Environ. Entomol. 2: 240–243.

Elsey, K. D., and Cheatham, J. S. 1976. Contact toxicity of insecticides to three natural enemies of the tobacco hornworm and tobacco budworm. Tobacco Int. 178(12): 76–77.

Emel'yanov, N. A., and Yakushev, B. S. 1981. Field edges for control of entomophagous insects. Zash. Rast. 9:29.

Endacott, C. J. 1983. Non-target organism mortality—a comparison of spraying techniques, p. 502. *In* Proc. 10th Intern. Cong. Plant Protec., Brighton, England, Nov. 20–25, 1983, vol. 2.

Englert, W. D. 1984. A standard method to test effects of pesticides on predaceous mites in vineyards, p. 732. *In* Abst. XVII Intern. Cong. Entomol., 960 pp.

Englert, W. D., and Kettner, J. 1983. Side-effects of plant protection materials on spider mites and predacious mites. Mitt. Deutsch. Ges. Allgem. Angew. Entomol. 4: 89–91.

EPA 1982. Pesticide Assessment Guidelines/Subdivision M/Biorational Pesticides, R. K. Hitch, Guideline Coordinator, OPP/USEPA, Washington, D.C., 284 pp.

Evans, F. C., and Murdoch, W. W. 1968. Taxonomic composition, trophic structure, and seasonal occurrence in a grassland insect community. J. Animal Ecol. 37: 259–273.

Eveleens, K. G., van den Bosch, R., and Ehler, L. E. 1973. Secondary outbreak induction of beet armyworm by experimental insecticide applications in cotton in California. Environ. Entomol. 2: 497–503.

Evenhuis, H. H. 1959. Effect van insecticiden op de bloedluis parasiet *Aphelinus mali* Hold. Meded. Dithuinb. 22: 306–311.

Everson, P. R., and Tonks, N. V. 1981. The effect of temperature on the toxicity of several pesticides to *Phytoseiulus persimilis* (Acarina: Phytoseiidae) and *Tetranychus urticae* (Acarina: Tetranychidae). Can. Entomol. 113: 333–336.

Fabellar, L. T., and Heinrichs, E. A. 1984: Toxicity of insecticides to predators of rice brown planthoppers, *Nilaparvata lugens* (Stal) (Homoptera: Delphacidae). Environ. Entomol. 13: 832–837.

Fagundes, A. C., Kesterke, R., and Arnt, T. 1985. Inseticidas no controle de pulgoes nas folhas de trigo (*Macrosiphum avenae*) e toxicidade a adultos de *Cycloneda sanguinea* (Col.: Coccinellidae). Agton. Sulriograndense, Porto Alegre 21: 73–85.

Falconer, D. S. 1981. Introduction to Quantitative Genetics, 2nd ed. Eangman Press, New York, 340 pp.

Fang, C. Y., Wen, S. G., Cui, S. Z., and Wang, Y. H. 1984. The role of natural enemies in the integrated control of insect pests on cotton. China Cotton (2): 42–43.

FAO 1967. Report of the first session of the FAO Working Party of Experts on Resistance of Pests to Pesticides. FAO, Rome, PL/1965/18, 125 pp.

FAO 1977. Pest resistance to pesticides and crop loss assessment. Rept. 1st FAO Panel of Experts. FAO Plant Protection Paper 6, 42 pp.

FAO 1979. Pest resistance to pesticides, Rept. 2nd Panel of Experts, Rome, Sept. 5–10, 1979. FAO Plant Protection Paper.

FAO 1984. Recommended method for the detection of resistance in agricultural pests to pesticides: methods for phytoseiid predatory mites. 8 pp.

Farghaly, H. T., Zhody, G. I., and Salama, A. 1974. Toxicity of certain acaricides on the two-spotted spider mite, *Tetranychus cinnabarinus* (Boisd.) and the predaceous mite, *Agistemus exsertus* Gonzalez, pp. 205–211. *In* Proc. Symp. Isotopes in Pesticides and Pest Control, Beirut, Mar. 1974.

Farlow, R. A., and Pitre, H. N. 1983. Bioactivity of the post-emergence herbicides aciflourfen and bentazon on *Geocoris punctipes*. J. Econ. Entomol. 76: 200–203.

Farrier, M. H., Rock, G. C., and Yeargan, R. 1980. Mite species in North Carolina apple orchard with notes on their abundance and distribution. Environ. Entomol. 9: 425–429.

Fashing, N. J., and Sagan, H. 1979. Effect of juvenile hormone analog Methoprene on *Nasonia vitripennis* when administered via a host, *Sarcophaga bullata*. Environ. Entomol. 8: 816–818.

Fast, P. G. 1964. Insect lipids: a review. Mem. Entomol. Soc. Can. 37, 50 pp.

Fast, P. B. 1970. Insect lipids. Prog. Chem. Fats Other Lipids. 11: 181–242.

Federal Register 1981. E. P. A., Certain pesticide products; intent to cancel registrations. Apr. 21, 1981, 46(76).

Feeney, A. M. 1983. The occurrence and effect of pesticides on aphid predators in Ireland, 1979–1982. *In* Aphid Antagonists, Proc. E. C. Experts Group, Portici, Italy. A. A. Balkena, Rotterdam, 1983.

Felton, G. W., and Dahlman, D. L. 1984. Nontarget effect of a fungicide: toxicity of maneb to the parasitoid *Microplitis croceipes* (Hymenoptera: Braconidae). J. Econ. Entomol. 77: 847–850.

Feng, H. T., and Wang, T. C. 1984. Selectivity of insecticide to *Plutella xylostella* (L.) and *Apanteles plutellae* Kurd. Plant Protec. Bull. Taiwan 26: 275–284.

Fernadez, B. R. I., and Clavijo, A. S. 1984. Effects of two insecticides (one chemical and the other biological) on the parasitism observed in larvae of *Spodoptera frugiperda* (S.) from experimental plots of maize. Revista Facultad Agron., Univ. Central Venezuela 13: 101–109.

Ferrari, R., and Tiberi, R. 1979. Efficiency of a control test with diflubenzuron against *Thaumetopoea pityocampa* (Den. & Schiff.) and first observation on the effect of the insecticide with respect to the hymenopterous chalcidoid egg parasitoid of the moth in Tuscany. Redia 72: 315–323.

Feyereisen, R. 1987. Chemical disruption of insect juvenile hormone biosynthesis, pp. 113–116. *In* Pesticide Science and Biotechnology. Eds. R. Greenhalgh and T. R. Roberts, Blackwell Sci. Publ.

Feyereisen, R., Koener, J. F., Farnsworth, D. E., and Nebert, D. W. 1989. Isolation and sequence of cDNA encoding a cytochrome P450 from an insecticide-resistant strain of the house fly, *Musca domestica*. Proc. Natl. Acad. Sci. USA, Vol. 86, pp. 1465–69 March 1989.

Ffrench-Constant, R. H., and Vickerman, G. P. 1985. Soil contact toxicity of insecticides to the European earwig *Forficula auricularia* (Dermaptera). Entomophaga 30: 271–278.

Field, R. P. 1974. Occurrence of an Australian strain of *Typhlodromus occidentalis* (Acarina: Phytoseiidae) tolerant to parathion. J. Aust. Entomol. Soc. 13: 255–256.

Field, R. P. 1976. Integrated pest control of Victorian peach orchards: the role of *Typhlodromus occidentalis* Nesbitt (Acarina: Phytoseiidae). Aust. J. Zool. 24: 565–572.

Field, R. P. 1978. Control of the two-spotted mite in a Victoria peach orchard with an introduced insecticide-resistant strain of the predatory mite *Typhlodromus occidentalis* Nesbitt (Acarina: Phytoseiidae). Aust. J. Zool. 26: 516–527.

Field, R. P. 1982. Development and implementation of integrated pest management in Victorian peach orchards, pp. 191–197. *In* Proc. Australasian Workshop on Devel. and Implemen. of IPM, Auckland, New Zealand.

Field, R. P., Webster, W. J., and Morris, D. S. 1979. Mass rearing *Typhlodromus occidentalis* Nesbitt (Acarina: Phytoseiidae) for release in orchards. J. Aust. Entomol. Soc. 18: 213–215.

Finlayson, D. G. 1979. Combined effects of soil-incorporated and foliar-applied insecticides in bed-system production of brassica crop. Can. J. Plant. Sci. 59: 399–410.

Finlayson, D. G., Mackenzie, J. R., and Campbell, C. J. 1980. Interactions of insecticides, a carabid predator, a staphylinid parasite, and cabbage maggots in cauliflower. Environ. Entomol. 9: 789–794.

Finney, D. J. 1978. Probit Analysis. Cambridge Univ. Press, Cambridge, UK, 318 pp.

Fisher, R. C. 1971. A study of insect multiparasitism. II. The mechanism and control of competition for possession of the host. J. Exp. Biol. 40: 531–540.

Fitzpatrick, G. E., and Dowell, R. V. 1981. Survival and emergence of citrus blackfly parasitoids after exposure to insecticides. Environ. Entomol. 10: 728–731.

Fix, L. A., and Plapp, F. W. 1983. Effect of parasitism on the susceptibility of the tobacco budworm (Lepidoptera: Noctuidae) to methyl parathion and permethrin. Environ. Entomol. 12: 976–978.

Flanders, R. V., Bledsoe, L. W., and Edwards, C. R. 1984. Effects of insecticides on *Pediobius foveolatus* (Hymenoptera: Eulophidae), a parasitoid of the Mexican bean bettle (Coleoptera: Coccinellidae). Environ. Entomol. 13: 902–906

Fleschner, C. A., and Scriven, G. T. 1957. Effect of soil-type and DDT on ovipositional responses of *Chrysopa californica* (Coq.) on lemon trees. J. Econ. Entomol. 50: 221–222.

Fleschner, C. A., Hall, J. C., and Ricker, D. W. 1955. Natural balance of mite pests in an avocado grove. Calif. Avocado Soc. Yearbook 39: 155–162.

Flexner, J. L. 1988. Organotin resistance in *Tetranychus urticae* Koch on pear: components and their integration for resistance management. Ph.D. dissertation, Oregon State University, Corvallis.

Flexner, J. L., Lighthart, B., and Croft, B. A. 1986. The effects of microbial pesticides on non-target, beneficial arthropods. Agric., Ecosys., Environ. 16: 203–254

Flexner, J. L., Croft, B. A., and Westigard, P. H. 1987. Effect of organotin formulations on organotin resistance of *Tetranychus urticae* Koch (Acarina: Tetranychidae). J. Econ. Entomol. 81: 766–769.

Flint, M. L., and van den Bosch, R. 1981. Introduction to Integrated Pest Management. Plenum Press, New York, 240 pp.

Ford, J. 1976. Setting a mite to catch a mite. Intern. Pest Contr. 18(5): 7–8.

Forsberg, C. W., Henderson, M., Henry, E., and Roberts, J. R. 1976. *Bacillus thuringiensis*: its effect on environmental quality. Environmental Secretariat, Natl. Res. Council Canada, NRCC 15385, 135 pp.

Fournier, D. 1981. Acquisition of resistance to methidathion with *Phytoseiulus persimilis*. Ph.D. Thesis, Univ. Paris, France, 44 pp.

Fournier, D., Pralavorio, M., Berge, J. B., and Cuany, A. 1985a. Pesticide resistance in Phytoseiidae, pp. 423–432. *In* Spider Mites—Their Biology, Natural Enemies and Control. Eds. W. Helle and M. W. Sabelis, Elsevier, Amsterdam, vol. 1B, 458 pp.

Fournier, D., Pralavorio, M., and Pourriere, O. 1985b. Studies on the use of the phytoseiid *Cydnodromus chilenensis* against *Tetranychus urticae* in glasshouses. Entomophaga 30: 113–120.

Fournier, D., Cuany, A., Pralavorio, M., Bride, J. M., and Berge, J. B. 1987a. Analysis of methidathion resistance mechanisms in *Phytoseiulus persimilis* A. M. Pestic. Biochem. Physiol. 28: 271–278.

Fournier, D., Pralavorio, M., Trottin-Caudal, Y., Coulon, J., Malezieux, S., and Berge, J. B. 1987b. Selection artificielle pour la resistance au méthidathion chez *Phytoseiulus persimilis* A. H. Entomophaga 32: 209–219.

Franz, J. M. 1974. Testing of side-effects of pesticides on beneficial arthropods in the laboratory—a review. Z. Pflkrankh. Pflschutz. 81: 141–174.

Franz, J. M. 1976. Introductory remarks, Symp. "Pesticides and Beneficial Arthropods" XV Int. Cong. Entomol., Washington, D.C., Aug. 25, 1976. Z. Pflkrankh. Pflschutz. 84: 129–131.

Franz, J. M., and Fabrietius, K. 1971. Testing the sensitivity to pesticides of entomophagous arthropods—trials using *Trichogramma*. Z. Angew. Entomol. 68: 278–288.

Franz, J. M., Hassan, S. A., and Bogenschutz, H. 1976. Some results of standardised laboratory tests on the effect of pesticides on entomophagous beneficial arthropods. Nachr. Pflschutz. 28(12): 181–183.

Franz, J. M., Bogenschutz, H., Hassan, S. A., Huang, P., Naton, E., Suter, H., and Viggiani, B. 1980. Results of a joint pesticide test programme by the working group: pesticides and beneficial arthropods. Entomophaga 25: 231–236.

Franzmann, B. A., and Rossiter, P. D. 1981. Toxicity of insecticides to *Trioxys complanatus* Quilis (Hymenoptera: Braconidae) in lucern. J. Aust. Entomol. Soc. 20: 313–315.

Frazer, B. D., and van den Bosch, R. 1973. Biological control of the walnut aphid in California: the interrelationship of the aphid and its parasite. Environ. Entomol. 2: 561–568.

Freitag, R., and Poulter, F. 1970. The effects of the insecticides sumithion and phosphamidon on populations of five species of carabid beetles and two species of lycosid spiders in northwestern Ontario. Can. Entomol. 102: 1307–1311.

Frisbie, R. F., and Adkisson, P. L., Eds. 1986. Integrated pest management on major agricultural systems. Texas Agric. Expt. Sta. Bull. MP-1616, 743 pp.

Frischknecht, M., and Jucker, W. 1978. Mode of action and practical possibilities of an insect growth regulator with juvenile hormone activity in pear psyllid control. Z. Pflkrankh. Pflschutz. 85: 334–340.

Fuhrer, E., and El-Sufty, R. 1979. Production of fungistatic metabolites by teratocytes of *Apanteles glomeratus* L. (Hym., Braconidae). Z. Parasitenk. 59: 21–25.

Fukami, J. 1980. Metabolism of several insecticides by glutathion S-transferase. Pharmac. Theor. 10: 473–514.

Fukuto, T. R. 1984. Propesticides, pp. 87–101. *In* Pesticide Synthesis through Rational Approaches. Eds. P. S. Mager and G. K. Kohn, Amer. Chem. Soc. Publ.

Fukuto, T. R., and Fahmy, M. A. H. 1981. Sulfur in propesticides action, pp. 35–49. *In* Sulfur in Pesticide Action and Metabolism. Eds. J. D. Rosen, P. S. Magee, and J. E. Casida. Amer. Chem. Soc. Symp. Ser. 158, Washington, D.C.

Gambaro, P. I. 1975. Selezione di popolazioni de Acari predatori resistenti ad alcumi insecticidi fosfonatiorganici. Estrat. Inform. Fitopath. 25(7): 21–25.

Gambaro, P. I. 1983. Biological control of mites on apple. 1. The harmful effect of some fungicides on predacious mites. Informatore Agrario 39(31): 26951–26953.

Gambaro, P. I. 1986. Validita del trasporta di popolazioni di *Amblyseius andersoni* Chant (Acarina: Phytoseiidae) OP resistenti su meli infestati da *Panonychus ulmi*. Estrat. Inform. Agrario 42(30): 43–45.

Garcia, A. C. 1975. Chemical control of the green lucerne aphid, *Acyrthosiphum pisum* (Harris) and its effect on the predator *Hippodamia convergens* G. Peruana Entomol. 17: 92–94.

Garcia, R. F. 1980. Action of biological and chemical agents in the reduction of populations of eggs of *Heliothis* spp. on cotton. Rev. Colombiana de Entomologia 6(1/2): 11–20.

Garg, A. K., and Sethi, G. R. 1984. Population buildup and effect of insecticidal treatments on *Brumoides suturalis* (Fabricius)—a predator of paddy pests. Indian J. Entomol. 46: 254–256.

Garg, D. O., and Taley, Y. M. 1977. Note on the effect of insecticides on sorghum midge and its parasite *Tetrastichus* spp. Indian J. Agric. Sci. 47: 313–314.

Gargov, V. P. 1968. A study of selectivity in the action of organophoshporus compounds on aphids that damage cotton and their coccinellid predators. Summary of Thesis Tashkent, sel'.-shoz. Inst., 16 pp.

Garrido, A., Tarancon, J., and Del Busto, T. 1982. Effect of some pesticides on the nymphal stages *of Cales noacki* How, a parasite of *Aleurothrixus floccosus* Mask. Anales Inst. Nac. Invest. Agrarias, Agricola 18: 73–96.

Garrido, A., Beitia, F., and Gruenholz, P. 1984. Effects of PP618 on immature stages on *Encarsia formosa* and *Cales noacki* (Hymenoptera: Aphelinidae), pp. 305–310. *In* Proc. 1984 Brit. Crop Protec. Conf. Pests and Diseases, Brighton Metropole, England, Nov. 19–22, 1984, vol. 1.

Garrido, A., Beitia, F., and Gruenholz, P. 1985. Incidence of the growth regulator NNI-750 on the immature stages of *Encarsia formosa* Gahan and *Cales noacki* How (Hym.: Aphelinidae). Anales Inst. Nac. Invest. Agrarias, Agricola 28(3): 137–145.

Genini, M., and Baillod, M. 1987. Introduction de souches résistantes de *Typhlodromus pyri* (Scheuten) et *Amblyseius andersoni* Chant (Acari: Phytoseiidae) en vergers de pommiers. Rev. Suisse Vitic., Arboric., Hortic. 19: 115–123.

Genini, M., Klay, A., Delucchi, V., Baillod, M., and Baumgartner, J. 1983. The species of phytoseiids (Acarina: Phytoseiidae) in apple orchards in Switzerland. Mitt. Schweiz. Entomol. Ges.56(1/2): 45–56.

Georghiou, G. P. 1967. Differential susceptibility and resistance to insecticides of coexisting populations of *Musca domestica, Fannia canicularis, F. femoralis*, and *Ophyra leucostoma*. J. Econ. Entomol. 60: 1338–1344.

Georghiou, G. P. 1972. The evolution of resistance to pesticides. Annu. Rev. Ecol. Sys. 3: 133–168.

Georghiou, G. P. 1980. Insecticide resistance and its prospects for management. Residue Rev. 76: 131–145.

Georghiou, G. P. 1983. Management of resistance in arthropods, pp. 769–792. *In* Pest Resistance to Pesticides. Eds. G. P. Georghiou and T. Saito, Plenum Press, New York, 809 pp.

Georghiou, G. P. 1986. The magnitude of the resistance problem, pp. 14–43. *In* Pesticide Resistance: Strategies and Tactics for Management. NAS/NRC Press, Washington, D.C., 471 pp.

Georghiou, G. P., and Saito, T., Eds. 1983. Pest Resistance to Pesticides. Plenum Press, New York/London, 809 pp.

Georghiou, G. P., and Taylor, C. E. 1976. Pesticide resistance as an evolutionary phenomenon. *In* Proc. XV Intern. Cong. Entomol. pp. 759–785.

Georghiou, G. P., and Taylor, C. E. 1977a. Genetic and biological influences in the evolution of insecticide resistance. J. Econ. Entomol. 70: 319.

Georghiou, G. P., and Taylor, C. E. 1977b. Operational influences in the evolution of insecticide resistance. J. Econ. Entomol 70: 653.

Gerolt, P. 1983. Insecticides: their route of entry, mechanism of transport and mode of action. Biol. Rev. 58: 233–274.

Getzin, L. W. 1960. Selective insecticides for vegetable leaf-miner control and parasite survival. J. Econ. Entomol. 53: 872–875.

Gholson, L. E., Beegle, C. C., Best, R. L., and Owens, J. C. 1978. Effects of several commonly used insecticides on cornfield carabids in Iowa. J. Econ. Entomol. 71: 416–418.

Gilbert, N., Gutierrez, A. P., Frazer, B. D., and Jones, R. E. 1976. Ecological Relationships. Freeman and Co., San Francisco, 156 pp.

Gilbert, L. I., Bollenbacher, W. E., Goddman, W., Smith, S. L., Ague, N., Granger, N., and Sedlak, B. J. 1980. Hormones controlling insect metamorphosis. Rec. Prog. Horm. Res.(36: 401–449.

Gilkeson, L. A., and Hill, S. B. 1986a. Genetic selection for and evaluation of nondiapause lines of predatory midge, *Aphidoletes aphidimyza* (Rondani) (Diptera: Cecidomyiidae). Can. Entomol. 118: 869–879.

Gilkeson, L. A., and Hill, S. B. 1986b. Diapause prevention in *Aphidoletes aphidimyza* (Diptera: Cecidomyiidae) by low-intensity light. Environ. Entomol. 15: 1067–1069.

Gilmore, J. U. 1938. Notes on *Apanteles congregatus* (Say) a parasite of tobacco hornworms. J. Econ. Entomol. 31: 712–715.

Gilyarov, M. S. 1977. The role of small arthropods in the decomposition of DDT. Zash. Rast. 5: 54.

Glass, E. H. 1960. Current status of pesticide resistance in insects and mites attacking orchard crops. Misc. Publ. Entomol. Soc. Amer. 2: 17–25.

Gloria, B. R. 1978. Chemical control of the woolly aphis *Eriosoma lanigerum* (Haus.) on apple. Rev. Peruana Entomol. 21: 118–120.

Godfrey, G. L., and Root, R. B. 1968. Emergence of parasites associated with the cabbage aphid during a chemical-control program. J. Econ. Entomol. 61:1762–1763.

Gong, H., Zhang, J. Z., and Zhai, Q. H. 1982. The synthesis of vitellogenin in *Coccinella septempunctata*. Sinozoologica 2: 175–181.

Gonzalez, R. H. 1981. The red spider mite of apple and pear. Rev. Fruticola 2: 3–9.

Gonzalez, R. H. 1983. La falsa aranita de la vid *Brevipalpus chilensis* (Baker) (Acarina, Tenuipalpidae). Rev. Fruticola 4: 61–65.

Gonzalez, V. M., Caballero Grande, R., and Acevedo, B. 1980. Effects of the sampling method and of chemical control on populations of *Heliothis virescens* and its natural enemy *Diadegma* sp. on tobacco under experimental conditions. Ciencias Agric. 7: 43–49.

Goodman, W., and Gilbert, L. I. 1978. The haemolymph titer of juvenile hormone binding protein and binding sites during the fourth larval instar of *Manduca sexta*. Gen. Comp. Endocrinol. 35: 27–34.

Goodwin, S., and Bowden, P. 1984. Laboratory testing of the side effects of pesticides against predatory mites, pp. 230–237. *In* Proc. 4th Aust. Appl. Entomol. Res. Conf., Adelaide, Sept., 24–28 1984.

Goos, A., and Goos, M. 1979. Experiments for the determination of side-effects of pesticides on parasites. Nachr. Pflschutz. 31(5): 65–69.

Goos, A., Goos, M., and Klein, K. 1974. Versuche zur Ermittlung der Nebenwirkungen von Pflanzenschutzmitteln. Nachr. Pflschutz. 26: 89–93.

Gopinath, P., Sarma, P. V., and Murthy, G. R. K. 1981. Efficacy of certain granular insecticides against paddy stem borer and gall midge. Pesticides 15: 43–44.

Gordon, H. T. 1961. Nutritional factors in insect resistance to chemicals. Annu. Rev. Entomol. 6: 27–54.

Gordon, R., and Cornect, M. 1986. Toxicity of the insect growth regulator diflubenzuron to the rove beetle *Aleochara bilineata*, a parasitoid and predator of the cabbage maggot *Delia radicum*. Entomol. Exp. Appl. 42: 179–185.

Gould, F. 1984. Role of behavior in the evolution of insect adaptation to insecticides and resistant host plants. Bull. Entomol. Soc. Amer. 30: 34–41.

Gould, H., Hussey, N. W., and Parr, W. J. 1969. Large-scale commercial control of *Tetranychus urticae* Koch in cucumbers by the predator *Phytoseiulus persimilis* A.-H., pp. 383–388. *In* Proc. 2nd Intern. Cong. Acarol. 1967.

Grafton-Cardwell, E. E. 1985. Intraspecific variability in response to pesticides in the common green lacewing, *Chrysoperla carnea* (Stephens), and selection for resistance to carbaryl: genetic improvement of a biological control agent. Ph.D. dissertation, Univ. Calif. Berkeley.

Grafton-Cardwell, E. E., and Hoy, M. A. 1983. Comparative toxicity of avermectin B1, to the predator *Metaseiulus occidentalis* (Nesbitt) (Acari: Phytoseiidae) and the spider mites *Tetranychus urticae* Kock and *Panonychus ulmi* (Koch) (Acari: Tetranychidae). J. Econ. Entomol. 762: 1216–1220.

Grafton-Cardwell, E. E., and Hoy, M. A. 1985. Short-term effects of permethrin and fenvalerate on oviposition by *Chrysoperla carnea* (Neuroptera: Chrysopidae). J. Econ. Entomol. 78: 955–959.

Grafton-Cardwell, E. E., and Hoy, M. A. 1986a. Selection of the common green lacewing for resistance to carbaryl. Calif. Agric. 40(9/10): 22–24.

Grafton-Cardwell, E. E., and Hoy, M. A. 1986b. Genetic improvement of the common green lacewing, *Chrysoperla carnea* (Neuroptera: Chrysopidae): selection for carbaryl resistance. Environ. Entomol. 15: 1130–1136.

Grafton-Cardwell, E. E., and Lighthart, B. 1986. Interim protocol for testing the effectiveness of microbial pathogens on the common green lacewing, *Chrysoperla carnea* (Neuroptera: Chrýopidae). Unpub. EPA Rep., US/EPA Corvallis OR, 16 pp.

Granett, J., and Weseloh, R. M. 1975. Dimilin toxicity to the gypsy moth larval parasitoid, *Apanteles melanoscelus*. J. Econ. Entomol. 68: 577–580.

Granett, J., Weseloh, R. M., and Helgert, E. 1975a. Activity of juvenile hormone analogues on hymenopterous parasitoids of the gypsy moth. Entomol. Exp. Appl. 18: 377–383.

Granett, J., Weseloh, R. M., and Dunbar, D. M. 1975b. Dimilin toxicity to *Apanteles melanoscelus* (Ratzeburg) (Hymenoptera: Braconidae) and effects on field populations. New York Entomol. Soc. 83: 242–243.

Granett, J., Dunbar, D. M., and Weseloh, R. M. 1976. Gypsy moth control with Dimilin sprays timed to minimize effects on the parasite *Apanteles melanoscelus*. J. Econ. Entomol. 69: 403–404.

Grapel, H. 1981. Influence of some insecticides upon the efficiency and fertility of *Coccinella septempunctata* L.; *Chrysopa carnea* Steph., and *Syrphus corollae* F. Mitt. Deutsch. Ges. Allg. Angew. Entomol. 3: 304–308.

Grapel, H. 1982. Investigations on the influence of some insecticides on natural enemies of aphids. Z. Pflkrankh. Pflschutz. 89:241–252.

Gratwick, M. 1957. The contamination of insects of different species exposed to dust deposits. Bull. Entomol. Res. 48: 741–753.

Gratwick, M. 1965. Laboratory studies of the relative toxicities of orchard insecticides to predatory insects. Rept. E. Malling. Res. Sta., pp. 171–176.

Graur, D. 1985. Gene diversity in Hymenoptera. Evolution 39: 190–199.

Gravena, S. 1984. Effect of insecticides used to control the coffee leaf-miner *Perileucoptera coffeella* (Guerin-Menevill, 1842) and *Bacillus thuringiensis* Berliner on the population of predacious ants. Anais Soc. Entomol. Brasil 13: 389–390.

Gravena, S., and Batiste, G. C. 1979a. Toxicity of insecticides to *Cycloneda sanguinea* (L.) (Coleoptera, coccinellidae). Cientifica 7: 267–272.

Gravena, S., and Batiste, G. C. 1979b. Selectivity of insecticides to natural enemies of the greenbug *Schizaphis graminum* (Rondani, 1852) (Homoptera, Aphididae) on grain sorghum under field conditions. Anais da Soc. Entomol. do Brasil 8: 335–344.

Gravena, S., Da Rocha, A. D., and Marconato, A. R. 1976. Influence of application methods of insecticides on the populations of predacious coccinellids and on the control of some cotton pests. Cientifica 4: 231–235.

Gray, M. E., and Coasts, J. R. 1983. Effects of an insecticide and a herbicide combination on nontarget arthropods in a cornfield. Environ. Entomol. 12:1171–1174.

Grebennikov, S. V. 1980. Microreserves. Zash. Rast. 10: 20.

Green, E. M. 1917. Note on the immunity of chalcid parasites to hydrocyanic acid gas. Ann. Appl. Biol. 4: 90.

Greene, G. L., Whitcomb, W. H., and Baker, R. 1974. Minimum rates of insecticide on soybeans: *Geocoris* and *Nabis* populations following treatment. Florida Entomol. 57: 114.

Greenhalgh, R., and Roberts, R. R., Eds. 1987. Pesticide Science and Biotechnology. Proc. 6th IUPAC Cong. Pesticide Chemistry, Ottawa, Ont., Canada, Aug. 1986. Blackwell Sci. Publ., Oxford/London.

Gregoire-Wibo, C. 1980. Standardized bioassay tests for detecting aldicarb by means of carabid beetles. Med. Fac. Landbouww. Rijkauniv. Gent 45: 675–690.

Gregoire-Wibo, C. 1983a. Incidences ecologiques des traitements phytosanitaires en culture de betterave sucrier, essais expérimentaux en champ. Pedobiologia 25: 37–48.

Gregoire-Wibo, C. 1983b. Incidences écologiques des traitements phytosanitaires en culture de betterave sucrier. II. Acariens, polydesmes, staphylins, et carabides. Pedobiologia 25: 93–108.

Gregoire-Wibo, C., and van Hoecke, A. 1979. Ecological effect of aldicarb on carabids from sugar beet fields. Med. Fac. Landbouww. Rijksuniv. Gent 44(1): 367–378.

Groner, A., Huber, J., Krieg, A., and Pinsdorf, W. 1978. Tests of two baculovirus preparations on honey bees. Nachr. Pflschutz. 30(3): 39–41.

Grosch, D. S. 1970. Reproductive performance of a braconid after heptachlor poisoning. J. Econ. Entomol. 63:1348–1349.

Grosch, D. S. 1972. The response of female arthropods' reproductive systems to radiation and chemical agents, pp. 217–227. *In* Proc. Symp. Sterility Principle for Insect Control or Eradication, IAEA/FAO, Athens, Greece, Sept. 14–18, 1972, 542 pp.

Grosch, D. S. 1975. Reproductive performance of *Bracon hebetor* after sublethal doses of carbaryl. J. Econ. Entomol. 68:659–662.

Grosch, D. S., and Hoffman, A. C. 1973. The vulnerability of specific cells in the oogenetic sequence of *Bracon hebetor* to some degradation products of carbamate pesticides. Environ. Entomol. 2:1029–1032.

Grosch, D. S., and Valcovic, L. R. 1967. Chlorinated hydrocarbon insecticides are not mutagenic in *Bracon hebetor* tests. J. Econ. Entomol. 60:1177–1179.

Gross, H. R., and Schuster, M. F. 1969. Responses of striped earwigs following applications of heptachlor and mirex and predator–prey relationships between imported fire ants and striped earwigs. J. Econ. Entomol. 62: 686–689.

Gruys, P. 1980a. Development of an integrated control program for orchards. Integrated Control of Insect Pests in Netherlands, pp. 5–10. PUDOC, Wageningen Neth.

Gruys, P. 1980b. Significance and practical application of selective pesticides, pp. 8–12. *In* Proc. Intern. Symp. on Integrated Control in Agriculture and Forestry, Vienna, Austria, 1979.

Gruys, P. 1980c. Solved and unsolved problems of integrated control in apple orchards, illustrated by examples from the Netherlands. pp. 359–364. *In* Proc. Intern. Symp. on Integrated Control in Agriculture and Forestry, Vienna, Austria, 1979.

Gruys, P. 1982. Hits and misses. The ecological approach to pest control in orchards. Entomol. Exp. Appl. 31: 70–87.

Guerra, A. A., Wolfenbarger, D. A., Lingren, P. D., and Garcia, R. D. 1977. Five experimental insect growth regulators: effect on populations of tobacco budworm and *Trichogramma* sp. in field cages. Ann. Entomol. Soc. Amer. 70: 771–774.

Gupta, M., Sultana, N., and Pawar, A. D. 1984. Toxicity of some insecticidal sprays on the egg parasitoid, *Trichogramma japonicum* Ashm. Plant Protec. Bull. India 36(1): 21–24.

Gusev, G. V., and Voroshilov, N. V. 1977. Selection of natural enemies. Zash. Rast. 6: 31.

Gusev, G. V., Svikle, M. J., Koval, Y. V., Zayats, Y. V., Lakhidov, A. I., and Sorokin, N. S. 1975. Prospects for using Colorado potato beetle entomophages in different geotraphical zones of the USSR. All-Union Plant Protec. Inst., Leningrad, USSR, pp. 34–38. *In* VIII International Plant Protec. Congr., Moscow, 1975, vol.3.

Gut, L. J. 1985. Arthropod community organization and development in pear. Ph.D. thesis, Oregon State Univ., Corvallis.

Gut, L. J., Jochums, C. E., Westigard, P. H., and Liss, W. J. 1982. Variations in pear psylla (*Psylla pyricola*) densities in southern Oregon orchards and its implications. Acta Hortic. 124: 101–111.

Guthrie, F. E., Rabb, R. L., Bowery, T. G., Lawson, F. R., and Baron, R. L. 1956. Control of hornworms and budworms on tobacco with reduced insecticide dosage. Tobacco Sci. 3: 65–68.

Habib, M. E. M., and Amaral, M. E. C. 1985. Aerial application of *Bacillus thuringiensis* against the velvetbean caterpillar, *Anticarsia gemmatalis* Huebner, in soybean fields. Rev. de Agric., Brazil, 60: 141–149.

Hadam, J. J., AliNiazee, M. T., and Croft, B. A. 1986. Phytoseiid mites of major crops in the Willamette Valley, OR, and pesticide resistance in *Typhlodromus pyri* Schulten. Environ. Entomol. 15: 1255–1263.

Hagen, K. S., and van den Bosch, R. 1968. Impact of pathogens, parasites and predators on aphids. Annu. Rev. Entomol. 13: 329–384.

Hagen, K. S., Sawall, E. F., and Tasson, R. L. 1970. The use of food sprays to increase effectiveness of entomophagous insects. Proc. Tall Timbers Conf. Ecol. Animal Contr. Habitat Mgmt. 1: 59–81.

Hagley, E. A. C., and Simpson, C. M. 1983. Effect of insecticides on predators of the pear psylla, *Psylla pyricola* (Hemiptera: Psyllidae), in Ontario. Can. Entomol. 115: 1409–1414.

Hagley, E. A. C., Trottier, R., Herne, D. H. C., Hikitchi, A., and Maitland, A. 1978. Pest management in Ontario orchards. Calif. Dept. Agric. Publ. A52–50, 23 pp.

Hagley, E. A. C., Pree, D. J., and Holliday, N. J. 1980. Toxicity of insecticides to some orchard carabids (Coleoptera: Carabidae). Can. Entomol. 112: 457–462.

Hagley, E. A. C., Pree, D. J., Simpson, C. M., and Hikichi, A. 1981. Toxicity of insecticides to parasites of the spotted tentiform leafminer (Lepidoptera: Gracillariidae). Can. Entomol. 113: 899–906.

Hagstrum, D. W. 1970. Laboratory studies on the effect of several insecticides on *Tarentula kochi*. J. Econ. Entomol. 63: 1844–1847.

Haidari, H. S., and Georgis, R. 1978. The toxicity of some pesticides to the woolly apple aphid parasite, *Aphelinus mali*. PANS 24: 109–110.

Hain, F. P., and Wallner, W. E. 1974. Control of *Exoteleia nepheos* with insecticides and their effect upon its parasitoids. J. Econ. Entomol. 67: 803.

Hall, F. 1986. Improved agrochemical and fertilizer application technology. ARI Newsl. spring issue, pp. 1–4.

Hall, F. R. 1987. Parameters governing dose transfer, pp. 259–264. *In* Pesticide Science and Biotechnology. Eds. R. Greenhalgh and T. R. Roberts, Blackwell Sci. Publ., Oxford/London, 604 pp.

Hall, R. A. 1982. Control of whitefly, *Trialeurodes vaporariorum*, and cotton aphid, *Aphis gossypii*, in glasshouses by two isolates of the fungus, *Verticillium lecanii*. Ann. Appl. Biol. 101: 1–11.

Hamed, A. R. 1979. Effects of *Bacillus thuringiensis* on parasites and predators of *Yponomeuta evonymellus* Lep., Yponomeutidae. Z. Angew. Entomol. 87: 294–311.

Hamel, D. R. 1977. The effects of *Bacillus thuringiensis* on parasitoids of the western spruce budworm, *Choristoneura occidentalis* (Lepidoptera: Tortricidae), spruce coneworm, *Dioryctria reniculelloides* (Lepidoptera: Pyralidae), in Montana. Can. Entomol. 109: 1409–1415.

Hamilton, E. W., and Kieckhefer, R. W. 1968. Integrated control of cereal aphids tests with malathion and parathion, pp.158–160. *In Proc. N. Central Branch Amer. Assoc. Econ. Entomol.*, vol.23.

Hamilton, E. W., and Kieckhefer, R. W. 1969. Toxicity of malathion and parathion to predators of the English grain aphid. J. Econ. Entomol. 62: 1190–1192.

Hamilton, J. T., and Attia, F. I. 1976. The susceptibility of the parasite *Apanteles glomeratus* (L.) (Hym.: Braconidae) to insecticides. J. Entomol. Soc. Aust. 9: 24–25.

Hamilton, J. T., and Attia, F. I. 1977. Effects of mixtures of *Bacillus thuringiensis* and pesticides on *Plutella xylostella* and the parasite *Thraella* collaris. J. Econ. Entomol. 70: 146–148.

Hamlen, R. A. 1975. Survival of hemispherical scale and an *Encyrtus* parasitoid after treatment with insect growth regulators and insecticides. Environ. Entomol. 4: 972–974.

Hamm, J. J., Nordlung, D. A., and Mullinix, B. G. 1983. Interaction of the microsporidium *Vairimorpha* sp. with *Microplitis croceipes* (Cresson) and *Cotesia marginiventris* (Cresson) (Hymenoptera: Braconidae), two parasitoids of *Heliothis zea* (Boddie) (Lepidoptera: Noctuidae). Environ. Entomol. 12: 1547–1550.

Hamm, J. J., Nordlung, D. A., and Marti, O. G. 1985. Effects of a nonoccluded virus of *Spodoptera frugiperda* (Lepidoptera: Noctuidae) on the development of a parasitoid, *Cotesia marginiventris* (Hymenoptera: Braconidae). Environ. Entomol. 14: 258–261.

Hammock, B. D., and Quistad, 1981. Metabolism and mode of action of juvenile hormone, juvenoids and other insect growth regulators, pp. 1–83. *In* Progress in Pesticide Biochemistry, Eds. D. H. Hutson and T. R. Roberts, Wiley, New York, vol. 1.

Hammock, B. D., Wing, K. D., McLaughlin, J., Lowell, V. M., and Sparks, T. C. 1982. Trifluoromethylketones as possible transition state analog inhibitors of juvenile hormone esterase. Pestic. Biochem. Physiol. 17: 76–88.

Hammock, B. D., Abdel-aal, Y. A. I., Mullin, C. A., Hanzlik, T. N., and Roe, R. M. 1984. Substituted thiotrifluoropropanines as potent selective inhibitors of juvenile hormone esterase. Pestic. Biochem. Physiol. 22: 209–223.

Hare, J. D., Logan, P. A., and Wright, R. J. 1983. Suppression of Colorado potato beetle, *Leptinotarsus decemlineata* (Say), (Coleoptera: Chrysomelidae) populations with antifeedant fungicides. Environ. Entomol. 12: 1470–1477.

Harper, A. M. 1978. Effect of insecticides on the pea aphid, *Acyrthosiphon pisum*, (Hemiptera: Aphididae), and associated fauna on alfalfa. Can. Entomol. 110: 891–894.

Harris, C. R., and Kinoshita, G. B. 1977. Influence of post-treatment temperature on the toxicity of pyrethroid insecticides. J. Econ. Entomol. 70: 215–218.

Hartl, D. L. 1971. Some'aspects of natural selection in arrhenotokous populations. Amer. Zool. 11: 309–325.

Hartley, G. S., and Graham-Bryce, I. J. 1980. Physical Principles of Pesticide Behavior. Academic Press, New York, vol. 1, pp. 1–518; vol. 2, pp. 519–1024.

Hassan, S. A. 1969. Observations on the effect of insecticides on Coleopterous predators of *Erioischia brassicae* (Diptera: Anthomyiidae). Entomol. Exp. Appl. 12: 157–168.

Hassan, S. A. 1973. The effects of insecticides on *Trybliographa rapae* West. (Hymenoptera: Cynipidae), a parasite of the cabbage root fly *Hylemya brassicae* (Bouche). Z. Angew. Entomol. 73: 93–102.

Hassan, S. A. 1974. Eine Methode zur Prüfung der Einwirkung von Pflanzenschutzmitteln auf Eiparasiten der Gattung *Trichogramma* (Hymenoptera: Trichogrammatidae). Ergebnisse einer Versuchsreihe mit Fungiziden. Z. Angew. Entomol. 76: 120–134.

Hassan, S. A. 1977. Standardized techniques for testing side-effects of pesticides on beneficial arthropods in the laboratory. Z. Pflkrankh. Pflschutz. 84: 158–163.

Hassan, S. A. 1980. A standard laboratory method to test the duration of harmful effects of pesticides on egg parasites of the genus *Trichogramma* (Hymenoptera, Trichogrammatidae). Z. Angew. Entomol. 89: 282–289.

Hassan, S. A. 1982a. Relative tolerance of three different strains of the predatory mite *Phytoseiulus persimilis* A.-H. (Acari, Phytoseiidae) to 11 pesticides used on glasshouse crops. Z. Angew. Entomol. 93: 55–63.

Hassan, S. A. 1982b. Comparison of the use of three different lines of *Phytoseiulus persimilis* in the control of *Tetranychus urticae* on cucumbers under glass. Z. Angew. Entomol. 93: 131–140.

Hassan, S. A. 1985. Standard methods to test the side-effects of pesticides on natural enemies of insects and mites developed by the IOBC/WPRS Working Group "Pesticides and Beneficial Organisms". Bull. OEPP/EPPO 15: 214–255.

Hassan, S. A., and Groner, A. 1977. The effect of the nuclear polyhedrosis virus (*Baculovirus* sp.) of *Mamestra brassicae* on *Trichogramma cacoeciae* (Hym., Trichogrammatidae) and *Chrysopa carnea* (Neur., Chrysopidae). Entomophaga 22: 291–288.

Hassan, S. A., and Krieg, A. 1975. *Bacillus thuringiensis* preparations harmless to the parasite *Trichogramma cacoeciae* (Hym., Trichogrammatidae). Z. Pflkrankh. Pflschutz. 82: 515–521.

Hassan, S. A., Bigler, F., Bogenschutz, H., Brown, J. U., Firth, S. I., et al. 1983. Results of the second joint pesticide testing programme by the IOBC/WPRS Working group "Pesticides and Beneficial Arthropods." Z. Angew. Entomol. 95: 151–158.

Hassan, S. A., Klingauf, F., and Shahin, F. 1985. Role of *Chrysopa carnea* as an aphid predator on sugar beet and the effect of pesticides. Z. Angew. Entomol. 100: 163–174.

Hassan, S. A., Albert, R., Bigler, F., Blaisinger, P., Bogenschutz, G., et al. 1987. Results of the third joint pesticide testing programme by the IOBC/WPRS Working Group "Pesticides and Beneficial Organisms." J. Appl. Entomol. 103; 92–107.

Hassanein, M. H., and Khalil, F. 1968. Effect of insecticides on predators of the cotton leaf worm. Bull. Entomol. Soc. Egypt 2: 247–264.

Hassell, M. P. 1978. The Dynamics of Arthropod Predator–Prey Systems. Princeton Univ. Press, Princeton, N.J., 237 pp.

Hassell, M. P. 1984. Insecticides in host-parasitoid interactions. Theor. Pop. Biol. 26: 378–386.

Hassell, M. P., and Waage, J. K. 1984. Host–parasitoid population interactions. Ann. Rev. Entomol. 24: 89–114.

Hassen, S. M., Zohdy, G. I., El-Badry, E. A., and Abo-Elghar, M. R. 1970. The effect of certain acaricides on *Agistemus exsertus* Gonzales (Acarina: Stigmaeidae). Bull. Entomol. Soc. Egypt, Econ. Ser. 4: 213–217.

Hatalane, Z. I., and Budai, C. 1982. Pesticide susceptibility of *Encarsia formosa* Gahan. Novenyvedelem 18: 25–26.

Hatzios, K. K., and Penner, D. 1982. Metabolism of Herbicides in Higher Plants. Cepco/Burgess Publ., Minneapolis, Minn, 142 pp.

Haug, G. W., and Peterson, A. 1938. The effect of insecticides on a beneficial coccinellid, *Hippodamia convergens* Guer. J. Econ. Entomol. 31: 87–92.

Haverty, M. I. 1982. Sensitivity of selected nontarget insects to the carrier of Dipel 4L in the laboratory Environ. Entomol. 11: 337–338.

Havron, A. 1983. Studies toward selection of *Aphytis* wasps for pesticide resistance. Ph.D. thesis. Hebrew Univ., Jerusalem, Israel.

Havron, A., and Rosen, D. 1984. Screening for pesticide resistance in *Aphytis.*, 10 pp. *In* Proc. 17th Intern. Cong. Entomol., Hamburg, Germany, 1984.

Havron, A., and Rosen, D. 1988. Selection for pesticide resistance in two *Aphytis* species, p. 315d. *In* Proc. 18th Intern. Cong. Entomol., Vancouver, B.C., Canada, July 3–9, 1988.

Havron, A., Rosen, D., and Rossler, Y. 1987a. A test method for pesticide tolerance in minute parasitic Hymenoptera. Entomophaga 32: 83–95.

Havron, A., Rosen, D., and Rossler, Y. 1987b. Selection on the male hemizygous genotype in arrhenotokous species. Entomophaga 32: 261–268.

Hawlitzky, N., and Boulag, C. 1974. Identification of the diets of the larva of an entomophagous insect *Phanerotoma flavitestacea* (Hym.: Braconidae). Entomophaga 19: 395–408.

Haynes, K. F. 1988. Sublethal effects of neurotoxic substances on the behavioral responses of insects. Annu. Rev. Entomol. 33: 149–168.

Haynes, K. F., Miller, T. A., Staten, R. T., Li, W. G., and Baker, T. C. 1986. Monitoring insecticide resistance with insect pheromones. Experientia 42: 1293–1295.

Head, R., Neel, W. W., Sartor, C. R., and Chambers, H. 1977. Methyl parathion and carbaryl resistance in *Chrysomela scripta* and *Coleomegilla maculata.* Bull. Environ. Contam. Tox. 17: 163–164.

Headley, J. C., and Hoy, M. A. 1986. The economics of integrated mite management in almonds. Calif. Agric. 40(1/2): 28–30.

Headley, J. C., and Hoy, M. A. 1987. Benefit/cost analysis of an integrated mite management program for almonds. J. Econ. Entomol. 80: 555–559.

Hegazi, E. M., Rawash, I. A., El-Gayar, F. H., and Kares, E. A. 1982. Effect of parasitism by *Microplitis rufiventris* Kok. on the susceptibility of *Spodoptera littoralis* (Boisd.) larvae to insecticides. Acta Phytopathol. Acad. Sci. Hungar. 17: 115–121.

Heinrichs, E. A., Basilio, R. P., and Valencia, S. 1984. Buprofezin, a selective insecticide for the management of rice planthoppers (Homoptera: Delphacidae) and leafhoppers (Homoptera: Cicadellidae). Environ. Entomol. 13: 515–521.

Helgesen, R. G., and Tauber, M. J. 1974. Pirimicarb, an aphicide nontoxic to three entomophagous arthropods. Environ. Entomol. 3: 99–104.

Helle, W. 1965. Resistance in Acarina: Mites. Advances in Acarol. 3: 71–93.

Helle, W., and Sabelis, M. W., Eds. 1986. Spider Mites. Their Biology, Natural Enemies and Control. Elsevier, Amsterdam, vol. 1B, 458 pp.

Helle, W., and van de Vrie, M. 1975. Problems with spider mites. Outlook on Agric. 8: 119.

Helle, W., Bolland, H. R., Arendonk, R. van, deBoer, R., Schulten, G. G. M., and Russel, V. M. 1978.

Genetic evidence for biparental males in haplo-diploid predatory mites (Acarina: Phytoseiidae). Genetica 49: 165–171.

Hellpap, C. 1982. Investigations of the effect of different insecticides on predators of cereal aphids under field conditions. Anz. Schädl., Pflschutz., Umweltschutz 55(9): 129–131.

Hellpap, C., and Schmutterer, H. 1982. Studies on the effect of lower concentrations of Pirimor on pea aphids *Acyrthosiphon pisum* (Harr.) and natural enemies. Z. Angew. Entomol. 94: 246–252.

Helyer, N. L. 1982. Laboratory screening of pesticides for use in integrated control programmes with *Phytoseiulus persimilis* (Acarina Phytoseiidae). Ann. Appl. Biol. 100(suppl): 64–65.

Henrick, C. A. 1982. Juvenile hormone analogs: Structure–activity relationships, pp. 315–402. *In* Insecticide Mode of Action. Ed. J. R. Coats, Academic Press, New York.

Herard, F. 1986. Annotated list of the entomophagous complex associated with pear psylla, *Psylla pyri*, in France. Agronomie 6: 1–34.

Herbert, D. A., and Harper, J. D. 1986. Bioassays of a beta-exotoxin of *Bacillus thuringiensis* against *Geocoris punctipes* (Hemiptera: Lygaeidae). J. Econ. Entomol. 79: 592–595.

Herne, D. H. C. 1963. Carabids collected in a DDT-sprayed peach orchard in Ontario (Coleoptera: Carabidae). Can. Entomol. 95: 357–362.

Herne, D. C., and Chant, D. A. 1965. Relative toxicity of parathion and kelthane to the predacious mite *Phytoseiulus persimilis* Athias-Henriot and its prey, *Tetranychus urticae* Koch, (Acarina: Phytoseiidae, Tetranychidae) in the lab. Can. Entomol. 97: 172–176.

Herne, D. H. C., Cranham, J. E., and Easterbrook, M. A. 1979. New acaricides to control resistant mites, pp. 95–104. *In* Recent Adv. Acarol. Academic Press, New York, vol. 1, 631 pp.

Herschbarger, J. C., and Forgash, A. J. 1964. Effect of lindane on the intracellular microorganisms of the American cockroach. J. Econ. Entomol. 57: 994–995.

Heymann, E. 1980. Carboxylesterases and amidases, pp. 291–323. *In* Enzymatic Basis of Detoxification. Ed. W. B. Jakoby, Academic Press, New York, vol. 2.

Heynen, C. 1985. Studies on the effect of diflubenzuron (Dimilin) on the host–parasite system *Spodoptera littoralis* Boisd. (Lep., Noctuidae)/*Microplitis rufiventris* Kok. (Hym., Braconidae). Z. Angew. Entomol. 100: 113–132.

Hislop, R. G., and Prokopy, R. J. 1979. Integrated management of phytophagous mites in Massachusetts (USA) apple orchards. 1. Foliage-inhabiting mite complexes in commercial and abandoned orchards. Prot. Ecol. 1: 279–290.

Hislop, R. G., and Prokopy, R. J. 1981. Integrated management of phytophagous mites in Massachusetts (USA) apple orchards. 2. Influences of pesticides on the mite predator *Amblyseius fallacis* under laboratory and field conditions. Prot. Ecol. 3: 157–172.

Hislop, R. G., Acker, A., Alves, N., and Prokopy, R. J. 1978. Laboratory toxicity of pesticides and growth regulators to *Amblyseius fallacis*, an important spider mite predator in Massachusetts apple orchards. Massachusetts Fruit Notes 43(5): 14–18.

Hislop, R. G., Auditore, P. J., Weeks, B. L., and Prokopy, R. J. 1981. Repellency of pesticides to the predatory mite *Amblyseius fallacis*. Protection Ecol. 3: 253–257.

Hoda, F. M., and Hafez, M. A. A. 1981. The effect of some acaricides on the population density of predacious insects and mites inhabiting cotton plants. Agric. Res. Rev. 59: 15–26.

Hodek, I. 1973. Biology of Coccinellidae. Academic Press, Prague, 260 pp.

Hodjat, S. H. 1971. Effects of sublethal doses of insecticides and of diet and crowding on *Dysdercus fasciatus* Sign. Bull. Entomol. Res. 60: 367–378.

Hofer, H. 1979. Pirimor, a selective compound for the control of aphids. Mitt. Schweiz. Landwirtsch. 27: 35–36.

Hoffman, A. C., and Grosch, D. S. 1971. The effects of ethyl methane sulfonate on the fecundity and fertility of *Bracon* (*Habrobracon*) females. 1. The influences of route of entry and physiological state. Pestic. Biochem. Physiol. 1: 319–326.

Holdsworth, R. P. 1968. Integrated control: Effect on European red mite and its more important predators. J. Econ. Entomol. 61: 1602–1607.

Holdsworth, R. P. 1974. Integrated control of apple pest insects in Ohio. In 1974 commercial fruit spray recommendations for Ohio. Bull. Ohio State Univ. Ext. Serv. 506: 20–26.

Holling, C. S. 1959. Some characteristics of simple types of predation and parasitization. Can. Entomol. 91: 386–398.

Hollingworth, R. M. 1975. Strategies in the design of selective insect toxicants, pp. 67–111. *In* Pesticide Selectivity. Ed. J. C. Street, Marcel Dekker, New York.

Hollingworth, R. M. 1976. The biochemical and physiological basis of selective toxicity, pp. 431–506. *In* Insecticide Biochemistry and Physiology. Ed. C. F. Wilkinson, Plenum Press, New York.

Holt, J. 1987. Simulation analysis of brown planthopper population dynamics on rice in the Philippines. J. Appl. Ecol. 24: 87–102.

Hoogcarspel, A. P., and Jobsen, J. A. 1984. Laboratory method for testing side-effects of pesticides on *Encarsia formosa* Gahan (Hym., Aphelinidae). Z. Angew. Entomol. 97: 268–78.

Horn, D. J. 1983. Selective mortality of parasitoids and predators of *Myzus persicae* on collards treated with malathion, carbaryl or *Bacillus thuringiensis*. Entomol. Exp. Appl. 34: 208–211.

Horton, D. L., Teague, T. G., Phillips, J. R., and Yearian, W. C. 1986. Fungicide interference with parasitization of *Heliothis zea* by *Microplitis croceipes*. J. Agric. Entomol. 3: 186–191.

Hotchkin, P. G., and Kaya, H. K. 1983a. Interactions between two baculoviruses and several insect parasites. Can. Entomol. 115: 841–846.

Hotchkin, P. G., and Kaya, H. K. 1983b. Pathological response of the parasitoid, *Glyptapanteles militaris*, to nuclear polyhedrosis virus-infected armyworm hosts. J. Invert. Pathol. 42: 51–61.

Hotchkin, P. G., and Kaya, H. K. 1985. Isolation of an agent affecting the development of an internal parasitoid. Ar. Ins. Biochem. Biophys. 2: 375–384.

Hough, W. S. 1928. Relative resistance to arsenical poisoning of two codling moth strains. J. Econ. Entomol. 21: 325–329.

House, G. J., All, J. N., Short, K. T., and Lay, S. E. 1983. Impact of synthetic pyrethroids using three spray application methods on beneficial arthropods in cotton. J. Agric. Entomol. 2: 161–166.

House, H. L. 1977. Nutrition of natural enemies, pp. 151–182. *In* Biological Control by Augmentation of Natural Enemies. Eds. R. L. Ridgway and S. B. Vinson, Plenum Press, New York.

Hower, A. A., and Davis, G. A. 1984. Selectivity of insecticides that kill the potato leafhopper (Homoptera: Cicadellidae) and alfalfa weevil (Coleoptera: Curculionidae) and protect the parasite *Microctonus aethiopoides* Loan (Hymenoptera: Braconidae). J. Econ. Entomol. 77: 1601–1607.

Hower, A. A., and Luke, J. 1979. Impact of methyl parathion spray program on the alfalfa weevil parasite, *Bathyplectes curculionis*, in Pennsylvania. Environ. Entomol. 8: 344–348.

Hoy, J. B., and Dahlsten, D. L. 1984. Effects of malathion and Staley's bait on the behavior and survival of parasitic Hymenoptera. Environ. Entomol. 13: 1483–1486.

Hoy, M. A. 1975. Effect of temperature and photoperiod on the induction of diapause in the mite *Metaseiulus occidentalis*. J. Insect Physiol. 21: 605–611.

Hoy, M. A. 1976. Genetic improvement of insects: fact or fantasy. Environ. Entomol. 5: 833–839.

Hoy, M. A. 1979. The potential for genetic improvement of predators for pest management programs, pp. 106–115. *In* Genetics in Relation to Insect Management. Eds. M. A. Hoy and J. J. McKelvey, Rockefeller Foundation, New York.

Hoy, M. A. 1982. Genetics and genetic improvement of the Phytoseiidae. *In* Recent Advances in Knowledge of the Phytoseiidae. Proc. Acarol. Soc. Amer., 1981, pp. 72–89. Ed. M. A. Hoy, Univ. Calif. Paid Publ. 3284, 92 pp.

Hoy, M. A. 1984a. Pesticide-resistant natural enemies in IPM systems. p. 832 (Abstr.). *In* Proc. 17th Intern. Cong. Entomol., Hamburg, Germany, 1984.

Hoy, M. A. 1984b. Genetic improvement of a biological control agent: multiple pesticide resistances and nondiapause in *Metaseiulus occidentalis* Nesbitt (Phytoseiidae), pp. 673–679. *In* Proc. 6th Intern. Cong. Acarol., Edinburgh, UK, 1982, Acarology VI, vol. 2. Eds. D. A. Griffiths and C. C. Bowman, Ellis Horwood Ltd., Halsted Press, New York.

Hoy, M. A. 1985. Recent advances in genetics and genetic improvement of the Phytoseiidae. Annu. Rev. Entomol. 30: 345–370.

Hoy, M. A. 1987. Developing insecticide resistance in insect and mite predators and opportunities for gene transfer, pp. 125–138. *In* Biotechnology in Agricultural Chemistry. Amer. Chem. Soc. Ser. 334.

Hoy, M. A., and Cave, F. E. 1986. Screening for thelytoky in the parahaploid phytoseiid, *Metaseiulus occidentalis* (Nesbitt). Exp. Appl. Acarol. 2: 273–276.

Hoy, M. A., and Cave, F. E. 1988. Guthion-resistant strain of walnut aphid parasite. Calif. Agric. 42(4): 4–5.

Hoy, M. A., and Flaherty, D. L. 1970. Photoperiodic induction of diapause in a predacious mite, *Metaseiulus occidentalis*. Ann. Entomol. Soc. Am. 63: 960–963.

Hoy, M. A., and Herzog, D. C., Eds. 1985. Biological Control in Agricultural IPM Systems. Academic Press, Orlando, 589 pp.

Hoy, M. A., and Knop, N. F. 1981. Selection for genetic analysis of permethrin resistance in *Metaseiulus occidentalis*: genetic improvement of a biological control agent. Entomol. Exp. Appl. 30: 10–18.

Hoy, M. A., and Lighthart, B. 1986. Interim protocol for testing the effects of microbial pathogens on parasitoid insects. Unpubl. EPA Rep., US/EPA, Corvallis, Ore.

Hoy, M. A., and Ouyang, Y. L. 1986. Selectivity of the acaricides clofentezine and hexythiazox to the predator *Metaseiulus occidentalis* (Acari: Phytoseiidae). J. Econ. Entomol. 79: 1377–1380.

Hoy, M. A., and Ouyang, Y. L. 1987. Toxicity of the beta-endotoxin of *Bacillus thuringiensis* to *Tetranychus pacificus* and *Metaseiulus occidentalis* (Acari: Tetranychidae and Phytoseiidae). J. Econ. Entomol. 80: 507–511.

Hoy, M. A., and Standow, K. A. 1981. Resistance to sulfur in a vineyard spider mite predator. Calif. Agric. 35: 8–10.

Hoy, M. A., and Standow, K. A. 1982. Inheritance of resistance to sulfur in the spider mite predator *Metaseiulus occidentalis*. Entomol. Exp. Appl. 31: 316–323.

Hoy, M. A., Roush, R. T., Smith, K. B., and Barclay, L. W. 1979. Spider mites and predators in San Joaquin Valley almond orchards. Calif. Agric. 33(10): 11–13.

Hoy, M. A., Knop, N. F., and Joos, J. L. 1980. Pyrethroid resistance persists in spider mite predator. Calif. Agric. 34: 11–12.

Hoy, M. A., Westigard, P. H., and Hoyt, S. C. 1983. Release and evaluation of laboratory selected, pyrethroid resistant strains of the predatory mite, *Typhlodromus occidentalis* (Acarina: Phytoseiidae) into southern Oregon pear orchards and Washington apple orchards. J. Econ. Entomol. 76: 383–388.

Hoy, M. A., van de, Bann, H. E. Groot, J. J. R., and Field, R. P. 1984. Aerial movement of mites in almonds: implications for pest management. Calif. Agric. 38(9): 21–23.

Hoy, M. A., Groot, J. J. R., and van de Bann, H. E. 1985. Influence of aerial dispersal on persistence and spread of pesticide-resistant *Metaseiulus occidentalis* in California almond orchards. Entomol. Exp. Appl. 37: 17–31.

Hoying, S. A., and Croft, B. A. 1977. Comparisons between populations of *Typhlodromus longipilus* Newbitt and *T. occidentalis* Newbitt: Taxonomy, distribution and hybridization. Ann. Entomol. Soc. Amer. 70: 150–159.

Hoyt, S. C. 1969a. Integrated chemical control of insects and biological control of mites on apple in Washington. J. Econ. Entomol. 62: 74–86.

Hoyt, S. C. 1969b. Population studies of five mite species on apple in Washington, pp. 117–133. *In* Proc. 2nd Intern. Cong. Acarol. Sutton Bonington, 1967.

Hoyt, S. C. 1972. Resistance to azinphosmethyl of *Typhlodromus pyri* (Acarina: Phytoseiidae) from New Zealand. New Zealand J. Sci. 15: 16–21.

Hoyt, S. C. 1973. Studies on integrated control of *Panonychus ulmi* in New Zealand apple orchards. New Zealand J. Exp. Agric. 1: 77–80.

Hoyt, S. C. 1982. Summary and recommendations for future research and implementation, ch. 5, pp. 90–92. *In* Recent Advances in Knowledge of the Phytoseiidae. Ed. M. A. Hoy, Univ. Calif. Publ. 3284, 92 pp.

Hoyt, S. C., and Caltagirone, L. E. 1971. The developing programs of integrated control of pests of apples in Washington and peaches in California, pp. 395–421, *In* Biological Control. Ed. C. B. Huffaker. Plenum, New York.

Hoyt, S. C., and Kinney, J. R. 1964. Field evaluation of acaricides for the control of the McDaniel spider mite. Wash. St. Res. Cir. 439, 13 pp.

Hoyt, S. C., and Retan, A. H. 1967. The appearance and location of the mite predator *Typhlodromus occidentalis* on apple trees. Wash. State Ext. Bull. 2744. 3 pp.

Hoyt, S. C., and Simpson, R. G. 1979. Economic thresholds and economic injury levels. *In* Biological Control and Insect Pest Management. Eds. D. W. Davis, S. C. Hoyt, J. A. McMurtry, and M. T. AliNiazee, Univ. of Calif. Priced Publ. 4096, 102 pp.

Hoyt, S. C., and Tanigoshi, L. K. 1983. Economic injury levels for apple insect and mite pests, pp. 203–218. *In* Integrated Management of Insect Pests of Pome and Stone Fruits, Eds. B. A. Croft and S. C. Hoyt, Wiley Interscience, New York, 454 pp.

Hoyt, S. C., Westigard, P. H., and Burts, E. C. 1978. Effects of two synthetic pyrethroids on the codling moth, pear psylla, and various mite species in northwest apple and pear orchards. J. Econ. Entomol. 71: 431–434.

Hoyt, S. C., Westigard, P. H., and Croft, B. A. 1985. Cyhexatin resistance in Oregon populations of *Tetranychus urticae*. J. Econ. Entomol. 78: 656–659.

Hsieh, C. Y. 1984. Effects of insecticides on *Diaeretiella rapae* (M'Intosh) with emphasis on bioassay techniques for aphid parasitoids. Ph.D. dissertation, Univ. Calif. Berkley, 69 pp.

Hsieh, C. Y., and Allen, W. W. 1986. Effects of insecticides on emergence, survival, longevity and fecundity of the parasitoid *Diaeretiella rapae* (Hymenoptera: Aphidiidae) from mummified *Myzus persicae* (Homoptera: Aphididae). J. Econ. Entomol. 79: 1599–1602.

Hsu, E. L., and Hsu, S. J. 1980. Studies of metabolism and fate of organochlorinated insecticides DDT, aldrin, dieldrin and y-HCH in the rice paddy model ecosystem. NTU Phytopathol. Entomol. 7: 44–65.

Hu, F. G. 1981. The efficacy of sumicidin in controlling the principal cotton pests. Zhiwu Baohu 7(3): 24.

Huang, M. D., Xiong, J. J., and Du, T. Y. 1987. The selection for and genetical analysis of phosmet resistance in *Amblyseius nicholsi*. Acta Entomol. Sinica 30: 133–139.

Huang, P. 1981. Zur Laborzucht von *Pales pavida* Meig. (Dipt., Tachinidae) am Ersatzwirt *Galleria mellonella L.* (Lep., Galleriidae). Z. Pflkrankh. Pflschutz. 88: 177–188.

Huckaba, R. M., Bradley, J. R., and van Duyn, J. W. 1983. Effects of herbicidal applications of toxaphene on the soybean thrips, certain predators and corn earworm in soybeans. J. Georgia Entomol. Soc. 18: 200–207.

Hudson, W. B., Rushmore, F. A., and Johnson, D. E. 1974. Mite counting in apple pest management. Wash. State Ext. Bull. 3886, 3 pp.

Huffaker, C. B. 1971. The ecology of pesticide interference with insect populations, pp. 92–107. *In* Agricultural Chemicals–Harmony or Discord for Food, People and the Environment. Ed. J. E. Swift, Univ. Calif. Div. Agric. Sci. Publ., 151 pp.

Huffaker, C. B., and Kennett, C. E. 1953. Differential tolerance to parathion in two *Typhlodromus* predators on cyclamen mite. J. Econ. Entomol. 46: 707–708.

Huffaker, C. B., and Kennett, C. E. 1956. Experimental studies on predation: predation and cyclamen mite populations on strawberries in California. Hilgardia 37: 283–335.

Huffaker, C. B., and Kennett, C. E. 1976. Studies of two parasites of olive scale, *Parlatoria oleae* (Colvee). IV. Biological control through the compensatory action of two introduced parasites. Hilgardia 37: 323–335.

Huffaker, C. B., and Kennett, C. E. 1969. Some aspects of assessing efficiency of natural enemies. Can. Entomol. 101: 425–447.

Huffaker, C. B., and Messenger, P. S., Eds. 1976. Theory and Practice of Biological Control. Academic Press, New York, 788 pp.

Huffaker, C. B., and Rabb, R. L. 1984. Ecological Entomology. Wiley Interscience, New York, 844 pp.

Huffaker, C. B., van de Vrie, M., and McMurtry, J. A. 1970. Ecology of tetranychid mites and their natural enemies; tetranychid populations and their possible control by predators: an evaluation. Hilgardia 40: 391–458.

Huger, A. M. 1984. Susceptibility of the egg parasitoid *Trichogramma evanescens* to the microsporidium *Nosema pyrausta* and its impact on fecundity. J. Invert. Pathol. 44: 228–229.

Huger, A. M., and Neuffer, G. 1978. Infection of the braconid parasite *Ascogaster quadridentata* (Hymenoptera: Braconidae) by a microsporidan of its host, *Laspeyresia pomonella*. *In* The Use of Integrated Control and the Sterile Insects Technique for Control of the Codling Moth. Summaries of papers presented at the Joint FAO/IAEA and IOBC/WPRS Research Coordination Meet at Heidelberg, Germany, Nov. 7–10, 1977. Ed. E. Dickler, Mitt. Biol. Bund. Land-Forst. Berlin-Dahlem, 1978, no. 180, 120 pp.

Hukusima, S., and Kondo, K. 1962. Further evaluation in the feeding potential of the predaceous insects and spiders in association with aphids harmful to apple and pear growing, and the effect of pesticides on predators. Jap. J. Appl. Entomol. Zool. 6: 274–280.

Hull, L. A. 1979a. Apple, tests of insecticides 1978. *In* Insecticide and Acaricide Tests. Entomol. Soc. Amer. Publ. 4: 20–22.

Hull, L. A. 1979b. Efficacy and cost analysis of insecticides and miticides in orchard spray programs. Pa. Fruit News 51: 1–4.

Hull, L. A., and Asquith, D. 1978. Apple integrated tests, 1977. *In* Insecticide and Acaricide Tests. Entomol. Soc. Amer. Publ. 3: 28–31.

Hull, L. A., and Beers, E. H. 1985. Ecological selectivity: modifying chemical control practices to preserve natural enemies, pp. 103–122. *In* Biological Control of Agricultural Integrated Pest Management Systems. Eds. M. A. Hoy and D. C. Herzog, Academic Press, New York, 589 pp.

Hull, L. A., and Starner, V. R. 1983. Impact of four synthetic pyrethroids on major natural enemies and pests of apple in Pennsylvania. J. Econ. Entomol. 76: 122–130.

Hull, L. A., Asquith, D., and Mowrey, P. D. 1978. Integrated control of the European red mite with and without the mite suppressant Dinocap. J. Econ. Entomol. 71: 880–885.

Hull, L. A., Beers, E. H., and Meagher, R. L. 1985a. Impact of selective use of the synthetic pyrethroid fenvalerate on apple pests and natural enemies in large-orchard trials. J. Econ. Entomol. 78: 163–168.

Hull, L. A., Beers, E. H., and Meagher, R. L. 1985b. Integration of biological and chemical control tactics for apple pests through selective timing and choice of synthetic pyrethroid insecticides. J. Econ. Entomol. 78: 714–721.

Humphrey, B. J., and Dahm, P. A. 1976. Chlorinated hydrocarbon insecticide residues in Carabidae and the toxicity of dieldrin to *Pterostichus chalcites*. Environ. Entomol. 5: 729–734.

Hunter, K. W., and Stoner, A. 1975. *Copidosoma truncatellum*: effect of parasitization on food consumption of larval *Trichoplusia ni*. Environ. Entomol. 4: 381–382.

Iacob, N., and Posoiu, V. 1981. The selectivity of several pesticides to the principal parasites used in the biological control in glasshouses. Res. Plant Prot. Inst., p. 435.

Iacob, N., Posoiu, V., and Manolescu, H. 1981. Selectivity of some pesticide products against the principal zoophagous arthropods used in integrated control in greenhouses. Analele Institutului de Cercetari pentru Protectia Plantelor 16: 357–373.

Ibrahim, M. M. 1962. An indication of the effect of the widespread use of pesticides on the population of some predators in cotton fields. Bull. Soc. Entomol. Egypt 46: 317–323.

Ilivicky, J. M., Dinamarca, L., and Agosin, M. 1964. Activity of NAD-kinase of nymph *Triatoma infestans* upon treatment with DDT and other compounds. Comp. Biochem. Physiol. 11: 291–301.

Iman, M., Soekarna, D., Situmorang, J., Adiputra, I. M. G., and Manti, I. 1986. Effect of insecticides on various field strains of diamondback moth and its parasitoid in Indonesia. *In* Diamondback Moth Mgmt., Proc. 1st Intern. Workshop, Tainan, Taiwan, Mar. 11–15, 1985.

Inglesfield, C. 1984. Field evaluation of the effects of a new pyrethroid insecticide, WL 85871, on the beneficial arthropod fauna of oilseed rape and wheat, pp. 325–330. *In* Proc. 1984 Brit. Crop Protection Conf. Pests and Diseases, Brighton Metropole, England, Nov 19–22, 1984, vol. 1.

Inserra, S. 1969. *Iridomyrmex humiis*, an undesirable enemy of *A. melinus*. Entomologica 5: 79–84.

Irabagon, T. A., and Brooks, W. M. 1974. Interaction of *Campoletis sonorensis* and a nuclear polyhedrosis virus in larvae of *Heliothis virescens*. J. Econ. Entomol. 67: 229–231.

Irving, S. N., and Wyatt, I. J. 1973. Effects of sublethal doses of pesticides on the oviposition behavior of *Encarsia formosa*. Ann. Appl. Biol. 75: 57–63.

Isaeva, L. I. 1983. The side effects of herbicides on beneficial insect fauna. Rastitelna Zashchita 31: 22–26.

Isaichev, V. V. 1978. Effects of pesticides on predacious carabids. Zash. Rast. 11: 35.

Ishaaya, I., and Casida, J. E. 1981. Pyrethroid esterase(s) may contribute to natural pyrethroid tolerance of larvae of the common green lacewing. Environ. Entomol. 10: 681–684.

Isley, D. 1926. Early summer dispersal of boll weevils. Ark. Agric. Expt. Sta. Bull. 204, 17 pp.

Issi, I. V., and Maslennikova, V. A. 1964. Effects of microsporidiosis on the diapause and survival of *Apanteles glomeratus* L. (Hymanoptera, Braconidae) and *Pieris brassicae* L. (Lepidoptera, Pieridae). Entomol. Rev. (USSR) 43: 56–58.

Jackson, J. G. 1984. Present trends in pesticide development regarding safety to beneficial organisms, pp. 387–394. *In* Proc. 1984 Brit. Crop Protec. Conf. Pests and Disease. Brighton Metropole, England, Nov. 19–22, 1984, vol. 1.

Jackson, G. J., and Ford, J. B. 1973. The feeding behaviour of *Phytoseiulus persimilis* (Acarina: Phytosiidae), particularly as affected by certain pesticides. Ann. Appl. Biol. 75: 165–171.

Jacobs, R. J., Kouskolekas, C. A., and Gross, H. R. 1984. Responses of *Trichogramma pretiosum* (Hymenoptera: Trichogrammatidae) to residues of permethrin and endosulfan. Environ. Entomol. 13: 355–358.

Jacobson, M. 1982. Plant, insects, and man—their interrelationships. Econ. Bot. 36: 346–354.

James, D. G. 1988. Reproductive diapause in *Typhlodromus occidentalis* Nesbitt (Acarina: Phytoseiidae) from southern New South Wales. J. Aust. Entomol. Soc. 27: 55–59.

Jao, L. T., and Casida, J. E. 1984. Insect pyrethroid-hydrolyzing esterases. Pestic. Biochem. Physiol. 4: 456–472.

Jaques, R. P. 1965. The effect of *Bacillus thuringiensis* Berliner on the fauna of an apple orchard. Can. Entomol. 97: 795–802.

Jeppson, L. R., McMurtry, J. A., Mead, D. W., Jesser, M. J., and Johnson, H. G. 1975. Toxicity of citrus pesticides to some predaceous phytoseiid mites. J. Econ. Entomol. 68: 707–710.

Johansen, C. A. 1957. History of biological control of insects in Washington. Northwest Sci. 31(2): 57–79.

Johansen, C. A., and Eves, J. D. 1973. Development of a pest management program on alfalfa grown for seed. Environ. Entomol. 2: 515–517.

Johnson, D. T., and Croft, B. A. 1981. Dispersal of *Amblyseius fallacis* in an apple tree ecosystem. Environ. Entomol. 10: 313–319.

Jones, D., Barlow, J. S., and Thompson, S. N. 1982. *Exeristes, Itoplectis, Aphaereta, Brachymeria*, and *Hyposoter* species: *in vitro* glyceride synthesis and regulation of fatty acid composition. Expl. Parasitol. 54: 340–351.

Jones, D., Snyder, M., and Granett, J. 1983. Can insecticides be integrated with biological control agents of *Trichoplusia ni* in celery? Entomol. Exp. Appl. 33: 290–296.

Jones, D. A. 1981. Cyanide and coevolution, pp. 509–516. *In* Cyanide in Biology, Eds. B. Vennesland et al., Academic Press, New York.

Jones, D. A., Parsons, J., and Rothschild, M. 1962. Release of hydrocyanic acid from crushed tissues of all stages in the life-cycle of species of the Zygaeninae (Lepidoptera). Nature 193: 52–53.

Jones, R. L., and Lewis, W. J. 1971. Physiology of the host parasite relationship between *Heliothis zea* and *Microplitis croceipes*. J. Insect Physiol. 17: 921–927.

Jones, V. P., and Parrella, M. P. 1983. Compatibility of six citrus pesticides with *Euseius stipulatus* (Acari: Phytoseiidae) populations in southern California. J. Econ. Entomol. 76: 942–944.

Juneja, P. S., Pearcy, S. C., Gholson, R. K., Burton, R. L., and Starks, K. J. 1975. Chemical bases for greenbug resistance in small grains. II. Identification of the major neutral metabolite of benzyl alcohol in barley. Plant Physiol. 56: 385–389.

Kabacik-Wasylik, D., and Jaworska, M. 1973. The effect of pesticides used to control the colorado beetle on the Carabidea (Coleoptera). Ekol. Polska 21: 369–375.

Kabacik-Wasylik, D., and Kmitowa, K. 1973. The effect of single and mixed infections of entomopathogenic fungi on the mortality of the Carabidae (Coleoptera). Ekol. Polska 21: 645–655.

Kalina, B. F. 1950. Development and viability of *Drosophila melanogaster* on a medium containing DDT. Science 111: 39–40.

Kalushkov, P. 1982. The effect of five insecticides on coccinellid predators (Coleoptera) of aphids *Phorodon humuli* and *Aphis fabae* (Homoptera). Acta Entomol. Bohemoslovaca 79: 167–180.

Kamenkova, K. V. 1971. The effects of insecticides on the hymenopterous egg parasites of the noxious pentatomid. Zash. Rast. 16(2): 8.

Kapetanakis, E. G., and Cranham, J. E. 1983. Laboratory evaluation of resistance to pesticides in the phytoseiid predator *Typhlodromus pyri* from English apple orchards. Ann. Appl. Biol. 103: 389–400.

Kapetanakis, E. G., Warman, T. M., and Cranham, J. E. 1986. Effects of permethrin sprays on the mite fauna of apple orchards. Ann. Appl. Biol. 108: 21–32.

Kapustina, O. V. 1975. The effect of certain pesticides on Trichogramma. Zash. Rast. 44: 33–47.

Karg, W. 1970. Studies on the acarifauna of apple orchards with regard to the transition from standard spray programmes to integrated treatment. Arch. Phytopathol. Pflschutz. 7: 243–279.

Karg, W. 1978. Investigations of selectivity of plant protectives as regards beneficial predatory mites destroying spider mites in orchards. Arch. Phytopathol. Pflschutz. 14: 41–55.

Kashio, T., and Tanaka, M. 1979. Effects of fungicides on the predacious mite, *Amblysieus deleoni* Muma and Denmark (Acarina: Phytoseiidae). Proc. Assoc. Plant Prot. of Kyushu 25: 153–156.

Kawahara, S., Kiritani, K., and Sasaba, T. 1971. The selective activity of rice-pest insecticides against the green rice leafhopper and spiders. Botyu-Kagaku 36: 121–128.

Kay, I. R. 1979. Toxicity of insecticides to *Coccinella repanda* Thunberg (Coleoptera: Coccinellidae). J. Aust. Entomol. Soc. 18: 233–234.

Kaya, H. K. 1970. Toxic factor produced by a granulosis virus in armyworm larvae: effect on *Apantales militaris*. Science 168: 251–253.

Kaya, H. K. 1979. Microsporidian spores: retention of infectivity after passage throughout the gut of the assasin bug, *Zelus exsanguis* (Stal). Proc. Hawaiian Entomol. Soc. 23: 91–94.

Kaya, H. K., and Dunbar, D. M. 1972. Effect of *Bacillus thuringiensis* and carbaryl on an elm spanworm egg parasite, *Telenomus alsophilae*. J. Econ. Entomol. 65: 1132–1134.

Kaya, H. K., and Hotchkin, P. G. 1981. The nematode *Neoaplectana carpocapsae* Weiser and its effect on selected ichneumonid and braconid parasites. Environ. Entomol. 10: 177–178.

Kaya, H. K., and Tanada, Y. 1972a. Response of *Apanteles militaris* to a toxin produced in a granulosis-virus infected host. J. Invert. Pathol. 19: 1–17.

Kaya, H. K., and Tanada, Y. 1972b. Pathology caused by a viral toxin in the parasitoid *Apanteles militaris*. J. Invert. Pathol. 19: 262–272.

Kaya, H. K., and Tanada, Y. 1973. Hemolymph factor in armyworm larvae infected with a nuclear-polyhedrosis virus toxic to *Apanteles militaris J. Invert. Pathol.* 21: 211–214.

Kazakova, S. B., and Dzhunusov, K. K. 1977. The effect of Bitoxibacillin-202 on certain orchard insects in the Issyk-kul' depression. Abs. Rev. Appl. Entomol. Ser. A, 65: 5987

Ke, L. 1987. Genetic improvement of pesticide resistance *Amblyseius pseudologispinousus (Acarina: Phytoseiidae)*. (pers. comm.)

Keever, D. W., Bradley, J. R., and Ganyard, M. C. 1977. Effects of diflubenzuron (Dimilin) on selected beneficial arthropods in cotton fields. Environ. Entomol. 6: 732–736.

Kehat, M., and Swirski, E. 1964. Chemical control of the date palm scale, *Parlatoria blanchardi*, and the effect of some insecticides on the lady beetle *Pharoscymnus* Aff. *Numidicus* Pic. Israel J. Agric. Res. 14(3): 101–110.

Kennedy, G. G., and Oatman, E. R. 1976. *Bacillus thuringiensis* and pirimicarb: selective insecticides for use in pest management on broccoli. J. Econ. Entomol. 69: 767–772.

Kennedy, G. G., Oatman, E. R., and Voth, V. 1976. Suitability of Plictran and Pirimor for use in a pest management program on strawberries in Southern California. J. Econ. Entomol. 69: 269–272.

Kennett, C. E. 1970. Resistance to parathion in the phytoseiid mite *Amblyseius hibisci*. J. Econ. Entomol. 63: 1999–2000.

Khalil, F. M., Maher Ali, A., Abdel Kawi, F., and Hafez, M. 1976. Effect of pesticides on population densities of predators of cotton pests. Agric. Res. Rev. 54: 63–70.

Kharizanov, A., and Babrikova, T. 1978. Toxicity of insecticides to certain species of chrysopids. Rastitelna Zashchita 26(5): 12–15.

Kharsun, A. L., and Karpenko, N. G. 1976. A study of the food chain of certain arthropods in relation to the use of pesticides. Parazity Teplokrovnykh Zhvotnykh, pp. 68–72.

Khukhrii, O. V. 1976. The effect of insecticides of *Aphidius ervi* Hold. a parasite of the pea aphid (*Acyrthosiphon pisum* Harr.). Zakhist Roslin 20: 25–27.

Kikkawa, H. 1964. Genetical studies on the resistance to parathion in *Drosophila melanogaster*. II. Induction of a resistance gene from its susceptible allele. Botyu-Kagaku 29: 37–42.

Kim, J. B., Cho, D. J., Hah, J. K., Chang, S. D., and Bark, Y. D. 1984. Effect of density variation on the natural enemies of rice pests by application method and time of carbofuran granules. Korean J. Plant Protec. 23(4): 233–236.

King, E. G., and Bell, J. V. 1978. Interactions between a braconid, *Microplitis croceipes*, and a fungus, *Nomuraea rileyi*, in laboratory-reared bollworm larvae. J. Invert. Pathol. 31: 337–340.

Kircher, H. W. 1982. Sterols and insects, pp. 1–50. *In* Cholesterol Systems in Insects and Animals. Ed. J. Dupont, CRC Press, Boca Raton, Fla.

Kiritani, K. 1977. Recent progress in pest management for rice in Japan. Jap. Agric. Res. Quart. 11: 40–49.

Kiritani, K., and Kawahara, S. 1973. Food-chain toxicity of granular formulations of insecticides to a predator, *Lycosa pseudoannulata*, of *Nephotettix cincticeps*. Botyu-Kagaku 38(2): 69–75.

Kirknel, E. 1974. Review of the effects of the relevant insecticides on insects parasitic or predatory on insects harmful to plants. Tidssk. Planteav. 78: 615–626.

Kirknel, E. 1975. The effects of various insecticides in laboratory experiments with two aphid predators, the seven spotted ladybird (*Coccinella septempunctata* (L.)) and larvae of hoverflies (*Metasyrphus corollae*). Tidssk. Planteav. 79: 393–404.

Kirknel, E. 1978. Influence of diazinon, trichloronate, carbofuran and chlorfenvinphos on the parasitization capacity of the rove-beetle *Aleochara bilineata* (Gyll.). Tidssk. Planteav. 82: 117–129.

Kirschbaum, J. B. 1985. Potential implications of genetic engineering and other biotechnologies to insect control. Annu. Rev. Entomol. 30: 51–70.

Kiselek, E. V. 1975. The effect of biopreparations on insect enemies. Zash. Rast. 12:23.

Kismali, S., and Erkin, E. 1984a. Effects of juvenile hormone analogues on the development of some useful insects. I. Effects on egg hatch in *Coccinella septempunctata* L. Turkiye Bitki Koruma Dergisi 8(2): 99–104.

Kismali, S., and Erkin, E. 1984b. Effects of juvenile hormone analogues on the development of some useful insects. II. Effects on larval development of *Coccinella septempunctata* L. Turkiye Bitki Koruma Dergisi 8(2): 231–236.

Kismir, A., and Sengonca, C. 1980. An investigation into the effect of chemical preparations applied in the Cukurova region against cotton pests on the insect predator *Anisochrysa carnea* (Stephens) (Neuroptera, Chrysopidae). Turkiye Bitki Koruma Dergisi 4: 243–250.

Klages, G. H., Emmerich, H., and Peter, M. G. 1980. High-affinity binding sites for juvenile hormone I in the larval integument of *Drosophila hydei*. Nature 286: 282–286.

Kmitowa, K., and Kabacik-Wasylik, D. 1971. An attempt at determining the pathogenicity of two species of entomopathogenic fungi in relation to Carabidae. Ekol. Polska 19(43): 727–733.

Knisley, C. B., and Swift, F. C. 1982. Qualitative study of mite fauna associated with apple foliage in New Jersey. J. Econ. Entomol. 65: 445–448.

Knowles, C. O. 1975. Basis for selectivity of acaricides, pp. 155–176. *In* Pesticide Selectivity. Ed. J. C. Street, Marcel Dekker, New York.

Kobayashi, T., Noguchi, Y., Hiwada, T., Kanayama, K., and Maruoka, N. 1978. Studies on the arthropod associations in paddyfields, with particular reference to insecticidal effects on them. III. Effect of insecticide applications on the faunistic composition of arthropods in paddy fields. Kontyu 46: 603–623.

Koebele, A. 1883. Studies of parasitic and predaceous insects in New Zealand, Australia and adjacent islands. USDA Bull., Washington D.C., 39 pp.

Koenig, K. 1983. Studies in the effects of insecticide usage on the epigeal fauna of sugar beet fields. Bayer. Landwirtsch. Jahrbuch 60: 235–312.

Koenig, K. 1985. Side effects of pesticides on the soil fauna. Nachr. Pflschutz. 37(1): 8–12.

Koenig, K., and Hassan, S. A. 1986. Resistance and cross-resistance of the predacious mite *Phytoseiulus persimilis* (Athias-Henriot) to organophosphates. J. Appl. Entomol. 101(3): 206–215.

Komives, T., and Casida, J. E. 1983. Acifluorfen increases the leaf content of phytoalexins and stress metabolites in several crops. J. Agric. Food Chem. 31: 751–755.

Koolman, J. 1982. Ecdysone metabolism. Insect Biochem. 12: 225–250.

Korostel, S. I., and Kapustina, O. V. 1975. Effect of the thermostable exotoxin of *Bacillus thuringiensis* on *Trichgramma* (*Trichogramma* sp.) and *Ageniaspis* (*Ageniaspis fusicollis* Dalm.). Tr. Vses. Nauch. Inst. Zash. Rast. 42: 102–109.

Korschgen, L. J. 1970. Soil–food chain–pesticide–wildlife relationships in aldrin-treated fields. J. Wildl. Mgmt. 34: 186–199.

Kostadinov, D. 1979. Susceptibility of single-sexed *Trichogramma* to some widely used pesticides. Rastitelna Zashchita 27(5): 28–30.

Kot, J., and Plewka, T. 1970. The influence of metasystox on different stages of the development of *Trichogramma evanescens*. Deutsch. Akad. Landwirtschaftwiss. Tagungsber. 110: 185–192.

Kot, J. T., Plewka, T., and Krukierek, T. 1971. Relationship in parallel development of host and parasite resistance to a common toxicant—Funal. Tech. Rept. PL-480, E21-Ent-19, F6-Po-203. Inst. Ecol. Polish Acad. Sci. 25, 66 pp.

Kot, J., Krukierek, T., and Plewka, T. 1975. Investigation on metasystox- and DDT resistance of five populations of *Trichogramma evanescens* Westw. (Hymenoptera. Trichogrammatidae). Polish Ecol. Stud. 1: 173–182.

Kovalick, G. E., and Koeppe, J. K. 1983. Assay and identification of juvenile hormone binding proteins in *Leucophala maderae*. Mol. Cell. Endocr. 31: 271–286.

Kowalska, T., and Pruszynski, S. 1969. Studies on the toxicity of some insecticides for the gold-eyed lacewing (*Chrysopa carnea Steph.*) (Neuroptera, Chrysopidae). Biol. Inst. Ochrony Roslin 45: 99–107.

Koyama, J. 1975. Studies on the diminution of insecticide application to the rice stem borer, *Chilo suppressalis* Walker, III. The effect of insecticide application on the density of larvae of the rice stem borer and spiders. Jap. J. Appl. Zool. 19: 125–130.

Kreig, A., Herfs, W. 1963. Über die Wirkung von *Bacillus thuringiensis* Einwirkungen auf Bienen (*Apis jellifera*). Entomol. Exp. Appl. 6: 1–9.

Kreig, A., and Herfs, W. 1964. Nebenwirken von *Bacillus thuringiensis* Einwirkungen auf Bienen. Entomophaga Mem. 2: 195–195.

Krieger, R. I., Feeny, P. P., and Wilkinson, C. F. 1971. Detoxication enzymes in the guts of caterpillars: an evolutionary answer to plant defenses. Science 172: 579–581.

Krishnamoorthy, A. 1983. Effect of some pesticides on the predatory mite, *Amblyseius tetranychivorus* (Gupta) (Acarina: Phytoseiidae). Entomon 8(3): 229–234.

Krukierek, T., Plewka, T., and Kot, J. 1975. Susceptibility of parasitoids of the genus *Trichogramma* (Hymenoptera, Trichogrammatidae) to metasystox in relation to their species, host species and ambient temperature. Polish Ecol. Stud. 1: 183–196.

Ku, T. Y., and Wang, S. C. 1981. Insecticidal resistance of the major insect rice pests, and the effect of insecticides on natural enemies and non-target animals. Phytopathol. Entomol. 8: 1–18.

Kubo, I., and Nakanishi, K. 1979. Some terpenoid insect antifeedants from tropical plants, pp. 284–294. *In* Advances in Pesticide Science, Part II. Ed. H. Geissbuhler. Pergamon Press, New York.

Kuhn, T. S. 1975. The Structure of Scientific Revolutions. Univ. Chicago Press, Chicago, Ill., 210 pp.

Kuhner, C., Klingauf, F., and Hassan, S. A. 1985. Development of laboratory and semi-field methods to test the side effect of pesticides on *Diaeretiella rapae* (Hym.: Aphidiidae). Meded. Facult. Landbouww., Rijks. Gent 50(2b): 531–538.

Kuhr, R. J., and Dorough, H. W. 1976. Carbamate Insecticides: Chemistry, Biochemistry and Toxicology. CRC Press, Cleveland, Ohio, 301 pp.

Kulkarni, A. P., and Hodgson, E. 1980. Metabolism of insecticides by mixed-function oxidase systems. Pharmac. Ther. 8: 379–475.

Kuo, J.-L. 1977. Auswirkungen zweier Wirtspflanzen von *Myzus persicae* (Sulz.) auf den räuberischen Blattlausfeind *Aphidoletes aphidimyza* (Rond.) (Diptera: Cecidomyiidae). Z. Angew. Entomol. 82: 229–303.

Kurdyukov, V. V. 1980. Pesticides and natural insect enemies. Zash. Rast. 4: 22.

Kuwahara, M. 1978. Toxicity of chlordimeform and analogues and MFO-inhibitors to three species of mites, improvement of toxicity by certain synergists. Appl. Entomol. Zool. 13: 296–303.

Lambert, G. A. 1981. Lucerne—a vital forage crop at Monto. Queensland Agric. J. 107: 293–299.

Lange, W. H., Agosta, G. G., Goh, K. S., and Kishiyama, J. S. 1980. Field effect of insecticides on chrysanthemum leafminer and a primary parasitoid, *Chrysocharis ainsliei* (Crawford), on artichokes in California. Environ. Entomol. 9: 561–562.

Larsen, L. L., Kenaga, E. E., and Morgan, R. W. 1985. Commercial and Experimental Organic Insecticides (Revision). Publ. Entomol. Soc. Amer. 105 pp.

Larson, H. 1970. Raising resistant predatory mites. Proc. Idaho Hort. Soc. (1969), pp. 32–37.

Larsson, R. 1979. Transmission of *Nosema mensili* (Paillot) (Microsporida, Nosematidae), a microsporidian parasite of *Pieris brassicae* L. (Lepidoptera, Pieridae) and its parasite *Apanteles glomeratus* L. (Hymenoptera, Braconidae). Zool. Anz. 203 (3/4): 151–157.

Laska, P. 1972. The effect of pirimicarb and some other insecticides on three aphid species and on ladybird beetle. Biol. Inst. Ochrony Roslin 8: 129–134.

Laska, P., Slovakova, J., and Bicik, V. 1980. Life cycle of *Trialeurodes vaporariorum* Westw. (Homoptera: Aleyrodidae) and its parasite *Encarsia formosa* Gah. (Hymenoptera: Aphelinidae) at constant temp. Sbornik Praci Prirodovedecke Fakulty University Palackeho v Olomouci, Biologie 20: 95–106.

Lasota, J. A., and Kok, L. T. 1986. Residual effects of methomyl, permethrin, and fenvalerate on *Pteromalus puparum* (Hymenoptera: Pteromalidae) adult parasites. J. Econ. Entomol. 79: 651–653.

Lawrence, P. O. 1974. Susceptibility of *Chrysopa rufilabris* to selected insecticides and miticides. Environ. Entomol. 3: 146–150.

Lawrence, P. O. 1981. Developmental and reproductive biologies of the parasitic wasp, *Biosteres longicaudatus*, reared on hosts treated with a chitin synthesis inhibitor. Ins. Sci. Appl. 1: 403–406.

Lawrence, P. O., Kerr, S. H., and Whitcomb, W. H. 1973. *Chrysopa rufilabris*: Effect of selected pesticides on duration of third larval stadium, pupal stage, and adult survival. Environ. Entomol. 2: 477–480.

Lawrence, P. O., Greany, P. D., Nation, J. L., and Oberlander, H. 1978. Influence of hydroprene on caribbean fruit fly suitability for parasite development. Florida Entomol. 61: 93–99.

Lawson, R. F., Rabb, R. L., Guthrie, F. E., and Bowry, T. G. 1961. Studies of integrated control systems for hornworms on tobacco. J. Econ. Entomol. 54: 93–97.

Lecrone, S., and Smilowitz, Z. 1980. Selective toxicity of pirimicarb, carbaryl and methamidophos to green peach aphid, (*Myzus persicae*) (Sulzer), *Coleomegilla maculata* lengi (Timberlake) and *Chrysopa oculata* Say. Environ. Entomol. 9: 752–755.

Ledieu, M. S. 1979. Laboratory and glasshouse screening of pesticides for adverse effects on the parasite *Encarsia formosa* Gahan. Pestic. Sci. 10(2): 23–32.

Legner, E. F. 1987a. Transfer of thelytoky to arrhenotokous *Muscidifurax raptor* Girault and Sanders (Hymenoptera: Pteromalidae). Can. Entomol. 119: 265–271.

Legner, E. F. 1986b. Inheritance of gregarious and solitary oviposition in *Muscidifurax raptorellus* Kogan and Legner (Hymenoptera: Pteromalidae). Can. Entomol. 119: 791–808.

Legner, E. F., and Warkentin, R. W. 1985. Genetic improvement and inbreeding effects in culture of beneficial arthropods. *In* Proc. and Papers 52nd Ann. Conf. Calif. Mosq. Vector Control Assoc., Inc., Long Beach, Calif., Jan. 29–Feb. 1, 1984.

Legotai, M. V. 1980. Effect of BTB on pests of cabbage and insect enemies. Zash. Rast. 8: 34–35.

Leighton, T., Marks, E., and Harghton, F. 1981. Pesticides: insecticides and fungicides are chitin synthesis inhibitors. Science 213: 905–907.

Lelièvre, D., and Chatenet, B. 1983. Phosalone selectivity and the environment. Défense Vegetaux 222: 195–202.

Lema, K. M., and Poe, S. L. 1978. Juvenile hormone analogues: effects of ZR-777 on *Liriomyza sativae* and its endoparasite. Florida Entomol. 61: 67–68.

Lentz, G. L., Chambers, A. Y., and Hayes, R. M. 1983. Effects of systemic insecticides-nematicides on mid-season pest and predator populations in soybean. J. Econ. Entomol. 76: 836–840.

Lesiewicz, D. S., van Duyn, J. W. and Bradley, J. R. 1984. Midseason response of three carabids to soil insecticides applied to field corn plots at planting. J. Georgia Entomol. Soc. 19: 271–275.

Levin, D. B., Laing, J. E., and Jacques, R. P. 1981. Interactions between *Apanteles glomeratus* (L.) (Hymenoptera: Braconidae) and granulosis virus in *Pieris rapae* (L.) (Lepidoptera: Pieridae). Environ. Entomol. 10: 65–68.

Levin, D. B., Laing, J. E., Jacques, R. P., and Corrigan, J. E. 1983. Transmission of the granulosis virus of *Pieris rapae* (Lepidoptera: Pieridae) by the parasitoid *Apanteles glomeratus* (Hymenoptera: Braconidae). Environ. Entomol. 12: 166–170.

Levins, R., and Wilson, M. 1980. Ecological theory and pest management. Annu. Rev. Entomol. 25: 287–308.

Lewontin, R. C. 1965. Selection for colonizing ability, pp. 77–94. *In* The Genetics of Colonizing Species. Eds. H. G. Baker and G. L. Stebbins, Academic Press, New York.

Li-ying, L., Xiong, X., and Zhang, M. 1988. Selecting *Trichogramma* spp. and *Anastatus japonicus* Ashmead for resistance to pesticides, p. 315f. *In* Proc. 18th Intern. Cong. Entomol., Vancouver, B.C., Canada, July 3–9, 1988.

Lienk, S. E., Watve, C., and Minns, J. 1976. Susceptibility of the European red mite and its predators to chemical treatments. Plant Sci. Entomol. 12, New York State Agric. Exp. Sta., 10 pp.

Lienk, S. E., Minns, J., and Labanowska, B. H. 1978. Evaluation of pesticides against the European red mite, apple rust mite, and two mite predators in 1976–1977. Plant Sci., Entomol. 71, New York State Agric. Exp. Sta., 10 pp.

Lighthart, B. 1986. Compilation of interim protocols to test the effects of microbial pest control agents on non-target beneficial arthropods. Terrestrial Microbial Ecology/Biotechnology Prog., Environ. Res. Lab., USEPA, Corvallis, Ore., Sept. 26, 1986, 80 pp.

Lim, G. S., Sivapragasam, A., and Ruwaida, M. 1986. Impact assessment of *Apanteles plutellae* on diamondback moth using an insecticide-check method, pp. 195–204. *In* Diamondback Moth Management. Proc. 1st Intern. Workshop, Tainan, Taiwan, Mar. 11–15, 1985.

Lindquist, R. K., and Wolgamott, M. L. 1980. Toxicity of acephate to *Phytoseiulus persimilis* and *Tetranychus urticae*. Environ. Entomol. 9: 389–392.

Lindquist, R. K., Frost, C., and Wolgamott, M. L. 1979. Alternative methods for control of greenhouse insect pests. Ohio Rept. Research and Develop. 64: 13–14.

Lingappa, S. S., Starks, K. J., and Eikenbary, R. D. 1972. Insecticidal effect on *Lysiphlebus testaceipes*, a parasite of the greenbug at three developmental stages. Environ. Entomol. 1: 520–521.

Lingren, P. D., and Ridgway, R. L. 1967. Toxicity of five insecticides to several insect predators. J. Econ. Entomol. 60: 1639–1641.

Lingren, P. D., Wolfenbarger, D. A., Nosky, J. B., and Diaz, M. 1972. Responses of *Campoletis perdistinctus* and *Apanteles marginiventris* to insecticides. J. Econ. Entomol. 63: 1295–1299.

Liotta, G. 1974. Secondary effects of the most common chemicals used against citrus diaspine coccids in Sicily on *Aspidiotiphagus citrinus* (Craw.) (Hym.-Aphilinidae). Boll. Inst. Entomol. Agraria Osserv. Fitopatol. Palermo 9: 187–194.

Liss, W. J., Gut, L. J., Westigard, P. H., and Warren, C. E. 1986. Perspectives on arthropod community structure, organization, and development in agricultural crops. Annu. Rev. Entomol. 31: 455–478.

Lloyd, M., and Ghelardi, R. J. 1964. A table for calculating the equitability component of species diversity. J. Animal Ecol. 33: 217–225.

Lo, P. K. 1986. Present status of biological control of mite pests in Taiwan. Plant Prot. Bull. Taiwan 28: 31–39.

Lo, P. K. C., and Chao, S. R. S. 1975. Preliminary study on the toxicity of 23 pesticides to natural enemies of red spider mites. J. Agric. Res. China 92: 81–86.

Lo, P. K. C., Wu, T. K., and Tseng, S. K. 1984. Studies on pesticide resistance in the phytoseiid mite, *Amblyseius longispinosus* (Evens), p. 816. *In* Proc. 17th Intern. Cong. Entomol., Hamburg, Germany, 1984.

Lo, P. K. C., Wu, T. K., and Lim, S. R. 1986. Studies on population dynamics and integrated control of spider mites on pear in temperate zone of Taiwan. Agric. Assoc. Taiwan, China, pp. 98–111.

Lockwood, J. A., Sparks, T. C., and Story, R. N. 1984. Evolution of insect resistance to insecticides: a re-evaluation of the roles of physiology and behavior. Bull. Entomol. Soc. Amer. 30: 41–51.

Logan, J. A. 1982. Recent advances and new directions in phytoseiid population models, ch. 3, pp. 49–71. *In* Recent Advances in Knowledge of the Phytoseiidae. Ed. M. A. Hoy, Div. Agric. Sci., Univ. Calif. Berkeley Paid Publ. 3294, 92 pp.

Long, K. Y., and Brattsten, L. B. 1982. Is rhodanese important in the detoxification of dietary cyanide in southern armyworm (*Spodoptera cridania* Cramer) larvae? Insect Biochem. 12: 367–375.

Lopez, J. D., and Morrison, R. K. 1985. Parasitization of *Heliothis* spp. eggs after augmentative releases of *Trichogramma pretiosum* Riley. Southwestern Entomol. (Suppl. 8): 110–137.

Lowery, D. T., and Sears, M. K. 1986. Stimulation of reproduction of the green peach aphid (Homoptera: Aphididae) by azinphosmethyl applied to potatoes. J. Econ. Entomol.79: 1530–1533.

Luck, R., Messenger, P. S., and Barbieri, J. 1981. The influence of hyperparasitism on the performance of biological control, pp. 43–39. *In* The Role of Hyperparasitism in Biological Control. A Symposium. Ed. D. Rosen, Div. Agric. Sci., Univ. Calif. Berkeley Paid Publ. 4103.

Luckey, T. D. 1968. Insect hormoligosis. J. Econ. Entomol. 61: 7–12.

Lyashova, L. V., and Evtushenko, I. I. 1983. The effect of trichothecin on coccinellid larvae. Zash. Rast. 4: 39.

Lykouressis, D. P., and van Emden, H. F. 1983. Movement away from the feeding site of the aphid *Sitobion avenae* when parasitized by *Aphelinus abdominalis*. Entomologia Hellenica 1: 59–63.

MacArthur, R. G., and Wilson, E. D. 1967. The Theory of Island Biogeography. Princeton Univ. Press., Princeton, N.J.

Maccolini, M. 1985. Effects of the juvenile-hormone analogue ZR-619 5E (Triprene) on *Chrysoperla carnea* (Stephens) (Neuroptera, Chrysopidae). Boll. Inst. Entomol. 'Guido Grandi' Univ. Stud. Bologna 39: 201–219.

MacDonald, D. R. 1959. Biological assessment of aerial forest spraying against spruce budworm in New Brunswick. III. Effect on two overwintering parasites. Can. Entomol. 91: 330–336.

MacDonald, D. R., and Webb, F. E. 1963. Insecticides and the spruce budworm. Mem. Entomol. Soc. Can. 31: 288–310.

Mackauer, M. 1976. Genetic problems in the production of biological control agents. Annu. Rev. Entomol. 21: 369–385.

Mackauer, M. 1986. Fecundity and host utilization of the aphid parasite *Aphelinus semiflavus* (Hymenoptera: Aphelinidae) at two host densities. Can. Entomol. 114: 721–726.

Mackauer, M., and van den Bosch, R. L. 1973. General applicability of evaluation results. J. Appl. Ecol. 10: 330–335.

MacPhee, A. W., and Sanford, K. N. 1961. The influence of spray programs on the fauna of apple orchards in Nova Scotia XII. Second supplement to VII. Effects on beneficial arthropods. Can. Entomol. 93: 671–673.

Madrid, F. J., and Stewart, R. K. 1981. Impact of diflubenzuron spray on gypsy moth parasitoids in the field. J. Econ. Entomol. 74: 1–2.

Madsen, H. F. 1968. Integrated control of deciduous tree fruit pests. World Crops 20: 20–23.

Madsen, H. F., and Carty, B. E. 1977. Fruit tree leafrollers: control of a population tolerant to diazinon. J. Econ. Entomol. 70: 615–616.

Makar, P. V., and Kadhav, J. 1980. Relative toxicity of some insecticides to the egg-larval parasite *Copidosoma Koehleri* Blanchard of potato tuberworm *Phthorimaea operculella* (Zell.). Indian J. Entomol. 42: 537–539.

Markar, P. V., and Kadhav, J. 1981. Toxicity of some insecticides to the aphid predator *Menochilus sexmaculatus* Fabricius. Indian J. Entomol. 43(2): 140–144.

Malley, C. W. 1916. On the selection and breeding of desirable strains of beneficial insects. S. Afr. J. Sci. 13: 191–195.

Malysheva, M. S., and Kartavtsev, N. I. 1977. Effect of chemical treatments by helicopter on the state of telenomines present within the eggs of their hosts. Zash. Rast. 44: 102–110.

Mangutova, S. A. 1971. The effect of aphicides on aphidophagous ladybird. Zash. Rast. 16(8): 34.

Mani, M., and Krishnamoorthy, A. 1984. Toxicity of some insecticides to *Apanteles plutellae*, a parasite of the diamondback moth. Trop. Pest Mgt. 30: 130–132.

Mani, M., and Krishnamoorthy, A. 1986. Susceptibility of *Telenomus remus* Nixon, an exotic parasitoid of *Spodoptera fitura* (F.), to some pesticides. Trop. Pest Mgt. 32: 49–51, 81, 84.

Mani, M., and Nagarkatti, S. 1983. Susceptibility of two braconid parasites *Apanteles angaleti* Muesebeck and *Bracon kirkpatricki* (Wilkinson) to several chemical pesticides. Entomon 8(1): 87–92.

Maniglia, G., 1978. Effects secondaires du carbaryl et du dimethoate sur *Opius conclolor* Szepl. (Hym. Baconidae) au laboratoire. Med. Fac. Landbouww. Rijksuniv. Gent 43: 487–491.

Manser, P. D., and Bennett, F. D. 1962. Possible effects of the application of malathion on the small moth borer, *Diatraea saccharalis* (F.) and its parasite *Lixophaga diatraeae* (Tns) in Jamaica. Bull. Entomol. Res. 53: 75–82.

Mansour, F., Rosen, D., Plaut, H. N., and Shulov, A. 1981. The effect of commonly used pesticides on *Chiracanthium mildei* and other spiders occurring on apple. Phytoparasitica 9: 139–144.

Manttari, J. 1980. Bioresmetriinia vastaan vastustuskykyisten ansaripetopunkkien kehittaminen. Biologisen torjunnan kehittaminen maataloudessa ja puutarhataloudessa. Seminaariraportti. Suomen Akatemian Julk. 15/1980, p. 26.

Marchal-Segault, D. 1975a. Larval development of the parasitic Hymenoptera *Apanteles glomeratus* L. and *Phanerotoma flavitestacea* F. in caterpillars infected by *Bacillus thuringiensis* Berliner. Ann. Parasitol. Hum. Comp. 50: 223–232.

Marchal-Segault, D. 1975b. Susceptibility of the hymenopterous braconids *Apanteles glomeratus* and *Phanerotoma glavitestacea* to the spore-crystal complex of *Bacillus thuringiensis* Berliner. Ann. Zool. Ecol. Anim. 6: 521–528.

Marchal-Segault, D. 1975c. Role of entomophagous larvae in the infection of the caterpillars of *Pieris brassicae* L. and *Anagasta kuehniella* Zell. by *Bacillus thuringiensis* Berliner. Rev. Zool. Agr. Path. Veg. 74: 68–84.

Mardzhanyan, G. M., and Ust'yan, A. K. 1966. Integrated control of the green peach aphid. *Myzus persicae* Sulz., on tobacco. Entomol. Obozr. 44: 441–448.

Margalef, R. 1957. La teoria de la informacion en ecologia. Mem. Real Acad. Ciencias y Artes de Barcelona 32: 373–449.

Margarit, G., and Hondru, N. 1981. Influence of some selective insecticides on the principal pests and their natural enemies in lucern crops. Analele Inst. Cercetari Pentru Protectia Plantelor 16: 159–166.

Markkula, M., and Tiittanen, K. 1976a. Mortality of *Aphidoletes aphidimyza* Rond. (Dipt., Itonididae) larvae treated with acaricides. Ann. Agric. Fenn. 15: 86–88.

Markkula, M., and Tiittanen, K. 1976b. "Pest in first" and "natural infestation" methods in the control of *Tetranychus urticae* Koch with *Phytoseiulus persimilis* A.-H. on glasshouse cucumbers. Ann. Agric. Fenn. 20: 28–31.

Markkula, M., and Tiittanen, K. 1980. Biological control of pests in glasshouses in Finland—the situation today and in the future. OILB SROP/WPRS Bull. 3: 127–133.

Markkula, M., Rimpilainen, M., and Tiittanen, K. 1979. Harmfulness of soil treatment with some fungicides and insecticides to the biological control agent *Aphidoletes aphidimyza* (Rond.) (Dipt., Cecidomyiidae). Ann. Agric. Fenn. 18: 168–170.

Markwick, N. P. 1984, p. 832. *In* Proc. 18th Intern. Cong. Entomol., Hamburg, Germany, Aug. 1984, vol. 8.

Markwick, N. P. 1986. Detecting variability and selecting for pesticide resistance in two species of phytoseiid mites. Entomophaga 31: 225–236.

Markwick, N. P. 1988. Evaluation of field- and laboratory-selected pyrethroid resistant predatory mites, p. 316a. *In* Proc. 18th Intern. Cong. Entomol., Vancouver, B.C., Canada, July 3–9, 1988.

Marsden, J. S., Martin, G. E., Parham, D. J., Ridsdill-Smith, T. J., and Johnston, B. G. 1980. Returns on Australian agricultural research. Joint Industries Assistance Comm., CSIRO, Div. Entomol., 107 pp.

Martin, N. A., and Workman, P. 1986. Buprofezin: a selective pesticide for greenhouse whitefly control, pp. 234–236. *In* Proc. 39th New Zealand Weed and Pest Control Conf., Palmerston North, Aug. 12–14, 1986.

Martin, W. R., and Brown, T. M. 1984. The action of acephate in *Pseudoplusia includens* (Lep.: Noctuidae) and *Pristhesanaus papnensis* (Hemiptera: Reduviidae). Entomol. Exp. Appl. 35: 3–9.

Matsumura, F., and Beeman, W. E. 1982. Toxic and behavioral effects of chlordimeform on the American cockroach *Periplaneta americana*, pp. 229–242. *In* Insecticide Mode of Action. Ed. J. R. Coats, Academic Press, New York.

Matsumura, F., and O'Brien, R. D. 1963. A comparative study of the modes of action of fluoroacetamide and fluoroacetate in the mouse and American cockroach. Biochem. Pharmacol. 12: 1201–1205.

Matvievskii, A. S. 1979. Integrated protection of orchards. Zash. Rast. 6: 26–27.

May, R. M. 1985. Evolution of pesticide resistance. Nature 315: 12–13.

May, R. M., and Dobson, A. P. 1986. Population dynamics and the rate of evolution of pesticide resistance, pp. 170–193. *In* Pesticide Resistance: Strategies and Tactics in Management. Nat. Acad. Sci. Press, Washington, D.C., 471 pp.

Mazzone, P., and Viggiani, G. 1980. Effects of diflubenzuron (Dimilin) on the larval instars of the predator *Cryptolaemus montrouzieri* Muls. (Col. Coccinellidae). Boll. Entomol. Agraria "Filippo Silvestri," Portici 37: 17–21.

Mazzone, P., Tranfaglia, A., and Viggiani, G. 1980. Effects of chemicals on the parasite *Leptomastix dactylopii* (How.) (Hymentopera: Encyrtidae). Boll. Entomol. Agraria "Filippo Silvestri," Portici 37: 13–15.

McClanahan, R. J. 1967. Food-chain toxicity of systemic acaricides to predaceous mites. Nature 215:1001.

McClure, M. S. 1977. Resurgence of the scale, *Fiorinia externa* (Homoptera: Diaspididae), on hemlock following insecticide application. Environ. Entomol. 6: 480–484.

McDonald, J. F. 1983. The molecular basis of adaptation: a critical review of relevant ideas and observations. Ann. Rev. Ecol. Sys. 14:77–102.

McDonald, S., and Harper, A. M. 1978. Laboratory evaluation of insecticides for control of *Acyrthosiphon pisum* (Hemiptera: Aphididae) in alfalfa. Can. Entomol. 110: 213–216.

McLaughlin, R. E., and Adams, C. B. 1966. Infection of *Bracon mellitor* (Hymenoptera: Braconidae) by *Mattesia grandis* (Protozoa: Neogregarinida). Ann. Entomol. Soc. Am. 59: 800–802.

McMullen, R. D., and Jong, C. 1967. The influence of three insecticides on predation of the pear psylla, *Psylla pyricola*. Can. Entomol. 99: 1292–1297.

McMurtry, J. A. 1977. Biological control of citrus mites. Proc. Intern. Soc. Citric. 2: 456–459.

McNeil, J. 1975. Juvenile hormone analogs: detrimental effects on the development of an endoparasitoid. Science 189: 640–642.

McNeil, J. M., and Brooks, W. M. 1974. Interactions of the hyperparasitoids *Catolaccus aeneoviridis* (Hymenoptera: Pteromalidae) and *Spilochalcis side* (Hymenoptera: Chalcididae) with the Microsporidans, *Nosema heliothidis* and *N. campoletidis*. Entomophaga 19: 195–204.

McWhorter, R., and Shepard, M. 1977. Response of Mexican bean beetle larvae and the parasitoid *Pediobius foveolatus* to Dimilin. Florida Entomol. 60: 55–56.

Meaghers, R. I., and Hull, L. A. 1986. Techniques to measure azinphosmethyl resistance in *Platynota idaeusalis* (Walker) (Lepidoptera: Tortricidae). unpubl.

Melander, A. L. 1914. Can insects become resistant to sprays? J. Econ. Entomol. 7: 167–173.

Mellini, E., and Boninsegni, G. 1983. Repercussions on the parasite *Gonia cinerascens* Rond. of hydroprene treatments effected on the host in the final phases of preimaginal development. Boll. Inst. Entomol. Univ. Studi Bologna 37: 171–191.

Mellini, E., and Cesari, R. 1982. Effects of the juvenoid ZR 512 4E (Hydroprene) on the host–parasite couple *Galleria mellonella* L.–*Gonia cinerascens* Rond. Boll. Inst. Entomol. Univ. Studi Bologna 36: 141–158.

Mellini, E., and Gironi, R. 1980. Effects of a juvenoid on the host–parasite couple *Galleria mellonella* L.–*Gonia cinerascens* Rond. Boll. Inst. Entomol. Univ. Studi Bologna 35: 189–213.

Mello Filho, A. de T., and Batista, G. C. de 1983. Thresholds of acceptance of some carbohydrates and of rejection of some salts by tarsel chemoreceptors of *Lixophaga diatraeae*, (Townsend, 1916) (Diptera, Tachinidae). Ciencia e Cultura 35: 491–495.

Menhinick, E. F. 1963. Insect species in the herb stratum of a *Sericae lespedeza* stand. US Atomic Energy Comm. Div. of Tech. Info. TID-19136.

Menn, J. J., and Henrick, C. A. 1981. Rational and biorational design of pesticides. Phil. Trans. R. Soc. Lond. B, 295: 57–71.

Messing, R. H. 1986. Biological control of the filbert aphid, *Myzocallis coryli*, in western Oregon. Ph.D. thesis, Oregon St Univ., Corvallis.

Metcalf, R. L. 1964. Selective toxicity of insecticides. World Rev. Pest Control 3: 28–43.

Metcalf, R. L. 1980. Changing role of insecticides in crop protection. Annu. Rev. Entomol. 25: 219–256.

Metcalf, R. L. 1982. Insecticides in pest management, pp. 217–278. *In* Introduction to Insect Pest Management, 2nd ed. Eds. R. L. Metcalf and W. H. Luckmann, Wiley Interscience, New York, 577 pp.

Metcalf, R. L., March, R. B., and Maxon, M. G. 1955. Substrate preferences of insect cholinesterases. Ann. Entomol. Soc. Amer. 48: 222–228.

Meyer, R. H. 1974. Management of phytophagous and predatory mites in Illinois orchards. Environ. Entomol. 3: 333–340.

Meyer, R. H. 1975. Release of carbaryl-resistant predatory mites in apple orchards. Environ. Entomol. 4:49.

Meyer, R. H. 1981. Integrated pest management. Proc. Ill. Hortic. Soc. 115: 30–32.

Michelbacher, A. E. 1962. Influence of natural factors on insect and spider mite populations, p. 694. *In* Proc. 11th Intern. Cong. Entomol. Vienna, Austria, 1960, vol. 2.

Michelbacher, A. E., and Hitchcock, S. 1958. Introduced increase of soft scales on walnut. J. Econ. Entomol. 51: 427–431.

Miller, J. C. 1980. Niche relationships among parasitic insects occurring in a temporary habitat. Ecology 6: 270–275.

Miller, J. C. 1983. Ecological relationships among parasites and the practice of biological control. Environ. Entomol. 12: 620–624.

Miller, J. R., Hendry, L. B., and Mumma, R. O. 1975. Norsesquiterpenes as defensive toxins of whirligig beetles (Coloeptera: Gyrinidae). J. Chem. Ecol. 1: 59–82.

Miller, L. K. 1983. A virus vector for genetic engineering in invertebrates, pp. 203–224. *In* Genetic Engineering in the Plant Sciences. Ed. N. J. Panopoulos, Praegen, New York.

Miller, M. C., White, R., and Ashley, J. 1973. Effect of granular applications of thimet on larvae and parasites of the alfalfa weevil in north Georgia. J. Georgia Entomol. Soc. 8(3): 213–216.

Miller, R. W., Croft, B. A., and Nelson, R. D. 1985. Effects of early season immigration on cyhexatin and formetanate resistance of *Tetranychus urticae* (Acari: Tetranychidae) on strawberry in central California. J. Econ. Entomol. 78: 1379–1388.

Mills, N. J. 1981. The mortality and fat content of *Adalia bipunctata* during hibernation. Entomol. Exp. Appl. 30: 265–268.

Mirianowski, J. A., and Carlson, G. A. 1986. Economic issues in public and private approaches to preserving pest susceptibility, pp. 436–448. *In* Pesticide Resistance: Strategies and Tactics for Management. NAS/NRC, Washington, D.C., 471 pp.

Mishra, U. S. 1979. Insect pest problems of rice in Chhattisgar. Pesticides 11: 24–26.

Mishra, N. C., and Satpathy, J. M. 1985. Selective toxicity of some insecticides against cabbage aphid, *Brevicoryne brassicae* L., and its coccinellid predator, *Coccinella repanda* Th. Indian J. Plant Protec. 12(1): 13–17.

Miszczak, M. 1975. Toxicity of several pesticides to the green lacewing *Chrysopa carnea* (Steph.) (Neuroptera: Chrysopidae). Roczniki Nauk Rolniczych 5: 31–41.

Miyata, T., and Saito, T. 1982. Mechanism of selective toxicity of malathion and pyridafenthion against insect pests of rice and their natural enemies, pp. 391–397. *In* Proc. Intern. Conf. Plant Protection in the Tropics, Mar. 1–4, 1982.

Moffitt, H. R., Anthon, E. W., and Smith, L. O. 1972. Toxicity of several commonly used orchard pesticides to adult *Hippodamia convergens*. Environ. Entomol. 1: 20–23.

Moffitt, H. R., Westigard, P. H., Mantey, K. D., and van de Baan, H. E. 1988. Resistance to diflubenzuron in the codling moth (Lepidoptera: Tortricidae). J. Econ. Entomol. 81: 1511–1515.

Moghaddam, H. R. 1978. Investigations into the residual effects, after treatment with insecticides, on the ladybird *Coccinella septempunctata* L. Entomol. Phytopathol. Appl. 46: 78–85.

Mohamad, R. B. 1974. Relative toxicity of selected insecticides to several populations of *Coleomegilla maculata* (De Geer), *Geocoris punctipes* (Say), and *Orius insidiosus* (Say). Thesis, Louisiana State Univ., 57 pp.

Monaco, R. 1969. The action taken against *D. oleae* by *O. concolor* distributed in Apulia in the Gargano olive groves, and by indigenous parasites in the same habitat. Studies of the working party of the C.N.R. for the Integrated control of animal pests of plants. Entomologica 5: 139–191.

Moosbeckhofer, R. 1983. Laboratory studies on the effect of diazinon, carbofuran and chlorfenvinphos on the locomotor activity of *Poecilus cupreus* L. (Col., Carabidae). Z. Angew. Entomol. 95: 15–21.

Morgan, C. V. C., and Madsen, H. F. 1976. Development of chemical, biological and physical methods for control of insects and mites. *In* History of Fruit Growing and Handling in United States of America and Canada 1860–1972. Regatta Press, Kelowna, B.C., Canada, 360 pp.

Mori, H., and Gotoh, T. 1986. Pesticide resistance in the Darmstadt strain of *Phytoseiulus persimilis* (Acarina: Phytoseiidae). Jap. J. Appl. Entomol. Zool. 30(1): 57–59.

Moriarty, F. 1969. The sublethal effects of synthetic insecticides on insects. Biol. Rev. 44: 321–357.

Morse, J. G., and Bellows, T. S. 1986. Toxicity of major citrus pesticides to *Aphytis melinus* (Hymenoptera: Aphelinidae) and *Cryptolaemus montrouzieri* (Coleoptera: Coccinellidae). J. Econ. Entomol. 79: 311–314.

Morse, J. G., and Croft, B. A. 1979. Development of resistance to azinphosmethyl through greenhouse selection in the predatory mite *Amblyseius fallacis* (Garman) and its prey, *Tetranychus urticae* (Koch). Recent Adv. in Acarology, Academic Press, New York 1: 397–399.

Morse, J. G., and Croft, B. A. 1981. Resistance to azinphosmethyl in a predator–prey system in greenhouse experiments. Entomophaga 26: 191–202.

Morse, J. G., and Croft, B. A. 1987. Biological control of *Aphis pomi* (Homoptera: Aphididae) by *Aphidoletes aphidimyza* (Diptera: Cecidomyiidae): A predator–prey model. Entomophaga 32: 339–356.

Morton, H. L., Moffett, J. O., and Stewart, F. D. 1975. Effect of alfalfa looper nuclear polyhedrosis virus on honeybees. J. Invert. Pathol. 26: 139–140.

Mosievskaya, L. M., and Makarov, E. M. 1974. The effect of bacterial preparations on the parasites of the codling moth. Zash. Rast. 11: 25.

Motoyama, N., and Dauterman, W. C. 1980. Glutathione S-transferases: their role in the metabolism of organophosphorus insecticides, pp. 49–69. *In* Review in Biochemical Toxicology. Eds. E. Hodgson, J. R. Bend, and R. M. Philpot. Elsevier/North-Holland, New York, vol. 2.

Motoyama, N., Rock, G. C., and Dauterman, W. C. 1970. Organophosphorous resistance in an apple orchard population of *Typhlodromus* (*Amblyseius*) *fallacis*. J. Econ. Entomol. 63: 1439–1442.

Motoyama, N., Rock, G. C., and Dauterman, W. C. 1972. Studies on the mechanisms of azinphosmethyl resistance in the predaceous mite *Neoseiulus* (*T.*) *fallacis*. Biochem. Physiol. 1: 205–215.

Motoyama, N., Rock, C. G., and Dauterman, W. C. 1977. Toxicity to O-alkyl analogues of azinphosmethyl to resistant and susceptible predaceous mite, *Amblyseius fallacis*. J. Econ. Entomol. 70: 475.

Mouches, C., Pasteur, N., Berge, J. B., Hyrien, O., Raymond, M., De Saint Vincent, B. R., De Silvestri, M., Georghiou, G. P. 1986. Amplification of an esterase gene is responsible for insecticide resistance in a California *Culex* mosquito. Science 233: 778–779.

Mowat, D. J., and Coaker, T. H. 1967. The toxicity of some soil insecticides to carabid predators of the cabbage root fly (*Erioischia brassicae* (Bouche)). Ann. Appl. Biol. 59: 349–354.

Mowrey, P. D., Asquith, D., and Hull, L. A. 1977. MITESIM, a computer predictive system for European red mite management. Penn. Fruit News 56: 64–67.

Muck, O., Hassan, S., Huger, A. M., and Krieg, A. 1981. The effect of *Bacillus thuringiensis* Berliner on the parasitic Hymenopterans *Apanteles glomeratus* L. (Braconidae) and *Pimpla turionella* (L.) (Ichneumonidae). Z. Angew. Entomol. 92: 303–314.

Mueller-Beilschmidt, D., and Hoy, M. A. 1987. Activity levels of genetically manipulated and wild strains of *Metaseiulus occidentalis* (Nesbitt) (Acarina: Phytoseiidae) compared as a method to assay quality. Hilgardia 55(6): 1–23.

Müller, E. W. 1960. Untersuchungen über den Einfluss chemischer Pflanzenschutzmittel auf den Populationsverlauf von Spinnmilben und Raubmilben im Obstbau. Pflanzenschutzdienst 11: 221–230.

Muller, G. 1974. Changes in the coleopteran fauna of the soil-surface of cultivated fields after herbicide treatment. Folia Entomol. Hung. 25(17): 297–305.

Mullin, C. A. 1985. Detoxification enzyme relationships in arthropods of differing feeding strategies, pp. 267–278. *In* Bioregulators for Pest Control. Ed. P. A. Hedin, Symp. Ser. 276, Amer. Chem. Soc.

Mullin, C. A., and Croft, B. A. 1983. Host-related alterations of detoxification enzymes in *Tetranychus urticae* (Acari: Tetranychidae). Environ. Entomol. 12: 1278–1282.

Mullin, C. A., and Croft, B. A. 1984. Trans-epoxide hydrolase: a key indicator enzyme for herbivory in arthropods. Experientia 40: 176–178.

Mullin, C. A., and Croft, B. A. 1985. An update on development of selective pesticides favoring arthropod natural enemies, pp. 123–150. *In* Biological Control of Agricultural Integrated Pest Management Systems. Eds. M. A. Hoy and D. C. Herzog, Academic Press, New York, 589 pp.

Mullin, C. A., Croft, B. A., Strickler, K., Matsumura, F., and Miller, J. R. 1982. Detoxification enzyme differences between a herbivorous and predatory mite. Science 217: 1270–1272.

Naixin, Z. 1983. Personal correspondence to B. A. Croft.

Nakashima, M. J., and Croft, B. A. 1974. Toxicity of benomyl to the life stages of *Amblyseius fallacis*. J. Econ. Entomol. 67: 657–677.

Nakasuji, F., Yamanaka, H., and Kiritani, K. 1973. Control of the tobacco cutworm, *Spodoptera litura* f., with polyphagous predators and ultra-low concentration of chlorophenamidine. Jap. J. Appl. Ent. Zool. 17: 171–180.

Nangia, N., and Channabasavanna, G. P. 1983. Effect of sulphur and dicofol on *Tetranychus ludeni* and its predator *Amblyseius tetranychivorus*, p. 1028. *In* Proc. 10th Intern. Cong. Plant Protec., Brighton, England, Nov. 20–25, 1983.

Naqui, S. M. Z. 1970. Comparative pesticide tolerances of selected freshwater invertebrates in Mississippi. Dissertation, Mississippi State Univ., State College, 55 pp.

Naqui, S. M., and de la Cruz, A. A. 1973. Mirex incorporation in the environment: residues in nontarget organisms—1972. Pestic. Monit. J. 7: 104–111.

NAS/NRC 1986. Pesticide Resistance: Strategies and Tactics for Management. Nat. Acad. Sci. Press, Washington, D.C., 471 pp.

NAS/NRC 1987. Balanced controls urged for engineered organism. NRC News Rept 37: 8–16.

Nasca, A. J., Fernandez, R. V., Herrero, A. J. de, and Manzur, B. E. 1983. Incidence of chemical treatment for control of fruit-flies (Trypetidae) on chrysopids and hemerobiids (Neuroptera) on citrus trees. CITRON 1(2): 47–73.

Nasseh, M. O. 1982. The effect of garlic extract on *Syrphus corollae* F., *Chrysopa carnea* (Steph.) and *Coccinella septempunctata* L. Z. Angew. Entomol. 94: 123–126.

Nath, D. K., and Sarkar, D. 1980. Predacious spiders in BPH endemic area of West Bengal, India. IRRI Newsl. 3(5): 15.

Naton, E. von 1978. On testing the side-effects of pesticides on beneficial arthropods on living plants. Test with *Phygadeuon trichops* Thomson (Ichneumonidae). Anz. Schädl., Pflschutz., Umweltschutz 51: 136–139.

Naton, E. von 1983. Testing of side effects of pesticides on *Phygadeuon trichops* Th. (Hym. Ichneumonidae). Anz. Schädl., Pflschutz., Umweltschutz 56: 82–91.

Natskova, V. 1974. The possibility of combining chemical control of the rose-leaf aphid, *Macrosiphum rosae* (L.) with control by its natural enemies. Rastitelna Zashchita 22(5): 10–13.

Navarajan, P. A. V., Dass, R., Ahmed, R., and Parshad, B. 1979. Effect of some insecticides on parasitism by the parasitoid *Trichogramma brasiliensis* (Ashmead) (Trichogrammatidae: Hymenoptera). Z. Angew. Entomol. 88: 399–403.

Nelson, E. E., Croft, B. A., Howitt, A. J., and Jones, A. L. 1973. Field and laboratory studies on the toxicity of pesticides to *Agistemus fleschneri*. Environ. Entomol. 2: 219–222.

Neuenschwander, P., Hagen, K. S., and Smith, R. F. 1975. Predation on aphids in California's alfalfa fields. Hilgardia 43: 53–78.

Newsom, L. D. 1967. Consequences of insecticide use on nontarget organisms. Annu. Rev. Entomol. 12: 257–286.

Newsom, L. D. 1974. Predator insecticide relationships. Entomophaga Mem. Ser. 7, 88 pp.

Newsom, L. D., Smith, R. F., and Whitcomb, W. H. 1976. Selective pesticides and selective use of pesticides, pp. 565–591. *In* Theory and Practice of Biological Control. Eds. C. B. Huffaker and P. S. Messenger, Academic Press, New York, 788 pp.

Niemczyk, E., Miszczak, M., and Nowakowski, Z. 1974. The influence of pesticides on some predacious insects. Biul. Inst. Ochrony Roslin (1972) 52, pp. 343–365.

Niemczyk, E., Misczyak, M., and Olszak, R. 1979. The toxicity of pyrethroids to predacious and parasitic insects. Roczniki Nauk Rolniczych, E 9: 105–115.

Niemczyk, E., Pruska, M., and Miszczak, M. 1985. Toxicity of diflubenzuron to predacious and parasitic insects. Roczniki Nauk Rolniczych, E. (Ochrona Roslin) 11: 181–191.

Nikusch, I., and Gernot, H. 1986a. Side effects on the beneficial fauna of some insecticides intended for integrated plant protection in apple cultivation. Bull. SROP 9(3): 12–14.

Nikusch, I., and Gernot, H. 1986b. Side-effects of some insecticides on the soil fauna (Carabidae etc.) of strawberries. Bull. SROP 9(3): 39–44.

Niwa, C. G. 1985. Effects of two isolates and dosages of *Bacillus thuringiensis* on parasites of the western spruce budworm, *Choristoneura occidentalis* (Lepidoptera: Tortricidae). Pacific Northwest Research Station, Forestry Sciences Laboratory, Corvallis, Ore. (unpubl.).

Noble, N. S. 1935. The woolly aphid parasite. Effect of orchard sprays on *Aphelinus mali*. Agric. Gaz. New South Wales, Oct. pp. 573–575.

Nohara, K., and Yasumatsu, K. 1968. Observations on the activity of spiders and the effect of insecticides on their populations in the citrus groves around Hagi City, Honshu, Japan. Sce. Bull. Fac. Agric. Kyushu Univ. 23: 151–167.

Norris, D. M. 1986. Anti-feeding compounds, pp. 97–146. *In* Biosynthesis Inhibitors and Antifeeding Compounds. Eds. H. Haug and H. Hoffman, Springer-Verlag, New York.

Nova Scotia Ministry of Agriculture 1982. Nova Scotia Tree Fruit Protection Guide—1982. N. S. Minist. Agric. Publ., 134 pp.

Novozhilov, K. V., Ed. 1984. Agrocoenotic aspects of plant protection. Vsesoyuznyi Nauchno-issledovatel'skii Inst. Zash. Rast., Leningrad, USSR, 102 pp.

Novozhilov, K. V., Kamenkova, K. V., and Smirnova, I. M. 1973. Development of the parasite *Trissolcus grandis* Thoms. (Hymenoptera, Scelionidae) where organophosphorus insecticides are in use against *Eurygaster integriceps* Put. (Hemiptera, Scutelleridae). Entomol. Obozr. 52: 20–28.

O'Brien, P. J., Elzen, G. W., and Vinson, S. B. 1985. Toxicity of azinphosmethyl and chlordimeform to parasitoid *Bracon mellitor* (Hymenoptera: Braconidae): lethal and reproductive effects. Environ. Entomol. 14: 891–894.

O'Brien, R. D. 1961. Selective toxicity of insecticides. Adv. Pest Control Res. 4: 75–116.

O'Brien, R. D. 1967. Insecticides, Action and Metabolism. Academic Press, New York, 255 pp.

O'Neill, W. J., and Hantsbarger, W. H. 1952. Apple and pear investigations at the Tree Fruit Experiment Station. Washington State Hort. Assn. Proc. 48: 170–173.

Oatman, E. R. 1966. Studies on integrated control of apple pests. J. Econ. Entomol. 59: 368–373.

Oatman, E. R., Badgley, M. E., and Platner, G. R. 1985. Predators of the two-spotted spider mite on strawberry. Calif. Agric. 39(1/2): 9–12.

Obadofin, A. A., and Finlayson, D. G. 1977. Interactions of several insecticides and a carabid predator (*Bembidion lampros* Hrbst.) and their effects on *Hylemya brassicae* (Bouche). Can. J. Plant Sci. 57: 1121–1126.

Odum, E. P. 1971. Fundamentals of Ecology, 3rd ed. W. B. Saunders, Philadelphia, Pa.

Oetting, R. D. 1985. Effects of insecticides applied to potting media on *Oenonogastra microrhopalae* (Ashmead) parasitization of *Liriomyza trifolii* (BUrgess). J. Entomol. Sci. 20: 405–410.

Olson, C. S. 1980. Effects of age and diet upon fat body volume and composition in adult female *habrobracon juglandis*. Ann. Entomol. Soc. Amer. 73: 427–431.

Ooi, P. A. C., and Sudderuddin, K. I. 1978. Control of diamondback moth in Cameron Highlands, Malaysia, pp. 214–227. *In* Proc. Conf. Plant Prot. 1978.

Ooi, P. A. C., Lim, G. S., and Koh, A. K. 1979. The feasibility of aerial spraying for control of the brown planthopper in Malaysia. Bull. Ministry Agric. Malaysia. 149, 18 pp.

Oomen, P. A., and Wiegers, G. L. 1984. Selective effects of a chitin synthesis inhibiting insecticide (CME 134–01) on the parasite *Encarsia formosa* and its white fly host *Trialeurodes vaporariorum*. Med. Fac. Landbouww. Rijksuniv. Gent 49(3a): 745–750.

Oonnithan, E. S., and Casida, J. E. 1968. Oxidation of methyl and dimethyl carbamate insecticide chemicals by microsomal enzymes and anticholinesterase activity of the metabolites. J. Agric. Food Chem. 16: 28–44.

Oppenoorth, F. J. 1984. Biochemistry of insecticide resistance. Pestic. Biochem. Physiol. 22: 187–193.

Oppenoorth, F. J., and Welling, W. 1979. Biochemistry and physiology of resistance, pp. 507–551. *In* Insecticide Biochemistry and Physiology. Ed. C. F. Wilkinson, Plenum Press, New York.

Orbtel, R. 1961. Effects of two insecticides on *Aphidius ervi* Hal. (Hym.: Braconidae), an internal parasite of *Acyrthosiphon onobrychis* (Boyer). Zool. Listy 24: 1–8.

Osborne, L. S. 1981. Utility of physiological time in integrating chemical and biological control of greenhouse whitefly. Environ. Entomol. 10: 885–888.

Osborne, L. S., and Petitt, F. L. 1985. Insecticidal soap and the predatory mite, *Phytoseiulus persimilis* (Acari: Phytoseiidae), used in management of the twospotted spider mite (Acari: Tetranychidae) on greenhouse grown foliage plants. J. Econ. Entomol. 78: 687–691.

Osman, A. A., and Keie, I. A. 1975. The effect of some granular pesticides on mites associated with cotton (Acarina). Bull. Entomol. Soc. Egypt, 9: 271–274.

Osman, A. A., and Zhody, G. I. 1976. Toxicity of some pesticides to the predaceous mite *Amblyseius gossipi* El-Badry in Egypt. Bull. Entomol. Soc. Egypt 10: 59–61.

Otvos, I. S., and Raske, A. G. 1980. The effects of fenitrothion, matachil, and *Bacillus thuringiensis* plus orthene on larval parasites of the spruce budworm, *Choristoneura fumiferana* (Lepidoptera: Tortricidae). Information Rep., Can. Forest. Service, 24 pp.

Outram, I. 1974. Influence of juvenile hormone on the development of some spruce budworm parasitoids. Environ. Entomol. 3: 361–363.

Ouyang, Y. L., and Hoy, M. A. 1988. Selection of *Metaseiulus occidentalis* (Acarina: Phytoseiidae) for resistance to permethrin and abamectin, p. 316c. *In* Proc. 18th Intern. Cong. Entomol., Vancouver, B.C., Canada, July 3–9, 1988.

Overmeer, W. P. J., and van Zon, A. Q. 1981. A comparative study of the effects of some pesticides on three predacious mite species: *Typhlodromus pyri, Amblyseius potentillae* and *A. bibens* (Acarina: Phytoseiidae). Entomophaga 26: 3–9.

Overmeer, W. P. J., and van Zon, A. Q. 1982. A standardized method for testing the side effects of pesticides on the predaceous mite, *Amblyseius potentillae* (Acarina: Phytoseiidae). Entomophaga 27: 357–364.

Overmeer, W. P. J., and van Zon, A. Q. 1983. Resistance to parathion in the predaceous mite *Typhlodromus pyri* Scheuten (Acarina: Phytoseiidae). Med. Fac. Landbouww. Rijksuniv. Gent 48: 247–252.

Page, M., Ryan, R. B., Rappaport, N., and Schmidt, F. 1982. Comparative toxicity of acephate, diflubenzuron, and malathion to larvae of the larch casebearer, *Coleophora joricella* (Lepidoptera: coleophoridae) and adults of its parasite, *Chrysocharis laricinellae*. Environ. Entomol. 11: 730–732.

Paine, R. T. 1966. Food web complexity and species diversity. Amer. Nat. 100: 65–75.

Panis, A. 1980. Damage caused by the Coccidae and Pseudoccoccidae (Homoptera, Coccoidea) of citrus in France and special effects of some pesticides on the orchard entomocoenosis. Fruits 35: 779–782.

Pape, D. J., and Crowder, L. A. 1981. Toxicity of methyl parathion and toxaphene to several insect predators in central Arizona. Southwest Entomol. 6: 44–48.

Paradis, R. O. 1981. Pest management approach in Quebec apple orchards. Res. Br. Agric. Can. Tech. Bull. 16, 31 pp.

Parent, B. 1961. Effects of certain pesticides products on *Typhlodromus rhenanus* (Oudms) and *Mediolata mali* (Ewing), two mite predators of the apple-tree red spider mites. Ann. Entomol. Soc. Quebec 6: 55–58.

Parent, B. 1967. Population studies of phytophagous mites and predators on apple in southwestern Quebec. Can. Entomol. 99: 771–778.

Parker, B. L., Ming, N. S., Peng, T. S., and Singh, G. 1976. The effect of malathion on fecundity, longevity, and geotropism of *Menochilus sexmaculatus*. Environ. Entomol. 5: 495–501.

Parker, F. D., Lawson, F. R., and Pinnell, R. E. 1971. Suppression of *Pieris rapae* using a new control system: mass releases of both the pest and its parasites. J. Econ. Entomol. 64: 721–735.

Parr, W. J., and Binns, E. S. 1971. Integrated control of red spider mite. Rept. Glasshouse Crops Res. Inst. 1970, pp. 119–121.

Parrella, M. P., McCaffrey, J. P., and Horsburgh, R. L. 1980. Compatibility of *Leptothrips mali* and *Stethorus punctum* and *Orius insidiosus*: Predators of *Panonychus ulmi*. Environ. Entomol. 9: 694–696.

Parrella, M. P., McCaffrey, J. P., and Horsburgh, R. L. 1981a. Comparison of two sampling methods for *Leptothrips mali* in Virginia apple orchards. J. N. Y. Entomol. Soc. 89: 166–169.

Parrella, M. P., McCaffrey, J. P., and Horsburgh, R. L. 1981b. Population trends of selected phytophagous arthropods and predators under different pesticide programs in Virginia apple orchards. J. Econ. Entomol. 74: 492–498.

Parrella, M. P., Rowe, D. L., and Horsburgh, R. L. 1982. Biology of *Leptothrips mali*, a common predator in Virginia apple orchards. Ann. Entomol. Soc. Am. 75: 130–135.

Pasqualini, E., and Malavolta, C. 1985. Possibility of natural limitation of *Panonychus ulmi* (Koch) (Acarina, Tetranychidae) on apple in Emilia-Romagna. Boll. Inst. Entomol. 'Guido Grandi' Univ. Studi Bologna 39: 221–230.

Pasteels, J. M. 1978. Apterous and brachypterous coccinellids at the end of the food chain, *Cionura erecta* (Asoepiadaceae)—*Aphis nerii*. Entomol. Exp. Appl. 24: 379–384.

Paul, A. V. N., Mohanasundaram, M., and Sobramanian, T. R. 1976. Effect of insecticides on the survival and emergence of the egg parasite, *Trichogramma* spp. Madras Agric. J., pp. 557–560.

Pavlova, G. A. 1975. *Stethorus*—a predator of the spider mite. Zash. Rast. 1: 23–24.

Pawlizki, K. H. 1984. Effects of graduated production intensities on the activity of field carabids (Coleoptera, Carabidae) and the self-regulation of agroecosystems. Bayer. Landwirtschaft. Jahrbuch, Sonderheft 61(2): 11–40.

Pedigo, L. P. 1967. Comparative biocoenotics of pond shore collembola. Ph.D. dissertation, Purdue Univ., Lafayette, Ind., 157 pp.

Peleg, B. A. 1983. Effect of a new insect growth regulator, RO 13–5223, on hymenopterous parasites of scale insects. Entomophaga 28: 367–372.

Peleg, B. A. 1988. Effect of a new phenoxy juvenile hormone analog on California red scale (Homoptera: Diaspididae), florida wax scale (Homoptera: Coccidae) and the ectoparasite *Aphytis holoxanthus* DeBach (Hymenoptera: Aphelinidae). J. Econ. Entomol. 81: 88–92.

Peleg, B. A., and Gothilf, S. 1980. Effect of the juvenoid Altosid on the development of three hymenopterous parasites. Entomophaga 25: 232–237.

Penman, D. R., and Chapman, R. B. 1980. Integrated control of apple pests in New Zealand. 17. Relationships of *Amblysieus fallacis* to phytophagous mites in an apple orchard. New Zealand J. Zool. 7: 281–287.

Penman, D. R., Ferro, D. N., and Wearing, C. H. 1976. Integrated control of apple pests in New Zealand. 7. Azinphosmethyl resistance in strains of *Typhlodromus pyri* from Nelson. New Zealand J. Exp. Agric. 4: 377–380.

Penman, D. R., Wearing, C. H., Collyer, E., and Thomas, W. P. 1979. The role of insecticide-resistant phytoseiids in integrated mite control in New Zealand. Rec. Adv. Acarol. 1: 59–69.

Penman, D. R., Chapman, R. B., and Jesson, K. S. 1981. Effects of fenvalerate and azinphosmethyl on two spotted spider mite and phytoseiid mites. Entomol. Exp. Appl. 30: 91–97.

Perera, P. A. C. R. 1982. Some effects of insecticide deposit patterns on the parasitism of *Trialeurodes vaporariorum* by *Encarsia formosa*. Ann. Appl. Biol. 101: 239–244.

Perkins, J. H. 1982. Insects, Experts and the Insecticide Crisis. Plenum Press, New York/London, 304 pp.

Peter, M. G., Dahm, K. H., and Roller, H. 1976. The juvenile hormones in blood of larvae and adults of *Manduca sexta* (Joh.). Z. Naturforsch. 31c: 129–131.

Peterson, R. C., Dunn, P. E., Seballos, H. L., Barbeau, B. K., Kein, P. S., Rilley, C. T., Heirickson, R. L., and Law, J. H. 1982. Juvenile hormone carrier protein of *Manduca sexta* haemolymph. Improved purification procedure, protein modification studies and sequence of the amino terminus of the protein. Insect Biochem. 12: 643–650.

Petrukka, O. I., and Gres', Y. A. 1974. *Thaumatomyia*—a predator of the beet root aphid. Zash. Rast. 11: 24–25.

Petrushov, A. Z., and Zelenkova, E. V. 1976. Methods of biological and ecological investigation with *Phytoseiulus*. Zash. Rast. 8: 19–20.

Pfeiffer, D. G. 1985. Toxicity of avermectin B_1 to San Jose scale (Homoptera: Diaspididae) crawler, and effects on orchard mites by crawler sprays compared with full-season applications. J. Econ. Entomol. 78: 1421–1424.

Phillips, J. 1981. Comparative physiology of insect renal function. Ann. J. Physiol. 241: R241-R257.

Pickett, A. D., Putman, W. L., and LeRoux, E. J. 1958. Progress in harmonizing biological and chemical control of orchard pests in eastern Canada, pp. 169–174. *In* Proc. X Intern. Cong. Entomol., vol. 3.

Pickford, R. J. J. 1983. A selective method of thrips control of cucumbers. Bull. SROP 6(3): 177–180.

Pielou, D. P. 1950. Selection for DDT tolerance in *Macrocentrus ancylivorus*. Rept. Entomol. Soc. Ontario 81: 44–45.

Pielou, D. P., and Glasser, R. F. 1951. Selection for DDT tolerance in a beneficial parasite, *Macroncentrus ancylivorus* 1. Some survival characteristics and the DDT resistance of the original laboratory strain. Can. J. Zool. 29: 90–101.

Pielou, D. P., and Glasser, R. F. 1952. Selection for DDT resistance in a beneficial insect parasite. Science 115: 117–118.

Pimentel, D. 1961. An ecological approach to the insecticide problem. J. Econ. Entomol. 52: 1103–1105.

Pimentel, D., and Wheeler, A. G. 1973. Influence of alfalfa resistance on a pea aphid population and its associated parasites, predators, and competitors. Environ. Entomol. 2: 1–11.

Pinsdorf, W. 1977. Vorläufige Richtlinie zur Prüfung der Wirkung von Pflanzenschutzmitteln auf *Coccinella septempunctata* L. Biol. Bundesanst. Lord-Forstwirtsch., Braunschweig, 17 pp.

Plapp, F. W. 1980. Ways and means of avoiding or ameliorating resistance to insecticides. Proc. 4th Intern. Cong. Plant Prot., Hamburg, Germany, 1957.

Plapp, F. W. 1981. The nature, modes of action, and toxicity of insecticides, pp. 3–16. *In* Handbook of Pest Management in Agriculture. Ed. D. Pimentel, CRC Press, Boca Raton, Fla., vol. 3, 656 pp.

Plapp, F. W., and Bull, D. L. 1978. Toxicity and selectivity of some insecticides to *Chrysopa carnea*, a predator of the tobacco budworm. Environ. Entomol. 7: 431–434.

Plapp, F. W., and Vinson, S. B. 1977. Comparative toxicities of some insecticides to the tobacco budworm and its ichneumonid parasite, *Campoletis sonorensis*. Environ. Entomol. 6: 381–384.

Plattner, H. C. 1979. Grundlagen und Methoden zur Prüfung der Nebenwirkungen von Pflanzenschutzmitteln auf den Parasiten *Phygadeuon trichops* Thomson. Dissertation, 188pp.

Plattner, H. C., and Naton, E. von, 1975. Testing the action of pesticides on beneficial arthropods. The state of preparatory work in Bavaria. Bayer. Landwirtschaft. Jahrbuch 52: 143–147.

Plewka, T., Kot, J., and Krukierek, T. 1975. Effect of insecticides on the longevity and fecundity of *Trichogramma evanescens* Westw. (Hymenoptera, Trichogrammatidae). Polish Ecol. Stud. 1: 197–210.

Pluthero, G. G., and Singh, R. S. 1981. Genetic differences in malathion avoidance and resistance in *Drosophila melanogaster*. J. Econ. Entomol. 74: 736–740.

Pluthero, F. G., and Threlkeld, S. G. H. 1984. Mutations in *Drosophila melanogaster* affecting physiological and behavioral response to malathion. Can. Entomol. 116: 411–418.

Poe, S. L. 1974a. Emergence of *Keiferia lycopersicella* (Lypidoptera: Gelechiidae) and *Apanteles* sp. (Hymenoptera: Braconidae) from pupae and soil treated with insect growth regulators. Entomophaga 19: 205–211.

Poe, S. L. 1974b. *Liriomyza munda* and parasite mortality from insect growth regulators. Florida Entomol. 57: 415–417.

Poe, S. L., and Enns, W. R. 1969. Predaceous mites associated with Missouri orchards. Trans. Mo. Acad. Sci. 3: 69–82.

Poehling, H. M., Dehne, H. W., and Sprick, P. 1985a. Investigations on the significance of carabids and staphylinids as natural enemies of aphids in winter wheat and on the adverse effects of insecticide products on them. Med. Fac. Landbouwwe, Rijksuniv. Gent 50(2b): 519–530.

Poehling, H. M., Dehne, H. W., and Picard, K. 1985b. Investigations on the use of fenvalerate for the control of cereal aphids in winter wheat, with special reference to side-effects on beneficial arthropods. Med. Fac. Landbouwwe, Rijksuniv. Gent 50(2b): 539–554.

Porres, M. A., McMurtry, J. A., and March, R. B. 1975. Investigations of leaf sap feeding by three species of phytoseiid mites by labelling with radioactive phophoric acid ($H_2^{32}PO_4$). Ann. Entomol. Soc. Amer. 68: 871–872.

Portillo, M. M. 1977. Action of the aphicide pirimicarb on the citrus aphid *Toxoptera citricidus* (Kirkalky), and on the coccinellid predator *Cycloneda sanguinea* (L.). Idia 331: 63–66.

Powell, J. E., and Scott, W. P. 1985. Effect of insecticide residues on survival of *Microplitis croceipes* adults (Hymenoptera: Braconidae) in cotton. Florida Entomol. 68: 692–693.

Powell, W. 1980. *Toxares deltiger* parasitizing the cereal aphid *Metopolophium dirhodum* in southern England: a new host parasitoid record. Bull. Entomol. Res. 70: 407–409.

Powell, W., Dean, G. J., and Bardner, R. 1986a. Effects of pirimicarb, dimethoate and benomyl on natural enemies of cereal aphids in winter wheat. Ann. Appl. Biol. 106: 235–242.

Powell, W., Wilding, N., Brobyn, P. J., and Clark, S. J. 1986b. Interference between parasitoids (Hym.: Aphidiidae) and fungi (Entomophthorales) attacking cereal aphids. Entomophaga 31: 293–302.

Pradhan, S., Jotwani, M. G., and Sarup, P. 1959. Effect of some important insecticides on *Coccinella septempunctata* a predator of mustard aphid (*Lipaphis erysimi* Kalt.). Indian Oilseeds J. 3(2): 121–124.

Pralavario, M., Fournier, D., and Berge, J. B. 1988. Analysis of methidathion resistance in *Phytoseiulus persimilis*, p. 316. *In* Proc. 18th Intern. Cong. Entomol., Vancouver, B.C., Canada, July 3–9, 1988.

Pree, D. J. 1979. Toxicity of phosmet, azinphosmethyl, and permethrin to the Oriental fruit moth and its parasite, *Macrocentrus ancylivorus*. Environ. Entomol. 8: 969–972.

Pree, D. J., and Hagley, E. A. C. 1985. Toxicity of pesticides to *Chrysopa oculata* Say (Neuroptera: Chrysopidae). J. Econ. Entomol. 78: 129–132.

Pree, D. J., Hagley, E. A. C., Simpson, C. M., and Hikichi, A. 1980. Resistance of the spotted tentiform leafminer, *Phyllonorycter blancardella* (Lepidoptera: Gracillariidae) to organophosphate insecticides in southern Ontario. Can. Entomol. 112: 469–474.

Pree, D. J., Marshall, D. B., and Archibald, D. E. 1987. Resistance to pyrethroid insecticides in the spotted tentiform leafminer, *Phyllonorycter blancardella* (Lepidoptera: Gracillariidae), in southern Ontario. J. Econ. Entomol. 79: 318–322.

Pree, D. J., Archibald, D. E., and Morrison, R. K. 1989. Resistance to insecticides in the common green lacewing *Chrysoperla carnea* (Stephens) (Neuroptera: Chrysopidae) in southern Ontario. J. Econ. Entomol. 82: 29–34.

Press, J. W., Flaherty, B. R., and McDonald, L. L. 1981. Survival and reproduction of *Bracon hebetor* on insecticide-treated *Ephestia cautella* larvae. J. Georgia Entomol. Soc. 16: 231–234.

Prestwich, G. D., Gayen, A. K., Phirwa, S., and Kline, T. B. 1983. 29-Fluorophytosterols: novel pro-insecticides which cause death by dealkylation. Bio/technol. 1: 62–65.

Prestwich, G. D., Eng, W. S., Roe, R. M., and Hammock, B. D. 1984. Synthesis and bioassay of isoprenoid 3-alkylthio-1, 1, 1-fluoro-2-propanones: potent, selective inhibitors of juvenile hormone esterase. Arch. Biochem. Biophys. 228: 639–645.

Price, J. F., and Shepard, M. 1978. *Calosoma sayi*. Seasonal history and response to insecticides in soybeans. Environ. Entomol. 7: 359–363.

Price, P. W. 1975. Insect Ecology. Wiley Interscience, New York, 514 pp.

Price, P. W. 1981. Semiochemicals in evolutionary time, pp. 251–279. *In* Semiochemicals: Their Role in Pest Control. Eds. D. A. Nordlund, R. L. Jones, and W. J. Lewis. Wiley, New York.

Price, P. W., and Waldbauer, G. P. 1975. Ecological aspects of pest management, pp. 37–73. *In* Introduction to Pest Management. Eds. W. Luckmann and R. L. Metcalf, Wiley Interscience, New York.

Price, P. W., Bouton, C. E., Gross, P., McPheron, B. A., Thompson, J. N., and Weis, A. E. 1980. Interactions among three trophic levels: influence of plants on interactions between insect herbivores and natural enemies. Annu. Rev. Ecol. Syst. 11: 41–65.

Pristavko, V. 1966. Processus pathologiques consécutifs à l'action de *Beauveria bassiana* associé à de faibles doses de DDT, chez *Leptinotarsa decimlineata*. Entomophaga 11: 311–324.

Prokopy, R. J., Coli, W. M., Hislop, R. G., and Hauschild, K. I. 1981. Integrated management of insect and mite pests in commercial apple orchards in Massachusetts. Environ. Entomol. 10: 529–535.

Pruszynski, S. 1984. Theoretical and practical principles for applying integrated methods in the protection of greenhouse crops against pests. Prace Naukowe Inst. Ochrony Roslin 26(1): 13–68.

Pruszynski, S., and Cone, W. W. 1973. Biological observations of *Typhlodromus occidentalis* (Acarina: Phytoseiidae) on Hops. Ann. Entomol. Soc. Amer. 66: 47–51.

Pruszynski, S., Siwek, L., Aumiller, P., and Konopinska, M. 1983. Results of biological control of spider mites (Tetranychidae) using the predacious mite *Phytoseiulus persimilis* in production glasshouses of the Naramowice state horticultural farm. Prace Naukowe Inst. Ochrony Roslin 24(2): 161–172.

Prutenskaya, M. D., Kryzhanovskaya, T. V., and Zhurba, L. N. 1984. Perspectives for the combined use of a predacious bug and some pesticides in glasshouses in botanical gardens. Dokl. Acad. Nauk Ukrainskoi SSR.B 9: 70–73.

Purcell, M., and Granett, J. 1985. Toxicity of benzoylphenyl ureas and thuringiensin to *Trioxys pallidus* (Hymenoptera: Braconidae) and the walnut aphid (Homoptera: Aphididae). J. Econ. Entomol. 78: 1133–1137.

Puritch, C. S., Tonks, N., and Downey, P. 1982. Effect of a commercial insecticidal soap on greenhouse whitefly (Hom.: Aleyrod.) and its parasitoid, *Encarsia formosa* (Hym.: Euloph.). J. Entomol. Soc. Brit. Columbia 79: 25–28.

Putman, W. L. 1956. Differences in susceptibility of two species of *Chrysopa* (Neuroptera: Chrysopidae) to DDT Can. Entomol. 88: 520.

Puttaswamy and Channabasavanna, G. P. 1976. Effect of some important acaricides on the predator *Stethorus pauperculus* Weise (Coleoptera: Coccinellide) and its host *Tetranychus cucurbitae* Rahman and Sapra. Mysore J. Agric. Sci. 10: 636–641.

Quayle, H. J. 1943. The increase in resistance in insects to insecticides. J. Econ. Entomol. 36: 493–500.

Quist, J. A. 1974. Apple insect population management of orchards in Colorado, New Mexico and Utah. Final Rept. of the Demonstration Grant, Four Corners Regional Commission, Colorado State Agric. Expt. Sta.

Rabb, R. L., and Kennedy, G. G. 1979. Movement of Highly Mobile Insects: Concepts and Methodology in Research. N. Carolina State Graph, 456 pp.

Rabb, R. L., Stinner, R. E., and van den Bosch, R. 1976. Conservation and augmentation of natural enemies, pp. 209–232. *In* Theory and Practice of Biological Control. Eds. C. B. Huffaker and P. S. Messenger, Academic Press, New York.

Rabbinge, R., and Hoy, M. A. 1980. A population model for two-spotted spider mite, *Tetranychus urticae* and its predator *Metaseiulus occidentalis*. Entomol. Exp. Appl. 28: 64–81.

Radke, S. G., and Barwad, W. L. 1978. New record of a parasite, *Aphidencyrtus aphidivorus* (Mayr) (Encyritidae: Hymenoptera) on *Rhopalosiphum maidis* (Fitch) and the efficacy of the various insecticides on the host and effect on parasitism. Indian J. Entomol. 40: 59–62.

Radwan, Z., and Lovei, G. L. 1982. Distribution and bionomics of ladybird beetles living in an apple orchard near Budapest, Hungar. Z. Ang. Entomol. 94: 169–175.

Raghunath, T. A. V. S. 1983. Effects of the applications of malathion on scale insect and its parasite *Anabrolepis mayurai* N & S (Hymenoptera: Encyrtidae). Coop. Sugar 14: 629–632.

Ragusa, S. 1977. Influence of azinphos-ethyl, endosulfan, mecarbam, omethoate, pyrethroids and unden on phytoseiid mites (Acarina: Mesostigmata). Atti Giornate Fitopatologicha, pp. 139–143.

Ragusa, S. 1983. Side-effects of chemicals on phytoseiid mites on hazel. Atti XIII Cong. Naz. Italiano Entomol., pp. 621–628.

Rai, B. K., Kanta, S., and Lal, R. 1964. Relative toxicity of some pesticides to *Eutetranychus banksi* (McGregor), (Tetranychidae: Acarina). Indian Oilseeds J. 8: 360–363.

Raj, D., and Ramakrishnan, N. 1983. Relative toxicity of menazon to mustard aphid, and its predator. Pesticides 17(6): 17–18.

Rajagopal, S., and Kareem, A. 1983. Studies on the toxic effects of some insecticides on parasites (*Aphelinus mali* Halt. and *Aphidius platensis* Breth.) and predator (*Menochilus sexmaculatus* F.) of chili aphid, *Myzus persicae* Sulzer, Pranikee 4: 308–315.

Rajakulendran, S. V., and Plapp, F. W. 1982a. Comparative toxicity of five synthetic pyrethroids to the tobacco budworm (Lepidoptera: Noctuidae) and a predator, *Chrysopa carnea*. J. Econ. Entomol. 75: 769–772.

Rajakulendran, S. V., and Plapp, F. W. 1982b. Synergism of five synthetic pyrethroids by chlordimeform against the tobacco budworm (Lepidoptera: Noctuidae) and a predator, *Chrysopa carnea* (Neuroptera: Chrysopidae). J. Econ. Entomol. 75: 1089–1092.

Ravensberg, W. J. 1981. The natural enemies of the woolly apple aphid *Eriosoma lanigerum* (Hausm.) (Homoptera: Aphididae), and their susceptibility to diflubenzuron. Med. Fac. Landbouww. Rijksuniv. Gent 46: 437–442.

Readshaw, J. L. 1973. Orchard mites, resistant predators, and dissemination. CSIRO, Div. Entomol. Rept. (1972–1973) pp. 95–96.

Readshaw, J. L. 1975. Biological control of orchard mites in Australia with an insecticide-resistant predator. J. Aust. Inst. Agric. Sci. 41: 213–214.

Readshaw, J. L. 1977. Routine technique for testing the effect of pesticides on predatory mites in the laboratory. Proc. XV Intern. Cong. Entomol., Washington, D.C., 1976, Z. Pflkrankh. Pflschutz. 84: 167.

Readshaw, J. L. 1979. Current status of biological control in deciduous fruit orchards in Australia. Aust. Appl. Ent. Research Conf., Canberra, 1979, pp. 240–246.

Readshaw, J. L., Helm, K. F., and Lee, B. 1982. Guidelines for controlling orchard mites using pesticide-resistant predators, pp. 173–176. *In* Proc. Australasian Workshop on Develop. and Implem. of IPM, Auckland, New Zealand.

Reboulet, J. N. 1986. Advances in a field methodology for defining the medium-term effects of pesticides on the beneficial fauna. Bull. SROP 9(3): 46–54.

Reboulet, J. N., Blanc, M., and Bony, D. 1984. Short-term effects of pesticides on the natural enemies of pear psyllids. Results of tests in pear orchards. Bull. SROP 7(5): 297–300.

Redmond, K. R., and Brazzel, J. R. 1968. Responses of the striped lynx spider *Oxyopes salticus* to two commonly used pesticides. J. Econ. Entomol. 61: 327–328.

Reede, R. H. de, Groendijk, R. F., and Wit, A. K. H. 1984. Field tests with the insect growth regulators, epofenonane and fenoxycarb, in apple orchards against leafrollers and side-effects on some leafroller parasites. Entomol. Exp. Appl. 35: 275–281.

Reeve, R. J., and French, J. V. 1978. Laboratory toxicity of pesticides to the brown lacewing, *Sympherobius barberi* (Banks). Southwest Entomol. 3: 121–123.

Reichstein, T. E., Euw, J. U., Parsons, J. A., and Rothschild, M. 1968. Heart poisons in the monarch butterfly. Science 161: 861–866.

Reissig, W. H. 1978. Biology and control of the obliquebanded leafroller on apples. J. Econ. Entomol. 71: 804–809.

Reissig, W. H., Heinrichs, E. A., and Valencia, S. L. 1982. Effects of insecticides on *Nilaparvata lugens* and its predators: spiders, *Microvelia atrolineata*, and *Cyrtorhinus lividipennis*. Environ. Entomol. 11: 193–199.

Reissig, W. H., Stanley, B. H., Valla, M. E., Seem, R. C., and Bourke, J. B. 1985. Effects of surface residues of azinphosmethyl in apple maggot behavior, oviposition and mortality. Environ. Entomol. 12: 815–822.

Respicio, N. C., and Forgash, A. J. 1984. Contact toxicity of six insecticides to the gypsy moth (Lepidoptera: Lymantriidae) and its parasites *Brachymeria intermedia* (Hymenoptera: chalcidiae) and *Compsilura concinnata* (Diptera: Tachinidae). Environ. Entomol. 13: 1357–1360.

Retnakaran, A., Granett, J., and Ennis, T. 1985. Insect growth regulators, pp. 529–601. *In* Comprehensive Insect Physiology, Biochemistry and Pharmacology. Eds. G. A. Kerkut and L. I. Gilbert, Pergamon Press, Oxford.

Reynolds, H. T., van den Bosch, R., and Dietrick, E. J. 1953. Systox on cotton, systemic insecticide successful in southern California control tests. Calif. Agric. 7:9.

Rice, R. E., Jones, R. A., and Hoffman, H. C. 1976. Seasonal fluctuations in phytophagous and predaceous mite populations on stone fruits in California. Environ. Entomol. 5: 557–564.

Riddiford, L. M. 1975. Host hormones and insect parasites, pp. 339–353. *In* Invertebrate Immunity. Eds. K. Maramorosch and R. E. Schope, Academic Press, New York.

Ridgway, R. L. 1969. Control of the bollworm and tobacco budworm through concentration and augmentation of predaceous insects. Proc. Tall Timbers Conf. Animal Contr. Habitat Mgmt. 1: 127–144.

Ridgway, R. L., and Vinson, S. B., Eds. 1977. Biological Control by Augmentation of Natural Enemies, pp. 281–450. Plenum Press, New York, 480 pp.

Ridgway, R. L., Lingren, P. D., Cowan, C. B., and Davis, J. W. 1967. Populations of arthropod predators and *Heliothis* spp. after applications of systemic insecticides to cotton. J. Econ. Entomol. 60: 1012–1016.

Ridland, P. M., Morris, D. S., Williams, D. G., and Tomkins, R. B. 1986. The occurrence of *Phytoseiulus persimilis* Athias-Henriot (Acarina: Phytoseiidae) in an orchard in Victoria. J. Aust. Entomol. Soc. 25: 79–80.

Riedl, H., and Hoying, S. A. 1980. Impact of fenvalerate and diflubenzuron on target and non-target arthropod species on Bartlett pears in northern California. J. Econ. Entomol. 73: 117–122.

Riedl, H., and Hoying, S. A. 1983. Toxicity and residual activity of fenvalerate to *Typhlodromus occidentalis* (Acari: Phytoseiidae) and its prey *Tetranychus urticae* (Acari: Tetranychidae) on pear. Can. Entomol. 115: 807–813.

Riedl, H., Seaman, A., and Henrie, F. 1985. Monitoring susceptibility to azinphosmethyl in field populations of the codling moth (Lepidoptera: Tortricidae) with pheromone traps. J. Econ. Entomol. 78: 697–699.

Ripper, W. E. 1956. Effects of pesticides on the balance of arthropod populations. Annu. Rev. Entomol. 1:403–438.

Ripper, W. E. 1957. Selective insecticides and the balance of Arthropod populations. Agric. Chem. 12(2): 36–7, 103, 105.

Ripper, W. E. 1959. Selective insect control and its application to the resistance problem. Publ. Entomol. Soc. Amer. 2: 153–156.

Ripper, W. E., Greenslade, R. M., Health, J., and Barker, K. 1948. New formulation of DDT with selective properties. Nature 161:484.

Ripper, W. E., Greenslade, R. M., and Lickerish, L. A. 1949. Combined chemical and biological control of insects by means of a systemic insecticide. Nature 163: 787–789.

Ripper, W. E., Greenslade, R. M., and Hartley, G. S. 1951. Selective insecticides and biological control. J. Econ. Entomol. 44: 448–459.

Riviere, J. L. 1975. The effect of a juvenile hormone analogue on the development of an entomophagous insect *Pales pavida* (Dipt.: Tachinidae). Entomophaga 20: 373–379.

Rizk, G. A., Sheta, I. B., and Ali, M. A. 1983a. Chemical control of the citrus brown mite, *Eutetranychus orientalis* (Klein), and relative effects on its predators. Bull. Entomol. Soc. Egypt (Econ.) 11: 97–104.

Rizk, G. A., Sheta, I. B., and Ali, M. A. 1983b. Chemical control of mites infesting grape-vine in Middle Egypt. Bull. Entomol. Soc. Egypt (Econ.) 11: 105–111.

Roberti, D., and Monaco, R. 1967. Observations carried out in Apulia in 1966 on the ectophagous parasites of the olive fly (*D. oleae*) in relation, also, to treatments with phosphoric esters. Entomologica 3: 237–275.

Robertson, J. G. 1957. Changes in resistance to DDT in *Macrocentrus ancylivorus* Rohw. (Hymenoptera: Braconidae). Can. J. Zool. 35: 629–633.

Robertson, J. L. 1972. Toxicity of zectran aerosol to the California oakworm, a primary parasite, and a hyperparasite. Environ. Entomol. 1: 115–117.

Robertson, L. N., Firth, A. C., and Davison, R. H. 1981. Predation on Australian soldiers fly, *Inopus rubriceps* (Diptera: Stratiomyidae), in pasture. New Zealand J. Zool. 8:431–439

Rock, G. C. 1972. Integrated control of mites. Agric. Expt. Serv., Fruit Insect Note A-1, N. Carolina State Univ. Publ., 18 pp.

Rock, G. C., and Apple, J. L. 1983. Organization and objectives of integrated pest and orchard management systems (IPOMS) project. N. Carolina State Univ. Publ., 17 pp.

Rock, G. C., and Yeargan, D. R. 1970. Relative toxicity of plictran to the European red mite, the two-spotted spider mite and the predaceous mite *Neoseiulus* (*Typhlodromus*) *fallacis*. Down to Earth 26: 1–4.

Rock, G. C., and Yeargan, D. R. 1971. Relative toxicity of pesticides to organophosphorus-resistant orchard populations of *Neoseiulus fallacis* and its prey. J. Econ. Entomol. 64: 350–352.

Rock, G. C., and Yeargan, D. R. 1972. Laboratory studies on toxicity of dinocap to *Neoseiulus fallacis* and its prey. J. Econ. Entomol. 65:932–933.

Rodriguez, E., and Levin, D. A. 1976. Biochemical parallelisms of repellents and attractants in higher plants and arthropods, pp. 214–270. *In* Recent Advances in Phytochemistry. Eds. J. W. Wallace and R. L. Mansell, Plenum Press, New York, vol. 10.

Root, R. B. 1973. Organization of a plant arthropod association in simple and diverse habitats: the fauna of collards. Ecol. Monographs 43:95–124.

Rosca, I., and Popov, C. 1983. Role of chemical treatments applied against cereal bugs on egg parasites. Studii Cercetari Biol., Biol. Animala 35(2): 148–152.

Rose, H. A., and Hooper, G. H. S. 1969. The susceptibility to insecticides of *Cydia pomonella* (L.) (Lepidoptera: Tortricidae) from Queensland. J. Aust. Entomol. Soc. 8: 79–86.

Rosen, D. 1967. Effect of commercial pesticides on the fecundity and survival of *Aphytis holoxanthus* (Hymenoptera: Aphelinidae). Israel J. Agric. Res. 17:47–52.

Rosenheim, J. A., and Hoy, M. A. 1986. Intraspecific variation in levels of pesticide resistance in field populations of a parasitoid, *Aphytis melinus* (Hymenoptera: Aphelinidae): the role of past selection pressures. J. Econ. Entomol. 79: 1161–1173.

Rosenheim, J. A., and Hoy, M. A. 1988. Genetic improvement of a parasitoid biological control agent: artificial selection for insecticide resistance in *Aphytis melinus* (Hymenoptera: Aphelinidae). J. Econ. Entomol. 81: 476–483.

Rosenstiel, R. G. 1950. Reactions of two-spotted mite and predator populations to acaricides. J. Econ. Entomol. 43:949–950.

Rosenthal, G. A., and Jansen, D. H., Eds. 1979. Herbivores, Their Interaction with Secondary Plant Metabolites. Academic Press, New York, 477 pp.

Roslavtseva, S. A. 1980. Pesticide resistance of natural enemies of arthropods. Khim Sel'skom Khoz. 18(5): 46–48.

Ross, D. C., and Brown, T. M. 1982. Inhibition of larval growth in *Spodoptera frugiperda* by sublethal dietary concentrations of insecticides. J. Agric. Food. Chem. 30: 193–196.

Rothschild, M. 1972. Secondary plant substances and warning coloration in insects, pp. 59–83. *In* Insect/Plant Relationships, Royal Entomol. Soc. London Symp. 6, Ed. H. F. van Emden, Blackwell Sci. Publ., Oxford.

Roush, R. T. 1979. Selection for insecticide resistance in *Metaseiulus occidentalis* (Nesbitt) (Acarina: Phytoseiidae): Genetic improvement of a spider mite predator. Ph.D. thesis. Univ. Calif. Berkeley, 87 pp.

Roush, R. T., and Croft, B. A. 1986. Experimental population genetics and ecology studies of resistance in arthropods, pp. 257–280. *In* Pesticide Resistance: Strategies and Tactics for management. NAS/NRC Press, Washington, D.C., 471 pp.

Roush, R. T., and Hoy, M. A. 1981a. Genetic improvement of *Metaseiulus occidentalis*: selection with methomyl, dimethoate, and carbaryl and genetic analysis of carbaryl resistance. J. Econ. Entomol. 74: 138–141.

Roush, R. T., and Hoy, M. A. 1981b. Laboratory, glasshouse, and field studies of artificially selected carbaryl resistance in *Metaseiulus occidentalis*. J. Econ. Entomol. 74: 142–147.

Roush, R. T., and Luttrell, R. G. 1987. The phenotypic expression of resistance in *Heliothis* and implications for resistance management, pp. 220–24. *In* Proc. 1987 Beltwide Cotton Res. Conf.

Roush, R. T., and McKenzie, J. 1987. Ecological genetics of insecticide and acaricide resistance. Annu. Rev. Entomol. 32: 361–380.

Roush, R. T., and Miller, G. L. 1986. Considerations for the design of insecticide resistance monitoring programs. J. Econ. Entomol. 79: 293–298.

Roush, R. T., and Plapp, F. W. 1982. Biochemical genetics of resistance to aryl carbamate insecticides in the predaceous mite, *Metaseiulus occidentalis*. J. Econ. Entomol. 75: 304–307.

Roush, R. T., Peacock, W. L., Flaherty, D. L., and Hoy, M. A. 1980. Dimethoate-resistant spider mite predator survives field tests. Calif. Agric. 34(5): 12–13.

Rudd, W., Ruesink, W. G., Newsom, L. D., Herzog, D. C., Jensen, R. L., and Marsolan, N. S. 1980. The systems approach to research and decision-making for soybean pest control, pp. 99–122. *In* New Technology of Pest Control. Ed. C. B. Huffaker, Wiley Interscience, New York, 500 pp.

Ruzicka, Z., Sehnal, F., and Cairo, V. G. 1974. The effects of juvenoids on the hover fly *Syrphus corollae* Fabr. (Dipt., Syrphidae). Z. Angew. Entomol. 76: 430–438.

Ruzicka, Z., Sehnal, F., and Holman, J. 1978. Effects of juvenoids on aphid predators. Acta Entomologica Bohemoslovace 75: 369–378.

Saba, F. 1978. Selection of pesticides for integrated pest control programmes in Moroccan citrus. Z. Angew. Entomol. 86: 443–446.

Saga, G. A., and Lee, W. R. 1970. A vacuum injection technique for obtaining uniform dosages in *Drosophila melanogaster*. Drosophila Inform. Ser. 45: 179.

Saharia, D. 1982. Field evaluation of some granular systemic insecticides on *Lipaphis erysimi* (Kltb.) and its predator *Coccinella repanda* Thnb. J. Res., Assam Agric. Univ. 3(2): 181–185.

Sailer, R. I. 1983. History of Insect Introductions, pp. 15–38. *In* Exotic Plant Pests and North American Agriculture. Eds. C. L. Wilson and C. L. Graham, Academic Press, New York.

Salama, H. S., and Zaki, F. N. 1983. Interaction between *Bacillus thuringiensis* Berliner and the parasites and predators of *Spodoptera littoralis* in Egypt. Z. Angew. Entomol. 95: 425–429.

Salama, H. S., and Zaki, F. N. 1985. Biological effects of *Bacillus thuringiensis* on the egg parasitoid, *Trichogramma evanescens*. Insect Sci. Appl. 6(2): 145–148.

Salama, H. S., Zaki, F. N., and Sharaby, A. F. 1982. Effect of *Bacillus thuringiensis* Berl. on parasites and predators of the cotton leafworm *Spodoptera littoralis* (Boisd.). Z. Angew. Entomol. 94: 498–504.

Salim, M., and Heinricks, E. A. 1985. Relative toxicity of insecticides to the whitebacked planthopper *Sogatella furcifera* (Horvath) (Homoptera: Delphacidae) and its predators. J. Plant Protec. Tropics 2(1): 45–47.

Salt, G 1961. Competition among insect parasitoids. Symp. Soc. Exp. Biol. 15: 96–119.

Samsoe-Petersen, L. 1983. Laboratory method for testing side effects of pesticides on juvenile stages of the predatory mite, *Phytoseiulus persimilis* (Acarina: Phytoseiidae) based on detached bean leaves. Entomophaga 28: 167–178.

Samsoe-Peterson, L. 1985. Laboratory tests to investigate the effects of pesticides on two beneficial arthropods: a predatory mite (*Phytoseiulus persimilis*) and a rove beetle (*Aleochara bilineata*). Pestic. Sci. 16: 321–324.

Samways, M. J. 1979. Integration of the predatory mite *Phytoseiulus persimilis* Athias-henriot, and the chemical dienochlor for the control of *Tetranychus urticae* (Koch, 1836) on glasshouse roses. Anais Soc. Entomol. Brasil 8: 149–153.

Sandburg, L. L., Kramer, K. J., Kezdy, F. J., Law, J. H., and Oberlander, H. 1975. Role of juvenile hormone esterases and carrier proteins in insect development. Nature 253: 266–267.

Sanford, K. H., and Herbert, H. J. 1970. The influence of spray programs on the fauna of apple orchards in Nova Scotia XX. Trends after altering levels of phytophagous mites or predators. Can. Entomol. 102: 592–601.

Santharam, G., and Kumaraswami, T. 1985. Effect of some insecticides on the emergence of the parasitoid, *Trichogramma chilonis* Ishi (Hymenoptera: Trichogrammatidae). Entomon 10: 47–48.

Saradamma, K., and Nair, M. R. G. K. 1968. Relative toxicity of insecticides to adults of *Microbracon brevicornis* (Wesmael) (Braconidae). Agric. Res. J. Kerala 6: 98–100.

Sarup, P., Singh, D. S., Srivastava, M. L., and Singh, R. N. 1972. Laboratory evaluation of pesticides against the adults of an important parasite, *Bracon brevicornis* Wesmael (Braconidae: Hymenoptera) of some crop pests. Indian J. Entomol. 33: 346–349.

Sasaba, T., Kiritani, K., and Urabe, T. 1973. A preliminary model to simulate the effects of insecticides on a spider-leafhopper system in the paddy field. Res. Pop. Ecol. 15: 9–22.

Sato, Y. 1968. Insecticidal action of phytoecdysones. Appl. Entomol. Zool. 3: 155–162.

Satpathy, J. M., Padhi, G. K., and Dutta, D. N. 1968. Toxicity of eight insecticides to the coccinellid predator *Chilomenes sexmaculata* Fabr. Indian J. Entomol. 30: 130–132.

Saxena, R. C. 1987. Neem seed oil—a potential antifeedant for insect pests of rise, pp. 139–144. *In* Proc. 6th IUPAC Cong. Pesticide Chemistry, Ottawa, Canada, Aug. 1986. Blackwell Sci. Ont. Publ., Oxford/London.

Sayedoleslami Esfahani, H. 1979. Aspects of the temporal and spatial coincidence of the white apple leafhopper (*Typhlocyba pomaria* McAtee, Cicadellidae: Homoptera) and two parasitic Hymenoptera. Ph.D. thesis, Michigan State Univ., 184 pp.

Scheurer, R., Fluck, V., and Ruzette, M. A. 1975. Experiments with insect growth regulators (IGRs) on lepidopterous pests and some of their parasitoids. Bull. Soc. Entomol. Suisse 3: 315–321.

Schicha, E. 1978. The morphology of three *Typhlodromus occidentalis* Nesbitt populations in Australia (Acarina: Phytoseiidae). Z. Angew. Zool. 65: 291–302.

Schimk, R. T., Kaufman, R. J., Alt, F. W., and Kellems, R. F. 1978. Gene amplification and drug resistance in cultured murine cells. Science 202: 1051–1054.

Schneider, H. 1958. Untersuchungen über den Einfluss neuzeitlicher Insektizide and Fungizide auf die Blutlauszehrwespe (*Aphelinus mali* Hald.). Z. Angew. Entomol. 43: 173–196.

Schoenbohm, R. B., and Turpin, F. T. 1977. Effect of parasitism by *Meteoris leviventris* on corn foliage consumption and corn seedling cutting by the black cutworm. J. Econ. Entomol. 70: 457–459.

Schooley, D. A., Judy, K. J., Bergot, B. J., Hall, M. S., and Hennings, R. C. 1976. Determination of the physiological levels of juvenile hormones in several insects and biosyntheses of the carbon skeletons of the juvenile hormones. *In* The Juvenile Hormone. Ed. G. L. Gilbert, Plenum Press, New York.

Schooley, D. A., Quistad, G. B., Skinner, W. S., and Adams, M. E. 1987. Insect neuropeptides and their physiological degradation: the basis for new insecticide discovery? pp. 97–100. *In* Pesticide Science and Biotechnology. Proc. 6th IUPAC Cong. Pesticide Chemistry, Ottawa, Ont., Canada, Aug. 1986. Blackwell Sci. Publ., Oxford/London.

Schoones, J., and Giliomee, J. H. 1982. The toxicity of methidathion to parasitoids of red scale, *Aonidiella aurantii* (Mask.) (Hemiptera: Diaspididae). J. Entomol. Soc. S. Africa 45: 261–273.

Schoonoven, L. M. 1982. Biological aspects of antifeedants. Entomol. Exp. Appl. 31: 57–69.

Schowalter, T. D. 1985. Adaptations of insects to disturbance, pp. 235–252. *In* The Ecology of Natural Disturbance and Patch Dynamics. Eds. T. A. Pickett and P. S. White, Academic Press, New York.

Schowalter, T. D., Hargrove, W. W., and Crossely, D. A. 1986. Herbivory in forested ecosystems. Annu. Rev. Entomol. 31: 177–196.

Schulten, G. G. M. 1985. Pseudo-arrhenotoky, pp. 64–72. *In* Spider Mites: Their Biology, Natural Enemies and Control. Eds. W. Helle and M. W. Sabelis, Elsevier, Amsterdam, 458 pp.

Schulten, G. G. M., and van de Klashorst, G. 1974. Genetics of resistance to parathion and demeton-*S*-methyl in *Phytoseiulus persimilis* A.-H. (Acari: Phytoseiidae), pp. 519–524. *In* Proc. 4th Intern. Cong. Acarol.

Schulten, G. G. M., van de Klashorst, G., and Russell, V. M. 1976. Resistance of *Phytoseiulus persimilis* A. H. (Acari: Phytoseiidae) to some insecticides. Z. Angew. Entomol. 80: 337–341.

Schultz, J. C. 1983. Impact of variable plant defensive chemistry on susceptibility of insects to natural enemies, pp. 37–54. *In* Plant Resistance to Insects, Symp. Ser. 208. Amer. Chem. Soc., Washington, D.C.

Schuster, D. J., Musgrave, C. A., and Jones, J. P. 1979. Vegetable leafminer and parasite emergence from tomato foliage sprayed with oxamyl. J. Econ. Entomol. 72: 208–210.

Scott, J. G., and Rutz, D. A. 1988. Comparative toxicity of seven insecticides to housefly (Diptera: Muscidae) and *Urolepis rufipes* (Ashmead) (Hymenoptera: Pteromalidae). J. Econ. Entomol. 81: 804–807.

Scott, J. G., Croft, B. A., and Wagner, S. W. 1983. Studies on the mechanism of permethrin resistance in *Amblyseius fallacis* (Acarina: Phytoseiidae) relative to previous insecticide use on apple. J. Econ. Entomol. 76: 6–10.

Searle, C. M. 1965. The susceptibility of *Pauridia peregrina* Timb. (Hymenoptera: Encyrtidae) to some pesticide formulations. J. Entomol. Soc. S. Africa 27: 239–249.

Sechser, B. 1981. An approach to integrated pest management from the chemical industry. Acta Phytopathol. Acad. Sci. Hungar. 16: 239–243.

Sechser, B., and Bathe, P. A. 1978. A new method for testing the selectivity of pesticides against beneficial insects in orchards. Z. Angew. Entomol. 77(3): 14–27.

Sechser, B., and Varty, I. W. 1978. Effect of an insect growth regulator on non-target arthropods in an aerial application against the sprucebudworm, *Choristoneura fumiferana* (Lepidoptera: Tortricidae), in New Brunswick, Canada. Can. Entomol. 110: 561–567.

Sekeroglu, E., and Uygun, N. 1980. Effect of some pesticides used for mite control in citrus orchards on *Sympherobius sanctus* Tjed (Neuroptera: Hemerobiidae) and *Cryptolaemus montrouzieri* (Muls.) (Coleoptera: Coccinellidae). Turkiye Bitki Koruma Dergisi 4: 251–256.

Sekita, N. 1986. Toxicity of pesticides commonly used in Japanese apple orchards to the predatory mite *Typhlodromus pyri* Scheuten (Acari: Phytoseiidae) from New Zealand. Appl. Entomol. Zool. 21: 173–175.

Sekulic, R., and Dedic, B. 1983. Species composition and side-effects of the insecticide lindane on ground beetles (Carabidae, Col.) in sugar beet fields in northeastern Yugoslavia. Ges. Allg. Angew. Entomol. 4: 80–82.

Self, L. S., Guthrie, F. E., and Hodgson, E. 1964. Metabolism of nicotine by tobacco-feeding insects. Nature 204: 300–301.

Selivanova, N. A. 1967. On the effects of systemic insecticides on the useful insect fauna. Nauch. Trudy Kursk. Sel'-kohz. Inst. 4: 240–242.

Sell, P. 1984a. Effects of pesticides on the efficiency of the aphidophagous larvae of *Aphidoletes aphidimyza* (Rond.) (Diptera, Cecidomyiidae). Z. Angew. Entomol. 98: 174–184.

Sell, P. 1984b. Studies on the effects of plant protection compounds on the efficiency of the predaceous gall midge *Aphidoletes aphidimyza* (Rond.) (Diptera, Cecidomyiidae) and its progeny. Z. Angew. Entomol. 98: 425–431.

Sell, P. 1985. Efficiency of the aphid predator *Aphidoletes aphidimyza* (Rond.) (Diptera: Cecidomyiidae) after exposure of 2- and 3-day old larvae to insecticide and fungicide deposits. Z. Pflkrankh. Pflschutz. 92: 157–163.

Sem'yanov, V. P. 1985. Population of seven-spotted ladybird in cotton fields depending on environmental conditions, pp. 170–172. *In* Ecological Principles of the Integrated Control of

Cotton Pests, Proc. All-Union Symp., Leningrad, Mar. 2–4, 1977. Ed. L. A. Skarlato, Amerind, New Delhi.

Sewall, D. K. 1986. Chemotherapeutic and non-target side-effects of benomyl to the orange tortrix *Argyrotaenia citrana* and the braconid endoparasite, *Apanteles aristoteliae*. MS thesis, Oregon State Univ., Corvallis, 69pp.

Sewall, D., and Croft, B. A. 1987. Chemotheraputic and nontarget side-effects of benomyl to the orange tortrix, *Argyrotaenia citrana* (Lepidoptera: Tortricidae), and braconid endoparasite *Apanteles aristoteliae* (Hymenoptera: Braconidae). Environ. Entomol. 16: 507–512.

Seymour, J. 1982a. Integrated control of orchard mites. Rural Res. 116: 15–19.

Seymour, J. 1982b. Spray-resistant mites to the rescue. Ecos 33: 3–7.

Shapiro, A. M. 1976. Beau geste? Amer. Natur. 110: 900–902.

Sharma, H. C., and Adlakha, R. L. 1981. Selective toxicity of some pesticides to the adults of ladybird beetle, *Coccinella septempunctata* L. and cabbage aphid. Indian J. Entomol. 43:92–99.

Shea, P. J., Johnson, D. R., and Nakamoto, R. 1984. Effects of five insecticides on two primary parasites of the western spruce budworm, *Choristoneura occidentalis* (Freeman) (Lepidoptera: Tortricidae). Protec. Ecol. 7: 259–268.

Shepard, M., Carner, G. R., and Turnipseed, S. G. 1977. Colonization and resurgence of insect pests of soybean in response to insecticides and field isolation. Environ. Entomol. 6: 501–506.

Shin, Y. H. 1970. On the bionomics of *Itoplectis narangae* (Ashmead) (Ichneumonidae, Hymenoptera). J. Fac. Agric. Kyushu Univ. 16: 1–75.

Shires, S. W. 1985. A comparison of the effect of cypermethrin, parathion-methyl and DDT on cereal aphids, predatory beetles, earthworms and litter decomposition in sprayed wheat. Crop Prot. 4: 177–193.

Shires, S. W., Inglesfield, C., and Murray, A. 1984. Laboratory studies on the effects of insecticides on *Trichogramma cacoeciae*, pp.349–354. *In* Proc. 1984 Brit. Metropole, England, Nov. 19–22, 1984.

Shishido, T. 1978. The role of glutathione S-transferases in pesticide metabolism. Nippon. Noyaku Gakkai 3: 465–473.

Shorey, H. H. 1963. Differential toxicity of insecticides to the cabbage aphid and two associated entomophagous insect species. J. Econ. Entomol. 56: 844–847.

Shour, M. H., and Crowder, L. A. 1980. Effects of pyrethroid insecticides on the common green lacewing. J. Econ. Entomol. 73: 306–309.

Simberloff, E. S., and Wilson, E. O. 1969. Experimental zoogeography of islands: A two year record. Ecology 51: 934–937.

Simmonds, S. P. 1973. More on biological pest control (use of the mite predator *Phytoseiulus persimilis* for the control of red spider mite *Tetranchus urticae* on violets growing out of doors in Devon.), J. Devon Trust Nature Conserv. 5(2): 70–73.

Simova, S. A. 1976. Effects of some pesticides on mites of the family Phytoseiidae on plum. Rastitelnozashitna Nauka 4: 111–115.

Singh, D. S., Dhingra; S., Srivastava, V. S., Sircar, P., and Lal, R. 1978. Relative susceptibility of *Aphis craccivora* to pesticides in relation to different hosts. Indian J. Entomol. 42: 746–756.

Singh, D. S., Dhingra, S., Saxena, V. S., Srivastava, V. S., Sircar, P., and Lal, R. 1979. Relative resistance of aphid predator, *Coccinella septempunctata* Linn. to insecticides. Indian J. Entomol. 41: 149–154.

Singh, H. N., and Rai, L. 1976. Properties of cholinesterase in herbivorous and carnivorous insects and its implication in developing specific insecticides. Indian J. Entomol. 38: 305–312.

Singh, O. P., and Rawat, R. R. 1981. Note on the safety of some insecticides for *Diaeretus rapae* M'Intosh, a parasite of the mustard aphid. Indian J. Entomol. 51: 204–205.

Singh, O. P., and Rawat, R. R. 1983. Seasonal incidence and toxicological studies on *Lipaphis erysimi* (Kalt.) and its parasite, *Aphidius* sp. (?) in Madhya Pradesh, India. Pranikee 4: 259–267.

Singh, P., and Moore, R. F., Eds. 1985. Handbook of Insect Rearing, Elsevier, New York, vol.1, 488pp.

Singh, P. P., and Varma, G. C. 1986. Comparative toxicities of some insecticides to *Chrysoperla carnea* (Chrysopidae: Neuroptera) and *Trichogramma brasiliensis* (Trichogrammatidae: Hymenoptera), two arthropod natural enemies of cotton pests. Agric., Ecosys. Environ. 15: 23–30.

Singh, R., and Malhotra, R. K. 1975a. Fatal effect of insecticide treated aphids on *Coccinella septempunctata*. Current Sci. 44: 325–326.

Singh, R., and Malhotra, R. K. 1975b. Susceptibility of *Coccinella septempunctata* Linn. (Coccinellidae: Coleoptera) to some insecticides. Haryana Agric. Univ. J. Res. 5: 29–34.

Sinha, R. N., Berck, B., and Wallace, H. A. H. 1967. Effect of phosphine on mites, insects, and microorganisms. J. Econ. Entomol. 60: 125–132.

Sithanantham, S. 1980a. Effects of insecticide applications on selected arthropod populations on sugarcane crops. Indian J. Plant Prot. 8: 85–88.

Sithanantham, S. 1980b. Inundative release of *Trichogramma* and integration with insecticidal methods. Biological Control of Sugar Cane Pests in India, Sugar Factories Ltd. pp.43–60.

Skrebtsov, M. F., and Skrebtsova, N. D. 1980. Pollinators of onion seed crops. Zash. Rast. 10: 22.

Slama, K., and Jarolim, V. 1981. Hydrolysis of 2(E) and 2(Z) isomers of juvenoid esters related to their juvenile hormone activity in insects. *In* Juvenile Hormone Biochemistry. Eds. G. E. Pratt and G. T. Brooks, Elsevier/North Holland, Biomedical Press.

Slansky, F. 1986. Nutritional ecology of endoparasitic insects and their hosts: an overview. J. Insect Physiol. 32: 255–261.

Slansky, F., and Scriber, J. M. 1982. Selected bibliograph and summary of quantitative food utilization by immature insects. Bull. Entomol. Soc. Amer. 28: 43–55.

Sluss, R. 1968. Behavioral and anatomic responses of the convergent lady beetle to parasitism by *Perilitus coccinellae*. J. Invert. Pathol. 10: 9–27.

Smilowitz, Z. 1973. Electrophoretic patterns in hemolymph protein of cabbage looper during development of the parasitoid *Hyposoter exiguae*. Ann. Entomol. Soc. Amer. 66: 93–99.

Smilowitz, Z., and Smith, C. L. 1977. Hemolymph proteins of developing *Pieris rapae* larvae parasitized by *Apanteles glomeratus*. Ann. Entomol. Soc. Amer. 70: 447–454.

Smilowitz, Z., Martinka, C. A., and Jowyk, E. A. 1976. The influence of a juvenile hormone mimic (JHM) on the growth and development of the cabbage looper, *Trichoplusia ni* (Lepidoptera: Noctuidae) and the endoparasite, *Hyposoter exiguae* (Hymenoptera: Ichneumonidae). Environ. Entomol. 5: 1178–1182.

Smith, R. F., and van den Bosch, R. 1967. Integrated control, pp.295–340. *In* Pest Control. Ed. R. L. Doutt, Academic Press, New York.

Smith, T. M., and Stratton, G. W. 1986. Effects of synthetic pyrethroid insecticides on nontarget organisms. Residue Reviews 97: 93–120.

Soderhall, K. 1982. Prophenoloxidase activating system and melanization—a recognition mechanism of arthropods? A Review. Devel. Comp. Immunol. 6: 601–611.

Soderlund, D. M., Saborn, J. R., and Lee, P. W. 1983. Metabolism of pyrethrins and pyrethroids in insects, pp.401–435. *In* Progress in Pesticides Biochemistry and Toxicology, Eds. D. J. Jutson and T. R. Roberts, Wiley, New York, vol.3.

Sohliesske, J. 1979. Effects of systemic insecticides via the prey to the population development of *Phytoseiulus persimilis* A-H (Acari: Phytoseiidae). Med. Fac. Landbouww. Rijksuniv. Gent 44: 153–157.

Sol, R. 1961. Über den Eingriff von Insektiziden in das Wechselspiel von *Aphis fabae* Scop. und einigen iherer Episiten. Entomophaga 1: 7–31.

Solomon, M. G., and Easterbrook, M. A. 1983. OP-resistant *Typhlodromus pyri* for apple orchard mite management, p. 999. *In* 10th Intern. Cong. Plant Protec., Brighton, England, Nov. 20–25, 1983, vol. 3.

Solomon, M. G., and Fitzgerald, J. D. 1984. The role of resistant *Typhlodromus pyri* in apple orchards, pp.1113–1116. *In* Proc. 1984 Brit. Crop Protec. Conf., Brighton Metropole, England, Nov. 19–22, 1984, vol.3.

Somchoudhury, A. K., and Dutt, N. 1980. Field bioecology of *Trichogramma perkinsi* Girault and *Trichogramma australicum* Girault (Hymenoptera: Trichogrammatidae) and their time of release for the control of *Chilo partellus* (Swinhoe) and *Heliothis armigera* Hubn. J. Entomol. Res. 4: 73–82.

Soultanopoulos, C., and Broumas, T. 1977. Comparative toxicity of insecticidal products to *Eupelmus urozonus* (Hym.: Eupelmidae) and *Dacus oleae* (Dipt.: Trypetidae). Entomophaga 22: 237–242.

Soultanopoulos, C., and Broumas, T. 1979. Comparative toxicity of various insecticides to *Habrobracon hebetor* and *Chelonus eleaphilus* (Hymenoptera: Braconidae). Med. Fac. Landbouww. Rijksuniv. Gent 44: 179–184.

Sparks, T. C., and Hammock, B. D. 1980. Comparative inhibition of the juvenile hormone esterases from *Trichoplusia ni, Tenebrio molitor* and *Musca domestica*. Pestic. Biochem. Physiol. 14: 290–302.

Spiller, D. 1958. Resistance of insects to insecticides. Entomologist 2(3): 1–18.

Spradling, A. C., and Rubin, G. M. 1982. Transposition of cloned P elements into *Drosophila* with transposable element vectors. Science 218: 341–347.

Sroka, P., and Vinson, S. B. 1978. Phenoloxidase activity in the haemolymph of parasitized and unparasitized *Heliothis virescens*. Insect Biochem. 3: 399–402.

Staal, G. B. 1972. Biological activity and bioassay of juvenile hormone analogs. *In* Insect Juvenile Hormones. Eds. J. J. Menn and M. Beroza, Academic Press, New York.

Staal, G. B. 1975. Insect growth regulators with juvenile hormone activity. Annu. Rev. Entomol. 20: 417–460.

Stacherska, B. 1963. Toxicity determination of several aphid-insecticides and their selectivity to some useful insects. Biol. Inst. Ochrony Roslin 23: 159–172.

Stam, P. A., Clower, D. F., Graves, J. B., and Schilling, P. E. 1978. Effects of certain herbicides on some insects and spiders found in Louisiana cotton fields. J. Econ. Entomol. 71: 477–480.

Stamenkovic, T., and Peric, P. 1984a. Susceptibility of the predator *Phytoseiulus persimilis* A-h. (Phytoseiidae) to some pesticides. Zastita Bilja 35: 317–321.

Stamenkovic, T., and Peric, P. 1984b. The effect of the acaricide bisclofentezin on *Encarsia formosa* G. and *Phytoseiulus persimilis* A.-H. Zastita Bilja 35: 135–139.

Stanley, B. H. 1987. Empirical and mathematical approaches to the management of apple maggot, *Rhagoletis pomonella* (Walsh) (Diptera: Tephritidae). Ph.D. dissertation, Cornell Univ., Ithaca, New York., 176pp.

Stanley-Samuelson, D. W., and Dadd, R. H. 1983. Long-chain polyunsaturated fatty acids: patterns of occurrence in insects. Insect Biochem. 13: 549–558.

Starks, K. J., Muniappan, R. , and Eikenbary, R. D. 1982. Interaction between plant resistance and parasitism against the greenbug on barley and sorghum. Ann. Entomol. Soc. Amer. 65: 650–655.

Staubli, A. 1986. Evaluation of the secondary effects of an insect growth regulator (IGR) on the beneficial fauna of orchards: the difficult choice of a method. Bull. SROP 9(3): 67–74.

Staubli, A., Hachler, M., Antonin, P., and Mittaz, C. 1984. Toxicity tests of various pesticides to the natural enemies of the principal pests of pear orchards in French-speaking Switzerland. Rev. Suisse Vitic., Arboric., Hortic. 16(5): 279–286.

Stein, W. 1961. The influence of different pest-control materials on egg-parasites of the genus *Trichogramma* (Hym. Trichogrammatidae). Anz. Schädl., Pflschutz., Umweltschutz 34(6): 87–89.

Steiner, H. 1977. Standardized field tests to measure side effects of pesticides in the tree level. Z. Pflkrankh. Pflschutz. 84: 164–166.

Stenseth, C. 1975. The effect of some fungicides and acaricides on the predacious mite *Phytoseiulus persimilis* Athias-Henriot (Acarina: Phytoseiidae). Forskning og Forsok i Landbrunek 26: 394–404.

Stenseth, C. 1979. Effect of some fungicides and insecticides on an organophosphorous-resistant strain of the predaceous mite, *Phytoseiulus persimilis* A-H. Forskning og Forsok i Landbrunek 30: 77–83.

Sterling, W. L., Dean, D. A., Fillman, D. A., & Jones, D. 1984. Naturally occurring biological control of the boll weevil (Col.: Curculionidae). Entomophaga 29: 1–9.

Stern, V. M., Smith, R. F., van den Bosch, R., and Hagen, K. S. 1959. The integrated control concept. Hilgardia 29: 81–101.

Stern, V. M. van den Bosch, R., and Reynolds, H. T. 1959. Effects of dylox and other insecticides on entomophagous insects attacking field crop pests in California. J. Econ. Entomol. 53: 67–72.

Sternburg, J., and Kearns, L. A. M. 1952. Metabolic fate of DDT when applied to certain naturally tolerant insects. J. Econ. Entomol. 45: 497–505.

Stevenson, J. H., and Walters, J. H. H. 1984. Evaluation of pesticides for use with biological control. Agric., Ecosys. Environ. 10: 201–215.

Stewart, J. G., and Philogene, J. R. 1983. Reproductive potential of laboratory reared *Manduca sexta* as affected by sex ratio. Can. Entomol. 115: 295–298.

Stinner, B. R., Krueger, H. R., and McCartney, D. A. 1986. Insecticide and tillage effects on pest and non-pest arthropods in corn agroecosystems. Agric., Ecosys. Environ. 15: 11–12.

Storozhkov, Y. V., Chabanovskii, A. G., Mozzhukin, Y. B., and Mereveli, N. P. 1977. The resistance of *Phytoseiulus* to pesticides. Zash. Rast. 10: 26.

Strawn, A. J. 1978. Organophosphate use and residues on citrus. M.S. Thesis, Univ. Cal. Riverside.

Streibert, H. P. 1981. A standardized laboratory rearing and testing method for the effects of pesticides on the predatory mite *Amblyseius fallacis* (Garman). Z. Angew. Entomol. 92: 121–127.

Strickler, K. and Croft, B. A. 1981. Variation in permethrin and azinphosmethyl resistance in populations of *Amblyseius fallacis* (Acarina: Phytoseiidae). Environ. Entomol. 10: 233–236.

Strickler, K., and Croft, B. A. 1982. Selection for permethrin resistance in the predatory mite *Amblyseius fallacis*. Entomol. Exp. Appl. 31: 339–345.

Strickler, K., and Croft, B. A. 1985. Comparative rotenone toxicity in the predator, *Amblyseius fallacis* (Acari: Phytoseidae), and the herbivore, *Tetranychus urticae* (Acari: Tetranychidae), grown on lima beans and cucumbers. Environ. Entomol. 14: 243–246.

Strickler, K., and Whalon, M. E. 1988. Selection for pyrethroid resistance in *Amblyseius fallacis* in Michigan: review and current status, p.316b. *In* Proc. 18th Intern. Cong. Entomol., Vancouver, B.C., Canada, July 3–9, 1988.

Strickler, K., Cushing, N., Whalon, M., and Croft, B. A. 1987. Mite (Acari) composition in Michigan apple orchards. Environ. Entomol. 16: 30–36.

Suckling, D. M., Wearing, C. H., and Thomas, W. P. 1984. Insecticide resistance in the lightbrown apple moth: a case for resistance management, pp.248–252. *In* Proc. 37th N. Z. Weed and Pest Control Conf.

Suckling, D. M., Penman, D. R., Chapman, R. B., and Wearing, C. H. 1985. Pheromone use in insecticide resistance surveys of lightbrown apple moths (Lepidoptera: Tortricidae). J. Econ. Entomol. 78: 204–207.

Sukhoruchenko, G. I., Niyazov, O. D., and Alekseev, Y. A. 1977. The effect of modern pesticides on the beneficial and injurious insect faunas of cotton. Entomol. Obozrenie 56: 3–14.

Sukhoruchenko, G. I., Smirnova, A. A., Vikar, Y. V., and Kapitan, A. I. 1982. The effect of pyrethroids on the arthropods of a cotton agrobiocoenosis. Rept. All-Union Pl. Protec. Res. Inst. 10pp.

Sullivan, W. N., Cawley, B., Hayes, D. K., Rosenthal, T., and Halberg, F. 1970. Circadian rhythm in susceptibility in houseflies and Madeira cockroaches to pyrethrum. J. Econ. Entomol. 63: 159–163.

Summers, C. G., Coviello, R. L., and Cothran, W. R. 1975. The effect of selected entomophagous insects of insecticides applied for pea aphid control in alfalfa. Environ. Entomol. 4: 612–614.

Sundaramurthy, V. T. 1980. Effects of diflubenzuron on a field population of the coconut black-headed caterpillar, *Nephantis serinopa* meyrick (Lepidoptera: Gelechiidae) and its parasite, *Parasierola nephantidis* (Nuesebeck). Bull. Entomol. Res. 70: 25–31.

Suss, L. 1983. Survival of pupal stage of '*Aphidius ervi*' Hal. in mummified '*Sitobion avenae*' F. to pesticide treatment, pp.129–134. *In* Aphid Antagonists. Proc. EC Experts' Group, Portici, Italy, Nov.23–24, 1982.

Sustek, Z. 1982. The effect of actellic EC 50 on the Carabidae and Staphylinidae in a Norway spruce forest in the Jizerske Hory Mountains. Zoologia 37: 131–139.

Suter, H. 1978. The effects of pesticides on the beneficial arthropod *Chrysopa carnea* Steph. (Neuroptera: Chrysopidae)—methods and results. Schweiz. Landwirtschaft. Forsch. 17: 37–44.

Suttman, C. E., and Barrett, G. W. 1979. Effects of sevin on arthropods in an agricultural and an old-field plant community. Ecology 60: 628–641.

Svoboda, J. A., and Lusby, W. R. 1986. Sterols of phytophagous and omnivorous species of hymenoptera. Arch. Insect Biochem. Physiol. 3: 13–18.

Svoboda, J. A., and Robbins, W. E. 1979. Comparison of sterols from a phytophagous and a predacious species of the family Coccinellidae. Experientia 35: 186–187.

Svoboda, J. A., Thompson, M. J., Robbins, W. E., and Kaplanis, J. N. 1978. Insect steroid metabolism. Lipids 13: 742–753.

Swezey, S. L., Page, M. L., and Dahlsten, D. L. 1982. Comparative toxicity of lindane, carbaryl, and chlorpyrifos to the western pine beetle (*Dendroctonus brevicomis*) (Coleoptera: Scolytidae) and two of its predators, *Enoclerus lecontei* and *Temnochila*. Can. Entomol. 114: 397–401.

Swift, F. C. 1970. Predation of *Typhlodromus* (*A.*) *fallacis* on the European red mite as measured by the insecticidal check method. J. Econ. Entomol. 53: 1617–1618.

Swirski, S., and Dorzia, N. 1968. Studies on the feeding, development and oviposition of the predaceous mite *Amblyseius limonicus* Garman and McGregor (Acarina: Phytoseiidae) on various kinds of food substances. Israel J. Agric. Res. 18: 71–75.

Swirski, E., Amitai, S., and Dorzia, N. 1967. Field and laboratory trials on the toxicity of some pesticides to predacious mites (Acarina: Phytoseiidae). Israel J. Agric. Res. 17: 149–159.

Syrett, P., and Penman, D. R. 1980. Studies of insecticide toxicity to lucerne aphids and their predators. New Zealand J. Agric. Res. 23: 575–580.

Tabashnik, B. E. 1986a. Computer simulation as a tool for pesticide resistance management, pp. 194–206. *In* Pesticide Resistance: Strategies and Tactics for Management. Eds. E. H. Glass etal., Nat. Acad. Sci., Washington D.C., 471pp.

Tabashnik, B. E. 1986b. Evolution of pesticide resistance in predator/prey systems. Bull. Entomol. Soc. Amer. 32: 156–161.

Tabashnik, B. E., and Croft, B. A. 1982. Managing pesticide resistance in crop-arthropod complexes: interactions between biological and operational factors. Environ. Entomol. 11: 1137–1144.

Tabashnik, B. E., and Croft, B. A. 1985. Evolution of pesticide resistance in apple pests and their natural enemies. Entomophaga 30: 37–49.

Tabashnik, B. E., and Cushing, N. L. 1989. Quantitative genetic analysis of insecticide resistance: variation in fenvalerate tolerance in a diamondback moth (Lepidoptera: Plutellidae) population. J. Econ. Entomol. 82: 5–10.

Tabashnik, B. E., and Johnson, M. W. 1988. Why nice guys finish last: pesticide impacts and pesticide resistance in arthropod natural enemies. *In* Principles and Applications of Biological Control. Ed. T. W. Fisher, Univ. Calif. Riverside (in press).

Tadic, M. 1979. The toxicity of some synthetic insecticides to the Hymenoptera *Ooencyrtus kuwanae* How. and *Telenomus terebrans* Ratz. Zastita Bilja 30: 205–210.

Takahashi, Y., and Kiritani, K. 1973. The selective toxicity of insecticides against insect pests of rice and their natural enemies. Appl. Entomol. Zool. 8: 220–226.

Takeda, S., Hukusima, S. and Yamada, H. 1965. Some physiological aspects of coccinellid beetles in relation to their tolerance against pesticide treatments. Foc. Agric. Bull. GIFU Univ., pp.83–91.

Tamaki, G., Halfhill, J. E., and Maitlen, J. C. 1969. The influence of UC-211 and the aphidiid parasite *Aphidius smithi* on populations of the pear aphid. J. Econ. Entomol. 62: 678–682.

Tamaki, G., Chavuin, R. L., Moffit, H. R., and Mantey, K. D. 1984. Diflubenzuron: differential toxicity to larvae of the Colorado potato beetle (Coleoptera: Chrysomelidae) and its internal parasite, *Doryphorophaga doryphorae* (Diptera: Tachinidae). Can. Entomol. 116: 197–202.

Tamashiro, M., and Sherman, M. 1955. Direct toxicity of insecticides to oriental fruit fly larvae and their internal parasites. J. Econ. Entomol. 48: 75–79.

Tanada, Y. 1964. Epizootiology of insect diseases, pp.548–578. *In* Biological Control of Insect Pests and Weeds. Ed. P. DeBach. Chapman and Hall, London.

Tanada, Y. 1984. *Bacillus thuringiensis*—integrated control, past, present, and future, pp.59–90. *In* Comparative Pathogen Biology, Ed. T. C. Cheng, Plenum Press, New York, vol.70.

Tanaka, K., and Ito, Y. 1982. Different response in respiration between predaceous and phytophagous lady beetles (Coleoptera: Coccinellidae) to starvation. Res. Popul. Ecol. 24: 132–141.

Tanigoshi, L. K., and Congdon, B. D. 1983. Laboratory toxicity of commonly-used pesticides in California citriculture to *Euseius hibisci* (Chant) (Acarina: Phytoseiidae). J. Econ. Entomol. 76: 247–250.

Tanigoshi, L. K., and Fargerlund, J. 1984. Implications of parathion resistance and toxicity of citricultural pesticides to a strain of *Euseius hibisci* (Chant) (Acarina: Phytoseiidae) from the San Joaquin Valley of California. J. Econ. Entomol. 77: 789–793.

Tanke, W., and Franz, J. M. 1977. Side-effects of some herbicides on the egg parasite *Trichogramma caoceciae* March. Z. Angew Entomol. 82: 288–293.

Taylor, C. E., and Georghiou, G. P. 1979. Suppression of insecticide resistance by alteration of dominance and migration. J. Econ. Entomol. 72: 105–109.

Taylor, E. A. 1954. Parasitization of the salt marsh caterpillar in Arizona. J. Econ. Entomol. 47: 525–530.

Teague, T. G., Horton, D. L., Yearian, W. C., and Phillips, J. R. 1985. Benomyl inhibition of *Cotesia marginiventris* in four lepidopterous hosts. J. Entomol. Sci. 20: 76–81.

Teetes, G. L. 1972. Differential toxicity of standard and reduced rates of insecticides to greenbugs and certain beneficial insects. Texas Agric. Exp. Stat. Prog. Rept., pp.1–9.

Teh, K. H., Ng, S. M., and Sudderuddin, K. I. 1983. The effect of malathion on the weight, fecundity and longevity of *Ischiodon scutellaris* Fabr. (Diptera: Syrphidae). Pertanika 6(1): 22–27.

Temerak, S. A. 1980. Detrimental effects of rearing a Braconid parasitoid on the pink borer larvae inoculated by different concentrations of the bacterium, *Bacillus thuringiensis* Berliner. Z. Angew. Entomol. 89: 315–319.

Temerak, S. A. 1982a. Transmission of two bacterial pathogens by means of the ovipositor of *Bracon brevicornis* Wesm. (Hym., Braconidae) into the body of *Sesamia cretica* Led. (Lep., Tortricidae). Anz. Schädl., Pflschutz., Umweltschutz 55(6): 89–92.

Temerak, S. A. 1982b. Interactions between *Bacillus thuringiensis* Berl. and the larvae of the braconid *Bracon brevicornis* Wesm. through the larvae of *Sesamia cretica* Led. at different temperatures. Anz. Schädl., Pflschutz., Umweltschutz 55(9): 137–140.

Teotia, T. P. S., and Tiwari, G. C. 1972. Toxicity of some important insecticides to the coccinellid predator, *Coccinella septempunctata* Linn. Labdev J. Sci. Tech. 10: 17–18.

Terriere, L. C. 1984. Induction of detoxification enzymes in insects. Ann. Rev. Entomol. 29: 71–88.

Terriere, L. C., and Yu, S. J. 1977. Juvenile hormone analogs: *in vitro* metabolism in relation to biological activity in blow flies and flesh flies. Pestic. Biochem. Physiol. 7: 161–168.

Tette, J. P. 1981. New York Apple Pest Management Manual. Cornell Univ. Publ., 86pp.

Tette, J. P., Brunno, D., Way, D., Baunister, R., Reynolds, E., and Cowles, R. 1982. An outline of the tree fruit integrated pest management program: 1979–1981. Ext. Publ. NY State Coll. Agric. Life Sci., 5pp.

Theiling, K. M. 1987. The SELCTV database: the susceptibility of arthropod natural enemies of agricultural pests to pesticdes. MS thesis, Oregon State University, Corvallis, 170pp.

Theiling, K. M., and Croft, B. A. 1988. Pesticide side-effects on arthropod natural enemies: a database summary. Agric., Ecosys. Environ. 21: 191–218.

Thomas, W. P., and Chapman, L. J. 1978. Integrated control of apple pests in New Zealand. 15. Introduction of two predaceous phytoseiid mites. Proc. 31st N.Z. Weed and Pest Control Conf. 31: 236–243.

Thompson, C. G., Neisess, J., and Batzer, D. O. 1977. Field tests of *Bacillus thuringiensis* and aerial application strategies on western mountainous terrain. USDA For. Serv., PNW For. Range Expt. Stn., Portland, Or. Res. Pap. PNW-230, 12pp.

Thompson, S. N. 1973. A review and comparative characterization of the fatty acid composition of seven insect orders. Comp. Biochem. Physiol. 45B: 467–482.

Thompson, S. N. 1977. Lipid nutrition during larval development of the parasitic wasp, *Exeristes*. J. Insect Physiol. 23: 579–583.

Thompson, S. N. 1981. Effects of dietary carbohydrate and lipid on nutrition and metabolism of metazoan parasites with special reference to parasitic Hymenoptera, pp.215–252. *In* Current Topics in Insect Endocrinology and Nutrition. Eds. G. Bhaskaran, S. Friedman and J. G. Rodriguez, Plenum Press, New York.

Thompson, S. N. 1983. Biochemical and physiological effects of metazoan parasites on their host species. Comp. Biochem. Physiol. 74: 183–211.

Thompson, S. N. 1986. Nutrition and *in vitro* culture of insect parasitoids. Annu. Rev. Entomol. 31: 197–220.

Thompson, S. N., and Barlow, J. S. 1972. Synthesis of fatty acids by the parasite *Exeristes comstockii* (Hymenop.) and two hosts, *Galleria mellonella* (Lep.) and *Lucilia sericata* (Dip.). Can. J. Zool. 50: 1105–1110.

Thompson, S. N., and Barlow, J. S. 1974. The fatty acid composition of parasitic Hymenoptera and its possible biological significance. Ann. Entomol. Soc. Amer. 67: 627–632.

Thompson, S. N., and Barlow, J. S. 1983. Metabolic determination and regulation of fatty acid composition in parasitic Hymenoptera and other animals, pp.73–106. *In* Metabolic Aspects of Lipid Nutrition in Insects. Ed. T. E. Mittler and R. E. Dadd, Westview Press, Boulder, Colo.

Thompson, S. N., and Johnson, J. 1978. Further studies on lipid metabolism in the insect parasite, *Exeristes roborator* (Fabricus). J. Parasitol. 64: 731–740.

Thoms, E. M., and Watson, T. F. 1986. Effect of Dipel (*Bacillus thuringiensis*) on the survival of immature and adult *Hyposoter exiguae* (Hymenoptera: Ichneumonidae). J. Invert. Pathol. 47: 178–183.

Thornhill, W. A., and Edwards, C. A. 1985. The effects of pesticides and crop rotation on the soil-inhabiting fauna of sugar-beet fields. Part I: The crop and macroinvertebrates. Crop Protec. 4: 409–422.

Thurling, D. J. 1980. Metabolic rate and life stage of the mite *Tetranychus cinnabarinus* Boised. (Prostigmata) and *Phytoseiulus persimilis* A-H. (Mesostigmata). Oecologia 46: 391–396.

Thurston, R., and Fox, P. M. 1972. Inhibition by nicotine of emergence of *Apanteles congregatus* from its host, the tobacco hornworm. Ann. Entomol. Soc. Amer. 65: 547–550.

Thwaite, G., and Bower, C. 1980. Predators spell doom for orchard mites. Agric. Gaz. New South Wales 91(4): 16–19.

Ticehurst, M., Fusco, R. A., and Blumenthal, E. M. 1982. Effects of reduced rates of Dipel 4L, Dylox 1.5 oil and Dimilin W-25 on *Lymantria dispar* (L.) (Lepidoptera: Lymantriidae), parasitism and defoliation. Environ. Entomol. 11: 1058–1062.

Tipping, P. W., and Burbutis, P. P. 1983. Some effects of pesticide residues on *Trichogramma nubilale* (Hymentoptera: Trichogrammatidae). J. Econ. Entomol. 76: 892–896.

Tiwari, G. C., and Moorthy, P. N. K. 1985. Selective toxicity of some synthetic pyrethroids and conventional insecticides to the aphid predator, *Menochilus sexmaculatus* Fabricius. Indian J. Agric. Sci. 55: 40–43.

Tolstova, Y. S., and Ionova, Z. A. 1976. Toxicity of pesticides to *Trichogramma*. Zash. Rast. 9: 21.

Tiwari, G. C., and Moorthy, P. N. K. 1985. Selective toxicity of some synthetic pyrethroids and conventional insecticides to the aphid predator. *Menochilus sexmaculatus* Fabricius. Indian J. Agric. Sci. 55: 40–43.

Tolstova, Y. S., and Ionova, Z. A. 1976. Toxicity of pesticides to *Trichogramma*. Zash. Rast. 9: 21.

Tomlin, A. D. 1975. The toxicity of insecticides by contact and soil treatment to two species of ground beetles (Coleoptera: Carabidae). Can. Entomol. 107: 529–532.

Toppozada, A., and O'Brien, R. D. 1967. Permeability of the ganglia of the willow aphid, *Tuberolachnus salignus*, to organic ions. J. Insect. Physiol 13: 941–954.

Torre-Bueno, J. R. 1962. A glossary of Entomology. Brooklyn Entomol. Soc., 336 pp.

Touzeau, J. 1986. The secondary effects of phytosanitary products and the Ecophyt database. Bull. SROP 9(3): 85–96.

Trammel, K. 1974. The white apple leafhopper—insecticide resistance and current control status. NY State Agric. Expt. Sta. Search Agriculture 4(8): 10 pp.

Trapman, M., and Blommers, L. 1985. The introduction of IPM in apple orchards. Med. Facult. Landbuww. Rijksuniv. Gent 50(2a): 425–430.

Travis, J. W., Hull, L. A., and Miller, J. D. 1978. Toxicity of insecticides to the aphid predator *Coccinella novemnotata*. Environ. Entomol. 7: 785–786.

Trevizoli, D., and Gravena, S. 1979. Efficiency and selectivity of insecticides for integrated control of citrus black aphid *Toxoptera citricidus* (Kirk, 1907). Cientifica 7: 115–120.

Trimble, R. M., and Pree, D. J. 1987. Relative toxicity of six insecticides to male and female *Pholetesor ornigis* (Weed) (Hymenoptera: Braconidae), a parasite of the spotted tentiform leafminer, *Phyllonorycter blancardella* (Fabr.) (Lepidoptera: Gracillariidae). Can. Entomol. 119: 153–157.

Trumble, J. T. 1984. Interfacing pesticides with successful biological control of *Liriomyza* spp., leafminers infesting *Apium* and *Lycopersicum* species. Abs. 17th Intern. Cong. Entomol. Hamburg, Germany, 1984.

Trumble, J. T. 1985. Integrated pest management of *Liriomyza trifolii*: influence of avermectin, cyromazine and methomyl on leafminer ecology in celery. Agric., Ecosys. Environ. 12: 181–188.

Tsankov, G., and Mirchev, P. 1983. The effect of some plant protection measures on the egg parasite complex of the pine processionary (*Thaumetopoea pityocampa*), Gorskostopanska Nauka 20(6): 84–89.

Tulisalo, U. 1984. Biological and integrated control by chrysopids, pp. 213–230. *In* Biology of Chrysopidae. Eds. M. Cunard, Y. Semeria, and T. R. New. D. W. Junk Publ., The Hague, The Netherlands.

Turnipseed, S. G., Todd, J. W., and Campbell, W. V. 1975. Field activity of selected foliar insecticides against geocorids, nabids and spiders on soybeans. Proc. Georgia Entomol. Soc. 10: 272–277.

Umarov, S. A., Nilova, G. N., and Davlyatov, T. D. 1975. The effect of Entobakterin and Dendrobacillin on beneficial arthropods. Zash. Rast. 3: 25–26.

Untersteuholer, G. 1970. Integrated pest control from the aspect of industrial research on crop protection chemicals. Nachr. Pflschutz. 23: 264–272.

Vaeck, M., Reynaerts, A., Hofte, H., Jansens, S., De Beuckeleer, M., Dean, C., Zabeau, M., van Montagu, M., and Leemans, J. 1987. Transgenic plants protected from insect attack. Nature 328: 33–37.

Vali, P. V. 1981. Cabbage looper nuclear polyhedrosis virus–parasitoid interactions. Environ. Entomol. 10: 517520.

van de Baan, H. E. 1988. Factors influencing pesticide resistance in pear psylla, *Psylla pyricola* Foerster, and susceptibility in its mirid predator, *Deraeocoris brevis* Knight. Ph. D. thesis, Oregon State Univ., Corvallis, 127 pp.

van de Baan, H. E., Kuijpers, L. A. M., Overmeer, W. P. J., and Oppenoorth, F. J. 1985. Organophosphorus and carbamate resistance in the predaceous mite *Typhlodromus pyri* due to insensitive acetylcholinesterase. Exp. Appl. Acarol. 1: 3–10.

van de Klashorst, G. 1984. Genetic improvement of an Australian strain of the mite predator *Typhlodromus occidentalis* Nesbitt, p. 817. *In* Proc. 17th Intern. Cong. Entomol., Hamburg, Germany, 1984, vol. 7.

van de Vrie, M. 1962. The influence of spray chemicals on predatory and phytophagous mites on apple trees in laboratory and field trials in the Netherlands. Entomophaga 7: 243.

van den Bosch, R., and Haramoto, F. H. 1951. *Opius oophilus* Fullaway, an egg-larval parasite of oriental fruit fly discovered in Hawaii. Proc. Hawaiian Entomol. Soc. 14: 251–255.

van den Bosch, R., and Stern, V. M. 1962. The integration of chemical and biological control of arthropod pests. Annu. Rev. Entomol. 7: 367–386.

van den Bosch, R., and Telford, A. D. 1964. Environmental modification and biological control, pp. 459–488. *In* Biological Control of Insect Pests and Weeds, Ed. P. DeBach, Reinhold Press, New York.

van den Bosch, R., Reynolds, H. T., and Dietrick, E. J. 1956. Toxicity of widely used insecticides to beneficial insects in California cotton and alfalfa fields. J. Econ. Entomol. 49: 359–363.

van den Bosch, R., Schlinger, E. I., and Hagen, K. S. 1962. Initial field observations in California on *Trioxys pallidus* (Haliday), a recently introduced parasite of the walnut aphid. J. Econ. Entomol. 55: 857–862.

van den Bosch, R., Hom, R. R., Matteson, R. R., Frazer, P., Messenger, P., and Dovis, S. 1979. Biological control of the walnut aphid in California: impact of the parasite *Trioxys pallidus*. Hilgardia 47: 1–13.

van Dinther, J. B. M. 1963. Residual effect of a number of insecticides on adults of the carabid *Pseudophonus rufipes* (Dej.). Entomophaga 8: 44–48.

van Driesche, R. G., Clark, J. M., Brooks, M. W., and Drummond, F. J. 1985. Comparative toxicity of orchard insecticides to the apple blotch leafminer, *Phyllonorycter crataegella* (Lepidoptera: Gracillariidae), and its eulophid parasitoid, *Sympiesis marylandensis* (Hymenoptera: Eulophidae). J. Econ. Entomol. 78: 926–932.

van Halteren, P. 1971. Preliminary investigations on deldrin accumulation in an insect food chain. J. Econ. Entomol. 64: 1055–1056.

van Zon, A. Q., and van der Geest, L. P. S. 1980. Effect of pesticides on predacious mites, pp. 227–230. *In* Integrated Control of Insect Pests in Netherlands. PUDDC, Wageningen, The Netherlands, 304 pp.

Varma, G. C., and Bindra, O. S. 1980. Studies on the chemical control of *Pyrilla perpusilla* (Walker) and effect of insecticides on its natural enemies. Pesticides 14(6): 9–11.

Vater, G. 1980. The effect of DDT/lindane on the brown-tail parasite *Eupteromalus peregrinus* (Pteromalidae). Angew. Parasitologie 21(3): 159–163.

Vazquez, G. M., and Carrillo, S. J. L. 1972. Effect of various insecticides on the beneficial fauna present on wheat in the Mexicali Valley during 1970–71. Agricultura Tecnica en Mexico 3(4): 145–149.

Velasco, L. R. I. 1985. Field parasitism of *Apanteles plutellae* Kurdj. (Braconidae, Hymenoptera) on the diamondback moth of cabbage. Philippine Entomol. 6(5/5): 539–553.

Verenini, M. 1984. Effects of a juvenoid ZR 515 4E (Methoprene) on the host–parasite couple *Galleria mellonella* L.–*Gonia cinerascens* Rond. Boll. Inst. Entomol. Univ. Studi Bologna 38: 95–115.

Versoi, P. L., and Yendol, W. G. 1982. Discrimination by the parasite, *Apanteles melanoscelus*, between health and virus-infected gypsy moth larvae. Environ. Entomol. 11: 42–45.

Via, S. 1986. Quantitative genetic models and the evolution of pesticide resistance, pp. 222–235. *In*

Pesticide Resistance: Strategies and Tactics for Management. NAS/NRC Press, Washington, D.C., 471 pp.

Vickerman, G. P., and Sunderland, K. D. 1977. Some effects of dimethoate on arthropods in winter wheat. J. Appl. Ecol. 14: 767–777.

Viggiani, G., and Tranfaglia, A. 1978. A method for laboratory testing of side-effects of pesticides on *Leptomastix dactylopii* (How.) (Hym. Encyrtidae). Boll. Entomol. Agraria "Filippo Silvestri," Portici 35: 8–15.

Viggiani, G., Castronuovo, N., and Borrelli, C. 1972. Secondary effects of 40 pesticides on *Leptomastiidea abnormis*-Grlt.—(Hym.: Encyrtidae) and *Scymnus includens* Kirsch (Col.: Coccinellidae), important natural enemies of *Planococcus citri* (Risso). Boll. Entomol. Agraria "Filippo Silvestri," Portici 30: 88–103.

Vigi, J., Boscheri, S., and Mantinger, H. 1985. Effect of various insecticides and acaricides on predatory mites. Obstbau Weinbau 22(4): 108–112.

Vijverberg, H. P. M., van de Zalm, J. M., van Kleef, R. G. D. M., and van der Bercken, J. 1983. Temperature- and structure-dependent interaction of pyrethroids with the sodium channels in frog node of Ranvier. Biochem. Biophys. Acta 728: 73–82.

Vinson, S. B. 1974. Effect of an insect growth regulator on two parasitoids developing from treated tobacco budworm larvae. J. Econ. Entomol. 67: 335–336.

Vinson, S. F. 1975. Biochemical coevolution between parasitoids and their hosts, pp. 14–48. *In* Evolutionary Strategies of Parasitic Insects and Mites. Ed. P. W. Price, Plenum Press, New York. 225 pp.

Vinson, S. B., and Barras, D. J. 1970. Effects of the parasitoid, *Cardiochiles nigricips*, on the growth, development, and tissues of *Heliothis virescens*. J. Insect Physiol. 16: 1329–1338.

Vinson, S. B., and Iwantsch, G. F. 1980. Host regulation by insect parasitoids. Quart. Rev. Biol. 55: 143–165.

Vogel, W., Masner, P., Graf, O., and Dorn, S. 1979. Types of response of insects on treatment with juvenile hormone active insect growth regulators. Experientia 35: 1254–1256.

Vorley, W. T. 1985. Spider mortality implicated in insecticide-induced resurgence of whitebacked planthopper (WBPH) and brown planthopper (BPH) in Kedah, Malaysia. IRRI Newsl. 10(5): 19–20.

Voroshilov, H. V. 1979. Heat resistant lines of the mite *Phytoseiulus persimilis* A.-H. Genetika 15: 70–76.

Voroshilov, H. V., and Kolmakova, L. I. 1977. Heritability of fertility in a hybrid population of *Phytoseiulus*. Genetika 13: 1496–1497.

Vybornov, R. V., Moiseev, V. A., and Kogan, V. S. 1982. On the question of the effect of viruses of the spotted cutworm, *Amanthes C-nigrum* L., on the *Trichogramma* (*Trichogramma euptoctidis* Gir., *Tr. evanescens* Westw.) Izv. Akad. Nauk Tadzh. SSR, Biol. Otd. Nauk 1: 77–81.

Waage, J. K., Hassell, M. P., and Godfray, H. C. J. 1985. The dynamics of pest-parasitoid-insecticide interactions. J. Appl. Ecol. 22: 825–838.

Waddill, V. H. 1978. Contact toxicity of four synthetic pyrethroids and methomyl to some adult insect parasites. Florida Entomol. 61: 27–30.

Waldner, W. 1985. Integrated control of red spider mite in south Tyrol fruit growing. Besseres Obst 3(1;2): 8–12, 43.

Walker, J. T. S., and Penman, D. R. 1978. Integrated control of apple pests in New Zealand 11. The influence of field application of azinphos-methyl on predation of European red mite by *Typhlodromus pyri*, pp. 208–213. *In* Proc. 31st New Zealand Weed and Pest Control Conf.

Walker, J. T., and Turnipseed, S. G. 1976. Predatory activity, reproductive potential and longevity of *Georcoris* spp. treated with insecticides. Proc. Georgia Entomol. Soc. 11: 266–271.

Walker, J. T. S., Wearing, C. H., Proffit, C. H., and Charles, J. G. 1981. Predator for two-spotted spider mite establishes in orchards. Orchardist of New Zealand 54: 340.

Walker, P. W., and Thurling, D. J. 1984. Insecticide resistance in *Encarsia formosa*, pp. 541–546. *In* 1984 Brit. Crop Protec. Conf. Pests and Diseases, Brighton Metropole, England, Nov. 19–22, 1984, vol. 2.

Waller, J. B. 1965. The effect of the venom of *Bracon hebetor* on the respiration of the wax moth, *Galleria mellonella*. J. Insect Physiol. 11: 1595–1599.

Wallner, W. E., Dubois, N. R., and Grinberg, P. S. 1983. Alteration of parasitism by *Rogas lymantriae* (Hymenoptera: Braconidae) in *Bacillus thuringiensis*-stressed gypsy moth (Lepidoptera: Lymantriidae) hosts. J. Econ. Entomol. 76: 275–277.

Walters, P. J. 1976a. Susceptibility of three *Stethorus* spp. (Coleoptera: Coccinellidae) to selected chemicals used in N.S.W. apple orchards. J. Aust. Entomol. Soc. 15: 49–52.

Walters, P. J. 1976b. Effect of five acaricides on *Tetranychus urticae* (Koch) and its predators *Stethorus* spp. (Coleoptera: Coccinellidae) in an apple orchard. J. Aust. Entomol. Soc. 15: 53–56.

Walters, P. J. 1976c. Chlordimeform: its prospects for the integrated control of apple pests at Bathurst. J. Aust. Entomol. Soc. 15: 57–61.

Ware, G. W., and McComb, M. 1970. Circadian susceptibility of pink bollworm to azinphosmethyl. J. Econ. Entomol. 63: 1941–1942.

Warner, L. A. 1981. Toxicities of azinphosmethyl and other orchard pesticides to the aphid predator, *Aphidoletes aphidimyza* (Rondani) (Diptera: Cecidomyiidae). M.S. thesis, Michigan State Univ., 72 pp.

Warner, L. A., and Croft, B. A. 1982. Selective toxicities of azinphosmethyl and selected orchard pesticides to the aphid predator, *Aphidoletes aphidimyza*. J. Econ. Entomol. 75: 410–415.

Warner, R. E., Peterson, K. K., and Borgman, L. 1966. Behavioral pathology in fish: a quantitative study of sublethal pesticide toxification. J. Appl. Ecol. 3(Suppl.): 223–247.

Washburn, J. A., Tassan, R. L., Grace, K., Bellis, E., Hagen, K. S., and Frankie, G. W. 1983. Effects of malathion sprays on the ice plant insect system. Calif. Agric. 37: 30–32.

Washizuka, Y., and Kuwana, S. 1959. Studies on the influence of agricultural chemicals on beneficial insect. L. Botyu-Kagaku 24: 137–140.

Waterhouse, D. E. 1973. Pest management in Australia. Nature 246: 269–271.

Watson, T. F. 1975. Practical considerations in the use of selective insecticides against major crop pests, pp. 47–65. *In* Pesticide Selectivity, Ed. J. C. Street, Marcel Dekker, New York.

Watve, C. M., and Lienk, S. E. 19?6. Toxicity of carbaryl and six organophosphorus insecticides to *Amblyseius fallacis* and *Typhlodromus pyri* from New York apple orchards. Environ. Entomol. 5: 368–370.

Way, M. J. 1949. Laboratory experiments on the effect of DDT and BHC on certain aphidophagous insects and their hosts. Bull. Entomol. Res. 40: 279–297.

Way, M. J. 1958. Effects of demeton-methyl on some aphid predators. Plant Pathol. 7: 9–13.

Way, M. J. 1963. Mutualism between ants and honeydew producing Homoptera. Annu. Rev. Entomol. 8: 307–344.

Way, M. J., Smith, P. M., and Potter, C. 1954. Studies on the bean aphid (*Aphis fabae* Scop.) and its control on field beans. Ann. Appl. Biol. 41: 117–131.

Wearing, C. H., and Ashley, E. 1982. A recommended spray programme for integrated mite control on apples. Orchardist of New Zealand Feb.: 27–29.

Wearing, C. H., and Proffitt, C. A. 1982. Integrated control of apple pests in New Zealand. 18. Introduction and establishment of the organo-phosphate-resistant predatory mite *Typhlodromus pyri* in the Waikato, pp. 22–26. *In* Proc. 35th New Zealand Weed and Pest Control Conf.

Wearing, C. J., Walker, J. T. S., Collyer, E., and Thomas, W. P. 1978a. Integrated control of apple pests in New Zealand. 8. Commercial assessment of an integrated control programme against European red mite using an insecticide-resistant predator. New Zealand J. Zool. 5: 823–837.

Wearing, C. H., Collyer, E., Thomas, W. P., and Cook, C. 1978b. Integrated control of apple pests in

New Zealand. 12. commercial implementation of integrated control of European red mite in Nelson, pp. 214–220. *In* Proc. 31st New Zealand Weed and Pest Control Conf.

Webb, B. A., and Dahlman, D. L. 1986. Ecdysteroid influence on the development of *Heliothis virescens* and its endoparasite *Microplitis croceipes*. J. Insect Physiol. 32: 339–345.

Wegorek, W., and Pruszynski, S. 1979. The toxicity of preparations applied for the control of the Colorado potato beetle to the two-spotted shield bug (*Perillus bioculatus*) (Heteroptera, Pentatomide). Biol. Inst. Ochrony Roslin 21: 119–127.

Weires, R. W. 1977. Control of *Phyllonorycter crateaegella* in eastern New York. J. Econ. Entomol. 70: 521–523.

Weires, R. 1985. Our problems with leafminers and leafrollers, and what we are doing about them (unpubl.).

Weires, R. W., and Chang, H. C. 1973. Integrated control prospects of major cabbage insect pests in Minnesota—based on the faunistic, host varietal, and trophic relationships. Univ. Minnesota Expt. Sta. Tech. Bull. 291, 42 pp.

Weires, R. W., Leeper, J. R., Reissig, W. H., and Lienk, S. E. 1982. Toxicity of several insecticides to the spotted tentiform leafminer (Lepidoptera: Gracillariidae) and its parasite, *Apanteles ornigis*. J. Econ. Entomol. 75: 680–684.

Weires, R. W., McNicholas, F. J., and Smith, G. L. 1979. Integrated mite control in Hudson and Champlain Valley apple orchards. Search Agric. 9: 1–11.

Welch, S. M. 1979. The application of simulation models to mite pest management, pp. 31–40. *In* Recent Advances in Acarology, Ed. J. G. Rodriguez, Academic Press, New York.

Welling, W. 1979. Dynamic aspects of insect–insecticide interactions. Annu. Rev. Entomol. 22: 53–78.

Welty, C., Reissig, W. H., Dennehy, T. J., and Weires, R. W. 1987. Cyhexatin resistance in New York populations of European red mite (Acari: Tetranychidae). J. Econ. Entomol. 80: 230–236.

Weseloh, R. M. 1984. Effect of the feeding inhibitor Plictran and low *Bacillus thuringiensis* Berliner doses on *Lymantria dispar* (L.) (Lepidoptera: Lymantriidae): implications for *Cotesia melanoscelus* (Ratzeburg) (Hymenoptera: Braconidae). Environ. Entomol. 13: 1371–1376.

Weseloh, R. M., and Andreadis, T. G. 1982. Possible mechanism for synergism between *Bacillus thuringiensis* and the gypsy moth (Leptidoptera: Lymantriidae) parasitoid, *Apanteles melanoscelus* (Hymenoptera: Braconidae). Ann. Entomol. Soc. Amer. 75: 435–438.

Weseloh, R. M., Andreadis, T. G., Moore, R. E. B., Anderson, J. F., Dubois, N. R., and Lewis, F. B. 1983. Field confirmation of a mechanism causing synergism between *Bacillus thuringiensis* and the gypsy moth parasitoid, *Apanteles melanoscelus*. J. Invert. Pathol. 41: 99–103.

West Virginia University Extension 1982. Integrated crop management program. West Virginia Univ. Ext. Mimeo, 7 pp.

Westigard, P. H. 1973. The biology of and effect of pesticides on *Deraeocors brevis piceatus* (Heteroptera: Miridae). Can. Entomol. 105: 1105–1111.

Westigard, P. H. 1974. Control of the pear psylla with insect growth regulators and preliminary effect on some non-target species. Environ. Entomol. 3: 256–258.

Westigard, P. H. 1979. Codling moth control on pear with diflubenzuron and effects on nontarget pest and beneficial species. J. Econ. Entomol. 72: 552–554.

Westigard, P. H., Gut, L. J., and Liss, W. J. 1986. Selective control program for the pear pest complex in southern Oregon. J. Econ. Entomol. 79: 250–257.

Whalon, M. E., and Croft, B. A. 1984. Apple IPM implementation in North America. Annu. Rev. Entomol. 29: 435–470.

Whalon, M. E., and Croft, B. A. 1985. Dispersal of apple pests and natural enemies into portable apple trees in Michigan. MSU Agric. Expt. Sta. Res. Rept. 467, 56 pp.

Whalon, M. E., and Croft, B. A. 1986. Immigration and colonization of portable apple trees by arthropod pests and their natural enemies. Crop Protec. 5: 376–384.

Whalon, M. E., Croft, B. A., and Mowry, T. M. 1982a. Establishment of a permethrin resistant predatory mite, *Amblyseius fallacis*, in a Michigan apple orchard. Environ. Entomol. 11:1096–1099.

Whalon, M. E., Berney, M. F., and Battenfield, S. L., Eds. 1982b. Biological Monitoring in Apple Orchards: An Instruction Manual. Michigan State Univ., Dept. Entomol. Manual 3, 356 pp.

White, E. B., DeBach, P., and Garber, M. J. 1970. Artificial selection for genetic adaptation to temperature extremes in *Aphytis lingnanensis* Compere (Hymenoptera: Aphelinidae). Hilgardia 40: 161–192.

Whittaker, R. H. 1952. A study of summer foliage insect communities in the Great Smokey Mountains. Ecol. Monographs 22:1–44.

Whitten, M. J. 1988. Regulatory issues in releasing genetically-manipulated arthropods, p. 317b. *In* Proc. 18th Intern. Cong. Entomol., Vancouver, B.C., Canada, July 3–9, 1988.

Whitten, M. J. McKenzie, J. A. 1982. The genetic basis for pesticide resistance, pp. 1–16. *In* Proc. 3rd Aust. Conf. Grassland Invert. Ecol., Ed. K. E. Lee, S. A. Govt. Printer, Adelaide.

WHO 1957. World Health Expert Committee on Insecticides. 7th Rept., WHO Tech. Rept., 88 pp.

Wiackowski, S. K. 1968. Laboratory investigations on the effect of insecticides on the larvae of *Chrysopa carnea* Steph. (Neuroptera, Chrysopidae). Polski Pismo Entomol. 38:601–609.

Wiackowski, S. K., and Dronka, K. 1968. Laboratory investigations on the effect of aphicides available in Poland on the most important natural enemies of aphids. Polski Pismo Entomol. 38:159–173.

Wiackowski, S. K., and Herman, E. 1968. Laboratory investigations on the effect of insecticides on adults of primary and secondary aphid parasites. Polskie Pismo Entomol. 38: 593–600.

Wiackowski, S. K., and Herman, E. 1971. Laboratory studies on the effect of chemical plant protection on *Encarsia formosa* Gah. (Hym.: Aphilinidae) brought from Canada for biological control of *Trialeurodes vaporariorum* Westw. (Hom.: Aleurodidae). Prac. Inst. Sadow. w Skiern. 15:181–183.

Wightman, J. A., and Whitford, D. N. J. 1982. Integrated control of pests of legume seed.crops. 1. Insecticides for mirid and aphid control. New Zealand J. Exp. Agric. 10:209–215.

Wiktelius, S. 1986. The effect of insecticides on some natural enemies of cereal aphids. Vaxtskydds-rapporter, Jordbruk 39:138–144.

Wilhelm, H. 1979. Control of aphids in sugar-beet plantings with respect to the sparing of natural enemies. Mitt. Biolog. Bundesanst. Land Forst. 191:200–205.

Wilkes, A. 1947. The effects of selective breeding on the laboratory propagation of insect parasites. Proc. R. Soc. London B 134:227–245.

Wilkes, A., Pielou, D. P., and Glasser, R. F. 1952. Selection for DDT tolerance in a beneficial insect, *In*, pp. 78–81. Conf. Insecticide Resistance and Ins. Physiol.

Wilkes, F. G., and Weiss, C. M. 1971. The accumulation of DDT by the dragonfly nymph, Tetragoneuria. Trans. Amer. Fish. Soc. 100:222–236.

Wilkinson, C. F. 1979. The use of insect subcellular components for studying the metabolism of xenobiotics, pp. 249–284. *In* Xenobiotic Metabolism: *In vitro* Methods, Symp. Ser. 97. Eds. G. D. Paulson, D. S. Frear, and E. P. Marks, Amer. Chem. Soc., Washington, D.C.

Wilkinson, J. D., and Ignoffo, C. M. 1973. Activity of a juvenile hormone analogue on a parasitoid, *Apanteles rubecula*, via its host, *Pieris rapae*. J. Econ. Entomol. 66:643–645.

Wilkinson, J. D., Biever, K. D., and Ignoffo, C. M. 1975. Contact toxicity of some chemical and biological pesticides to several insect parasitoids and predators. Entomophaga 20:113–120.

Wilkinson, J. D., Biever, K. D., and Ignoffo, C. M. 1979. Synthetic pyrethroid and organophosphate insecticides against the parasitoid *Apanteles marginiventris* and the predators *Geocoris punctipes, Hippodamia convergens*, and *Podiusus maculiventris*. J. Econ. Entomol. 72:473–475.

Williams, C. B., Walton, G. S., and Tierman, C. F. 1969. Zectran and naled affect incidence of parasitism of the budworm, *Choristoneura occidentalis* in Montana. J. Econ. Entomol. 62:310–312.

Williams, M. A. 1978. Integrated control of orchard mites in Tasmania. Tasmanian J. Agric. 49:292–232

Williams, M. C., and James, L. F. 1983. Effects of herbicides on the concentration of poisonous compounds in plants: a review. Amer. J. Vet. Res. 44:2420–2422.

Williams, M. W., and Batjer, L. P. 1964. Site and mode of action of 1-napthyl *N*-methylcarbamate (Sevin) in thinning apples. Proc. Amer. Soc. Hortic. Sci. 85:1–10.

Wilson, T. G., and Fabian, J. 1986. A *Drosophila melanogaster* mutant resistant to a chemical analog of juvenile hormone. Develop. Biol. 118:190–201.

Wilton, B. E., and Klowden, M. J. 1985. Solubilized crystals of *Bacillus thuringiensis* subsp. *israelensis*: effects of adult houseflies, stable flies (Diptera: Muscidae) and green lacewings (Neuroptera: Chrysopidae). J. Amer. Mosq. Control Assoc. 1:97–98.

Winteringham, E. P. W. 1969. Mechanisms of selective insecticidal action. Annu. Rev. Entomol. 14: 409–442.

Wissinger, W. L., and Grosch, D. S. 1975. Influence of juvenile hormone analogues on reproductive performance in the wasp, *Habrobracon juglandis*. J. Insect Physiol. 21:1559–1564.

Wolfenbarger, D. A., and Getzin, L. W. 1963. Selective toxicants and toxicant-surfactant combinations for leaf miner, *Liriomyza munda* Frick, control and parasite survival. Florida Entomol. 46:251–265.

Wong, S. W., and Chapman, R. B. 1979. Toxicity of synthetic pyrethroid insecticides to predaceous phytoseiid mites and their prey. Aust. J. Agric. Res. 30:497–501.

Wright, J. E., and Spates, G. E. 1972. A new approach in integrated control: insect juvenile hormone plus a hymenopteran parasite against the stable fly. Science 178:1292–1293.

Wu, T. K., Lin, S. R., and Lo, K. C. 1985. Selectivity of acaricides to two-spotted spider mite, *Tetranychus urticae* Koch and the predatory mite, *Amblyseius californicus* (McGregor). J. Agric. Res. China 34:469–476.

Wustner, D. A., Smith, C., and Fukuto, T. R. 1978. Aryl *N*-methoxy-*N*-methyl-carbamates: potent competitive reversible inhibitors of cholinesterases. Pestic. Biochem. Physiol. 9:281–292.

Xie, D. L., Li, T. J., Chen, W. R., Xie, Y. Y., and He, S. F. 1984. Effects of pesticides on egg-parasitic wasps. Kunchong Zhishi 21(1):17–19.

Yabar, L. E. 1980. Insecticides against the "bean looper" and their effect on populations of spiders. Rev. Peruana Entomol. 23:149–150.

Yamada, M., and Kawashima, K. 1983. Selective effect of insecticides for the control of apple golden leaf miner, *Phyllonorycter ringoniella* Matsumura. Ann. Rept. Soc. Plant Protec. North Japan 34:55–58.

Yokoyama, V. Y., and Pritchard, J. 1984. Effects of pesticides on mortality, fecundity, and egg viability of *Geocoris pallens* (Hemiptera: Lygaeidae). J. Econ. Entomol. 77:876–879.

Yokoyama, V. Y., Pritchard, J., and Dowell, R. M. 1984. Laboratory toxicity of pesticides to *Geocoris pallens* (Hemiptera: Lygaeidae), a predator in California cotton. J. Econ. Entomol. 77:10–15.

Yoo, J. K., Kwon, U. W., Park, H. M., and Lee, H. R. 1984. Studies on the selective toxicity of insecticides for rice insect pests between some dominant rice insect pests and a predacious spider, *Pirata subpiraticus*. Korean J. Plant Protec. 23(3):166–171.

Young, O. P., and Hamm, J. J. 1984. Compatibility of two fall armyworm pathogens with the predaceous beetle, *Calosoma sayi* (Coleoptera: Carabidae). J. Entomol. Sci. 20(2):212–218.

Yousten, A. A. 1973. Effect of the *Bacillus thuringiensis* delta-endotoxin on an insect predator which has consumed intoxicated cabbage looper larvae. J. Invert. Pathol. 21:312–314.

Yu, S. J. 1982. Induction of microsomal oxidases by host plants in the fall armyworm. *Spodoptera frugiperda* (J. E. Smith). Pestic. Biochem. Physiol. 17:59–67.

Yu, S. J. 1983. Induction of detoxifying enzymes by allelochemicals and host plants in the fall armyworm. Pestic. Biochem. Physiol. 19:330–336.

Yu, S. J. 1987. Biochemical defense capacity in the spined soldier bug (*Podisus maculiventris*) and its lepidopterous prey. Insect Physiol. Biochem. 28:216–223.

Yu, S. J. 1988. Selectivity of insecticides to the spined soldier bug (Hemiptera: Pentatomidae) and its lepidopterous prey. J. Econ. Entomol. 81:119–122.

Yu, S. J., and Terriere, L. C. 1973. Phenobarbitol induction of detoxification enzymes in resistant and susceptible houseflies. Pestic. Biochem. Physiol. 3:141–148.

Yu, S. J., and Terriere, L. C. 1978. Metabolism of juvenile hormone I by microsomal oxidase, esterase and epoxide hydrase of *Musca domestica* and some comparisons with *Phormia regina* and *Sarcophaga bullata*. Pestic. Biochem. Physiol. 9:237–246.

Yu, S. J., Berry, R. E., and Terriere, L. C. 1979. Host plant stimulation of detoxification enzymes in a phytophagous insect. Pestic. Biochem. Physiol. 12:280–284.

Yun, Y. M., and Ruppel, R. F. 1964. Toxicity of insecticides to a coccinellid predator of the cereal leaf beetle. J. Econ. Entomol. 57:835–837.

Zeleny, J. 1965. The effects of insecticides (Fosfotion, Intration, Soldep) on some predators and parasites of aphids (*Aphis craccivora* Koch, *Aphis fabae* Scop.). Rozpr. Col. Acad. Ved. 75: 3–73.

Zeleny, J. 1969. A biological and toxicological study of *Cycloneda limbifer* Casey (Coleoptera, Coccinellidae). Acta Entomol. Bohemoslovaca 66:333–344.

Zhang, N. X., and Kong, J. A. 1985. Responses of *Amblyseius fallacis* Garman to various relative humidity regimes. Chinese J. Biol. Contr. 1(3):6–9.

Zhang, N. X., and Kong, J. A. 1986. Studies on the feeding habit of *Amblyseius fallacis*. Chinese J. Biol. Control 2(1):10–13.

Zhao, F., Liu, F., and Chen, W. 1981. Studies on the biological characteristics of *Clubiona japonicola* Boes.-Str. (Acarina: Clubionidae). Zool. Res. 2:125–135.

Zhody, G. I., Osman, A. A., and Momen, F. M. 1979. Toxicity of some pyrethroid compounds to the predatory mite, *Amblyseius gossipi* El-Badry, pp. 507–512. *In* Proc. 3rd Pesticide Conf., Tanta Univ., Sept. 1979, vol. 1.

Zhody, N. Z. M. 1976. On the effect of the food of *Myzus persicae* Sulz. on the hymenopterous parasite *Aphelinus asychis* Walker. Oecologia 26:185–191.

Zhuravskaya, S. A., Bobyreva, T. V., Akramov, S. A., and Mamatkazina, A. 1976. Use of radio-active isotopes for studying the selective toxicity of phthalophos for the cotton (cucurbit) aphid and the common lacewing, Uzbekskii Biologicheskii Zhurnal 2:52–55.

Zilbermints, I. V., and Petrushov, A. Z. 1980. Dominant resistance to organophosphorus compounds and its practical use. Biology 15:882–885.

Zseller, H., and Budai, C. 1982. Studies on pesticide sensitivity of *Encarsia formosa* Gahan. Novenyvedelem 18:25–26.

Zungoli, P. A., Steinhauer, A. L., and Linduska, J. J. 1983. Evaluation of diflubenzuron for Mexican bean beetle (Coleoptera: Coccinellidae) control and impact on *Pediobius foveolatus* (Hymenoptera: Eulophidae). J. Econ. Entomol. 76:188–191.

Zwick, R. W. 1972. Studies on the integrated control of spider mites on apples in Oregon's Hood River Valley. Environ. Entomol. 1:169–176.

SUBJECT INDEX

SPECIES INDEX

PESTICIDE INDEX